INTRODUCTION TO THE

Human Body

The Essentials of
Anatomy and Physiology

SECOND EDITION

Gerard J. Tortora

Bergen Community College

HarperCollins*Publishers*

Sponsoring Editor: Bonnie Roesch
Senior Development Editor: Robert Ginsberg
Project Editor: Thomas R. Farrell
Art Direction: Teresa J. Delgado
Art Coordination: Claudia DePolo
Text and Cover Design: D.D.I.
Cover and title page photo: David Stoecklein/The Stock Market
Photo Research: Mira Schachne/Kelly Mountain
Production: Kewal K. Sharma
Compositor, Printer, and Binder: Arcata Graphics/Kingsport
Cover Printer: The Lehigh Press, Inc.
Chapter-opening photos: *Chapter 1:* ScottForesman; *Chapter 2:* Four by Five;
Chapter 3: Four by Five; *Chapter 4:* Four by Five; *Chapter 5:* Walter Iooss,
Jr./The Image Bank; *Chapter 6:* Weinberg-Clark/The Image Bank; *Chapter 7:*
David Brownell/The Image Bank; *Chapter 8:* David Stoecklein/The Stock Market;
Chapter 9: U. Wiesmeier/The Image Bank; *Chapter 10:* Brooks Dodge; *Chapter
11:* Jean Kepler/Rainbow; *Chapter 12:* Janeart/The Image Bank; *Chapter 13:*
Michael Kevin Daly/The Stock Market; *Chapter 14:* The Image Bank; *Chapter
15:* CO Rentmeester/The Image Bank; *Chapter 16:* Superstock; *Chapter 17:*
Walter Iooss, Jr./The Image Bank; *Chapter 18:* Dean Hulse/Rainbow; *Chapter
19:* Sam Abell; *Chapter 20:* Bob Daemmrich; *Chapter 21:* Mitchell B. Reibel/
Sports Chrome, Inc.; *Chapter 22:* Four by Five; *Chapter 23:* Elyse Lewin/The
Image Bank: *Chapter 24:* From *The Incredible Machine*/Lennart Nilsson/Bonnier
Fakta.

Introduction to the Human Body: The Essentials of Anatomy and Physiology, *Second Edition*

Library of Congress Cataloging-in-Publication Data

Tortora, Gerard J.
    Introduction to the human body : the essentials of anatomy and
physiology / Gerard J. Tortora. —2nd ed.
        p.   cm.
    Includes index.
    ISBN 0-06-046697-9 (Student Edition)
    ISBN 0-06-500033-1 (Teacher Edition)
    1. Human physiology. 2. Body, Human. I. Title.
    QP36.T67   1991
    612—dc20   90-4794
                                                            CIP

92 93 9 8 7 6 5 4

To Angelina M. Tortora,
my mother,
whose love, support, guidance, and patience
have been such an important part of my
personal and professional life

Contents in Brief

Preface   xix
Note to the Student   xxiii

1	Organization of the Human Body	2
2	Introductory Chemistry	18
3	Cells	35
4	Tissues	60
5	The Integumentary System	84
6	The Skeletal System	95
7	Articulations	130
8	The Muscular System	141
9	Nervous Tissue	187
10	Central and Somatic Nervous System	203
11	Autonomic Nervous System	230
12	Sensations	240
13	The Endocrine System	268
14	The Cardiovascular System: Blood	297
15	The Cardiovascular System: Heart	316
16	The Cardiovascular System: Blood Vessels	333
17	The Lymphatic System and Immunity	352
18	The Respiratory System	375
19	The Digestive System	400
20	Metabolism	426
21	The Urinary System	448
22	Fluid, Electrolyte, and Acid–Base Balance	470
23	The Reproductive Systems	484
24	Development and Inheritance	510

Bibliography   B-1
Glossary of Combining Forms, Word Roots, Prefixes,
and Suffixes   G-1
Glossary of Terms   G-7
Index   I-1

Contents in Detail

Preface xix
Note to the Student xxiii

1

Organization of the Human Body 2

Anatomy and Physiology Defined 3
Levels of Organization 3
Life Processes 5
Anatomical Position 6
Directional Terms 6
Planes and Sections 6
Body Cavities 8
Abdominopelvic Regions and Quadrants 9
Homeostasis: Maintaining Physiological Limits 10
Stress and Homeostasis 11
Homeostasis of Blood Pressure (BP) 12
Homeostasis of Blood Sugar (BS) Level 13
Wellness: Life-Style, Health, and Homeostasis 14

2

Introductory Chemistry 18

Introduction to Basic Chemistry 19
Chemical Elements 19
Structure of Atoms 19

Atoms and Molecules 21
    *Ionic Bonds, 22 · Covalent Bonds, 23 · Hydrogen
    Bonds, 23*
Chemical Reactions 23
    *Synthesis Reactions—Anabolism, 23 · Decomposition
    Reactions—Catabolism, 24 · Energy and Chemical
    Reactions, 25*
Chemical Compounds and Life Processes 25
Inorganic Compounds 25
    *Water, 25 · Inorganic Acids, Bases, and Salts, 25 · Acid-
    Base Balance: The Concept of pH, 25 · Maintaining pH:
    Buffer Systems, 26*
Organic Compounds 27
    *Carbohydrates, 27 · Lipids, 27 · Proteins, 31 · Nucleic
    Acids: Deoxyribonucleic Acid (DNA) and Ribonucleic
    Acid (RNA), 31 · Adenosine Triphosphate (ATP), 32*
Wellness: Vegetarian Diets: Alphabet Soup 28

3

Cells 35

Generalized Animal Cell 36
Plasma (Cell) Membrane 36
Chemistry and Structure 36
Functions 36
Movement of Materials Across Plasma Membranes 38
    Passive Processes, 39 · Active Processes, 41
Cytoplasm 44
Organelles 44
Nucleus 44
Ribosomes 44
Endoplasmic Reticulum (ER) 45
Golgi Complex 46
Lysosomes 46
Mitochondria 47
The Cytoskeleton 47

Centrosome and Centrioles   47
Flagella and Cilia   47
Cell Inclusions   48
Gene Action   48
Protein Synthesis   48
    Transcription, 48 · Translation, 48
Normal Cell Division   51
Somatic Cell Division   51
    Mitosis, 52 · Cytokinesis, 53
Abnormal Cell Division: Cancer (CA) 53
Definition   53
Spread   53
Types   56
Possible Causes   56
Treatment   56
Medical Terminology and Conditions   56
Wellness: The Cancer-Prevention Life-Style 54

5

The Integumentary System   84

Skin   85
Functions   85
Structure   86
Epidermis   86
Dermis   87
Skin Color   87
Accessory Organs of the Skin   90
Hair   90
Glands   91
    Sebaceous (Oil) Glands, 91 · Sudoriferous (Sweat)
    Glands, 91 · Ceruminous Glands, 91
Nails   91
Homeostasis of Body Temperature   92
Common Disorders   92
Medical Terminology and Conditions   93
Wellness: More Than Skin Deep   88

4

Tissues   60

Extracellular Materials   61
Types of Tissues   61
Epithelial Tissue   62
Covering and Lining Epithelium   62
    Classification by Cell Shape, 62 · Classification by
    Arrangement of Layers, 62 · Types and Functions, 62
Glandular Epithelium   69
Connective Tissue   69
Classification   69
Embryonic Connective Tissue   69
Adult Connective Tissue   76
    Connective Tissue Proper, 76 · Cartilage, 76 · Osseous
    Tissue (Bone), 77 · Vascular Tissue (Blood), 77
Muscle Tissue   77
Nervous Tissue   77
Membranes   81
Mucous Membranes   81
Serous Membranes   81
Cutaneous Membrane   81
Synovial Membranes   81
Wellness: Fat Tissue Issues   78

6

The Skeletal System   95

Functions   96
Types of Bones   96
Histology   97
Compact Bone   98
Spongy Bone   98
Ossification: Bone Formation   98
Intramembranous Ossification   99
Endochondral Ossification   99
Homeostasis   101
Bone Growth and Maintenance   101
Mineral Storage   101
Surface Markings   102
Divisions of the Skeletal System   103
Skull   103
Sutures   103
Cranial Bones   105

Facial Bones   107
Fontanels   110
Foramina   111
Hyoid Bone   111
Vertebral Column   111
Divisions   111
Normal Curves   111
Typical Vertebra   112
Cervical Region   113
Thoracic Region   113
Lumbar Region   113
Sacrum and Coccyx   113
Thorax   114
Sternum   114
Ribs   114
Pectoral (Shoulder) Girdle   115
Clavicle   115
Scapula   115
Upper Extremity   116
Humerus   116
Ulna and Radius   116
Carpals, Metacarpals, and Phalanges   116
Pelvic (Hip) Girdle   117
Lower Extremity   117
Femur   118
Patella   120
Tibia and Fibula   120
Tarsals, Metatarsals, and Phalanges   121
Arches of the Foot   122
Female and Male Skeletons   123
Common Disorders   126
Medical Terminology and Conditions   127
Wellness: Osteoporosis: A Question of Calcium Balance   124

7

Articulations   **130**

Classification   131
Functional   131
Structural   131
Synarthroses (Immovable)   131
Suture   131
Gomphosis   132
Synchondrosis   132
Amphiarthroses (Slightly Movable)   132

Syndesmosis   132
Symphysis   132
Diarthroses (Freely Movable)   132
Types of Synovial Joints   133
Gliding   133
Hinge   134
Pivot   134
Ellipsoidal   134
Saddle   136
Ball-and-Socket   136
Special Movements at Synovial Joints   136
Common Disorders   136
Medical Terminology and Conditions   140
Wellness: Living with Arthritis   138

8

The Muscular System   **141**

Characteristics   142
Functions   142
Types   142
Skeletal Muscle Tissue   142
Connective Tissue Components   143
Nerve and Blood Supply   143
Histology   144
Contraction   144
Sliding-Filament Mechanism   144
Neuromuscular Junction   144
Physiology of Contraction   146
Energy for Contraction   148
Homeostasis   149
    *Oxygen Debt, 150 · Muscle Fatigue, 150 · Heat
    Production, 150*
Motor Unit   150
All-or-None Principle   150
Kinds of Contraction   150
    *Twitch, 150 · Tetanus, 151 · Treppe, 151 · Isotonic and
    Isometric, 151*
Muscle Tone   151
Muscular Atrophy and Hypertrophy   154
How Skeletal Muscles Produce Movement   154
Origin and Insertion   154
Lever Systems   154
Group Actions   154
Naming Skeletal Muscles   154

Principal Skeletal Muscles 155
Cardiac Muscle Tissue 180
Smooth Muscle Tissue 180
Common Disorders 183
Medical Terminology and Conditions 184
Wellness: Survival of the Fit 152

9

Nervous Tissue 187

Organization 188
Histology 188
Neuroglia 188
Neurons 189
   Structure, 189 · Classification, 191
Functions 194
Nerve Impulses 194
   *Ion Channels in Plasma Membranes, 194 · Membrane
   Potentials, 194 · Excitability, 195 · All-or-None
   Principle, 195 · Saltatory Conduction, 195 · Speed of
   Nerve Impulses, 197*
Conduction Across Synapses 197
   *Excitatory Transmission, 199 · Inhibitory Transmission,
   199 · Integration at Synapses, 199 · Alteration of
   Synaptic Conduction, 199*
Graded Potentials 200
Regeneration 200
Wellness: Biofeedback: Expanding the Nervous System 192

10

Central and Somatic Nervous Systems 203

Grouping of Neural Tissue 204
Spinal Cord 204
Protection and Coverings 204
   Vertebral Canal, 204 · Meninges, 204
General Features 205
Structure in Cross Section 206

Functions 206
   *Impulse Conduction, 206 · Reflex Center, 207 · Reflex
   Arc and Homeostasis, 207*
Spinal Nerves 208
Names 208
Composition and Coverings 209
Distribution 209
   *Branches, 209 · Plexuses, 209 · Intercostal (Thoracic)
   Nerves, 209*
Brain 209
Principal Parts 209
Protection and Coverings 209
Cerebrospinal Fluid (CSF) 209
Blood Supply 210
Brain Stem 212
   Medulla Oblongata, 212 · Pons, 213 · Midbrain, 213
Diencephalon 214
   Thalamus, 214 · Hypothalamus, 214
Reticular Activating System (RAS), Consciousness,
and Sleep 215
Cerebrum 215
   *Lobes, 215 · Brain Lateralization (Split-Brain Concept),
   216 · White Matter, 217 · Basal Ganglia (Cerebral
   Nuclei), 217 · Limbic System, 217 · Functional Areas
   of Cerebral Cortex, 217 · Electroencephalogram (EEG),
   221*
Cerebellum 221
Neurotransmitters 221
Cranial Nerves 221
Common Disorders 224
Medical Terminology and Conditions 226
Wellness: In Search of Morpheus 218

11

Autonomic Nervous System 230

**Comparison of Somatic and Autonomic Nervous
Systems** 231
Structure of the Autonomic Nervous System 232
Visceral Efferent Pathways 232
   *Preganglionic Neurons, 232 · Autonomic Ganglia,
   234 · Postganglionic Neurons, 235*
Sympathetic Division 235
Parasympathetic Division 235
Functions of the Autonomic Nervous System 235
Neurotransmitters 235
Activities 235

Wellness: Stress Management: When You Can't Fight or Flee 236

Wellness: Flexibility: Taking It to the Limit 244

12

13

Sensations 240

Sensations	**241**
Definition	241
Characteristics	241
Classification of Receptors	242
Location, 242 · Stimulus Detected, 242 · Simplicity or Complexity, 242	
General Senses	**242**
Cutaneous Sensations	242
Tactile Sensations, 242 · Thermoreceptive Sensations, 243 · Pain Sensations, 243	
Proprioceptive Sensations	246
Receptors, 246	
Special Senses	**246**
Olfactory Sensations	**246**
Structure of Receptors	246
Stimulation of Receptors	247
Olfactory Pathway	247
Gustatory Sensations	**247**
Structure of Receptors	247
Stimulation of Receptors	248
Gustatory Pathway	248
Visual Sensations	**248**
Accessory Structures of Eye	249
Structure of Eyeball	249
Fibrous Tunic, 249 · Vascular Tunic, 249 · Retina (Nervous Tunic), 249 · Lens, 251 · Interior, 251	
Physiology of Vision	251
Retinal Image Formation, 251 · Stimulation of Photoreceptors, 253 · Visual Pathway, 254	
Auditory Sensations and Equilibrium	**254**
External Ear	254
Middle Ear	254
Internal Ear	254
Sound Waves	260
Physiology of Hearing	260
Physiology of Equilibrium	261
Static Equilibrium, 261 · Dynamic Equilibrium, 262	
Common Disorders	**264**
Medical Terminology and Conditions	**265**

The Endocrine System 268

Endocrine Glands	**270**
Chemistry of Hormones	**270**
Mechanism of Hormonal Action	**270**
Overview	270
Receptors	270
Interaction with Plasma Membrane Receptors	272
Interaction with Intracellular Receptors	272
Prostaglandins (PGs) and Hormones	272
Control of Hormonal Secretions: Feedback Control	**272**
Pituitary	**274**
Anterior Pituitary	274
Human Growth Hormone (hGH), 274 · Thyroid-Stimulating Hormone (TSH), 275 · Adrenocorticotropic Hormone (ACTH), 275 · Follicle-Stimulating Hormone (FSH), 275 · Luteinizing Hormone (LH), 276 · Prolactin (PRL), 277 · Melanocyte-Stimulating Hormone (MSH), 277	
Posterior Pituitary	277
Oxytocin (OT), 277 · Antidiuretic Hormone (ADH), 277	
Thyroid	**278**
Function and Control of Thyroid Hormones	280
Calcitonin (CT)	281
Parathyroids	**281**
Parathyroid Hormone (PTH)	281
Adrenals	**281**
Adrenal Cortex	281
Mineralocorticoids, 282 · Glucocorticoids, 283 · Gonadocorticoids, 284	
Adrenal Medulla	284
Epinephrine and Norepinephrine (NE), 284	
Pancreas	**285**
Glucagon	285
Insulin	286
Ovaries and Testes	**286**
Pineal	**286**
Thymus	**286**
Other Endocrine Tissues	**286**
Stress and the General Adaptation Syndrome (GAS)	**287**
Stressors	287
Alarm Reaction	287

Resistance Reaction   287
Exhaustion   290
Stress and Disease   291
Common Disorders   291
Medical Terminology and Conditions   293
Wellness: Diabetes and the Wellness Life-Style   288

14

The Cardiovascular System: Blood   297

Functions of Blood   298
Physical Characteristics of Blood   298
Components of Blood   299
Formed Elements   299
    Origin, 299 · Erythrocytes (Red Blood Cells), 301 ·
    Leucocytes (White Blood Cells), 302 · Thrombocytes
    (Platelets) 304
Plasma   304
Hemostasis of Blood   304
Vascular Spasm   304
Platelet Plug Formation   305
Coagulation   305
    Extrinsic Pathway, 308 · Intrinsic Pathway, 308 ·
    Retraction and Fibrinolysis, 308 · Hemostatic Control
    Mechanisms, 308 · Intravascular Clotting, 310
Grouping (Typing) of Blood   310
ABO   310
Rh   311
Common Disorders   312
Medical Terminology and Conditions   313
Wellness: Cholesterol, Lipoproteins, and Life-Style   306

15

The Cardiovascular System: Heart   316

Location of Heart   317
Pericardium   317

Heart Wall   317
Chambers of the Heart   319
Great Vessels of the Heart   319
Valves of the Heart   319
Atrioventricular (AV) Valves (Cuspid Valves)   319
Semilunar Valves   319
Blood Supply   322
Conduction System   322
Electrocardiogram (ECG or EKG)   322
Blood Flow Through the Heart   323
Cardiac Cycle   324
Phases   324
Timing   324
Sounds   324
Cardiac Output (CO)   325
Heart Rate   325
Autonomic Control   325
Chemicals   325
Temperature   325
Emotions   329
Sex and Age   329
Risk Factors in Heart Disease   329
Common Disorders   329
Medical Terminology and Conditions   330
*Wellness: Diet and Heart Disease: Redefining
the Good Life*   326

16

The Cardiovascular System: Blood Vessels   333

Arteries   334
Arterioles   334
Capillaries   334
Venules   335
Veins   336
Blood Reservoirs   336
Physiology of Circulation   336
Blood Flow   336
    Blood Pressure, 336 · Resistance, 337
Factors That Affect Arterial Blood Pressure   337
    Cardiac Output (CO), 337 · Blood Volume, 337 ·
    Peripheral Resistance, 337
Homeostasis of Blood Pressure Regulation   337
    Vasomotor Center, 337 · Baroreceptors, 338 ·

Chemoreceptors, 338 · Regulation by Higher Brain Centers, 338 · Chemicals, 338 · Autoregulation, 338

Capillary Exchange	339
Factors That Aid Venous Return	339

Pumping Action of the Heart, 339 · Velocity of Blood Flow, 339 · Skeletal Muscle Contractions and Valves, 339 · Breathing, 339

Checking Circulation	339
Pulse	339
Measurement of Blood Pressure (BP)	340
Shock and Homeostasis	340
Circulatory Routes	341
Systemic Circulation	341
Pulmonary Circulation	342
Cerebral Circulation	343
Hepatic Portal Circulation	348
Fetal Circulation	348
Exercise and the Cardiovascular System	348
Common Disorders	348
Medical Terminology and Conditions	349
Wellness: Hypertension Prevention	346

17

The Lymphatic System and Immunity 352

Functions	353
Lymph and Interstitial Fluid	353
Lymphatic Vessels	354
Lymphatic Tissue	356
Lymph Nodes	356
Tonsils	357
Spleen	357
Thymus Gland	357
Lymph Circulation	357
Route	357
Maintenance	358
Nonspecific Resistance to Disease	358
Skin and Mucous Membranes	358

Mechanical Factors, 358 · Chemical Factors, 359

Antimicrobial Substances	359

Interferon (IFN), 359 · Complement, 359 · Properdin, 359

Phagocytosis	359

Kinds of Phagocytes, 359 · Mechanism, 359

Inflammation	360

Symptoms, 360 · Stages, 360

Fever	362

Immunity (Specific Resistance to Disease)	362
Antigens (Ags)	363
Antibodies (Abs)	363
Cellular and Humoral Immunity	363

Formation of T Cells and B Cells, 364 · T Cells and Cellular Immunity, 364 · B Cells and Humoral Immunity, 365 · Actions of Antibodies, 368

The Skin and Immunity	368
Immunology and Cancer	369
Common Disorders	369
Medical Terminology and Conditions	373
Wellness: Giving Viruses a Cold Reception	366

18

The Respiratory System 375

Organs	376
Nose	376

Anatomy, 376 · Functions, 376

Pharynx	376
Larynx	377

Anatomy, 377 · Voice Production, 378

Trachea	379
Bronchi	380
Lungs	380

Gross Anatomy, 380 · Lobes and Fissures, 382 · Lobules, 382 · Alveolar-Capillary (Respiratory) Membrane, 382 · Blood Supply, 383

Respiration	383
Pulmonary Ventilation	383

Inspiration, 383 · Expiration, 384 · Collapsed Lung, 384

Pulmonary Air Volumes and Capacities	385

Pulmonary Volumes, 385 · Pulmonary Capacities, 386

Exchange of Respiratory Gases	386
External (Pulmonary) Respiration	387
Internal (Tissue) Respiration	388
Transportation of Respiratory Gases	388

Oxygen, 388 · Carbon Dioxide, 390

Control of Respiration	390
Nervous Control	390

Medullary Rhythmicity Area, 390 · Pneumotaxic Area, 391 · Apneustic Area, 391

Regulation of Respiratory Center Activity 391
   Cortical Influences, 391 · Inflation Reflex, 391 ·
   Chemical Stimuli, 391 · Other Influences, 394
Common Disorders 395
Medical Terminology and Conditions 396

Wellness: Living with Asthma: Learning to Breathe
Easy 392

19

The Digestive System 400

Digestive Processes 400
Organization 400
General Histology 400
   Mucosa, 400 · Submucosa, 401 · Muscularis, 401 ·
   Serosa, 401
Mouth (Oral Cavity) 401
Tongue 402
Salivary Glands 402
   Composition of Saliva, 403 · Secretion of Saliva,
   403
Teeth 404
Digestion in the Mouth 405
   Mechanical, 405 · Chemical, 405
Pharynx 405
Esophagus 406
Deglutition 406
Stomach 407
Anatomy 407
Digestion in the Stomach 407
   Mechanical, 407 · Chemical, 408
Regulation of Gastric Secretion 408
   Stimulation, 408 · Inhibition, 409
Regulation of Gastric Emptying 409
Absorption 410
Pancreas 410
Anatomy 410
Pancreatic Juice 410
Regulation of Pancreatic Secretions 410
Liver 410
Anatomy 410
Blood Supply 410
Bile 411
Regulation of Bile Secretion 414

Functions of the Liver 414
Gallbladder (GB) 414
Functions 414
Emptying of the Gallbladder 415
Small Intestine 415
Anatomy 415
Intestinal Juice 416
Digestion in the Small Intestine 416
   Mechanical, 416 · Chemical, 416
Regulation of Intestinal Secretion 416
Absorption 417
   Carbohydrates, 417 · Proteins, 417 · Lipids,
   417 · Water, 419 · Electrolytes, 419 · Vitamins,
   419
Large Intestine 419
Anatomy 419
Digestion in the Large Intestine 420
   Mechanical, 420 · Chemical, 421
Absorption and Feces Formation 421
Defecation 421
Common Disorders 421
Medical Terminology and Conditions 422
Wellness: Alcohol: Use or Abuse? 412

20

Metabolism 426

Regulation of Food Intake 427
Nutrients 427
Metabolism 427
Anabolism 428
Catabolism 428
Metabolism and Enzymes 428
Oxidation-Reduction Reactions 428
Carbohydrate Metabolism 429
Fate of Carbohydrates 429
Glucose Catabolism 429
   Glycolysis, 429 · Krebs Cycle, 430 · Electron Transport
   Chain, 430
Glucose Anabolism 430
   Glucose Storage: Glycogenesis, 430 · Formation of
   Glucose from Proteins and Fats: Gluconeogenesis,
   430

Lipid Metabolism 430
Fate of Lipids 430
Fat Storage 431
Lipid Catabolism 431
    Glycerol, 433 ∙ Fatty Acids, 433
Lipid Anabolism: Lipogenesis 434
Protein Metabolism 434
Fate of Proteins 434
Protein Catabolism 435
Protein Anabolism 435
Regulation of Metabolism 435
Minerals 436
Vitamins 437
Metabolism and Body Heat 437
Measuring Heat 437
Production of Body Heat 437
Basal Metabolic Rate (BMR) 442
Loss of Body Heat 442
    Radiation, 442 ∙ Conduction, 442 ∙ Convection, 442 ∙
    Evaporation, 442
Body Temperature Regulation 442
    Hypothalamic Thermostat, 442 ∙ Mechanisms of Heat
    Production, 442 ∙ Mechanisms of Heat Loss, 443 ∙
    Fever, 443
Common Disorders 444
Wellness: The Metabolic Realities of Weight Control 440

Tubular Reabsorption 457
Tubular Secretion 457
Hemodialysis Therapy 458
Homeostasis 459
Ureters 459
Structure 460
Functions 460
Urinary Bladder 460
Structure 461
Functions 461
Urethra 464
Functions 464
Urine 464
Volume 464
Physical Characteristics 464
Chemical Composition 464
Abnormal Constituents 464
Common Disorders 466
Medical Terminology and Conditions 467
Wellness: Urinary Tract Infections 462

22

Fluid, Electrolyte, and Acid-Base Balance 470

Fluid Compartments and Fluid Balance 471
Water 471
Fluid Intake and Output 471
Regulation of Intake 471
Regulation of Output 472
Electrolytes 472
Concentration 472
Distribution 473
Functions and Regulation 474
    Sodium, 474 ∙ Potassium, 474 ∙ Calcium, 474 ∙
    Magnesium, 475 ∙ Chloride, 475 ∙ Phosphate,
    475
Movement of Body Fluids 475
Between Plasma and Interstitial Compartments 475
Between Interstitial and Intracellular Compartments 478
Acid–Base Balance 478
Buffer Systems 479
    Carbonic Acid–Bicarbonate, 479 ∙ Phosphate, 479 ∙
    Hemoglobin, 479 ∙ Protein, 480

21

The Urinary System 448

Kidneys 449
External Anatomy 449
Internal Anatomy 449
Blood Supply 450
Nephron 450
Juxtaglomerular Apparatus (JGA) 450
Functions 453
Glomerular Filtration 453
    Production of Filtrate, 453 ∙ Net Filtration Pressure
    (NFP), 454 ∙ Glomerular Filtration Rate (GFR), 455 ∙
    Regulation of GFR, 455

Respiration 480
Kidney Excretion 480
Acid–Base Imbalances 480
Wellness: Endurance Exercise: A Challenge to Fluid and Electrolyte Balance 476

23

The Reproductive Systems 484

Male Reproductive System 485
Scrotum 485
Testes 485
    Spermatogenesis, 486 · Spermatozoa, 489 · Testosterone and Inhibin, 489
Male Puberty 489
Ducts 490
    Ducts of the Testis, 490 · Epididymis, 490 · Ductus (Vas) Deferens, 490 · Ejaculatory Duct, 490 · Urethra, 490
Accessory Sex Glands 490
Semen (Seminal Fluid) 491
Penis 491
Female Reproductive System 492
Ovaries 493
    Oogenesis, 493
Uterine (Fallopian) Tubes 494
Uterus 495
Vagina 496
Vulva 496
Perineum 496
Mammary Glands 497
Female Puberty 498
Female Reproductive Cycle (FRC) 498
Hormonal Regulation 499
Menstrual Phase (Menstruation) 499
Preovulatory Phase 501
Ovulation 501
Postovulatory Phase 501
Menopause 501
Common Disorders 502
Medical Terminology and Conditions 507
Wellness: Life-Style and the Menstrual Cycle 504

24

Development and Inheritance 510

Sexual Intercourse 511
Male Sexual Act 511
    Erection, 511 · Lubrication, 511 · Orgasm, 511
Female Sexual Act 511
    Erection, 511 · Lubrication, 512 · Orgasm (Climax), 512
Development During Pregnancy 512
Fertilization 512
Formation of the Morula 512
Development of the Blastocyst 512
Implantation 512
Embryonic Development 513
Beginnings of Organ Systems 514
    Embryonic Membranes, 514 · Placenta and Umbilical Cord, 514 · Fetal Circulation, 514
Fetal Growth 516
Hormones of Pregnancy 516
Gestation 517
Prenatal Diagnostic Techniques 519
Amniocentesis 519
Chorionic Villus Sampling (CVS) 520
Parturition and Labor 520
Lactation 524
Birth Control (BC) 524
Sterilization 524
Hormonal 524
Intrauterine Devices (IUDs) 524
Barrier 525
Chemical 525
Physiologic (Natural) 525
Coitus Interruptus (Withdrawal) 525
Induced Abortion 525
Inheritance 525
Genotype and Phenotype 525
Genes and the Environment 528
Inheritance of Sex 528
Color Blindness and X-Linked Inheritance 528
Medical Terminology and Conditions 529
Wellness: Eating for Two 522

Bibliography   B-1
Glossary of Combining Forms, Word Roots, Prefixes, and Suffixes   G-1
Glossary of Terms   G-7
Index   I-1

Preface

Introduction to the Human Body: The Essentials of Anatomy and Physiology, Second Edition, is designed for use in a one-semester course in human anatomy and physiology or in human biology. It assumes no previous study of the human body. The successful approach of the first edition—to provide students with a basic understanding of the structure and functions of the human body with an emphasis on homeostasis—has been maintained. The development of the second edition focused on improving the acknowledged strengths of the text as well as the introduction of several new and innovative features.

NEW TO THIS EDITION

WELLNESS BOXES

The importance of maintaining homeostasis is one of the keys to understanding anatomy and physiology. It is also one of the keys to good health. In order to increase the students' appreciation of the relevancy of the concepts in this text, we have added innovative, new, boxed material to each chapter on wellness and self-care topics.

The wellness philosophy supports the notion that the life-style choices made by individuals throughout the years have an important influence on their mental and physical well-being. An understanding of the anatomy and physiology of the human body increases one's ability to understand how life-style factors such as diet, exercise, and stress management affect the maintenance of health.

The wellness boxes have been written especially for this text. In each chapter students can relate the topics covered to recognized health issues such as cancer prevention, healthy skin, obesity and weight control, osteoporosis, physical fitness, stress management, cholesterol, diet and heart disease, hypertension, alcohol abuse, and pregnancy diets.

REINFORCEMENT QUESTIONS

Another new and helpful feature of this edition is the questions placed with the illustrations in the text. These reinforcement questions encourage students to look at the art carefully and get the maximum benefit from each drawing. They enable students to review or apply a newly learned concept before moving on to another one. Placing the questions with the illustrations also means that the reading of the narrative is not interrupted for the student. They truly help the student integrate the text and the art into a clearer understanding of the material. *Introduction to the Human Body,* Second Edition, is the first text to offer this valuable learning tool in an extensive and consistent way.

REVISED ART PROGRAM

The choice to extensively develop questions to go along with the art program reveals our understanding of the importance of the drawings and photographs in the book. Although reviews praised the first edition's visual program, the second edition is even better. Many illustrations have been redrawn, and many new ones have been added to this edition to complement and amplify the narrative and provide students with an additional pedagogical aid to assist their learning efforts. Examples of revised or newly drawn art may be found in Chapter 3 (cells), Chapter 7 (joints), Chapter 8 (muscles), Chapter 12 (eye and ear), Chapter 13 (heart), and Chapter 23 (reproductive organs). New art that specifically emphasizes physiological concepts has been added throughout the book. See Figures 6.4, 14.5, and 21.7 as examples. Many of the photomicrographs in Chapter 4 have been replaced to provide students with clearer images of the microscopic level of structure.

LEVEL

In response to feedback from instructors and students, the entire book has been carefully scrutinized to ensure that the level of presentation and detail is appropriate for users of the text. Throughout this edition, discussions have been simplified and the level of detail lightened.

ORGANIZATION

As in the first edition, the book is divided into 24 chapters and follows a systemic approach to a study of the human body. Chapter 1, Organization of the Human Body, presents students with a broad overview of the human body with emphasis on levels of structural organization, life processes, anatomical position, directional terms, planes, body cavities, and homeostasis.

In Chapter 2, Introductory Chemistry, atomic structure, chemical bonding, chemical reactions and the importance of inorganic and organic compounds to the human body are discussed.

Chapter 3, Cells, now contains a section on extracellular materials that was in Chapter 2 in the first edition. The chapter also describes the basic components and functions of cells, normal cell division (mitosis and meiosis), and abnormal cell division (cancer).

In Chapter 4, Tissues, there is an extensive discussion of epithelial and connective tissues and general discussions of muscle tissue and nervous tissue. Membranes and tissue repair are also presented.

The structure and functions of the skin and its accessory organs are treated in Chapter 5, The Integumentary System, along with the role of the skin in homeostasis.

In Chapter 6, The Skeletal System, there is emphasis on the histology of bone tissue, ossification, homeostasis of bone (remodeling, growth, maintenance, mineral storage), types of bones, surface markings, and bones of the axial and appendicular skeletons.

Chapter 7, Articulations, discusses the various types of joints based on movement and structure and the various kinds of movements at joints.

Chapter 8, The Muscular System, deals with the types of muscle tissue, the histology of muscle tissue, the physiology of muscular contraction, the homeostasis of muscle tissue, types of contractions, how skeletal muscles produce movement, how skeletal muscles are named, and the various muscles of the body in terms of origin, insertion, and action.

In Chapter 9, Nervous Tissue, attention is given to the histology of nervous tissue and neuroglia, the physiology of the nerve impulse, conduction across synapses, neurotransmitters, and regeneration.

The structure and functions of the spinal cord, spinal nerves, the structure of the brain and its functions, cranial nerves and neurotransmitters are considered in Chapter 10, Central and Somatic Nervous Systems. The discussion of sleep, memory, and sensory and motor pathways have been moved from Chapter 12 to Chapter 10.

In Chapter 11, Autonomic Nervous System, the structure and physiology of the autonomic nervous system are considered with emphasis on the roles of the sympathetic and parasympathetic divisions in helping to maintain homeostasis.

In Chapter 12, Sensations, attention is turned to the characteristics of sensations, the classification of receptors, general senses (cutaneous and proprioceptive), and special senses (olfactory, gustatory, visual, auditory, and equilibrium).

Chapter 13, The Endocrine System, deals with the chemistry of hormones, the mechanisms of hormone action, the control of hormone secretions, the structure and physiology of endocrine glands and tissues, and stress and the general adaptations syndrome.

Chapter 14, The Cardiovascular System: Blood, examines the components of blood and its functions. There is also discussion of grouping (typing) of blood.

In Chapter 15, The Cardiovascular System: Heart, consideration is given to the structure and functions of the heart, the elctrocardiogram, factors that affect heart rate, and risk factors in heart disease.

The structure and functions of blood vessels, the physiology of circulation, pulse, blood pressure, circulatory routes, and exercise and the cardiovascular system are discussed in Chapter 16, The Cardiovascular System: Blood Vessels. The section on shock and homeostasis is now in Chapter 16.

In Chapter 17, the Lymphatic System and Immunity, attention is turned to the structure of the lymphatic system, lymph circulation, components of nonspecific resistance to disease, immunity, and immunology and cancer.

Chapter 18, The Respiratory System, includes the structure of the respiratory system, the physiology of respiration, and the mechanisms that control respiration.

Chapter 19, The Digestive System, deals with the structure and functions of the digestive system, the regulation of digestion, and the absorption of digested nutrients.

The metabolism of foods, the importance of minerals and vitamins, the importance and regulation of metabolism, and the relationships of metabolism to body heat are discussed in Chapter 20, Metabolism.

Chapter 21, The Urinary System, discusses the anatomy and physiology of the urinary organs, physiology of the nephron, hemodialysis, the role of the kidneys in homeostasis, and the normal and abnormal characteristics of urine.

In Chapter 22, Fluid, Electrolyte, and Acid–Base Balance, students are introduced to fluid compartments and fluid balance, avenues of fluid intake and output, the functions of electrolytes, movements of body fluids, acid–base balance, and acid–base imbalances.

In Chapter 23, The Reproductive Systems, there is a discussion of the structure and functions of the male and female organs of reproduction, male and female puberty, and the female reproductive cycle.

Sexual intercourse, fertilization, embryonic and fetal growth, gestation, parturition and labor, lactation, birth control, and the principles of inheritance are discussed in Development and Inheritance, Chapter 24.

LEARNING AIDS

In order to present the material in the book at an appropriate level and to make it understandable and enjoyable, numerous learning aids that proved successful in the first edition have been maintained.

STUDENT OBJECTIVES

Each chapter opens with a comprehensive list of Student Objectives. Each objective describes a knowledge or skill students should acquire while studying the chapter. (See Note to the Student for an explanation of how the objectives can be used.)

A LOOK AHEAD

Each chapter contains a topical outline of the material covered.

EXHIBITS

Health-science students are generally expected to learn a great deal about the anatomy of certain organ systems. In order to avoid interrupting the discussion of concepts and to organize the data, anatomical details have been presented in tabular form in exhibits, most of which are accompanied by illustrations. There are also summary exhibits in relation to physiological principles. The number of exhibits has been increased in the second edition.

PHONETIC PRONUNCIATIONS

Throughout the text, phonetic pronunciations are provided in parentheses for selected anatomical and physiological terms. These pronunciations are given at the point where the terms are introduced and are repeated in the Glossary of Terms. The Note to the Student on page xxiii explains the pronunciation key.

COMMON DISORDERS

Abnormalities of structure or function are grouped at the end of appropriate chapters. These sections provide a review of normal body processes as well as demonstrate the importance of the study of anatomy and physiology to a career in any of the health fields.

MEDICAL TERMINOLOGY AND CONDITIONS

Glossaries of selected medical terms and conditions appear at the end of appropriate chapters.

STUDY OUTLINE

A Study Outline at the end of each chapter provides a brief summary of major topics. This section consolidates the essential points covered in the chapter so that students can recall and relate the points to one another. Page numbers in the outlines make it easier to locate topics within chapters.

REVIEW QUESTIONS

Review Questions at the end of each chapter provide a check to see if the objectives stated at the beginning of the chapter have been met. New to the second edition is the addition of page numbers to the review questions. This will help students locate the responses more quickly.

GLOSSARY

Two glossaries appear at the end of the book. The first deals with prefixes, suffixes, and combining forms. The second is a comprehensive glossary of terms.

SELECTED READINGS

The bibliography lists current suggested readings at the end of the book.

INSIDE COVER MATERIAL

Symbols and Abbreviations is an alphabetical list of commonly encountered medical symbols and abbreviations. Eponyms contains a listing of eponyms and their current terminology used in the textbook. Both elements are now inside the covers, rather than appendixes as they were in the first edition.

SUPPLEMENTS

The following supplementary items are avilable to accompany *Introduction to the Human Body,* Second Edition.

TEST BANK

The complimentary Test Bank contains numerous questions for each of the 24 chapters in the book. The test bank is available in standard printed format as well as on HarperTest for use with IBM, Apple, or Macintosh computers.

INSTRUCTOR'S MANUAL

Each chapter in the complimentary Instructor's Manual consists of a chapter overview, a list of instruction concepts relating to the chapter, suggested problem-solving essay questions, and lists of audiovisual materials relating to the chapter topic.

TRANSPARENCIES

Seventy-five full-color transparencies are available to adopters. Illustrations from the text that are often shown and discussed in class are the subjects for the transparencies, and special care has been taken to make them clear and usable as overhead projections.

STUDENT STUDY GUIDE

New for the second edition, this valuable supplement contains study reviews, activities, and self-tests corresponding to each chapter of the text.

EXPLORING THE HUMAN BODY

A Laboratory Manual by Patricia J. Donnelly and George A. Wistreich, both of East Los Angeles College, is available for use in a coordinating laboratory course. An Instructor's Manual and Test Bank for the laboratory manual are also available.

SOFTWARE

Several software packages are available to amplify the text material. Please contact your local HarperCollins representative for information.

ACKNOWLEDGMENTS

I wish to thank the following people for their helpful contributions to this edition. Harriett Prentiss read the entire manuscript and recommended cuts and simplifications throughout the text. Jean Edwards, of Centennial College of Applied Arts and Technology

(Ontario), read and evaluated the first draft of the manuscript and advised on the appropriateness of the coverage for nonmajors and those fulfilling a general-science requirement. A special thanks must, of course, go to Barbara Brehm of Smith College, who contributed the Wellness boxes and whose work considerably enhances the practical aspect of the text.

The following reviewers commented on the entire manuscript and made many helpful suggestions:

Roger B. Corzine
Odessa College

Jean A. Edwards
Centennial College of Applied Arts and Technology

Dale M. Featherston
Germanna Community College

Alice W. Glover
Germanna Community College

Ann M. Gunkel
College of Mount Saint Joseph

David Houghton
Cambrian College

Gary G. Kwiecinski
University of Scranton

Kathleen P. Lee
The University of Vermont

Michael C. Mahaney
University of Guelph

Michael J. Power
Sir Sandford Fleming College

Leba Sarkis
Aims Community College

Jane Schneider
Westchester Community College

Finally, for typing and reading all drafts of the manuscript and numerous other duties associated with the task of putting together a textbook, thanks to Geraldine C. Tortora.

I should like to invite all readers and users of the book to send their reactions and suggestions to me so that plans can be formulated for subsequent editions.

Gerard J. Tortora
Natural Sciences and Mathematics, S229
Bergen Community College
400 Paramus Road
Paramus, NJ 07652

Note to the Student

At the beginning of each chapter is a listing of Student Objectives. Before you read the chapter, please read the objectives carefully. Each objective is a statement of a skill or knowledge that you should acquire. To meet these objectives, you will have to perform several activities. Obviously, you must read the chapter carefully. If there are sections of the chapter that you do not understand after one reading, you should reread those sections before continuing. In conjunction with your reading, pay particular attention to the figures and exhibits; they have been carefully coordinated with the textual narrative. A Look Ahead, also found at the beginning of each chapter, provides a brief overview of the material to be covered.

The illustrations in the text have been carefully developed to help you understand the text. With nearly every figure you will find some reinforcement questions. As you stop to look at the illustrations, you should carefully consider each of the questions asked. If you are able to answer them, you should feel confident that you understand the concepts being illustrated and discussed in the text. If not, a closer look at the illustration and another reading of the text will be helpful before moving on to the next part of the chapter.

At the end of each chapter are two, and sometimes three, other learning guides that you may find useful. The first, Study Outline, is a concise summary of important topics discussed in the chapter. This section is designed to consolidate the essential points covered in the chapter, so that you may recall and relate them to one another. The second guide, Review Questions, is a series of questions designed specifically to help you master the objectives. After you have answered the review questions, you should return to the beginning of the chapter and reread the objectives to determine whether you have achieved your goals. Both the Study Outline and the Review Questions are page referenced so that you can quickly find the topic in the chapter for further study, if needed. A third aid, Medical Terminology and Conditions, appears in some chapters. This is a listing of terms or conditions designed to build your medical vocabulary.

As a further aid, we have included pronunciations for many terms that may be new to you. These appear in parentheses immediately following the new words, and they are repeated in the glossary of terms at the back of the book. (Of course, since there will always be some disagreement among medical personnel and dictionaries about pronunciation, you will come across variations in different sources.) Look at the words carefully and say them out loud several times. Learning to pronounce a new word will help you remember it and make it a useful part of your medical vocabulary. Take a few minutes now to read the following pronunciation key, so it will be familiar as you encounter new words. The key is repeated at the beginning of the glossary of terms.

PRONUNCIATION KEY

1. The strongest accented syllable appears in capital letters, for example, bilateral (bī-LAT-er-al) and diagnosis (dī-ag-NŌ-sis).
2. If there is a secondary accent, it is noted by a single quote mark ('), for example, constitution (kon'-sti-TOO-shun) and physiology (fiz'-ē-OL-ō-jē). Any additional secondary accents are also noted by a single quote mark, for example, decarboxylation (dē'-kar-bok'-si-LĀ-shun).
3. Vowels marked with a line above the letter are pronounced with the long sound as in the following common words.

   ā as in *māke*
   ē as in *bē*
   ī as in *īvy*
   ō as in *pōle*

4. Vowels not so marked are pronounced with the short sound as in the following words.

   e as in *bet*
   i as in *sip*
   o as in *not*
   u as in *bud*

5. Other phonetic symbols are used to indicate the following sounds:

   a as in *above*
   oo as in *sue*
   yoo as in *cute*
   oy as in *oil*

Introduction to the Human Body

Organization of the Human Body

STUDENT OBJECTIVES

1. Describe how the human body is organized.
2. List and define the principal systems that make up the human body.
3. Define the life processes of humans.
4. Compare the common and anatomical names for various regions of the human body.
5. List by name and location the principal body cavities and the organs contained within them.
6. Define homeostasis and describe its importance in health and disease.
7. Explain how the regulation of blood pressure and blood sugar are examples of homeostasis.

A LOOK AHEAD

ANATOMY AND PHYSIOLOGY
  DEFINED
LEVELS OF ORGANIZATION
LIFE PROCESSES
ANATOMICAL POSITION
DIRECTIONAL TERMS
PLANES AND SECTIONS
BODY CAVITIES
ABDOMINOPELVIC REGIONS AND
  QUADRANTS
HOMEOSTASIS: MAINTAINING
  PHYSIOLOGICAL LIMITS
    Stress and Homeostasis
    Homeostasis of Blood Pressure (BP)
    Homeostasis of Blood Sugar (BS)
    Level

Y ou are about to begin a study of the human body so that you can learn how it is organized and how it functions. In order to understand what happens to the body when it is injured, diseased, or placed under extreme stress, you must first have a basic understanding of how the body is organized, how its different parts normally work, and the various conditions that will affect the operation of its different parts in order to maintain health and life.

In this chapter, you will be introduced to the various systems that compose the human body. You will also learn how the various systems generally cooperate with each other to maintain the health of the body as a whole. In later chapters, when you have studied body systems in some detail, it will be pointed out how body systems interact to keep you healthy.

ANATOMY AND PHYSIOLOGY DEFINED

To understand the structures and functions of the human body, we study the sciences of anatomy and physiology. *Anatomy* (a-NAT-o-mē; *anatome* = to dissect) refers to the study of *structure* and the relationships among structures. *Physiology* (fiz'-ē-OL-ō-jē) deals with *functions* of the body parts, that is, how the body parts work. Since function can never be separated completely from structure, you will learn about the human body by studying anatomy and physiology together. You will see how each structure of the body is designed to carry out a particular function and how the structure of a part often determines the functions it performs. For example, the hairs lining the nose filter air that we inhale. The fused bones of the skull protect the brain, whereas the spaces between long bones in the limbs permit various types of movement. The external ear is shaped in such a way as to collect sound waves, which results in hearing. The air sacs of the lungs are so thin that oxygen and carbon dioxide readily pass between the sacs and the blood. Body functions, in turn, often influence the size, shape, and health of the structures.

LEVELS OF ORGANIZATION

The human body consists of several levels of structural organization (Figure 1-1). The lowest level of organization, the *chemical level,* includes all chemical substances essential for maintaining life. Chemicals are made up of atoms, and certain of these, such as carbon (C), hydrogen (H), oxygen (O), nitrogen (N), calcium (Ca), potassium (K), and sodium (Na), are essential for maintaining life. Atoms combine to form molecules. Familiar examples of molecules are proteins, carbohydrates, fats, and vitamins.

Chemicals, in turn, combine to form the next higher level of organization: the *cellular level. Cells* are the basic structural and functional units of an organism. Among the many kinds of cells in your body are muscle cells, nerve cells, and blood cells. Figure 1-1 shows several cells from the lining of the stomach. Each has a different structure, and each performs a different function.

The third higher level of organization is the *tissue level. Tissues* are groups of similar cells that, together, perform a particular function. The cells in Figure 1-1 form a tissue called epithelium which lines the stomach. Each cell has a specific function. Mucous cells produce mucus to protect the stomach lining; parietal cells produce acid; and zymogenic cells produce enzymes to digest food. Other examples of tissues in your body are muscle tissue, connective tissue, and nervous tissue.

When different kinds of tissues join together, they form the next higher level of organization: the *organ level. Organs* usually have a recognizable shape, specific functions, and are composed of two or more different tissues. Examples of organs are the heart, liver, lungs, brain, and stomach. Figure 1-1 shows several tissues that make up the stomach. The serosa is a layer of connective tissue and epithelial

3

FIGURE 1-1 Levels of organization in the human body.

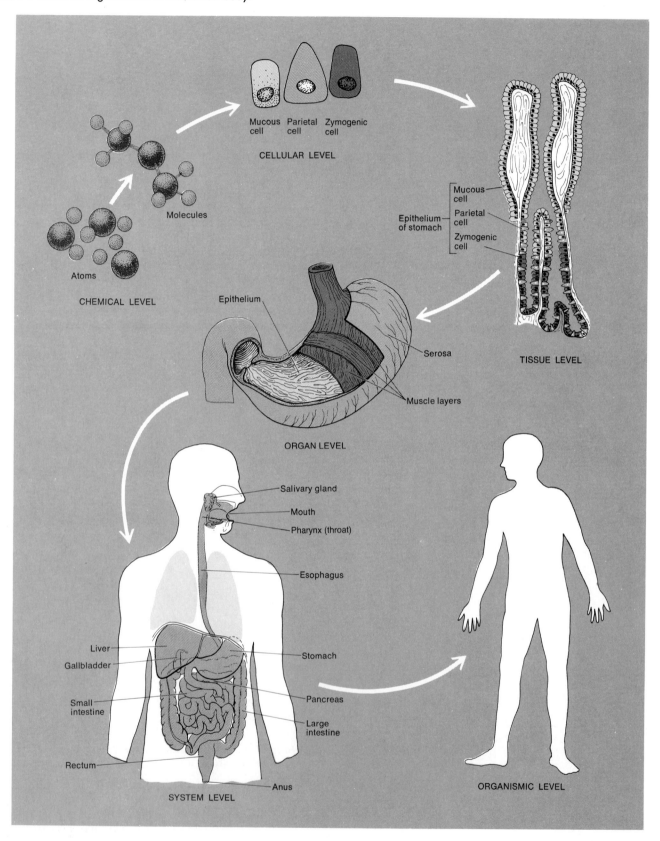

Which level of structural organization is composed of two or more tissues and has a usually recognizable shape?

tissue around the outside of the stomach which protects it. The muscle tissue layers of the stomach are under the serosa and contract to mix food and pass it on. The epithelial tissue layer lining the stomach produces mucus, acid, and enzymes.

The fifth highest level of organization is the *system level.* A *system* consists of related organs that have a common function. The digestive system, which functions in the breakdown and absorption of food, is composed of these organs: mouth, salivary glands, pharynx (throat), esophagus, stomach, small intestine, large intestine, rectum, liver, gallbladder, and pancreas.

The highest level of organization is the *organismic level.* All the systems of the body functioning with one another make up the total *organism*—one living individual.

The chapters that follow take up the anatomy and physiology of each of the body systems. Exhibit 1-1 lists these systems.

LIFE PROCESSES

All living organisms have certain characteristics that set them apart from nonliving things. Following are the important life processes of humans:

1. **Metabolism.** Metabolism is the sum of all the chemical reactions that occur in the body. One phase of metabolism, called *catabolism,* provides the energy needed to sustain life by breaking down substances, food molecules, for example. The other phase of metabolism, called *anabolism,* uses the energy from catabolism to make various substances that form body structures and enable them to function.
2. **Excitability.** Excitability is the ability to sense change within

---

EXHIBIT 1-1
Principal Systems of Human Body,
Representative Organs, and Functions

1. INTEGUMENTARY
Definition: Skin and structures derived from it, such as hair, nails, and sweat and oil glands.
Function: Helps regulate body temperature, protects the body, eliminates wastes, helps make vitamin D, receives certain stimuli such as temperature, pressure, and pain.
Reference: See Figure 5-1.

2. SKELETAL
Definition: All the bones of the body, their associated cartilages and joints.
Function: Supports and protects the body, assists with body movements, houses cells that produce blood cells, stores minerals.
Reference: See Figure 6-5.

3. MUSCULAR
Definition: Specifically refers to skeletal muscle tissue (other muscle tissues include visceral and cardiac).
Function: Participates in bringing about movement, maintains posture, produces heat.
Reference: See Figure 8-8.

4. NERVOUS
Definition: Brain, spinal cord, nerves, and sense organs, such as the eye and ear.
Function: Regulates body activities through nerve impulses.
Reference: See Figures 10-1 and 10-5.

5. ENDOCRINE
Definition: All glands and tissues that produce hormones.
Function: Regulates body activities through hormones transported by the blood of the cardiovascular system.
Reference: See Figure 13-1.

6. CARDIOVASCULAR
Definition: Blood, heart, and blood vessels.
Function: Distributes oxygen and nutrients to cells, carries carbon dioxide and wastes from cells, helps maintain the acid–base balance of the body, protects against disease, prevents hemorrhage by forming blood clots, helps regulate body temperature.
Reference: See Figures 15-1 and 16-8.

7. LYMPHATIC AND IMMUNE
Definition: Lymph, lymphatic vessels, and structures or organs containing lymphatic tissue (large numbers of white blood cells called lymphocytes), such as the spleen, thymus gland, lymph nodes, and tonsils.
Function: Returns proteins and plasma to the cardiovascular system, transports fats from the gastrointestinal tract to the cardiovascular system, filters body fluid, produces white blood cells, protects against disease.
Reference: See Figure 17-3.

8. RESPIRATORY
Definition: Lungs and associated passageways such as the pharynx (throat), larynx (voice box), trachea (windpipe), and bronchial tubes leading into and out of them.
Function: Supplies oxygen, eliminates carbon dioxide, helps regulate the acid–base balance of the body.
Reference: See Figure 18-1.

9. DIGESTIVE
Definition: Gastrointestinal tract and associated organs such as the salivary glands, liver, gallbladder, and pancreas.
Function: Breaks down and absorbs food for use by cells, eliminates solid and other wastes.
Reference: See Figure 19-1.

10. URINARY
Definition: Kidneys, ureters, urinary bladder, and urethra that, together, produce, collect, and eliminate urine.
Function: Regulates the chemical composition of blood, eliminates wastes, regulates fluid and electrolyte balance and volume, helps maintain the acid–base balance of the body, helps regulate red blood cell count.
Reference: See Figure 21-1.

11. REPRODUCTIVE
Definition: Organs (testes and ovaries) that produce reproductive cells (sperm and ova) and other organs that transport and store reproductive cells (vagina, uterine tubes, uterus, penis).
Function: Reproduces the organism.
Reference: See Figures 23-1 and 23-7.

and around us. We continually respond to stimuli such as light, heat, chemicals, and pain in order to make adjustments that maintain health.

3. **Conductivity.** Conductivity refers to the ability of cells to communicate the effect of a stimulus from one part of a cell to another. Nerve and muscle cells are highly conductive.

4. **Contractility.** Contractility is the capacity of cells to generate force to shorten (contract) for movement. Muscle fibers (cells) exhibit a high degree of contractility.

5. **Growth.** Growth refers to an increase in size. It may be due to an increase in the size of existing cells, the number of cells, or the amount of substance surrounding cells.

6. **Differentiation.** Differentiation is the process whereby unspecialized cells become specialized cells. Specialized cells differ in structure and function from the cells from which they originated. For example, following the union of a sperm and ovum, the fertilized egg undergoes tremendous differentiation and progresses through various stages into a unique individual who is similar to, yet quite different from, either of the parents (Chapter 24).

7. **Reproduction.** Reproduction refers to either the formation of new cells for growth, repair, or replacement or the production of a new individual.

ANATOMICAL POSITION

In anatomy, there is universal agreement that descriptions of the human body assume the body is in a specific position, called the *anatomical position*. In the anatomical position, the subject is standing upright, facing the observer, with the upper extremities (limbs) placed at the sides, the palms turned forward, and the feet flat on the floor (Figure 1-2). The common and anatomical terms for several body regions are also presented in Figure 1-2.

DIRECTIONAL TERMS

To locate various body structures in relationship to each other, anatomists use certain *directional terms*. Directional terms are defined in Exhibit 1-2; the examples given there are shown in Figures 1-3 and 1-7. Study the exhibit and figures together.

PLANES AND SECTIONS

The human body may also be described in terms of the *planes* (imaginary flat surfaces) that pass through it (Figure 1-4). A *sagittal* (SAJ-i-tal) *plane* is a vertical plane that divides the body into right and left sides. A *midsagittal plane* passes through the midline of the body and divides the body into *equal* right and left sides. A *parasagittal* (*para* = near) *plane* does not pass through the midline of the body and divides the body into *unequal* left and right portions. A *frontal* (*coronal;* kō-RŌ-nal) *plane* is a vertical plane that divides the body into anterior (front) and posterior (back) portions. Finally, a *horizontal* (*transverse*) *plane* divides the body into superior (upper) and inferior (lower) portions.

EXHIBIT 1-2
Directional Terms. Study This Exhibit with Figures 1-3 and 1-7. Locate the Structures Mentioned in Each Example.

Term	Definition	Example
Superior (Cephalic or Cranial)	Toward the head or the upper part of a structure.	The heart is superior to the liver.
Inferior (Caudal)	Away from the head or toward the lower part of a structure.	The stomach is inferior to the lungs.
Anterior (Ventral)	Nearer to or at the front of the body. In the *prone* position, the body lies anterior side down. In the *supine* position, the body lies anterior side up.	The sternum is anterior to the heart.
Posterior (Dorsal)	Nearer to or at the back of the body.	The esophagus is posterior to the trachea.
Medial	Nearer to the midline of the body or a structure. The *midline* is an imaginary vertical line that divides the body into equal left and right sides.	The ulna is on the medial side of the forearm.
Lateral	Farther from the midline of the body or a structure.	The ascending colon of the large intestine is lateral to the urinary bladder.
Intermediate	Between two structures.	The ring finger is intermediate between the little and middle fingers.
Proximal	Nearer to the attachment of an extremity (limb) to the trunk or a structure; nearer to the point of origin.	The humerus is proximal to the radius.
Distal	Farther from the attachment of an extremity (limb) to the trunk or a structure; farther from the point of origin.	The phalanges are distal to the carpals (wrist bones).
Superficial	Toward or on the surface of the body.	The muscles of the thoracic wall are superficial to the viscera in the thoracic cavity (see Figure 1-7).
Deep	Away from the surface of the body	The ribs are deep to the skin of the chest (see Figure 1-7).

When you study a body structure, you will often view it in section, meaning that you look at only one surface of the three-dimensional structure. Figure 1-5 shows three different sections through the brain.

FIGURE 1-2 Anatomical position. Most body parts have both common and anatomical terms (in parentheses). (a) Anterior view. (b) Posterior view.

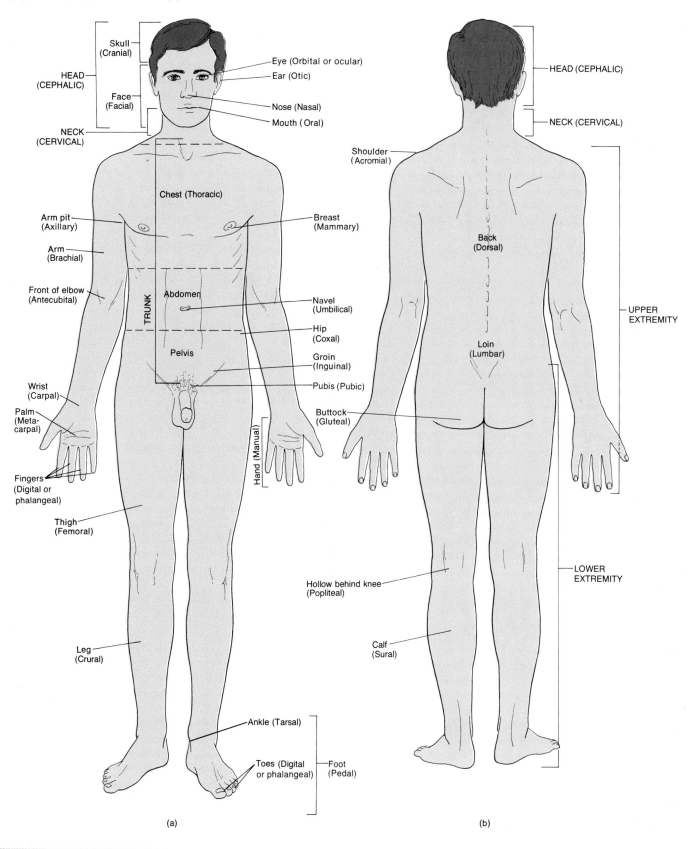

(a)

(b)

What are the four characteristics of the anatomical position?

FIGURE 1-3 Anatomical and directional terms. Study Exhibit 1-2 with this figure to understand the directional terms: *superior, inferior, anterior, posterior, medial, lateral, intermediate, proximal,* and *distal.*

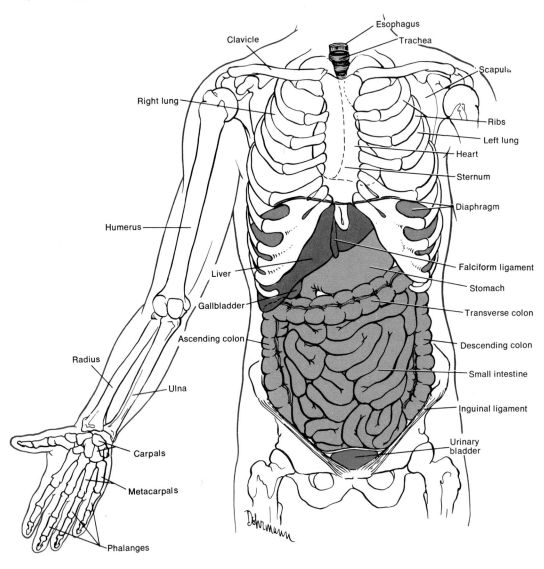

Is the radius proximal to the humerus? Is the falciform ligament closer to the midline than the gallbladder? Is the esophagus anterior to the trachea? Are the ribs superficial to the liver? Is the urinary bladder inferior to the transverse colon?

BODY CAVITIES

Spaces within the body that contain internal organs are called **body cavities.** Figure 1-6 shows the two principal body cavities: dorsal and ventral. The **dorsal body cavity** is located near the posterior (back) surface of the body. It is composed of a **cranial cavity,** which is formed by the cranial (skull) bones and contains the brain, and a **vertebral (spinal) canal,** which is formed by the vertebrae of the backbone and contains the spinal cord and the beginnings of spinal nerves.

The **ventral body cavity** is located on the anterior (front) aspect of the body and contains organs collectively called **viscera** (VIS-er-a). Like the dorsal body cavity, the ventral body cavity has two principal subdivisions—an upper portion, called the **thoracic** (thō-RAS-ik) **cavity** (or chest cavity), and a lower portion, called the **abdominopelvic** (ab-dom'-i-nō-PEL-vik) **cavity.** The diaphragm (DĪ-a-fram; *diaphragma* = partition or wall), a dome-shaped sheet of muscle, divides the ventral body cavity into the thoracic and abdominopelvic cavities.

The thoracic cavity contains several divisions. There are two **pleural cavities,** one surrounding each lung (Figure 1-7). Each pleural cavity is a small potential space (not an actual space) between

FIGURE 1-4 Planes of the human body.

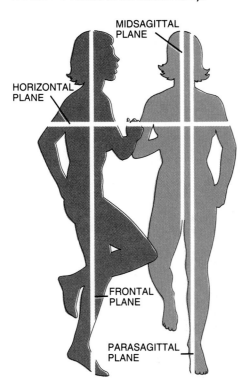

MIDSAGITTAL PLANE

HORIZONTAL PLANE

FRONTAL PLANE

PARASAGITTAL PLANE

FIGURE 1-5 Sections through different parts of the brain. (a) Cross (horizontal or transverse) section. (b) Frontal section. (c) Midsagittal section.

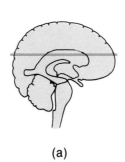
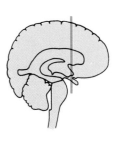
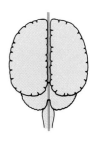

(a)   (b)   (c)

> Which plane would divide the brain into an anterior and posterior portion?

the visceral pleura, which is a membrane that covers the lungs, and the parietal pleura, which is a membrane that lines the wall of the thoracic cavity. The pleural cavities contain a small amount of fluid. The **pericardial** (per'-i-KAR-dē-al; *peri* = around; *cardi* = heart) **cavity** is a small potential space between the visceral pericardium, which is a membrane that covers the heart, and the parietal pericardium, which is a membrane that lines the wall of the thoracic cavity (Figure 1-7). It also contains a fluid. The **mediastinum** (mē'-dē-as-TĪ-num; *medias* = middle; *stare* = stand in) is the mass of tissues between the lungs that extends from the sternum (breastbone) to the vertebral column (Figure 1-7). The mediastinum includes all the structures in the thoracic cavity except the lungs themselves. Among the structures in the mediastinum are the heart, thymus gland, esophagus, and many large blood and lymphatic vessels.

The abdominopelvic cavity, as the name suggests, is divided into two portions, although no specific structure separates them (see Figure 1-6). The membrane lining the cavity and covering the organs within it is called the **peritoneum**. The upper portion, the **abdominal cavity,** contains the stomach, spleen, liver, gallbladder, pancreas, small intestine, and most of the large intestine. The lower portion, the **pelvic cavity,** contains the urinary bladder, sigmoid colon, rectum, and the internal male or female reproductive organs. The pelvic cavity is located between two imaginary planes, which are indicated by broken lines in Figure 1-6a.

A summary of the body cavities is presented in Exhibit 1-3.

EXHIBIT 1-3
Summary of Body Cavities

Cavity	Comments
DORSAL	
Cranial	Formed by cranial bones and contains brain.
Vertebral	Formed by vertebrae and contains spinal cord and beginnings of spinal nerves.
VENTRAL	
Thoracic	Chest cavity; separated from abdominal cavity by diaphragm.
Pleural	Contains lungs.
Pericardial	Contains heart.
Mediastinum	Region between the lungs from the breastbone to backbone that contains heart, thymus gland, esophagus, trachea, bronchi, and many large blood and lymphatic vessels.
Abdominopelvic	Subdivided into abdominal and pelvic cavities.
Abdominal	Contains stomach, spleen, liver, gallbladder, pancreas, small intestine, and most of large intestine.
Pelvic	Contains urinary bladder, cecum, appendix, sigmoid colon, rectum, and internal female and male reproductive organs.

ABDOMINOPELVIC REGIONS AND QUADRANTS

To locate organs easily, the abdominopelvic cavity is divided into the *nine regions* shown in Figure 1-8. The abdominopelvic cavity may also be divided into **quadrants** (*quad* = four). These are shown in Figure 1-9. A horizontal line and a vertical line are passed through the umbilicus (navel). These two lines divide the abdomen into a **right upper quadrant (RUQ), left upper quadrant**

FIGURE 1-6 Body cavities. (a) Location of the dorsal and ventral body cavities in right lateral view. (b) Subdivisions of the ventral body cavity in anterior view.

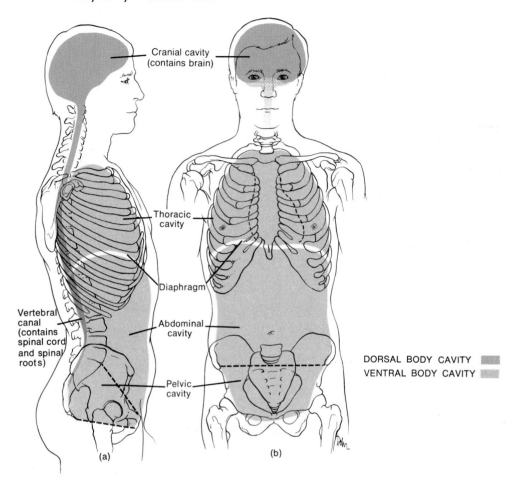

In which cavity (thoracic, abdominal, or pelvic) is each of the following located: Urinary bladder? Stomach? Sigmoid colon? Heart? Pancreas? Small intestine? Lungs? Internal female reproductive organs? Mediastinum? Spleen? Rectum? Liver?

(LUQ), right lower quadrant (RLQ), and left lower quadrant (LLQ). Whereas the nine-region division is more widely used for anatomical studies, clinicians find the four-quadrant division is better suited for locating the site of an abdominopelvic pain, tumor, or other abnormality.

HOMEOSTASIS: MAINTAINING PHYSIOLOGICAL LIMITS

As we have seen, the human body is composed of various systems and organs, each of which consists of millions of cells. These cells need relatively stable conditions in order to function effectively and contribute to the survival of the body as a whole. The maintenance of stable conditions for its cells is an essential function of

the human body which physiologists call homeostasis, one of the major themes of this textbook.

Homeostasis (hō′-mē-ō-STĀ-sis) is a condition in which the body's internal environment remains within certain physiological limits (*homeo* = same; *stasis* = standing still). The internal environment refers to the fluid in which body cells live, called extracellular fluid, discussed in more detail in Chapter 3.

For the body's cells to survive, the composition of the extracellular fluid must be properly maintained at all times. Among the substances in extracellular fluid are gases, nutrients, and electrically charged chemical particles called ions (electrolytes), such as sodium (Na^{2+}) and chloride (Cl^-), each of which has its own normal range that must be maintained as part of the body's overall homeostasis. An organism is said to be in homeostasis when its internal environment contains the appropriate concentration of chemicals,

FIGURE 1-7 **Mediastinum seen in a cross section of the thorax. Some of the structures shown and labeled may be unfamiliar to you now, but they will be discussed in later chapters.**

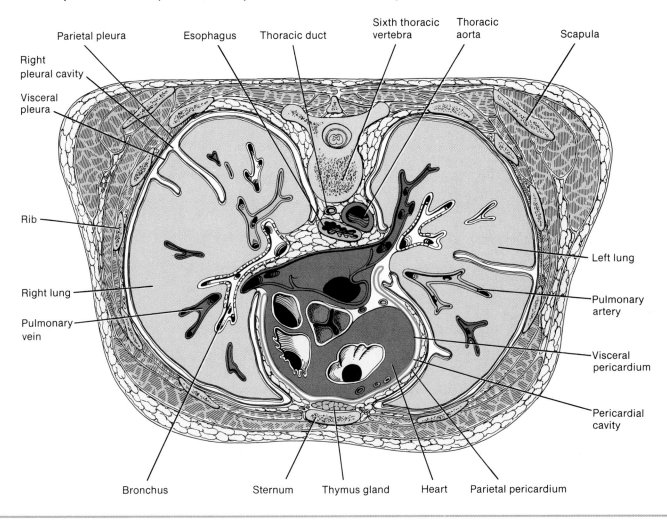

Which of the following structures is contained in the mediastinum: Thymus gland? Right lung? Heart? Esophagus? Scapula?

is at an appropriate temperature, and is at an appropriate pressure. When homeostasis is disturbed, ill health may result. If body fluids are not eventually brought back into homeostasis, death may occur.

STRESS AND HOMEOSTASIS

Homeostasis may be disturbed by *stress,* which is any stimulus that creates an imbalance in the internal environment. The stress may come from the external environment in the form of stimuli such as heat, cold, or lack of oxygen. Or the stress may originate within the body in the form of stimuli such as high blood pressure, tumors, or unpleasant thoughts. Most stresses are mild and routine. Extreme stress might be caused by poisoning, overexposure to temperature extremes, and surgical operations.

Fortunately, the body has many regulating (homeostatic) devices that may bring the internal environment back into balance. Every body structure, from the cellular to the system level, attempts to keep the internal environment within normal physiological limits.

The homeostatic mechanisms of the body are under the control of the nervous system and the endocrine system. The nervous system regulates homeostasis by detecting when the body deviates from its balanced state and then sending messages (nerve impulses) to the proper organs to counteract the stress. The endocrine system is a group of glands that secrete chemical messengers, called hormones, into the blood. Whereas nerve impulses coordinate homeostasis rapidly, hormones work more slowly. Following are two examples of how the nervous and endocrine systems exert homeostatic control.

FIGURE 1-8 Abdominopelvic cavity. (a) The nine regions. (b) Anterior view showing the most superficial organs in the various regions.

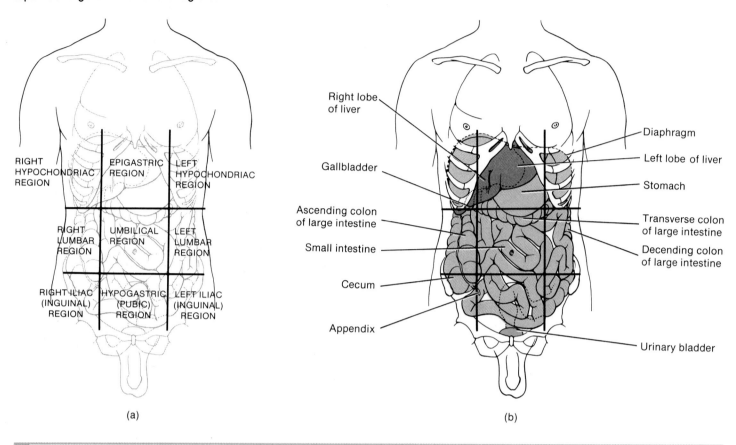

(a)

(b)

In which abdominopelvic region is each of the following found: Most of the liver? Cecum? Transverse colon? Urinary bladder? Gallbladder?

FIGURE 1-9 Quadrants of the abdominopelvic cavity. The two lines intersect at right angles at the umbilicus (navel).

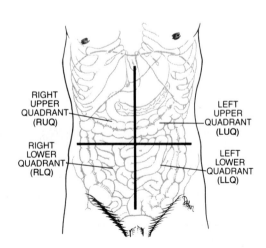

HOMEOSTASIS OF BLOOD PRESSURE (BP)

Blood pressure (BP) is the force of blood as it passes through blood vessels, especially arteries. In order to sustain life, blood must not only be kept circulating, it must also circulate at an appropriate pressure. For example, if blood pressure is too low, organs of the body, such as the brain, will not receive adequate oxygen and nutrients to function properly. High blood pressure, on the other hand, has adverse effects on organs such as the heart, kidneys, and brain. High blood pressure contributes to the development of heart attacks and stroke. Among other factors, blood pressure depends on the rate and strength of the heartbeat. If some stress causes the heartbeat to speed up, the following sequence occurs (Figure 1-10). As the heart pumps faster, it pushes more blood into the arteries, increasing blood pressure. The higher pressure is detected by pressure-sensitive nerve cells in the walls of certain arteries, which respond by sending nerve impulses to the brain. The brain, in turn, responds by sending impulses to the heart to slow the heart rate, thus decreasing blood pressure. The continual monitoring of blood pressure by the nervous system is an attempt to maintain a normal blood pressure and involves what is called a feedback system.

A *feedback system* is any circular situation in which information about the status of something is continually reported (fed back) to a central control region. In the case of regulating blood pressure, the *input* (*stimulus*) is the information (high blood pressure) picked up by the pressure-sensitive nerve cells, and the *output* (*response*) is the return toward normal blood pressure due to decreased heartbeat. Figure 1-10 shows that the system runs in a circle. The pressure-sensitive nerve cells continue to monitor pressure and to feed this information to the brain, even after the return to homeosta-

sis has begun and blood pressure has begun to normalize. In other words, if the pressure is still too high, the brain can continue to send out impulses to slow the heartbeat.

This type of feedback system reverses the direction of the initial condition (here, from a rising to a falling blood pressure) and is called a *negative feedback system.* If, instead, the brain had signaled the heart to beat even faster and the blood pressure had kept on rising, the system would have been a *positive feedback system.* In a positive feedback system, the output (reaction of the body) *intensifies* the input (the stress). Most of the feedback systems of the body are negative, but a few positive ones do occur, such as the increasing rate of uterine contractions in childbirth. Most positive feedback systems are destructive, however, as you will see later.

FIGURE 1-10 Homeostasis of blood pressure. Note that the output is fed back into the system, and the system continues to lower blood pressure until there is a return to homeostasis.

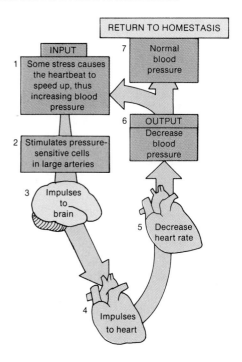

Is this a positive or negative feedback system?

HOMEOSTASIS OF BLOOD SUGAR (BS) LEVEL

An example of homeostatic regulation by hormones is the maintenance of blood sugar (BS) level. Glucose, the sugar found in blood, is not only one of the body's principal sources of energy but is also required in a specific concentration in order to prevent brain damage. Under normal circumstances, the concentration of sugar in the blood averages about 90 milligrams/100 milliliters of blood. This level is maintained primarily by two hormones secreted by the pancreas: insulin and glucagon. Suppose you have just eaten some candy. The sugar in the candy is broken down in the digestive tract and moves into the blood. The sugar then becomes a stress because it raises the blood sugar level above normal. In response to this stress, the cells of the pancreas are stimulated to secrete insulin (see Figure 13-19). When insulin enters the blood, it increases the sugar uptake by cells, which lowers the blood sugar level, and it stimulates the liver and muscles to store more sugar. Thus, even more sugar is removed from the blood.

The other hormone produced by the pancreas, glucagon, has the opposite effect of insulin. Suppose you have not eaten for several hours and your blood sugar level is steadily decreasing. Lack of sugar is now the stress, and under this condition, the pancreas secretes glucagon (see Figure 13-19). Glucagon accelerates the process by which sugar stored in the liver is sent back into the bloodstream. The blood sugar level is thus increased until it returns to normal.

S TUDY OUTLINE

Anatomy and Physiology Defined (p. 3)

1. Anatomy is the study of structure and the relationship among structures.
2. Physiology is the study of how body structures function.

Levels of Structural Organization (p. 3)

1. The human body consists of several levels of organization: chemical, cellular, tissue, organ, system, and organismic.

2. Cells are the basic structural and functional units of an organism.
3. Tissues consist of groups of similarly specialized cells.
4. Organs are structures of definite shape composed of two or more different tissues that have specific functions.
5. Systems consist of associations of organs that have a common function.
6. The human organism is a collection of structurally and functionally integrated systems.

Life-Style, Health, and Homeostasis

Homeostasis is the basis of good health. Knowledge of the physiological processes responsible for the maintenance of homeostasis is essential for understanding the nature of health and disease.

One might view the human body as a large orchestra, and its life processes as the music the orchestra produces. The nervous and endocrine systems act together as the conductor to direct the various orchestral sections to keep all parts blended into a harmonious balance. The quality of the music suffers if even one of the instruments is off key or plays at the wrong time. The study of anatomy and physiology is like learning about the instruments in the orchestra and how they all play together.

The term *homeostasis* does not imply that the body stays the same, that it exists in a static state. Rather, homeostasis is a dynamic balance, a balance among the components of the body and between the body and its environment. Internal and external events demand continuous physiological adjustments. The body is in a continuous state of flux.

Health Behavior

The body's ability to maintain homeostasis gives it tremendous healing power and a remarkable resistance to abuse. But lifelong good health is not something that just happens. Two important factors in this balance we call health are the environment and behavior of the organism in question. Our homeostasis is affected by the air we breathe, the food we eat, and even the thoughts we think.

Disease can result from a disruption of homeostasis, which may be brought on by one's behavior and interaction with the environment. The way we live our lives can either support or interfere with the body's ability to maintain homeostasis. Many diseases are the result of years of poor health behavior. An obvious example is smoking. Smoking tobacco exposes sensitive lung tissue to a multitude of chemicals that cause cancer and damage the lung's ability to repair itself. Since lung cancer is difficult to treat and very rarely cured, it is much wiser to quit

smoking (or never start) than to hope the doctor can fix you up once you are diagnosed with this disease.

The health-care system's power is limited when it comes to treating many life-style diseases such as lung cancer, emphysema, obesity, and heart disease. Health-care professionals face frustration daily in their attempts to treat patients who resist treatment or are unable to change the life-style habits necessary to promote effective healing.

An Ounce of Prevention . . .

The treatment of disease is increasingly costly, and as a result, the United States and Canada are approaching a crisis in escalating health-care costs. We all pay the price of this gigantic health-care bill, both in terms of our own health-care and insurance costs as well as in the higher prices of goods and services as businesses pass on their increased health-care costs to the consumer. Public health officials are beginning to take a long, hard look at how health-care dollars are spent.

Many people equate good health with access to medical technology and expertise. This is symptomatic of the "fix it once it's broken" approach. But as with love, money can't buy health. There is no doubt that modern medicine has saved many lives, and relieved much pain and suffering. However, it is interesting that while the United States spends more per capita on health care than any other country except Sweden, our life expectancy falls below that of 18 other countries. It takes more than money to produce a nation of healthy citizens.

Life-expectancy figures represent an average length of life based on the life expectancy of each person in a defined group. Excess rates of death at any age will lower the average. And excess deaths in infancy and early childhood have an especially strong impact on this statistic. Our rate of infant mortality is relatively high for a developed country, especially among low-income groups. This is one of the most important factors that accounts for our lower life expectancy.

But life expectancy alone isn't the whole health-care story. Our high infant death rate is due partly to high rates of premature birth and poor prenatal health care. While some babies die, others are born with complications that cost enormous amounts both in terms of expensive neonatal health care and emotional trauma.

Consider the value of prenatal care. Counseling pregnant women about the impact of their life-style choices on the developing fetus can help prevent premature births and possible lifelong medical complications for the child. The cost savings of preventing a single premature birth can be equivalent to the cost of prenatal care for hundreds of pregnant women.

Heart Disease

Heart disease provides one of the best illustrations of the connection between life-style and disease. Blood flow to the heart is threatened when the arteries of the heart become clogged with plaque, a process called atherosclerosis, which will be discussed in Chapter 15. Heart attacks and their complications are still the leading cause of premature death in Canada and the United States. But mortality rates from heart disease have declined dramatically over the past 20 years. Experts believe this is due to national improvements in life-style behaviors.

Many factors contribute to the process of atherosclerosis, and some of them are under an individual's control. The three most important controllable factors are smoking, blood cholesterol levels, and blood pressure. People can substantially reduce their risk of heart disease by not smoking, and controlling blood cholesterol and blood pressure levels through appropriate diet, exercise, weight control, and, if necessary, medication. Statistics show that we have improved our control of these variables.

Fewer North Americans are smoking. We're eating less fat, exercising more, and keeping an eye on cholesterol and blood pressure. There is still plenty of room for improvement, but we've shown that life-style can make a difference.

Beyond Health

For many years, the word *health* was used to mean simply an absence of disease. If you were free of medical symptoms requiring a physician's intervention, you were healthy. Many people were uncomfortable with this notion, and felt that there was more to health than simply not being sick. The term *wellness* came into use to indicate that health means more than the presence or absence of disease. And just as life-style can make us sick, it can also help make us well.

Cardiologist George Sheehan, a devout believer in the values of health behavior, once said that when he thought about life expectancy, he preferred to think about what he expected from life. Health is more than the absence of disease. The word *wellness* is often used to mean optimal health and living the sort of life-style that engenders it.

Wellness means taking responsibility for one's health, preventing accidents and illness, knowing when to consult a health-care professional, and working with health-care providers when necessary. Wellness encourages consumer awareness and promotes the establishment of social systems and environments conducive to health-promoting behavior. Paradoxically, disease and disability do not prevent a wellness life-style, for wellness simply means maximizing one's potential for physical, psychological, and social well-being.

7. The systems of the human body are the integumentary, skeletal, muscular, nervous, endocrine, cardiovascular, lymphatic, respiratory, digestive, urinary, and reproductive (see Exhibit 1-1).

Life Processes (p. 5)

1. All living forms have certain characteristics that distinguish them from nonliving things.
2. Among the life processes in humans are metabolism, excitability, conductivity, contractility, growth, differentiation, and reproduction.

Anatomical Position (p. 6)

1. When in the anatomical position, the subject stands erect facing the observer, with arms at the sides and palms turned forward and the feet are flat on the floor.
2. Parts of the body have both anatomical and common names for different regions. Examples include cranial (skull), thoracic (chest), brachial (arm), patellar (knee), cephalic (head), and gluteal (buttock).

Directional Terms (p. 6)

1. Directional terms indicate the relationship of one part of the body to another.
2. Commonly used directional terms are superior (toward the head or upper part of a structure), inferior (away from the head or toward the lower part of a structure), anterior (near or at the front of the body), posterior (near or at the back of the body), medial (nearer the midline of the body or a structure), intermediate (between a medial and lateral structure), proximal (nearer the attachment of an extremity to the trunk or a structure), distal (farther from the attachment of an extremity to the trunk or a structure), superficial (toward or on the surface of the body), and deep (away from the surface of the body).

Planes and Sections (p. 6)

1. Planes are imaginary flat surfaces that are used to divide the body or organs into definite areas. A midsagittal plane is a vertical plane through the midline of the body that divides the body (or an organ) into equal right and left sides. A parasagittal plane is a plane that does not pass through the midline of the body and divides the body or organs into unequal right and left sides. A frontal (coronal) plane divides the body into anterior and posterior portions. A horizontal (transverse) plane divides the body into superior and inferior portions.
2. Sections result from cuts through body structures. They are named according to the plane on which the cut is made and include cross sections, frontal sections, and midsagittal sections.

Body Cavities (p. 8)

1. Spaces in the body that contain internal organs are called cavities.
2. The dorsal and ventral cavities are the two principal body cavities.
3. The dorsal cavity is subdivided into the cranial cavity, which contains the brain, and the vertebral (spinal) canal, which contains the spinal cord and beginnings of spinal nerves.
4. The ventral body cavity is subdivided by the diaphragm into an upper thoracic cavity and a lower abdominopelvic cavity.
5. The thoracic cavity contains two pleural cavities and a mediastinum, which includes the pericardial cavity.
6. The mediastinum is a mass of tissues between the lungs that extends from the sternum to the vertebral column. It contains all the structures of the thoracic cavity, except the lungs.
7. The abdominopelvic cavity is divided into a superior abdominal and an inferior pelvic cavity. No specific structure divides them.
8. Viscera (organs) of the abdominal cavity include the stomach, spleen, pancreas, liver, gallbladder, small intestine, and most of the large intestine.
9. Viscera of the pelvic cavity include the urinary bladder, sigmoid colon, rectum, and internal female and male reproductive structures.

Abdominopelvic Regions and Quadrants (p. 9)

1. To describe the location of organs easily, the abdominopelvic cavity may be divided into nine regions.
2. The names of the nine abdominopelvic regions are epigastric, right hypochondriac, left hypochondriac, umbilical, right lumbar, left lumbar, hypogastric (pubic), right iliac (inguinal), and left iliac (inguinal).
3. The abdominopelvic cavity may also be divided into quadrants by passing an imaginary horizontal and vertical line through the umbilicus.
4. The names of the abdominopelvic quadrants are right upper quadrant (RUQ), left upper quadrant (LUQ), right lower quadrant (RLQ), and left lower quadrant (LLQ).

Homeostasis: Maintaining Physiological Limits (p. 10)

1. Homeostasis is a condition in which the internal environment (extracellular fluid) of the body remains within certain physiological limits in terms of chemical composition, temperature, and pressure.
2. All body systems attempt to maintain homeostasis.
3. Homeostasis is controlled mainly by the nervous and endocrine systems.

Stress and Homeostasis (p. 11)

1. Stress is an external or internal stimulus that creates a change in the internal environment.
2. If a stress acts on the body, homeostatic mechanisms attempt to counteract the effects of the stress and bring the condition back to normal.

Homeostasis of Blood Pressure (BP) (p. 12)

1. Blood pressure (BP) is the force exerted by blood as it passes through blood vessels, especially arteries.
2. If a stress causes the heartbeat to increase, blood pressure also increases. Pressure-sensitive nerve cells in certain arteries inform the brain, and the brain responds by sending impulses that decrease heartbeat, thus decreasing blood pressure back to normal.
3. Any circular situation in which information about the status

of something is continually fed back to a control region is called a feedback system.

4. A negative feedback system is one in which the reaction of the body (output) counteracts the stress (input) in order to maintain homeostasis; most feedback systems of the body are negative.

5. A positive feedback system is one in which the output intensifies the input; a positive feedback system is usually destructive, but a few are beneficial, such as contractions in childbirth.

Homeostasis of Blood Sugar (BS) Level (p. 13)

1. A normal blood sugar (BS) level is maintained by the actions of two different pancreatic hormones: insulin and glucagon.

2. Insulin lowers blood sugar level by increasing sugar uptake by cells and accelerating sugar storage as glycogen in the liver and skeletal muscles.

3. Glucagon raises blood sugar level by accelerating the rate of sugar released from glycogen by the liver.

REVIEW QUESTIONS

1. Define anatomy and physiology. (p. 3)
2. Give several examples of how structure and function are related. (p. 3)
3. Define each of the following terms: cell, tissue, organ, system, and organism. (pp. 3, 5)
4. Using Exhibit 1-1 as a guide, outline the functions of each system of the body, and list several organs that compose each system. (p. 5)
5. List and define the life processes of humans. (pp. 5, 6)
6. Define the anatomical position. Why is the anatomical position used? (p. 6)
7. Review Figure 1-2. Locate each region on your own body, and name each by its common and anatomical terms. (p. 7)
8. What is a directional term? Why are these terms important? (p. 6) Use each of the directional terms listed in Exhibit 1-2 in a complete sentence without using the same examples given in the exhibit. (p. 6)
9. Using Figure 1-3 as a guide, answer the following:
   a. Is the stomach superior to the urinary bladder?
   b. Are the carpals distal to the radius?
   c. Are the ribs superficial to the heart?
   d. Are the lungs lateral to the heart? (p. 8)
10. Define the various planes that may be passed through the body. Explain how each plane divides the body. Describe the meaning of cross section, frontal section, and midsagittal section. (p. 6)
11. Define a body cavity. List the body cavities discussed, and tell which major organs are located in each. What landmarks separate the various body cavities from one another? What is the mediastinum? (p. 8)
12. Describe how the abdominopelvic area is subdivided into nine regions. Name and locate each region and list the organs, or parts of organs, in each, using Figure 1-8 as a guide. (p. 9)
13. Describe how the abdominopelvic cavity is divided into quadrants, and name each quadrant. (pp. 9–10)
14. Define homeostasis. What is extracellular fluid (ECF)? (p. 10)
15. Under what conditions is the internal environment said to be in homeostasis? (p. 10)
16. What is a stress? Give several examples. How is stress related to homeostasis? (p. 11)
17. Discuss briefly how the regulation of blood pressure (BP) and blood sugar (BS) level are examples of homeostasis. (pp. 12–13)
18. Define a feedback system. Distinguish between a negative and a positive feedback system. (p. 13)

2

Introductory Chemistry

A LOOK AHEAD

INTRODUCTION TO BASIC CHEMISTRY
  Chemical Elements
  Structure of Atoms
  Atoms and Molecules
    Ionic Bonds
    Covalent Bonds
    Hydrogen Bonds
  Chemical Reactions
    Synthesis Reactions—Anabolism
    Decomposition Reactions—
      Catabolism
    Energy and Chemical Reactions
CHEMICAL COMPOUNDS AND LIFE
  PROCESSES
  Inorganic Compounds
    Water
    Inorganic Acids, Bases, and Salts
    Acid–Base Balance: The Concept
      of pH
    Maintaining pH: Buffer Systems
  Organic Compounds
    Carbohydrates
    Lipids
    Proteins
    Nucleic Acids:
      Deoxyribonucleic Acid (DNA)
      and Ribonucleic Acid (RNA)
    Adenosine Triphosphate (ATP)

Many common substances we eat and drink—water, sugar, table salt, cooking oil—play vital roles in keeping us alive. In this chapter, you will learn how these substances function in your body. That requires a knowledge of basic chemistry and chemical processes since your body is composed of chemicals and all body activities are chemical in nature.

INTRODUCTION TO BASIC CHEMISTRY

CHEMICAL ELEMENTS

All living and nonliving things consist of **matter,** which is anything that occupies space and has mass. Matter may exist as a solid, liquid, or gas. All forms of matter are made up of building units called **chemical elements,** substances that cannot be broken down into simpler substances by ordinary chemical reactions. At present, scientists recognize 109 different elements. Elements are designated by letter abbreviations called **chemical symbols:** H (hydrogen), C (carbon), O (oxygen), N (nitrogen), Na (sodium), K (potassium), Fe (iron), and Ca (calcium).

Approximately 26 elements are found in the human organism. Oxygen, carbon, hydrogen, and nitrogen make up about 96 percent of the body's weight. These four elements together with calcium and phosphorus constitute approximately 99 percent of the total body weight. Twenty other chemical elements found in the human body are called **trace elements** because they are found in low concentrations and compose the remaining 1 percent (Exhibit 2-1).

STRUCTURE OF ATOMS

Each element is made up of **atoms,** the smallest units of matter. An element is composed of all the same type of atoms. A sample of the element carbon, such as pure coal, contains only carbon atoms. A tank of oxygen contains only oxygen atoms.

An atom consists of two basic parts: the nucleus and the electrons (Figure 2-1). The centrally located **nucleus** contains positively charged particles called **protons** (p^+) and uncharged (neutral) particles called **neutrons** (n^0). Because each proton has one positive charge, the nucleus itself is positively charged. **Electrons** (e^-) are negatively charged particles that move around the nucleus. The number of

FIGURE 2-1 Structure of an atom. In this highly simplified version of a carbon atom, note the centrally located nucleus. The nucleus contains six neutrons and six protons, although all are not visible in this view. The six electrons move about the nucleus at varying distances from its center.

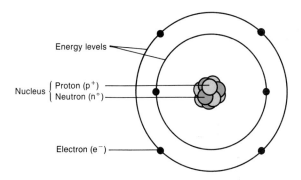

Energy levels

Nucleus { Proton (p^+) / Neutron (n^0)

Electron (e^-)

What is the atomic number of this atom?

19

EXHIBIT 2-1

Representative Chemical Elements Found in the Body

Chemical Element (Symbol)	Percentage of Total Body Weight	Comment
Oxygen (O)	65.0	Constituent of water and organic molecules; functions in cellular respiration.
Carbon (C)	18.5	Found in every organic molecule.
Hydrogen (H)	9.5	Constituent of water, all foods, and organic molecules; as an ion (H^+), it contributes to acidity.
Nitrogen (N)	3.2	Component of all protein and nucleic acid molecules.
Calcium (Ca)	1.5	Constituent of bone and teeth; required for blood clotting, intake (endocytosis) and output (exocytosis) of substances through plasma membranes, motility of cells, movement of chromosomes prior to cell division, glycogen metabolism, synthesis and release of neurotransmitters, and contraction of muscle.
Phosphorus (P)	1.0	Component of many proteins, nucleic acids, ATP, and cyclic AMP; required for normal bone and tooth structure; found in nerve tissue.
Potassium (K)	0.4	Required for growth and important in conduction of nerve impulses and muscle contraction.
Sulfur (S)	0.3	Component of many proteins, especially the contractile proteins of muscle.
Sodium (Na)	0.2	A cation (positively charged ion) of NaCl; structural component of bone; essential in blood to maintain water balance; needed for conduction of nerve impulses and muscle contraction.
Chlorine (Cl)	0.2	An anion (negatively charged ion) of NaCl, a salt important in water movement between cells.
Magnesium (Mg)	0.1	Component of many enzymes.
Iodine (I)	0.1	Vital to production of hormones by thyroid gland.
Iron (Fe)	0.1	Essential component of hemoglobin and some respiratory enzymes.

Chromium (Cr)	Fluorine (F)	Selenium (Se)	Vanadium (V)
Cobalt (Co)	Manganese (Mn)	Silicon (Si)	Zinc (Zn)
Copper (Cu)	Molybdenum (Mo)	Tin (Sn)	

These elements, as well as a few others, are called trace elements because they are required in low concentrations.

☐ compose about 96 percent of total body weight
☐ compose about 3 percent of total body weight
☐ compose about 1 percent of total body weight

electrons in an atom always equals the number of protons. Since each electron carries one negative charge, the negatively charged electrons and the positively charged protons balance each other, and the atom is electrically neutral.

What makes the atoms of one element different from those of another? The answer is the number of protons. Figure 2-2 shows that a hydrogen atom contains one proton. A helium atom contains two. A carbon atom has six, and so on. Each different kind of atom has a different number of protons in its nucleus. The number of protons in an atom is called the atom's **atomic number.** Therefore, each kind of atom, or element, has a different atomic number. The total number of protons and neutrons in an atom is its **atomic weight.** Thus, an atom of sodium has an atomic weight of 23 because there are 11 protons and 12 neutrons in its nucleus.

ATOMS AND MOLECULES

When atoms combine with or break apart from other atoms, a **chemical reaction** occurs. In the process, new products with different properties are formed. Chemical reactions are the foundation of all life processes.

It is the electrons of an atom that participate in chemical reactions. The electrons move around the nucleus in regions (shown in Figure 2-2 as concentric circles) called **energy levels.** Each energy level has a maximum number of electrons it can hold. For instance, the energy level nearest the nucleus never holds more than two electrons, no matter what the element. This energy level is referred to as the first energy level. The second energy level never holds

more than eight electrons. The third energy level can hold a maximum of eight electrons if the atom's atomic number is less than 20. If the atom's atomic number is greater than 20, the third energy level can hold a maximum of 18 electrons.

An atom always attempts to fill its outermost energy level with the maximum number of electrons it can hold. To do this, the atom may give up an electron, take on an electron, or share an electron with another atom—whichever is easier. The **valence** (combining capacity) is the number of extra or deficient electrons in the outermost energy level. Look at the chlorine atom in Figure 2-2. Its outermost energy level, which happens to be the third level, has seven electrons. Since the third level of an atom can hold a maximum of eight electrons, chlorine can be described as having a shortage of one electron. In fact, chlorine usually does try to pick up an extra electron (see Figure 2-3b). Sodium, by contrast, has only one electron in its outer level (Figure 2-2). This again happens to be the third energy level. It is much easier for sodium to get rid of the one electron than to fill the third level by taking on seven more electrons (see Figure 2-3a). Atoms of a few elements, like helium, have completely filled outer energy levels and do not need to gain or lose electrons. These are called **inert elements** and do not enter into chemical reactions.

Atoms with incompletely filled outer energy levels, like sodium and chlorine, tend to combine with other atoms in a chemical reaction. During the reaction, the atoms can trade off or share electrons and thereby fill their outer energy levels.

When two or more atoms combine in a chemical reaction, the resulting combination is called a **molecule** (MOL-e-kyool). A mole-

FIGURE 2-2 **Atomic structures of some representative atoms.**

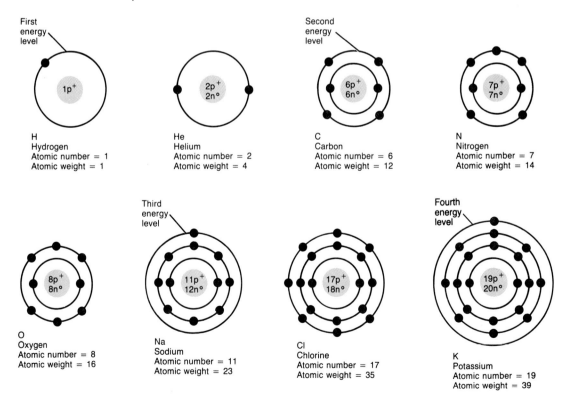

H
Hydrogen
Atomic number = 1
Atomic weight = 1

He
Helium
Atomic number = 2
Atomic weight = 4

C
Carbon
Atomic number = 6
Atomic weight = 12

N
Nitrogen
Atomic number = 7
Atomic weight = 14

O
Oxygen
Atomic number = 8
Atomic weight = 16

Na
Sodium
Atomic number = 11
Atomic weight = 23

Cl
Chlorine
Atomic number = 17
Atomic weight = 35

K
Potassium
Atomic number = 19
Atomic weight = 39

Which of these atoms is inert?

cule may contain two atoms of the same kind, as in the hydrogen molecule: H_2. The subscript 2 indicates that there are two hydrogen atoms in the molecule. Molecules may also contain two or more different kinds of atoms, as in the hydrochloric acid molecule: HCl. Here an atom of hydrogen is attached to an atom of chlorine.

A *compound* is a substance composed of two or more *different* elements. Compounds can be broken down into their constituent elements by chemical means. Whereas hydrochloric acid (HCl) is a compound, a molecule of hydrogen (H_2) is not. The atoms in molecules and compounds are held together by electrical forces of attraction called **chemical bonds**. Energy is needed to form these bonds but, when they are broken, energy is released.

Ionic Bonds

Atoms are electrically neutral because the number of positively charged protons equals the number of negatively charged electrons. But when an atom gains or loses electrons, this balance is upset. If the atom gains electrons, it acquires an overall negative charge. If the atom loses electrons, it acquires an overall positive charge. Such a negatively or positively charged particle is called an **ion** (Ī-on) or **electrolyte**. As you will see shortly, an ion is always symbolized by writing the chemical abbreviation followed by the number of positive (+) or negative (−) charges the ion acquires.

Consider a sodium ion (Figure 2-3a). A sodium atom (Na) has 11 protons and 11 electrons, with 1 electron in its outer energy

FIGURE 2-3 Formation of an ionic bond. (a) An atom of sodium attains stability by passing a single electron to an electron acceptor. The loss of this single electron results in the formation of a sodium ion (Na^+). (b) An atom of chlorine attains stability by accepting a single electron from an electron donor. The gain of this single electron results in the formation of a chloride ion (Cl^-). (c) When the Na^+ and the Cl^- ions are combined, they are held together by the attraction of opposite charges, which is known as an ionic bond, and a molecule of sodium chloride (NaCl) is formed.

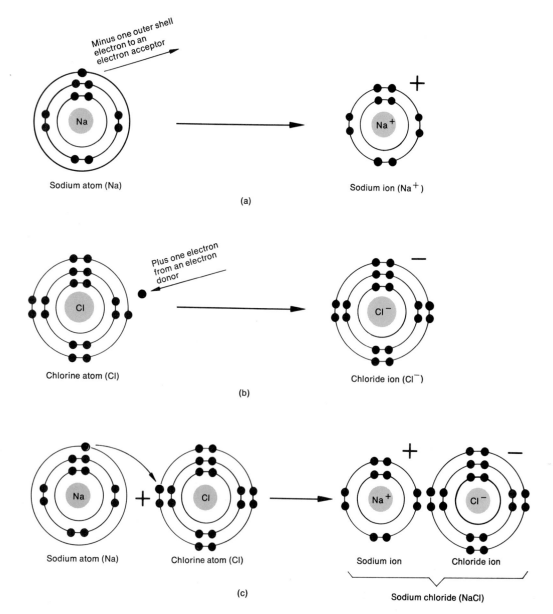

level. When sodium gives up the single electron in its outer level, it is left with 11 protons and only 10 electrons. It is an *electron donor* because it gives up electrons. The atom now has an overall positive charge of one (+1). This positively charged sodium atom is called a sodium ion (written Na^+). Generally, atoms whose outer energy level is less than half-filled lose electrons and form positively charged ions called *cations* (KAT-ī-ons). Other examples of cations are the potassium ion (K^+), calcium ion (Ca^{2+}), and iron ion (Fe^{2+}).

Another example is the formation of a chloride ion (Figure 2-3b). Chlorine has a total of 17 electrons, 7 of them in the outer energy level. Since this energy level can hold 8 electrons, chlorine tends to pick up an electron that has been lost by another atom. Chlorine is an *electron acceptor* because it picks up electrons. By accepting an electron, chlorine acquires a total of 18 electrons. However, it still has only 17 protons in its nucleus. The chloride ion therefore has a negative charge of one (−1) and is written Cl^-. Atoms whose outer energy level is more than half-filled tend to gain electrons and form negatively charged ions called *anions* (AN-ī-ons). Other examples of anions include the iodine ion (I^-) and sulfur ion (S^{2-}).

The positively charged sodium ion (Na^+) and the negatively charged chloride ion (Cl^-) attract each other—opposite charges attract each other. The attraction, called an *ionic bond,* holds the two ions together, and a molecule is formed (Figure 2-3c). The formation of this molecule, sodium chloride (NaCl) or table salt, is one of the most common examples of ionic bonding. Thus, an ionic bond is an attraction between ions in which one atom loses electrons and another atom gains electrons.

Hydrogen is an example of an atom whose outer level is exactly half-filled. The first energy level can hold two electrons, but in hydrogen atoms it contains only one. Hydrogen may lose its electron and become a positive ion (H^+). This is precisely what happens when hydrogen combines with chlorine to form hydrochloric acid (H^+Cl^- or HCl). However, hydrogen is equally capable of forming another kind of bond altogether, called a covalent bond.

Covalent Bonds

The second type of chemical bond to be considered is the *covalent bond.* This is a more common bond in the human body than an ionic bond and is more stable. When a covalent bond is formed, neither of the combining atoms loses or gains an electron. Instead, the two atoms share one, two, or three electron pairs. Look at the hydrogen atom again. One way a hydrogen atom can fill its outer energy level is to combine with another hydrogen atom to form the molecule H_2 (Figure 2-4a). In the H_2 molecule, the two atoms share a pair of electrons. Each hydrogen atom has its own electron plus one electron from the other atom. When one pair of electrons is shared between atoms, as in the H_2 molecule, a *single covalent bond* is formed. When two pairs of electrons are shared between two atoms, a *double covalent bond* is formed (Figure 2-4b). A *triple covalent bond* occurs when three pairs of electrons are shared (Figure 2-4c).

The same principles that apply to covalent bonding between atoms of the same element also apply to atoms of different elements.

Methane (CH_4), also known as marsh gas, is an example of covalent bonding between atoms of different elements (Figure 2-4d). The outer energy level of the carbon atom can hold eight electrons but has only four of its own. Each hydrogen atom can hold two electrons but has only one of its own. In the methane molecule, the carbon atom shares four pairs of electrons. One pair is shared with each hydrogen atom. Each of the four carbon electrons orbits around both the carbon nucleus and a hydrogen nucleus. Each hydrogen electron circles around its own nucleus and the carbon nucleus.

Elements whose outer energy levels are half-filled, such as hydrogen and carbon, form covalent bonds quite easily. In fact, carbon always forms covalent bonds. It never becomes an ion. However, many atoms whose outer energy levels are more than half-filled may also form covalent bonds. An example is oxygen.

Hydrogen Bonds

A *hydrogen bond* consists of a hydrogen atom covalently bonded to one oxygen atom or one nitrogen atom but also attracted to another oxygen or nitrogen atom. Because hydrogen bonds are weak, only about 5 percent as strong as covalent bonds, they do not bind atoms into molecules. However, they do serve as bridges between different molecules or between various parts of the same molecule. As weak bonds, they may be formed and broken fairly easily. Hydrogen bonds are found in large complex molecules such as proteins and nucleic acids (see Figure 2-10). It should be noted that even though hydrogen bonds are relatively weak, large molecules may contain several hundred of these bonds, resulting in considerable strength and stability.

CHEMICAL REACTIONS

Chemical reactions are nothing more than the making or breaking of bonds between atoms. These reactions occur continually in all cells of your body. Chemical reactions are the means by which body structures are built and body functions carried out. After a chemical reaction, the total number of atoms remains the same, but because they are rearranged, there are new molecules with new properties. This section looks at the basic chemical reactions common to all living cells.

Synthesis Reactions—Anabolism

When two or more atoms, ions, or molecules combine to form new and larger molecules, the process is called a *synthesis* or *combination reaction.* The word *synthesis* means "combination," and synthesis reactions involve the *forming of new bonds:*

$$
\underset{\substack{\text{Atom, ion, or}\\\text{molecule A}}}{A} \quad + \quad \underset{\substack{\text{Atom, ion, or}\\\text{molecule B}}}{B} \quad \overset{\substack{\text{Combine}\\\text{to form}}}{\longrightarrow} \quad \underset{\substack{\text{New molecule}\\\text{or compound AB}}}{AB}
$$

FIGURE 2-4 Formation of a covalent bond between atoms of the same element and between atoms of different elements. (a) A single covalent bond between two hydrogen atoms. (b) A double covalent bond between two oxygen atoms. (c) A triple covalent bond between two nitrogen atoms. (d) Single covalent bonds between a carbon atom and four hydrogen atoms. In the boxed representation, called structural formulas, each covalent bond is represented by a straight line between atoms. The molecular formulas to the right of the structural formulas show the number of atoms in each molecule.

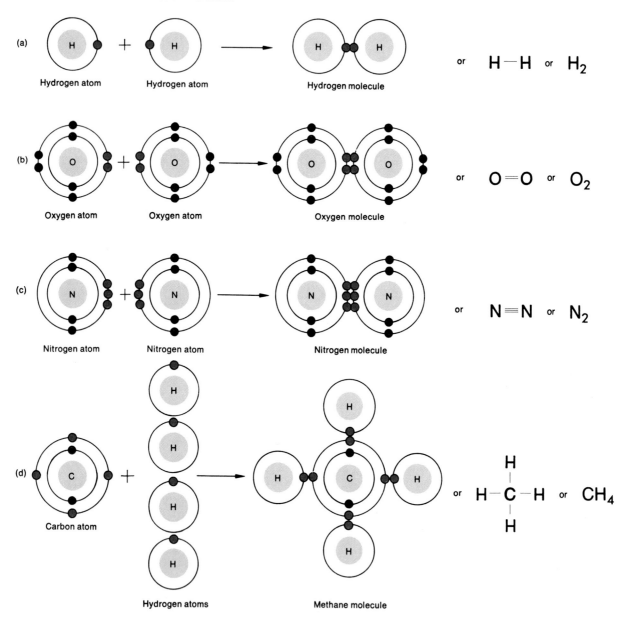

Draw a diagram similar to (d) for covalent bonds in a molecule of carbon tetrachloride (CCl_4). Refer to Figure 2-2.

An example of a synthesis reaction is:

$$N + 3H \longrightarrow NH_3$$

Nitrogen atom   Hydrogen atoms   Ammonia molecule

All the synthesis reactions that occur in your body are collectively termed **anabolism** (a-NAB-ō-lizm). Combining glucose molecules to form glycogen and combining amino acids to form proteins are two examples of anabolism. The importance of anabolism is considered in detail in Chapter 20.

Decomposition Reactions—Catabolism

The reverse of a synthesis reaction is a **decomposition reaction**. The word *decompose* means to break down into smaller parts. In a decomposition reaction, the *bonds are broken*.

$$AB \xrightarrow{\text{Breaks down into}} A + B$$

Molecule or compound AB   Atom, ion, or molecule A   Atom, ion, or molecule B

Under the proper conditions, methane can decompose into carbon and hydrogen:

$$CH_4 \longrightarrow C + 4H$$

Methane   Carbon   Hydrogen
molecule   atom   atoms

All the decomposition reactions that occur in your body are collectively termed *catabolism* (ka-TAB-ō-lizm). The digestion and oxidation of food molecules are examples of catabolism. The importance of catabolism is also considered in detail in Chapter 20.

Energy and Chemical Reactions

Chemical energy is the energy released or absorbed in the breaking or forming of chemical bonds. When a chemical bond is formed, energy is required. When a bond is broken, energy is released. In other words, synthesis reactions need energy and decomposition reactions give off energy. The building processes of the body—the construction of bones, the growth of hair and nails, the replacement of injured cells—occur basically through synthesis reactions. The breakdown of foods, on the other hand, occurs through decomposition reactions that release energy that can be used in the body's building processes. The energy from decomposition reactions is stored in a molecule called ATP (adenosine triphosphate) until needed by the body at some future point. Then, when energy is needed for synthesis reactions, it is provided by the breakdown of ATP. These reactions are covered in detail later in the chapter.

CHEMICAL COMPOUNDS AND LIFE PROCESSES

Most of the chemicals in the body exist in the form of compounds. Chemists divide these compounds into two principal classes: inorganic and organic. *Inorganic compounds* usually lack carbon. They are usually small, ionically bonded molecules. They include water, oxygen, carbon dioxide, and many salts, acids, and bases. *Organic compounds* always contain at least carbon and hydrogen and are held together mostly or entirely by covalent bonds. Organic compounds in the body include carbohydrates, lipids, proteins, nucleic acids, and adenosine triphosphate (ATP). Both organic and inorganic compounds are vital to life functions.

INORGANIC COMPOUNDS

Water

One of the most abundant, inorganic substances in the human organism is *water*. About 60 percent of red blood cells, 75 percent of muscle tissue, and 92 percent of blood plasma are water (plasma is the liquid portion of blood). The following functions of water explain why it is such a vital compound:

1. Water is an excellent solvent and suspending medium. A *solvent* is a liquid or gas in which some other material, called a *solute,* has been dissolved. The combination of solvent plus solute is called a *solution.* Water is the solvent that carries nutrients into and wastes out of body cells. Oxygen dissolves in water in the blood to be carried to body cells and carbon dioxide dissolves to be carried away for exhalation. Furthermore, if lung air sacs are not moist, oxygen cannot move into the blood in the first place.

   As a suspending medium, water is also vital to survival. Many large organic molecules are suspended in the water of body cells, which allows them to come in contact with other chemicals, leading to essential chemical reactions.

2. Water can participate in chemical reactions. In digestion, water helps break down large nutrient molecules. Water molecules are also part of the synthesis reactions that produce hormones and enzymes.

3. Water absorbs and releases heat very slowly. In comparison with other substances, water requires a large input of heat to increase its temperature and a great loss of heat to decrease its temperature. Thus, the large amount of water in the body moderates the effects of fluctuating environmental temperatures, thereby helping maintain a homeostatic body temperature.

4. Water requires a large amount of heat to change from a liquid to a gas. When water (perspiration) evaporates from the skin, it takes with it large quantities of heat, providing an excellent body cooling mechanism.

5. Water serves as a lubricant. It is a major part of mucus and other lubricating fluids, in the chest and abdomen where internal organs touch and slide over each other, and at joints where bones, ligaments, and tendons rub against each other. In the gastrointestinal tract, water in saliva moistens foods for smooth passage.

Inorganic Acids, Bases, and Salts

When molecules of inorganic acids, bases, or salts dissolve in the water of the body cells, they undergo *ionization* (ī-on-i-ZĀ-shun) or *dissociation* (dis'-sō-sē-Ā-shun), that is, they break apart into ions. An *acid* dissociates into *hydrogen ions* (H^+). A *base,* by contrast, dissociates into *hydroxyl ions* (OH^-). A *salt,* when dissolved in water, dissociates into cations and anions, neither of which is H^+ or OH^- (Figure 2-5). Acids and bases react with one another to form salts. For example, the combination of hydrochloric acid (HCl), an acid, and sodium hydroxide (NaOH), a base, produces sodium chloride (NaCl), a salt, and water (H_2O).

Many salts are found in cells and body fluids, such as lymph, blood, and the extracellular fluid of tissues. The ions of salts provide many essential chemical elements. Sodium (Na^+) and chloride (Cl^-) ions are present in high concentrations in extracellular body fluids. Inside cells, phosphate (HPO_4^{2-}) and potassium (K^+) ions are abundant.

Acid–Base Balance: The Concept of pH

Since acids dissociate into hydrogen ions (H^+) and bases dissociate into hydroxyl ions (OH^-), it follows that the more hydrogen ions in a solution, the more acid it is. Conversely, the more hydroxyl ions, the more basic (alkaline) the solution. The term *pH* is used to describe the degree of *acidity* or *alkalinity (basicity)* of a solution.

A solution's acidity or alkalinity is expressed on a *pH scale*

FIGURE 2-5 Acids, bases, and salts. (a) When placed in water, hydrochloric acid (HCl) dissociates into H⁺ ions and Cl⁻ ions. (b) When the base sodium hydroxide (NaOH) is placed in water, it dissociates into OH⁻ ions and Na⁺ ions. (c) When table salt (NaCl) is placed in water, it dissociates into positive and negative ions (Na⁺ and Cl⁻), neither of which is H⁺ or OH⁻.

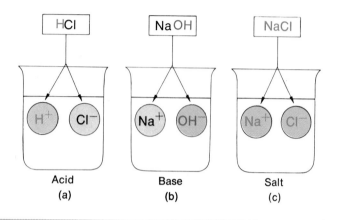

Acid
(a)

Base
(b)

Salt
(c)

If HCl dissociates as HCl → H⁺ + Cl⁻, write the dissociation of potassium chloride (KCl) and calcium chloride ($CaCl_2$).

that runs from 0 to 14 (Figure 2-6). The pH scale is based on the number of H^+ ions in a solution (expressed in certain chemical units called moles per liter). A solution that is 0 on the pH scale has many H^+ ions and few OH^- ions. A solution that rates 14, by contrast, has many OH^- ions and few H^+ ions. The midpoint is 7, where the concentration of H^+ and OH^- ions is equal. A

pH of 7 (pure water) is neutral. A solution with more H^+ ions than OH^- is *acid* and has a pH below 7. A solution with more OH^- ions than H^+ is *basic (alkaline)* and has a pH above 7. A change of one whole number on the pH scale represents a 10-fold change from the previous concentration. That is, a pH of 2 indicates 10 times fewer H^+ ions than a pH of 1. A pH of 3 indicates 10 times fewer H^+ ions than a pH of 2 and 100 times fewer H^+ ions than a pH of 1.

Your body fluids must maintain a fairly constant balance of acids and bases because the biochemical reactions that occur in living systems are extremely sensitive to even small changes in the acidity or alkalinity of the environment. Moreover, since H^+ and OH^- ions are involved in practically all biochemical processes, any departure from normal H^+ and OH^- concentrations affects a cell's functioning.

Maintaining pH: Buffer Systems

Although the pH of various body fluids may differ, the normal limits for each are quite specific and narrow. Exhibit 2-2 shows the pH values for certain body fluids compared with common substances. Even though strong acids and bases are continually taken into the body as foods and beverages, the pH levels of body fluids remain relatively constant because of the body's *buffer systems*. Buffers are weak acids or bases found in the body's fluids. They replace strong acids or bases to prevent drastic changes in pH and help maintain homeostasis. (Although the preceding discussion of pH has focused on inorganic acids and bases, there are also organic acids and bases and they too can cause pH problems. More will be said about buffers in Chapter 22.)

FIGURE 2-6 pH scale. At pH 7 (neutrality), the concentrations of H⁺ and OH⁻ ions are equal. A pH value below 7 indicates an acid solution. A pH value above 7 indicates an alkaline (basic) solution. A change in one whole number on the pH scale represents a 10-fold change from the previous concentration.

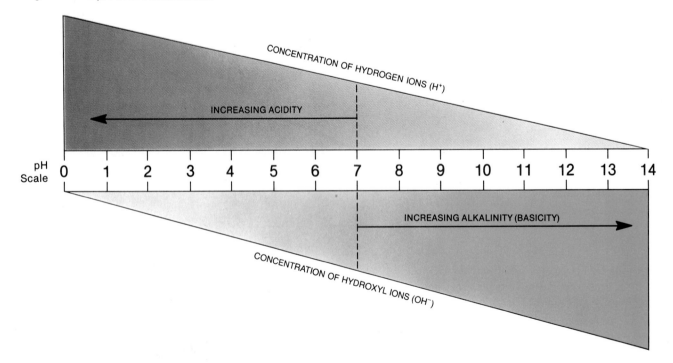

Which pH is more acid? 6.82 or 6.91? Which pH is closer to neutrality? 8.42 or 5.59?

EXHIBIT 2-2
Normal pH Values of Representative Substances

Substance	pH Value
Gastric juice (digestive juice of the stomach)	1.2–3.0
Lemon Juice	2.2–2.4
Grapefruit juice, vinegar, beer, wine	3.0
Cider	2.8–3.3
Pineapple juice, orange juice	3.5
Tomato juice	4.2
Coffee	5.0
Clam chowder	5.7
Urine	4.6–8.0
Saliva	6.35–6.85
Milk	6.6–6.9
Pure (distilled) water	7.0
Blood	7.35–7.45
Semen (fluid containing sperm)	7.20–7.60
Cerebrospinal fluid (fluid associated with nervous system)	7.4
Pancreatic juice (digestive juice of the pancreas)	7.1–8.2
Eggs	7.6–8.0
Bile (liver secretion that aids in fat digestion)	7.6–8.6
Milk of magnesia	10.0–11.0
Limewater	12.3

ORGANIC COMPOUNDS

Carbohydrates

Carbohydrates are sugars and starches. They are the body's most readily available source of energy. They also provide energy reserves. For example, glycogen, which is stored glucose, is found in the liver and skeletal muscles for emergency energy needs. Some carbohydrates are used in building cell structures. Also, some carbohydrates are components of deoxyribonucleic acid (DNA), the molecule that carries hereditary information, and ribonucleic acid (RNA), the molecule involved in protein synthesis.

Carbohydrates are composed of carbon, hydrogen, and oxygen. The ratio of hydrogen to oxygen atoms is typically 2 to 1: ribose ($C_5H_{10}O_5$), glucose ($C_6H_{12}O_6$), sucrose ($C_{12}H_{22}O_{11}$), and so on. Carbohydrates are divided into three major groups on the basis of size: monosaccharides, disaccharides, and polysaccharides.

1. **Monosaccharides.** *Monosaccharides* (mon-ō-SAK-a-rīds), or simple sugars, are compounds containing from three to seven carbon atoms. Pentoses (five-carbon sugars) and hexoses (six-carbon sugars) are exceedingly important monosaccharides. The pentose called deoxyribose is a component of genes. The hexose called glucose is the main energy-supplying molecule of the body.
2. **Disaccharides.** *Disaccharides* (dī-SAK-a-rīds) are sugars that consist of two small monosaccharides joined chemically into a large, more complex molecule. When two monosaccharides combine to form a disaccharide, a molecule of water is always lost. This reaction is known as *dehydration synthesis* (*dehydration* = loss of water) and is shown in Figure 2-7. Glucose and fructose are two monosaccharides that form the disaccharide, sucrose. Similarly, glucose and galactose unite to form lactose (milk sugar) and two molecules of glucose produce maltose.

   Disaccharides can be broken down into their smaller, simpler molecules by adding water. This reverse chemical reaction is called *digestion* **(hydrolysis),** which means to split by using water. Sucrose, for example, may be digested into glucose and fructose by adding water. See Figure 2-7.
3. **Polysaccharides.** *Polysaccharides* (pol'-ē-SAK-a-rīds) consist of several monosaccharides joined together through dehydration synthesis. Like disaccharides, polysaccharides can be broken down into their constituent sugars through hydrolysis reactions. Unlike monosaccharides or disaccharides, however, they usually lack the characteristic sweetness of sugars like sucrose and are usually not soluble in water. Glycogen is a polysaccharide.

Lipids

Like carbohydrates, *lipids* are also composed of carbon, hydrogen, and oxygen, but they do not have a 2:1 ratio of hydrogen to oxygen. There is less oxygen in lipids than in carbohydrates. Most

FIGURE 2-7 Dehydration synthesis and hydrolysis of a molecule of sucrose. In the dehydration synthesis reaction (read from left to right), the two smaller molecules, glucose and fructose, are joined to form a larger molecule of sucrose. Note the loss of a water molecule. In hydrolysis (read from right to left), the sucrose molecule is broken down into the two smaller molecules, glucose and fructose. Here, a molecule of water is added to sucrose for the reaction to occur.

Which chemical reaction is anabolic?

Vegetarian Diets: Alphabet Soup

Once upon a time, and not so far away, meat was regarded as one of the most nutritious foods around, and it was not uncommon to find meat at every meal: bacon, ham, or sausage at the breakfast table, sandwich meats for lunch, and meat as the main dinner entree. High meat consumption was a symbol of American prosperity.

To discover why Americans have such high rates of atherosclerosis (artery disease), researchers began to look closely at the American diet and to compare it to diets in countries where rates of artery disease are very low. They found that one of the biggest differences is our fat intake, and one of the major sources of fat in our diet is meat.

Research has shown that vegetarians are generally healthier than their meat-eating friends. Not only do they have less artery disease, they also have lower rates of high blood pressure, diabetes, and certain types of cancer; a more healthily functioning digestive system; and lower cholesterol and body fat levels. Scientists attribute much of this difference to the vegetarian's low-fat, high-fiber diet. A well-planned vegetarian diet is also high in vitamins and minerals and low in cholesterol.

So while 30 years ago most Americans thought vegetarianism was only for "health nuts," the current quest for a low-fat life-style has promoted widespread interest in the vegetarian style of eating among all kinds of people. Some become lacto-ovo vegetarians, people who do not eat meat but continue to eat eggs and dairy products. Others become vegans, people who eat no animal products at all. Many people continue to eat meat, but are choosing to eat less of it and to have several meatless meals each week. Vegetarian meals have become part of mainstream America.

People new to vegetarian meal planning sometimes wonder if there is anything special they should know to be sure they get all the nutrients they used to get from meat. Since meat is a concentrated source of protein, getting enough protein from nonmeat sources is often their first concern.

Protein is found in many foods other than meat. The average adult male needs about 63 g of protein daily. Adult females need about 50 g. A cup of rice contains 6 g protein; an egg has 6 g; a serving of oatmeal contains 5 g. Most vegetarians have no problem getting the protein they need. But is it the right kind of protein?

Dietary protein provides the amino acids used to build proteins in the body. Our bodies use at least 20 different amino acids to manufacture the wide variety of proteins needed. Given an adequate protein intake, we can make all but 10 of the amino acids we need. These 10 are called essential amino acids, which means we must obtain them from our diet.

Many people have heard that the quality of plant protein is inferior to animal protein. Nutritionists use the word *quality* to refer to the amino acid composition of protein. Animal products, including meat, eggs, and dairy products, are considered high-quality protein sources because they contain all 10 essential amino acids in the relative proportions needed by the body. Plant sources are often low on one or two essential amino acids.

"Low on one or two essential amino acids" might not sound too bad, but consider the analogy of proteins to words made from amino acid letters. Even if you had plenty of *Y*s and *S*s, you still couldn't spell *YES* if you were missing the letter *E*.

Is it difficult to get all the amino acid letters without eating meat? Fortunately, it is easy to obtain enough of the right kinds of amino acids from plant sources as long as a wide variety are consumed. For example, the amino acids lacking in rice are found abundantly in beans. Rice and beans are an example of complementary proteins. Consume the two together, or at some time during the day, and you will get all the amino acids necessary for good health.

To ensure an adequate intake of all 10 essential amino acids from plant sources, it is helpful to think of three groups of protein foods:

1. Legumes (peas and beans) and their products. Examples are lentils and split peas; lima, kidney, pinto,

and other dried beans; soybeans and soybean products such as tofu, tempeh, and soy milk; hommous (a spread made from garbanzo beans or chick peas); and peanut butter. Meat analogues, such as textured vegetable protein, are usually made from legumes.
2. Grains, such as rice, wheat, barley, oats, corn, and their products, such as bread, pasta, and cereals.
3. Nuts and seeds, such as sesame and sunflower seeds.

While foods in each group provide incomplete protein, the missing amino acids will be provided if a food from one of the other groups or dairy products or eggs are eaten as well. Rice and bean casseroles, peanut butter and whole wheat bread, and chili and cornbread are examples of food combinations that provide high-quality protein.

The more restricted a person's diet, the more careful he or she needs to be about menu planning. A person who simply forgoes meat two or three days a week and has a generally healthful diet need not be more concerned about nutritional deficiencies than any other person. Lacto-ovo vegetarians can obtain all the nutrients found in a meat-eater's diet. They must be careful, however, not to rely solely on eggs and dairy products for their protein, or they may consume too much fat and cholesterol.

Vegans must be especially careful to include a wide variety of foods to be sure they get all the amino acids they need, as well as the vitamins and minerals others get from eggs, dairy products, and meat. Population groups with special nutritional needs and considerations, such as pregnant and lactating women, infants and children, are especially vulnerable to possible nutritional deficiencies imposed by a restricted diet.

Vitamin B_{12} is found only in animal foods and in nutritional supplements, such as yeast grown in a medium enriched with vitamin B_{12}. Vegans can obtain vitamin B_{12} from nutritional yeast, or from vitamin B_{12}-fortified products such as soy milk. Approximately a four years' supply of this vitamin is stored in the body, so it takes years for a vitamin B_{12} deficiency to develop in healthy adults. Pregnant and lactating women, however, must be aware that their infants can develop this damaging deficiency even though the mother may remain symptom-free.

Vitamin D, riboflavin, and calcium are supplied in abundance by milk and dairy products. Plant sources do not provide vitamin D, although regular exposure to the sun will prevent a deficiency. Vitamin D supplements may be necessary for vegans who live in cloudy climates or don't get outdoors much. Too much of this

vitamin is toxic, so minimum recommended dosages should not be exceeded.

Riboflavin is present in dark green vegetables, so vegans who consume good amounts of these foods are able to meet their needs for this B vitamin. Dark green vegetables also supply some calcium, as do sesame seeds, legumes, and grains, but since daily requirements for this mineral are fairly high, it is often difficult to meet calcium needs through plant sources alone. For this reason, many vegans rely on calcium-fortified soy milk. An adequate calcium intake is especially crucial for children and pregnant and lactating women.

It's important to note that simply eliminating meat from one's diet does not automatically make it a healthful one. There are plenty of fatty foods free of animal products, so you can give up hamburgers and still get plenty of fat from french-fried potatoes, milkshakes, and grilled cheese sandwiches.

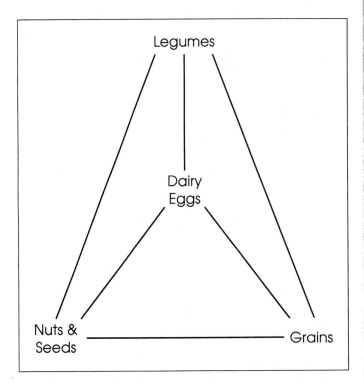

While legumes, grains, nuts, and seeds eaten alone do not provide all the essential amino acids needed by humans, when eaten with each other or with dairy products or eggs, they provide high-quality protein, along with vitamins, minerals, carbohydrates, and fiber. This diagram illustrates that foods connected by lines may be consumed together to provide complete protein.

lipids are insoluble in water, but they readily dissolve in solvents such as chloroform and ether. Functionally, lipids, such as fats, protect, insulate, and serve as a source of energy. Fats represent the body's most highly concentrated source of energy. They provide more than twice as much energy per weight as either carbohydrates or proteins. However, fats are less efficient as body fuels than are carbohydrates, because they are more difficult to break down. Other lipids, such as phospholipids, make up parts of plasma (cell) membranes. Still other lipids are components of cholesterol, bile salts, hormones, and certain vitamins. Among the classes of lipids are fats, phospholipids (lipids that contain phosphorus), steroids, carotenes, vitamins E and K, and prostaglandins (PGs). Since lipids are a large and diverse group of compounds, we discuss only fats in detail at this point.

A molecule of **fat** (triglyceride) consists of two basic components: **glycerol** and **fatty acids** (Figure 2-8a,b). A single molecule of fat is formed when a molecule of glycerol combines with three

molecules of fatty acids. This reaction, like the one described for disaccharide formation, is a dehydration synthesis reaction. During digestion, a single molecule of fat is broken down into fatty acids and glycerol.

Because of their relationship to cardiovascular disease, fats are very important in our daily lives. At this point, it will be helpful to compare three types of dietary fat and how they relate to cholesterol. A **saturated fat** has no double bonds between its carbon atoms. There are only single covalent bonds and since all the carbon atoms are bonded to the maximum number of hydrogen atoms, such a fat is said to be *saturated* with hydrogen atoms (Figure 2-8c). Saturated fats (and some cholesterol) occur mostly in animal foods such as beef, pork, butter, whole milk, eggs, and cheese. They also occur in some plant products such as cocoa butter, palm oil, and coconut oil. Since the liver produces cholesterol from some breakdown products of saturated fats, eating of these fats is discouraged for individuals with high cholesterol levels. A

FIGURE 2-8 Structure and reactions of glycerol and fatty acids. (a) Structure of glycerol. (b) Structure of a fatty acid. When glycerol and the fatty acid are joined in dehydration synthesis, a molecule of water is lost. (c) Fats consist of one molecule of glycerol joined to three molecules of fatty acids, which vary in the number of double bonds between carbon atoms (C=C). Shown here is a molecule of a fat that contains three different fatty acids: a saturated fatty acid, a polyunsaturated fatty acid, and a monounsaturated fatty acid. Note that when a molecule of glycerol combines with three fatty acids in dehydration synthesis to form a molecule of fat, three molecules of water are lost.

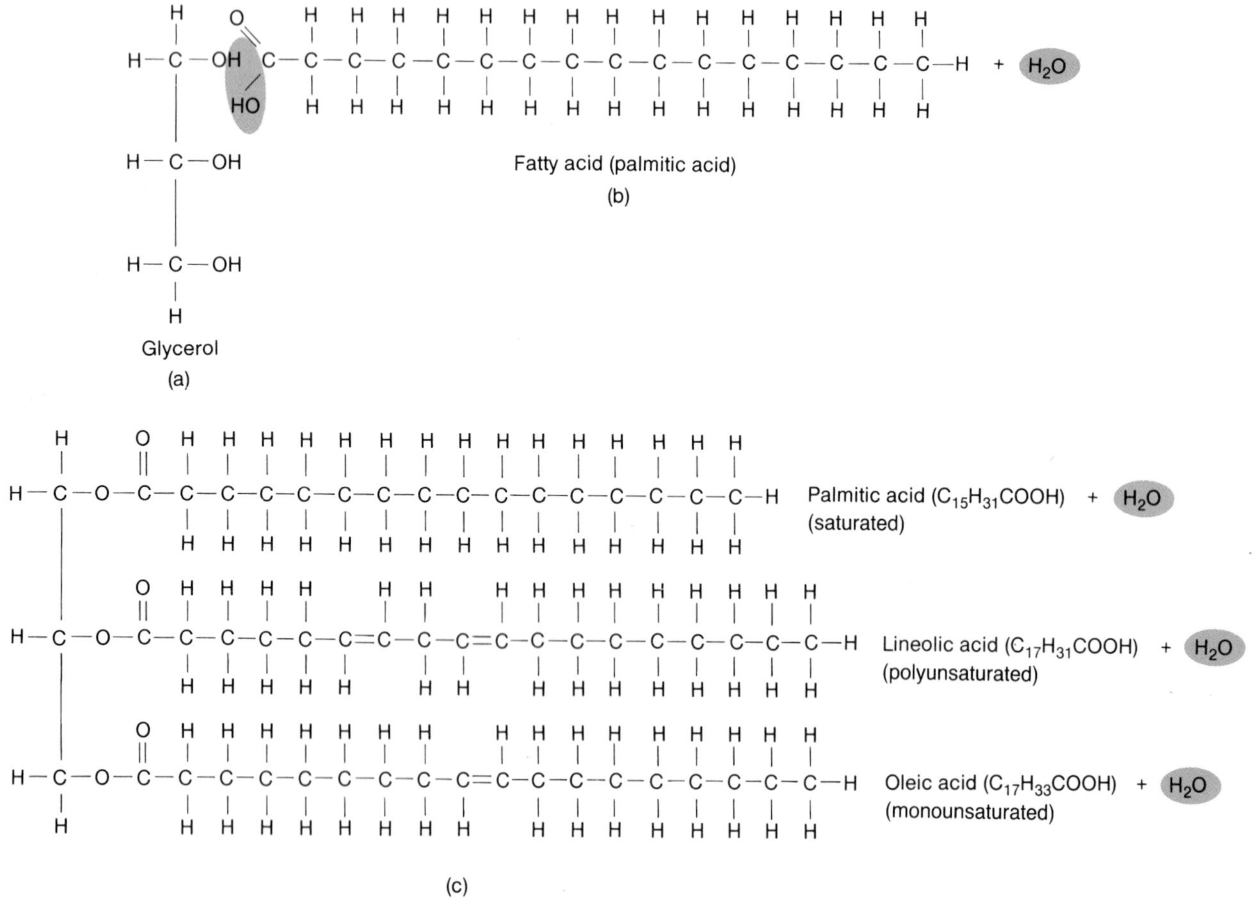

Fatty acid (palmitic acid)
(b)

Glycerol
(a)

Palmitic acid ($C_{15}H_{31}COOH$) + H_2O
(saturated)

Lineolic acid ($C_{17}H_{31}COOH$) + H_2O
(polyunsaturated)

Oleic acid ($C_{17}H_{33}COOH$) + H_2O
(monounsaturated)

(c)

Which type of fat is associated with raising cholesterol level?

monounsaturated fat contains only one double covalent bond between its carbon atoms; it is not completely saturated with hydrogen atoms (Figure 2-8c). Examples are olive oil and peanut oil, which are believed to help reduce cholesterol levels. A *polyunsaturated fat* contains more than one double covalent bond between its carbon atoms (Figure 2-8c). Corn oil, safflower oil, sunflower oil, cottonseed oil, sesame oil, and soybean oil are examples of polyunsaturated fats, which researchers believe also help to reduce cholesterol in the blood.

Proteins

Proteins are much more complex in structure than carbohydrates or lipids and are involved in numerous physiological activities. Proteins are largely responsible for the structure of body cells. Some proteins in the form of enzymes function as catalysts to speed up certain chemical reactions. Enzymes are very important in regulating chemical reactions in cells to help maintain homeostasis. Other proteins assume an important role in muscular contraction. Antibodies are proteins that defend the body against invading microbes. Some types of hormones are proteins.

Chemically, proteins always contain carbon, hydrogen, oxygen, and nitrogen, and sometimes sulfur. *Amino acids* are the building blocks of proteins. There are at least 20 different amino acids. In protein formation, amino acids combine to form more complex molecules; the bonds formed between amino acids are called *peptide bonds* (Figure 2-9).

When two amino acids combine, a *dipeptide* results. Adding another amino acid to a dipeptide produces a *tripeptide*. Further additions of amino acids result in the formation of *polypeptides*, which are large protein molecules. All have the same basic composition, but each also has additional atoms arranged in a specific way. A great variety of proteins is possible therefore, because each variation in the number or sequence of amino acids produces a different protein. The situation is similar to using an alphabet of 20 letters to form words. Each letter would be equivalent to an amino acid, and each word would be a different protein.

Nucleic Acids: Deoxyribonucleic Acid (DNA) and Ribonucleic Acid (RNA)

Nucleic (noo-KLĒ-ic) *acids* are exceedingly large organic molecules containing carbon, hydrogen, oxygen, nitrogen, and phospho-

rus. There are two principal kinds: *deoxyribonucleic* (dē-ok′-sē-rī-bō-noo-KLĒ-ik) *acid (DNA)* and *ribonucleic acid (RNA).*

Whereas the basic structural units of proteins are amino acids, the basic units of nucleic acids are *nucleotides.* A molecule of DNA is a chain composed of repeating nucleotide units. Each nucleotide of DNA consists of three basic parts (Figure 2-10a):

1. One of four possible *nitrogenous bases,* which are ring-shaped structures containing atoms of C, H, O, and N. The nitrogenous bases found in DNA are adenine (A), thymine (T), cytosine (C), and guanine (G).
2. A pentose sugar called *deoxyribose.*
3. *Phosphate groups.*

Nucleotides are named according to their nitrogenous bases; for example, a nucleotide containing thymine is called a *thymine nucleotide.*

Figure 2-10b shows the following structural characteristics of the DNA molecule:

1. The molecule consists of two strands with crossbars. The strands twist about each other in the form of a *double helix* so that the shape resembles a twisted ladder.
2. The uprights of the DNA ladder consist of alternating phosphate groups and the deoxyribose portions of the nucleotides.
3. The rungs of the ladder contain paired nitrogenous bases. Adenine always pairs with thymine and cytosine always pairs with guanine.

About 1000 rungs of DNA comprise a *gene,* the hereditary material in cells. Humans have about 100,000 functional genes that direct the production of body proteins. This is discussed in detail in Chapter 3. Genes determine which traits we inherit, and they control all the activities that take place in our cells throughout a lifetime. When a cell divides, its hereditary information is passed on to the next generation of cells. The passing of information is possible because of DNA's unique structure. Its structure is also responsible for the almost endless number of variations among humans.

RNA, the other kind of nucleic acid, differs from DNA in several respects. RNA is single stranded; DNA is double stranded. The sugar in the RNA nucleotide is the pentose ribose. And RNA does not contain the nitrogenous base thymine. Instead of thymine,

FIGURE 2-9 Protein formation. When two or more amino acids are chemically united, the resulting bond between them is called a peptide bond. Here, the amino acids glycine and alanine are joined to form the dipeptide glycylalanine. The peptide bond is formed at the point where water is lost.

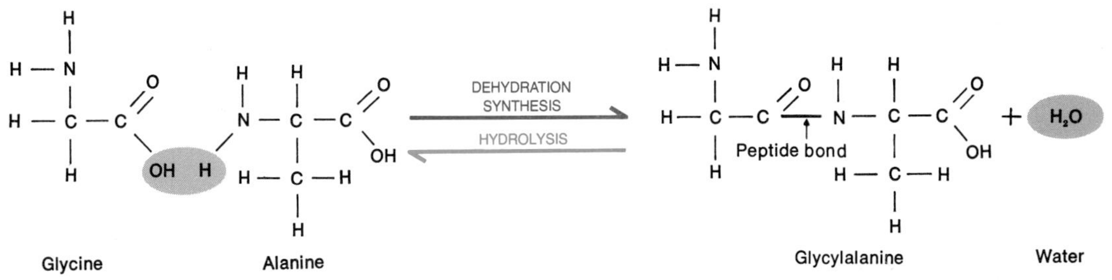

FIGURE 2-10 **DNA molecule. (a) Adenine nucleotide. (b) Portion of an assembled DNA molecule.**

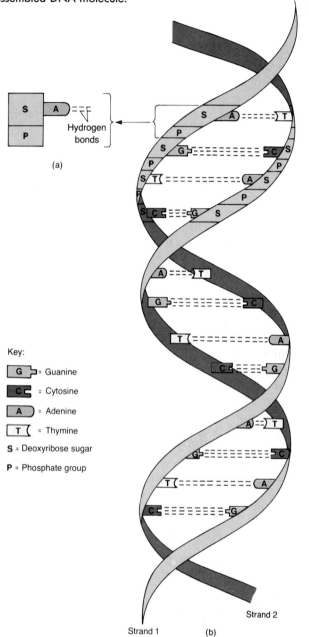

Key:

G = Guanine
C = Cytosine
A = Adenine
T = Thymine
S = Deoxyribose sugar
P = Phosphate group

Strand 1   (b)   Strand 2

Which nitrogenous base is not present in RNA?

RNA has the nitrogenous base uracil. Three different kinds of RNA have been identified in cells. Each type has a specific role to perform with DNA in protein synthesis reactions (see Chapter 3).

Adenosine Triphosphate (ATP)

A molecule that is indispensable to the life of a cell is *adenosine* (a-DEN-ō-sēn) *triphosphate* (*ATP*). ATP is found universally in living systems. Its essential function is to store energy for the cell's basic life activities. Structurally, ATP consists of three phosphate groups (PO_4^{3-}) and an adenosine unit composed of adenine and the five-carbon sugar ribose (Figure 2-11). ATP releases a

great amount of usable energy when it is broken down by the addition of a water molecule (hydrolysis).

When the terminal phosphate group (here symbolized as P) is hydrolyzed and therefore removed, the changed molecule is called *adenosine diphosphate* (*ADP*). This reaction is represented as follows:

$$\text{ATP} \rightleftharpoons \text{ADP} + \text{P} + \text{E}$$

Adenosine triphosphate   Adenosine diphosphate   Phosphate   Energy

The energy supplied by the catabolism of ATP into ADP is constantly being used by the cell. Since the supply of ATP at any given time is limited, a mechanism exists to replenish it: a phosphate group is added to ADP to manufacture more ATP. The reaction is represented as follows:

$$\text{ADP} + \text{P} + \text{E} \rightleftharpoons \text{ATP}$$

Adenosine diphosphate   Phosphate   Energy   Adenosine triphosphate

Energy is required to make ATP. This energy is required to attach a phosphate group to ADP and is supplied primarily by the breakdown of glucose in the cell in a process called cellular respiration. Cellular respiration has two phases:

1. **Anaerobic.** In the absence of oxygen, glucose is partially broken down into a substance called pyruvic acid, a process called *glycolysis*. Since glucose breakdown is incomplete, only a small amount of energy is produced to form ATP.
2. **Aerobic.** In the presence of oxygen, glucose is completely broken down into carbon dioxide and water. A large amount of energy is produced to form ATP and generate heat. The details of cellular respiration will be discussed in Chapters 8 and 20.

FIGURE 2-11 **Structure of ATP and ADP. The high-energy bonds are indicated by ~.**

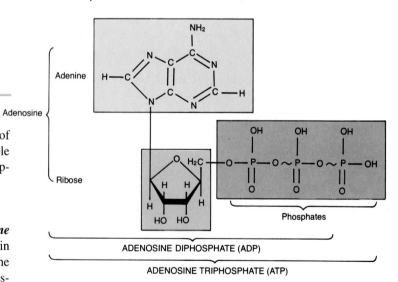

Is there more energy in ATP or ADP?

STUDY OUTLINE

Introduction to Basic Chemistry (p. 19)

Chemical Elements (p. 19)
1. Matter is anything that occupies space and has mass. It is made up of building units called chemical elements.
2. Oxygen, carbon, hydrogen, and nitrogen make up 96 percent of the body's weight. These elements together with calcium and phosphorus make up 99 percent of the total body weight.

Structure of Atoms (p. 19)
1. Units of chemical elements are called atoms.
2. Atoms consist of a nucleus, which contains protons and neutrons, and electrons that move about the nucleus in energy levels.
3. The total number of protons in an atom is its atomic number. This number is equal to the number of electrons in the atom.
4. The combined total of protons and neutrons of an atom is its atomic weight.

Atoms and Molecules (p. 21)
1. Electrons are the parts of an atom that participate in chemical reactions.
2. A molecule is two or more combined atoms. A molecule containing different kinds of atoms is a compound.
3. In an ionic bond, outer-energy-level electrons are transferred from one atom to another. The transfer forms ions, whose unlike charges attract each other and form ionic bonds.
4. In a covalent bond, there is a sharing of pairs of outer-energy-level electrons.
5. Hydrogen bonding provides temporary bonding between certain atoms within large complex molecules such as proteins and nucleic acids.

Chemical Reactions (p. 23)
1. Synthesis reactions produce a new molecule. The reactions are anabolic; bonds are formed.
2. In decomposition reactions, a substance breaks down into other substances. The reactions are catabolic; bonds are broken.
3. When chemical bonds are formed, energy is usually needed. When bonds are broken, energy is released.

Chemical Compounds and Life Processes (p. 25)

1. Inorganic substances usually lack carbon, contain ionic bonds, resist decomposition, and dissolve readily in water.
2. Organic substances always contain carbon and hydrogen. Most organic substances contain covalent bonds and many are insoluble in water.

Inorganic Compounds (p. 25)
1. Water is the most abundant substance in the body. It is an excellent solvent and suspending medium, participates in chemical reactions, absorbs and releases heat slowly, and lubricates.
2. Inorganic acids, bases, and salts dissociate into ions in water. Cations are positively charged ions; anions are negatively charged ions. An acid ionizes into H^+ ions; a base ionizes into OH^- ions. A salt ionizes into neither H^+ nor OH^- ions.
3. The pH of different parts of the body must remain fairly constant for the body to remain healthy. On the pH scale, 7 represents neutrality. Values below 7 indicate acid solutions, and values above 7 indicate alkaline solutions.
4. The pH values of different parts of the body are maintained by buffer systems which consist of weak acids and weak bases. Buffer systems eliminate excess H^+ ions and excess OH^- ions in order to maintain pH homeostasis.

Organic Compounds (p. 27)
1. Carbohydrates are sugars or starches and are the most common sources of the energy that is needed for life. They may be monosaccharides, disaccharides, or polysaccharides. Carbohydrates, and other organic molecules, are joined together to form larger molecules with the loss of water (dehydration synthesis). In the reverse process, called digestion (hydrolysis), large molecules are broken down into smaller ones upon the addition of water.
2. Lipids are a diverse group of compounds that includes fats, phospholipids, steroids, carotenes, vitamins E and K, and prostaglandins (PGs). Fats protect, insulate, and provide energy.
3. Proteins are constructed from amino acids. They give structure to the body, regulate processes as enzymes, provide protection, and help muscles to contract.
4. Deoxyribonucleic acid (DNA) and ribonucleic acid (RNA) are nucleic acids consisting of nitrogenous bases, sugar, and phosphate groups. DNA is a double helix and is the primary chemical in genes. RNA differs in structure and chemical composition from DNA and is mainly concerned with protein synthesis reactions.
5. The principal energy-storing molecule in the body is adenosine triphosphate (ATP). When an ATP molecule is decomposed, energy is liberated and the changed molecule is known as adenosine diphosphate (ADP). ATP is manufactured from ADP and P primarily by using the energy supplied by the decomposition of glucose in a process called cellular respiration.
6. Cellular respiration may be anaerobic (without oxygen) or aerobic (in the presence of oxygen). Aerobic respiration provides more energy to generate ATP.

REVIEW QUESTIONS

1. What is matter? (p. 19)
2. Define a chemical element. List the chemical symbols for several different chemical elements. Which chemical elements make up the bulk of the human organism? (p. 19)
3. What is an atom? Diagram the positions of the nucleus, protons, neutrons, and electrons in an atom of oxygen and an atom of nitrogen. What is an atomic number? (pp. 19–21)
4. What is meant by an energy level? (p. 21)
5. How are chemical bonds formed? Distinguish between an ionic bond and a covalent bond. Give at least one example of each. (p. 22)
6. Can you determine how a molecule of $MgCl_2$ is ionically bonded? Magnesium has two electrons in its outer energy level. Construct a diagram to illustrate your answer.
7. Refer to Figure 2-4b and c. See if you can determine why there is a double covalent bond between atoms in an oxygen molecule (O_2) and a triple covalent bond between atoms in a nitrogen molecule (N_2).
8. Define a hydrogen bond. Why are hydrogen bonds important? (p. 23)
9. How are anabolism and catabolism related to synthesis and decomposition reactions, respectively? How is energy related to chemical reactions? (pp. 23–25)
10. How do inorganic compounds differ from organic compounds? List and define the principal inorganic and organic compounds that are important to the human body. (p. 25)
11. What are the essential functions of water in the body? (p. 25)
12. Define an inorganic acid, a base, and a salt. How does the body acquire some of these substances? (p. 25)
13. What is pH? Why is it important to maintain a relatively constant pH? What is the pH scale? If there are 100 OH^- ions at a pH of 8.5, how many OH^- ions are there at a pH of 9.5? (pp. 25–26)
14. List the normal pH values of some common fluids, biological solutions, and foods. Refer to Exhibit 2-2, and select the two substances whose pH values are closest to neutrality. Is the pH of milk or of cerebrospinal fluid closer to neutrality? Is the pH of bile or of urine farther from neutrality?
15. What are the components of a buffer system? What is the function of a buffer? How is this an example of homeostasis? (p. 26)
16. Define a carbohydrate. Why are carbohydrates essential to the body? How are carbohydrates classified? (p. 27)
17. Compare dehydration synthesis and hydrolysis. Why are they significant? (p. 27)
18. How do lipids differ from carbohydrates? What are some types of lipids in the body? (p. 27)
19. Define a protein. What is a peptide bond? Discuss the importance of proteins to the body. (p. 31)
20. What is a nucleic acid? How do deoxyribonucleic acid (DNA) and ribonucleic acid (RNA) differ with regard to chemical composition, structure, and function? (p. 31)
21. What is adenosine triphosphate (ATP)? What is the essential function of ATP in the human body? How is this function accomplished? (p. 32)
22. How is energy supplied to produce ATP? (p. 32)

Cells

1. Define a cell and list its generalized parts.
2. Explain the structure and functions of the plasma membrane.
3. Describe how materials move across plasma membranes.
4. Explain the chemical composition and functions of cytoplasm.
5. Describe the structure and functions of the various cellular organelles.
6. Define a cell inclusion and give several examples.
7. Define a gene and explain the sequence of events involved in protein synthesis.
8. Discuss the stages, events, and significance of cell division.
9. Describe cancer as a homeostatic imbalance of cells.
10. Define medical terminology and conditions associated with cells.

GENERALIZED ANIMAL CELL
PLASMA (CELL) MEMBRANE
    Chemistry and Structure
    Functions
    Movement of Materials Across
        Plasma Membranes
        Passive Processes
        Active Processes
CYTOPLASM
ORGANELLES
    Nucleus
    Ribosomes
    Endoplasmic Reticulum (ER)
    Golgi Complex
    Lysosomes
    Mitochondria
    The Cytoskeleton
    Centrosome and Centrioles
    Flagella and Cilia
CELL INCLUSIONS
GENE ACTION
    Protein Synthesis
        Transcription
        Translation
NORMAL CELL DIVISION
    Somatic Cell Division
        Mitosis
        Cytokinesis
ABNORMAL CELL DIVISION: CANCER
(CA)
    Definition
    Spread
    Types
    Possible Causes
    Treatment
MEDICAL TERMINOLOGY AND
    CONDITIONS

I t is at the cellular level of organization that activities essential to life occur and disease processes originate. A *cell* is the basic, living, structural and functional unit of the body and, in fact, of all organisms. *Cytology* (sī-TOL-ō-jē; *cyto* = cell; *logos* = study of) is the branch of science concerned with the study of cells. This chapter examines the structure, functions, and reproduction of cells.

GENERALIZED ANIMAL CELL

A *generalized animal cell* is a composite of many different cells in the body. Examine the generalized cell in Figure 3-1, but keep in mind that no such single cell actually exists.

For convenience, we divide the generalized cell into four principal parts:

1. **Plasma (cell) membrane.** The outer, limiting membrane separating the cell's internal parts from the extracellular materials and external environment. (Extracellular materials will be examined in Chapter 4.)
2. **Cytoplasm.** The intracellular fluid between the nucleus and the plasma membrane in which organelles and inclusions are suspended and solutes are dissolved.
3. **Organelles.** Structures that are specialized for specific cellular activities.
4. **Inclusions.** The secretions and storage products of cells.

PLASMA (CELL) MEMBRANE

The exceedingly thin structure that separates one cell from other cells and from the external environment is called the *plasma (cell) membrane.*

CHEMISTRY AND STRUCTURE

The plasma membrane consists primarily of phospholipids (lipids that contain phosphorus) and proteins. Other chemicals in lesser amounts include cholesterol (a lipid), glycolipids (combinations of carbohydrates and lipids), and carbohydrates called oligosaccharides (Figure 3-2). The phospholipid molecules are arranged in two parallel rows, forming a *phospholipid bilayer.*

There are two kinds of membrane proteins: integral and peripheral. *Integral proteins* penetrate through the phospholipid bilayer. *Peripheral proteins* are loosely attached to the exterior or interior surface of the membrane (Figure 3-2).

FUNCTIONS

Following are the functions of the plasma membrane:

1. It provides a flexible boundary that *gives shape* to a cell and *encloses* and *protects* the cellular contents. Some peripheral proteins help to support the membrane.
2. It *separates* a cell from the external environment.
3. Some integral proteins and oligosaccharides provide *receptor sites,* that is, binding sites, that enable a cell to recognize cells of its own kind to form a tissue, respond to foreign cells that might be potentially dangerous, and attach to hormones, nutrients, antibodies, neurotransmitters, and other chemicals.
4. Some peripheral proteins function as *enzymes* that catalyze various chemical reactions.
5. Some peripheral proteins help *change membrane shape* during processes such as cell division, locomotion, and ingestion.

FIGURE 3-1 Generalized animal cell.

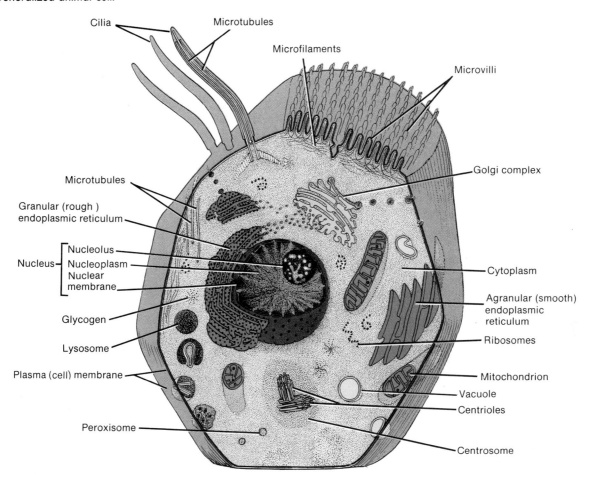

What are four factors that affect the permeability of plasma membranes?

FIGURE 3-2 Plasma membrane. The phospholipids form a bilayer (double layer) and the integral proteins penetrate through it. Peripheral proteins are attached to the inside of the membrane.

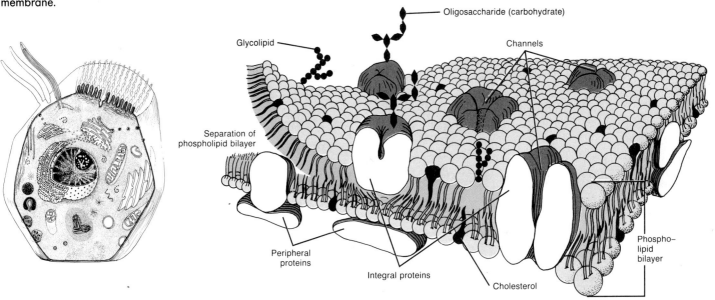

What are the four principal parts of a generalized cell?

6. It regulates the entrance and exit of materials. The ability of a plasma membrane to permit certain substances to enter and exit, but to restrict the passage of others, is called **selective permeability.** Let us look at this function in more detail.

A membrane is said to be *permeable* to a substance if it allows the free passage of that substance. Plasma membranes are not actually freely permeable to any substance, but they do permit some substances to pass more readily than others. Selective permeability is a function of several factors.

1. **Size of molecules.** Large molecules, including most proteins, cannot pass through the plasma membrane. Water and amino acids are small molecules and can generally enter and exit easily. Many scientists believe that giant molecules do not enter the cell because they are larger than the minute channels in the integral proteins and cannot dissolve in the phospholipid bilayer.
2. **Solubility in lipids.** Substances that dissolve easily in lipids pass through the membrane more readily, since a major part of the plasma membrane consists of phospholipid molecules. Lipid-soluble substances include oxygen, carbon dioxide, and steroid hormones.
3. **Charge on ions.** The charge of an ion attempting to cross the plasma membrane can determine how easily the ion enters or leaves the cell. If an ion has a charge opposite that of the protein portion of the membrane, it is attracted to the membrane and passes through more readily. If the ion attempting to cross has the same charge as the membrane, it is repelled.
4. **Presence of carrier molecules.** Some integral proteins function as carriers to transport substances across the membrane regardless of size, ability to dissolve in lipids, or membrane charge. This mechanism will be described shortly.

MOVEMENT OF MATERIALS ACROSS PLASMA MEMBRANES

Before actually discussing how materials move into and out of a cell, we will examine the various fluids through which they move. Fluid outside body cells is called **extracellular** (*extra* = outside) **fluid (ECF).** There are two types, found in two principal places. The fluid filling the microscopic spaces between the cells of tissues is called **interstitial** (in′-ter-STISH-al) **fluid** (*inter* = between) or **intercellular fluid.** The extracellular fluid in blood vessels is called **plasma** (Figure 3-3) and in lymphatic vessels is called **lymph.** Fluid within cells is called **intracellular** (*intra* = within, inside) **fluid (ICF)** or **cytoplasm.** Extracellular fluid contains gases, nutrients, and ions, all needed for maintaining life. Extracellular fluid circulates through the blood and lymphatic vessels and from there moves into the spaces between tissue cells. Thus, it is in constant motion throughout the body in order to maintain homeostasis. Essentially, all body cells are surrounded by the same fluid environment. For this reason, extracellular fluid is often called the body's **internal environment.**

The processes involved in the movement of substances across plasma membranes are classified as either passive or active. In **passive processes,** substances move across plasma membranes with-

FIGURE 3-3 Body fluid. (a) Extracellular fluid (ECF) is found in two principal places: in blood vessels as plasma and between cells as interstitial (intercellular) fluid. Plasma circulates through arteries and arterioles and then into microscopic blood vessels called capillaries. From there it moves into the spaces between body cells, where it is called interstitial fluid. This fluid then returns to capillaries as plasma and passes through the venules and then into the veins. Intracellular fluid (ICF) is found inside body cells. (b) Enlarged detail.

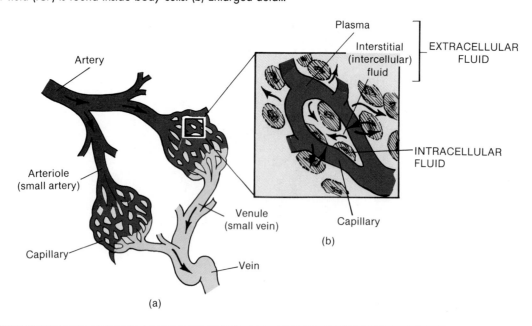

Why is extracellular fluid called the body's internal environment?

out the use of energy (from the splitting of ATP) by the cell. They move by other means that will be described shortly. In *active processes,* the cell uses energy (from the splitting of ATP).

Passive Processes

Molecules in solutions are constantly moving as a result of the energy of motion of molecules called *kinetic energy.* This motion is affected by several factors, including (1) concentration of the molecules of solute and solvent (review this on p. 25), (2) temperature—as temperature increases, the rate of molecular motion increases, and (3) pressures exerted by gravity and other forces.

■ **Diffusion** *Diffusion* occurs where there is a *net* (greater) movement of molecules or ions from a region of higher concentration to a region of lower concentration; that is, they move from an area where there are more of them to an area where there are fewer of them. The movement from high to low continues until the molecules are evenly distributed, a condition called *equilibrium.* The difference between high and low concentrations is callee the *concentration gradient.* Molecules moving from the high-concentration area to the low-concentration area are said to move *down* or *with* the concentration gradient (Figure 3-4). The movement involves only the kinetic energy of individual molecules. The substances move on their own down the concentration gradient.

If a dye pellet is placed in a beaker filled with water, the color of the dye is seen immediately around the pellet. At increasing distances from the pellet, the color becomes lighter. Later, however, the water solution will be a uniform color. The dye molecules possess kinetic energy and move about at random and they move down the concentration gradient from an area of high dye concentration to an area of low dye concentration. The water molecules also move from their high-concentration to their low-concentration area. When dye molecules and water molecules are evenly distributed among themselves, equilibrium is reached and diffusion ceases, even though molecular movements continue. If the water in the beaker were boiled, the rate of diffusion of the dye molecules would increase tremendously since an increase in temperature speeds up diffusion.

Diffusion across selectively permeable membranes follows the same principles as seen in the example of the dye pellet in water. In the human body, diffusion is important in activities such as the movement of oxygen and carbon dioxide between blood and body cells and blood and lungs during respiration, the absorption of nutrients by body cells, the excretion of wastes by body cells, and electrolyte balance—all activities that contribute to homeostasis.

■ **Facilitated Diffusion** In a type of diffusion called *facilitated diffusion,* large and lipid-insoluble substances move across plasma

FIGURE 3-4 Principle of diffusion. Small lipid-insoluble substances may diffuse through channels formed by integral proteins, whereas larger, lipid-soluble substances may diffuse through the phospholipid bilayer.

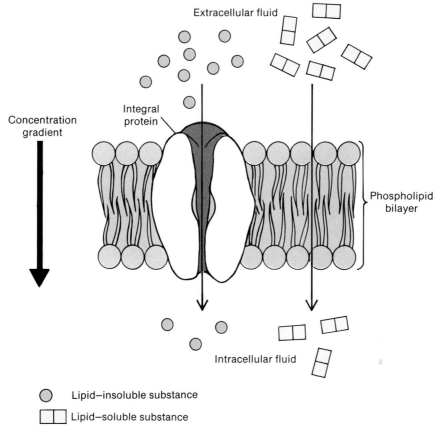

How would fever affect body processes that involve diffusion?

membranes with the assistance of integral proteins that function as carriers. Most sugars, including glucose, are transported via facilitated diffusion. As an example of facilitated diffusion, we will use the transport of glucose across a plasma membrane (Figure 3-5). An integral protein (carrier) attaches to glucose, making it more soluble. Then, by altering its shape, the integral protein transports the glucose across the membrane. In facilitated diffusion, the cell does not use energy from the splitting of ATP, and the movement of large and lipid-insoluble substances is from a region of their higher concentration to a region of their lower concentration.

■ **Osmosis** *Osmosis* is the net movement of water molecules through a selectively permeable membrane from an area of higher water concentration to an area of lower water concentration. Osmosis refers only to water. The water molecules pass through channels in integral proteins in the membrane. Here again, the only energy involved is the kinetic energy of motion.

The apparatus in Figure 3-6 demonstrates the process. It consists of a sac constructed from cellophane, a selectively permeable membrane. The cellophane sac is filled with a colored, 20 percent sugar (sucrose) solution (20 parts sugar and 80 parts water). The upper portion of the cellophane sac is plugged with a rubber stopper through which a glass tube is fitted. The cellophane sac is placed into a beaker containing distilled (pure) water (0 parts sugar and 100 parts water). The concentrations of water on either side of the selectively permeable membrane are different. There is a lower concentration of water inside the cellophane sac than outside. Because of this difference, water begins to move from the beaker into the cellophane sac. This passage of water through a selectively permeable membrane produces a pressure called osmotic pressure. *Osmotic pressure* is the pressure required to prevent the movement of pure water (no solutes) into a solution containing solutes when the two solutions are separated by a selectively permeable mem-

brane. In other words, osmotic pressure is the pressure needed to stop the flow of water across the membrane. The greater the solute concentration of the solution, the greater its osmotic pressure. There is no movement of sugar from the cellophane sac inside the beaker, since the cellophane is impermeable to molecules of sugar: sugar molecules are too large to go through the pores of the membrane. As water moves into the cellophane sac, the sugar solution becomes increasingly diluted and the increased volume and pressure force the mixture up the glass tubing. In time, the weight of the water that has accumulated in the cellophane sac and the glass tube will be equal to the osmotic pressure, and this keeps the water volume from increasing in the tube. When water molecules leave and enter the cellophane sac at the same rate, equilibrium is reached. Osmotic pressure is an important force in the movement of water between various compartments of the body in order to maintain homeostasis.

■ **Isotonic, Hypotonic, and Hypertonic Solutions** Osmosis may also be understood by considering the effects of different water concentrations on red blood cells. If the normal shape of a red blood cell is to be maintained, the cell must be in an *isotonic* (*iso* = same) **solution,** one where the concentrations of water molecules (solvent) and solute (solid) molecules are the same on both sides of the selectively permeable cell membrane. Under ordinary circumstances, a 0.90 percent NaCl (salt) solution is isotonic for red blood cells. In this condition, water molecules enter and exit the cell at the same rate, allowing the cell to maintain its normal shape (Figure 3-7a).

A different situation results if red blood cells are placed in a solution that has a lower concentration of solutes and therefore a higher concentration of water. This is called a **hypotonic** (*hypo* = less than) **solution.** Water molecules then enter the cells faster than they can leave, causing the red blood cells to swell and eventu-

FIGURE 3-5 Facilitated diffusion. An integral protein (carrier) picks up glucose while it is in one shape and transports the glucose across the membrane while it is in another shape.

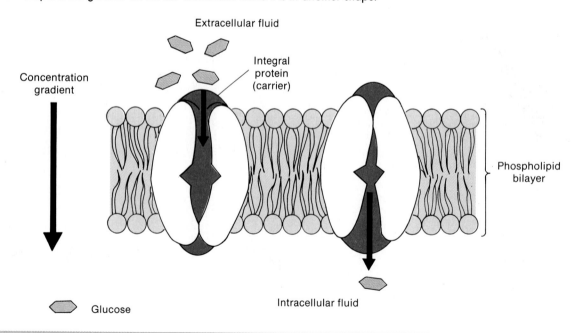

Is energy from ATP required in this process? Why or why not?

FIGURE 3-6 Principle of osmosis. (a) Apparatus at the start of the experiment. (b) Apparatus at equilibrium. In (a), the cellophane sac (a selectively permeable membrane) contains a 20 percent sugar solution and is immersed in a beaker of distilled water. The 20 percent sugar (sucrose) solution contains 20 parts sugar and 80 parts water, whereas the distilled water contains 0 parts sugar and 100 parts water. The arrows indicate that water molecules can pass freely into the sac, but that sugar (sucrose) molecules are held back by the selectively permeable membrane. As water moves into the sac by osmosis, the sugar solution is diluted and the volume of the solution in the cellophane sac increases. This increased volume is shown in (b), with the sugar solution moving up the glass tubing. The final height (FH) reached occurs at equilibrium and represents the osmotic pressure. At this point, the number of water molecules leaving the cellophane sac is equal to the number of water molecules entering the sac.

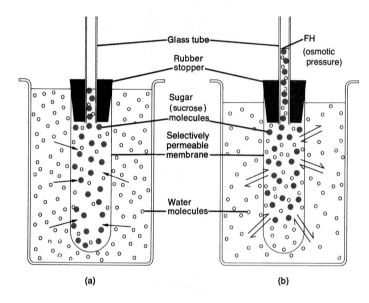

What is osmotic pressure?

ally burst. The rupture of red blood cells is called **hemolysis** (hē-MOL-i-sis) (Figure 3-7b). Distilled water is a strongly hypotonic solution.

A **hypertonic** (*hyper* = greater than) **solution** has a higher concentration of solutes and a lower concentration of water than the red blood cells. A hypertonic solution might be 10 percent NaCl. In such a solution, water molecules move out of the cells faster than they can enter, causing the cells to shrink. This shrinkage is called **crenation** (krē-NĀ-shun) (Figure 3-7c).

■ **Filtration** *Filtration* involves the movement of solvents such as water and dissolved substances such as sugar across a selectively permeable membrane by gravity or mechanical pressure, usually hydrostatic (water) pressure. Such movement is always from an area of higher pressure to an area of lower pressure and continues as long as a pressure difference exists. Most small to medium-sized molecules can be pushed through a cell membrane by filtration pressure.

Filtration is a very important process in which water and nutrients in the blood are pushed into interstitial fluid for use by body cells. It is also the primary force that begins the process of urine formation (Chapter 21).

Active Processes

When cells actively participate in moving substances across membranes, they must expend energy by splitting ATP. The active processes are active transport and endocytosis (phagocytosis, pinocytosis, and receptor-mediated endocytosis).

■ **Active Transport** The process by which substances are transported across plasma membranes, from an area of low concentration

FIGURE 3-7 Principle of osmosis applied to red blood cells. Shown here are the effects on red blood cells when placed in an (a) isotonic solution, in which they maintain normal shape; (b) a hypotonic solution, in which they undergo hemolysis; and (c) a hypertonic solution, in which they undergo crenation.

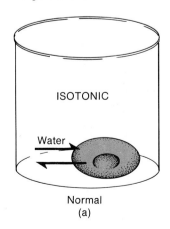
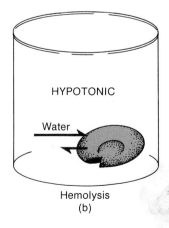
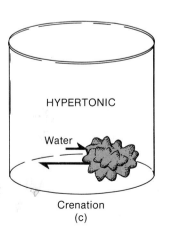

What are hemolysis and crenation?

to an area of high concentration, is called **active transport.** Movement is against the concentration gradient (Figure 3-8). A typical body cell is estimated to use up to 40 percent of its ATP for active transport.

In active transport, the substance being moved enters a channel in an integral protein and makes contact with an active site in the channel. (An **active site** is a particular region on the surface of a cell or molecule—here, the integral protein—that binds a particular kind of molecule or ion that ''fits'' it to cause a chemical reaction.) At contact, ATP splits (the chemical reaction) and releases energy that changes the shape of the integral protein, which, in turn, expels the substance (into or out of the cell).

Active transport is vitally important in maintaining the concentrations of some ions inside body cells and other ions outside body cells, especially potassium and sodium ions. For example, before a nerve cell can conduct an action potential (nerve impulse) the concentration of potassium (K^+) ions must be considerably higher inside the nerve cell than outside. This is accomplished by a ''potassium pump,'' which consists of an integral protein serving as the ''pump'' that is powered by ATP (see Figure 9-4a). At the same time, the nerve cells must have a higher concentration of sodium (Na^+) ions outside than inside. This is accomplished by a ''sodium pump,'' which is also an integral protein powered by ATP (see Figure 9-4a). The ''potassium pump'' and ''sodium pump'' are components of the same pump, referred to as the ''sodium–potassium pump.''

■ **Endocytosis** Large molecules and particles pass through plasma membranes by **endocytosis,** in which a portion of the plasma membrane surrounds the substance, encloses it, and brings it into the cell. The export of substances from the cell occurs by a reverse process called **exocytosis,** a very important mechanism for cells that secrete various substances. There are three kinds of endocytosis: phagocytosis, pinocytosis, and receptor-mediated endocytosis.

In **phagocytosis** (fag'-ō-sī-TŌ-sis), or ''cell eating,'' projections of cytoplasm, called **pseudopodia** (soo'-dō-PŌ-dē-a), surround a large solid particle outside the cell (Figure 3-9a,b). Once the particle is surrounded, the membrane folds inward, forming a membrane sac around the particle. This newly formed sac, called a **phagocytic vesicle,** breaks off from the plasma membrane, and the solid material inside the vesicle is digested by enzymes from organelles called lysosomes (discussed shortly). Phagocytosis is a vital body defense mechanism. Through phagocytosis, certain white blood cells engulf and destroy bacteria and other foreign substances.

In **pinocytosis** (pi'-nō-sī-TŌ-sis), or ''cell drinking,'' the engulfed material is an extracellular liquid rather than a solid. Moreover, no cytoplasmic projections are formed. Instead, a minute droplet of liquid is attracted to the surface of the membrane. The membrane folds inward, forms a **pinocytic vesicle** that surrounds the liquid, and detaches from the membrane. Indigestible particles and cell products are removed from the cell by exocytosis. Few cells, such as endothelial (lining) cells of the wall of blood vessels, especially capillaries, are capable of phagocytosis, but many carry out pinocytosis, especially in the kidneys and urinary bladder.

Receptor-mediated endocytosis applies to large molecules or particles, but it is a highly selective process, only transporting certain substances. The interstitial fluid that bathes cells contains many chemicals, most of which are in concentrations lower than in the cells themselves. These chemicals are called **ligands.** Ligands include nutrients (amino acids, iron, and vitamins), hormones, and waste products or poisonous materials that would interfere with normal cell functioning if not brought in and digested.

Receptor-mediated endocytosis is similar to phagocytosis but here the substance attaches to a receptor protein on the plasma membrane prior to the phagocytic process (Figure 3-9c). Afterward, the receptor protein detaches from the substance and returns to the membrane for reuse. Once inside the cell the ligands are usually broken down by powerful enzymes provided by lysosomes.

FIGURE 3-8 Active transport. The substance transported moves from lower to higher concentration and the cell uses energy from the splitting of ATP.

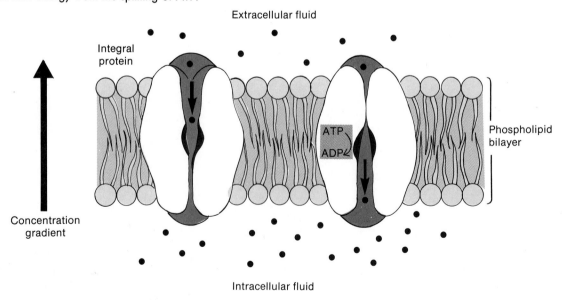

Why is this considered an active process?

FIGURE 3-9 Endocytosis. (a) Diagram of phagocytosis. (b) Photomicrographs of phagocytosis. On the left, a white blood cell engulfs a microbe. The right shows a later stage of phagocytosis in which the engulfed microbe is being destroyed. (Courtesy of Abbott Laboratories.) (c) Receptor-mediated endocytosis.

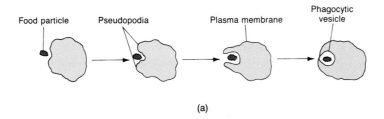

(a)

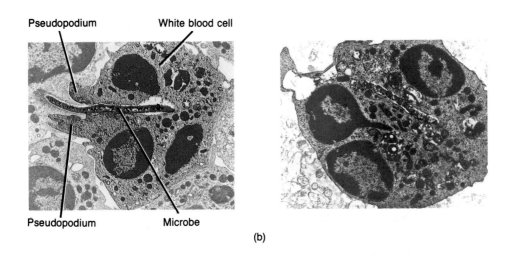

(b)

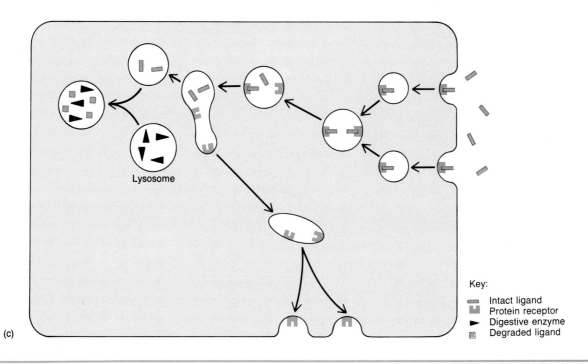

(c)

Key:

- Intact ligand
- Protein receptor
- Digestive enzyme
- Degraded ligand

What is the opposite process of endocytosis called?

The various passive and active processes by which substances move across plasma membranes are summarized in Exhibit 3-1.

EXHIBIT 3-1
Summary of Processes by Which Substances Move Across Membranes

Process	Description
PASSIVE PROCESSES	Substances move on their own down a concentration gradient from an area of higher to lower concentration or pressure; cell does not expend energy (ATP).
Diffusion	Net movement of molecules or ions due to their kinetic energy from an area of higher to lower concentration until an equilibrium is reached.
Facilitated Diffusion	Diffusion of larger molecules across a selectively permeable membrane with the assistance of integral proteins in the membrane that serve as carriers.
Osmosis	Net movement of *water* molecules due to kinetic energy across a selectively permeable membrane from an area of higher to lower concentration of water until an equilibrium is reached. The solute concentration produces osmotic pressure.
Filtration	Movement of solvents (such as water) and solutes (such as glucose) across a selectively permeable membrane as a result of gravity or hydrostatic (water) pressure from an area of higher to lower pressure.
ACTIVE PROCESSES	Substances move against a concentration gradient from an area of lower to higher concentration; cell must expend energy (ATP).
Active Transport	Movement of substances across a selectively permeable membrane from a region of lower to higher concentration with integral proteins as carriers; the process requires energy expenditure.
Endocytosis	Movement of large molecules and particles through plasma membranes in which the membrane surrounds the substance, encloses it, and brings it into the cell. Examples include phagocytosis (''cell eating''), pinocytosis (''cell drinking''), and receptor-mediated endocytosis.
Exocytosis	Export of substances from the cell by reverse endocytosis.

CYTOPLASM

The substance inside the cell's plasma membrane and external to the nucleus that contains organelles and inclusions is called *cytoplasm* (see Figure 3-1). Cytoplasm is a thick, semitransparent, elastic fluid. It contains organelles, a cytoskeleton (which will be described shortly), and various organic and inorganic particles dissolved or suspended in water which makes up 75 to 90 percent of cytoplasm.

Some chemical reactions occur in the cytoplasm. It receives raw materials from the external environment and converts them into usable energy by decomposition reactions. Cytoplasm is also the site where new substances are synthesized for cellular use. It packages chemicals for transport to other parts of the cell or body and facilitates the excretion of waste materials.

ORGANELLES

Despite the numerous chemical activities occurring simultaneously in the cell, there is little interference of one reaction with another. This is because the cell has a system of compartmentalization provided by its *organelles* (''little organs''). These are specialized structures that have a characteristic appearance and specific roles in growth, maintenance, repair, and control. The numbers and types of organelles vary in different kinds of cells, depending on their functions.

NUCLEUS

The *nucleus* is generally spherical or oval. It is the largest structure in the cell (Figure 3-10). The nucleus is the control center of the cell because it contains the body's genes, which carry the inherited directions for cell structure and cellular activities. Most body cells contain a single nucleus, although some, such as mature red blood cells, do not have a nucleus. Skeletal muscle fibers (cells) and a few other cells contain several nuclei.

The nucleus is bounded by a membranous sac called the *nuclear membrane (envelope)*. The surface of the nuclear membrane is studded with ribosomes (described next) and is continuous at certain points with granular (rough) endoplasmic reticulum (also described shortly). Throughout the nuclear membrane are openings called *nuclear pores* that provide a means of communication between the nucleus and cytoplasm, specifically for the passage of large molecules such as RNA and various proteins.

Inside the nucleus are *nucleoplasm,* like cytoplasm, and one or more spherical bodies called *nucleoli.* They are composed of protein, DNA, and RNA, and they are not membrane-bound. They can disperse and disappear, which they do during cell division, re-forming once new cells are formed. A type of RNA called ribosomal RNA (rRNA), which is very important in protein synthesis, is produced in nucleoli. Finally, DNA (the hereditary material) and protein form a loose mass known as *chromatin* (which coils into rod-shaped bodies called *chromosomes* during cell division).

RIBOSOMES

Ribosomes are tiny granules composed of ribosomal RNA (rRNA) and protein. Structurally, ribosomes consist of two subunits, one

FIGURE 3-10 Diagram of a nucleus with one nucleolus.

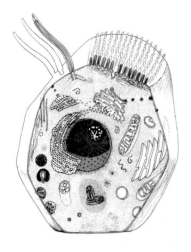

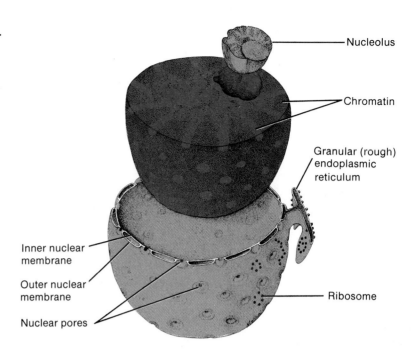

Why is the nucleus called the control center of the cell?

FIGURE 3-11 Endoplasmic reticulum (ER) and ribosomes. (a) Diagram of the endoplasmic reticulum and ribosomes. (b) Diagram of the three-dimensional structure of ribosomes.

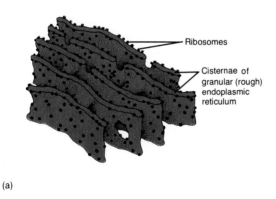

(a)

What is the difference between smooth ER and granular ER?

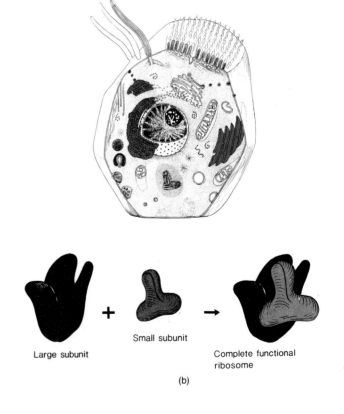

about half the size of the other (Figure 3-11b). Ribosomes are the sites of protein synthesis (which is discussed later in the chapter).

Some ribosomes are scattered in the cytoplasm; others are attached to an organelle called the endoplasmic reticulum (ER).

ENDOPLASMIC RETICULUM (ER)

Throughout the cytoplasm winds a system of double membranous channels (*cisternae*) of varying shapes called the **endoplasmic reticulum** or **ER** (Figure 3-11a). The channels are continuous with the nuclear membrane. There are two types of ER. **Granular (rough) ER** is studded with ribosomes; **agranular (smooth) ER** is free of ribosomes.

Endoplasmic reticulum has numerous functions. Both types help support and distribute the cytoplasm. Both provide a transportation system for the intracellular exchange of materials with the cytoplasm

FIGURE 3-12 Diagram of a Golgi complex.

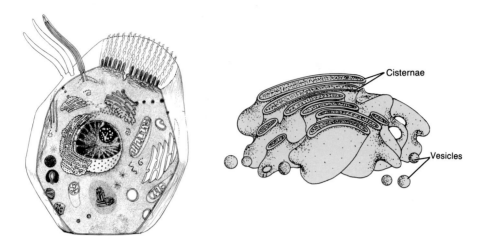

What is the main function of the Golgi complex?

and a surface area for chemical reactions. Smooth ER is involved in the production of lipids, particularly steroid hormones. Granular ER stores the proteins synthesized by ribosomes, packages them, and transports them to another organelle, the Golgi complex.

GOLGI COMPLEX

The **Golgi** (GOL-jē) **complex** is generally near the nucleus. It consists of four to eight flattened membranous sacs (cisternae) that are stacked on each other like a pile of dishes with expanded areas at their ends (vesicles) (Figure 3-12).

The Golgi complex acts as a processing and packaging plant. It processes proteins received from the granular ER and packages them into vesicles. Some vesicles may become lysosomes; others (secretory vesicles) move to the cell membrane to release various secretions from the cells, such as digestive tract enzymes. Lipids, such as steroids, received from agranular ER are similarly packaged and released within the cell or by the cell membrane.

LYSOSOMES

Lysosomes (*lysis* = dissolution; *soma* = body) are membrane-enclosed sacs that contain powerful digestive enzymes capable of breaking down many kinds of molecules (Figure 3-13). These enzymes are also capable of digesting bacteria and other substances that enter the cell in phagocytic vesicles (see Figure 17-6). White blood cells, which ingest bacteria by phagocytosis, contain large numbers of lysosomes.

Lysosomes not only function in phagocytosis, pinocytosis, and receptor-mediated endocytosis, they also use their enzymes to recycle the cell's own molecules. A lysosome can engulf another organelle, digest it, and return the digested components to the cytoplasm for reuse. In this way, worn-out cellular structures are continually renewed. The process by which worn-out organelles are digested is called **autophagy** (aw-TOF-a-jē; *auto* = self; *phagio* = to eat). Normally, the lysosomal membrane is impermeable to the passage of enzymes into cytoplasm. This prevents digestion of the cytoplasm. However, under certain conditions, *au-*

FIGURE 3-13 Diagram of a lysosome.

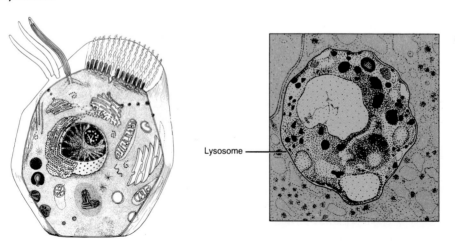

What is inside a lysosome?

tolysis or programmed self-destruction by lysosomes occurs. For example, autolysis is responsible for muscle shrinkage during prolonged inactivity and reduction of the size of the uterus following delivery.

Lysosomes also function in extracellular digestion. Lysosomal enzymes released at sites of injury help digest away cellular debris, which prepares the injured area for repair.

MITOCHONDRIA

Small, rod-shaped structures called *mitochondria* (mī-tō-KON-drē-a) appear throughout the cytoplasm. Because of their function in generating energy, they are referred to as "powerhouses" of the cell (Figure 3-14). A mitochondrion consists of two membranes, each of which is similar in structure to the plasma membrane. The outer mitochondrial membrane is smooth, but the inner membrane is arranged in a series of folds called *cristae*. The center of a mitochondrion is called the *matrix.*

The folds of the cristae provide an enormous surface area for chemical reactions, and it is here that ATP is produced during the aerobic phase of cellular respiration. The enzymes that catalyze ADP into ATP are located on the cristae. Cells that expend a lot of energy, such as muscle, liver, and kidney tubule cells, have a large number of mitochondria.

THE CYTOSKELETON

Cytoplasm has an internal structure consisting of microfilaments, microtubules, and intermediate filaments, together referred to as the *cytoskeleton* (see Figure 3-1).

Microfilaments are rodlike structures made of the proteins *actin* and *myosin*. Microfilaments are involved in the contraction of muscle fibers (cells). In nonmuscle cells, microfilaments provide support and shape and assist in locomotion of entire cells (as in the movement of phagocytes) and movements within cells.

Microtubules are straight, slender tubes made of the protein *tubulin*. Besides providing support and shape for cells, microtubules serve as channels for substances to move through the cytoplasm. As you will see shortly, other cell structures are also composed of microtubules. These include flagella and cilia, centrioles, and the mitotic spindle.

Intermediate filaments are made up of various proteins. The functions of intermediate filaments are not completely understood, but it appears they provide structural reinforcement in some cells and assist in contraction in others.

CENTROSOME AND CENTRIOLES

A *centrosome* is a dense area of cytoplasm near the nucleus. In the centrosome is a pair of cylindrical structures called *centrioles* (see Figure 3-1). Centrioles are composed of microtubules and they are at right angles to each other. Centrioles are involved in cell division and reproduction, to be described shortly. Certain cells, such as nerve cells, do not have centrioles and so do not reproduce, which is why they cannot be replaced if destroyed.

FLAGELLA AND CILIA

Some body cells possess projections for moving the entire cell or for moving substances along the surface of the cell. These projections contain cytoplasm and are covered by the plasma membrane. If the projections are few (usually occurring singly or in pairs) and long, they are called *flagella*. The only example of a flagellum in the human body is the tail of a sperm cell, used for locomotion (see Figure 23-4). If the projections are numerous and short, resembling many hairs, they are called *cilia*. Ciliated cells of the respiratory tract move mucus that has trapped foreign particles (see Figure 18-4). Both cilia and flagella consist of microtubules.

FIGURE 3-14 **Diagram of a mitochondrion.**

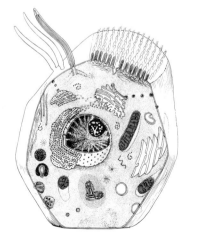
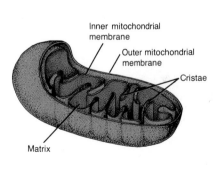

Inner mitochondrial membrane
Outer mitochondrial membrane
Cristae
Matrix

How does the multifold inner structure of a mitochondrion contribute to its energy-producing function?

CELL INCLUSIONS

Cell inclusions are chemical substances produced by cells. Some have recognizable shapes. They are mostly organic and may appear or disappear at various times. *Melanin,* a pigment stored in certain cells of the skin, hair, and eyes, protects the body by screening out ultraviolet rays from the sun. *Glycogen,* a polysaccharide stored in the liver and skeletal muscle cells, provides quick energy in these cells when broken down into glucose. *Lipids,* which are stored in adipocytes (fat cells), can also be decomposed for energy.

The major parts of the cell and their functions are summarized in Exhibit 3-2.

EXHIBIT 3-2
Cell Parts and Their Functions

Part	Functions
Plasma Membrane	Protects cellular contents; makes contact with other cells; provides receptors for hormones, enzymes, and antibodies; regulates entrance and exit of materials.
Cytoplasm	Serves as the substance in which chemical reactions occur.
Organelles	
Nucleus	Contains genes and therefore controls cellular activities.
Ribosomes	Sites of protein synthesis.
Endoplasmic Reticulum (ER)	Supports cytoplasm; provides transportation system for intracellular exchange of materials; provides surface area for chemical reactions; smooth ER helps produce lipids such as steroid hormones; granular ER stores, packages, transports protein to Golgi complex.
Golgi Complex	Packages synthesized proteins for secretion; produces lysosomes; secretes lipids.
Mitochondria	Sites for production of ATP.
Lysosomes	Digest cell substances, bacteria, and foreign microbes.
Cytoskeleton	
Microfilaments	Involved in muscle fiber (cell) contraction; provide support and shape; assist in cellular and intracellular movement.
Microtubules	Provide support and shape; form intracellular conducting channels; assist in cellular movement; form flagella, cilia, centrioles, and mitotic spindle.
Intermediate Filaments	Provide structural reinforcement in some cells.
Centrioles	Help organize mitotic spindle during cell division.
Flagella and Cilia	Allow movement of entire cell (flagella) or movement of particles along surface of cell (cilia).
Inclusions	Melanin (pigment in skin, hair, eyes) screens out ultraviolet rays; glycogen (stored glucose) decomposes to provide energy; lipids (stored in fat cells) decompose to produce energy.

GENE ACTION

PROTEIN SYNTHESIS

Basically, cells are protein factories that constantly produce large numbers of different proteins. Some proteins are structural, helping to form plasma membranes, microfilaments, microtubules, centrioles, flagella, cilia, the mitotic spindle, and other parts of cells. Others serve as hormones, antibodies, and contractile elements in muscle tissue. Still others serve as enzymes that regulate the chemical reactions that occur in cells.

The instructions for making proteins are in DNA. Cells make proteins by translating the genetic information encoded in DNA into specific proteins. In the process, the genetic information in DNA is copied to produce a molecule of RNA. Then the information in RNA is translated into a new protein molecule. Let us take a look at how DNA directs protein synthesis. There are two principal steps, transcription and translation.

Transcription

Transcription is the process by which genetic information in DNA is transferred to a type of RNA called *messenger RNA (mRNA).* It is called transcription because it resembles the transcription, or copying, of a sequence of words from one tape to another. In transcription, a segment of DNA uncoils and free nitrogenous bases attach to one-half of the segment, forming a strand of RNA, which now has the blueprint or pattern for a particular type of protein. Thus, the genetic information stored in the nitrogenous bases of one side of that segment is copied to the nitrogenous bases of the mRNA strand (Figure 3-15). Specifically, the "rewriting" works this way: cytosine (C) in the DNA template or pattern dictates a guanine (G) in the mRNA strand being made; a G in the DNA template dictates a C in the mRNA strand; and a thymine (T) in the DNA template dictates an adenine (A) in the mRNA. Since RNA contains uracil (U) instead of T, an A in the DNA template dictates a U in the mRNA. For example, if the DNA template has the nitrogenous base sequence of ATGCAT, the transcribed mRNA strand would have the complementary base sequence UACGUA.

DNA synthesizes mRNA through transcription. DNA also synthesizes two other kinds of RNA. One is *ribosomal RNA (rRNA)* that, with protein, makes up ribosomes. The other is called *transfer RNA (tRNA),* to be explained next. Once synthesized, mRNA, rRNA, and tRNA leave the nucleus of the cell. In the cytoplasm, they participate in the next step in protein synthesis—translation.

Translation

Just as DNA provides the template for mRNA to be made, so mRNA provides a template for a protein to be made. *Translation* is the process by which the information in the nitrogenous bases of mRNA "direct" the arrangement of the amino acids of a protein (Figure 3-16). It is the arrangement of the amino acids that defines a protein, that is, whether it is an enzyme, or plasma membrane carrier, or contractile element in a muscle, and so on. Translation then is the key part of protein synthesis.

FIGURE 3-15 Transcription. In the process, the genetic information in DNA is copied to mRNA. Once completed, mRNA leaves the nucleus and enters the cytoplasm where translation occurs. DNA also synthesizes rRNA and tRNA.

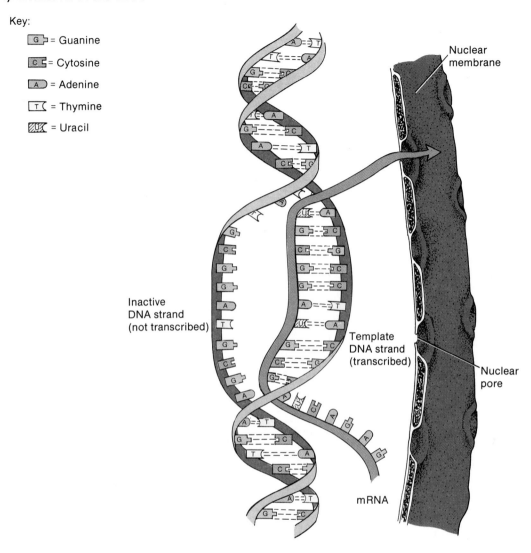

Key:

G ⊐ = Guanine

C Ϲ = Cytosine

A ⊃ = Adenine

T ⊂ = Thymine

U = Uracil

Nuclear membrane

Inactive DNA strand (not transcribed)

Template DNA strand (transcribed)

Nuclear pore

mRNA

If the DNA template had the nitrogenous base sequence AGCT, what would the mRNA base sequence be?

Translation, and all of protein synthesis, takes place on a ribosome. One protein molecule is made per ribosome. The process begins when an mRNA strand attaches to a ribosome. Now amino acids must be brought to the ribosome so the actual protein can be formed, and that is the job of tRNA.

The tRNA picks up whichever of the 20 different amino acids in the cytoplasm that may participate in protein synthesis. For each different amino acid, there is a different tRNA (so there are 20 kinds of tRNA too). One end of a tRNA molecule attaches to an amino acid. The other end has three nitrogenous bases (triplet) known as an **anticodon.** The tRNA, with the attached amino acid, travels to the ribosome. There, the anticodon recognizes a corresponding triplet, or **codon,** on the mRNA and attaches to it. If the tRNA anticodon is UAC, the mRNA codon would be AUG (Figure 3-16c).

Once the tRNA attaches to mRNA, the ribosome moves along the mRNA strand and the next tRNA with its amino acid attaches. These two amino acids are joined by a peptide bond through dehydration synthesis and the first tRNA is released and can pick up another molecule of the same amino acid if necessary.

Amino acids are attached one by one until the protein is complete and further synthesis is stopped by a special *termination codon.* The newly formed protein molecule is then released from the ribosome.

As the ribosome moves along the mRNA strand, it "reads" the information coded in mRNA and synthesizes a protein according to that information; the ribosome synthesizes the protein by translating the codon sequences into an amino acid sequence (Figure 3-16i).

Based on this description of protein synthesis, we can now

FIGURE 3-16 Translation. Note that during protein synthesis the ribosomal subunits join together. When not involved in protein synthesis, the two subunits separate.

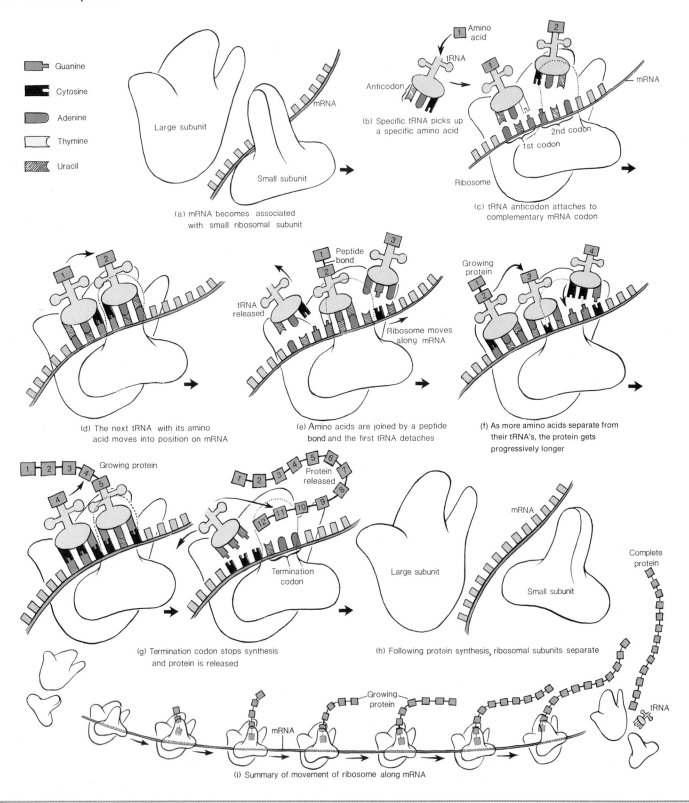

Guanine

Cytosine

Adenine

Thymine

Uracil

(a) mRNA becomes associated with small ribosomal subunit

(b) Specific tRNA picks up a specific amino acid

(c) tRNA anticodon attaches to complementary mRNA codon

(d) The next tRNA with its amino acid moves into position on mRNA

(e) Amino acids are joined by a peptide bond and the first tRNA detaches

(f) As more amino acids separate from their tRNA's, the protein gets progressively longer

(g) Termination codon stops synthesis and protein is released

(h) Following protein synthesis, ribosomal subunits separate

(i) Summary of movement of ribosome along mRNA

In a sentence, what is translation?

define a *gene* as a group of nucleotides on a DNA molecule that serves as the master pattern for manufacturing a specific protein. Genes average about 1000 pairs of nucleotides, which appear in a specific sequence on the DNA molecule. No two genes have exactly the same sequence of nucleotides and this is the key to heredity.

Remember that the nitrogenous base sequence of the gene determines the sequence of the bases in the mRNA. The sequence of the bases in the mRNA then determines the sequence and kind of amino acids that will form the protein. Thus, each gene is responsible for making a particular protein as follows:

$$\text{DNA} \xrightarrow{\text{transcription}} \text{RNA} \xrightarrow{\text{translation}} \text{protein}$$

NORMAL CELL DIVISION

Most of the cell activities discussed thus far maintain the cell on a day-to-day basis. However, cells become damaged, diseased, or worn out and die. New cells must be produced as replacements and for growth. In a 24-hour period, the average adult loses billions of cells from different parts of the body. Obviously, these cells must be replaced. Cells that have a short life span, like those of the outer layer of skin, are continually replaced. In addition, sperm and egg cells must be produced.

Cell division is the process by which cells reproduce themselves. It consists of a nuclear division (mitosis) and a cytoplasmic division (cytokinesis) which will be described shortly. Because there are two kinds of nuclear division, two kinds of cell division are recognized.

In the first kind of division, *somatic cell division,* a single starting cell called a *parent cell* divides to produce two identical cells called *daughter cells*. This ensures that each daughter cell has the same *number* and *kind* of chromosomes as the original parent cell and thus the same hereditary material and genetic potential as the parent cell. Somatic cell division results in an increase in the number of body cells and is the means by which dead or injured cells are replaced and new ones are added for body growth.

The second type of cell division is called *reproductive cell division.* It is the mechanism by which sperm and egg cells are produced, the cells required to form a new organism. This process consists of a nuclear division called *meiosis* plus *cytokinesis*. Here we will discuss somatic cell division; reproductive cell division will be discussed in Chapter 23.

SOMATIC CELL DIVISION

Human cells contain 23 pairs of chromosomes. A *chromosome* is a tightly coiled DNA molecule that stores hereditary information in units called *genes.* Following somatic cell division, both daughter cells must also contain 23 pairs of chromosomes identical to those of the parent cell DNA. Therefore, prior to mitosis, the parent cell must produce a duplicate set of chromosomes by replicating (producing duplicates of) its DNA.

DNA replication takes place when a cell is between divisions, when it is said to be in *interphase.* Also during this stage, the

RNA and protein needed to produce all the structures doubled in cell division are manufactured. When DNA replicates, its helix partially uncoils (Figure 3-17). Those portions of DNA that remain coiled stain darker than the uncoiled portions. This unequal distribution of stain causes the DNA to appear as a granular mass called *chromatin* (see Figure 3-18a). To uncoil, the DNA separates where its nitrogenous bases are connected. Each exposed nitrogenous base then picks up a complementary nitrogenous base. Uncoiling and complementary base pairing continue until each of the two original DNA strands is matched and joined with two newly formed DNA strands. The original DNA molecule has become two DNA molecules. Also during interphase, the paired centrioles in the cell also replicate so that two pairs are present. Now mitosis begins.

FIGURE 3-17 Replication of DNA. The two strands of the double helix separate by breaking the bonds between nucleotides. New nucleotides attach at the proper sites, and a new strand of DNA is paired off with each of the original strands. After replication, the two DNA molecules, each consisting of a new and an old strand, return to their helical structure.

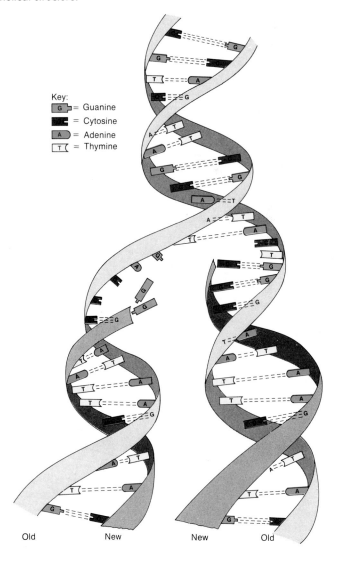

Key:
- G = Guanine
- = Cytosine
- A = Adenine
- T = Thymine

Old   New   New   Old

Why is DNA replication a key step that occurs prior to somatic cell division?

Mitosis

Mitosis is the distribution of the two sets of chromosomes into two separate and equal nuclei following the replication of the chromosomes of the parent nucleus. Biologists divide the process into four stages: prophase, metaphase, anaphase, and telophase. These are arbitrary classifications, but mitosis is actually a continuous process, with one stage merging imperceptibly into the next.

▪ **Prophase** During *prophase* (*pro* = before) (Figure 3-18b), the first stage of mitosis, chromatin shortens and coils into chromosomes. The nucleoli become less distinct and the nuclear membrane

FIGURE 3-18 **Cell division: mitosis and cytokinesis.** Photomicrographs and diagrammatic representations of the various stages of cell division in whitefish eggs. Read the sequence starting at (a) and move clockwise until you complete the cycle. (Photographs by Carolina Biological Supply Company.)

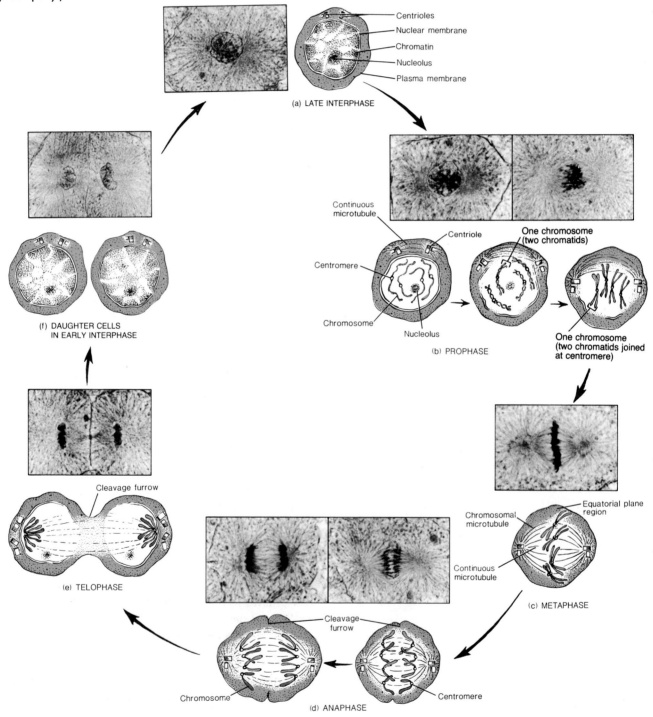

When does cytokinesis begin and end?

disappears. Each prophase "chromosome" is actually composed of a pair of structures called *chromatids.* A chromatid is a complete chromosome consisting of a double-stranded DNA molecule, and each is attached to its chromatid partner by a small spherical body called a *centromere.* During prophase, the chromatid pairs assemble across the center of the cell, which is called the *equatorial plane region* (*equator*).

Also during prophase, the two pairs of centrioles separate and each pair moves to an opposite pole (end) of the cell. Then, a group of *continuous microtubules* in the vicinity of each pair of centrioles begins to grow toward each other. Thus, they extend from one pole of the cell to another. As they grow toward each other, a second group of microtubules develops, called *chromosomal microtubules,* apparently growing out of the centromeres. They extend from a centromere to a pole of the cell. Together, the continuous and chromosomal microtubules constitute the *mitotic spindle.*

■ **Metaphase** During *metaphase* (*meta* = after) (Figure 3-18c), the second stage of mitosis, the centromeres of the chromatid pairs line up on the equatorial plane of the cell. The centromeres of each chromatid pair form a chromosomal microtubule that attaches the centromere to a pole of the cell.

■ **Anaphase** In the third stage of mitosis, *anaphase* (*ana* = upward) (Figure 3-18d), the centromeres divide and the complete identical sets of chromatids, now called chromosomes, move to opposite poles of the cell. During this movement, the centromeres attached to the chromosomal microtubules seem to drag the trailing parts of the chromosomes toward opposite poles.

■ **Telophase** *Telophase* (*telos* = far or end) (Figure 3-18e), the final stage of mitosis, consists of a series of events that are nearly the reverse of prophase. By this time, two identical sets of chromosomes have reached opposite poles. In telophase, new nuclear membranes enclose the chromosomes, the chromosomes start to assume their chromatin form, nucleoli reappear, and the mitotic spindles disappear. The formation of two nuclei identical to those of cells in interphase terminates telophase. A mitotic cycle has thus been completed (Figure 3-18f).

■ **Time Required** The time required for mitosis, and for each stage, varies with the kind of cell, its location, and such factors as temperature. Prophase is usually the longest stage, lasting from one to several hours. Metaphase ranges from 5 to 15 minutes. Anaphase is the shortest stage, lasting from 2 to 10 minutes. Telophase lasts from 10 to 30 minutes. Overall, mitosis represents only a small part of the life cycle of the cell.

Cytokinesis

Division of the parent cell's cytoplasm and organelles is called *cytokinesis* (sī'-tō-ki-NĒ-sis). It usually begins during late anaphase and terminates at the same time as telophase. Cytokinesis begins with the formation of a *cleavage furrow* that extends around the cell's equatorial plane region. The furrow progresses inward, like a constricting ring, finally cutting completely through the cell to form two separate portions of cytoplasm and organelles (Figure 3-18d–f).

ABNORMAL CELL DIVISION: CANCER (CA)

DEFINITION

When cells in some area of the body duplicate without control, the excess tissue that develops is called a *tumor, growth,* or *neoplasm.* The study of tumors is called *oncology* (*onco* = swelling or mass; *logos* = study of) and a physician who specializes in this field is called an *oncologist.* Tumors may be cancerous and sometimes fatal or they may be quite harmless. A cancerous growth is called a *malignant tumor* or *malignancy.* A noncancerous growth is called a *benign tumor.* Benign tumors do not spread to other parts of the body, but they may be removed if they interfere with a normal body function or are disfiguring.

SPREAD

Cells of malignant tumors duplicate continuously and very often quickly and without control. They are also capable of *metastasis* (me-TAS-ta-sis), the spread of cancerous cells to other parts of the body. Cancer cells secrete a protein called *autocrine motility factor* (*AMF*) that enables them to metastasize. Metastatic groups of cells are more difficult to detect and eliminate than primary tumors.

In the process of metastasis, there is an initial invasion of the malignant cells into surrounding tissues. As the cancer grows, it begins to compete with normal tissues for space and nutrients. Eventually, the normal tissue atrophies (wastes away) and dies. When nonmalignant cells of the body divide and migrate (e.g., skin cells that multiply to heal a superficial cut), their further migration is inhibited by contact on all sides with other skin cells. This is called *contact inhibition.* Unfortunately, malignant cells do not conform to the rules of contact inhibition and thus invade healthy body tissues.

Following invasion, some of the malignant cells may detach from the initial (primary) tumor and invade a body cavity (abdominal or thoracic) or enter the blood or lymph. This can lead to widespread metastasis, the next step in metastasis. Malignant cells now invade adjacent body tissues and establish additional (secondary) tumors. It is believed that additional tumor cells have certain properties that even enhance metastasis. In the final stage of metastasis, the additional tumors establish networks of blood vessels that provide nutrients for their further growth. Proteins that serve as chemical triggers for blood vessel growth are called *tumor angiogenesis factors* (*TAFs*), which have been isolated from human colon tumors. In all stages of metastasis, the malignant cells resist the antitumor defenses of the body. The pain associated with cancer develops when the growth puts pressure on nerves or blocks a passageway so that secretions build up pressure.

The Cancer-Prevention Life-Style

While the process involved in the development of cancer is only partially understood, it is known that people often get cancer through contact with one or more carcinogens that cause the genetic changes that allow cells to grow out of control.

It is believed that there are two types of carcinogens involved in causing most cancers. Initiators start the cellular damage that can lead to cancer, and promoters allow genetically damaged cells to proliferate at a greater rate than the normal cells. For example, alcohol promotes cancer of the mouth and esophagus when combined with an initiator such as tobacco. Cancers are more common in tissues that already have a high rate of proliferation, such as the skin, gastrointestinal tract lining, and the uterine lining.

Cancer rates vary considerably among population groups. Scientists try to draw links between life-style and cancer risk by comparing groups with high and low rates of various types of cancer. It is not easy to isolate life-style factors. Once scientists have strong evidence that a particular factor is associated with cancer risk, they then test it on animals in the laboratory, or follow more groups of people to see if indeed that factor is consistently associated with higher cancer rates.

Scientists at the National Cancer Institute estimate that about 80 percent of all cancer cases are related to life-style. Life-style factors can be a source of carcinogens that initiate and/or promote cancer. The good news is that many cancers are preventable by avoiding carcinogens and following recommendations for a cancer-prevention life-style.

Tobacco Use

Former U.S. Surgeon General C. Everett Koop called tobacco use America's single most important preventable cause of death. Tobacco use contributes to the three leading causes of death in North America: heart disease, cancer, and stroke.

Cigarette smoking causes 30 percent of all cancer deaths. About 90 percent of lung-cancer patients are smokers. Cigarette smoking also causes cancers of the larynx, esophagus, pancreas, bladder, kidney, and mouth. Low-tar and nicotine cigarettes are no solution, because they actually increase the smoker's risk of cancers of the mouth and throat. Pipe smoking increases the smoker's risk of cancers of the mouth, tongue, and throat, and chewing tobacco increases the user's risk of cancer of the mouth.

Tobacco smoke contains hundreds of damaging chemicals, and includes both cancer initiators and promoters. For example, the phenols found in tobacco tar greatly increase the carcinogenic potency of benzopyrene, another substance found in cigarette tar. (Cigarette tar refers collectively to several hundred different chemicals in cigarettes that, when condensed, form a brown, sticky substance.)

Alcohol

Heavy alcohol use (more than two drinks a day) is associated with cancers of the mouth, throat, esophagus, and liver. (One drink contains about an ounce of alcohol, and is comparable to 1½ ounces of liquor, a 5-ounce glass of wine, or a 12-ounce glass of beer.) As already mentioned, people who drink and smoke have a much greater risk of getting cancers of the mouth and esophagus. Alcohol may also contribute to the development of breast cancer.

Diet

Several dietary factors are associated with a person's risk of developing cancer. Carcinogens are found in foods that are smoked, cured, and pickled. Consumption of pesticide residues on fruits and vegetables and in meats may increase an individual's cancer risk.

Fat intake may be the major and most controllable dietary carcinogen in North America. Fat appears to act as a cancer promoter, especially in cancers of the breast and colon. While for the population of North America as a whole, lung cancer is the leading cause of cancer deaths for both men and women, nonsmoking men are more likely to develop colon cancer, the second leading cause of cancer deaths for men. For nonsmoking women, breast and colon cancers are the first and second leading causes of death.

Fat may act through the action of hormones such as estrogen, or by suppressing the immune response. Foods high in fat include meats with marbling, cold cuts, fried foods, whole milk and cream, creamed soups and sauces, rich desserts, and any food prepared with butter, oils, or margarine.

Some dietary chemicals may block promotion: retinoids, vitamins C and E, and the mineral selenium. Researchers have theorized that some carcinogens cause cancer by producing highly reactive oxygen atoms called free radicals, which cause damage to cell components, such as DNA. Retinoids, vitamins C and E, and selenium appear to act as antioxidants, neutralizing these reactive chemicals and thus blocking their carcinogenic effect.

Retinoids are metabolic by-products of many dark green or yellow vegetables. Beta-carotene, the most studied of the retinoids, is responsible for the yellow/orange color of plants. (In dark green vegetables like broccoli, it is masked by the darker green pigment chlorophyll.) Retinoids can be converted to vitamin A in the body. Animals fed a high-retinoid diet are very resistant to the carcinogenic effect of promoters. Preliminary studies show that people whose diets are rich in retinoids have fewer cancers of the lung, colon, stomach, prostate, and cervix.

While too much vitamin A is toxic, retinoids in quantity appear to be safe. Foods are preferable to supplements, since foods high in retinoids are also good sources of other vitamins, minerals, and fiber. Good sources of retinoids include dark green vegetables such as spinach, beet greens, broccoli, and kale and yellow fruits and vegetables like cantaloupe, apricots, carrots, yams, and winter squashes.

Vitamin C is found in many fruits and vegetables, such as citrus fruits, strawberries, and potatoes. Vitamin E, a normal component of cell membranes, is found in vegetable oils, and selenium is found in seafood, whole grains, and organ meats.

Studies on both humans and animals suggest that fiber may help prevent cancers of the colon and rectum. Fiber is plant material that people can't digest. These materials may be divided into two categories: water soluble and water insoluble. It is the insoluble type, which comes from the structural components of plants, that is linked with a healthy digestive tract. Insoluble fiber is found in whole-grain cereals, wheat bran, breads, and many vegetables.

Dietary fiber appears to help block the carcinogenic effects of bile acids. Although bile acids aid fat digestion, they also promote tumor growth in the colon. Fiber may help prevent cancer in several ways. It is believed to (1) help increase the bulk of the stool, thus diluting the concentration of carcinogens; (2) hasten the passage of the stool through the colon, thus decreasing the length of time the colon is exposed to carcinogens; and (3) bind to and inactivate carcinogens. Some researchers have pointed out that a diet high in fiber tends to be low in fat, so an increase in dietary fiber may help prevent cancer by decreasing dietary fat.

A family of plants known as cruciferous vegetables also contains antioxidants, and has received some attention as possible cancer preventers. These vegetables include bok choy, broccoli, Brussels sprouts, cabbage, cauliflower, collards, kale, and turnips.

Weight Control and Physical Activity

Life insurance data indicate that people who are more than 20 percent above their recommended weight have a higher than average risk of many types of cancer. Cancer risk increases with the amount of extra weight.

Several studies have found an association between low levels of occupational and/or recreational physical activity and risk of colon cancer. Sedentary participants in one study had two to three times the risk of colon cancer as more active participants. The intensity of the activity does not appear to be an important variable: doing something is better than doing nothing. People who performed even mild physical activity, such as walking, had a lower cancer risk than sedentary folks.

Preliminary evidence also suggests that active women may have fewer cancers of the breast and reproductive system. Physical activity may exert its protective effect by preventing obesity, and/or decreasing stool transit time by stimulating the movement of substances through the gastrointestinal tract.

Exposure to Carcinogens

Carcinogens are present in many industrial and household products. Regulatory agencies, industries, and organized labor have developed procedures for reducing hazardous exposure to carcinogens in the workplace. Household and garden products should always be used as directed to reduce exposure to fumes and contact with skin.

Ultraviolet light is a potent carcinogen, especially for fair-skinned people. Protective clothing and sunscreens can prevent much of the carcinogenic effect of sunlight.

Personality and Stress

Some studies have found that cancer is more common among people who keep their anger and other emotions bottled up and who cope poorly with internal turmoil and stress. A poor stress response has been associated with a decreased immune function, which may be the link between stress and cancer. The association between stress and disease will be explored in greater detail in the Wellness Box in Chapter 11.

TYPES

The name of the cancer is derived from the type of tissue in which it develops. *Carcinoma* (*carc* = cancer; *omo* = tumor) refers to a malignant tumor consisting of epithelial cells. A tumor that develops from a gland is called an *adenosarcoma* (*adeno* = gland). *Sarcoma* is a general term for any cancer arising from connective tissue. *Osteogenic sarcomas* (*osteo* = bone; *genic* = origin), the most frequent type of childhood cancer, destroy normal bone tissue and eventually spread to other areas of the body. *Myelomas* (*myelos* = marrow) are malignant tumors, occurring in middle-aged and older people, that interfere with the blood-cell-producing function of bone marrow and cause anemia. *Chondrosarcomas* (*chondro* = cartilage) are cancerous growths of cartilage.

POSSIBLE CAUSES

What triggers a perfectly normal cell to lose control and become abnormal? Scientists are uncertain. First, there are environmental agents: substances in our air, water, and food. A chemical or other environmental agent that produces cancer is called a *carcinogen.* The World Health Organization estimates that carcinogens may be associated with 60 to 90 percent of all human cancers. The hydrocarbons found in cigarette tar are carcinogens. Ninety percent of all lung cancer patients are smokers. Another environmental factor is radiation. Ultraviolet (UV) light from the sun, for example, may cause genetic mutations in exposed skin cells and lead to skin cancer, especially among light-skinned people.

Viruses are a second cause of cancer, at least in animals. These agents are tiny packages of nucleic acids, either DNA or RNA, that are capable of infecting cells and converting them to virus producers. With over 100 separate viruses identified as carcinogens in many species and tissues of animals, it is also probable that at least some cancers in humans are due to viruses.

A great deal of cancer research is now directed toward studying *oncogenes* (ONG-kō-jēnz)—genes found in every human cell that have the ability to transform a normal cell into a cancerous cell when they are inappropriately *activated*. Oncogenes are derived from normal genes that regulate growth and development, called *proto-oncogenes.* It is believed that some oncogenes cause extra production of growth factors, chemicals that stimulate cell growth. Other oncogenes may cause changes in a surface receptor, causing it to send signals as though it were being activated by a growth factor. As a result, the growth pattern of the cell becomes abnormal.

It appears that some proto-oncogenes are activated to oncogenes by various types of mutation in which the DNA of the proto-oncogenes is altered by carcinogens, viruses, or a rearrangement of a cell's chromosomes in which segments of DNA are exchanged. Burkitt's lymphoma, malignant tumors of the colon and rectum, and one type of lung cancer are linked to oncogenes.

Researchers have also determined that some cancers are not caused by oncogenes but may be caused by genes called *anti-oncogenes.* These genes can cause cancer when they are inappropriately *inactivated*. A cancer caused by an anti-oncogene is a childhood cancer of the eye called *retinoblastoma.* Such inactivation may also be involved in some types of breast cancer and one type of lung cancer.

Currently, scientists are also trying to establish a relationship between stress and cancer. Some believe that stress may play a role not only in the development but also in the metastasis of cancer.

There is also great interest in determining the effects of alterations of the immune system and nutrition in the development of cancer. In 1982 the National Research Council issued a series of guidelines related to diet and cancer. The main recommendations were (1) to reduce fat intake; (2) to increase consumption of fiber, fruits, and vegetables; (3) to increase intake of complex carbohydrates (potatoes, pasta); and (4) to reduce consumption of salted, smoked, and pickled foods and simple carbohydrates (refined sugars).

TREATMENT

Treating cancer is difficult because it is not a single disease and because all the cells in a single population (tumor) do not behave in the same way. The same cancer may contain a diverse population of cells by the time it reaches a clinically detectable size. Although tumor cells look alike under the microscope, they do not necessarily behave in the same manner in the body. For example, some metastasize and others do not. Some are sensitive to drugs and some are resistant. Consequently, a single chemotherapeutic drug may destroy susceptible cells but resistant cells will proliferate. This is one reason combination chemotherapy is usually more successful. Also, radiation therapy, surgery, and hyperthermia (abnormally high temperatures) may be used alone or in combination.

Scientists are moving closer to developing a vaccine for cancer that would stimulate certain cells of the body into marshaling a successful attack against the cancer cells.

MEDICAL TERMINOLOGY AND CONDITIONS

NOTE TO THE STUDENT

Each chapter in this text that discusses a major system of the body is followed by a glossary of *medical terminology and conditions.* Both normal and pathological conditions of the system are included in these glossaries.

Some of these disorders, as well as disorders discussed in the text, are referred to as local or systemic. A *local disease* is one that affects one part of a limited area of the body. A *systemic*

disease affects either the entire body or several parts.

The science that deals with why, when, and where diseases occur and how they are transmitted in a human community is known as *epidemiology* (ep′-i-dē′-mē-OL-ō-jē; *epidemios* = prevalent; *logos* = study of). The science that deals with the effects and uses of drugs in the treatment of disease is called *pharmacology* (far′-ma-KOL-ō-jē; *pharmakon* = medicine; *logos* = study of).

Atrophy (AT-rō-fē; *a* = without; *tropho* = nourish) A decrease in the size of cells with subsequent decrease in the size of the affected tissue or organ; wasting away.

Biopsy (BĪ-op-sē; *bio* = life; *opsis* = vision) The removal and microscopic examination of tissue from the living body for diagnosis.

Dysplasia (dis-PLĀ-zē-a; *dys* = abnormal; *plas* = to grow) Alteration in the size, shape, and organization of cells due to chronic irritation or inflammation; may progress to neoplasia (tumor formation, usually malignant) or revert to normal if the stress is removed.

Hyperplasia (hī'-per-PLĀ-zē-a; *hyper* = over) Increase in the number of cells due to an increase in the frequency of cell division.

Hypertrophy (hī-PER-trō-fē) Increase in the size of cells without cell division.

Metaplasia (met'-a-PLĀ-zē-a; *meta* = change) The transformation of one cell into another.

Metastasis (me-TAS-ta-sis; *stasis* = standing still) The spread of cancer to surrounding tissues (local metastasis) or to other body tissues (distant metastasis).

Necrosis (ne-KRŌ-sis; *necros* = death; *osis* = condition) Death of a group of cells.

Neoplasm (NĒ-ō-plazm; *neo* = new) New growth that may be benign or malignant.

STUDY OUTLINE

Generalized Animal Cell (p. 36)

1. A cell is the basic living, structural, and functional unit of the body.
2. A generalized cell is a composite that represents various cells of the body.
3. Cytology is the science concerned with the study of cells.
4. The principal parts of a cell are the plasma (cell) membrane, cytoplasm, organelles, and inclusions.

Plasma (Cell) Membrane (p. 36)

Chemistry and Structure (p. 36)
1. The plasma (cell) membrane surrounds the cell and separates it from other cells and the external environment.
2. It is composed primarily of phospholipids and proteins.

Functions (p. 36)
1. Functionally, the plasma membrane facilitates contact with other cells, provides receptors, and regulates the passage of materials.
2. The membrane's selectively permeable nature restricts the passage of certain substances. Substances can pass through the membrane depending on their molecular size, lipid solubility, electrical charges, and the presence of carriers.

Movement of Materials Across Plasma Membranes (p. 38)
1. Fluid outside cells is called extracellular fluid (interstitial, plasma, lymph); fluid inside cells is called intracellular.
2. Substances move across plasma membranes between extracellular and intracellular fluids.
3. Passive processes involve the kinetic energy of individual molecules; no ATP is required.

4. Diffusion is the net movement of molecules or ions from an area of higher concentration to an area of lower concentration until an equilibrium is reached.
5. In facilitated diffusion, certain molecules, such as glucose, combine with a carrier to become soluble in the phospholipid portion of the membrane.
6. Osmosis is the movement of water through a selectively permeable membrane from an area of higher water concentration to an area of lower water concentration until equilibrium is reached.
7. In an isotonic solution, red blood cells maintain their normal shape; in a hypotonic solution, they undergo hemolysis; in a hypertonic solution, they undergo crenation.
8. Filtration is the movement of water and dissolved substances across a selectively permeable membrane by pressure.
9. Active processes involve the use of ATP by the cell.
10. Active transport is the movement of ions across a cell membrane from lower to higher concentration.
11. Endocytosis is the movement of substances through plasma membranes in which the membrane surrounds the substance, encloses it, and brings it into the cell.
12. Phagocytosis is the ingestion of solid particles by pseudopodia. It is the process used by white blood cells to destroy bacteria.
13. Pinocytosis is the ingestion of a liquid by the plasma membrane. In this process, the liquid becomes surrounded by a vacuole.
14. Receptor-mediated endocytosis is the selective uptake of large molecules and particles by cells.

Cytoplasm (p. 44)

1. Cytoplasm is the intracellular material inside the cell between the plasma membrane and nucleus that contains organelles and inclusions.

2. It is composed of mostly water plus proteins, carbohydrates, lipids, and inorganic substances.
3. Functionally, cytoplasm is the medium in which some chemical reactions occur.

Organelles (p. 44)

1. Organelles are specialized structures in the cell that carry on specific activities.
2. They assume specific roles in cellular growth, maintenance, repair, and control.

Nucleus (p. 44)

1. Usually the largest organelle, the nucleus controls cellular activities and contains the genetic information.
2. Most body cells have a single nucleus; some (red blood cells) have none; others (skeletal muscle fibers) have several.
3. The parts of the nucleus include the nuclear membrane, nucleoplasm, nucleoli, and genetic material (DNA), which comprises the chromosomes.

Ribosomes (p. 44)

1. Ribosomes are granular structures consisting of ribosomal RNA and ribosomal proteins.
2. They occur free or attached to endoplasmic reticulum.
3. Functionally, ribosomes are the sites of protein synthesis.

Endoplasmic Reticulum (ER) (p. 45)

1. The ER is a network of parallel membranes continuous with the plasma membrane and nuclear membrane.
2. Granular or rough ER has ribosomes attached to it. Agranular or smooth ER does not contain ribosomes.
3. The ER provides mechanical support, exchanges materials with cytoplasm, and transports substances within cells. Granular ER stores the protein synthesized by ribosomes. Smooth ER is involved in the production of lipids, especially steroids.

Golgi Complex (p. 46)

1. The Golgi complex consists of four to eight stacked, flattened membranous sacs (cisternae).
2. The principal function of the Golgi complex is to process, sort, and deliver proteins within the cell. It also secretes lipids and forms lysosomes.

Lysosomes (p. 46)

1. Lysosomes are spherical structures that contain digestive enzymes. They are formed from Golgi complexes.
2. They are found in large numbers in white blood cells, which carry on phagocytosis.
3. Digestion by lysosomes of worn-out cell parts is called autophagy; programmed self-destruction by lysosomes is called autolysis.
4. Lysosomes also function in extracellular digestion.

Mitochondria (p. 47)

1. Mitochondria consist of a smooth outer membrane and a folded inner membrane surrounding the interior matrix. The inner folds are called cristae.
2. The mitochondria are called ''powerhouses of the cell'' because ATP is produced in them.

The Cytoskeleton (p. 47)

1. Microfilaments, microtubules, and intermediate filaments form the cytoskeleton.
2. Microfilaments are rodlike structures consisting of the protein actin or myosin. They are involved in muscular contraction, support, and movement.
3. Microtubules are slender tubes made of the protein tubulin. They support, provide movement, and form flagella, cilia, centrioles, and the mitotic spindle.
4. Intermediate filaments appear to provide structural reinforcement in some cells.

Centrosome and Centrioles (p. 47)

1. The dense area of cytoplasm containing the centrioles is called a centrosome.
2. Centrioles are paired cylinders arranged at right angles to one another. They are important in cell reproduction.

Flagella and Cilia (p. 47)

1. These cellular projections have the same basic structure and are used in movement.
2. If projections are few (usually occurring singly or in pairs) and long, they are called flagella. If they are numerous and hairlike, they are called cilia.
3. Flagella move an entire cell (e.g., the flagellum on a sperm cell). Cilia move objects along a cell surface (the cilia in mucus toward the throat for elimination).

Cell Inclusions (p. 48)

1. Cell inclusions are chemical substances produced by cells. They are usually organic and may have recognizable shapes.
2. Examples are melanin, glycogen, and lipids.

Gene Action (p. 48)

1. Cells are basically protein factories.
2. Cells make proteins by translating the genetic information encoded in DNA into specific proteins. This involves transcription and translation.
3. In transcription, genetic information encoded in DNA is passed to a strand of messenger RNA (mRNA).
4. DNA also synthesizes ribosomal RNA (rRNA) and transfer RNA (tRNA).
5. In translation, the information in the nitrogenous base sequence of mRNA dictates the amino acid sequence of a protein.
6. Protein synthesis occurs on ribosomes where a strand of mRNA is attached.
7. tRNA transports specific amino acids to the mRNA at the ribosome. A portion of the tRNA has a triplet of bases called an anticodon; it matches a codon of three bases on the mRNA.
8. The ribosome moves along an mRNA strand as amino acids are joined to form the protein molecule.
9. Peptide bonds hold the amino acids together.

Normal Cell Division (p. 51)

1. Cell division is the process by which cells reproduce themselves. It consists of nuclear division and cytoplasmic and organelle division (cytokinesis).

2. Cell division that results in an increase in body cells is called somatic cell division and involves a nuclear division called mitosis and cytokinesis.

3. Cell division that results in the production of sperm and eggs is called reproductive cell division and consists of a nuclear division called meiosis and cytokinesis (to be examined in Chapter 23).

Somatic Cell Division (p. 51)

1. Prior to mitosis and cytokinesis, the DNA molecules, or chromosomes, replicate themselves so the same hereditary traits are passed on to future generations of cells.

2. DNA replication occurs during interphase.

3. Mitosis is the distribution of two sets of chromosomes into separate and equal nuclei following their replication.

4. It consists of prophase, metaphase, anaphase, and telophase.

5. Cytokinesis usually begins in late anaphase and terminates in telophase.

6. A cleavage furrow forms at the cell's equatorial plane and progresses inward, cutting through the cell to form two separate portions of cytoplasm.

Abnormal Cell Division: Cancer (p. 53)

1. Cancerous tumors are referred to as malignant; noncancerous tumors are called benign; the study of tumors is called oncology.

2. The spread of cancer from its primary site is called metastasis.

3. Carcinogens include environmental agents and viruses.

4. Treating cancer is difficult because all the cells in a single population do not behave the same way.

REVIEW QUESTIONS

1. Define a cell. What are the four principal portions of a cell? (p. 36)

2. Discuss the chemistry and structure of the plasma membrane. (p. 36)

3. Describe the various functions of the plasma membrane. What determines selective permeability? (pp. 36–38)

4. Distinguish between extracellular and intracellular fluid. (p. 38)

5. What are the major differences between active processes and passive processes in moving substances across plasma membranes? (pp. 38–39)

6. Define and give an example of each of the following: diffusion, facilitated diffusion, osmosis, filtration, active transport, phagocytosis, pinocytosis, and receptor-mediated endocytosis. (pp. 39–42)

7. Compare the effect on red blood cells of an isotonic, hypertonic, and hypotonic solution. What is osmotic pressure? (pp. 40–41)

8. Discuss the chemical composition of cytoplasm. What is its function? (p. 44)

9. What is an organelle? By means of a labeled diagram, indicate the parts of a generalized animal cell. (p. 44)

10. Describe the structure and functions of the nucleus of a cell. (p. 44)

11. What is the function of ribosomes? (pp. 44–45)

12. What are the functions of the endoplasmic reticulum (ER)? (pp. 45–46)

13. Describe the structure and functions of the Golgi complex. (p. 46)

14. List and describe the various functions of lysosomes. (pp. 46–47)

15. Why are mitochondria referred to as "powerhouses of the cell"? (p. 47)

16. Contrast the structure and functions of microfilaments, microtubules, and intermediate filaments. (p. 47)

17. Describe the structure and function of centrioles. (p. 47)

18. How are cilia and flagella different in structure and function? (p. 47)

19. Define a cell inclusion. Provide examples and indicate their functions. (p. 48)

20. Summarize the steps involving gene action in protein synthesis. (pp. 48–51)

21. Distinguish between the two types of cell division. Why is each important? (p. 51)

22. Describe DNA replication. Why is it important? (p. 51)

23. Describe the principal events of each stage of mitosis. (pp. 52–53)

24. What is a tumor? Distinguish between malignant and benign tumors. Describe the principal types of malignant tumor. (p. 53)

25. Define metastasis. What factors contribute to metastasis? (p. 53)

26. Discuss several possible causes of cancer. (p. 56)

27. What are some of the problems with respect to treating cancer? (p. 56)

28. Refer to the glossary of medical terminology and conditions associated with cells. Be sure you can define each term. (pp. 56–57)

Tissues ^{4}

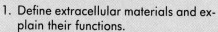

STUDENT OBJECTIVES

1. Define extracellular materials and explain their functions.
2. Define a tissue and list the four basic types of tissues.
3. Compare the characteristics and functions of epithelial, connective, muscle, and nervous tissues.
4. Define a gland and distinguish the two principal types of glands.
5. Describe the principal membranes in the human body.

A LOOK AHEAD

EXTRACELLULAR MATERIALS
TYPES OF TISSUES
EPITHELIAL TISSUE
    Covering and Lining
        Epithelium
        Classification by Cell Shape
        Classification by
            Arrangement of Layers
        Types and Functions
    Glandular Epithelium
CONNECTIVE TISSUE
    Classification
    Embryonic Connective Tissue
    Adult Connective Tissue
        Connective Tissue Proper
        Cartilage
        Osseous Tissue (Bone)
        Vascular Tissue (Blood)
MUSCLE TISSUE
NERVOUS TISSUE
MEMBRANES
    Mucous Membranes
    Serous Membranes
    Cutaneous Membrane
    Synovial Membranes

Cells are highly organized units, but they do not function in isolation. They work together in groups of similar cells called tissues.

Before we look at the basic types of tissues that compose the body, we will examine extracellular materials which, among other functions, support and bind cells together in a tissue.

EXTRACELLULAR MATERIALS

The substances that lie outside cells are called ***extracellular materials.*** Body fluids (interstitial fluid and plasma), discussed in Chapter 3 (p. 38), are one type of extracellular material. Another is secretions, such as mucus, enzymes, or hormones. A third type particularly important to the study of tissues is ***matrix.*** Matrix is a ground substance in which cells of a tissue called connective tissue are embedded. Matrix materials are produced by certain cells and deposited outside their plasma membranes. Matrix supports cells, binds them together, and gives strength and elasticity to the tissue. Some matrix materials have no specific shape. ***Hyaluronic*** (hī′-a-loo-RON-ik) ***acid*** is a viscous, fluidlike substance that lubricates joints and maintains the shape of the eyeballs. ***Chondroitin*** (kon-DROY-tin) ***sulfate*** is a jellylike substance that provides support and adhesiveness in cartilage, bone, heart valves, the cornea of the eye, and the umbilical cord.

Other matrix materials contain fibers. Among these are ***collagenous*** (*kolla* = glue) ***fibers,*** which are tough, white fibers consisting of the protein *collagen*. Collagenous fibers are found in tissues such as cartilage, tendons, and ligaments. ***Reticular*** (*rete* = net) ***fibers*** consist of the protein collagen and a coating of glycoprotein; they form a network around fat cells, nerve fibers, muscle fibers (cells), and blood vessels. They also form the framework for many soft organs of the body such as the spleen. ***Elastic fibers,*** which are yellow in color, consist of the protein *elastin* and give elasticity to skin and to tissues forming the walls of blood vessels.

TYPES OF TISSUES

A ***tissue*** is a group of similar cells and their intercellular substance that function together to perform a specialized activity. The science that deals with the study of tissues is called ***histology*** (his′-TOL-ō-jē; *histio* = tissue; *logos* = study of). There are four types of tissue in the body:

1. **Epithelial** (ep′-i-THĒ-lē-al) **tissue,** which covers body surfaces, lines body cavities and tubes and forms glands.
2. **Connective tissue,** which protects and supports the body and its organs, binds organs together and stores energy reserves.
3. **Muscular tissue,** which is responsible for movement.
4. **Nervous tissue,** which initiates and transmits nerve impulses that coordinate body activities.

Epithelial tissue and connective tissue, except for bone and blood, will be discussed in detail in this chapter. Detailed discussions of bone tissue and blood occurs later in the book. Similarly, the detailed discussions of muscle tissue and nervous tissue are taken up later.

Muscle and nervous tissue are so highly differentiated (specialized) that they have lost their capacity for mitosis. Epithelial tissue, by contrast, is constantly being regenerated. Connective tissues with a good blood supply, such as bone, regenerate better than those with a poorer blood supply, such as cartilage and fibrous connective tissue. This can affect the rate at which some injuries will heal.

EPITHELIAL TISSUE

Epithelial tissue, or more simply *epithelium,* may be divided into two subtypes: (1) *covering and lining epithelium* and (2) *glandular epithelium.* Covering and lining epithelium forms the outer covering of external body surfaces and the outer covering of some internal organs. It lines body cavities and the interiors of the respiratory and gastrointestinal tracts, blood vessels, and ducts. It makes up, along with nervous tissue, the parts of the sense organs that produce sensations such as smell and hearing. And, it is the tissue from which gametes (sperm and eggs) develop. Glandular epithelium constitutes the secreting portion of glands.

Both types of epithelium consist of closely packed cells with little or no intercellular material between them. The points of attachment between adjacent epithelial cells are called **cell junctions**. Epithelial cells are arranged in continuous sheets that may be either single or multilayered. Nerves may extend through these sheets, but blood vessels do not. They are **avascular** (*a* = without; *vasculum* = blood vessels). The vessels that supply nutrients and remove wastes are located in underlying connective tissue.

Both types of epithelium overlie and adhere firmly to connective tissue, which holds the epithelium in position and prevents it from being torn. The attachment between the epithelium and the connective tissue is called the **basement membrane**. This structure provides physical support for epithelium, provides for cell attachment, serves as a filter in the kidneys, and guides cell migration during development and repair.

COVERING AND LINING EPITHELIUM

Classification by Cell Shape

Epithelial tissue, which covers or lines various parts of the body, contains cells with *four basic shapes*:

1. *Squamous* (SKWĀ-mus) cells are flat or scalelike and attach to each other like mosaic tiles. Their thinness allows for the active and passive movement of substances through them when only a single layer exists (Chapter 3).
2. *Cuboidal* cells are thicker, being cube or hexagon shaped. They produce several important body *secretions* (fluids that are produced and released by cells, such as sweat and enzymes). They may also be used for *absorption* of fluids and other substances, such as digested foods in the intestines.
3. *Columnar* cells are tall and cylindrical, thereby protecting underlying tissues. They may also be specialized for secretion and absorption. Some may also have cilia.
4. *Transitional* cells range in shape from flat to columnar and often change shape due to distention, expansion, or movement of body parts.

Classification by Arrangement of Layers

Epithelial cells tend to be arranged in one or more layers depending on the function of the particular body part. The terms used to refer to their classification by arrangement of layers include the following:

1. *Simple epithelium* is a single layer of cells found in areas where passive and active movement of molecules is needed.
2. *Stratified epithelium* contains two or more layers of cells used for protection of underlying tissues in areas where there is considerable wear and tear.
3. *Pseudostratified epithelium* contains one layer of a mixture of cell shapes. The tissue looks multilayered but not all cells reach the surface; those that do are either ciliated or secrete mucus (see Exhibit 4-1 Pseudostratified Columnar).

In summary, epithelium may be classified by both shape and arrangement of layers of cells as indicated as follows:

Simple
1. Squamous
2. Cuboidal
3. Columnar

Stratified
1. Squamous
2. Cuboidal
3. Columnar
4. Transitional

Pseudostratified

Types and Functions

Each of the epithelial tissues described in the following section is illustrated in Exhibit 4-1.

▪ **Simple Squamous Epithelium** This tissue lines the air sacs of the lungs, the heart, blood and lymphatic vessels, and the abdominal cavity. It makes up the walls of microscopic blood vessels (capillaries) throughout the body including those in the kidneys (glomeruli). The flat, single layer of cells allows for the diffusion of gases in the lungs and the movement of fluid and dissolved substances between the blood and tissue cells by osmosis and filtration.

▪ **Simple Cuboidal Epithelium** This tissue covers the ovaries and lines kidney tubules, ducts of some glands, and a part of the lens of the eye. It makes up secretory units of some glands such as the thyroid and forms part of the retina (inner lining) of the eye. Its single layer of cells is important in secretion of substances, such as tears and saliva, and absorption, such as the reabsorption of water by the kidney tubule cells.

▪ **Simple Columnar Epithelium** The surface view of simple columnar epithelium is similar to that of simple cuboidal tissue. In cross sections, however, the cells appear rectangular. The nucleus is located near the base of the cell. Simple columnar epithelium lines the gastrointestinal tract from the stomach to the anus, as well as the gallbladder, and excretory ducts of many glands. Some of these cells are modified to aid in food-related activities. In the small intestine, for example, the plasma membranes of the cells are folded into **microvilli** (see Figure 3-1). Microvilli increase the surface area of the plasma membrane and thereby allow larger amounts of digested nutrients and fluids to be absorbed into the body.

Other modified columnar cells, called **goblet cells**, secrete mucus and are so named because the mucus accumulates in the upper

EXHIBIT 4-1
Epithelial Tissues

COVERING AND LINING EPITHELIUM
Simple Squamous
Description: Single layer of flat, scalelike cells; centrally located nuclei.
Location: Lines air sacs of lungs, heart, blood lymphatic vessels, and abdominal cavity and makes up walls of blood capillaries throughout the body, including the kidney (glomeruli).
Function: Filtration, absorption, exchange, and secretion in serous membranes.

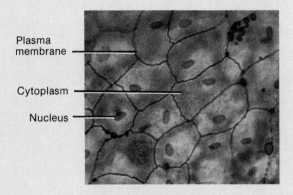

Plasma membrane

Cytoplasm

Nucleus

**Surface view of mesothelial lining of
peritoneal cavity (243 x)**

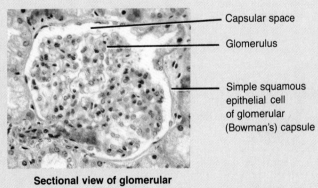

Capsular space

Glomerulus

Simple squamous epithelial cell of glomerular (Bowman's) capsule

**Sectional view of glomerular
capsule of kidney (300x)**

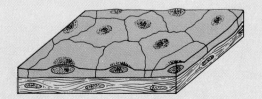

EXHIBIT 4-1 (*Continued*)
Epithelial Tissues

Simple Cuboidal
Description: Single layer of cube-shaped cells; centrally located nuclei.
Location: Covers surface of ovary; lines kidney tubules, ducts of some glands, and part of lens of eye; forms part of retina of eye; and makes up secreting portion of some glands, such as the thyroid.
Function: Secretion and absorption.

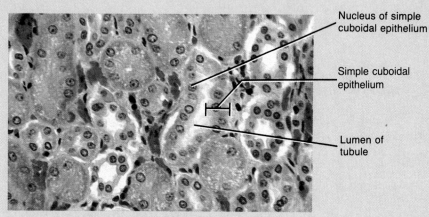

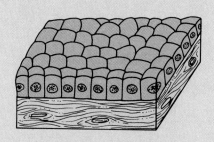

Nucleus of simple cuboidal epithelium

Simple cuboidal epithelium

Lumen of tubule

Sectional view of kidney (450x)

Simple Columnar (Nonciliated)
Description: Single layer of nonciliated rectangular cells; contains goblet cells in some locations; nuclei at bases of cells.
Location: Lines the gastrointestinal tract from the stomach to the anus, excretory ducts of many glands, and gallbladder.
Function: Secretion and absorption.

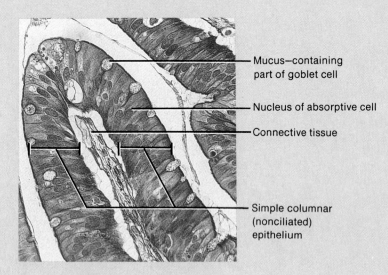

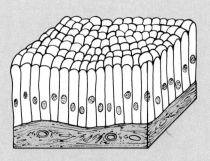

Mucus–containing part of goblet cell

Nucleus of absorptive cell

Connective tissue

Simple columnar (nonciliated) epithelium

Sectional view of epithelium from the small intestine (140x)

Simple Columnar (Ciliated)

Description: Single layer of ciliated columnar cells; contains goblet cells in some locations; nuclei at bases of cells.

Location: Lines portions of upper respiratory tract, uterine (Fallopian) tubes, and uterus.

Function: Moves mucus and other substances by ciliary action.

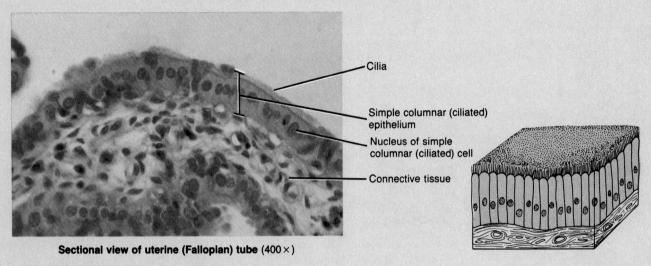

Cilia

Simple columnar (ciliated) epithelium

Nucleus of simple columnar (ciliated) cell

Connective tissue

Sectional view of uterine (Fallopian) tube (400×)

Stratified Squamous

Description: Several layers of cells; squamous cells in superficial layers; basal cells replace surface cells as they are lost.

Location: Nonkeratinized variety lines wet surfaces such as lining of the mouth, tongue, esophagus, and vagina; keratinized variety forms outer layer of skin.

Function: Protection.

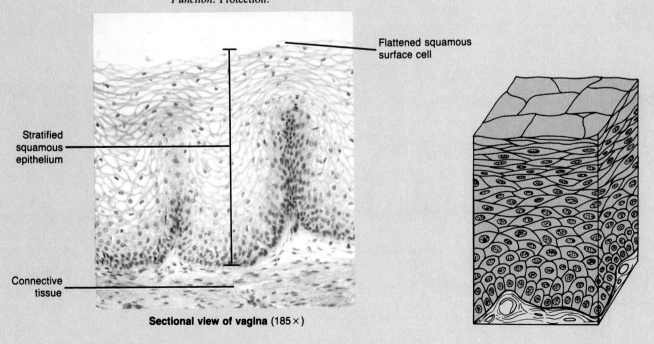

Flattened squamous surface cell

Stratified squamous epithelium

Connective tissue

Sectional view of vagina (185×)

EXHIBIT 4-1 (*Continued*)
Epithelial Tissues

Stratified Cuboidal
Description: Two or more layers of cells in which the surface cells are cube-shaped.
Location: Ducts of adult sweat glands, conjunctiva of eye, male urethra, and pharynx.
Function: Protection.

Stratified cuboidal epithelium

Nucleus of stratified cuboidal cell

Lumen of duct of sweat gland

Connective tissue

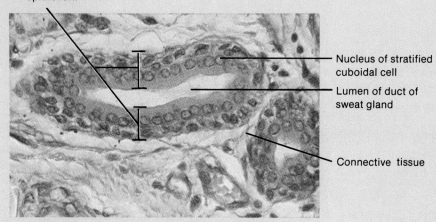
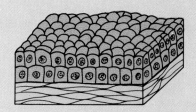

Sectional view of the duct of a sweat gland (450x)

Stratified Columnar
Description: Several layers of cells; columnar cells in superficial layer.
Location: Lines part of male urethra, large excretory ducts of some glands, and small areas in anal mucous membrane.
Function: Protection and secretion.

Nuclei of stratified columnar cells

Lumen of duct

Connective tissue

Stratified columnar epithelium

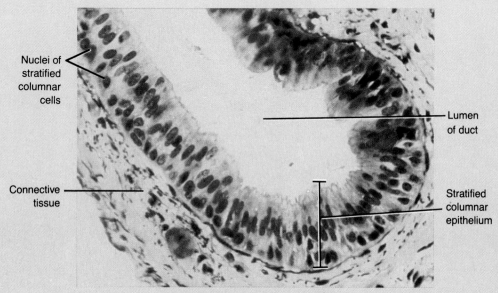
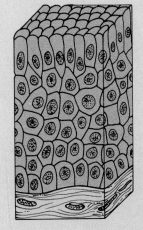

Sectional view of the duct of the
submandibular salivary gland (375×)

Transitional
Description: Superficial cells are large with rounded free surface, resemble squamous cells.
Location: Lines urinary bladder and portions of ureters and urethra.
Function: Permits distention.

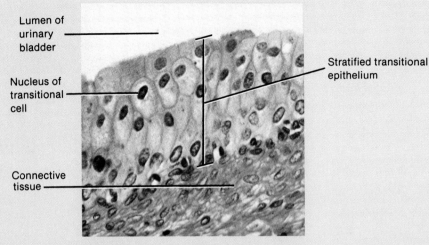
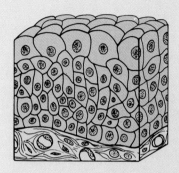

Lumen of urinary bladder

Nucleus of transitional cell

Connective tissue

Stratified transitional epithelium

Sectional view of urinary bladder in relaxed state (240x)

Pseudostratified Columnar
Description: Not a true stratified tissue; nuclei of cells at different levels; all cells attached to basement membrane, but not all reach surface.
Location: Lines larger excretory ducts of many large glands, epididymis, male urethra, and auditory (Eustachian) tubes; ciliated variety with goblet cells lines most of the upper respiratory tract and some ducts of male reproductive system.
Function: Secretion and movement of mucus and sperm cells by ciliary action.

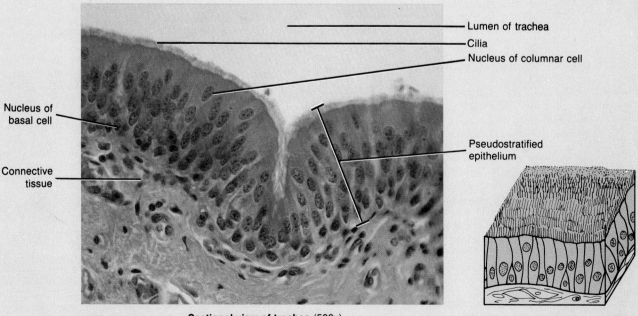

Nucleus of basal cell

Connective tissue

Lumen of trachea

Cilia

Nucleus of columnar cell

Pseudostratified epithelium

Sectional view of trachea (500x)

EXHIBIT 4-1 (Continued)
Epithelial Tissues

GLANDULAR EPITHELIUM
Exocrine Gland
Description: Secretes products into ducts or directly onto external or internal body surface.
Location: Sweat, oil, wax, and mammary glands of the skin; digestive glands such as salivary glands that secrete into mouth cavity and pancreas that secretes into the small intestine.
Function: Produces mucus, perspiration, oil, wax, milk, or digestive enzymes.

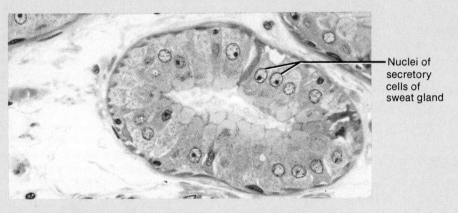

Nuclei of secretory cells of sweat gland

Sectional view of the secretory portion of a sweat gland (1532x)

Endocrine Gland
Description: Secretes hormones into blood.
Location: Pituitary, thyroid, parathyroids, adrenals (suprarenals), ovaries, testes, pineal, and thymus.
Function: Produces hormones that regulate various body activities.

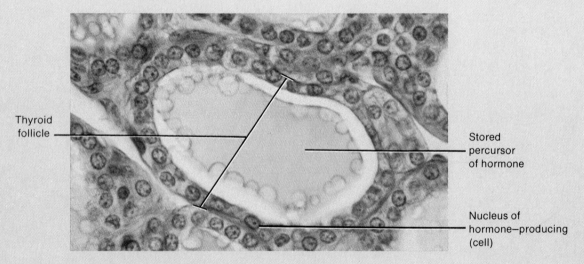

Thyroid follicle

Stored percursor of hormone

Nucleus of hormone–producing (cell)

Sectional view of thyroid gland (744x)

Photomicrographs of mesothelium and exocrine, © Biophoto, Photo Researchers; glomerular capsule, simple cuboidal, simple columnar (ciliated), stratified squamous, stratified columnar, and pseudostratified columnar, courtesy of Andrew J. Kuntzman; simple columnar (nonciliated) and endocrine, Lester Bergman and Associates; stratified cuboidal, Biological Photo Service; exocrine, © Bruce Iverson.

half of the cell, causing the area to bulge out so the whole cell resembles a goblet or wine glass. The secreted mucus serves as a lubricant and protective film between the food and the walls of the gastrointestinal tract.

A third modification of columnar epithelium is the addition of hairlike processes called *cilia*. In the lining of the upper respiratory tract, ciliated cells help with the removal of mucus and foreign particles. In the female reproductive tract cilia help with the transportation of ova and sperm.

■ **Stratified Epithelium** Stratified epithelium is classified according to the shape of the *surface* cells.

■ **Stratified Squamous Epithelium** This tissue is found in the top layers of the skin and lines the mouth, tongue, esophagus, and vagina. The basal (bottom) layer of cells is constantly reproducing and pushing new cells upward. As the upper layers move away from the underlying blood vessels, the lack of oxygen and nutrients causes them to shrink and eventually die. Friction then causes them to be sloughed off. Wet surfaces may absorb substances in contact with them but absorption is minimal through the skin due to the formation of *keratin*. This protein is produced by epithelial skin cells called *keratinocytes* as they are pushed upward. Keratin protects skin from injury and microbial invasion and makes it waterproof.

■ **Stratified Cuboidal Epithelium** This is a relatively rare type of epithelium found in the ducts of sweat glands, the conjunctiva of the eye, the male urethra, and the pharynx. Its function is mainly protective.

■ **Stratified Columnar Epithelium** This type is also uncommon in the body. It lines part of the male urethra, some larger excretory ducts such as lactiferous (milk) ducts in the mammary glands, and small areas in the anal mucous membrane. It functions in protection and secretion.

■ **Transitional Epithelium** The cells in the outer layer in transitional epithelium tend to be large and rounded, which allows the tissue to be stretched (distended) without the cells breaking apart. Because of this arrangement, transitional epithelium lines hollow structures that are subjected to expansion from within, such as the urinary bladder. Its function is to help prevent a rupture of the organ.

■ **Pseudostratified Columnar Epithelium** This tissue lines the auditory (Eustachian) tubes, the tubes that connect the middle ear cavity and upper part of the throat, larger excretory ducts of many glands, and parts of the male urethra. Pseudostratified columnar epithelium that is ciliated and contains goblet cells lines most of the upper respiratory tract and certain ducts of the male reproductive system.

GLANDULAR EPITHELIUM

The function of glandular epithelium is secretion, which is accomplished by cells that lie in clusters deep below the covering and lining epithelium. A *gland* may consist of one cell or a group of highly specialized epithelial cells that secrete substances into ducts, onto a surface, or into the blood. Glands of the body are classified as exocrine or endocrine.

Exocrine glands secrete their products into ducts (tubes), which, in turn, empty onto an external or internal surface. Or an exocrine gland may release its product directly at the skin surface or into the lumen (cavity) of a hollow organ. Examples of exocrine glands include sweat glands, salivary and other digestive tract glands, and goblet cells. Exocrine secretions may be watery such as saliva, sweat, tears, and serous fluid or more viscous like mucus, oil (sebum), or wax (cerumen). The functions of these substances will be discussed later in the text. *Endocrine glands* are ductless and secrete their products into the blood. Endocrine glands only secrete hormones, chemicals that regulate various physiological activities. The pituitary, thyroid, and adrenal (suprarenal) glands are examples of endocrine glands.

CONNECTIVE TISSUE

The most abundant tissue in the body is *connective tissue*. It is highly vascular and thus has a rich blood supply, with the exception of cartilage, which is avascular. Cells in connective tissue are widely scattered, rather than closely packed, and there is considerable intercellular material (matrix). The general functions of connective tissues are protection, support, binding together various organs, separating structures such as skeletal muscles, and storage of reserve energy.

CLASSIFICATION

Connective tissue is classified as follows:

I. Embryonic connective tissue
   A. Mesenchyme
   B. Mucous connective tissue
II. Adult connective tissue
   A. Connective tissue proper
      1. Loose (areolar) connective tissue
      2. Adipose tissue
      3. Dense (collagenous) connective tissue
      4. Elastic connective tissue
      5. Reticular connective tissue
   B. Cartilage
      1. Hyaline cartilage
      2. Fibrocartilage
      3. Elastic cartilage
   C. Osseous (bone) tissue
   D. Vascular (blood) tissue

Each of the connective tissues described in the following sections is illustrated in Exhibit 4-2.

EMBRYONIC CONNECTIVE TISSUE

Connective tissue that is present primarily in the embryo or fetus is called *embryonic connective tissue*. An *embryo* refers to a developing human from fertilization through the first two months of pregnancy; a *fetus* refers to a developing human from the third month of pregnancy to birth.

An embryonic connective tissue found almost exclusively in

EXHIBIT 4-2
Connective Tissues

EMBRYONIC CONNECTIVE TISSUE
Mesenchyme
Description: Consists of highly branched mesenchymal cells embedded in a fluid matrix.
Location: In embryo primarily and in adult.
Function: Forms all other kinds of connective tissue.

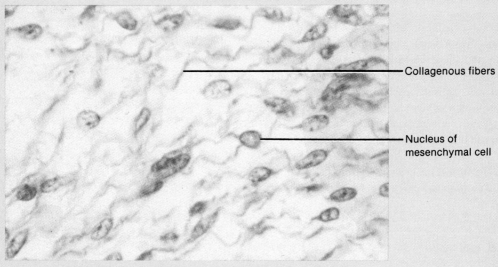

Collagenous fibers

Nucleus of mesenchymal cell

Sectional view of mesenchyme from a developing fetus (1008x)

Mucous
Description: Consists of flattened or spindle-shaped cells embedded in a mucuslike matrix containing fine collagenous fibers.
Location: Umbilical cord of fetus.
Function: Support.

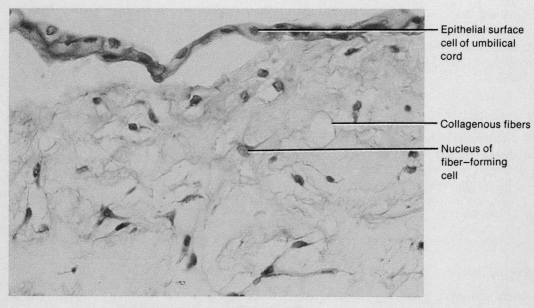

Epithelial surface cell of umbilical cord

Collagenous fibers

Nucleus of fiber−forming cell

Sectional view of the umbilical cord (457x)

ADULT CONNECTIVE TISSUE
Loose (Areolar)
Description: Consists of fibers (collagenous, elastic, and reticular) and several kinds of cells (fibroblasts, macrophages, plasma cells, and mast cells) embedded in a viscous matrix.
Location: Around organs, dermis of skin, and subcutaneous layer.
Function: Strength, elasticity, and support.

Collagenous fibers

Elastic fibers

Sectional view of subcutaneous tissue (224x)

Adipose
Description: Consists of adipocytes, "signet ring"-shaped cells with peripheral nuclei, that are specialized for fat storage.
Location: Subcutaneous layer of skin, around heart and kidneys, marrow of long bones, padding around joints, under eyeballs.
Function: Reduces heat loss through skin, serves as an energy reserve, supports, and protects.

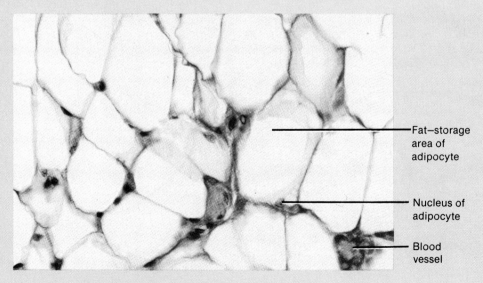

Fat–storage area of adipocyte

Nucleus of adipocyte

Blood vessel

Sectional view of adipocytes of the pancreas (1720x)

EXHIBIT 4-2 (Continued)
Connective Tissues

Dense (Collagenous)
Description: Consists of densely packed collagenous fibers that may be regularly or irregularly arranged; fibroblasts are present.
Location: Regularly arranged fibers are found in tendons and ligaments; irregularly arranged fibers are found around the heart, eyeballs, bone and cartilage, liver, kidneys, testes, and lymph nodes.
Function: Provides protection, support, or attachment.

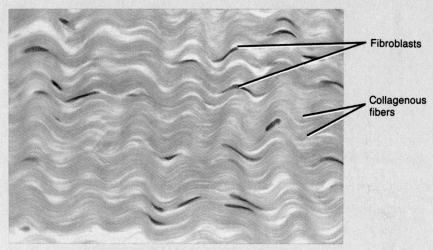

Sectional view of a tendon (250×)

Elastic
Description: Consists of predominantly freely branching elastic fibers and relatively few fibroblasts.
Location: Lung tissue, wall of arteries, trachea, bronchial tubes, and true vocal cords.
Function: Allows stretching of various organs.

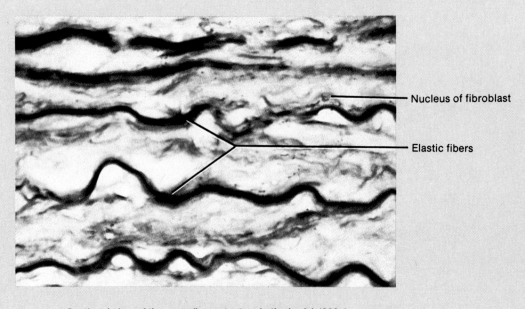

Sectional view of the aorta (largest artery in the body) (300x)

Reticular

Description: Consists of a network of interlacing reticular fibers with cells wrapped around fibers.

Location: Liver, spleen, lymph nodes, and basal lamina underlying epithelia.

Function: Forms stroma of organs; binds together smooth muscle tissue cells.

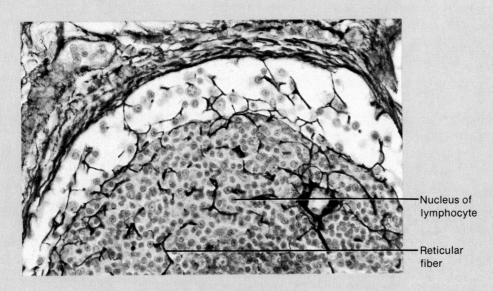

Nucleus of lymphocyte

Reticular fiber

Sectional view of lymph node (496x)

Hyaline Cartilage

Description: Also called gristle; appears as a bluish white, glossy mass; contains numerous chondrocytes; is the most abundant type of cartilage.

Location: Ends of long bones and ribs, nose, parts of larynx, trachea, bronchial tubes, embryonic skeleton.

Function: Provides movement at joints, flexibility, and support.

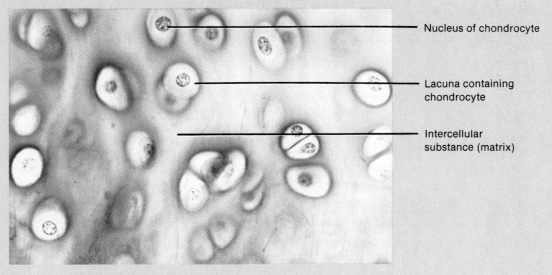

Nucleus of chondrocyte

Lacuna containing chondrocyte

Intercellular substance (matrix)

Sectional view of hyaline cartilage from trachea (512 x)

EXHIBIT 4-2 (*Continued*)
Connective Tissues

Fibrocartilage
Description: Consists of chondrocytes scattered among bundles of
collagenous fibers.
Location: Joint between hip bones, intervertebral discs, knees.
Function: Support and fusion.

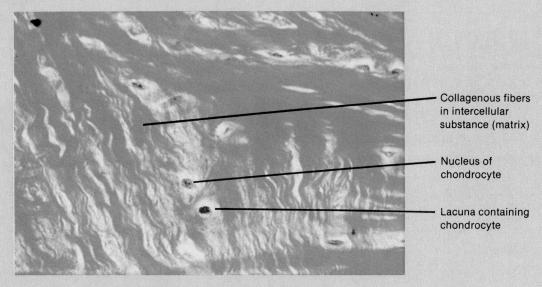

Collagenous fibers
in intercellular
substance (matrix)

Nucleus of
chondrocyte

Lacuna containing
chondrocyte

Sectional view of fibrocartilage from the patellar tendon insertion (288×).

Elastic Cartilage
Description: Consists of chondrocytes located in a threadlike network
of elastic fibers.
Location: Epiglottis of larynx, external ear, and auditory (Eusta-
chian) tubes.
Function: Gives support and maintains shape.

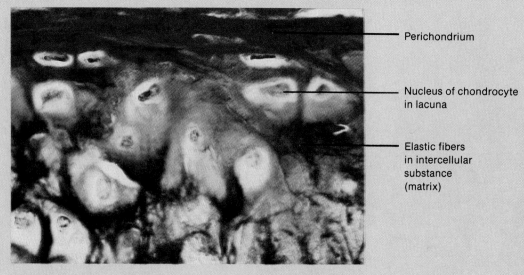

Perichondrium

Nucleus of chondrocyte
in lacuna

Elastic fibers
in intercellular
substance
(matrix)

Sectional view of elastic cartilage from pinna of ear (288x)

Osseous (Bone)

Description: Compact bone consists of osteons (Haversian systems) that contain lamellae, lacunae, osteocytes, canaliculi, and central (Haversian) canals. See also Figure 6-2.

Location: Both compact and spongy bone comprise the various bones of the body.

Function: Support, protection, storage, houses blood-forming tissue, serves as levers that, with muscle tissue, provide movement.

Osteocyte in lacuna

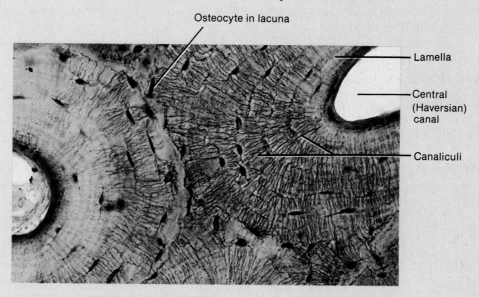

Lamella

Central (Haversian) canal

Canaliculi

Sectional view of portions of two osteons (Haversian systems) from the femur (150x)

Vascular (Blood)

Description: Consists of plasma (liquid matrix) and formed elements (erythrocytes, leucocytes, and thrombocytes).

Location: Within blood vessels (arteries, arterioles, capillaries, venules, and veins).

Function: Erythrocytes transport oxygen and carbon dioxide; leucocytes carry on phagocytosis and are involved in allergic reactions and immunity; and thrombocytes are essential to the clotting of blood.

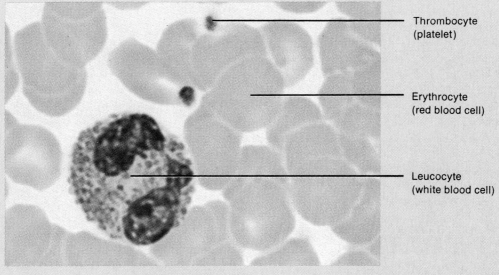

Thrombocyte (platelet)

Erythrocyte (red blood cell)

Leucocyte (white blood cell)

Blood smear (2350x)

Photomicrographs of mesenchyme and dense collagenous, courtesy of Andrew J. Kuntzman; mucous and vascular (blood), Lester Bergman and Associates; loose areolar, elastic, reticular, hyaline cartilage, fibrocartilage, elastic cartilage, and osseous (bone), Biophoto, Photo Researchers; adipose, © Bruce Iverson.

the embryo is **mesenchyme** (MEZ-en-kīm)—the tissue from which all other connective tissues are produced. Mesenchymal cells are branched and found in a fluid matrix. Mesenchyme is located beneath the skin and along the developing bones of the embryo. Some mesenchymal cells are found in adult connective tissue where they assist in wound healing.

Mucous connective tissue (Wharton's jelly) is an embryonic connective tissue in which flattened or spindle-shaped cells are in a mucuslike matrix with collagenous fibers. It is found primarily in the fetus in the umbilical cord where it supports the wall of the cord.

ADULT CONNECTIVE TISSUE

Adult connective tissue is connective tissue that exists in the newborn and that does not change after birth. It is subdivided into several kinds.

Connective Tissue Proper

There is variation among the types of connective tissue proper, but it can be said that the most characteristic cell is the **fibroblast,** a large, flat cell with branching processes, and that the matrix varies from viscous to semisolid and contains some type of fiber.

▪ **Loose (Areolar) Connective Tissue** Loose or areolar (a- RĒ-ō-lar) connective tissue is one of the most widely distributed connective tissues in the body. It occurs around body organs and in the upper region of the dermis of the skin. Combined with adipose tissue, it forms the **subcutaneous** (sub'-kyoo-TĀ-nē-us; *sub* = under; *cut* = skin) **layer**—the layer of tissue that attaches the skin to underlying tissues and organs.

The term ''loose'' refers to the loosely woven arrangement of fibers in the matrix. There are three kinds of fibers: collagenous, elastic, and reticular. The matrix is viscous primarily because of **hyaluronic acid**.

The cells in loose connective tissue are many and varied. They include fibroblasts, macrophages, plasma cells, and mast cells.

Fibroblasts are the predominant cells. They produce new fibers and matrix if the tissue is injured. When a fibroblast becomes mature and no longer produces fibers or matrix it is called a **fibrocyte**.

Macrophages (MAK-rō-fa-jez; *macro* = large; *phagein* = to eat) engulf bacteria and cellular debris by phagocytosis, thus providing a vital defense for the body. As you will see in Chapter 17, macrophages are important in the inflammatory response.

Plasma cells give rise to antibodies and, accordingly, provide another defensive mechanism.

Mast cells are found in abundance along blood vessels and form heparin, an anticoagulant that prevents blood from clotting, and produce histamine and serotonin, chemicals that dilate (widen) blood vessels during the inflammatory response.

Other cells in loose connective tissue include **melanocytes (pigment cells)**, **adipocytes (fat cells),** and **leucocytes (white blood cells).**

▪ **Adipose Tissue** Adipose tissue is a loose connective tissue in which the cells, called **adipocytes**, are specialized for fat storage. Adipocytes are derived from fibroblasts and the cells have the shape of a ''signet ring'' because the cytoplasm and nucleus are pushed to the edge of the cell by a large droplet of fat. Adipose tissue is found wherever loose connective tissue is located. Specifically, it is in the subcutaneous layer below the skin, around the kidneys, at the base and on the surface of the heart, in the marrow of long bones, as a padding around joints, and behind the eyeballs. Adipose tissue is a poor conductor of heat and therefore reduces heat loss through the skin. It is also a major energy reserve and generally supports and protects various organs.

▪ **Dense (Collagenous) Connective Tissue** This tissue is found in and around many organs to provide protection and support or attachment. Collagenous fibers, the predominant fibers, are densely packed. Fibroblasts are present between fibers. The collagenous fibers are organized in regularly arranged parallel bundles in areas that need great strength and some flexibility such as in **tendons,** which attach muscles to bones, and **ligaments,** which hold bones together. Collagenous fibers may also be arranged irregularly with bundles interwoven to provide protective coverings around the heart, eyeballs, bones and cartilage, liver, kidneys, testes, and lymph nodes. This tissue has a silvery white appearance. It has relatively few fibroblasts and elastic fibers.

▪ **Elastic Connective Tissue** Elastic connective tissue has a predominance of freely branching elastic fibers that are a yellowish color. There are relatively few fibroblasts that are present between the fibers. Elastic connective tissue can be stretched and will snap back into shape. It is a component of the walls of arteries, the true vocal cords, the trachea, the bronchial tubes in the lungs, and the lungs themselves, all structures that need to stretch to function efficiently.

▪ **Reticular Connective Tissue** Reticular connective tissue consists of interlacing reticular fibers with cells wrapped around them. It helps to form a delicate supporting stroma (framework) for many organs, including the liver, spleen, and lymph nodes. Reticular connective tissue also helps to bind together the cells of smooth muscle tissue.

Cartilage

Cartilage is capable of enduring considerably more stress than the tissues just discussed. Unlike other connective tissues, cartilage has no blood vessels or nerves, except for those in the **perichondrium** (per'-i-KON-drē-um; *peri* = around; *chondro* = cartilage), the dense connective tissue membrane around the surface of cartilage. Cartilage consists of a dense network of collagenous fibers and elastic fibers embedded in a firm matrix of chondroitin sulfate, a jellylike substance. Whereas the strength of cartilage is due to its collagenous fibers, its resilience (ability to assume its original shape after deformation) is due to chondroitin sulfate. The cells of mature cartilage, called **chondrocytes** (KON-drō-sīts), occur singly or in groups within spaces called **lacunae** (la-KOO-nē) in the matrix. There are three kinds of cartilage: hyaline, fibrocartilage, and elastic. (See Exhibit 4-2.)

▪ **Hyaline Cartilage** This cartilage, also called gristle, appears in the body as a bluish-white, shiny substance. It is the most abundant kind of cartilage in the body. Hyaline cartilage affords flexibility and support. It is found at joints over the ends of long bones (where it is called *articular cartilage)* and forms the *costal cartilages* where the ribs join to the sternum. Hyaline cartilage also helps form the nose, larynx, trachea, bronchi, and bronchial tubes of the lungs. Most of the embryonic skeleton consists of hyaline cartilage. The role of hyaline cartilage in bone formation is discussed in Chapter 6.

■ **Fibrocartilage** This tissue combines strength and rigidity. Fibrocartilage is found between the hip bones, in the intervertebral discs between vertebrae, and in the knee joint.

■ **Elastic Cartilage** Elastic cartilage provides strength and maintains the shape of certain organs—the epiglottis, a lid on top of the larynx (voice box), the external part of the ear, and the auditory (Eustachian) tubes.

Osseous Tissue (Bone)

Together, cartilage, joints, and *osseous* (OS-ē-us) *tissue (bone)* comprise the skeletal system. Mature bone cells are called *osteocytes*. The matrix consists of collagenous fibers and mineral salts containing calcium and phosphorus. The salts are responsible for the hardness of bone.

Bone tissue is classified as either compact (dense) or spongy (cancellous), depending on how the matrix and cells are organized. At this point, we will discuss compact bone only. The basic unit of compact bone is called an *osteon (Haversian system)*. Each consists of *lamellae*, rings of hard matrix; *lacunae*, small spaces between lamellae that contain osteocytes; *canaliculi*, minute canals that project from lacunae and provide numerous routes for nutrients to reach osteocytes and for wastes to be removed from them; and a *central (Haversian) canal* that contains blood vessels and nerves. Note that osseous tissue is vascular, while cartilage is not. Also, the lacunae of osseous tissue are interconnected by canaliculi; those of cartilage are not.

Functionally, the skeletal system supports soft tissues, protects delicate structures, works with skeletal muscles to produce movement, stores calcium and phosphorus, and houses red marrow, which produces several kinds of blood cells, and yellow marrow, which contains lipids as an energy source.

The details of compact and spongy bone are discussed in Chapter 6.

Vascular Tissue (Blood)

Vascular tissue (blood) is a liquid connective tissue that consists of a matrix called plasma and formed elements (cells and cell-like structures). *Plasma* is a straw-colored liquid that consists mostly of water plus some dissolved substances (nutrients, enzymes, hormones, respiratory gases, and ions). The formed elements are erythrocytes (red blood cells), leucocytes (white blood cells), and thrombocytes (platelets). *Erythrocytes* transport oxygen to body cells and carbon dioxide from them. *Leucocytes* (also spelled *leukocytes*) are involved in phagocytosis, immunity, and allergic reactions. *Thrombocytes* function in blood clotting.

The details of blood are considered in Chapter 14.

MUSCLE TISSUE

Muscle tissue is highly specialized for contraction, which allows motion, maintains posture, and produces heat. Muscle tissue is classified into three types: skeletal, cardiac, and smooth (Exhibit 4-3).

Skeletal muscle tissue is named for its location—attached to bones. Skeletal muscle tissue is also *voluntary* because it can be made to contract by conscious control. A single skeletal muscle fiber (cell) is cylindrical and appears *striated* (striped) under a microscope; when organized in a tissue, the fibers are parallel to each other. Each muscle fiber has a plasma membrane, the *sarcolemma,* surrounding the cytoplasm, or *sarcoplasm.* Skeletal muscle fibers are multinucleate (more than one nucleus) and the nuclei are near the sarcolemma.

Cardiac muscle tissue forms the bulk of the wall of the heart and its contraction results in pumping blood to all parts of the body. It is *involuntary* because contraction is not under conscious control. Cardiac muscle fibers are roughly quadrangular and branch to form networks throughout the tissue. They also are striated, but usually have only one centrally located nucleus. Cardiac muscle fibers are separated from each other by thickenings of the sarcolemma called *intercalated discs.* These are unique to cardiac muscle and serve to strengthen the tissue and aid in regulation of the heart rate.

Smooth muscle tissue is located in the walls of hollow internal structures such as blood vessels, the stomach, intestines, gallbladder, and urinary bladder. Its contraction helps break down food, move food and fluids through the body, and evacuate wastes. Smooth muscle fibers are involuntary. The fibers are also spindle-shaped, wide at the center with tapering ends, and nonstriated. They contain a single, centrally located nucleus.

A more detailed discussion of muscle tissue is considered in Chapter 8.

NERVOUS TISSUE

Despite the tremendous complexity of the nervous system, it consists of only two kinds of cells: neurons and neuroglia. *Neurons,* or nerve cells, are highly specialized cells capable of picking up stimuli, converting the stimuli to nerve impulses, and conducting the nerve impulses to other neurons, muscle fibers, or glands. Neurons have three parts: a cell body and two kinds of processes called dendrites and axons (Exhibit 4-4).

The *cell body* contains the nucleus and other organelles. *Dendrites* are highly branched processes that conduct nerve impulses toward the cell body. *Axons* are single, long processes that conduct nerve impulses away from the cell body.

Neuroglia are cells that protect and support neurons. They are of clinical interest because they are the sites of tumors of the nervous system.

The detailed structure and function of neurons and neuroglia are considered in Chapter 9.

Fat Tissue Issues

Adipose tissue: no other body tissue has come close to achieving such popular attention, fame, and notoriety. It has been featured in almost every major newspaper and magazine, and provides a recurring theme in women's magazines as a focus of fashion and fascination. No other body tissue has inspired the creation of more products claiming to compress, massage, shrink, and melt it.

Adipose tissue has even sparked political and philosophical controversy. Fat people have rallied to expose the prejudice and discrimination they face in our culture. Feminists argue that women are judged too frequently by how much fat they have and its anatomical location. Physiological gender differences (females have about 50 percent more fat than males) in body composition often conflict with fashion ideals, leading young girls to develop negative body images, low self-esteem, and eating disorders. This has led some women to say that fat is a feminist issue (since, as a physiology professor once pointed out, fat is a feminine tissue).

Fashion aside, fat performs several vital physiological functions. The fat incorporated into such organs and tissues as mammary glands, nerves, brain, and lungs and the fat in adipose tissue that cushions and supports vital organs is known as essential fat. Storage fat is considered nonessential, though very important as it provides a concentrated source of energy.

Too Much of a Good Thing

Obesity refers to excess body fat and is usually defined as being 10 to 20 percent or more overweight. About 30 percent of the people in the United States fall into this category.

Life insurance data have shown a negative association between excessive fatness and mortality, and the desirable weight for height charts have been derived from these data. People falling into the weight ranges for a given gender and height on these charts generally have a longer life expectancy.

Obesity alone may not pose a risk of premature death, but the frequency of several serious health problems does increase with excess weight. Obesity increases the risk of hypertension, type II diabetes mellitus, and hyperlipidemia (high serum cholesterol and/or triglycerides), all three of which increase a person's risk of heart disease and stroke. Gallstones, gallbladder disease, cirrhosis

of the liver, kidney disease, and some cancers are more common in obese individuals. In addition, extra fat results in increased stress on the muscles and bones, with higher rates of osteoarthritis, back pain, and other orthopedic complications.

How Much Is Too Much?

Height–weight charts give a useful range for healthful body weights, but weight alone can be misleading. While the scale does not lie, it doesn't tell the whole truth either. Weight is not the same thing as fat. For most sedentary people, a weight gain does mean an increase in body fat. But it is also possible to get fatter over the years and see no change in weight. Sedentary adults may weigh the "correct" amount but be overly fat because their muscles have atrophied from disuse. Likewise, people who embark on a vigorous exercise program may lose fat yet see no change or even an increase in body weight. Bodybuilders and other strength athletes may produce a hefty number on the scale, and weigh more than they "should" for their height, yet be relatively lean.

To get around the limitations of the height–weight tables, various methods for assessing fatness, or body composition, have been devised, based on the physiological properties of adipose tissue. These methods divide body composition into two components: fat and everything else or fat-free mass (FFM). Body composition is often expressed as the percentage of the body's volume that is comprised of fat. The desirable body fat range for men is 12 to 20 percent, with athletes as low as 5 percent. Women should be about 20 to 30 percent fat, with female athletes at 16 percent.

Most people know whether they have too much adipose tissue without a body composition assessment. But such assessments can help an obese person set a realistic weight (fat) loss goal and help monitor body composition change in obesity research. Many fitness centers offer some sort of body composition assessment to help motivate clients to exercise.

Methods of Estimating Body Composition

All methods of estimating body composition are just that: estimates. The only way actually to measure how much fat is contained in a body is cadaver dissection. Since water is not present in adipocytes, and since water comprises a known fraction of the FFM, percent fat can be determined from the amount of total body water, estimated from isotope dilution. Estimating total body potassium works in a similar fashion, since the intracellular cation potassium is not found in stored fat.

Differences in the density between fat and FFM provide the basis for hydrostatic densitometry, also known as underwater weighing. Since adipose tissue is less dense than other body components, the more easily a person floats, the fatter he is. Percent fat is estimated from density, using a standard equation.

Anthropometric measures, such as girths and skinfolds, are also used to estimate body composition, especially outside the laboratory. To measure skinfold thickness, the skin and subcutaneous fat are pulled (pinched) away from the underlying muscle and the thickness of the fold is measured with calipers.

The accuracy of this method varies considerably, depending on both the skill of the technician and the group of people being measured. The skinfold method assumes that the thickness of subcutaneous fat can be used to estimate total body fat, and that the measurement sites represent the average thickness of subcutaneous fat. These assumptions are not true for everyone, and introduce a substantial degree of error into the method.

When skinfolds or girths are used to measure change in body fatness, they are more accurate than when they are used to predict percent fat. This is because when they are used to predict, they are put into a regression equation that has been derived from a specific group of people. That equation will generally predict a good average for the group from which it was derived. However, it may not predict accurately for any given individual in that or any other group. In the long run, on the average, it may predict fairly well, but it cannot be guaranteed for any given individual. How do you know which individuals will fit the equation? You can't tell.

Another method of estimating body composition is bioelectric impedance analysis. It is based on the principle that fat tissue is a poorer conductor of an electric current than lean tissue. Small electrodes are placed on the skin to measure the conductance of a very weak electric current. The easier the conductance, the leaner the person.

Haunch Versus Paunch

All fat stores are not created equal. Research has found that extra fat on the torso is associated with a greater health risk than extra fat on the hips and thighs. Central adiposity is more common among males, although it occurs in women as well, and may be one of the reasons that males have a higher risk of cardiovascular disease.

The health risks associated with having an "apple" versus a "pear" shape include insulin resistance and type II diabetes mellitus, high blood cholesterol, and high blood pressure, all of which increase a person's risk of heart disease.

Body shape, or fat pattern distribution, can be assessed very simply by comparing waist and hip girths. A waist–hip ratio (WHR) greater than 1.0 for men or 0.8 for women is considered to put an individual at increased risk for the above disorders.

Adipocytes in different regions appear to be metabolically different. Fat cells in the abdominal region release directly into the vein that goes to the liver, the site of a great deal of fat, protein, and carbohydrate metabolism, and the place where cholesterol-containing lipoproteins are produced. Abdominal fat cells may thus have a stronger impact on fat and cholesterol metabolism than cells in more peripheral locations. Studies have also shown that abdominal adipocytes also respond more readily to several hormones, including adrenalin.

Fat pattern distribution appears to be genetically determined, and is not influenced by diet or exercise. If an obese person loses weight, fat is generally lost from all areas of the body. As one researcher put it, "An individual continues to resemble himself in relative fat pattern, despite weight loss."

Many people persist in performing calisthenics in a misguided attempt to influence fat deposits in specific body areas. "Spot reduction" does not occur, however, Sit-ups will strengthen abdominal muscles, but will not reduce the amount of adipose tissue stored in that area.

EXHIBIT 4-3
Muscle Tissue

Skeletal Muscle Tissue
Description: Cylindrical, striated fibers with several nuclei near the sarcolemma; voluntary.
Location: Attached to bones.
Function: Motion, posture, heat production.

Cardiac Muscle Tissue
Description: Quadrangular, branching, striated fibers with one centrally located nucleus; contains intercalated discs; involuntary.
Location: Heart wall.
Function: Motion (contraction of heart that pumps blood to all parts of the body).

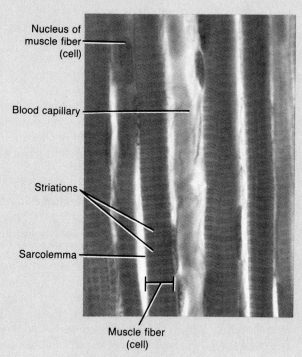

Nucleus of muscle fiber (cell)

Blood capillary

Striations

Sarcolemma

Muscle fiber (cell)

Section of skeletal muscle (800×)

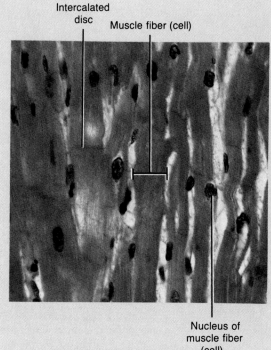

Intercalated disc

Muscle fiber (cell)

Nucleus of muscle fiber (cell)

Section of cardiac muscle (400×)

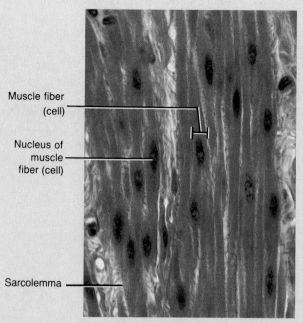

Muscle fiber (cell)

Nucleus of muscle fiber (cell)

Sarcolemma

Section of smooth muscle (840x)

Smooth Muscle Tissue
Description: Spindle-shaped, nonstriated fibers with one centrally located nucleus; involuntary.
Location: Walls of hollow internal structures such as blood vessels, stomach, intestines, gallbladder, and urinary bladder.
Function: Motion (constriction of blood vessels, propulsion of foods through gastrointestinal tract; contraction of gallbladder).

Photomicrographs courtesy of Andrew J. Kuntzman.

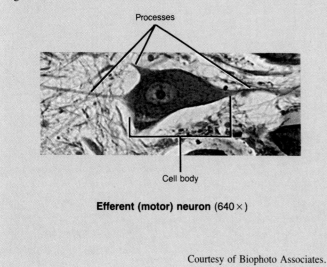
MEMBRANES

The combination of an epithelial layer and an underlying connective tissue layer constitutes an *epithelial membrane*. The three types of epithelial membranes are mucous, serous, and cutaneous. The fourth kind of membrane in the body, *synovial*, does not contain epithelium.

MUCOUS MEMBRANES

A *mucous membrane*, or *mucosa*, lines a body cavity that opens directly to the exterior. Mucous membranes line the entire gastrointestinal, respiratory, excretory, and reproductive tracts.

The epithelial layer of a mucous membrane secretes mucus, which prevents the cavities from drying out. It also traps particles in the respiratory passageways, lubricates and absorbs food as it moves through the gastrointestinal tract, and secretes digestive enzymes.

The connective tissue layer binds the epithelium to the underlying structures. It also provides the epithelium with oxygen and nutrients and removes wastes since the blood vessels are located here (see Figure 19-2).

SEROUS MEMBRANES

A *serous membrane*, or *serosa*, lines a body cavity that does not open directly to the exterior, and it covers the organs that lie within the cavity. Serous membranes consist of two portions. The part attached to the cavity wall is called the *parietal* (pa-RĪ-e-tal) *portion*; the part that covers the organs is the *visceral portion*. The serous membrane lining the thoracic cavity and covering the lungs is called the *pleura* (see Figure 18-5). The one lining the heart cavity and covering the heart is the *pericardium* (*cardio* = heart). The one lining the abdominal cavity and covering the abdominal organs and some pelvic organs is called the *peritoneum*.

The epithelial layer, called *mesothelium*, secretes a lubricating serous fluid that allows the organs to glide against one another or against the cavity walls. The connective tissue layer is a thin layer of loose connective tissue.

CUTANEOUS MEMBRANE

The *cutaneous membrane*, or skin, constitutes an organ of the integumentary system and is discussed in the next chapter.

SYNOVIAL MEMBRANES

Synovial membranes line the cavities of the joints (see Figure 7-1). They are composed of loose connective tissue with elastic fibers and varying amounts of fat; they do not have an epithelial layer. Synovial membranes secrete *synovial fluid*, which lubricates the ends of bones as they move at joints and nourishes the articular cartilage covering the bones.

STUDY OUTLINE

Extracellular Materials (p. 61)

1. These are substances that lie outside the plasma membrane.
2. They include body fluids (interstitial fluid and plasma), secretions (mucus, hormones, etc.), and matrix.

3. Matrix is a ground substance in which connective tissue cells and fibers are embedded. Matrix binds cells together into tissues and gives strength and elasticity to the tissue.

Types of Tissues (p. 61)

1. A tissue is a group of similar cells and their intercellular substance that are specialized for a particular function.
2. The tissues of the body are classified into four principal types: epithelial, connective, muscular, and nervous.

Epithelial Tissue (p. 62)

1. Epithelium has many cells, little intercellular material, and no blood vessels (avascular). It is attached to connective tissue by a basement membrane. It can replace itself.
2. The subtypes of epithelium include covering and lining epithelium and glandular epithelium.

Covering and Lining Epithelium (p. 62)

1. Cell shapes include squamous (flat), cuboidal (cubelike), columnar (rectangular), and transitional (variable). Layers are arranged as simple (one layer), stratified (several layers), and pseudostratified (one layer that appears as several).
2. Simple squamous epithelium is adapted for diffusion and filtration and is found in lungs and kidneys.
3. Simple cuboidal epithelium is adapted for secretion and absorption. It is found covering ovaries, in kidneys and eyes, and lining some glandular ducts.
4. Nonciliated simple columnar epithelium lines most of the gastrointestinal tract. Specialized cells containing microvilli perform absorption. Goblet cells secrete mucus. In a few portions of the respiratory tract, the cells are ciliated to move foreign particles trapped in mucus out of the body.
5. Stratified squamous epithelium is protective. It lines the upper gastrointestinal tract and vagina and forms the outer layer of skin.
6. Stratified cuboidal epithelium is found in sweat glands, the pharynx, and portions of the urethra.
7. Stratified columnar epithelium protects and secretes. It is found in the male urethra and large excretory ducts.
8. Transitional epithelium lines the urinary bladder and is capable of stretching.
9. Pseudostratified columnar epithelium has only one layer but gives the appearance of many. It lines larger excretory ducts, parts of the urethra, auditory (Eustachian) tubes, and most upper respiratory structures, where it protects and secretes.

Glandular Epithelium (p. 69)

1. A gland is a single cell or a mass of epithelial cells adapted for secretion.
2. Exocrine glands (sweat, oil, and digestive glands) secrete into ducts or directly onto a free surface.
3. Endocrine glands secrete hormones into the blood.

Connective Tissue (p. 69)

1. Connective tissue is the most abundant body tissue. It has few cells, extensive intercellular substance (matrix), and a rich blood supply (vascular), except for cartilage.
2. The matrix determines the tissue's qualities.
3. Connective tissue protects, supports, and binds organs together.
4. Connective tissue is classified into two principal types: embryonic and adult.

Embryonic Connective Tissue (p. 69)

1. Mesenchyme forms all other connective tissues.
2. Mucous connective tissue is found in the umbilical cord of the fetus, where it gives support.

Adult Connective Tissue (p. 76)

1. Adult connective tissue is connective tissue that exists in the newborn and does not change after birth. It is subdivided into connective tissue proper, cartilage, bone tissue, and vascular tissue.
2. Connective tissue proper has a viscous to semisolid matrix; a typical cell is the fibroblast. Five examples of such tissues may be distinguished.
3. Loose (areolar) connective tissue is a widely distributed connective tissue in the body. Its matrix (hyaluronic acid) contains fibers (collagenous, elastic, and reticular) and various cells (fibroblasts, macrophages, plasma cells, mast cells, and melanocytes). Loose connective tissue is found around body organs and in the subcutaneous layer.
4. Adipose tissue is a form of loose connective tissue in which the cells, called adipocytes, are specialized for fat storage. It is found in the subcutaneous layer and around various organs.
5. Dense (collagenous) connective tissue has a close packing of fibers (regularly or irregularly arranged). It is found in and around many organs and in tendons and ligaments.
6. Elastic connective tissue has a predominance of freely branching elastic fibers that give it a yellow color. It is found in arteries, the trachea, bronchial tubes, and true vocal cords.
7. Reticular connective tissue consists of interlacing reticular fibers and forms the stroma of the liver, spleen, and lymph nodes.
8. Cartilage has a firm jellylike matrix containing collagenous and elastic fibers and chondrocytes.
9. Hyaline cartilage is found in the embryonic skeleton, at the ends of bones, in the nose, and in respiratory structures. It is flexible, allows movement, and provides support.
10. Fibrocartilage connects the hip bones and the vertebrae. It provides strength.
11. Elastic cartilage maintains the shape of organs such as the larynx, auditory (Eustachian) tubes, and external ear.
12. Osseous tissue (bone) consists of a matrix of mineral salts and collagenous fibers that contribute to the hardness of bone and cells called osteocytes. It supports, protects, helps provide movement, stores minerals, and houses red marrow.
13. Bone tissue is classified as either compact (dense) or spongy (cancellous), depending on how the matrix and cells are organized.
14. Vascular tissue (blood) is a liquid connective tissue. It consists of plasma and formed elements (erythrocytes, leucocytes, and thrombocytes). Functionally, its cells transport, carry on phagocytosis, participate in allergic reactions, provide immunity, and bring about blood clotting.

Muscle Tissue (p. 77)

1. Muscle tissue is modified for contraction and thus provides motion, maintains posture, and produces heat.
2. Skeletal muscle tissue is attached to bones, is striated, and is voluntary.
3. Cardiac muscle tissue forms most of the heart wall, is striated, and is involuntary.
4. Smooth muscle tissue is found in the walls of hollow internal structures (blood vessels and viscera), is nonstriated, and is involuntary.

Nervous Tissue (p. 77)

1. The nervous system is composed of neurons (nerve cells) and neuroglia (protective and supporting cells).
2. Neurons consist of a cell body and two types of processes called dendrites and axons.

3. Neurons are specialized to pick up stimuli, convert stimuli into nerve impulses, and conduct nerve impulses.

Membranes (p. 81)

1. An epithelial membrane is an epithelial layer overlying a connective tissue layer. Examples are mucous, serous, and cutaneous membranes.
2. Mucous membranes line cavities that open to the exterior, such as the gastrointestinal tract.
3. Serous membranes (pleura, pericardium, peritoneum) line closed cavities and cover the organs in the cavities. These membranes consist of parietal and visceral portions.
4. The cutaneous membrane is the skin.
5. Synovial membranes line joint cavities and do not contain epithelium.

REVIEW QUESTIONS

1. What are extracellular materials? Give examples and the functions of each. (p. 61)
2. Define a tissue. What are the four basic kinds of human tissue? (p. 61)
3. What characteristics are common to all epithelium? (p. 62)
4. Describe the various layering arrangements and cell shapes of epithelium. (p. 62)
5. How is epithelium classified? List the various types. (p. 62)
6. For each of the following kinds of epithelium, briefly describe the appearance, location in the body, and functions: simple squamous, simple cuboidal, simple columnar (nonciliated and ciliated), stratified squamous, stratified cuboidal, stratified columnar, transitional, and pseudostratified columnar. (pp. 62–69)
7. Define the following terms: secretion, absorption, goblet cell, and keratin. (pp. 62–69)
8. What is a gland? (p. 69)
9. How does connective tissue differ from epithelium? (p. 69)
10. How are connective tissues classified? List the various types. (p. 69)
11. Describe the following connective tissues with regard to appearance, location in the body, and function: loose (areolar), adipose, dense (collagenous), elastic, reticular, hyaline cartilage, fibrocartilage, elastic cartilage, osseous tissue (bone), and vascular tissue (blood). (pp. 69–77)
12. Define the following terms: hyaluronic acid, collagenous fiber, elastic fiber, reticular fiber, fibroblast, macrophage, plasma cell, mast cell, melanocyte, adipocyte, chondrocyte, lacuna, osteocyte, lamella, and canaliculi. (p. 76–77)

13. Describe muscle tissue. How is it classified? What are its functions? (p. 77)
14. Distinguish between neurons and neuroglia. Describe the structure and function of neurons. (p. 77)
15. Following are some descriptions of various tissues of the body. For each description, name the tissue described.
    a. An epithelium that permits distention (stretching).
    b. A single layer of flat cells concerned with filtration and absorption.
    c. Forms all other kinds of connective tissue.
    d. Specialized for fat storage.
    e. An epithelium with waterproofing qualities.
    f. Forms the framework of many organs.
    g. Produces perspiration, wax, oil, or digestive enzymes.
    h. Cartilage that shapes the external ear.
    i. Contains goblet cells and lines the intestines.
    j. Most widely distributed connective tissue.
    k. Forms tendons and ligaments.
    l. Specialized for the secretion of hormones.
    m. Provides support in the umbilical cord.
    n. Lines kidney tubules and is specialized for absorption and secretion.
    o. Permits extensibility of lung tissue.
    p. Stores red marrow, protects, supports.
    q. Nonstriated, usually involuntary muscle tissue.
    r. Composed of a cell body, dendrites, and axon.
16. Define the following kinds of membranes: mucous, serous, cutaneous, and synovial. Where is each located in the body? What are their functions? (p. 81)

5

The Integumentary System

STUDENT OBJECTIVES

1. Describe the structure and functions of the skin.
2. Explain the basis for skin color.
3. Describe the functions of hair and skin glands.
4. Explain how the skin helps to regulate normal body temperature.
5. Define a burn and describe its effects on the body.
6. Define common disorders and medical terminology and conditions associated with the integumentary system.

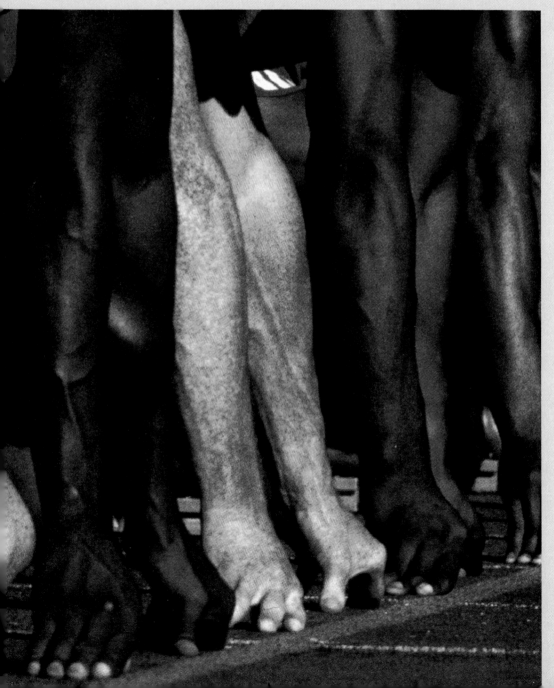

A LOOK AHEAD

SKIN
  Functions
  Structure
  Epidermis
  Dermis
  Skin Color
ACCESSORY ORGANS OF THE SKIN
  Hair
  Glands
    Sebaceous (Oil) Glands
    Sudoriferous (Sweat) Glands
    Ceruminous Glands
  Nails
HOMEOSTASIS OF BODY
    TEMPERATURE
COMMON DISORDERS
MEDICAL TERMINOLOGY AND
    CONDITIONS

A group of tissues that performs a specific function is an *organ*. The next higher level of organization is a *system*—a group of organs operating together to perform specialized functions. The skin and its accessory organs, such as hair, nails, glands, and several specialized receptors, constitute the *integumentary* (in'-teg-yoo-MEN-tar-ē) *system* of the body.

Of all the body's organs, none is more exposed to inspection, disease, and injury than the skin. Because of its visibility, skin is a reflection of emotions as evidenced by frowning, blushing, and perspiring. Diseases of internal organs may be revealed by changes in the skin such as color changes (paleness, redness, yellow coloration, or bluish coloration) or abnormal eruptions or rashes (chickenpox, cold sores, or measles). The skin is also subject to a number of disorders that involve just itself, such as warts, age spots, or pimples. The skin's exposure to the environment makes it susceptible to damage from trauma, sunlight, and microbes.

Many interrelated factors may affect both the appearance and health of the skin, including nutrition, hygiene, circulation, age, immunity, genetic factors, psychological state, and drugs. So important is the skin to body image that people spend a great deal of time and money, and may even undergo surgery, to restore skin to a more normal or youthful appearance.

SKIN

The *skin* is an organ because it consists of the cutaneous membrane and accessory organs such as sweat and oil glands, hair follicles, nails, nerve receptors, and blood vessels joined together to perform specific activities. It is one of the larger organs in terms of surface area. For the average adult, the skin occupies a surface area of approximately 2 square meters (3000 square inches). *Dermatology* (der'-ma-TOL-ō-jē; *dermato* = skin; *logos* = study of) is the medical specialty that deals with the diagnosis and treatment of skin disorders.

FUNCTIONS

The numerous functions of the skin are as follows:

1. **Regulation of body temperature.** Perspiration produced by sweat glands and changes in the flow of blood to the skin help regulate body temperature (described in detail later in the chapter).
2. **Protection.** The skin provides a physical barrier against abrasion, microbial invasion, dehydration, and ultraviolet (UV) radiation. Hair and nails are also protective, as described later.
3. **Reception of stimuli.** Numerous nerve endings and receptors detect stimuli related to temperature, touch, pressure, and pain (see Chapter 12).
4. **Excretion.** Small amounts of water, salts, and several organic compounds, components of perspiration, are excreted by sweat glands.
5. **Synthesis of vitamin D.** Exposure of the skin to ultraviolet (UV) radiation helps with the synthesis of vitamin D, a substance that aids in the absorption of calcium and phosphorus. Significant amounts of vitamin D are synthesized by the skin in spring, summer, and fall. In a city like Boston, for example, a 10- to 15-minute exposure of the hands, arms, and face per day for 3 days per week is sufficient to produce the body's need for vitamin D. (However, the amount produced is insignificant, especially in temperate climates, and the risk of skin cancer outweighs the possible benefits.)
6. **Immunity.** As you will see in Chapter 17, certain cells of the skin assume a role in bolstering immunity, your ability to fight disease by producing antibodies.

STRUCTURE

Structurally, the skin consists of two principal parts (Figure 5-1). The outer, thinner portion, which is composed of epithelium, is called the *epidermis*. The inner, thicker, connective tissue part is called the *dermis*. Beneath the dermis is a *subcutaneous (SC) layer,* which attaches the skin to underlying structures.

EPIDERMIS

The *epidermis* is composed of stratified squamous epithelium and contains four types of cells (Figure 5-2). The most numerous is known as a *keratinocyte*, a cell that undergoes keratinization. (This process will be described shortly.) Keratinocytes produce keratin, which helps waterproof and protect the skin, and they participate in immunity. The second type of cell is called a *melanocyte*, which can also be found in the dermis. It produces melanin, one of the pigments responsible for skin color, and absorbs ultraviolet (UV) radiation. The third and fourth types of cells are both called *nonpigmented granular dendrocytes*, but they are two distinct cell types, formerly known as *Langerhans' cells* and *Granstein cells*. These cells are involved in the immune response.

The keratinocytes of the epidermis are organized into either four or five cell layers. Where exposure to friction is greatest, such as in the palms and soles, the epidermis has five layers. In all other parts of the body it has four. The names of the five layers from the deepest to the most superficial are as follows:

1. **Stratum basale.** This single layer of cells produces melanin and is capable of continued cell division. As the cells multiply, they push up toward the surface and become part of the layers to be described next. Eventually the cells are shed from the top layer of the epidermis. The stratum basale is sometimes

FIGURE 5-1 Skin. Structure of the skin and underlying subcutaneous layer.

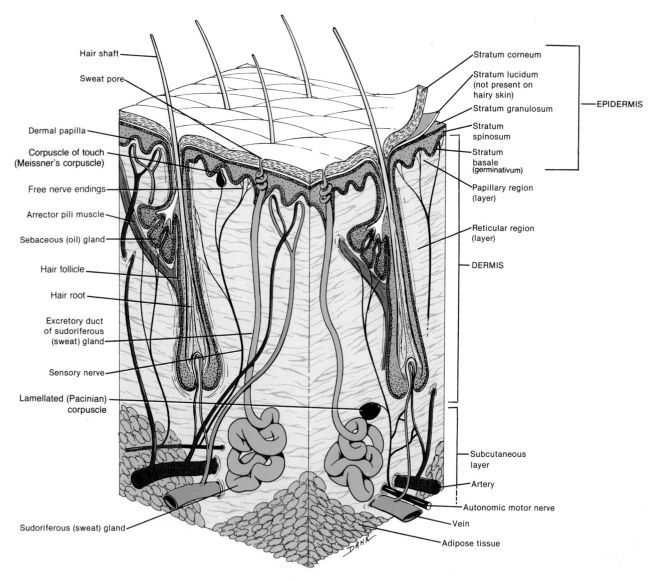

What stratum of the epidermis is continually shed? Capable of cell division?

referred to as the *stratum germinativum* (jer'-mi-na-TĒ-vum) to indicate its role in germinating new cells. The stratum basale of hairless skin contains nerve endings sensitive to touch called *tactile (Merkel's) discs.*

2. **Stratum spinosum.** This layer has eight to ten rows of cells with spinelike projections (*spinosum* = prickly) that join the cells together. Melanin is also found in this layer.
3. **Stratum granulosum.** The third layer consists of three to five rows of flattened cells that contain *keratohyalin* (ker'-a-tō-HĪ-a-lin), a substance involved in keratin formation. Keratin protects the skin from injury and microbial invasion and makes it waterproof.
4. **Stratum lucidum.** This is the layer found only in the palms and soles. It consists of several rows of clear, flat, dead cells that are translucent (*lucidum* = clear).
5. **Stratum corneum.** This layer consists of 25 to 30 rows of flat, dead cells completely filled with keratin. These cells are continuously shed and replaced.

In the process of *keratinization*, newly formed cells produced in the basal layers are pushed to the surface. As the cells move upward, the cytoplasm, nucleus, and other organelles are replaced by keratin and the cells die. Eventually, the keratinized cells are sloughed off and replaced by underlying cells that, in turn, become keratinized. The whole process takes about 2 weeks.

DERMIS

The second part of the skin, the *dermis* (see Figure 5-1), is very thick in the palms and soles and very thin in the eyelids, penis, and scrotum.

The upper region of the dermis consists of loose connective tissue containing fine elastic fibers and its surface area is greatly increased by small, fingerlike projections called *dermal papillae* (pa-PIL-ē). Dermal papillae cause ridges in the epidermis, which produce fingerprints and help us to grip objects. Some dermal papillae contain tactile receptors called *corpuscles of touch*, also known as *Meissner's (MĪS-nerz) corpuscles*, nerve endings sensitive to touch. Others contain blood capillaries.

The lower region of the dermis consists of dense, irregular connective tissue containing collagenous and elastic fibers, adipose tissue, hair follicles, nerves, oil glands, and the ducts of sweat glands. The combination of collagenous and elastic fibers gives the skin its strength, extensibility, and elasticity. (*Extensibility* is the ability to stretch; *elasticity* is the ability to return to original shape after extension or contraction.) The ability of the skin to stretch can readily be seen during pregnancy, obesity, and tissue swelling (edema). Small tears in the skin due to extensive stretching that remain visible as silvery white streaks are called *striae* (STRĪ-ē). The lower region is attached to underlying bone and muscle by the subcutaneous layer. The subcutaneous layer contains nerve endings called *lamellated (Pacinian) corpuscles*, which are sensitive to pressure.

Cold receptors, which are probably free nerve endings, are found in and just below the dermis, whereas warmth receptors, which are also probably free nerve endings, are found in the upper and middle dermis.

SKIN COLOR

Skin color is due to *melanin*, a pigment in the epidermis; carotene, a pigment mostly in the dermis; and blood in capillaries in the dermis. The amount of melanin varies the skin color from pale

FIGURE 5-2 Structure of the epidermis. (a) Photomicrograph at a magnification of 496×. (Courtesy of Lester Bergman and Associates.) (b) Diagram.

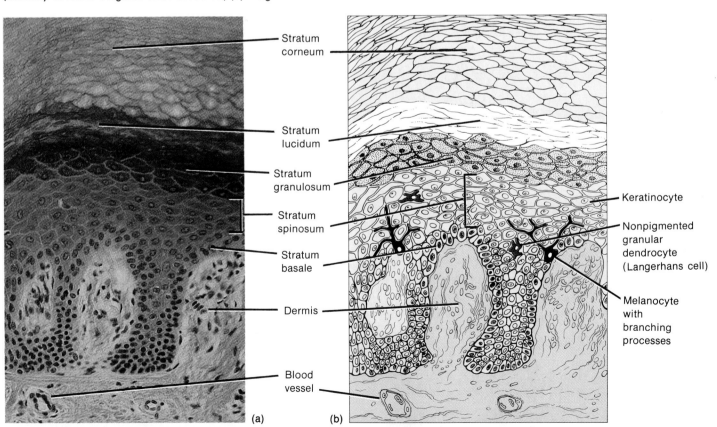

Stratum corneum

Stratum lucidum

Stratum granulosum

Stratum spinosum

Stratum basale

Dermis

Blood vessel

Keratinocyte

Nonpigmented granular dendrocyte (Langerhans cell)

Melanocyte with branching processes

(a)　　(b)

More Than Skin Deep

The skin reflects a person's general health and well-being. What's good for your health is often good for your skin. And the health benefits of skin care are more than skin deep, since what affects the skin can affect other parts of the body and a person's self-image as well. A healthy life-style can help keep skin healthy, reduce the severity of many dermatological disorders, and slow skin aging.

Nutrition

The best nutritional recommendation for healthy skin is not magic, just plain good sense, and the same phrase you've heard a hundred times before: Eat a well-balanced diet. Vitamin and mineral deficiencies are often associated with skin disorders. For example, scurvy, caused by vitamin C deficiency, is marked by bleeding skin and gums. Pellagra, caused by niacin deficiency, is characterized by a skin rash. Other vitamin and mineral deficiencies may cause dermatitis, mouth ulcers, and redness and cracking at the corners of the mouth.

Getting more than the recommended daily allowance of vitamins and minerals does not make above-average skin. Supplements will only improve the skin if they are correcting an existing deficiency. In fact, too much of some vitamins can actually cause skin problems. For example, too much vitamin A is toxic and can make the skin rough and dry. Niacin can cause flushing and itching. And vitamin E applied topically can cause acne and allergic reactions in some people.

Gelatin and protein supplements have been promoted as useful for improving the strength of nails and hair. However, increasing protein consumption will improve hair and nails in a malnourished person only. While it's true that nails and hair are made from protein, most Americans consume far more protein than they need already. Extra protein is not packed into the nails, but transformed into fat.

Food Sensitivity

Some skin disorders are linked to food sensitivities. One person in five experiences hives at some time, which may be brought on by many things, including certain foods. Eggs, nuts, beans, chocolate, strawberries, tomatoes, citrus fruit, seafood, corn, and pork are the most common problematic foods. People who believe their hives may be caused by foods must undergo some sort of controlled diet, eliminating then reintroducing possible offenders.

Several food additives may also cause hives. Tartrazine (yellow dye #5), the preservative sodium benzoate, and sulfites are common triggers. Tartrazine and sodium benzoate must be listed on labels; sulfites are more difficult to avoid. They are used as a preservative in many processed foods, on prepared foods in restaurants and salad bars, and in wine and cider. Salicylates also cause hives in some people. Salicylates are found in all aspirin-containing medications. They also occur naturally in almonds, apples, peaches, potatoes, and other foods.

Some evidence suggests that some foods exacerbate atopic eczema, or dermatitis. In one study of 36 children with eczema, 12 experienced significant improvement when eggs and cow's milk were eliminated from their diets.

Diet and Acne

Popular opinion to the contrary, most research has failed to establish a clear link between diet and acne. At one time it was a commonly held belief that foods such as chocolate, cola drinks, nuts, and dairy products could make acne worse. Some acne patients (and their physicians) claim to have isolated foods that do affect acne severity, so it remains possible that diet, especially one high in fat and refined carbohydrates, may be a contributing factor in some cases.

Some of the most effective acne drugs, Retin-A and Accutane, are synthetic derivatives of vitamin A. They work as drugs, however, and must be used under medical supervision. Accutane produces side effects in some patients, and must not be taken during pregnancy, since it causes birth defects.

Stress, Fatigue, and Emotional Problems

Why is it that students (and teachers) look so much better in September than in December? Stress may be one of the factors responsible for this difference. Stress tends to worsen many preexisting skin problems, especially herpes, acne, eczema, hives, psoriasis, and warts. Stress management techniques, including hypnosis, biofeedback, and other relaxation exercises, have been shown to improve many skin problems. These techniques may work by changing hormone levels and nervous system activity. (The stress response will be discussed in more detail in Chapter 13.)

Herpes is caused by a virus. Managing stress and reducing fatigue can improve immune system function and help the body resist the virus's effects. Acne can worsen when the adrenal glands become overstimulated during times of stress and produce too many of the hormones responsible for acne outbreaks. Relaxation exercises can decrease the itching of those who suffer from eczema.

Noted scientist Lewis Thomas has written that it's "one of the great mystifications of science: warts can be ordered off the skin by hypnotic suggestion." The skin responds to a person's beliefs. In one study, subjects were told their skin was being exposed to poison ivy, and many broke out even though the poison ivy was only in their imaginations.

A person's behavior often makes a skin problem worse. Squeezing pimples and scratching irritated skin are two common examples. Relaxation techniques can be used to help lessen the picking and scratching, and also to decrease the itching associated with conditions such as eczema.

Exercise

During exercise, the body shunts blood to the skin to help release excess heat produced by the contracting muscles. This increased blood flow provides the skin with nutrients and gets rid of wastes. One study found that regular exercisers had thicker skin than sedentary individuals. Thicker skin ages more gracefully as it develops wrinkles later than thinner skin.

Outdoor exercisers must take care to protect their skin from the elements. Sun can cause skin cancer, and wind and cold can damage the skin and cause dryness and frostbite. Swimmers often experience dry skin and hair that can be treated with moisturizing products.

Protection from the Sun

Protecting skin from the sun's damaging rays will help prevent premature aging and cancers of the skin, whose rates have been rising rapidly. One in seven Americans, over 500,000 people per year, will develop skin cancer during their lifetime.

The sun's ultraviolet rays are the source of skin damage. Until recently, scientists and consumers were only concerned about UV-B rays, since they are the ones that cause sunburn and skin cancer. UV-A rays cause tanning, and were once thought to be harmless. But UV-A rays actually penetrate the skin more deeply and can damage the skin's connective tissue, causing sagging and wrinkling of the skin. UV-A seems to increase the cancer-causing effects of UV-B rays. While glass effectively blocks UV-B rays, UV-A can penetrate glass, which is why your arm can get tan on a long car ride. Tanning salons promise "harmless UV-A" light, but there is no such thing.

Skin protection is the way to go outdoors. The most effective skin protection is some form of sun block. Tightly woven clothing (hold it up to a light and see how much shines through) helps keep the sun's rays from reaching the skin, and wide-brimmed hats provide some protection.

When a sun block is not practical, a sunscreen should be used. These do not shield the skin completely, but they do reduce the damaging effects of the ultraviolet rays. A screen with two or more ultraviolet-absorbing ingredients offers the best protection against both kinds of ultraviolet rays. When selecting a sunscreen look for PABA (para-aminobenzoic acid), Padimate O, Padimate A, or cinnamates, which protect you from UV-B rays, and Parsol 1789, benzophenones, oxybenzone, methoxybenzone, or sufisobenzone, which prevent UV-A damage.

To be effective, sunscreens must be applied at least 15 minutes before sun exposure, and then again every 60–90 minutes, and after swimming or sweating. Sunscreens are marked with an SPF (sun-protection factor) rating; the higher the factor, the stronger the sunscreen. A sunscreen with SPF-15 is generally recommended. Sunscreens and blocks should be used conscientiously whenever outdoors, even on cloudy days since some radiation penetrates the cloud cover.

Skin Self-Care

Daily skin self-care includes hygiene and the application of lotions and medications when required. Hygiene is especially important in the prevention and treatment of acne.

An enormous variety of over-the-counter skin care products are available. Some are effective in moisturizing the skin, and in reducing inflammation and itching. But some products make unsubstantiated claims, and it is up to the consumer to become educated about skin care products.

A monthly skin self-exam is recommended to catch early signs of skin cancer. A physician should be consulted if unusual changes in the skin are observed, such as unusual moles or changes in existing moles, irregularly shaped spots with different coloration, crusted sores that do not heal, and red, rough patches or bumps.

yellow to black. Since the number of melanocytes is about the same in all races, differences in skin color are due to the amount of pigment the melanocytes produce and distribute. An inherited inability of an individual in any race to produce melanin results in **albinism** (AL-bin-izm); the pigment is absent in the hair and eyes as well as the skin. In some people, melanin tends to form in patches called **freckles**.

When the skin is repeatedly exposed to ultraviolet radiation, the amount and darkness of melanin increase, which tans and protects the body against radiation. Overexposure to ultraviolet light, however, may lead to skin cancer. Among the most serious skin cancers is **malignant melanoma** (*melano* = dark colored; *oma* = tumor), cancer of the melanocytes. Fortunately, most skin cancers involve basal and squamous cells that can be removed surgically.

The pigment, **carotene** (KAR-o-tēn), is found in the stratum corneum of people of Asian origin. Together, carotene and melanin account for the yellowish hue of their skin.

Lesser amounts of melanin in Caucasian skin makes the epidermis translucent. Their skin will appear pink to red depending on the amount and quality of the blood moving through vessels in the dermal and subcutaneous layers.

ACCESSORY ORGANS OF THE SKIN

Accessory organs of the skin that develop from the epidermis of an embryo—hair, glands, nails—perform vital functions. Hair and nails protect the body. Sweat glands help regulate body temperature.

HAIR

Hairs (pili) are variously distributed over the body. Their primary function is protection. Hair on the head guards the scalp from injury and the sun's rays; eyebrows and eyelashes protect the eyes from foreign particles; hair in the nostrils protects against inhaling insects and foreign particles.

Each hair is a thread of fused, keratinized cells that consists of a shaft and a root (Figure 5-3). The **shaft** is the superficial portion, most of which projects above the surface of the skin. The **root** is the portion below the surface that penetrates into the dermis and even into the subcutaneous layer. Surrounding the root is the **hair follicle**, which is composed of two layers of epidermal cells: external and internal root sheaths surrounded by a connective tissue sheath.

The base of each follicle is enlarged into an onion-shaped structure, the **bulb**. This structure contains an indentation, the **papilla of the hair**, which contains many blood vessels and provides nourishment for the growing hair. The bulb also contains a region of cells called the **matrix**, which produce new hairs by cell division when older hairs are shed.

Normal hair loss in the adult scalp is about 70 to 100 hairs per day. Both the rate of growth and the replacement cycle may be altered by any of the following factors: illness, diet, age, genetics, gender, drugs, radiation therapy, and psychological stress.

FIGURE 5-3 **Principal parts of a hair root and associated structures.**

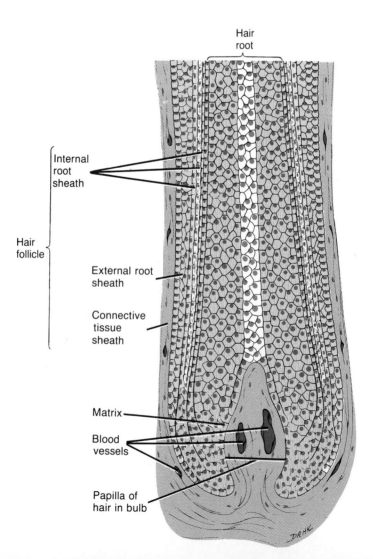

Hair root

Internal root sheath

Hair follicle

External root sheath

Connective tissue sheath

Matrix

Blood vessels

Papilla of hair in bulb

Why does the papilla of the hair contain blood vessels?

Hair color is due to melanin (brown, black, and yellow), which produces the various shades, ranging from blond to black. Decreased melanin production and air in the hair shaft produce gray or white hair.

Sebaceous (oil) glands (to be described shortly) are also associated with hair. There is also a bundle of smooth muscle called **arrector pili** along the side of the hair follicle (see Figure 5-1). These muscles contract under stresses of fright and cold, pulling the hairs into a vertical position and resulting in ''goosebumps'' or ''gooseflesh.''

Around hair follicles are nerve endings, called **hair root plexuses**, that are sensitive to movements of the hair (see Figure 12-1).

GLANDS

Three kinds of glands associated with the skin are sebaceous, sudoriferous, and ceruminous.

Sebaceous (Oil) Glands

Sebaceous (se-BĀ-shus) or **oil glands**, with few exceptions, are connected to hair follicles (see Figure 5-1). The secreting portions of the glands lie in the dermis and open into the necks of hair follicles or directly onto a skin surface (lips, glans penis, labia minora, and tarsal glans of the eyelids). There are no sebaceous glands in the palms and soles.

Sebaceous glands secrete an oily substance called **sebum** (SĒ-bum), a mixture of fats, cholesterol, proteins, and inorganic salts. Sebum keeps hair from drying out, prevents excessive evaporation of water from the skin, keeps the skin soft, and inhibits the growth of certain bacteria.

When sebaceous glands of the face become enlarged because of accumulated sebum, **blackheads** develop. Since sebum is nutritive to certain bacteria, **pimples** or **boils** often result. The color of blackheads is due to melanin and oxidized oil, not dirt. Although sebaceous gland activity increases during adolescence, it decreases with aging. This contributes to wrinkles and increased fragility of the skin.

Sudoriferous (Sweat) Glands

Sudoriferous (soo'-dor-IF-er-us; *sudor* = sweat; *ferre* = to bear) or **sweat glands** are divided into two types. **Apocrine sweat glands** are found in the skin of the axilla (armpit), pubic region, and pigmented areas (areolae) of the breasts. Apocrine sweat gland ducts open into hair follicles. They begin to function at puberty and produce a viscous (sticky) secretion. They are responsive during times of emotional stress.

Eccrine sweat glands are distributed throughout the skin except for the margins of the lips, nail beds of the fingers and toes, glans penis, glans clitoris, labia minora, and eardrums. Eccrine sweat glands are most numerous in the skin of the palms and the soles. Eccrine sweat gland ducts terminate at a pore at the surface of the epidermis (see Figure 5-1) and function throughout life and produce a more watery secretion than that of apocrine sweat glands.

Perspiration, or **sweat**, is the substance produced by sudoriferous glands. It is a mixture of water, salts, urea, uric acid, amino acids, ammonia, sugar, lactic acid, and ascorbic acid. Its principal function is to help regulate body temperature. It also helps eliminate wastes.

Since the mammary glands are actually modified sudoriferous glands, they could be discussed here. However, because of their relationship to the reproductive system, they will be considered in Chapter 23.

Ceruminous Glands

In certain parts of the skin, sudoriferous glands are modified as **ceruminous** (se-ROO-mi-nus) **glands**. Such modified glands are present in the external auditory meatus (canal), the outer ear canal. Ceruminous gland ducts open either directly onto the surface of the external auditory meatus or into ducts of sebaceous glands. The combined secretion of the ceruminous and sebaceous glands is called **cerumen** (*cera* = wax). Cerumen and the hairs in the external auditory meatus provide a sticky barrier against foreign bodies.

NAILS

Plates of tightly packed, hard, keratinized cells of the epidermis are referred to as **nails**. Each nail (Figure 5-4) consists of a nail body, a free edge, and a nail root. The **nail body** is the portion of the nail that is visible; the **free edge** is the part that extends past the end of the finger or toe; the **nail root** is the portion that is not visible. Most of the nail body is pink because of the underlying vascular tissue. The whitish semilunar area near the nail root is

FIGURE 5-4 Structure of nails. (a) Fingernail viewed from above. (b) Sagittal section of a fingernail and nail bed.

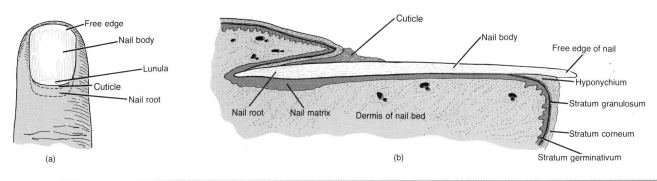

Why is a nail so hard?

called the *lunula* (LOO-nyoo-la). It appears whitish because the vascular tissue underneath does not show through due to the thickened stratum basale in the area.

Essentially, nail growth occurs by the transformation of superficial cells of the *nail matrix* into nail cells. The average growth of fingernails is about 1 mm (0.04 inch) per week; growth is somewhat slower in toenails. For some reason, the longer the digit, the faster the nail grows. Adding supplements such as gelatin to an otherwise healthy diet has no effect on making nails grow faster or stronger. The *cuticle* consists of stratum corneum.

Functionally, nails help us to grasp and manipulate small objects and provide protection against trauma to the ends of the digits.

HOMEOSTASIS OF BODY TEMPERATURE

One of the best examples of homeostasis in humans is the regulation of body temperature by the skin. As warm-blooded animals, we are able to maintain a remarkably constant body temperature of 37°C (98.6°F) even though the environmental temperature varies greatly.

Suppose you are in an environment where the temperature is 37.8°C (100°F). A sequence of events is set into operation to counteract this above-normal temperature, which may be considered a stress. Sensing structures in the skin called *thermoreceptors* pick up the stimulus—in this case, heat—and activate nerve cells that send a message (nerve impulse) to your brain. A temperature-regulating area of the brain then sends nerve impulses to the sudoriferous glands, which produce more perspiration. As the perspira-

FIGURE 5-5 **One role of the skin in regulating the homeostasis of body temperature.**

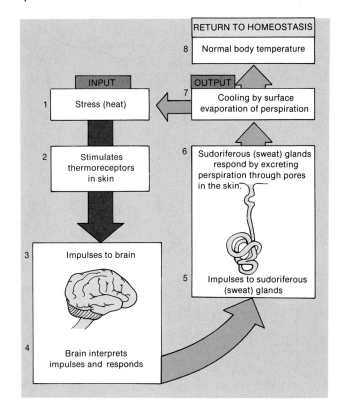

Why is this a negative feedback cycle?

■ COMMON DISORDERS ■

Burns
Tissue damage from excessive heat, electricity, radioactivity, or corrosive chemicals that destroy the protein in cells is called a *burn*. A *first-degree burn* involves only the surface epidermis. It is characterized by mild pain and redness, but no blisters. A *second-degree burn* involves the entire epidermis and possibly part of the dermis. Some skin functions are lost. Redness, blister formation, edema, and pain are characteristic of such a burn. A *third-degree burn* destroys the epidermis, dermis, and the epidermal organs, and skin functions are lost. Such burns vary in appearance from marble-white to mahogany-colored to charred, dry wounds.

Acne
Acne is an inflammation of sebaceous (oil) glands that usually begins at puberty.

Systemic Lupus Erythematosus
Systemic lupus erythematosus (er-i'-thēm-a-TŌ-sus), *SLE*, or *lupus* is an autoimmune, inflammatory disease, occurring mostly in young women in their reproductive years. (An *autoimmune disease* is one in which the body attacks its own tissues.)

Psoriasis
Psoriasis (sō-RĪ-a-sis) is a chronic, occasionally acute, noncon-

tagious relapsing skin disease that ordinarily involves the scalp, elbows and knees, back, and buttocks.

Decubitus Ulcers
Decubitus (dē-KYOO-bi-tus) *ulcers,* also known as *bedsores* or *pressure sores,* are caused by a constant deficiency of blood to tissues over a bony projection that has been subjected to prolonged pressure against an object like a bed, cast, or splint.

Skin Cancer
Excessive sun exposure can result in *skin cancer*, the most common cancer in white people. There are three common forms of skin cancer. *Basal cell carcinomas (BCCs)* account for over 75 percent of all skin cancers. The tumors arise from the epidermis and rarely spread. *Squamous cell carcinomas (SCCs)* also arise from the epidermis but less common than BCCs and have a variable tendency to spread. Most SCCs arise from preexisting lesions on sun-exposed skin. *Malignant melanomas* arise from melanocytes and are the leading cause of death from all skin diseases since they spread rapidly. Malignant melanomas account for only about 3 percent of all skin cancers.

Sunburn
Sunburn is injury to the skin as a result of acute, prolonged exposure to the ultraviolet (UV) rays of sunlight.

tion evaporates from the surface of your skin, it is cooled and your body temperature is lowered (Figure 5-5).

Note that this temperature regulation involves a feedback system. After the sudoriferous glands are activated, the skin receptors keep the brain informed about the external temperature. The brain, in turn, continues to send nerve impulses to the sudoriferous glands until body temperature returns to 37°C (98.6°F). This is a negative feedback system because the output, cooling, is the opposite of the original condition, overheating.

Other mechanisms for lowering body temperature include adjusting blood flow to the skin (dilation of blood vessels in the skin results in the release of more heat), regulating metabolic rate (a slower metabolic rate reduces heat production), and regulating skeletal muscle contractions (decreased muscle tone results in less heat production). These mechanisms are discussed in detail in Chapter 20.

In response to a lower-than-normal body temperature, production of perspiration is decreased, blood vessels in the skin constrict so that less heat is released, metabolic rate is increased to produce more heat, and increased muscle tone and shivering of skeletal muscles result in increased heat production.

MEDICAL TERMINOLOGY AND CONDITIONS

Abrasion (a-BRĀ-shun; *ab* = away; *rasion* = scraped) A portion of the skin that has been scraped away.

Athlete's (ATH-lēts) *foot* A superficial fungus infection of the skin of the foot.

Callus (KAL-lus) An area of hardened and thickened skin that is usually seen in palms and soles and is due to pressure and friction.

Carbuncle (KAR-bung-kl; *carbunculus* = little coal) A hard, round, deep, painful inflammation of the subcutaneous tissue that causes necrosis (death) and pus formation (abscess).

Cold sore (KŌLD sor) A lesion, usually in oral mucous membrane, caused by type 1 herpes simplex virus (HSV), transmitted by oral or respiratory routes. Triggering factors include ultraviolet (UV) radiation, hormonal changes, and emotional stress. Also called a *fever blister*.

Comedo (KOM-ē-dō; *comedo* = to eat up) A collection of sebaceous material and dead cells in the hair follicle and excretory duct of the sebaceous (oil) gland. Found over the face, chest, and back, and more commonly during adolescence. Also called *blackhead* or *whitehead*.

Corn (KORN) A painful conical thickening of the skin that may be hard or soft, depending on location. Hard corns are usually found over toe joints, and soft corns are usually found between the fourth and fifth toes.

Cyst (SIST; *cyst* = sac containing fluid) A sac with a distinct connective tissue wall, containing a fluid or other material.

Dermabrasion (der-ma-BRĀ-shun; *derm* = skin) Removal of acne, scars, tattoos, or moles by sandpaper or a high-speed brush.

Detritus (de-TRĪ-tus; *deterere* = to rub away) Particulate matter produced by or remaining after the wearing away or disintegration of a substance or tissue; scales, crusts, and loosened skin.

Eczema (EK-ze-ma; *ekzein* = to boil out) An acute or chronic superficial inflammation of the skin, characterized by redness, oozing, crusting, and scaling. Also called *chronic dermatitis*.

Erythema (er'-e-THĒ-ma; *erythema* = redness) Redness of the skin caused by an engorgement of capillaries in lower layers of the skin. Erythema occurs with any skin injury, infection, or inflammation.

Furuncle (FYOOR-ung-kul) A boil; an abscess resulting from infection of a hair follicle.

Hemangioma (hē-man'-jē-Ō-ma; *hemo* = blood; *angio* = blood vessel; *oma* = tumor) Localized tumor of the skin and subcutaneous layer that results from an abnormal increase in blood vessels; one type is a *port-wine stain*, a flat, pink, red, or purple lesion present at birth, usually at the nape of the neck.

Hives (HĪVZ) Condition of the skin marked by reddened elevated patches that are often itchy. Most commonly caused by infections, physical trauma, medications, emotional stress, food additives, and certain foods. Also called *urticaria* (yoor-ti-KAR-ē-a).

Hypodermic (hī'-pō-DER-mik; *hypo* = under) Relating to the area beneath the skin. Also called *subcutaneous*.

Impetigo (im'-pe-TĪ-go) Superficial skin infection caused by staphylococci or streptococci; most common in children.

Intradermal (in'-tra-DER-mal; *intra* = within) Within the skin. Also called *intracutaneous*.

Keratosis (ker'-a-TŌ-sis; *kera* = horn) Formation of a hardened growth of tissue.

Laceration (las'-er-Ā-shun; *lacerare* = to tear) Wound or irregular tear of the skin.

Nevus (NE-vus) A round, pigmented, flat, or raised skin area that may be present at birth or develop later. Varying in color from yellow-brown to black. Also called a *mole* or *birthmark*.

Nodule (NOD-yool; *nodulus* = little knot) A large cluster of cells raised above the skin but extending deep into the tissues.

Papule (PAP-yool) A small, round, skin elevation varying in size from a pinpoint to that of a split pea. One example is a pimple.

Polyp (POL-ip) A tumor on a stem found especially on mucous membranes.

Pruritus (proo-RĪ-tus; *pruire* = to itch) Itching, one of the most common dermatological disorders. It may be caused by skin disorders (infections), systemic disorders (cancer, kidney failure), or psychogenic factors (emotional stress).

Pustule (PUS-tyool) A small, round elevation of the skin containing pus.

Topical (TOP-i-kal) Pertaining to a definite area; local. Also in reference to a medication, applied to the surface rather than ingested or injected.

Wart (WORT) Mass produced by uncontrolled growth of epithelial skin cells; caused by a virus (papillomavirus). Most warts are noncancerous.

STUDY OUTLINE

Skin (p. 85)

1. The skin and its accessory organs (hair, glands, and nails) constitute the integumentary system.
2. The skin functions to help regulate body temperature, protect, receive stimuli, excrete water, salts, and several organic compounds, synthesize vitamin D, and provide immunity.
3. The principal parts of the skin are the outer epidermis and inner dermis. The dermis overlies the subcutaneous layer.
4. The epidermal layers, from deepest to most superficial, are the strata basale, spinosum, granulosum, lucidum, and corneum. The basale undergoes continuous cell division and produces all other layers. Epidermal cells include keratinocytes, malanocytes, and nonpigmented granular dendrocytes (Langerhans' and Granstein cells).
5. The dermis consists of two regions. The upper region is loose connective tissue containing blood vessels, nerves, hair follicles, dermal papillae, and corpuscles of touch (Meissner's corpuscles). The lower region is dense, irregularly arranged connective tissue containing adipose tissue, hair follicles, nerves, sebaceous (oil) glands, and ducts of sudoriferous (sweat) glands.
6. Skin color is due to melanin, carotene, and blood in capillaries in the dermis.

Accessory Organs of the Skin (p. 90)

1. Accessory organs of the skin are structures developed from the epidermis of an embryo.
2. They include hair, skin glands (sebaceous, sudoriferous, and ceruminous), and nails.

Hair (p. 90)

1. Hairs are threads of fused, keratinized cells that function in protection.
2. Hairs consist of a shaft above the surface, a root that penetrates the dermis and subcutaneous layer, and a hair follicle.
3. Associated with hairs are sebaceous (oil) glands, arrectores pilorum muscles, and hair root plexuses.
4. New hairs develop from cell division of the matrix in the bulb; hair replacement and growth occurs in a cyclic pattern.

Glands (p. 91)

1. Sebaceous (oil) glands are usually connected to hair follicles; they are absent in the palms and soles. Sebaceous glands produce sebum, which moistens hairs and waterproofs the skin.
2. Sudoriferous (sweat) glands are divided into apocrine and eccrine. Apocrine sweat glands are only found in the skin of the axilla, pubis, and areolae; their ducts open into hair follicles. Eccrine sweat glands have an extensive distribution; their ducts terminate at pores at the surface of the epidermis. Sudoriferous glands produce perspiration, which carries small amounts of wastes to the surface and assists in maintaining body temperature.
3. Ceruminous glands are modified sudoriferous glands that secrete cerumen. They are found in the external auditory meatus.

Nails (p. 91)

1. Nails are hard, keratinized epidermal cells covering the terminal portions of the fingers and toes.
2. The principal parts of a nail are the body, free edge, root, lunula, cuticle, and matrix. Cell division of the matrix cells produces new nails.

Homeostasis of Body Temperature (p. 92)

1. One of the functions of the skin is the regulation of normal body temperature of 37°C (98.6°F).
2. If environmental temperature is high, skin receptors sense the stimulus (heat) and generate impulses that are transmitted to the brain. The brain then causes the sweat glands to produce perspiration. As the perspiration evaporates, the skin is cooled.
3. The skin-cooling response is a negative feedback mechanism.
4. Temperature regulation is also accomplished by adjusting blood flow to the skin, regulating metabolic rate, and regulating skeletal muscle contractions.

REVIEW QUESTIONS

1. What is the integumentary system? (p. 85)
2. List the principal functions of the skin. (p. 85)
3. Compare the structure of the epidermis and the dermis. What is the subcutaneous layer? (pp. 86–87)
4. List and describe the epidermal layers from the deepest outward. What is the importance of each layer? (pp. 86–87)
5. Contrast the differences between the two regions of the dermis. (p. 87)
6. Explain the factors that produce skin color. What is an albino? (pp. 87–88)
7. Describe the structure of a hair. How are hairs moistened? What produces "goosebumps" or "gooseflesh"? (p. 90)
8. Contrast the locations and functions of sebaceous (oil) glands, sudoriferous (sweat) glands, and ceruminous glands. What are the names and chemical components of the secretions of each? (p. 91)
9. Explain with a labeled diagram how the skin helps regulate normal body temperature. (p. 92–93)
10. Define the following: burn, acne, systemic lupus erythematosus (SLE), psoriasis, decubitus ulcers, skin cancer, and sunburn. (p. 92)
11. Refer to the glossary of medical terminology and conditions associated with the integumentary system. Be sure that you can define each term. (p. 93)

6

The Skeletal System

STUDENT OBJECTIVES

1. Discuss the components and functions of the skeletal system.
2. Describe the microscopic structure of bone tissue.
3. Explain the steps involved in bone formation.
4. Describe the processes of bone construction and destruction as an example of homeostasis.
5. Describe the conditions necessary for normal bone growth and replacement.
6. Describe the bones of the various regions of the body.
7. Compare the principal differences between female and male bones.
8. Define common disorders and medical terminology and conditions associated with the skeletal system.

A LOOK AHEAD

FUNCTIONS
TYPES OF BONES
HISTOLOGY
   Compact Bone
   Spongy Bone
OSSIFICATION: BONE FORMATION
   Intramembranous
      Ossification
   Endochondral Ossification
HOMEOSTASIS
   Bone Growth and
      Maintenance
   Mineral Storage
SURFACE MARKINGS
DIVISIONS OF THE SKELETAL SYSTEM
SKULL
   Sutures
   Cranial Bones
      Frontal Bone
      Parietal Bones
      Temporal Bones
      Occipital Bone
      Sphenoid Bone
      Ethmoid Bone
   Facial Bones
      Nasal Bones
      Maxillae
      Paranasal Sinuses
      Zygomatic Bones
      Mandible
      Lacrimal Bones
      Palatine Bones
      Inferior Nasal Conchae
      Vomer
   Fontanels
   Foramina
HYOID BONE
VERTEBRAL COLUMN
   Divisions
   Normal Curves
   Typical Vertebra
   Cervical Region
   Thoracic Region
   Lumbar Region
   Sacrum and Coccyx
THORAX
   Sternum
   Ribs
PECTORAL (SHOULDER) GIRDLE
   Clavicle
   Scapula
UPPER EXTREMITY
   Humerus
   Ulna and Radius
   Carpals, Metacarpals, and Phalanges
PELVIC (HIP) GIRDLE
LOWER EXTREMITY
   Femur
   Patella
   Tibia and Fibula
   Tarsals, Metatarsals, and Phalanges
   Arches of the Foot
FEMALE AND MALE SKELETONS
COMMON DISORDERS
MEDICAL TERMINOLOGY AND
      CONDITIONS

T he framework of bones and cartilage that protects our organs and us to move is called the *skeletal system*. The specialized b   al medicine that deals with the preservation and restoration of the s   al system, articulations (joints), and associated structures is called *orthopedics* (or'-thō-PĒ-diks; *ortho* = correct or straighten; *pais* = child).

FUNCTIONS

The skeletal system performs the following functions:

1. **Support**. The skeleton provides a framework for the body and, as such, supports soft tissues and provides points of attachment for skeletal muscles.
2. **Protection.** Internal organs are protected from injury by the skeleton. For example, the brain is protected by the cranial bones while the heart and lungs are protected by the rib cage.
3. **Movement**. Skeletal muscles are attached to bones. When muscles contract, they pull on bones and together they produce movement. This function is discussed in detail in Chapter 8.
4. **Mineral storage**. Bones store several minerals, especially calcium and phosphorus, that can be distributed to other parts of the body upon demand. The homeostatic mechanism that deposits and removes calcium and phosphorus is discussed in detail on pages 101–102.
5. **Storage of blood cell–producing cells**. In certain bones, a connective tissue called red marrow produces blood cells, a process called **hemopoiesis** (hēm'-ō-poy-Ē-sis). Red marrow is found in developing bones and adult bones such as the pelvis, ribs, breastbone, backbones, skull, and ends of the arm bone and thigh bone. Hemopoiesis is discussed in detail in Chapter 14.
6. **Storage of energy**. Lipids stored in cells of yellow marrow are an important source of chemical energy.

TYPES OF BONES

The bones of the body may be classified into four principal types on the basis of shape: long, short, flat, and irregular. **Long bones** have greater length than width and consist of a shaft and extremities (ends). They are slightly curved for strength. Long bones consist mostly of compact bone (dense bone with few spaces) but also contain considerable amounts of spongy bone (bone with large spaces). The details of compact and spongy bone are discussed shortly. Long bones include bones of the thighs, legs, toes, arms, forearms, and fingers. Figure 6-1a shows the parts of a long bone.

Short bones are somewhat cube-shaped and nearly equal in length and width. They are spongy except at the surface where there is a thin layer of compact bone. Short bones include the wrist and ankle bones.

Flat bones are generally thin and composed of two more or less parallel plates of compact bone enclosing a layer of spongy bone. Flat bones afford considerable protection and provide extensive areas for muscle attachment. Flat bones include the cranial bones, the sternum (breastbone), ribs, and the scapulas (shoulder blades).

Irregular bones have complex shapes and cannot be grouped into any of the three categories just described. They also vary in the amount of spongy and compact bone present. Such bones include the vertebrae (backbones) and certain facial bones.

There are two additional types of bones not included in this classification by shape; they are classified by location. **Sutural** (SOO-chur-al) **bones** are small bones between the joints of certain cranial bones. Their number varies greatly from person to person. **Sesamoid bones** are small bones in tendons where considerable pressure develops, for instance, in the wrist. These, like sutural bones, also vary in number. Two sesamoid bones, the patellas (kneecaps) are present in all individuals.

FIGURE 6-1 Osseous tissue. (a) Macroscopic appearance of an adult long bone that has been partially sectioned. (b) Histological structure of bone.

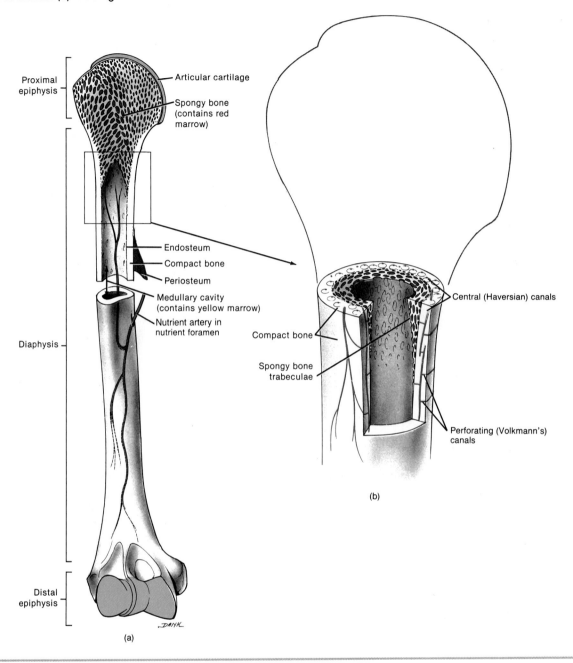

Which part of a bone reduces friction at joints? Produces blood cells? Is the shaft? Lines the medullary cavity?

HISTOLOGY

The skeletal system consists of four types of connective tissue: cartilage, bone, bone marrow, and the periosteum, the membrane around bones. We described the microscopic structure of cartilage in Chapter 4. Here, we will discuss the microscopic structure of bone tissue.

Like other connective tissues, **bone** or **osseous** (OS-ē-us) **tissue** contains a great deal of intercellular substance (matrix) surrounding widely separated cells. The matrix consists of mineral salts (67%) and collagenous fibers (33%). There are four types of cells in bone tissue: osteoprogenitor (osteogenic) cells, osteoblasts, osteocytes, and osteoclasts. **Osteoprogenitor** (os'-tē-ō-prō-JEN-i-tor; *osteo* = bone; *pro* = precursor; *gen* = to produce) *cells* undergo mitosis to become osteoblasts. **Osteoblasts** (OS-tē-ō-blasts'; *blast* = germ or bud) are the cells that form bone. They do not have mitotic potential. Osteoblasts initially form bone, but once they become isolated in the bony matrix, they are called *osteocytes* (OS-tē-ō-sīts'; *cyte* = cell), or mature bone cells. They are the principal cells of bone tissue. Like osteoblasts, osteocytes have no mitotic potential. Osteocytes maintain daily cellular activities of bone tissue. *Osteoclasts* (OS-tē-ō-clasts'; *clast* = to break) develop from a type of white blood cell called a monocyte. Osteoclasts function in bone resorption (degradation), which is important in the development, growth, maintenance, and repair of bone.

Unlike other connective tissues, the matrix of bone contains abundant mineral salts, primarily calcium phosphate and some calcium carbonate. As these salts are deposited by osteoblasts in the framework formed by the collagenous fibers of the matrix, the tissue hardens, that is, becomes calcified.

The gross structure of a long bone, in this case, the humerus (arm bone), will be examined before considering the microscopic structure of bone. A typical long bone consists of the following parts (Figure 6-1):

1. *Diaphysis* (dī-AF-i-sis). The shaft or long, main portion of the bone.
2. *Epiphyses* (ē-PIF-i-sēz). The extremities or ends of the bone. (Singular is *epiphysis*.)
3. *Articular cartilage*. A thin layer of hyaline cartilage covering the epiphysis where the bone forms a joint with another bone. The cartilage reduces friction and absorbs shock at freely movable joints.
4. *Periosteum* (per'-ē-OS-tē-um). The periosteum (*peri* = around; *osteo* = bone) is a dense, white, fibrous covering around the surface of the bone not covered by articular cartilage. It consists of two layers. The outer layer contains blood vessels and nerves that pass into the bone. The inner layer contains osteoprogenitor cells, osteoclasts, and osteoblasts. The periosteum is necessary for the protection, nutrition, growth, and repair of bones and is the site for attachment for ligaments and tendons.
5. *Medullary* (MED-yoo-lar'-ē) or *marrow cavity*. The space within the diaphysis that contains the fatty *yellow marrow* in adults. Yellow marrow consists primarily of fat cells and a few scattered blood cells. Thus, yellow marrow functions in energy storage.
6. *Endosteum* (end-OS-tē-um). The lining of the medullary cavity that consists of osteoprogenitor cells, osteoblasts, and scattered osteoclasts.

Bone is not completely solid. In fact, all bone has some spaces between its hard components. The spaces provide channels for blood vessels that supply bone cells with nutrients. The spaces also make bones lighter. Depending on the size and location of the spaces, bone may be classified as compact or spongy (see Figure 6-1).

Compact (dense) bone tissue contains few spaces. It covers spongy bone tissue. It provides protection and support and helps long bones resist the stress of weight placed on them. *Spongy (cancellous) bone tissue*, by contrast, contains many large spaces filled with red marrow. It makes up most of the bone tissue of short, flat, and irregularly shaped bones and most of the epiphyses of long bones. Spongy bone in bones such as the pelvis, ribs, breastbone, backbones, skull, and ends of certain long bones is the only site of red marrow in adults.

COMPACT BONE

Compare the differences between spongy and compact bone tissue by looking at the highly magnified sections in Figure 6-2. One main difference is that adult compact bone has a concentric-ring structure, whereas spongy bone appears as an irregular latticework. Nutrient arteries and nerves from the periosteum penetrate compact bone through *perforating (Volkmann's) canals.* These blood vessels connect with blood vessels and nerves of the medullary cavity and those of the *central (Haversian) canals.* The central canals run lengthwise through the bone. Around the canals are *concentric lamellae* (la-MEL-ē)—rings of hard, calcified, intercellular substance (matrix). Between the lamellae are small spaces called *lacunae* (la-KOO-nē; *lacuna* = little lake), which contain osteocytes. Projecting outward in all directions from the lacunae are minute canals called *canaliculi* (kan'-a-LIK-yoo-lī), which contain slender processes of osteocytes. The canaliculi connect with those of other lacunae and, eventually, with the central canals. Thus, an intricate branching network of canaliculi is formed throughout the bone to provide numerous routes for nutrients and oxygen to reach osteocytes and remove wastes. Each central canal, with its surrounding lamellae, lacunae, osteocytes, and canaliculi, is called an *osteon (Haversian system).*

SPONGY BONE

In contrast to compact bone, spongy bone usually does not contain true osteons. It consists of an irregular latticework of thin plates of bone called *trabeculae* (tra-BEK-yoo-lē). See Figure 6-2. The spaces between the trabeculae of some bones are filled with red marrow. Within the trabeculae lie lacunae, which contain osteocytes. Blood vessels from the periosteum penetrate through to the spongy bone, and osteocytes in the trabeculae are nourished directly from the blood circulating through the marrow cavities.

Most people think all bone is hard and rigid, but the bones of an infant are quite soft and become completely rigid only after growth stops during late adolescence. Even then, bone is constantly broken down and rebuilt. It is a dynamic, living tissue. Let us now see how bones are formed and how they grow.

OSSIFICATION: BONE FORMATION

The process by which bone forms is called *ossification* (os'-i-fi-KĀ-shun). The "skeleton" of a human embryo is composed of fibrous membranes and hyaline cartilage. Both are shaped like bones and provide the sites for ossification. Ossification begins around the sixth or seventh week of embryonic life and continues throughout adulthood.

Two methods of bone formation occur. The first is called intramembranous (in'-tra-MEM-bra-nus) ossification. This refers to the formation of bone directly on or within fibrous membranes (*intra* = within; *membranous* = membrane). The second kind, endochondral (en'-dō-KON-dral) ossification, refers to the formation of bone within a cartilage model (*endo* = within; *chondro* = cartilage). These two methods of ossification do *not* lead to differences in the structure of mature bones. They are simply different methods of bone development. Both mechanisms involve the replacement of a preexisting connective tissue with bone.

FIGURE 6-2 Enlarged aspect of osteons (Haversian systems) in compact bone and trabeculae in spongy bone.

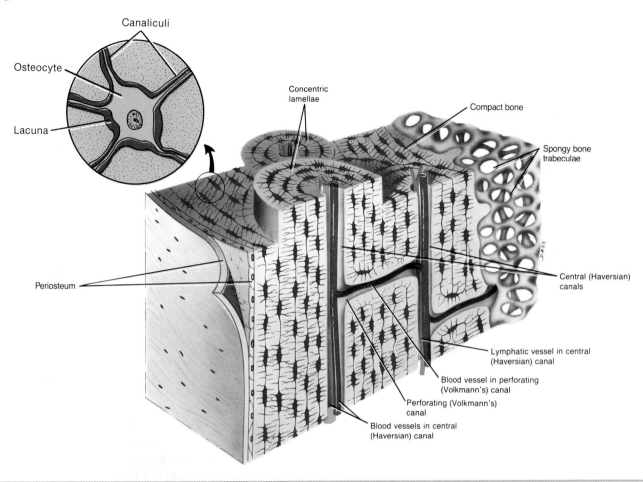

As people age, some central (Haversian) canals may become blocked. What effect would this have on osteocytes, osteoblasts, and bone formation?

The first stage in the development of bone is the appearance of osteoprogenitor cells which undergo mitosis to produce osteoblasts which will produce bony matrix by either intramembranous or endochondral ossification.

INTRAMEMBRANOUS OSSIFICATION

Of the two methods of bone formation, *intramembranous ossification* is easier to understand. Most of the surface skull bones and the clavicles (collarbones) are formed in this way. The process occurs as follows.

Osteoblasts formed from osteoprogenitor cells cluster in a fibrous membrane and begin to secrete intercellular substances partly composed of collagenous fibers that form a framework, or matrix, in which calcium salts are quickly deposited. The deposition of calcium salts is called *calcification.* When a cluster of osteoblasts is completely surrounded by the calcified matrix, it is called a *trabecula.* As trabeculae form in various ossification centers, they fuse with one another to create the open latticework appearance of spongy bone. With the formation of successive layers of bone, some osteoblasts become trapped in the lacunae. The entrapped osteoblasts

lose their ability to form bone and are called osteocytes. The space between the trabeculae is filled with red marrow. The original connective tissue that surrounds the growing mass of bone becomes the periosteum. Eventually, the surface layers of the spongy bone will be reconstructed into compact bone. Also, much of this newly formed bone will be remodeled (destroyed and re-formed) so the bone may reach its final adult size and shape.

ENDOCHONDRAL OSSIFICATION

The replacement of cartilage by bone is called *endochondral ossification.* Most bones of the body are formed this way, but the process is best observed in a long bone (Figure 6-3).

Early in embryonic life, a cartilage model of the future bone is laid down. This model is covered by a membrane called the *perichondrium* (per-i-KON-drē-um). Once the perichondrium starts to form bone, it is called the *periosteum.* Midway along the shaft of the cartilage model, a nutrient artery penetrates the perichondrium, stimulating the cells in its inner layer to enlarge and become osteoblasts. The cells begin to form a periosteal collar of compact bone around the middle of the diaphysis.

FIGURE 6-3 Endochondral ossification of the tibia. (a) Cartilage model. (b) Periosteal collar formation. (c) Development of primary ossification center. (d) Entrance of blood vessels (nutrient arteries). (e) Medullary (marrow) cavity formation. (f) Thickening and lengthening of collar. (g) Formation of secondary ossification centers. (h) Remains of cartilage as articular cartilage and epiphyseal plate. (i) Formation of epiphyseal lines.

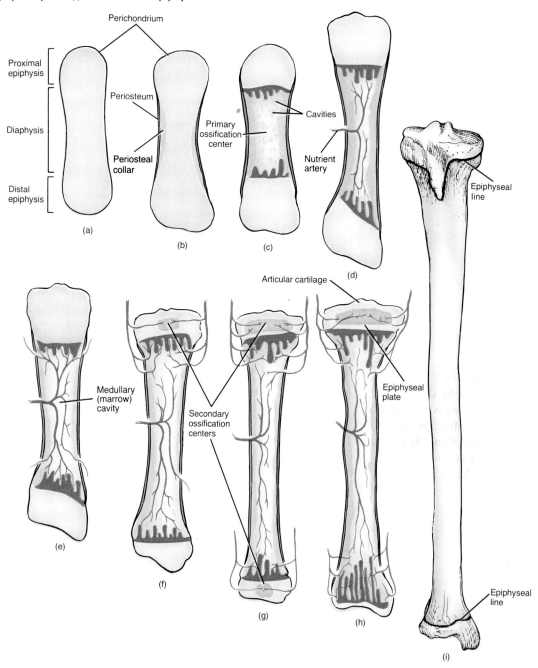

Which structure signals that bone growth in length has stopped?

As these internal changes occur, the cartilage cells degenerate and are replaced by osteoblasts from the periosteum, thus forming the **primary ossification center**. Osteoblasts produce bone around this center, which gradually expands toward the epiphyses. Osteoclast activity then breaks down bone in the middle and produces the medullary (marrow) cavity.

As these internal changes occur, the osteoblasts of the periosteum deposit successive layers of bone on the outer surface so that the periosteal collar thickens, becoming thickest in the diaphysis. The

cartilage model continues to grow at its ends, steadily increasing in length. Eventually, blood vessels enter the epiphyses and **secondary ossification centers** appear and lay down spongy bone.

After the two secondary ossification centers have formed, bone tissue completely replaces cartilage, except in two regions. Cartilage continues to cover the articular surfaces of the epiphyses, where it is called **articular cartilage**. It also remains as a region between the epiphysis and diaphysis, called the **epiphyseal plate**.

The epiphyseal plate allows the diaphysis of the bone to increase in length until early adulthood. The rate of growth is controlled by hormones such as human growth hormone (hGH). As the child grows, cartilage cells are produced by mitosis on the epiphyseal side of the plate. They are then destroyed and the cartilage is replaced by bone on the diaphyseal side of the plate. In this way, the thickness of the epiphyseal plate remains fairly constant, but the bone on the diaphyseal side increases in length.

Growth in diameter occurs along with growth in length. In this process, the bone lining the medullary cavity is destroyed by osteoclasts so that the cavity increases in diameter. At the same time, osteoblasts from the periosteum add new osseous tissue around the outer surface of the bone. Initially, diaphyseal and epiphyseal ossification produce only spongy bone. Later, by remodeling, the outer region of spongy bone is reorganized into compact bone.

The epiphyseal cartilage cells stop dividing and the cartilage is eventually replaced by bone. The new structure is called the *epiphyseal line*, a remnant of the once active epiphyseal plate. With the appearance of the epiphyseal line, bone growth in length stops. Ossification of most bones is usually completed by age 25.

HOMEOSTASIS

BONE GROWTH AND MAINTENANCE

Bone, like skin, continually replaces itself throughout adult life. *Remodeling* is the replacement of old bone tissue by new bone tissue. Compact bone is formed by the transformation of spongy bone. The diameter of a long bone is increased by the destruction of bone internally and the construction of new bone externally. Even after bones have reached their adult shapes and sizes, old bone is continually destroyed and new bone tissue is formed in its place. Bone is never metabolically at rest; it constantly remodels. Remodeling removes worn and injured bone, replacing it with new tissue. It also allows bone to serve as the body's storage area for calcium (to be discussed shortly). Many other tissues in the body need calcium to perform their functions. The blood continually trades off calcium with the bones, removing calcium when it and other tissues are not receiving enough and resupplying the bones to keep them from losing mass.

Osteoclasts are responsible for the resorption (destruction) of bone tissue. A delicate homeostasis is maintained between the action of the osteoclasts in removing calcium and collagen and the action of the bone-making osteoblasts in depositing calcium and collagen. Should too much new tissue be formed, the bones become abnormally thick and heavy. If too much calcium is deposited, the surplus may form thick bumps (spurs) that interfere with movement at joints. A loss of too much tissue or calcium makes bones breakable or too flexible.

Bone growth in the young and bone replacement in the adult depend on several factors. First, sufficient quantities of calcium and phosphorus, components of the primary salt that makes the matrix of bone hard, must be included in the diet.

Second, vitamin D is required for the absorption of calcium from the gastrointestinal tract into the blood, proper bone mineralization, calcium removal from bone, and kidney reabsorption of calcium that might otherwise be lost in urine. Vitamin C is required

to maintain the matrix of bone (and other connective tissues). Deficiency leads to retardation of bone growth and delayed healing of fractures. Vitamin A helps regulate osteoblast and osteoclast differentiation. Its deficiency results in a decreased rate of growth. Vitamin B_{12} may also play a role in osteoblast activity.

Third, the body must manufacture the proper amounts of hormones responsible for bone tissue activity (Chapter 13). Human growth hormone (hGH), secreted by the pituitary gland, is responsible for the general growth of bones. Too much or too little hGH during childhood makes the adult abnormally tall or short. The sex hormones (estrogen and testosterone) aid osteoblastic activity and thus promote the growth of new bone. They also bring about the degeneration of cartilage cells in epiphyseal plates. The typical adolescent experiences a spurt of growth during puberty, when sex hormone levels start to increase, and quickly completes the growth process as the epiphyseal cartilage disappears. Premature puberty can actually prevent one from reaching an average adult height because of the simultaneous premature degeneration of the plates. Insulin and thyroid hormones are also important for normal bone growth and maturity.

As you will see shortly, several hormones are also involved in mineral storage.

Within limits, bone has the ability to alter its strength in response to mechanical stress. Bone tissue deposits more mineral salts and produces more collagen fibers under stress; lack of stress results in the loss of salts and collagen, in other words, in decalcification of the matrix. One of the chief stresses to which bone is subjected is the pull of skeletal muscles caused by weight-bearing and exercise, for example. Bones of astronauts show some degree of loss of mass due to the weightless environment.

In response to mechanical stress, bone produces minute currents of electricity, called the *piezoelectric* (pē-ā-zō-e-LEK-trik) *effect*, which stimulates formation of more osteoblasts which make additional bone matrix. Regular exercise in which bones are stressed, such as walking or running, stimulates bone growth. It also increases the production of calcitonin (CT) by the thyroid gland, which inhibits osteoclasts and, thus, resorption.

With aging comes a loss of calcium from bones. In females, it begins after age 30, accelerates greatly around age 40–45 as estrogen levels decrease, and continues until as much as 30 percent of bone calcium is lost by age 70. In males, calcium loss typically does not begin until after age 60. The loss of calcium from bones causes osteoporosis, which will be described shortly.

The second principal effect of aging on the skeletal system is a decrease in the rate of protein formation, which results in a decreased ability to produce the organic portion of bone matrix. Consequently, bone accumulates less organic matrix and more inorganic. In some elderly individuals, this process causes their bones to become brittle and more susceptible to fracture.

MINERAL STORAGE

Bones contain more calcium than any other organs, but calcium is also needed for muscle and nerve function and blood clotting. Adequate levels of calcium are needed in the blood but excessive amounts may also be harmful and the bones serve as a useful storage sites when blood calcium increases beyond its normal homeostatic level. Bones also store more phosphate (phosphorus) than any other organs. Appropriate levels of phosphate are required to produce nucleic acids (DNA and RNA) and ATP. Several hormones are involved in mineral storage, especially storage of calcium and phosphate. Calcitonin (CT), produced by the thyroid gland, is released in response to high blood levels of calcium. It inhibits osteoclastic activity, speeds up calcium absorption, and increases deposition of calcium salts. The net result is that CT promotes bone formation and decreases blood calcium level (Figure 6-4). In response to low blood levels of calcium, the parathyroid glands secrete a hormone called parathyroid hormone (PTH). This hormone increases the number and activity of osteoclasts, which release calcium (and phosphate) from bones into blood. (PTH also causes the transport of calcium from fluid that will become urine into blood.) The net result is that PTH promotes bone resorption and increases blood levels of calcium (Figure 6-4). In response to high blood levels of calcium, the thyroid gland secretes CT.

Overall, CT and PTH help regulate blood levels of calcium by promoting deposition or removal of calcium from bones. Blood calcium regulation is another example of how the body uses negative feedback cycles to maintain homeostasis.

SURFACE MARKINGS

The surfaces of bones have various structural features adapted to specific functions. These features are called *surface markings*. Long bones that bear a great deal of weight have large, rounded ends that can form sturdy joints, for example. Other bones have depressions that receive the rounded ends.

Following are examples of surface markings:

I. *Depressions and Openings*
  A. *Foramen* (fō-RĀ-men; *foramen* = hole) An opening through which blood vessels, nerves, or ligaments pass, such as the foramen magnum of the occipital bone (see Figure 6-7).
  B. *Meatus* (mē-Ā-tus; *meatus* = canal) A tubelike passageway running within a bone, such as the external auditory meatus of the temporal bone (see Figure 6-6b).

FIGURE 6-4 Actions of calcitonin (CT) and parathyroid hormone (PTH) in the maintenance of blood levels of calcium and storage and release of calcium by bones.

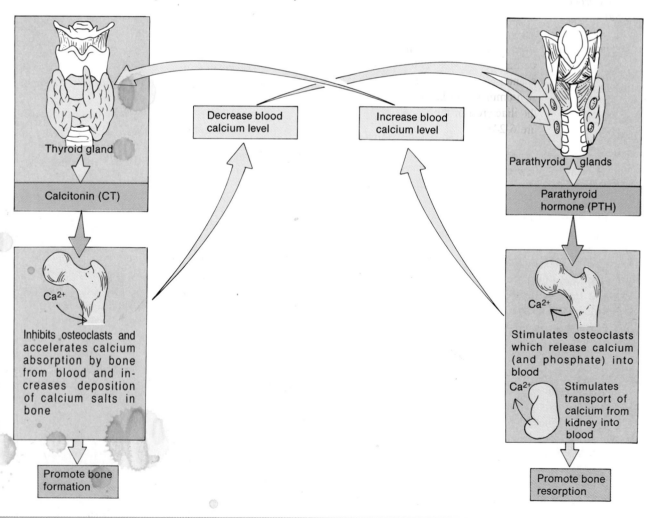

Use a red pencil to trace one negative feedback cycle and a blue pencil to trace the other negative feedback cycle.

C. *Paranasal sinus* (*sin* = cavity) — An air-filled cavity within a bone connected to the nasal cavity, such as the frontal sinus in the frontal bone (see Figure 6-6c).

D. *Fossa* (*fossa* = basinlike depression) — A depression in or on a bone, such as the mandibular fossa of the temporal bone (see Figure 6-7).

II. *Processes that Form Joints*

A. *Condyle* (KON-dīl; *condylus* = knuckle-like process) — A large, rounded articular prominence, such as the medial condyle of the femur (see Figure 6-27).

B. *Head* — A rounded articular projection supported on the constricted portion (neck) of a bone, such as the head of the femur (see Figure 6-27).

C. *Facet* — A smooth, flat surface such as the facet on a vertebra (see Figure 6-15).

III. *Processes to Which Tendons, Ligaments, and Other Connective Tissues Attach*

A. *Tuberosity* — A large, rounded, usually roughened process, such as the deltoid tuberosity of the humerus (see Figure 6-21).

B. *Spinous process* (*spine*) — A sharp, slender projection, such as the spinous process of a vertebra (see Figure 6-16).

C. *Trochanter* (trō-KAN-ter) — A large, blunt projection found only on the femur, such as the greater trochanter (see Figure 6-27).

D. *Crest* — A prominent border or ridge, such as the iliac crest of the hip bone (see Figure 6-24).

DIVISIONS OF THE SKELETAL SYSTEM

The adult human skeleton consists of 206 bones grouped as the *axial skeleton* and the *appendicular skeleton*. The axial division consists of the bones of the skull, auditory ossicles, hyoid bone, ribs, breastbone, and backbone. The appendicular division consists of the bones of the upper and lower extremities (limbs), plus the bones called girdles, which connect the extremities to the axial skeleton. There are 80 bones in the axial division and 126 in the appendicular (Exhibit 6-1). Refer to Figure 6-5 to see how the two divisions are joined to form the complete skeleton.

SKULL

The *skull* rests on top of the vertebral column and is composed of two sets of bones: cranial bones and facial bones. The eight *cranial bones* enclose and protect the brain. They are the frontal bone, parietal bones (2), temporal bones (2), occipital bone, sphenoid bone, and ethmoid bone. There are 14 *facial bones*: nasal bones (2), maxillae (2), zygomatic bones (2), mandible, lacrimal bones (2), palatine bones (2), inferior nasal conchae (2), and vomer. Locate the skull bones in the various views of the skull (Figure 6-6).

EXHIBIT 6-1
Divisions of the Skeletal System

Regions of the Skeleton	Number of Bones
AXIAL SKELETON	
Skull	
Cranium	8
Face	14
Hyoid	1
Auditory Ossicles[a]	
(3 in each ear)	6
Vertebral Column	26
Thorax	
Sternum	1
Ribs	24
	80
APPENDICULAR SKELETON	
Pectoral (Shoulder) Girdles	
Clavicle	2
Scapula	2
Upper Extremities	
Humerus	2
Ulna	2
Radius	2
Carpals	16
Metacarpals	10
Phalanges	28
Pelvic (Hip) Girdle	
Coxal (Hip) Bone	2
Lower Extremities	
Femur	2
Fibula	2
Tibia	2
Patella	2
Tarsals	14
Metatarsals	10
Phalanges	28
	126
	Total = 206

[a] Although the auditory ossicles are not considered part of the axial or appendicular skeleton, but rather as a separate group of bones, they are placed with the axial skeleton for convenience. The auditory ossicles are exceedingly small bones in the middle portion of each ear and are named for their shapes— malleus, incus, and stapes (commonly called hammer, anvil, and stirrup). They are described in detail in Chapter 12.

SUTURES

A *suture* (SOO-chur), meaning seam or stitch, is an immovable joint found only between skull bones (Figure 6-6). Examples are the following:

1. **Coronal suture** between the frontal bone and the two parietal bones.
2. **Sagittal suture** between the two parietal bones.
3. **Lambdoidal** (lam-DOY-dal) **suture** between the parietal bones and the occipital bone.
4. **Squamosal** (skwa-MŌ-sal) **suture** between the parietal bones and the temporal bones.

FIGURE 6-5 Divisions of the skeletal system. The axial skeleton is indicated in gold.

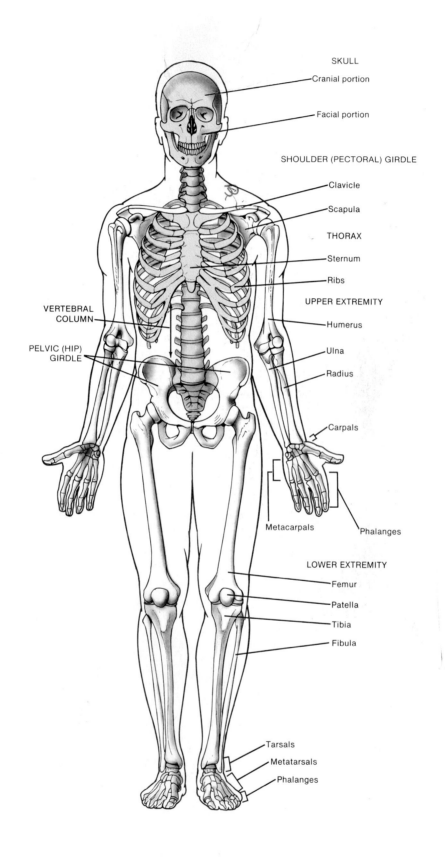

SKULL
Cranial portion
Facial portion

SHOULDER (PECTORAL) GIRDLE
Clavicle
Scapula

THORAX
Sternum
Ribs

UPPER EXTREMITY
Humerus
Ulna
Radius
Carpals

VERTEBRAL COLUMN

PELVIC (HIP) GIRDLE

Metacarpals   Phalanges

LOWER EXTREMITY
Femur
Patella
Tibia
Fibula

Tarsals
Metatarsals
Phalanges

Identify an example of each of the following bones: long, short, flat, irregular, and sesamoid.

FIGURE 6-6 Skull. (a) Anterior view.

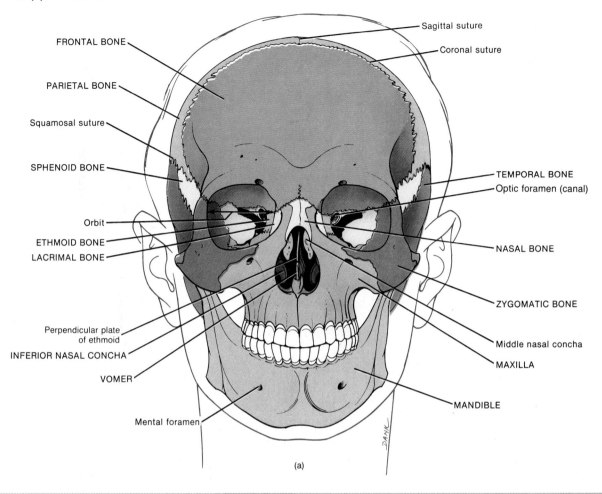

Sagittal suture
Coronal suture
FRONTAL BONE
PARIETAL BONE
Squamosal suture
SPHENOID BONE
TEMPORAL BONE
Optic foramen (canal)
Orbit
ETHMOID BONE
LACRIMAL BONE
NASAL BONE
Perpendicular plate of ethmoid
ZYGOMATIC BONE
INFERIOR NASAL CONCHA
Middle nasal concha
MAXILLA
VOMER
MANDIBLE
Mental foramen

(a)

Which skull bones are unpaired?

CRANIAL BONES

Frontal Bone

The *frontal bone* forms the forehead (the anterior part of the cranium), part of the *orbits* (eye sockets), and most of the anterior (front) part of the cranial floor. The *frontal sinuses* lie deep within the frontal bone (see Figure 6-6c). These mucous-membrane-lined cavities act as sound chambers that give the voice resonance.

Parietal Bones

The two *parietal bones* (pa-RĪ-e-tal; *paries* = wall) form the greater portion of the sides and roof of the cranial cavity (Figure 6-6).

Temporal Bones

The two *temporal bones* form the inferior (lower) sides of the cranium and part of the cranial floor.

In the lateral (side) view of the skull in Figure 6-6b, note that

the temporal and zygomatic bones join to form the *zygomatic arch*.

The *mandibular fossa* forms a joint with a projection on the mandible (lower jawbone) called the condylar process to form the temporomandibular joint (TMJ). The mandibular fossa is seen best in Figure 6-7.

The *external auditory (acoustic) meatus* is the canal in the temporal bone that leads to the middle ear. The *mastoid process* is a rounded projection of the temporal bone posterior to (behind) the external auditory meatus. It serves as a point of attachment for several neck muscles. The *styloid process* projects downward from the undersurface of the temporal bone and serves as a point of attachment for muscles and ligaments of the tongue and neck.

Occipital Bone

The *occipital* (ok-SIP-i-tal) *bone* forms the posterior (back) part and a prominent portion of the base of the cranium (Figure 6-7).

FIGURE 6-6 (*Continued*) (b) Right lateral view.

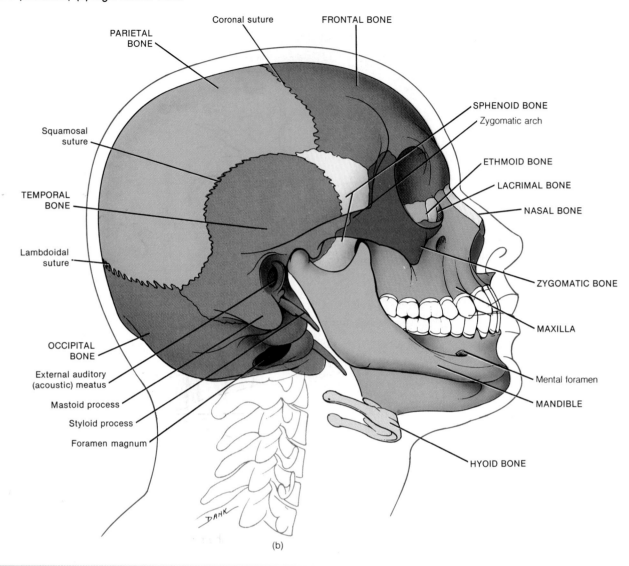

Coronal suture

FRONTAL BONE

PARIETAL
BONE

SPHENOID BONE

Zygomatic arch

Squamosal
suture

ETHMOID BONE

LACRIMAL BONE

NASAL BONE

TEMPORAL
BONE

Lambdoidal
suture

ZYGOMATIC BONE

MAXILLA

OCCIPITAL
BONE

External auditory
(acoustic) meatus

Mastoid process

Mental foramen

Styloid process

MANDIBLE

Foramen magnum

DAHK

HYOID BONE

(b)

What are the names of the facial bones? Refer to views (b) right lateral and (c) median.

The *foramen magnum* is a large hole in the bone through which the medulla oblongata (part of the brain stem that is continuous with the spinal cord) and the vertebral arteries and nerves pass.

The *occipital condyles* are oval projections, one on either side of the foramen magnum, that articulate (form a joint) with the first cervical vertebra.

Sphenoid Bone

The *sphenoid* (SFĒ-noyd) *bone* is situated at the middle part of the base of the skull (Figure 6-8). *Spheno* means wedge. This bone is referred to as the keystone of the cranial floor because it attaches to all the other cranial bones. The shape of the sphenoid is frequently described as a bat with outstretched wings.

The cubelike central portion of the sphenoid bone contains the *sphenoidal sinuses*, which drain into the nasal cavity (see Figure 6-11). On the superior surface of the sphenoid body is a depression called the *sella turcica* (SEL-a TUR-si-ka = Turk's saddle). This depression contains the pituitary gland.

Ethmoid Bone

The *ethmoid bone* is a light, spongy bone located in the front part of the floor of the cranium between the orbits (Figure 6-9).

The ethmoid contains several air spaces, or "cells," ranging in number from 3 to 18. It is from these "cells" that the bone derives its name (*ethmos* = sieve). The ethmoid "cells" together form the *ethmoidal sinuses*. The sinuses are shown in Figure 6-11. The *perpendicular plate* forms the upper portion of the nasal septum (see Figure 6-10). The *cribriform plate* forms the roof of

FIGURE 6-6 (*Continued*) (c) Median view.

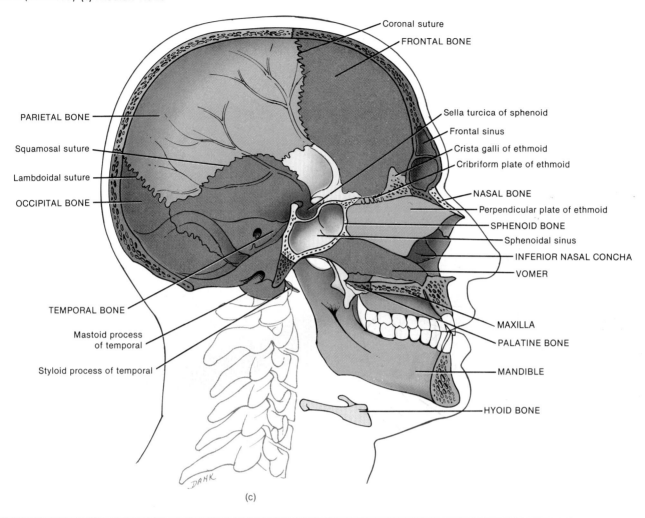

(c)

What features of the skull help to reduce its weight and serve as resonating chambers for the voice?

the nasal cavity. Projecting upward from the cribriform plate is a triangular process called the ***crista galli*** (= cock's comb), which serves as a point of attachment for the membranes (meninges) that cover the brain.

The ethmoid bone also contains two thin, scroll-shaped bones on either side of the nasal septum. These are called the ***superior nasal concha*** (KONG-ka; *concha* = shell) and the ***middle nasal concha***. The conchae filter air before it passes into the trachea, bronchi, and lungs.

FACIAL BONES

Nasal Bones

The paired ***nasal bones*** form part of the bridge of the nose (see Figures 6-6 and 6-10). The lower portion of the nose, indeed the major portion, consists of cartilage.

Maxillae

The ***maxillae*** (mak-SIL-ē) unite to form the upper jawbone and articulate with every bone of the face except the mandible, or lower jawbone (Figure 6-10).

Each maxillary bone contains a ***maxillary sinus*** that empties into the nasal cavity (see Figure 6-11). The ***alveolar*** (al-VĒ-ō-lar; *alveolus* = hollow) ***process*** contains the ***alveoli*** (bony sockets) into which the maxillary (upper) teeth are set. The maxilla forms the anterior three-fourths of the hard palate. The left and right sides of the maxillary bones fuse before birth. If the fusion does not occur, a condition called ***cleft palate*** results. Incomplete fusion of the sides of the palatine bones may also occur (see Figure 6-7). A ***cleft lip***, which involves a split upper lip, may also be associated with cleft palate. Depending on the extent and position of the cleft, speech and swallowing may be affected. A surgical procedure can sometimes improve the condition.

FIGURE 6-7 **Skull in inferior view.**

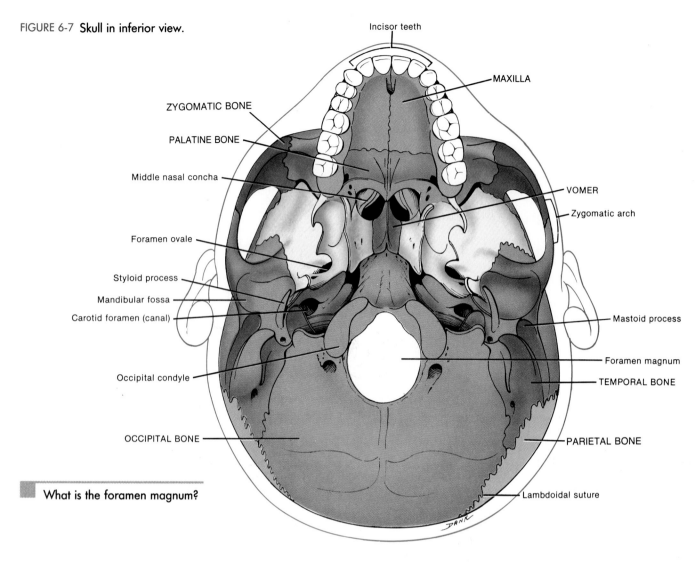

Incisor teeth

MAXILLA

ZYGOMATIC BONE

PALATINE BONE

Middle nasal concha

VOMER

Zygomatic arch

Foramen ovale

Styloid process

Mandibular fossa

Carotid foramen (canal)

Mastoid process

Occipital condyle

Foramen magnum

TEMPORAL BONE

OCCIPITAL BONE

PARIETAL BONE

Lambdoidal suture

What is the foramen magnum?

FIGURE 6-8 **Sphenoid bone viewed in the floor of the cranium from above.**

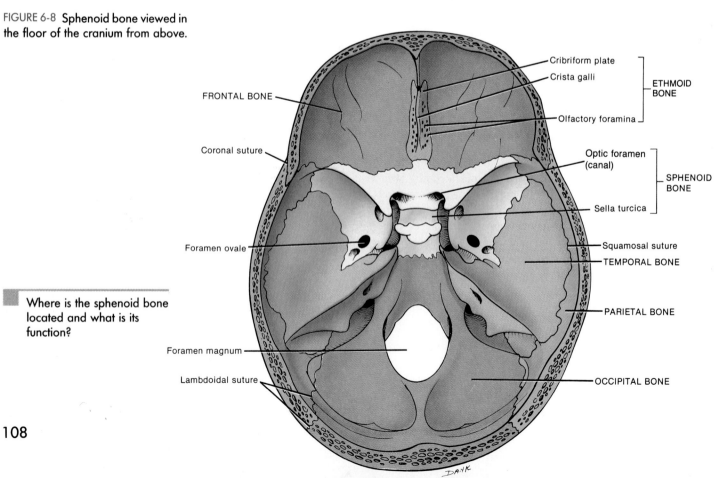

Cribriform plate

Crista galli

ETHMOID BONE

FRONTAL BONE

Olfactory foramina

Coronal suture

Optic foramen (canal)

SPHENOID BONE

Sella turcica

Foramen ovale

Squamosal suture

TEMPORAL BONE

PARIETAL BONE

Where is the sphenoid bone located and what is its function?

Foramen magnum

Lambdoidal suture

OCCIPITAL BONE

FIGURE 6-9 Ethmoid bone. Median view showing the ethmoid bone on the inner aspect of the left part of the skull.

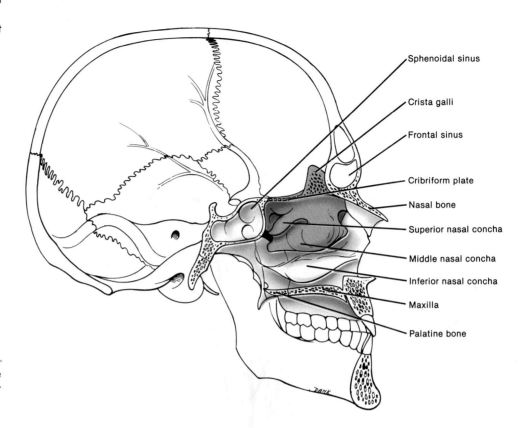

- Sphenoidal sinus
- Crista galli
- Frontal sinus
- Cribriform plate
- Nasal bone
- Superior nasal concha
- Middle nasal concha
- Inferior nasal concha
- Maxilla
- Palatine bone

Name three distinct parts of the ethmoid bone and identify their functions.

FIGURE 6-10 Maxillae. (a) Median view of the left maxilla. (b) Inferior view of the skull, showing the maxillae.

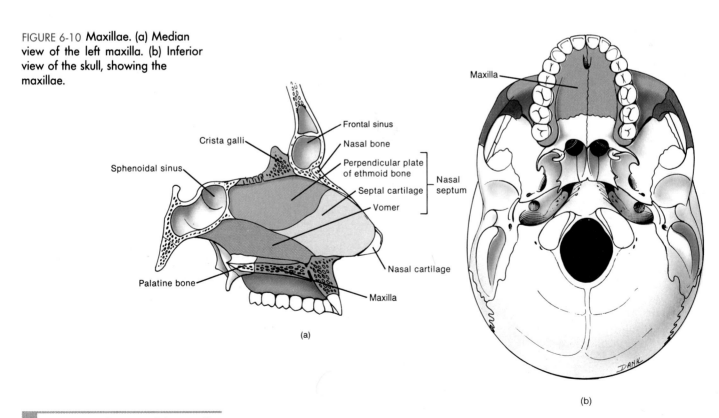

Sphenoidal sinus

Crista galli

Frontal sinus

Nasal bone

Perpendicular plate of ethmoid bone

Septal cartilage

Vomer

Nasal septum

Nasal cartilage

Palatine bone

Maxilla

(a)

Maxilla

(b)

What is the function of the maxilla?

Paranasal Sinuses

Paired cavities, called *paranasal sinuses*, are located in certain bones near the nasal cavity (Figure 6-11). The paranasal sinuses are lined with mucous membranes that are continuous with the lining of the nasal cavity. Cranial bones containing paranasal sinuses are the frontal, sphenoid, ethmoid, and maxillae. (The paranasal sinuses were described in the discussion of each of these bones.) Besides producing mucus, the paranasal sinuses lighten the skull bones and serve as resonant chambers for sound as we speak.

Secretions of the mucous membrane of a paranasal sinus drain into the nasal cavity. An inflammation of the membrane due to an allergy or infection is called *sinusitis*. If the membrane swells enough to block drainage into the nasal cavity, fluid pressure builds up in the paranasal sinus and a sinus headache results.

FIGURE 6-11 Paranasal sinuses seen in median view.

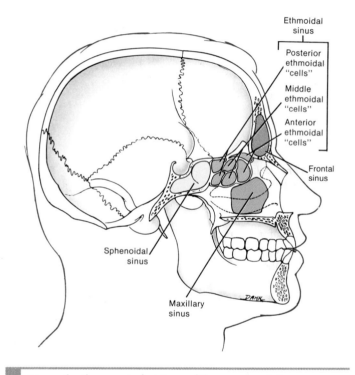

Ethmoidal sinus

Posterior ethmoidal "cells"

Middle ethmoidal "cells"

Anterior ethmoidal "cells"

Frontal sinus

Sphenoidal sinus

Maxillary sinus

Name the three main functions of the paranasal sinuses.

Zygomatic Bones

The two *zygomatic bones*, commonly referred to as the cheekbones, form the prominences of the cheeks (Figure 6-6b).

Mandible

The *mandible*, or lower jawbone, is the largest, strongest facial bone (Figure 6-12). It is the only movable skull bone. The mandible has a *condylar* (KON-di-lar) *process* that forms a joint with the mandibular fossa of the temporal bone to form the temporomandibular joint (TMJ). The *alveolar process* is an arch containing the *alveoli* (sockets) for the mandibular (lower) teeth.

Lacrimal Bones

The paired *lacrimal* (LAK-ri-mal; *lacrima* = tear) *bones* are thin bones roughly resembling a fingernail in size and shape. They are the smallest bones of the face. They can be seen in the anterior and lateral views of the skull in Figure 6-6.

Palatine Bones

The two *palatine* (PAL-a-tīn) *bones* are L-shaped bones that form the posterior portion of the hard palate. These can be seen in Figure 6-7.

Inferior Nasal Conchae

Refer to the views of the skull in Figures 6-6a and 6-9. The two *inferior nasal conchae* (KONG-kē) are scroll-like bones that project into the nasal cavity inferior to the superior and middle nasal conchae of the ethmoid bone. They serve the same function as the superior and middle nasal conchae, that is, filtration of air before it passes into the lungs.

Vomer

The *vomer* (= plowshare) is a roughly triangular bone that forms the lower and back part of the nasal septum. It is clearly seen in the anterior view of the skull in Figure 6-6a and the inferior view in Figure 6-7.

The lower border of the vomer articulates with the cartilage septum that divides the nose into a right and left nostril. Its upper border articulates with the perpendicular plate of the ethmoid bone. Thus, the structures that form the *nasal septum*, or partition, are the perpendicular plate of the ethmoid, the septal cartilage, and the vomer (Figure 6-10a). A *deviated nasal septum (DNS)* projects sideways from the midline of the nose. If the deviation is severe, it may entirely block the nasal passageway. Even an incomplete blockage may lead to nasal congestion, blocked paranasal sinuses, chronic sinusitis, headaches, or nosebleeds.

FONTANELS

The "skeleton" of a newly formed embryo consists of cartilage or fibrous membrane structures shaped like bones. Gradually, the cartilage or fibrous membrane is replaced by bone, but at birth, *fontanels* (fon'-ta-NELZ; = little fountains) are still found between cranial bones. These "soft spots" are areas of dense connective tissue that primarily enable the fetal skull to compress as it passes through the birth canal and permit rapid growth of the brain during infancy. Although an infant may have many fontanels at birth, the form and location of six are fairly constant.

The *anterior (frontal) fontanel* is located between the two parietal bones and the two segments of the frontal bone. It is roughly diamond-shaped and is the largest of the fontanels. It usually closes 18 to 24 months after birth.

FIGURE 6-12 **Mandible in right lateral view.**

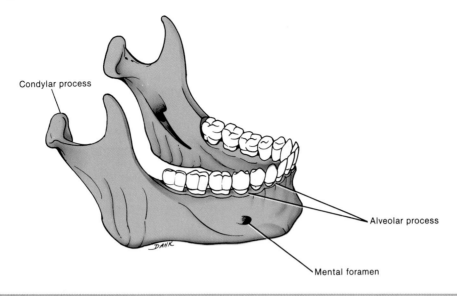

Condylar process

Alveolar process

Mental foramen

DAHK

What is the distinctive feature of the mandible among all the skull bones?

The *posterior (occipital;* ok-SIP-i-tal) *fontanel* is situated between the two parietal bones and the occipital bone. It is also diamond-shaped, but considerably smaller than the anterior fontanel. It generally closes about 2 months after birth.

The *anterolateral (sphenoidal;* sfē-NOY-dal) *fontanels* are paired. One is located on each side of the skull at the junction of the frontal, parietal, temporal, and sphenoid bones. These fontanels are quite small and irregular in shape. They normally close about 3 months after birth.

The *posterolateral (mastoid) fontanels* are also paired. One is situated on each side of the skull at the junction of the parietal, occipital, and temporal bones. They are irregularly shaped. They begin to close 1 or 2 months after birth, but closure is not generally complete until 12 months.

FORAMINA

The names and locations of certain foramina (singular = foramen) are included in the illustrations of the bones of the skull. In later chapters, when reference is made to the foramina, you can refer back to the illustrations to locate them.

HYOID BONE

The single *hyoid* (*hyoedes* = U-shaped) *bone* is a unique component of the axial skeleton because it does not attach to (articulate with) any other bone. Rather, it is suspended from the styloid process of the temporal bone by ligaments and muscles. The hyoid is located in the neck between the mandible and larynx. It supports the tongue and provides attachment for some of its muscles. Refer to Figure 6-6 to see the position of the hyoid bone.

The hyoid bone is frequently fractured during strangulation. As a result, it is carefully examined in an autopsy when strangulation is suspected.

VERTEBRAL COLUMN

DIVISIONS

The *vertebral column (spine)* is composed of a series of bones called *vertebrae*. The vertebral column is a strong, flexible rod that moves anteriorly, posteriorly, and laterally. It encloses and protects the spinal cord, supports the head, and serves as a point of attachment for the ribs and the muscles of the back. Between vertebrae are openings called *intervertebral foramina*. The nerves that connect the spinal cord to various parts of the body pass through these openings.

The adult vertebral column contains 26 vertebrae (Figure 6-13): 7 *cervical* (*cervix* = neck) *vertebrae* in the neck region; 12 *thoracic vertebrae* posterior to the thoracic cavity; 5 *lumbar* (*lumbus* = loin) *vertebrae* supporting the lower back; 5 *sacral vertebrae* fused into one bone called the *sacrum*; and 4 *coccygeal* (kok-SIJ-ē-al) *vertebrae* fused into the *coccyx* (KOK-six). Prior to the fusion of the sacral and coccygeal vertebrae, the total number of vertebrae is 33.

Between adjacent vertebrae from the first vertebra (atlas) to the sacrum are fibrocartilaginous *intervertebral discs*. Each disc is composed of an outer fibrous ring consisting of fibrocartilage called the *annulus fibrosus* and an inner soft, pulpy, highly elastic structure called the *nucleus pulposus*. The discs form strong joints, permit various movements of the vertebral column, and absorb vertical shock.

NORMAL CURVES

When viewed from the side, the vertebral column shows four *curves* (Figure 6-13). Two of them, called cervical and lumbar curves, are convex ( )), and two, called thoracic and sacral, are

FIGURE 6-13 **Vertebral column in right lateral view.**

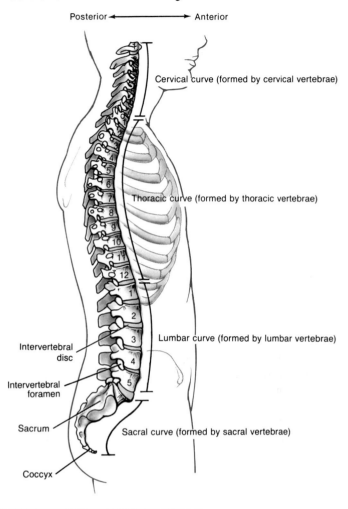

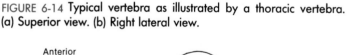

What is the function of the curves of the vertebral column?

concave (**(**). The curves of the column, like the curves in a long bone, increase its strength, help maintain balance in the upright position, absorb shocks from walking, and help protect the column from fracture.

In the fetus, there is only a single anteriorly concave curve. At approximately the third postnatal month, when an infant begins to hold its head erect, the *cervical curve* develops. Later, when the child stands and walks, the *lumbar curve* develops.

TYPICAL VERTEBRA

Vertebrae in different regions of the column vary in size, shape, and detail, but they are similar enough that we can discuss a typical vertebra (Figure 6-14).

1. The *body* is the thick, disc-shaped front portion; it is the weight-bearing part of a vertebra.
2. The *vertebral (neural) arch* extends behind the body. It is formed by two short, thick processes, the *pedicles* (PED-i-kuls), which unite with the *laminae* (LAM-i-nē), flat parts that end in a single spinous process, a sharp slender projection. The space between the vertebral arch and body contains the spinal cord and is known as the *vertebral foramen*. The vertebral foramina of all vertebrae together form the *vertebral (spinal) canal*. The pedicles are notched in such a way that when they are arranged in the column there is an opening between vertebrae, the *intervertebral foramen*, each of which permits the passage of a single spinal nerve.
3. Seven *processes* arise from the vertebral arch. At the point where a lamina and pedicle join, a *transverse process* extends laterally. There is the single *spinous process (spine)* behind and below the junction of the laminae. These three processes serve as points of muscular attachment. The remaining four processes form joints with other vertebrae. The two *superior*

FIGURE 6-14 **Typical vertebra as illustrated by a thoracic vertebra. (a) Superior view. (b) Right lateral view.**

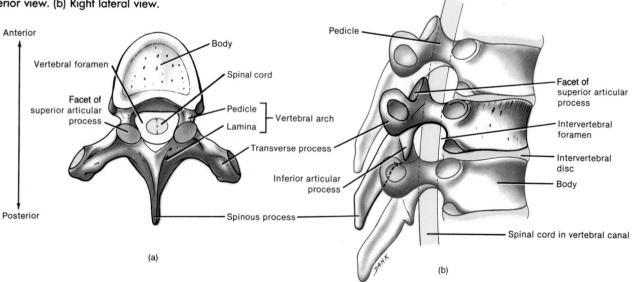

What is the purpose of the (a) vertebral foramen; (b) intervertebral foramina?

FIGURE 6-15 Cervical vertebrae. (a) Superior view of a cervical vertebra. (b) Superior view of the atlas. (c) Anterior view of the axis.

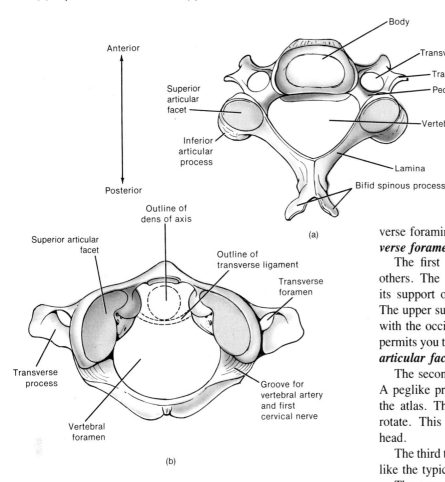

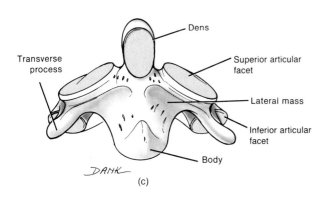

How many cervical vertebrae are there?

verse foramina. Each cervical transverse process contains a *transverse foramen* through which blood vessels and nerves pass.

The first two cervical vertebrae differ considerably from the others. The first cervical vertebra (C1), the *atlas*, is named for its support of the head. It lacks a body and a spinous process. The upper surface contains *superior articular facets* that articulate with the occipital condyles of the occipital bone. This articulation permits you to nod your head. The inferior surface contains *inferior articular facets* that articulate with the second cervical vertebra.

The second cervical vertebra (C2), the *axis*, does have a body. A peglike process called the *dens* projects up through the ring of the atlas. The dens makes a pivot on which the atlas and head rotate. This arrangement permits side-to-side movement of the head.

The third through sixth cervical vertebrae (C3–C6) are structured like the typical cervical vertebra described above.

The seventh cervical vertebra (C7), called the *vertebra prominens*, is somewhat different. It is marked by a large, nonbifid spinous process that may be seen and felt at the base of the neck.

THORACIC REGION

Viewing a typical *thoracic vertebra* from above, you can see that it is considerably larger and stronger than a cervical vertebra (see Figure 6-14). Except for the eleventh and twelfth thoracic vertebrae, the transverse processes have *facets* for articulating with the ribs.

LUMBAR REGION

The *lumbar vertebrae* (L1–L5) are the largest and strongest in the column (Figure 6-16). The spinous processes are well adapted for the attachment of the large back muscles.

SACRUM AND COCCYX

The *sacrum* is a triangular bone formed by the union of five sacral vertebrae. These are indicated in Figure 6-17 as S1 through S5. Fusion begins between 16 and 18 years of age and is usually completed by the mid-twenties. The sacrum serves as a strong foundation for the pelvic girdle. It is positioned at the posterior portion of the pelvic cavity between the two hip bones.

The anterior and posterior sides of the sacrum contain four pairs of *sacral foramina*. Nerves and blood vessels pass through

articular processes of a vertebra articulate with the vertebra immediately above, and the two *inferior articular processes* articulate with the vertebra below them. The articulating surfaces of the articular processes are referred to as *facets*.

CERVICAL REGION

The spinous processes of the second through sixth *cervical vertebrae* are often *bifid*, that is, with a cleft (Figure 6-15). All cervical vertebrae have three foramina: the vertebral foramen and two trans-

FIGURE 6-16 Lumbar vertebra seen in superior view.

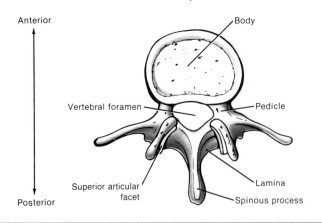

> Why are the lumbar vertebrae the largest and strongest in the vertebral column?

the foramina. The *sacral canal* is a continuation of the vertebral canal. The lower entrance is called the *sacral hiatus* (hī-A-tus).

The *coccyx* is also triangular in shape and is formed by the fusion of the coccygeal vertebrae, usually the last four. These are indicated in Figure 6-17 as Co1 through Co4. The coccyx articulates superiorly with the sacrum.

THORAX

Anatomically, the term *thorax* refers to the entire chest. The skeletal portion of the thorax is a bony cage formed by the sternum, costal cartilage, ribs, and the bodies of the thoracic vertebrae (Figure 6-18).

The thoracic cage encloses and protects the organs in the thoracic cavity and supports the bones of the shoulder girdle and upper extremities.

STERNUM

The *sternum*, or breastbone, is a flat, narrow bone located in the center of the anterior thoracic wall.

The sternum (see Figure 6-18) is divided into the *manubrium* (ma-NOO-brē-um), the upper portion; the *body*, the middle, largest portion; and the *xiphoid* (ZĪ-foyd) *process*, the lower, smallest portion. The manubrium articulates with the clavicles and the first and second ribs. The body of the sternum articulates directly or indirectly with the second through tenth ribs. The xiphoid process has no ribs attached to it but provides attachment for some abdominal muscles. If the hands of a rescuer are mispositioned during cardiopulmonary resuscitation (CPR), there is the danger of fracturing the xiphoid process and driving it into the liver.

Since the sternum possesses red bone marrow throughout life and because it is readily accessible and has thin compact bone, it is a common site for withdrawal of marrow for biopsy (*marrow aspiration*).

RIBS

Twelve pairs of *ribs* make up the sides of the thoracic cavity (see Figure 6-18). The ribs increase in length from the first through

FIGURE 6-17 Sacrum and coccyx. (a) Anterior view. (b) Posterior view.

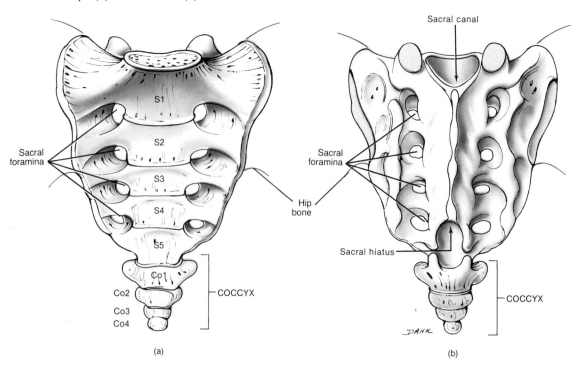

(a)

(b)

> What is the function of the foramina?

FIGURE 6-18 **Skeleton of the thorax in anterior view.**

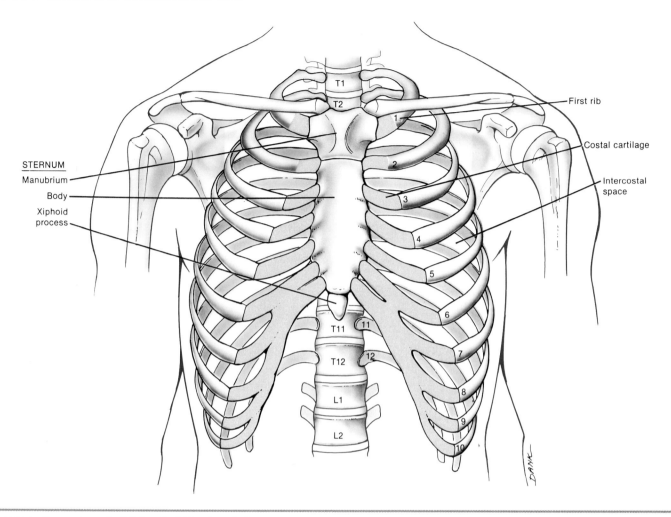

Which ribs are true ribs? False ribs? Floating ribs?

seventh. Then they decrease in length to the twelfth rib. Each rib articulates posteriorly with its corresponding thoracic vertebra.

The first through seventh ribs have a direct anterior attachment to the sternum by a strip of hyaline cartilage, called *costal cartilage* (*costa* = rib). These ribs are called *true ribs*. The remaining five pairs of ribs are referred to as *false ribs* because their costal cartilages do not attach to the sternum indirectly or do not attach at all. The cartilages of the eighth, ninth, and tenth ribs attach to each other and then to the cartilage of the seventh rib. The eleventh and twelfth false ribs are also called *floating ribs* because their anterior ends do not attach at all; they attach only posteriorly to the thoracic vertebrae. Spaces between ribs, called *intercostal spaces*, are occupied by intercostal muscles, blood vessels, and nerves.

PECTORAL (SHOULDER) GIRDLE

The *pectoral* (PEK-tō-ral) or *shoulder girdles* attach the bones of the upper extremities to the axial skeleton (Figure 6-19). Each of the two pectoral girdles consists of two bones: a clavicle and a scapula. The clavicle is the anterior component and articulates with the sternum at the sternoclavicular joint. The posterior component, the scapula, articulates with the clavicle and humerus. The pectoral girdles do not articulate with the vertebral column. Although the shoulder joints are not very stable, they are freely movable and thus allow movement in many directions.

CLAVICLE

The *clavicles* (KLAV-i-kuls), or collarbones, are long, slender bones with a double curvature. They lie horizontally and are superior to the first rib.

The rounded, medial end articulates with the sternum. The broad, flat, lateral end articulates with the acromion of the scapula. (See Figure 6-19.)

SCAPULA

The *scapulae* (SCAP-yoo-lē), or shoulder blades, are large, triangular, flat bones situated in the posterior part of the thorax between the second and seventh ribs (Figure 6-20).

A sharp ridge, the **spine**, runs diagonally across the posterior surface of the triangular **body**. The end of the spine projects as a flattened, expanded process called the **acromion** (a-KRŌ-mē-on). This process articulates with the clavicle. Inferior to the acromion is a depression called the **glenoid cavity**. This cavity articulates with the head of the humerus to form the shoulder joint.

FIGURE 6-19 **Right pectoral (shoulder) girdle and upper extremity seen in anterior view.**

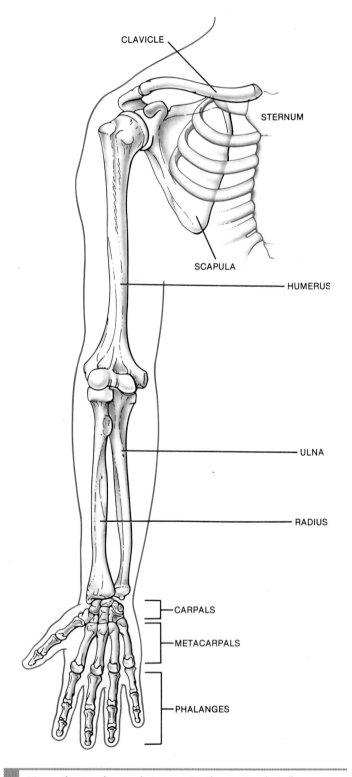

CLAVICLE

STERNUM

SCAPULA

HUMERUS

ULNA

RADIUS

CARPALS

METACARPALS

PHALANGES

Name the two bones that comprise the pectoral girdle.

UPPER EXTREMITY

The **upper extremities** consist of 60 bones. The skeleton of the right upper extremity is shown in Figure 6-21. Each upper extremity includes a humerus in the arm, ulna and radius in the forearm, carpals (wrist bones), metacarpals (palm bones), and phalanges in the fingers of the hand.

HUMERUS

The **humerus** (HYOO-mer-us) is the longest and largest bone of the upper extremity (Figure 6-21). It articulates with the scapula and at the elbow with both ulna and radius.

The proximal end of the humerus consists of a **head** that articulates with the glenoid cavity of the scapula. It also has an **anatomical neck**, which is a groove just below the head.

The **body (shaft)** of the humerus contains a roughened area called the **deltoid tuberosity**, which serves as a point of attachment for the deltoid muscle.

The following parts are found at the distal end of the humerus. The **capitulum** (ka-PIT-yoo-lum) is a rounded knob that articulates with the head of the radius. The **radial fossa** is a depression that receives the head of the radius when the forearm is flexed. The **trochlea** (TRŌK-lē-a) is a surface that articulates with the ulna. The **coronoid fossa** is a depression that receives part of the ulna when the forearm is flexed. The **olecranon** (ō-LEK-ra-non) **fossa** is a depression that receives the olecranon of the ulna when the forearm is extended.

ULNA AND RADIUS

The **ulna** is the medial bone of the forearm (Figure 6-22). In other words, it is located at the little finger side. The proximal end of the ulna has an **olecranon**, which forms the elbow. The **coronoid process,** together with the olecranon, receives the trochlea of the humerus. The **trochlear notch** is a curved area between the olecranon and the coronoid process. The trochlea of the humerus fits into this notch. The **radial notch** is a depression for the head of the radius. A **styloid process** is at the most distal end.

The **radius** is the lateral bone of the forearm, that is, situated on the thumb side. The proximal end of the radius articulates with the capitulum of the humerus and radial notch of the ulna. It has a raised, roughened area called the **radial tuberosity** that provides a point of attachment for the biceps brachii muscle. The shaft of the radius widens distally to articulate with two bones (carpals) of the wrist. Also at the distal end is a **styloid process** and a concave **ulnar notch** that articulates with the ulna.

CARPALS, METACARPALS, AND PHALANGES

The **carpus (wrist)** consists of eight small bones, the **carpals,** connected to each other by ligaments (Figure 6-23). The bones are arranged in two rows, four bones in each row. In the anatomical position, the top row of carpals consists of the **scaphoid, lunate,**

FIGURE 6-20 **Right scapula (a) Anterior view. (b) Posterior view.**

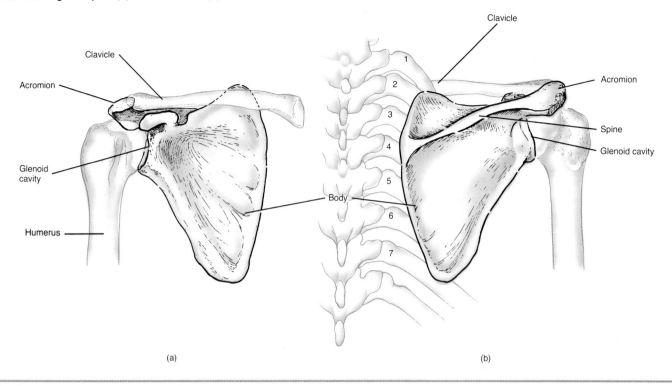

(a)

(b)

What is the common term we use for the scapula?

triquetral, and pisiform. The bottom row of carpals consists of the *trapezium, trapezoid, capitate*, and *hamate*.

The five bones of the *metacarpus* constitute the palm of the hand. Each metacarpal bone consists of a proximal *base, a shaft*, and a distal *head*. The metacarpal bones are numbered I to V, starting with the lateral bone in the thumb. The heads of the metacarpals are commonly called the "knuckles" and are readily visible when the fist is clenched.

The *phalanges* (fa-LAN-jēz), or bones of the fingers, number 14 in each hand. A single bone of the finger (or toe) is referred to as a *phalanx* (FĀ-lanks). Each phalanx consists of a proximal *base*, a *shaft*, and a distal *head*. There are two phalanges (proximal and distal) in the first digit, called the thumb *(pollex)*, and three phalanges (proximal, middle, and distal) in each of the remaining four digits. These digits are commonly referred to as the index finger, middle finger, ring finger, and little finger.

PELVIC (HIP) GIRDLE

The *pelvic (hip) girdle* consists of the two *hip bones* (Figure 6-24). The pelvic girdle provides a strong and stable support for the lower extremities on which the weight of the body is carried. The hip bones are united to each other anteriorly at the symphysis (SIM-fi-sis) pubis. They unite posteriorly to the sacrum.

Together with the sacrum and coccyx, the two hip bones of the pelvic girdle form the basinlike structure called the *pelvis*. The pelvis is divided into a greater pelvis and a lesser pelvis.

The *greater (false) pelvis* is the portion situated superior to the brim of the pelvis. The *lesser (true) pelvis* is inferior to the brim. The lesser pelvis contains a superior opening called the *pelvic inlet* and an inferior opening called the *pelvic outlet*.

Pelvimetry is the measurement of the size of the inlet and outlet of the birth canal. The size of the pelvic cavity in females is important because the fetus must pass through the narrower opening of the lesser pelvis at birth.

Each of the two *hip bones* of a newborn consists of a superior *ilium*, an inferior and anterior *pubis*, and an inferior and posterior *ischium* (Figure 6-25). Eventually, the three separate bones fuse into one. The area of fusion is a deep, lateral fossa (depression) called the *acetabulum* (as'-e-TAB-yoo-lum), which serves as the socket for the head of the femur. Although the adult hip bones are single bones, it is common to discuss them as if they still consisted of three portions.

The ilium is the largest of the three subdivisions. Its upper border is called the *iliac crest* (border or ridge). On the lower surface is the *greater sciatic* (sī-AT-ik) *notch*. The ischium joins with the pubis and together they surround the *obturator* (OB-too-rā'-tor) *foramen*. The pubis contributes to the formation of the symphysis pubis that consists of fibrocartilage.

LOWER EXTREMITY

The *lower extremities* are composed of 60 bones (Figure 6-26). Each extremity includes the femur in the thigh, patella (kneecap),

FIGURE 6-21 **Right humerus seen in anterior view.**

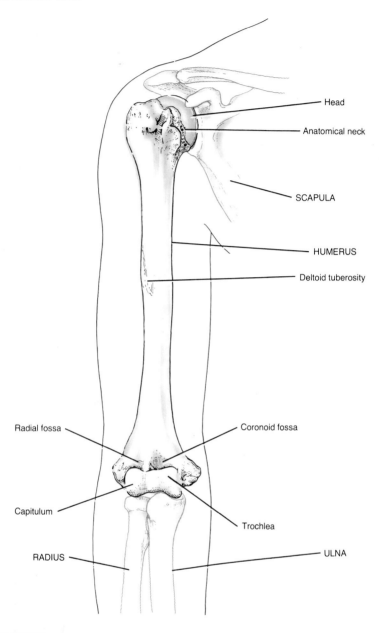

Head

Anatomical neck

SCAPULA

HUMERUS

Deltoid tuberosity

Radial fossa

Coronoid fossa

Capitulum

Trochlea

RADIUS

ULNA

With what part of the scapula does the humerus articulate?

fibula and tibia in the leg, tarsals (ankle bones), metatarsals, and phalanges in the toes.

FEMUR

The *femur*, or thighbone, is the longest and heaviest bone in the body (Figure 6-27). Its proximal end articulates with the coxal bone. Its distal end articulates with the tibia. The *body (shaft)* of the femur bows medially so that the knee joints are brought nearer to the body's line of gravity. The convergence is greater in the female because the female pelvis is broader.

The *head* of the femur articulates with the acetabulum of the hip bone. The *neck* of the femur is a constricted region below the head. A fairly common fracture in the elderly occurs at the neck of the femur. The neck may become so weak it fails to support the body.

The body of the femur contains a rough vertical ridge called the *linea aspera*, which serves to attach several thigh muscles.

The distal end of the femur expands into the *medial condyle* and *lateral condyle*, which articulate with the tibia. The *patellar surface* is located between the condyles on the anterior surface.

FIGURE 6-22 **Right ulna and radius. (a) Anterior view. (b) Details of proximal end of ulna.**

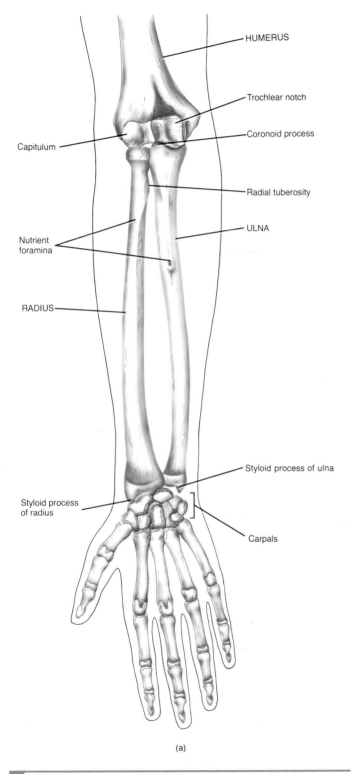

HUMERUS

Trochlear notch

Coronoid process

Capitulum

Radial tuberosity

ULNA

Nutrient foramina

RADIUS

Styloid process of ulna

Styloid process of radius

Carpals

(a)

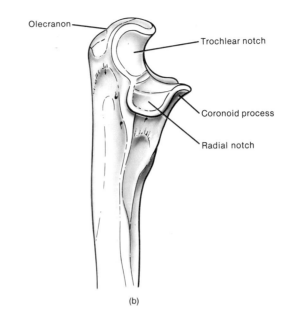

Olecranon

Trochlear notch

Coronoid process

Radial notch

(b)

FIGURE 6-23 **Right wrist and hand seen in anterior view.**

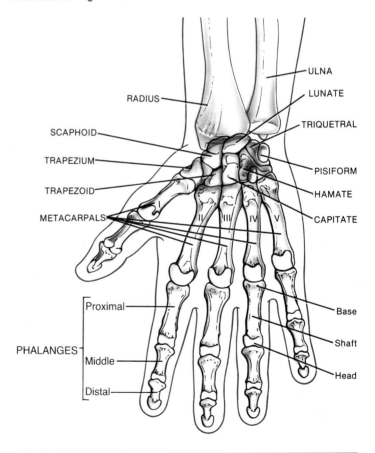

ULNA

LUNATE

TRIQUETRAL

RADIUS

SCAPHOID

PISIFORM

TRAPEZIUM

HAMATE

TRAPEZOID

CAPITATE

METACARPALS

I   II   III IV   V

Base

Proximal

Shaft

PHALANGES

Middle

Head

Distal

What part of the ulna is called the elbow?

How many bones make up the wrist and hand? Why are there so many bones in this area?

FIGURE 6-24 **Pelvic (hip) girdle of a female in anterior view.**

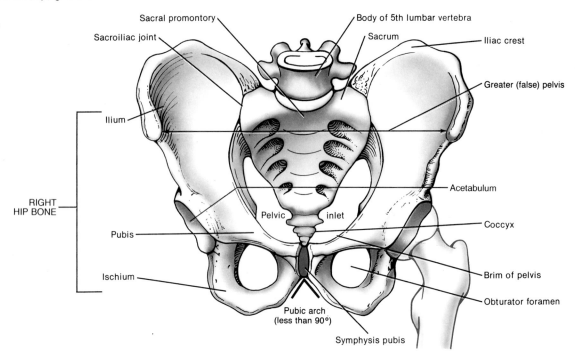

Name the two bones that comprise the pelvic girdle. With what two additional bones do they form the pelvis?

FIGURE 6-25 **Right hip bone seen in lateral view. The lines of fusion of the ilium, ischium, and pubis are not always visible in an adult bone.**

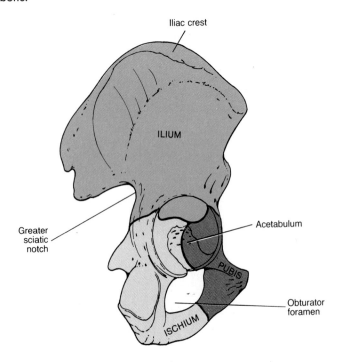

The acetabulum is the socket for which bone?

PATELLA

The *patella*, or kneecap, is a small, triangular bone in front of the knee joint (see Figure 6-26). It is a sesamoid bone that develops in the tendon of the quadriceps femoris muscle.

TIBIA AND FIBULA

The *tibia*, or shinbone, is the larger medial bone of the leg (Figure 6-28). It bears the major portion of the weight of the leg. The tibia articulates at its proximal end with the femur and fibula and at its distal end with the fibula and talus bone of the ankle.

The proximal end of the tibia expands into a *lateral condyle* and a *medial condyle*, which articulate with the condyles of the femur. The *tibial tuberosity* is a point of attachment for the patellar ligament.

The medial surface of the distal end of the tibia forms the *medial malleolus* (mal-LĒ-ō-lus), which articulates with the talus bone of the ankle and forms the bump that can be felt on the medial surface of your ankle. The *fibular notch* articulates with the fibula.

Shinsplints is the name given to soreness or pain along the tibia. It is probably caused by inflammation of the periosteum brought on by repeated tugging of the muscles and tendons attached to the periosteum, often the result of walking or running up and down hills or vigorous activity followed by relative inactivity.

FIGURE 6-26 Right pelvic (hip) girdle and lower extremity seen in anterior view.

FIGURE 6-27 Right femur seen in anterior view.

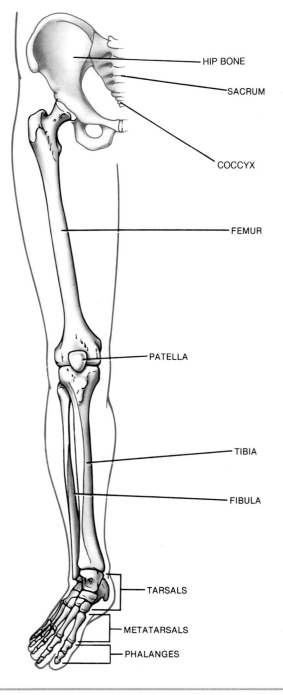

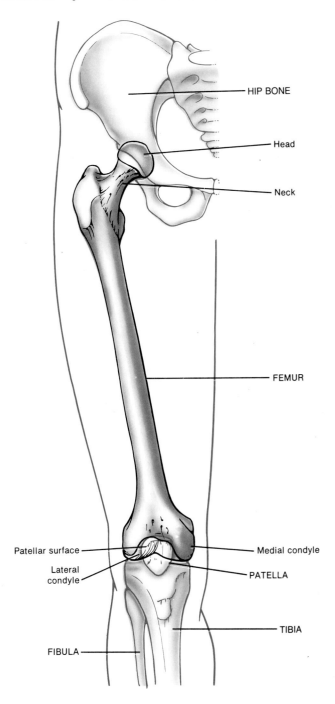

What kind of bone is the patella?

With which bone does the distal end of the femur articulate?

The *fibula* is parallel to the tibia (Figure 6-28) and considerably smaller. The proximal end of the fibula articulates with the lateral condyle of the tibia just below the knee joint. The distal end has a projection called the *lateral malleolus* that articulates with the talus bone of the ankle. This forms the prominence on the lateral surface of the ankle. The inferior portion of the fibula also articulates with the tibia at the fibular notch.

TARSALS, METATARSALS, AND PHALANGES

The *tarsus* is a collective term for the seven bones of the ankle called *tarsals* (Figure 6-29). The *talus* and *calcaneus* (kal-KĀ-nē-us) are located on the posterior part of the foot. The anterior part contains the *cuboid, navicular,* and three *cuneiform bones* called the *first (medial), second (intermediate),* and *third (lateral)*

FIGURE 6-28 **Right tibia and fibula seen in anterior view.**

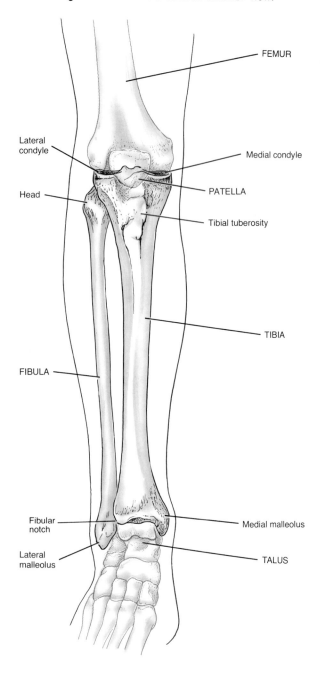

FIGURE 6-29 **Right foot seen in superior view.**

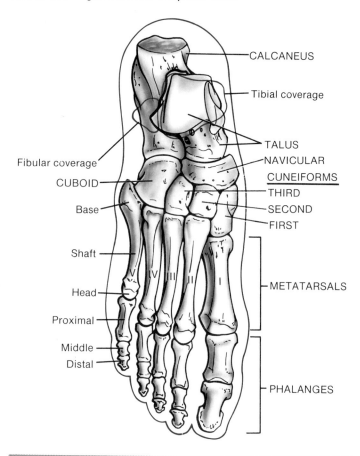

What is the purpose of the arches of the foot?

With which bone is shinsplints associated?

cuneiform. The talus is the only bone of the foot that articulates with the fibula and tibia. It is surrounded by the medial malleolus of the tibia and the lateral malleolus of the fibula. During walking, the talus initially bears the entire weight of the body. About half the weight is then transmitted to the calcaneus, and the remainder is distributed over the other five tarsal bones. The calcaneus, or heel bone, is the largest and strongest tarsal bone.

The ***metatarsus*** consists of five metatarsal bones numbered I to V from the medial to lateral position. Like the metacarpals of the palm of the hand, each metatarsal consists of a proximal ***base***, a ***shaft***, and a distal ***head***. The first metatarsal is thicker than the others because it bears more weight.

The ***phalanges*** of the foot resemble those of the hand both in number and arrangement. Each also consists of a proximal ***base***, a middle ***shaft***, and a distal ***head***. The great (big) toe, or ***hallux***, has two large, heavy phalanges (proximal and distal) while the other four toes each have three phalanges (proximal, middle, and distal).

ARCHES OF THE FOOT

The bones of the foot are arranged in two ***arches***. These arches enable the foot to support the weight of the body and provide leverage while walking. The arches are not rigid. They yield as weight is applied and spring back when the weight is lifted.

The ***longitudinal arch*** runs from the front to the back of the foot, from the calcaneus to the distal ends of the metatarsals. The ***transverse arch*** is formed by the navicular, three cuneiforms, cuboid, and the proximal ends of the five metatarsals.

The bones composing the arches are held in position by ligaments and tendons. If these ligaments and tendons are weakened, the height of the longitudinal arch may "fall." The result is *flatfoot*.

Clawfoot is a condition in which the longitudinal arch is abnormally elevated. It is frequently caused by muscle imbalance, such as may result from poliomyelitis.

A *bunion* is a deformity of the great toe. Although the condition may be inherited, it is typically caused by wearing tightly fitting shoes. The condition produces inflammation of bursae (fluid-filled sacs at the joint), bone spurs, and calluses.

FEMALE AND MALE SKELETONS

The bones of the male are generally larger and heavier than those of the female. In addition, since certain muscles of the male are larger than those of the female, the points of attachment—tuberosities, lines, ridges—are larger in the male skeleton.

Also, significant structural differences between female and male skeletons exist in the pelvis; most are related to pregnancy and childbirth. See Exhibit 6-2 and Figure 6-30.

FIGURE 6-30 Comparison of (a) female and (b) male pelvis in anterior view.

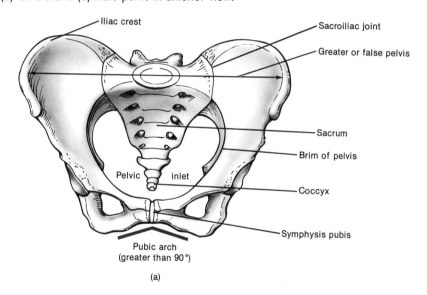

(a)

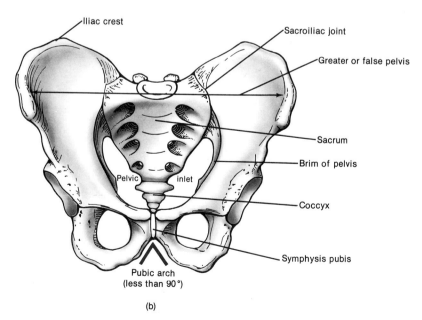

(b)

What are five ways a female pelvis differs from a male pelvis?

Osteoporosis: A Question of Calcium Balance

As the number of elderly people in our population increases, so does the public health impact of osteoporosis. Public health policymakers and health-care professionals have become increasingly concerned about the treatment and prevention of this disease.

Though the loss of bone minerals may begin as early as the fourth decade of life, osteoporosis is often not diagnosed until so much bone mass has been lost that the patient develops a fracture, which may have been caused by something no more traumatic than a sneeze. In the United States, there are about 1.2 million fractures associated with osteoporosis each year.

One of the most noticeable signs of osteoporosis is the rounded back posture known as "dowager's hump," so often depicted in the advertisements for calcium supplements. Spongy bone is more sensitive than compact bone to metabolic influences and is thus more readily lost during the development of osteoporosis. When the vertebrae lose spongy bone, the spinal column loses strength, and individual vertebrae collapse because of what are known as crush fractures. When these fractures accumulate in several vertebrae, the spine develops a hunchback that is often accompanied by back pain and loss of height.

Of women who live to be 90 years old, one-third will experience a hip fracture. Hip fractures are the most costly and debilitating of fractures associated with osteoporosis. Long hospital stays often accelerate bone loss and muscle weakness and can lead to institutionalization.

It has been found that age-associated bone loss occurs to some extent in all populations and races studied and thus appears to be an inevitable part of the aging process. Its rate of progression does vary greatly from group to group and among individuals within groups. These variations have allowed scientists to describe several life-style risk factors for osteoporosis.

Calcium Intake

Calcium intake has been the most publicized risk factor, and the one with which Americans are probably most familiar. Several scientific and government groups have espoused increasing calcium intake to prevent osteoporosis, and the calcium supplement industry has not been slow to jump on the bandwagon.

A great deal of controversy surrounds the calcium supplementation issue. Studies of whether calcium supplements help prevent or treat osteoporosis have yielded mixed results. Some studies have failed to show a clear association between either childhood or adulthood calcium intake and the development of osteoporosis. In fact, it appears that the body can adapt to fairly low levels of dietary calcium. People in some countries maintain calcium balance on intakes of 400 to 500 mg per day. (The U.S. RDA is 1000 mg.) Calcium balance means that there is neither a net loss nor a gain in bone tissue. As you learned in the section on bone tissue homeostasis, the process of bone remodeling is finely regulated by a number of hormones, and a high calcium intake may be countered by homeostatic adjustments.

Nutritionists recommend foods over supplements for several reasons. Some studies suggest that many supplements are not absorbed as well as the calcium from foods. Calcium supplements also interfere with the body's absorption of iron consumed at the same time. Possible overdoses are always a problem with supplements, since well-intentioned people sometimes figure that if some is good, more is better. Vitamin D is often included in calcium supplements and is one of the most toxic vitamins in high doses.

Despite the confusion over the importance of calcium intake, nutritionists still agree that an adequate intake of dietary calcium is important for good health, which includes the growth and maintenance of bone tissue. Good calcium sources include milk and milk products, dark green vegetables like broccoli, sardines, and canned salmon (eat the bones, too). Tofu and soy milk are often fortified with calcium for people who avoid dairy products.

Other Foods and Nutrients

Consumption of excess protein may contribute somewhat to loss of bone mineral. Doubling protein intake increases urinary calcium excretion 50 percent. This appears to be due to decreased uptake of calcium by the kidney. Epidemiological studies have found lower rates of osteoporosis in British vegetarians and female Seventh-Day Adventists in the United States than in meat-eating individuals. However, these differences could also be due to other factors, such as activity or heredity. But since most Americans consume more protein than they need, decreasing their intake of concentrated protein sources such as meat and eggs could be beneficial.

Some researchers have expressed concern that a very high-fiber diet may decrease calcium absorption from the intestine, since various plant components and dietary fiber may bind with the calcium. Most fiber studies have used fiber amounts much higher than those consumed by most people, so health professionals continue to recommend a moderate increase in fiber intake since such a diet is associated with many health benefits.

Vitamin D improves the efficiency of intestinal absorption of calcium. Vitamin D is difficult to obtain from natural dietary sources, so it is added to most dairy products.

Other minerals, such as boron, are involved in bone maintenance but are readily received from a well-balanced diet. Boron is found in pears, apples, grapes, leafy vegetables, and legumes.

Frequent dieting can decrease bone density, as the body draws calcium from bone to make up for calcium missing from the diet. Although on the surface it sounds like this goes against evidence that the body adapts to a low intake, adaptation takes time, which binge dieting doesn't give it. Therefore, it seems that frequent periods of calcium deficiency contribute to the development of osteoporosis.

Alcohol

The association between a high level of alcohol consumption and osteoporosis may be a consequence of poor nutrition, lower body weight, liver disease, or other illness, or it may be a result of a toxic effect of alcohol. A moderate intake of alcohol does not appear to cause bone loss.

Cigarette Smoking

People who smoke cigarettes have a higher risk of osteoporosis than nonsmokers. This risk may be explained by the fact that smokers tend to be thinner, which is a risk factor. Female smokers have lower estrogen levels and undergo menopause at an earlier age than nonsmokers, so they experience less of estrogen's protective effect.

Physical Activity

People confined to bedrest and astronauts in a weightless environment lose up to 1 percent of their spongy bone per week. Bone mass is restored once they resume normal activity. Active people tend to have denser bones. Bone density in a tennis player's dominant arm is greater than that of the other arm. Weight-bearing activity is strongly recommended to help increase bone mass during adolescence and young adulthood and to delay its loss thereafter. Weight-bearing refers to activities in which the skeletal system must support body weight, for example, walking, hiking, and cross-country skiing as opposed to swimming. Physical exercise places demands for strength on the bone, and homeostatic mechanisms for regulation of bone density respond. The adage "use it or lose it" certainly applies to bone density.

"Moderation in all things" applies as well. Excessive exercise may lead to loss of bone mineral, possibly through an alteration of sex hormone levels. Several studies have documented osteoporosis in young women athletes who have stopped menstruating, a condition known as amenorrhea. An absence of the menstrual cycle often signals low levels of the female hormone estrogen, which helps bones retain calcium. Bone-density measurements have revealed that many of these young athletes had the bone density of women in their fifties. In some cases, the bone damage sustained was considered irreversible.

Prevention and Treatment

Maintaining an adequate calcium intake by eating a well-balanced diet; avoiding frequent dieting, heavy use of alcohol, and smoking; and participating in regular weight-bearing physical activity are recommendations that bestow many health benefits in addition to preservation of bone mass. People in high-risk groups, such as fair-skinned, small, postmenopausal women, may benefit from additional medical treatment, such as hormone replacement therapy.

EXHIBIT 6-2 *Comparison of a Typical Female and Male Pelvis (see Figure 6-30)*		
Point of Comparison	**Female**	**Male**
General Structure	Light and thin.	Heavy and thick.
Greater Pelvis	Shallow.	Deep.
Pelvic Inlet	Larger and more oval.	Heart-shaped.
Pelvic Outlet	Comparatively large.	Comparatively small.
Sacrum	Short, wide, flat, curving forward in lower part.	Long, narrow, with smooth concavity.
Pubic Arch	Greater than a 90° angle.	Less than a 90° angle.
Symphysis Pubis	Less deep.	More deep.
Ilium	Less vertical.	More vertical.
Iliac Crest	Less curved.	More curved.
Acetabulum	Small.	Large.
Obturator Foramen	Oval.	Round.

▪ COMMON DISORDERS ▪

Osteoporosis

Osteoporosis (os′-tē-ō-pō-RŌ-sis) is an age-related disorder characterized by decreased bone mass and increased susceptibility to fractures as a result of decreased levels of estrogens.

The disorder primarily affects middle-aged and elderly people—especially women (white more than blacks). Estrogens maintain osseous tissue by stimulating osteoblasts to form new bone.

Women produce smaller amounts of sex hormones, especially estrogens, after menopause, and both men and women produce smaller amounts during old age. As a result, the osteoblasts become less active, and there is a decrease in bone mass.

When bone mass is so depleted that the skeleton can no longer withstand the mechanical stresses of everyday living, fractures result. Osteoporosis also causes shrinkage of the backbone and height loss and hunched backs. Osteoporosis affects the entire skeletal system, especially the vertebral bodies, ribs, proximal femur (hip), humerus, and distal radius. It can also lead to loss of teeth as the bone mass around them decreases.

Other factors implicated in osteoporosis, besides race and sex, are (1) body build (short females are at greater risk since they have less total bone mass), (2) weight (thin females are at greater risk since adipose tissue is a great source of estrone, an estrogen that retards bone loss), (3) smoking (smoking decreases blood estrogen levels), (4) calcium deficiency and malabsorption, (5) vitamin D deficiency, (6) exercise (sedentary people are more likely to develop bone loss), (7) certain drugs (alcohol, some diuretics, cortisone, and tetracycline), and (8) premature menopause. Estrogen replacement therapy (ERT), calcium supplements, and weight-bearing exercise are prescribed to prevent or retard the development of osteoporosis in postmenopausal women.

Rickets

Rickets is a vitamin D deficiency in children in which the body does not absorb calcium and phosphorus. As a result, when the child walks, the weight of the body causes the bones in the legs to bow.

Osteomalacia

Demineralization (loss of excessive calcium and phosphorus) caused by vitamin D deficiency is called *osteomalacia* (os′-tē-ō-ma-LĀ-shē-a; *malacia* = soft). With demineralization, the weight of the body produces a bowing of the leg bones, a shortening of the backbone, and a flattening of the pelvic bones.

Paget's Disease

Paget's disease is an irregular thickening and softening of the bones, especially among bones of the skull, pelvis, and extremities.

Osteomyelitis

The term *osteomyelitis* (os′-tē-ō-mī-i-LĪ-tis) refers to all infectious diseases of bone.

Fractures (Fxs)

A *fracture (Fx)* is any break in a bone. Several types of fractures include:

1. **Partial (incomplete).** The break across the bone is incomplete.
2. **Complete.** The break across the bone is complete.
3. **Closed (simple).** The bone does not break through the skin.
4. **Open (compound).** The broken ends of the bone protrude through the skin.
5. **Stress.** A partial fracture resulting from inability to withstand repeated stress due to a change in training, harder surfaces, longer distances, or greater speed.

6. **Pathologic.** A fracture due to weakening of a bone caused by disease processes such as osteomyelitis, osteoporosis, or osteomalacia.

The healing process for bone varies with several factors such as the extent of the fracture, age, nutrition, the general health of the individual, and the type of treatment received.

Herniated (Slipped) Disc

A *herniated (slipped) disc* is characterized by a protrusion of the nucleus pulposus of an intervertebral disc.

Abnormal Curves

Scoliosis (skō′-lē-Ō-sis; *scolio* = bent) is a lateral bending of the vertebral column, usually in the thoracic region. *Kyphosis* (kī-FŌ- sis; *kypho* = hunchback) is an exaggeration of the thoracic curve of the vertebral column. Mild kyphosis is termed ''round-shouldered.'' *Lordosis* (lor-DŌ-sis; *lordo* = sway-back) is an exaggeration of the lumbar curve of the vertebral column.

Spina Bifida

Spina bifida (SPĪ-na BIF-i-da) is a congenital defect of the vertebral column in which laminae fail to unite at the midline.

MEDICAL TERMINOLOGY AND CONDITIONS

Achondroplasia (a-kon′-drō-PLĀ-zē-a; *a* = without; *chondro* = cartilage; *plasia* = growth) Imperfect ossification within cartilage of long bones during fetal life; also called *fetal rickets*. It causes a form of dwarfism.

Craniotomy (krā-nē-OT-ō-mē; *cranium* = skull; *tome* = a cutting) Any surgery that requires cutting through the bones surrounding the brain.

Necrosis (ne-KRŌ-sis; *necros* = death; *osis* = condition) Death of tissues or organs; in the case of bone, death results from deprivation of blood supply.

Osteitis (os′-tē-Ī-tis; *osteo* = bone) Inflammation or infection of bone.

Osteoblastoma (os′-tē-ō-blas-TŌ-ma; *oma* = tumor) A benign tumor of the osteoblasts.

Osteochondroma (os′-tē-ō-kon-DRŌ-ma; *chondro* = cartilage) A benign tumor of the bone and cartilage.

Osteoma (os′-tē-Ō-ma) A benign bone tumor.

Osteosarcoma (os′-tē-ō-sar-KŌ-ma; *sarcoma* = connective tissue tumor) A malignant tumor composed of osseous tissue.

Pott's (POTS) *disease* Inflammation of the backbone, caused by the microorganism that produces tuberculosis.

STUDY OUTLINE

Functions (p. 96)

1. The skeletal system consists of all bones attached at joints and cartilage between joints.
2. The functions of the skeletal system include support, protection, movement, mineral storage, housing blood-forming tissue, and storage of energy.

Types of Bones (p. 96)

1. On the basis of shape, bones are classified as long, short, flat, or irregular.
2. Sutural bones are found between the sutures of certain cranial bones. Sesamoid bones develop in tendons or ligaments. The patellas are sesamoid bones.

Histology (p. 97)

1. Osseous tissue consists of widely separated cells surrounded by large amounts of intercellular substance (matrix). The four principal types of cells are osteoprogenitor cells, osteoblasts, osteocytes, and osteoclasts. The matrix contains collagenous fibers and mineral salts, consisting mainly of calcium phosphate.
2. Parts of a typical long bone are the diaphysis (shaft), epiphyses (ends), articular cartilage, periosteum, medullary (marrow) cavity, and endosteum.
3. Compact (dense) bone consists of osteons (Haversian systems) with little space between them. Compact bone lies over spongy bone and composes most of the bone tissue of the diaphysis. Functionally, compact bone protects, supports, and resists stress.
4. Spongy (cancellous) bone consists of trabeculae surrounding many red and yellow marrow-filled spaces. It forms most of the structure of short, flat, and irregular bones and the epiphyses of long bones. Functionally, spongy bone stores some red and yellow marrow and provides some support.

Ossification: Bone Formation (p. 98)

1. Bone forms by a process called ossification.
2. The two types of ossification, intramembranous and endochon-

dral, involve the replacement of a preexisting connective tissue with bone.

3. Intramembranous ossification occurs within fibrous membranes of the embryo and the adult.

4. Endochondral ossification occurs within a cartilage model. The primary ossification center of a long bone is in the diaphysis. Cartilage degenerates, leaving cavities that merge to form the medullary (marrow) cavity. Osteoblasts lay down bone. Next, ossification occurs in the epiphyses, where bone replaces cartilage, except for the epiphyseal plate.

5. Because of the activity of the epiphyseal plate, the diaphysis of a bone increases in length.

6. Bone grows in diameter as a result of the addition of new bone tissue around the outer surface of the bone.

Homeostasis (p. 101)

Bone Growth and Maintenance (p. 101)

1. The homeostasis of bone growth and development depends on a balance between bone formation and resorption.

2. Old bone is constantly destroyed by osteoclasts, while new bone is constructed by osteoblasts. This process is called remodeling.

3. Normal growth depends on minerals (calcium and phosphorus), vitamins (D, C, A, B_{12}), and hormones (human growth hormone, sex hormones).

4. Under mechanical stress, bone increases in mass.

5. With aging, bone loses calcium and organic matrix.

Mineral Storage (p. 101).

1. Bones store and release calcium and phosphate.

2. This is controlled by calcitonin (CT), parathyroid hormone (PTH), and vitamin D.

Surface Markings (p. 102)

1. Surface markings are structural features visible on the surfaces of bones.

2. Each marking has a specific function—joint formation, muscle attachment, or passage of nerves and blood vessels.

Divisions of the Skeletal System (p. 103)

1. The axial skeleton consists of bones arranged along the longitudinal axis. The parts of the axial skeleton are the skull, hyoid bone, auditory ossicles, vertebral column, sternum, and ribs.

2. The appendicular skeleton consists of the bones of the girdles and the upper and lower extremities. The parts of the appendicular skeleton are the pectoral (shoulder) girdles, bones of the upper extremities, pelvic (hip) girdle, and bones of the lower extremities.

Skull (p. 103)

1. The skull consists of the cranium and the face.

2. Sutures are immovable joints between bones of the skull. Examples are the coronal, sagittal, lambdoidal, and squamosal sutures.

3. The 8 cranial bones include the frontal, parietal (2), temporal (2), occipital, sphenoid, and ethmoid.

4. The 14 facial bones are the nasal (2), maxillae (2), zygomatic (2), mandible, lacrimal (2), palatine (2), inferior nasal conchae (2), and vomer.

5. Paranasal sinuses are cavities in bones of the skull that communicate with the nasal cavity. They are lined by mucous membranes. Cranial bones containing paranasal sinuses are the frontal, sphenoid, ethmoid, and maxillae.

6. Fontanels are membrane-filled spaces between the cranial bones of fetuses and infants. The major fontanels are the anterior, posterior, anterolaterals, and posterolaterals.

7. The foramina of the skull bones provide passages for nerves and blood vessels.

Hyoid Bone (p. 111)

1. The hyoid bone is a U-shaped bone that does not articulate with any other bone.

2. It supports the tongue and provides attachment for some of its muscles as well as some neck muscles.

Vertebral Column (p. 111)

1. The bones of the adult vertebral column are the cervical vertebrae (7), thoracic vertebrae (12), lumbar vertebrae (5), the sacrum (5, fused), and the coccyx (4, fused).

2. The vertebral column contains normal curves that give strength, support, and balance.

3. The vertebrae are similar in structure, each consisting of a body, vertebral arch, and seven processes. Vertebrae in the different regions of the column vary in size, shape, and detail.

Thorax (p. 114)

1. The thoracic skeleton consists of the sternum, ribs, costal cartilages, and thoracic vertebrae.

2. The thoracic cage protects vital organs in the chest area.

Pectoral (Shoulder) Girdle (p. 115)

1. Each pectoral (shoulder) girdle consists of a clavicle and scapula.

2. Each attaches an upper extremity to the trunk.

Upper Extremity (p. 116)

1. The bones of each upper extremity include the humerus, ulna, radius, carpals, metacarpals, and phalanges.

Pelvic (Hip) Girdle (p. 117)

1. The pelvic (hip) girdle consists of two hip bones.

2. It attaches the lower extremities to the trunk at the sacrum.

3. Each hip bone consists of three fused components— ilium, pubis, and ischium.

Lower Extremity (p. 117)

1. The bones of each lower extremity include the femur, tibia, fibula, tarsals, metatarsals, and phalanges.
2. The bones of the foot are arranged in two arches, the longitudinal arch and the transverse arch, to provide support and leverage.

Female and Male Skeletons (p. 123)

1. The female pelvis is adapted for pregnancy and childbirth. Differences in pelvic structure are listed in Exhibit 6-2.
2. Male bones are generally larger and heavier than female bones and have more prominent markings for muscle attachment.

REVIEW QUESTIONS

1. Define the skeletal system. What are its six principal functions? (p. 96)
2. What are the four principal types of bones? Give an example of each. Distinguish between a sutural and a sesamoid bone. (p. 96)
3. Diagram the parts of a long bone and list the functions of each part. (p. 98)
4. Distinguish between compact and spongy bone in terms of appearance, location, and function. (p. 98)
5. Outline the major events involved in intramembranous and endochondral ossification and explain the principal differences. (p. 99)
6. What is the significance of the epiphyseal line? (p. 101)
7. Define remodeling. How does the balance between osteoblast activity and osteoclast activity demonstrate the homeostasis of bone? (p. 101)
8. List the primary factors involved in bone growth and replacement. (p. 101)
9. Distinguish between the axial and appendicular skeletons. What subdivisions and bones are contained in each? (p. 103)
10. What are the bones that compose the skull? (p. 103)
11. Define a suture. What are the four prominent sutures of the skull? Where are they located? (p. 103)
12. What is a fontanel? Describe the location of the six common fontanels. (p. 110)
13. What is a paranasal sinus? What cranial bones contain paranasal sinuses? (p. 110)
14. Distinguish between the number of vertebrae found in the adult vertebral column and that of a child. (p. 111)
15. What are the normal curves in the vertebral column? What are the functions of the curves? (p. 111)
16. What are the principal distinguishing characteristics of the bones of the various regions of the vertebral column? (p. 113)
17. What bones form the skeleton of the thorax? What are the functions of the thoracic skeleton? (p. 114)
18. How are ribs classified on the basis of their attachment to the sternum? (p. 115)
19. What is the pectoral (shoulder) girdle? Why is it important? (p. 115)
20. What are the bones of the upper extremity? (p. 116)
21. What is the pelvic (hip) girdle? Why is it important? (p. 117)
22. What are the bones of the lower extremity? (p. 117)
23. What are the two arches of the foot? What is the function of an arch? (p. 122)
24. What are the principal structural differences between typical female and male skeletons? Use Exhibit 6-2 as a guide in formulating your response. (p. 126)
25. Define each of the following disorders: osteoporosis, rickets, osteomalacia, Paget's disease, osteomyelitis, fracture (Fx), herniated (slipped) disc, abnormal curves, and spina bifida. (p. 126)
26. Refer to the glossary of medical terminology and conditions associated with the skeletal system. Be sure that you can define each term. (p. 127)

7

*A*rticulations

1. Define an articulation (joint) and classify articulations by function and structure.
2. Explain the various movements that occur at articulations.
3. Define common disorders and medical terminology and conditions associated with articulations.

A LOOK AHEAD

CLASSIFICATION
  Functional
  Structural
SYNARTHROSES (IMMOVABLE)
  Suture
  Gomphosis
  Synchondrosis
AMPHIARTHROSES (SLIGHTLY
      MOVABLE)
  Syndesmosis
  Symphysis
DIARTHROSES (FREELY MOVABLE)
TYPES OF SYNOVIAL JOINTS
  Gliding
  Hinge
  Pivot
  Ellipsoidal
  Saddle
  Ball-and-Socket
SPECIAL MOVEMENTS AT
      SYNOVIAL JOINTS
COMMON DISORDERS
MEDICAL TERMINOLOGY
      AND CONDITIONS

Bones are too rigid to bend without damage. Fortunately, the skeletal system consists of many separate bones held together at joints by flexible connective tissue. All body movement occurs at joints. You can understand the importance of joints if you imagine how a knee cast or a finger splint affect movements, large and small, that you take for granted.

An *articulation (joint)* is a point of contact between bones, between cartilage and bones, or between teeth and bones. The scientific study of joints is referred to as *arthrology* (ar-THROL-ō-jē; *arthro* = joint; *logos* = study of).

The joint's structure determines how it functions. Some joints permit no movement, others permit slight movement, and still others afford fairly free movement. In general, the closer the fit at the point of contact, the stronger the joint. At tightly fitted joints, however, movement is restricted. The looser the fit, the greater the movement, but loosely fitted joints are prone to dislocation. Movement at joints is also determined by the structure (shape) of articulating bones, the flexibility (tension or tautness) of the connective tissue that binds the bones together, and the position of associated ligaments, muscles, and tendons. Joint flexibility may also be affected by hormones. For example, relaxin, a hormone produced by the placenta and ovaries, relaxes the symphysis pubis and ligaments between the sacrum, hip bone, and coccyx toward the end of pregnancy, which assists in delivery.

CLASSIFICATION

FUNCTIONAL

Functionally, joints are classified as *synarthroses* (sin′-ar-THRŌ-sēz), immovable joints; *amphiarthroses* (am′-fē-ar-THRŌ-sēz), slightly movable joints; and *diarthroses* (dī-ar-THRŌ-sēz), freely movable joints.

STRUCTURAL

The structural classification of joints is based on the presence or absence of a synovial (joint) cavity (a space between the articulating bones) and the kind of connective tissue that binds the bones together. Structurally, joints are classified as *fibrous*, in which there is no cavity and the bones are held together by fibrous connective tissue; *cartilaginous,* in which there is no cavity and the bones are held together by cartilage; and *synovial*, in which there is a cavity and the bones forming the joint are united by a surrounding articular capsule and frequently accessory ligaments (described in detail later). We will discuss joints in terms of their functions, but referring to their structural characteristics as well.

SYNARTHROSES (IMMOVABLE)

There are three kinds of synarthroses (immovable joints)—sutures, gomphoses, and synchondroses.

SUTURE

Sutures (SOO-cherz) are fibrous joints found between bones of the skull. In a suture, the bones are united by dense fibrous connective tissue. Their irregular structure gives them added strength and decreases the chance of fractures. An example is the coronal suture between the parietal and temporal bones (see Figure 6-6b).

GOMPHOSIS

A **gomphosis** (gom-FŌ-sis) is a type of fibrous joint in which a cone-shaped peg fits into a socket. Examples are the articulations of the roots of the teeth with the alveoli (sockets) of the maxillae and mandible.

SYNCHONDROSIS

A **synchondrosis** (sin'-kon-DRŌ-sis) is a hyaline cartilage joint which is eventually replaced by bone. Examples include the epiphyseal plate (see Figure 6-3) and the joint between the first rib and the sternum (see Figure 6-18).

AMPHIARTHROSES (SLIGHTLY MOVABLE)

Amphiarthroses (slightly movable joints) include syndesmoses and symphyses.

SYNDESMOSIS

In a **syndesmosis** (sin'-dez-MŌ-sis) joint, there is a much more dense fibrous connective tissue than in a suture, but the fit between bones is not quite as tight, which permits some flexibility. An example is the distal articulation of the tibia and fibula (see Figure 6-28).

SYMPHYSIS

A **symphysis** (SIM-fi-sis) is a cartilaginous joint in which the connecting material is a broad, flat disc of fibrocartilage. Examples include the intervertebral joints (see Figure 6-13) and the symphysis pubis (see Figure 6-24).

DIARTHROSES (FREELY MOVABLE)

Diarthroses, or freely movable joints, are synovial joints. At this point, we will discuss the general structure of a synovial joint and later consider the various types.

A joint in which there is a space between articulating bones is called a **synovial** (si-NŌ-vē-al) **joint**. The space is called the **synovial (joint) cavity** (Figure 7-1). Synovial joints are freely movable because of this cavity, its surrounding capsule, and ligaments.

The ends of the bones in a synovial joint are covered by a layer of hyaline cartilage called **articular cartilage**.

Synovial joints are surrounded by a sleevelike **articular capsule** that encloses the synovial cavity and unites the articulating bones. The outer layer of the articular capsule, the **fibrous capsule**, is composed of ligaments. The fibers in ligaments run in one direction, which gives them tensile strength to resist dislocation but allows flexibility. The inner layer of the capsule is the **synovial membrane**. It secretes **synovial fluid (SF)**, which has the consistency of uncooked egg white and lubricates and nourishes the articular cartilage.

Inside some synovial joints are pads of fibrocartilage called **articular discs (menisci)**. They lie between two bones with different shapes and help stabilize the joint by providing a tighter fit (Figure 7-1b). A tearing of articular discs in the knee, commonly called **torn cartilage**, occurs frequently among athletes.

The various movements of the body create friction between moving parts. To reduce this friction, saclike structures called **bursae** are situated in body tissues. They are filled with a fluid

FIGURE 7-1 Synovial joint. (a) Generalized synovial joint in frontal section. (b) Sagittal section of the knee joint showing articular discs (menisci) and bursae.

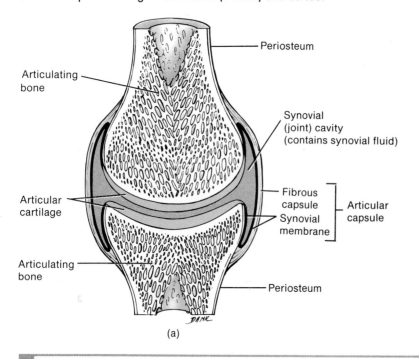

(a)

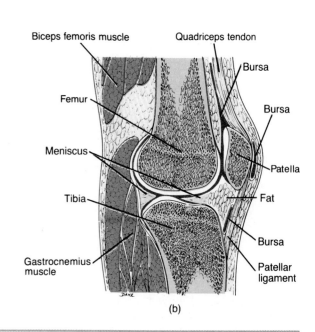

(b)

What is the functional classification of this joint?

similar to synovial fluid. Bursae are located between the skin and bone in places where skin rubs over bone, between tendons and bones, muscles and bones, and ligaments and bones. As fluid-filled sacs, they cushion the movement of one part of the body over another. An inflammation of a bursa is called *bursitis*.

TYPES OF SYNOVIAL JOINTS

Though all synovial joints are similar in structure, variations exist in the shape of the articulating surfaces. Accordingly, synovial joints are divided into six subtypes: gliding, hinge, pivot, ellipsoidal, saddle, and ball-and-socket joints.

GLIDING

The articulating surfaces of bones in *gliding joints* are usually flat. These joints permit a *gliding movement* in which one surface moves back-and-forth and from side-to-side over another surface without any angular or rotary motion (Figure 7-2a). Examples of gliding joints are those between the carpals, tarsals, sternum and clavicle, and scapula and clavicle.

FIGURE 7-2 **Simplified representations of types of synovial joints.**

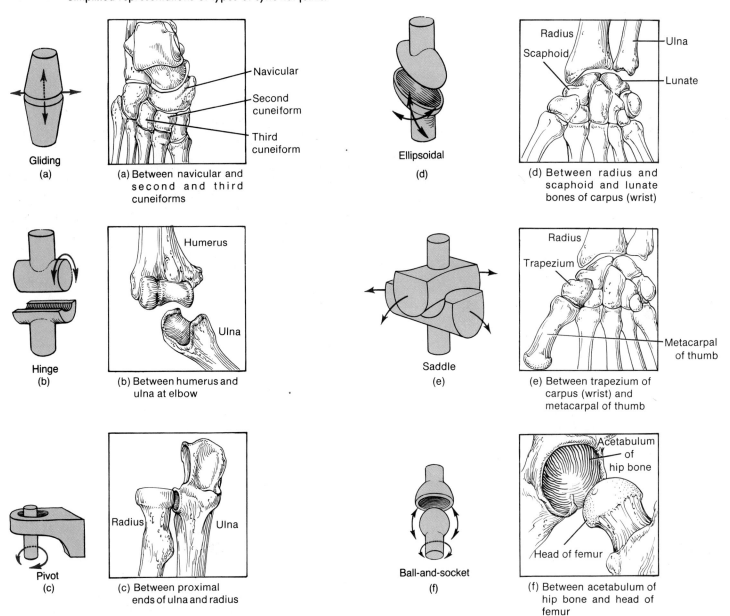

Gliding
(a)

(a) Between navicular and second and third cuneiforms

Hinge
(b)

(b) Between humerus and ulna at elbow

Pivot
(c)

(c) Between proximal ends of ulna and radius

Ellipsoidal
(d)

(d) Between radius and scaphoid and lunate bones of carpus (wrist)

Saddle
(e)

(e) Between trapezium of carpus (wrist) and metacarpal of thumb

Ball-and-socket
(f)

(f) Between acetabulum of hip bone and head of femur

Which joint usually permits flexion and extension only? Which joint is found between the atlas and axis? What kind of joint is formed by the trapezium bone of the wrist and the thumb? Which joint allows movement in three planes? Which joint is formed by the radius and carpals at the wrist? Which joint is found between carpals?

HINGE

In a **hinge joint**, the convex surface of one bone fits into the concave surface of another bone (Figure 7-2b). Movement is primarily in a single plane, similar to that of a hinged door. Movement is usually flexion and extension. **Flexion** decreases the angle between articulating bones and occurs when you bend your knee or elbow (Figure 7-3c,f). **Extension** increases the angle between articulating bones to restore a body part to its anatomical position after it has been flexed (Figure 7-3c,f). Other hinge joints are the ankle joints, interphalangeal joints, and joint between the occipital bone and atlas. Some hinge joints are capable of **hyperextension**, continuation of extension beyond the anatomical position, such as when the head bends backward (Figure 7-3a).

PIVOT

In a **pivot joint,** a rounded or pointed surface of one bone articulates within a ring formed partly by bone and partly by a ligament (Figure 7-2c). The primary movement is **rotation**, movement of a bone around its own axis. We rotate the atlas around the axis when we move the head from side-to-side (Figure 7-3k). Another pivot joint is found between the proximal ends of the ulna and radius and it allows us to turn the palms forward (or upward) and downward (or backward).

ELLIPSOIDAL

In an **ellipsoidal joint**, an oval-shaped condyle of one bone fits into an elliptical cavity of another bone (see Figure 7-2d). The

FIGURE 7-3 Representative movements at synovial joints. (Copyright © 1983 by Gerard J. Tortora. Courtesy of Lynne and James Borghesi.)

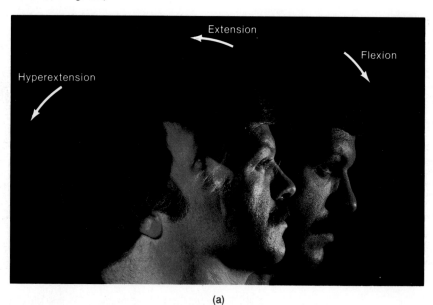

(a)

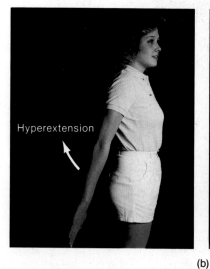
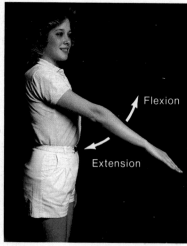
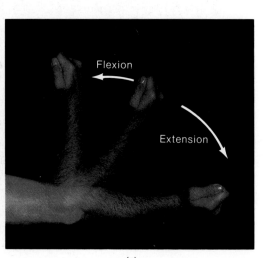

(b)   (c)

Which movement decreases the angle between bones? Moves a bone toward the midline? Moves a bone around its axis?

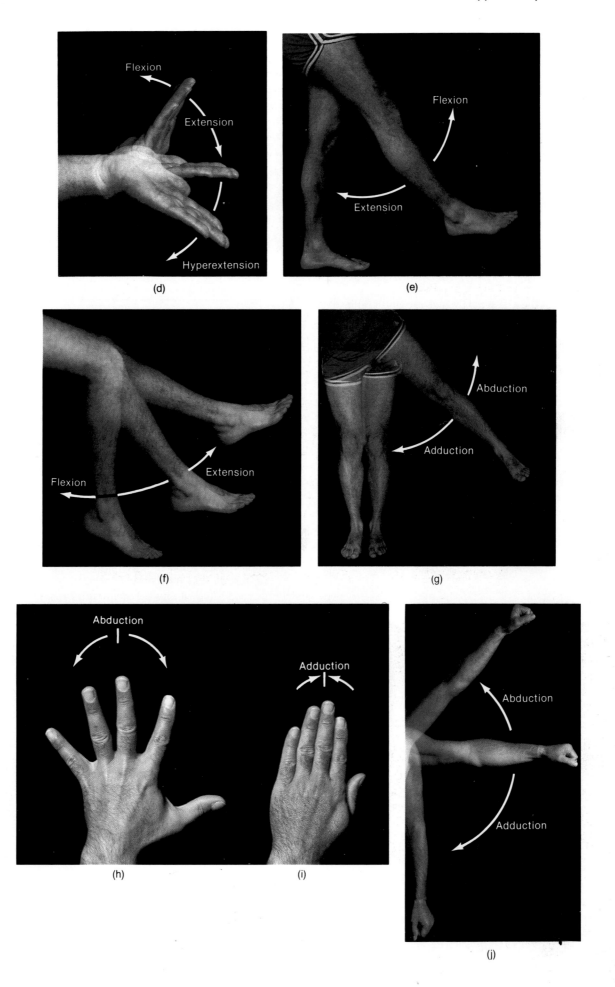

(d)

(e)

(f)

(g)

(h)

(i)

(j)

FIGURE 7-3 (*Continued*)

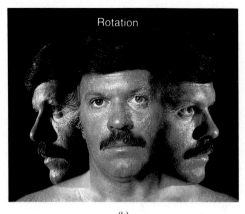

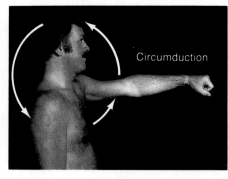

joint at the wrist between the radius and carpals is ellipsoidal. The movement permitted by such a joint is side-to-side and back-and-forth, as when you flex and extend, abduct and adduct, and circumduct the wrist. ***Abduction*** refers to movement away from the midline of the body (see Figure 7-3g–j). ***Adduction*** refers to movement toward the midline of the body (see Figure 7-3g–j). ***Circumduction*** is a movement in which the distal end of a bone moves in a circle, while the proximal end remains stable (see Figure 7-3l).

SADDLE

In a ***saddle joint***, the articular surface of one bone is saddle-shaped and the articular surface of the other is shaped like a rider sitting in a saddle. Movements at a saddle joint are side-to-side and back-and-forth (see Figure 7-2e). The joint between the trapezium of the carpus and metacarpal of the thumb is a saddle joint.

BALL-AND-SOCKET

A ***ball-and-socket joint*** consists of a ball-like surface of one bone fitted into a cuplike depression of another bone (see Figure 7-2f). Such a joint permits movement in three planes: flexion–extension, abduction–adduction, and rotation–circumduction (see Figure 7-2f). Examples are the shoulder joint and hip joint. The range of movement at a ball-and-socket joint is illustrated by circumduction of the arm (Figure 7-3l).

SPECIAL MOVEMENTS AT SYNOVIAL JOINTS

In addition to gliding movements, flexion, extension, hyperextension, rotation, abduction, adduction, and circumduction, several other movements also occur at synovial joints. These are called ***special movements*** and occur only at particular joints (Figure 7-4).

Inversion is moving the sole of the foot inward so the soles face toward each other. ***Eversion*** is moving the sole outward so the soles face away from each other. ***Dorsiflexion*** involves bending the foot up. ***Plantar flexion*** involves bending the foot down.

Protraction is moving the mandible or shoulder girdle forward. Thrusting the jaw outward is protraction of the mandible. Bringing your arms forward until the elbows touch requires protraction of the clavicle or shoulder girdle. ***Retraction*** is moving back the protracted part of the body. Pulling the lower jaw back in line with the upper jaw is retraction of the mandible.

Supination is a movement of the forearm in which the palm of the hand is turned forward or upward. ***Pronation*** is a movement of the forearm in which the palm is turned backward or downward.

Elevation is an upward movement of a body part. You elevate your mandible when you close your mouth. ***Depression*** is a downward movement of a body part. You depress your mandible when you open your mouth. The shoulders can also be elevated and depressed.

▪ COMMON DISORDERS ▪

Rheumatism

Rheumatism (*rheumat* = subject to flux) refers to any painful state of the supporting structures of the body—its bones, ligaments, joints, tendons, or muscles. Arthritis is a form of rheumatism in which the joints are inflamed.

Osteoarthritis

Osteoarthritis (os′-tē-ō-ar-THRĪ-tis) is a degenerative joint disease characterized by deterioration of articular cartilage and spur (new bone) formation.

Gouty Arthritis

In *gouty* (GOW-tē) *arthritis,* sodium urate crystals are deposited in the soft tissues such as the kidneys and cartilage of joints. The crystals eventually destroy all the joint tissues.

Rheumatoid Arthritis (RA)

Rheumatoid (ROO-ma-toyd) *arthritis* (*RA*) is an autoimmune disease in which an individual's antibodies attack the joint tissues. Although RA is a less common type of arthritis, it is a more severe and potentially crippling disease. The primary symptom is inflammation of the synovial membrane.

Lyme Disease

Lyme disease is caused by a bacterium (*Borrelia burgdorferi*) transmitted to humans by deer ticks and other ticks. Within a few weeks of the tick bite, a rash may appear at the site and may be accompanied by joint stiffness, fever and chills, headache, stiff neck, nausea, and low back pain.

Bursitis

An acute chronic inflammation of a bursa is called *bursitis.*

Dislocation

A *dislocation* or *luxation* (luks-Ā-shun) is the displacement of a bone from a joint with tearing of ligaments, tendons, and articular capsules. A partial or incomplete dislocation is called a *subluxation.*

Sprain and Strain

A *sprain* is the forcible wrenching or twisting of a joint with partial rupture or other injury to its attachments without luxation. A sprain is more serious than a *strain*, which is the overstretching of a muscle.

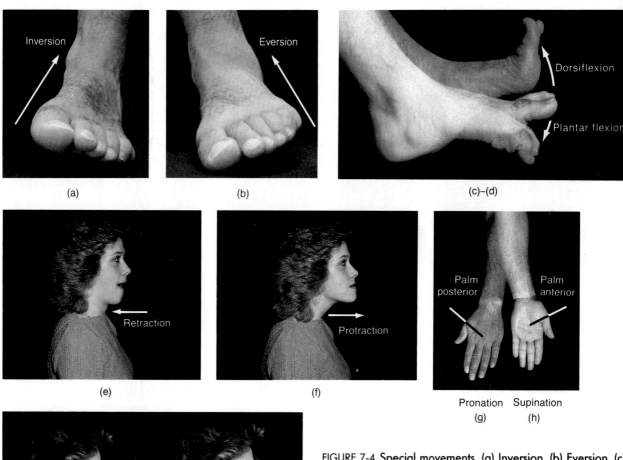

FIGURE 7-4 Special movements. (a) Inversion. (b) Eversion. (c) Dorsiflexion. (d) Plantar flexion. (e) Retraction. (f) Protraction. (g) Pronation. (h) Supination. (i) Elevation. (j) Depression. (Copyright © 1983 by Gerard J. Tortora. Courtesy of Lynne and James Borghesi.)

Why are these movements termed "special"?

Living with Arthritis

A wellness life-style is not just for people who are already healthy. It's also for people with chronic diseases and disabilities. Limitation is a fact of life, and something each of us learns to live with at some level. Limitation makes it even more important to live in a way that maximizes our potential for health and quality of life. The limitations imposed by arthritis provide an excellent illustration of this point.

There are over 100 different kinds of arthritis, and the disease affects one in seven people in the United States. It has been estimated that about 85 percent of people age 70 and over have some degree of osteoarthritis, the most common form of arthritis. Arthritis is not just a disease of the elderly; it also occurs in children and adults of all ages.

Except for infectious arthritis, which is caused by a specific disease agent, the causes of arthritis are not known. The symptoms vary in severity from day to day and even within a given day. People with arthritis often experience periods of remission. Since no cures are available, arthritis treatment means figuring out the best way to live with it. Education and self-care form the basis of arthritis treatment.

Self-Care Means Working with Health-Care Professionals

Self-care doesn't mean a person must "go it alone." On the contrary, self-care means knowing when to get professional help and working with health-care professionals to understand the nature of arthritis and treatment methods. The earlier the arthritis diagnosis the better, since although there is no cure, medical and life-style treatment can still significantly reduce or delay its progression. The goal of arthritis medical care is to relieve pain, reduce inflammation, and prevent deformity.

Inflammation is the body's normal way of responding to injury or disease agents. White blood cells invade the site of injury to clean up damaged cells and repair tissues. But in many forms of arthritis, the inflammation becomes part of the problem, and damage is caused in healthy joint tissues. Inflammation causes swelling, redness, pain, and loss of motion in the joint. A positive feedback cycle is created, in which inflammation leads to damage, which in turn accelerates inflammation and further damage.

Various medications are prescribed to reduce inflammation and relieve pain. Nonsteroidal antiinflammatory drugs (NSAIDs) such as aspirin and ibuprofen are the medications most frequently prescribed. Other drugs, such as corticosteroids and immunosuppressive drugs, are sometimes prescribed during severe flares (when the disease symptoms are most troublesome), but the serious side effects of these drugs must be weighed against the treatment benefits. Self-care means understanding and following medication recommendations and consulting the physician when questions arise.

Surgery may occasionally be indicated for severe arthritis and can decrease pain and improve joint function.

Pain-Relief Techniques

Physical therapists play an important role in arthritis treatment. They educate the arthritis patient about various pain-relief techniques and therapeutic exercise programs.

Pain-relief techniques include hot and/or cold treatments, joint protection, and rest. Hot treatments include hot baths and showers, hot packs, heat lamps, electric heating pads and mitts, and paraffin wax. Cold treatments include ice packs and compresses. Both can decrease pain and improve joint mobility. Joint protection includes education about joint-sparing body mechanics. For example, people with arthritis in the wrist learn to push open doors with the side of their body rather than the hand. Joint protection also includes appropriate use of orthopedic devices such as splints, walkers, and canes to reduce joint stress.

Rest is an important component of arthritis self-care. Prescribed rest may include complete bedrest, or periodic resting of affected joints, and/or stress-management relaxation techniques. Emotional rest includes participation in social groups and recreation. Arthritis can be a challenge to quality of life, and living with it requires a good attitude that can transform a potential invalid into an active family and community member.

Exercise

For many years, medical scientists believed that osteoarthritis and other forms of arthritis were the result of "wear and tear" on joints and that exercise accelerated joint degeneration. Studies have shown that exercise does not appear to cause arthritis in healthy joints. Repetitive, high-impact movements such as running can speed the progression of arthritis in already damaged joints, however. Unfortunately, because of the confusion that has surrounded the arthritis–exercise issue, many people with arthritis have avoided all but the very mildest forms of exercise. Many people with less severe arthritis have unnecessarily restricted aerobic activity for fear it would worsen the disease.

Here comes another positive feedback cycle. Sedentary life-styles lead to loss of muscle strength and low fitness levels, which make movement even more painful and difficult, leading to further restrictions in activity, and an even greater decline in fitness. When muscles and joints atrophy, the resulting weakness makes joints even more unstable. Recent studies have demonstrated that aerobic activities, especially those that support body weight, such as cycling and swimming, can be appropriate for people with osteoarthritis and rheumatoid arthritis who have fairly good joint mobility, can increase aerobic capacity, muscle strength, and functional status; and can improve pain tolerance, mood, and quality of life.

Physical therapists help educate arthritis patients about therapeutic exercises for the maintenance of joint function. These typically include range-of-motion (ROM) exercises to increase joint mobility and resistance exercises to increase the strength of muscles, tendons, ligaments, and other tissues that are part of the joint structure. Flexibility and strength help protect joints from stress.

Arthritis self-care means learning to balance exercise and rest. Too much exercise can lead to pain and inflammation, while too much rest can cause joint stiffness.

Exercise must often be done in very small amounts several times during the day if arthritis is severe. If pain persists for an hour or more after exercise, the patient has overdone it and must reduce activity to a lower level.

Nutrition and Weight Control

Being overweight is a risk factor for the development of arthritis, and extra weight increases the stress on arthritic joints. Weight control achieved through a well-balanced diet and a moderate exercise program can significantly slow the development of arthritis. Good nutrition is also important for maintaining general good health and well-being, which is as vital for the person with arthritis as it is for anyone else.

Living with Arthritis

The life-style modifications required by people with arthritis vary with the severity of their condition. Many people with arthritis continue to lead very active lives. Tennis player Billie Jean King is a good example. She has maintained a successful career as a professional player despite five bouts of knee surgery for osteoarthritis.

For people with more severe arthritis, coping with such simple tasks as preparing a meal or cleaning the house can provide real challenges. Some problems have simple solutions, like moving dishes to lower cupboards, using an electric can opener, or eating out more often. Relocating items and rearranging rooms can improve the comfort and safety of the home. People confined to a wheelchair will need to make more elaborate changes in their home environments. Self-care means managing one's resources in order to maximize quality of life.

MEDICAL TERMINOLOGY AND CONDITIONS

Ankylosis (ang'-ki-LŌ-sus; *ankyle* = stiff joint; *osis* = condition) Severe or complete loss of a movement at a joint.

Arthralgia (ar-THRAL-jē-a; *arth* = joint; *algia* = pain) Pain in a joint.

Arthrosis (ar-THRŌ-sis) Refers to an articulation; also a disease of a joint.

Bursectomy (bur-SEK-tō-mē; *ectomy* = removal of) Removal of a bursa.

Chondritis (kon-DRĪ-tis; *chondro* = cartilage) Inflammation of cartilage.

Rheumatology (roo'-ma-TOL-ō-jē; *rheumat* = subject to flux) The study of joints; the field of medicine devoted to joint diseases and related conditions.

Synovitis (sin'-ō-VĪ-tis; *synov* = joint) Inflammation of a synovial membrane in a joint.

STUDY OUTLINE

Classification (p. 131)

1. An articulation (joint) is a point of contact between two or more bones.
2. Functional classification of joints is based on the degree of movement permitted. Joints may be synarthroses, amphiarthroses, or diarthroses.
3. Structural classification is based on the presence of a synovial (joint) cavity and type of connecting tissue. Structurally, joints are classified as fibrous, cartilaginous, or synovial.

Synarthroses (Immovable Joints) (p. 131)

1. Synarthroses are immovable joints.
2. These joints include sutures (found in the skull), gomphoses (roots of teeth in alveoli of mandible and maxilla), and synchondroses (temporary cartilage between diaphysis and epiphyses).

Amphiarthroses (Slightly Movable Joints) (p. 132)

1. Amphiarthroses are slightly movable joints.
2. These joints include syndesmoses (such as the tibiofibular articulation) and symphyses (the symphysis pubis).

Diarthroses (Freely Movable Joints) (p. 133)

1. Diarthroses are freely movable joints (synovial).
2. Synovial joints contain a synovial (joint) cavity, articular cartilage, and a synovial membrane; some also contain ligaments, articular discs, and bursae.

Types of Synovial Joints (p. 133)

1. Synovial joints can be classified according to the shape of the articulating surfaces.
2. Types of synovial joints include gliding joints (wrist bones), hinge joints (elbow), pivot joints (between radius and ulna), ellipsoidal joints (between radius and wrist), saddle joints (between wrist and thumb), and ball-and-socket joints (shoulder).

Special Movements at Synovial Joints (p. 137)

1. In addition to gliding movements, flexion, extension, hyperextension, rotation, abduction, adduction, and circumduction, special movements also occur at synovial joints.
2. Special movements include inversion, eversion, dorsiflexion, plantar flexion, protraction, retraction, supination, pronation, elevation, and depression.

REVIEW QUESTIONS

1. Distinguish among the three kinds of joints on the basis of function and structure. (p. 131)
2. Describe the structure of the following synarthroses: sutures, gomphoses, and synchondroses. (p. 131)
3. Compare the structure of syndesmoses and symphyses, both amphiarthroses. (p. 132)
4. Explain the components of a synovial joint. Indicate the relationship of ligaments and tendons to the strength of the joint and restrictions on movement. (p. 132)
5. Explain how the articulating bones in a synovial joint are held together. (p. 132)
6. What is an articular disc? Why is it important? (p. 132)
7. What are bursae? What is their function? (p. 132)
8. Define each of the following types of synovial joints: gliding, hinge, pivot, ellipsoidal, saddle, and ball-and-socket. (p. 132)
9. Define the following principal movements: gliding, flexion, extension, hyperextension, rotation, abduction, adduction, and circumduction. (p. 133)
10. Define a special movement and give several examples. (p. 136)
11. Have another person assume the anatomical position and execute each of the movements at joints discussed in the text. Reverse roles, and see if you can execute the same movements.
12. Define each of the following disorders: rheumatism, osteoarthritis, gouty arthritis, rheumatoid arthritis (RA), Lyme disease, bursitis, dislocation, sprain, and strain. (p. 136)
13. Refer to the glossary of medical terminology and conditions associated with articulations. Be sure you can define each term. (p. 140)

8

The Muscular System

STUDENT OBJECTIVES

1. Explain the characteristics and functions of muscle tissue.
2. Compare the three kinds of muscle tissue in terms of location, structure, and function.
3. Describe the importance of the blood and nerve supply to muscles.
4. Explain how muscle fibers (cells) contract.
5. Explain the ways in which muscle tissue exhibits homeostasis.
6. Describe the different types of muscle contractions.
7. Describe how bones and muscles cooperate to bring about body movements.
8. Describe how most body movements are the result of several muscles acting together.
9. List and describe several ways that skeletal muscles get their names.
10. For various regions of the body, describe the attachment of skeletal muscles and identify their functions.
11. Define common disorders and medical terminology and conditions associated with the muscular system.

A LOOK AHEAD

CHARACTERISTICS
FUNCTIONS
TYPES
SKELETAL MUSCLE TISSUE
   Connective Tissue Components
   Nerve and Blood Supply
   Histology
CONTRACTION
   Sliding-Filament Mechanism
   Neuromuscular Junction
   Physiology of Contraction
   Energy for Contraction
   Homeostasis
      Oxygen Debt
      Muscle Fatigue
      Heat Production
   Motor Unit
   All-or-None Principle
   Kinds of Contractions
      Twitch
      Tetanus
      Treppe
      Isotonic and Isometric
   Muscle Tone
   Muscular Atrophy and
      Hypertrophy
HOW SKELETAL MUSCLES
      PRODUCE MOVEMENT
   Origin and Insertion
   Lever Systems
   Group Actions
NAMING SKELETAL MUSCLES
PRINCIPAL SKELETAL MUSCLES
CARDIAC MUSCLE TISSUE
SMOOTH MUSCLE TISSUE
COMMON DISORDERS
MEDICAL TERMINOLOGY
      AND CONDITIONS

B ones and joints form the framework of the body, but they cannot move the body by themselves. Motion results from the contraction and relaxation of muscles. Muscle tissue constitutes about 40 to 50 percent of the total body weight and is composed of highly specialized cells. The scientific study of muscles is known as *myology* (mī-OL-ō-jē; *myo* = muscle; *logos* = study of).

CHARACTERISTICS

Muscle tissue has four principal characteristics that are important in understanding its functions:

1. *Excitability* is the ability of muscle tissue to receive and respond to stimuli.
2. *Contractility* is the ability to shorten and thicken, or contract.
3. *Extensibility* is the ability of muscle tissue to stretch or extend.
4. *Elasticity* is the ability of muscle tissue to return to its original shape after contraction or extension.

FUNCTIONS

Through contraction, muscle performs three important functions:

1. Motion.
2. Maintenance of posture.
3. Heat production.

Motion is obvious in movements involving the whole body, such as walking and running, and in localized movements, such as grasping a pencil or nodding the head. Less noticeable kinds of motion produced by muscles are the beating of the heart, churning of food in the stomach, and contraction of the urinary bladder to expel urine.

Muscle tissue also enables the body to maintain posture. The contraction of skeletal muscles holds the body in stationary positions, such as standing and sitting.

The third function of muscle tissue is heat production. Skeletal muscle contractions produce heat and thereby help maintain normal body temperature. As much as 85 percent of all body heat is generated by muscle contractions.

TYPES

There are three types of muscle tissue.

Skeletal muscle tissue, which is named for its location, is attached to bones and moves the skeleton. It is *striated*; that is, striations, or alternating light and dark bands, are visible under a microscope. It is *voluntary* because it can be made to contract and relax by conscious control.

Cardiac muscle tissue forms the bulk of the wall of the heart. It is *striated* and *involuntary*; that is, its contractions are not under conscious control.

Smooth muscle tissue is involved with internal processes. It is located in the walls of hollow internal structures, such as blood vessels, the stomach, and the intestines. It is *nonstriated*, because it lacks striations, and *involuntary*.

SKELETAL MUSCLE TISSUE

To understand how muscles move, you need some knowledge of their connective tissue coverings, nerve and blood supply, and the structure of individual muscle fibers (cells).

CONNECTIVE TISSUE COMPONENTS

The term *fascia* (FASH-ē-a) is applied to a sheet or broad band of fibrous connective tissue beneath the skin or around muscles and other organs of the body. There are two types of fascia—*superficial fascia*, immediately under the skin, and more important to the study of muscles, *deep fascia*, which holds muscles together, separating them into functioning groups. Functionally, deep fascia allows free movement of muscles, carries nerves and blood and lymphatic vessels, fills spaces between muscles, and sometimes provides the origin (one point of attachment to a bone) for muscles.

Skeletal muscles are further protected, strengthened, and attached to other structures by several other connective tissue coverings (Figure 8-1). The entire muscle is wrapped in fibrous connective tissue called the *epimysium* (ep´-i-MĪZ-ē-um), which is deep fascia. Bundles of fibers (cells) called *fasciculi* (fa-SIK-yoo-lī) or *fascicles* (FAS-i-kuls) are covered by a fibrous connective tissue called the *perimysium* (per´-i-MĪZ-ē-um). Finally, a fibrous connective tissue called the *endomysium* (en´-dō-MĪZ-ē-um) wraps each individual muscle fiber (cell).

Epimysium, perimysium, and endomysium extend beyond the muscle as a *tendon* (*tendere* = to stretch out)—a cord of connective tissue that attaches a muscle to a bone. When a tendon extends as a broad, flat layer, it is called an *aponeurosis*. An example of an aponeurosis is the galea aponeurotica (see Figure 8-9). Certain tendons, especially those of the wrist and ankle, are enclosed in tubes of fibrous connective tissue called *tendon sheaths*, which permit tendons to slide easily and prevent them from slipping out of place.

NERVE AND BLOOD SUPPLY

Skeletal muscles are well supplied with nerves and blood vessels, both of which are directly related to contraction, the chief characteristic of muscle. For a skeletal muscle fiber (cell) to contract, it must first be stimulated by an electric current called an action potential. An action potential that is associated with a nerve cell is called a *nerve action potential* or *nerve impulse*. One associated with a muscle cell is called a *muscle action potential*. Muscle contraction also requires a good deal of energy and therefore large amounts of nutrients and oxygen. Moreover, the waste products of these energy-producing reactions must be eliminated. Thus,

FIGURE 8-1 Relationships of connective tissue to skeletal muscle depicted by a three-dimensional drawing of a muscle indicating the relative positions of the epimysium, perimysium, and endomysium.

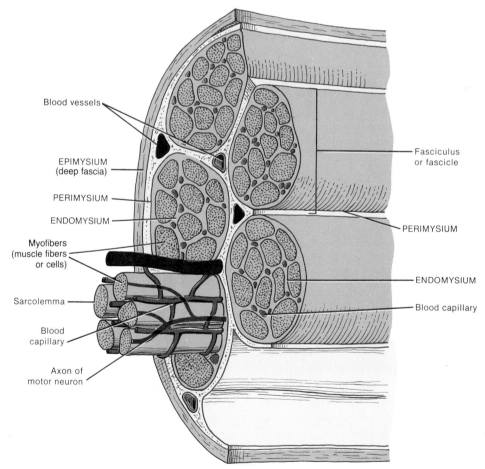

Blood vessels

EPIMYSIUM (deep fascia)

PERIMYSIUM

ENDOMYSIUM

Myofibers (muscle fibers or cells)

Sarcolemma

Blood capillary

Axon of motor neuron

Fasciculus or fascicle

PERIMYSIUM

ENDOMYSIUM

Blood capillary

What are the functions of deep fascia?

prolonged muscle action depends on a rich blood supply to deliver nutrients and oxygen and remove wastes.

Generally, an artery and one or two veins accompany each nerve that penetrates a skeletal muscle. Microscopic blood vessels called capillaries are distributed within the endomysium. Each muscle fiber is thus in close contact with one or more capillaries. Each skeletal muscle fiber also makes contact with a portion of a nerve called a synaptic end bulb.

HISTOLOGY

If you tease apart a skeletal muscle and view it under a microscope, you will see it consists of thousands of elongated, cylindrical cells called *muscle fibers* or *myofibers* (see Exhibit 4-3). Each fiber is covered by a plasma membrane called the *sarcolemma* (*sarco* = flesh; *lemma* = sheath). The cytoplasm, called *sarcoplasm*, of a muscle fiber contains many nuclei (multinucleate). In the sarcoplasm are numerous mitochondria. The large numbers are related to the large amounts of energy (ATP) that muscle tissue must generate in order to contract. The sarcoplasm also contains myofibrils (described shortly), special high-energy molecules (also to be described shortly), enzymes, and *sarcoplasmic reticulum* (sar′-kō-PLAZ-mik re-TIK-yoo-lum), a network of membrane-enclosed tubules comparable to smooth endoplasmic reticulum (Figure 8-2a). Perpendicular to the sarcoplasmic reticulum are *transverse tubules* (*T tubules*). The tubules are tunnellike extensions of the sarcolemma that pass through the muscle fiber at right angles to the sarcoplasmic reticulum and also open to the exterior of the muscle fiber. Skeletal muscle fibers also contain *myoglobin*, a reddish pigment similar to hemoglobin in blood. Myoglobin stores oxygen until needed by mitochondria to generate ATP.

Each skeletal muscle fiber is composed of cylindrical structures called *myofibrils* which run longitudinally through the muscle fiber. They, in turn, consist of two kinds of even smaller structures called *thin myofilaments* and *thick myofilaments*.

The myofilaments do not extend the entire length of a muscle fiber: they are arranged in compartments called *sarcomeres* (Figure 8-2b). Sarcomeres are separated from one another by narrow zones of dense material called *Z lines*. Within a sarcomere is a dark area, called the *A band*, composed of the thick myofilaments. The ends of the A band are darker because of overlapping thick and thin myofilaments. The length of darkening depends on the extent of overlapping. A light-colored area called the *I band* is composed of thin myofilaments. This combination of alternating dark A bands and light I bands gives the muscle fiber its striated (striped) appearance.

Thin myofilaments are anchored to the Z lines and project in both directions. They are composed mostly of the protein *actin* arranged in two single strands that entwine like a rope (Figure 8-3a). Each actin molecule contains a *myosin-binding site* for a myosin molecule (described shortly). Besides actin, the thin myofilaments contain two other protein molecules, *tropomyosin* and *troponin*, that help regulate muscle contraction.

Thick myofilaments are composed mostly of the protein *myosin*, which is shaped like a golf club. The tails (handles of the golf club) are arranged parallel to each other, forming the shaft of the thick myofilament. The heads of the golf clubs project outward on the surface of the shaft. These projecting heads are referred to as *cross bridges* and contain an *actin-binding site* and an *ATP-binding site* (Figure 8-3b).

In recent years, attention has been focused on the use of *muscle-building anabolic steroids* by amateur athletes. These steroids, a derivative of the hormone testosterone, are used to build muscle proteins and therefore increase strength and endurance. However, anabolic steroids have a number of side effects, including liver cancer, kidney damage, increased risk of heart disease, irritability and aggressive behavior, psychotic symptoms (including hallucinations), and mood swings. In females, additional side effects include sterility, the development of facial hair, deepening of the voice, atrophy of the breasts and uterus, enlargement of the clitoris, and irregularities of menstruation. In males, additional side effects include testicular atrophy, baldness, excessive development of breast glands, and diminished hormone secretion and sperm production by the testes. Steroids may also be addictive.

CONTRACTION

SLIDING-FILAMENT MECHANISM

During muscle contraction, thin myofilaments slide inward toward the center of a sarcomere (see Figure 8-5d). The sarcomere shortens, but the lengths of the thin and thick myofilaments themselves do not change. The myosin cross bridges of the thick myofilaments connect with portions of actin of the thin myofilaments. The myosin cross bridges move like the oars of a boat on the surface of the thin myofilaments, and the thin and thick myofilaments slide past each other. The myosin cross bridges may pull the thin myofilaments of each sarcomere so far inward toward the center of a sarcomere that their ends overlap. As the thin myofilaments slide inward, the Z lines are drawn toward each other and the sarcomere is shortened. The sliding of myofilaments and shortening of sarcomeres cause the shortening of the muscle fibers. This is the *sliding-filament mechanism* of muscle contraction. This process occurs only when there are sufficient calcium ions (Ca^{2+}) and an adequate supply of energy.

NEUROMUSCULAR JUNCTION

For a skeletal muscle fiber to contract, it must be stimulated by a nerve cell, or *neuron*. The particular type of neuron that stimulates muscle tissue is called a *motor neuron*.

When the axon of a motor neuron enters a skeletal muscle, it branches into axon terminals that approach—but do not touch—the sarcolemma of a muscle fiber. This region of the sarcolemma near the axon terminal is known as the *motor end plate*. The term *neuromuscular junction* refers to the axon terminal of a motor neuron together with the motor end plate (Figure 8-4). The ends of the axon terminals are expanded into bulblike structures called *synaptic end bulbs* (see Figure 9-7). These bulbs contain sacs called *synaptic vesicles* filled with chemicals called *neurotransmitters*. The space between the axon terminal and sarcolemma is known as a *synaptic cleft*.

When a nerve impulse (nerve action potential) reaches an axon terminal, Ca^{2+} ions from interstitial fluid enter the synaptic end

FIGURE 8-2 Histology of skeletal muscle tissue. (a) Enlarged aspect of several myofibrils of a muscle fiber (cell) based on an electron micrograph. (b) Enlarged aspect of a sarcomere showing thin and thick myofilaments.

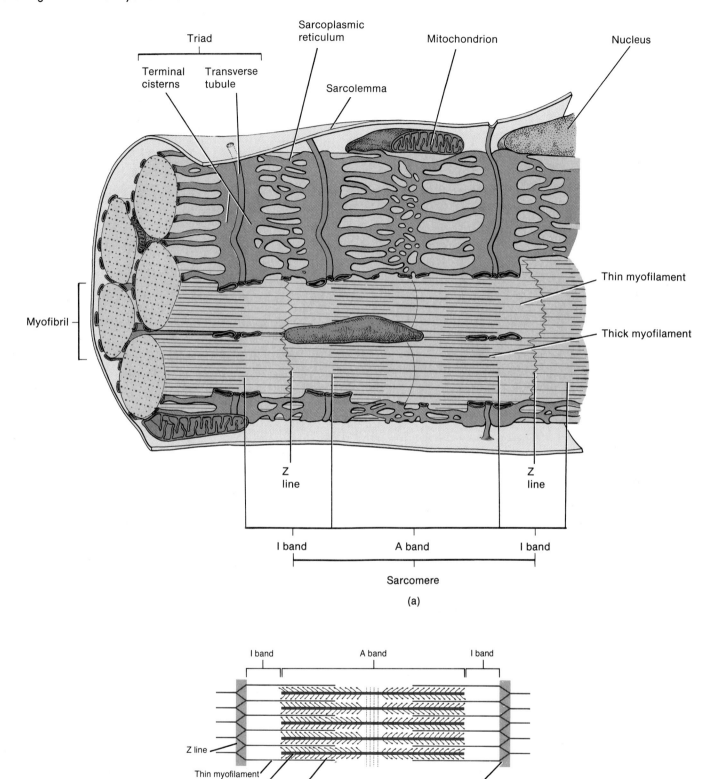

(a)

(b)

Of what does the A band consist? Of what does the I band consist? Where are calcium (Ca^{2+}) ions stored?

FIGURE 8-3 Detailed structure of portions of myofilaments. (a) Thin myofilament. (b) Thick myofilament.

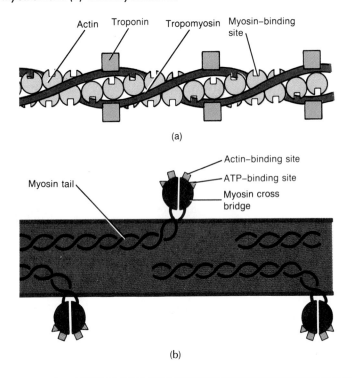

(a)

(b)

FIGURE 8-4 Neuromuscular junction. (a) Diagram based on a photomicrograph. (b) Enlarged aspect based on an electron micrograph.

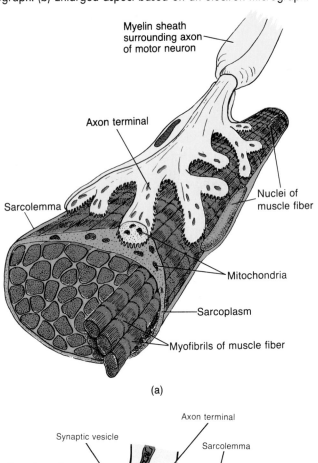

(a)

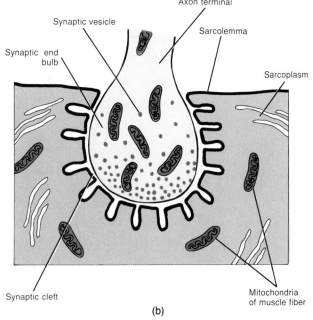

(b)

Thin myofilaments are composed of two strands of molecules of which substance? Thick myofilaments are composed of molecules of which substance?

---

bulb, causing synaptic vesicles to release a neurotransmitter (Figure 8-4c). The neurotransmitter released at neuromuscular junctions is *acetylcholine* (as'-ē-til-KŌ-lēn), or *ACh*. The ACh diffuses across the synaptic cleft to combine with receptors on the sarcolemma. This combination alters the permeability of the sarcolemma to sodium and potassium ions and ultimately produces a muscle action potential that travels along the sarcolemma and results in contraction. The details of nerve impulse generation and muscle action potential are discussed in Chapter 9.

As long as ACh is present in the neuromuscular junction, it will stimulate the muscle fiber. Continuous stimulation by ACh is prevented by an enzyme called *acetylcholinesterase (AChE)*, or simply *cholinesterase* (kō'-lin-ES-ter-ās). AChE is found on the sarcolemma surface. AChE inactivates ACh within ¹/₅₀₀ second by breaking it down into its components, acetate and choline. This permits the muscle fiber to prepare itself so another muscle action potential may be generated. When the next nerve impulse comes through, the synaptic vesicles release more ACh, another muscle action potential is generated, and AChE again inactivates ACh. This cycle is repeated over and over again. Eventually, choline reenters the synaptic end bulb, where it combines with acetate produced in the end bulb to form more ACh that is distributed into synaptic vesicles.

PHYSIOLOGY OF CONTRACTION

Let us look more closely at the neuromuscular junction and how the molecular events there cause a muscle fiber to contract. Ca^{2+}

What is the motor end plate?

---

ions and energy, in the form of ATP, are required for a muscle to contract. When a muscle fiber is relaxed (not contracting), there is a low concentration of Ca^{2+} ions in the sarcoplasm. Also, in a relaxed muscle, the concentration of ATP is high and attached to the ATP-binding sites on myosin cross bridges (Figure 8-5a). The myosin cross bridges cannot combine with actin of the thin myofila-

FIGURE 8-4 (*Continued*) (c) Release, action, breakdown, and synthesis of acetylcholine (ACh). (1) A nerve action potential causes (2) calcium ions (Ca^{2+}) to (3) enter a synaptic end bulb and (4) induce the release of ACh from a synaptic vesicle. (5) ACh diffuses across the synaptic cleft and two molecules of ACh bind to ACh receptors on the sarcolemma. In the absence of ACh, the channel remains closed. When two molecules of ACh bind to the receptor, (6) a change occurs, permitting positive ions, such as Na^+ ions, to pass through. (7) This produces a muscle action potential, causing a muscle fiber to contract. Once it has accomplished its function, ACh is (8) broken down by AChE into acetate and choline to prevent continuous generation of action potentials. Within the synaptic end bulb, (9) acetate and choline combine to form (10) ACh.

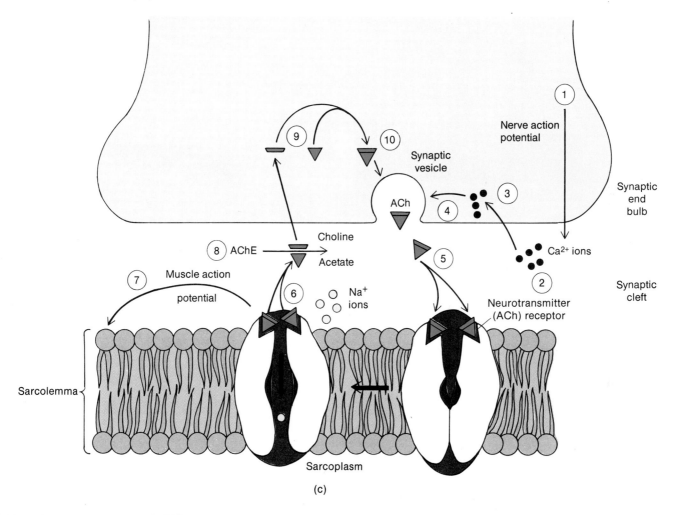

(c)

What is ACh and how does it function in a contraction? What is AChE and its function?

ments as long as tropomyosin and troponin are attached to actin and the ATP is attached to the myosin cross bridges.

What causes the concentration of Ca^{2+} ions to increase enough for contraction to occur? To understand, we need to back up to when ACh diffuses across the synaptic cleft.

When ACh diffuses across the synaptic cleft and combines with receptors of the sarcolemma, it causes a muscle action potential to develop and spread over the surface of the sarcolemma, into the transverse tubules, and to the sarcoplasmic reticulum. The reticulum then releases Ca^{2+} ions into the sarcoplasm.

The Ca^{2+} ions cause the troponin on the thin myofilaments to change structurally in such a way that the troponin and the attached tropomyosin move into a groove between actin strands, thus exposing the myosin-binding sites on actin (Figure 8-5b).

The energy for muscle contraction is supplied by ATP. ATP is found on myosin cross bridges. When a muscle action potential stimulates a muscle fiber, the myosin cross bridges split ATP into ADP + P, releasing energy. The myosin cross bridges, activated (energized) by the energy from the splitting of ATP, combine with myosin-binding sites on actin (Figure 8-5c). As a result, the ADP + P is released, and the myosin cross bridge moves toward the center of the sarcomere. This movement of the myosin cross bridges, called the ***power stroke***, is the force that causes the thin actin myofilaments to slide past the thick myosin myofilaments.

FIGURE 8-5 Mechanism of muscle contraction. (a) Relaxed state of a muscle fiber in which tropomyosin and troponin cover the myosin-binding site on actin and the ATP-binding site of the myosin cross bridge is occupied. (b) Calcium ions combine with troponin, and troponin and the attached tropomyosin move, exposing the myosin-binding site. (c) A myosin cross bridge splits ATP into ADP + P. The energy from the splitting of ATP activates the myosin cross bridge and it combines with the myosin-binding site on actin. As a result, the myosin cross bridge moves toward the center of the sarcomere (power stroke), which slides the actin myofilament past the myosin myofilament.

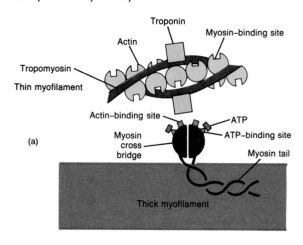

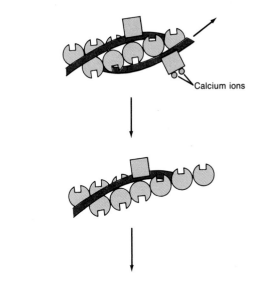

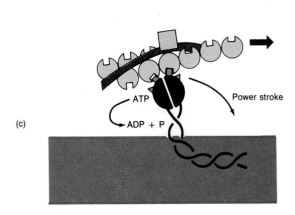

What is the immediate source of energy for contraction? What substance is used to generate ATP if contraction continues for more than a few seconds?

Once the power stroke is complete, the myosin cross bridge detaches from actin. A given myosin cross bridge then combines with another myosin-binding site further along the actin strand. Again, ATP is split and the cycle repeats itself. The myosin cross bridges keep moving back and forth like the cogs of a ratchet with each power stroke, moving the thin actin myofilaments toward the center of a sarcomere. The Z lines are drawn toward each other, the sarcomere shortens, the muscle fibers contract, and thus, the muscle itself contracts (Figure 8-5d).

What happens when a muscle fiber goes from a contracted state back to a relaxed state? The process simply reverses itself from the point at which AChE destroys the ACh that initiated the events leading to contraction. Once ACh is inactivated, no more muscle action potentials are generated. Ca^{2+} ions are pumped back into the sarcoplasmic reticulum for storage, myosin cross bridges detach from actin, ADP is resynthesized to ATP, and the thin myofilaments slide back to their relaxed positions. Thus, the muscle resumes its original length.

Following death, muscles become rigid, unable to contract or stretch, a condition called *rigor mortis*. Without ATP, myosin cross bridges remain attached to the actin myofilaments, thus preventing relaxation. The time elapsing between death and the onset of rigor mortis varies greatly among individuals. Those who have had long, wasting illnesses undergo rigor mortis more quickly.

ENERGY FOR CONTRACTION

Skeletal muscle fibers, unlike other body cells, alternate between virtual inactivity and continuous activity. Although ATP is the immediate source of energy for muscular contraction, muscle fibers usually contain only enough ATP to sustain activity for about 5 to 6 seconds. Skeletal muscle fibers contain the enzyme ATPase and a high-energy molecule called *phosphocreatine* (fos'-fō-KRĒ-a-tin), both of which can be used to quickly produce more ATP during prolonged exercise.

In the presence of ATPase, ATP is broken down into ADP + P and energy is released as follows:

$$ATP \rightarrow ADP + P + Energy$$

Phosphocreatine breaks down into creatine and phosphate, and in

FIGURE 8-5 (*Continued*) (d) Sliding-filament mechanism of muscle contraction. Shown are the positions of the various parts of two sarcomeres in relaxed, contracting, and maximally contracted states. Note the movement of the thin myofilaments and the relative size of the sarcomeres.

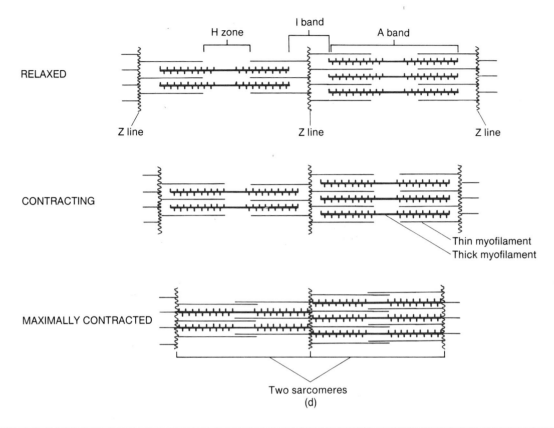

(d)

In the sliding-filament mechanism of contraction, which filaments slide?

the process, large amounts of energy are released as follows:

$$Phosphocreatine \rightarrow Creatine + Phosphate + Energy$$

The released energy is used to convert ADP to ATP and the ATP is then used as a source of energy for contraction. This mechanism provides enough energy for muscles to contract maximally for about 15 seconds, as in a 100-meter dash. In other words, it is good for maximal short bursts of activity.

When muscle activity continues so that even the supply of phosphocreatine is depleted, glucose must be broken down. Skeletal muscles store glucose in the form of glycogen. During exercise, glycogen is converted back to glucose. Once this occurs, glucose is split into two molecules of pyruvic acid in the sarcoplasm, a process called **glycolysis**. In the process, energy is released and used to form ATP. Glycolysis does not require oxygen, so it is an **anaerobic process**. For this reason, glycolysis is referred to as **anaerobic respiration**. It may be summarized as follows:

$$1\ Glucose \rightarrow 2\ Pyruvic\ Acid + Energy\ (ATP)$$

Normally, muscle fibers contain oxygen (O_2). This allows the pyruvic acid formed by glycolysis to enter mitochondria, where it is completely catabolized to carbon dioxide and water. Since this breakdown requires O_2, it is referred to as **aerobic respiration**

or **cellular respiration**. The process is assisted by the presence of myoglobin, which helps store O_2 for use during vigorous exercise. The complete catabolism of pyruvic acid also yields energy that is used to generate most of a muscle fiber's ATP:

$$Pyruvic\ acid + O_2 \rightarrow CO_2 + H_2O + Energy\ (ATP)$$

If there is insufficient oxygen for the complete catabolism of pyruvic acid, most of the pyruvic acid is converted to lactic acid, some of which diffuses out of the muscle fibers and into blood. The production of lactic acid also releases energy from glucose that can be used to produce ATP, and it occurs anaerobically. Anaerobic respiration (glycolysis) and aerobic respiration (cellular respiration), together, allow for prolonged muscular activity, such as jogging, and will continue as long as nutrients and adequate oxygen last. Anaerobic respiration can provide sufficient energy for 30 to 40 seconds of maximal muscle activity, such as a 400-meter dash.

HOMEOSTASIS

Muscle tissue plays a vital role in maintaining the body's homeostasis. Three examples are the relationship of muscle tissue to oxygen, to fatigue, and to heat production.

Oxygen Debt

During exercise, blood vessels in muscles dilate and blood flow is increased in order to increase the available oxygen supply. But when muscular exertion is very great, oxygen cannot be supplied to muscle fibers fast enough, and the aerobic breakdown of pyruvic acid cannot produce all the ATP required for further muscle contraction. Additional ATP is then generated by anaerobic glycolysis. In the process, however, most of the pyruvic acid is converted to lactic acid. About 80 percent of this lactic acid diffuses from the skeletal muscles and is transported to the liver for conversion back to glucose or glycogen, but some lactic acid accumulates in muscle tissue.

Ultimately, this lactic acid must be broken down completely into carbon dioxide and water. After exercise has stopped, extra oxygen is required to metabolize the lactic acid; replenish ATP, phosphocreatine, and glycogen; and pay back any oxygen that has been borrowed from hemoglobin, myoglobin, air in the lungs, and body fluids. The additional oxygen that must be taken into the body after vigorous exercise to restore all systems to their normal states is called *oxygen debt*. The debt is paid back by labored breathing that continues after exercise has stopped. Thus, accumulated lactic acid causes hard breathing and sufficient discomfort to stop muscle activity until homeostasis is restored.

The maximum rate of oxygen consumption during the aerobic catabolism of pyruvic acid is called *maximal oxygen uptake*. Highly trained athletes can have maximal oxygen uptakes that are twice that of average people, probably owing to a combination of genetics and training. As a result, they are capable of greater activity without increasing their lactic acid production, and their oxygen debts are less. For these reasons, they do not become short of breath as readily as untrained individuals.

Muscle Fatigue

If a skeletal muscle or group of skeletal muscles is continuously stimulated for an extended period of time, the contraction becomes progressively weaker until the muscles no longer respond. The inability of a muscle to maintain its strength of contraction is called *muscle fatigue*. It is related to an inability of muscle to produce sufficient energy to meet its needs. Although its exact mechanism is not completely understood, it may be related to insufficient oxygen, depletion of glycogen, and/or lactic acid buildup. Increased lactic acid would cause a decrease in the pH of the cells' environment. Muscle fatigue may, therefore, be viewed as a homeostatic mechanism that prevents pH levels from dropping below the normal acceptable range for the homeostasis of cells.

Heat Production

Homeostatic mechanisms are used to regulate temperature (as previously described in Chapter 5). Of the total energy released during muscular contraction, only a small amount is used for mechanical work (contraction). As much as 85 percent is released as heat to help maintain normal body temperature. Excessive heat loss by the body results in shivering, an increase in muscle tone, which increases the rate of heat production by several hundred percent as an effort to raise body temperature back to normal.

MOTOR UNIT

A *motor unit* is composed of a motor neuron and all the muscle fibers it stimulates. A single motor neuron connects to many muscle fibers. However, muscles that control precise movements, such as the external eye muscles, have fewer than 10 muscle fibers in each motor unit. Muscles of the body that are responsible for gross movements, such as the biceps brachii in the arm and gastrocnemius in the leg, may have as many as 2000 muscle fibers in each motor unit.

Stimulation of one motor neuron causes all the muscle fibers in that motor unit to contract simultaneously. However, the number of motor units that are activated at any one time in a muscle will vary with the amount of strength needed for a given action. The process of increasing the number of active motor units is called *recruitment* and is determined by the needs of the body at a given time. While some motor units are active, others are inactive; all the motor units are not contracting at the same time. This pattern of firing various motor neurons prevents fatigue by allowing a brief rest for the inactive units. Yet, the alternating motor units relieve one another so smoothly the contraction can be sustained. A state of partial contraction in a muscle is called muscle tone (described later in the chapter).

ALL-OR-NONE PRINCIPLE

The weakest stimulus from a neuron that can still initiate a contraction is called a *threshold stimulus*. According to the *all-or-none principle*, when a threshold, or greater, stimulus is applied, individual muscle fibers of a motor unit will contract to their fullest extent or will not contract at all. In other words, *individual muscle fibers* do not partly contract. This does not mean the entire muscle must be either fully relaxed or fully contracted because, of the many motor units that comprise the entire muscle, some are contracting and some are relaxing. Thus, the muscle as a whole can contract to a greater or lesser degree. Strength of contraction may be decreased by fatigue, lack of nutrients, or lack of oxygen.

KINDS OF CONTRACTIONS

Skeletal muscles produce different kinds of contractions, depending on how frequently they are stimulated.

Twitch

The *twitch contraction* is a rapid, jerky response to a single threshold or greater stimulus. Figure 8-6 is a graph *(myogram)* of a twitch contraction. Note that a brief period exists between application of the stimulus and the beginning of contraction, the *latent period*. During this time, Ca^{2+} ions are released from the sarcoplasmic reticulum and myosin cross-bridge activity begins. The second phase is the *contraction period*, and the third phase, the *relaxation period*. The durations of these periods vary with the muscle involved. All periods are very short for muscles that move the eyes, but longer periods are required for large leg muscles.

If two stimuli are applied one immediately after the other, the muscle will respond to the first stimulus but not to the second.

FIGURE 8-6 Myogram of a twitch contraction. The red arrow indicates the point at which the stimulus is applied.

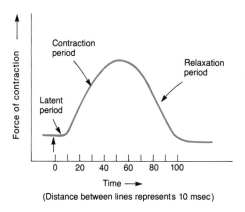

During which period do sarcomeres get smaller?

When a muscle fiber receives enough stimulation to contract, it temporarily loses its excitability and cannot contract again until its responsiveness is regained. This period of lost excitability is the **refractory period**. Its duration also varies with the muscle involved. Skeletal muscle has a short refractory period. Cardiac muscle has a long refractory period.

Tetanus

When two stimuli are applied but the second is delayed until the refractory period is over, the skeletal muscle will respond to both stimuli. In fact, if the second stimulus is applied after the refractory period, but before the muscle has finished relaxing, the second contraction will be stronger than the first. This phenomenon, in which stimuli arrive at different times, is called **wave summation**.

If a human muscle is accidentally shocked (electrocuted) or a frog muscle is electrically stimulated in a laboratory at a rate of 20 to 30 stimuli per second, the muscle can only partly relax between stimuli. As a result, the muscle maintains a sustained contraction called **incomplete tetanus**. Stimulation at an increased rate (35 to 50 stimuli per second) results in **complete tetanus**, a sustained contraction that lacks even partial relaxation between stimuli.

Both kinds of tetanus result from the additional Ca^{2+} ions released at the second stimulus while Ca^{2+} ions are still in the sarcoplasm from the first stimulus. This causes the rapid succession of separate twitches. Relaxation is either partial or does not occur at all. Voluntary contractions, such as contraction of the biceps brachii muscle in order to flex the forearm, are tetanic contractions. In fact, most of our muscular contractions are short-term tetanic contractions and are thus smooth sustained contractions.

Treppe

Treppe is a condition in which a skeletal muscle contracts more forcefully to the same strength of stimulus after it has contracted several times. It is demonstrated by stimulating an isolated muscle with a series of stimuli at the same frequency and intensity, but not at a rate fast enough to produce tetanus. Time is allowed between stimuli to permit relaxation. In this case, the first few tracings on the myogram will show an increasing height with each contraction. This is treppe—the staircase phenomenon. It is the principle athletes use when warming up. After the first few stimuli, the muscle reaches its peak of performance and undergoes its strongest contraction. Treppe is thought to result from increased availability of Ca^{2+} ions after several contractions.

Isotonic and Isometric

In an **isotonic** (*iso* = equal; *tonos* = tension) **contraction,** the muscle shortens and pulls on another structure, such as a bone, to produce movement. During such a contraction, the tension remains constant and energy is expended.

In an **isometric contraction**, there is a minimal shortening of the muscle, but the *tension* on the muscle increases greatly. Although isometric contractions do not result in body movement, energy is still expended.

Both isotonic and isometric training methods increase muscular strength in relatively short periods, but studies where direct comparisons are made tend to favor isotonic methods. The greatest advantage of isotonic exercise is that it works all the involved muscles over the entire range of a particular movement. Isometric exercise requires several separate and different maneuvers to work all the same muscles.

It is well documented that blood pressure increases considerably during isometric maneuvers. Therefore, they are a potentially dangerous form of exercise for rehabilitating cardiac patients, older adults, or anyone with hypertension.

MUSCLE TONE

A muscle may be in a state of partial contraction even though the muscle fibers operate on an all-or-none basis. Under normal conditions, at any given time, some fibers in a muscle are contracted while others are relaxed. This contraction tightens a muscle, but there may not be enough fibers contracting at the time to produce movement. Fibers contracting at different times (recruitment) allow the contraction to be sustained for long periods.

A sustained partial contraction of portions of a skeletal muscle results in **muscle tone**. Tone is essential for maintaining posture. For example, when the muscles in the back of the neck are in tonic contraction, they keep the head from slumping forward, but they do not apply enough force to pull the head all the way back.

Abnormalities of muscle tone are referred to as hypotonia or hypertonia. **Hypotonia** means decreased or lost muscle tone. Such muscles are said to be **flaccid** (FLAK-sid or FLAS-sid). Flaccid muscles are loose, their normal rounded contour is replaced by a flattened appearance, and the affected limbs are hyperextended.

Hypertonia means increased muscle tone and is expressed in two ways: spasticity or rigidity. **Spasticity** is characterized by increased muscle tone (stiffness) associated with a change in normal reflexes. **Rigidity** also refers to increased muscle tone, but reflexes are not affected.

Survival of the Fit

It used to be that physical fitness meant you could do lots of sit-ups or push-ups, maybe lift heavy weights, or play sports. These accomplishments were regarded by some people as valuable in and of themselves, but most people figured that you only needed fitness if it were an occupational requirement, for instance, if you were a soldier or a firefighter. Otherwise, fitness was for kids and athletes, and exercise was regarded as an inappropriate activity for adults, an unproductive waste of time.

Times have changed, and research has shown that some level of physical fitness is important for everyone, because it means cleaner arteries, denser bones, fewer back problems, and many other health benefits. Hundreds of studies have demonstrated that regular exercise reduces one's risk of developing many debilitating diseases such as heart disease, hypertension, and type II diabetes mellitus.

The human body responds to regular physical activity by making a number of physiological adaptations. These adaptations are called *training effects.* By understanding training effects and how regular exercise influences the physiological systems that maintain homeostasis, it is easy to see how fitness contributes to good health and helps to prevent disease.

Specificity of Training

There are many types of exercise, and the specific training effects vary according to the physiological systems stressed. For example, many people do sit-ups in a misguided attempt to reduce abdominal fat. This doesn't work, of course, because while sit-ups strengthen abdominal muscles, they do not require much energy, so they do not stress the body's energy production systems, which use fat as fuel.

Fitness is often defined in terms of four components: cardiovascular (or aerobic) endurance, muscle strength, muscle endurance, and joint flexibility. Cardiovascular fitness is achieved through regular aerobic exercise, that is, activities that involve repetitive movement of large muscle groups for extended periods of time. Examples of aerobic activities include walking, jogging, swimming, bicycling, rowing, cross-country skiing, and sports such as basketball, racquet sports, soccer, and field hockey: in other words, anything that keeps you moving continuously.

Muscle strength is measured by the amount of weight that can be moved by a muscle group, while muscular endurance refers to the ability of a muscle to repeat a movement over a period of time. In real life, these two components are related. Both can be improved by applying muscular force against a resistance. Calisthenics use body parts as the resistance. For example, push-ups force the muscles of the upper body to move against the body's weight. Lifting weights uses the resistance of a machine or the weight itself. Isometrics involve no muscle movement, for example, when pushing one hand against the other at chest level. But the muscles are still applying a force against a resistance.

Flexibility refers to a joint's range of motion and is limited by a joint's structure. When a muscle is relaxed, it is the connective tissue structures that limit how far that muscle can be stretched. These structures include the muscle coverings (epimysium, perimysium, and endomysium) and tendons. Flexibility is increased by regular stretching.

Athletes use the specificity of training principle to train specific energy production systems. (You will learn more about these systems in Chapter 20.) Sprinters practice short bursts of high energy production, while endurance athletes focus more on the aerobic energy systems that provide long-term energy production. Specificity of training also applies to the specific muscles being used for activity. Bicycling, for example, will not do much to increase the muscular endurance of your arms.

Overload

Fitness will improve only if the body is asked to do more than it is already doing; an overload must be applied. The overload must be appropriate to a person's fitness level. An elderly, sedentary man might find walking around the block each day enough of an overload to stimulate cardiovascular improvements. As he improves, however, the overload must gradually be increased if further improvement is to occur.

Healthy adults will generally experience or maintain a cardiovascular training effect if they exercise three to five times a week for 15 to 60 minutes per session at a moderate intensity. Intensity is often measured by heart rate, since heart rate and metabolic rate are closely related. Healthy individuals are usually advised to exercise at an exercise, or target, heart rate of approximately

60 to 90 percent of their maximal heart rate, depending on their current fitness level. People who are already quite fit will need to exercise at the higher end of their target heart rate range to achieve sufficient overload. Intensities lower than 60 percent may not lead to an aerobic training effect for most people, while intensities higher than this produce fatigue and discomfort. Maximal heart rate may be measured by having a person exercise at maximal levels. This is often not practical, so maximal heart rate is usually estimated by the formula 220 − age. Maximal heart rate varies considerably among individuals of the same age. Target heart rate ranges are quite broad, however, and are usually sufficient for providing a general exercise intensity guideline.

Improvements in muscle strength and endurance occur if the muscles are experiencing a moderate resistance several times a week and if the resistance or the number of repetitions is increased as the body adapts.

Reversibility

Here's the depressing part. Although lifelong exercise is a good investment in your health, you can't bank your exercise hours or your improvements. When less is required of the body, the training effects gradually disappear. Just as bone is lost during bedrest, muscle tone is lost during desk work. The body is a master of energy conservation. Why waste energy maintaining large muscles if small ones will do? The extra energy is instead stored as fat in case of future famine. Once again: Use it or lose it.

Training Effects

Aerobic exercise improves the body's oxygen delivery and energy production systems. It causes an increase in blood volume and in the contractility of cardiac muscle. These mean that the heart can pump more blood per beat (stroke volume) and per minute (cardiac output). A large stroke volume means that the heart can pump a given amount of blood with fewer beats. The heart will then be a more efficient pump both during exercise and at rest. Highly trained endurance athletes sometimes have resting heart rates of 40 beats per minute or less. Exercise is an important component of cardiac rehabilitation because it improves cardiac function.

Part of the training effect is an increase in the capillary-to-fiber (cell) ratio in both cardiac and skeletal muscle. Blood vessels increase in both size and number. A better blood supply means better delivery of oxygen and nutrients and better removal of wastes.

Energy production systems improve because of an increase in the size and number of mitochondria in the

Age-Predicted Maximal Heart Rates and Target Zones		
Age	Approximate Maximal Heart Rate	Target Zone
20	200	120–180
25	195	117–176
30	190	114–171
35	185	108–162
40	180	105–158
45	175	102–153
50	170	99–149
55	165	96–144
60	160	93–140
65	155	90–135

working muscles. The concentration of the enzymes that regulate both aerobic and anaerobic energy production also increases. During exercise, the working muscles of a trained person are able to extract more oxygen from the blood. The higher maximal oxygen uptake observed in athletes is a function of these training effects: greater oxygen-carrying capacity of the blood, better oxygen delivery systems, and an increase in the ability to use the oxygen for energy production.

Regular aerobic exercise improves the body's ability to mobilize and use fatty acids for fuel. It prevents obesity by preserving lean body mass and burning calories. Exercise helps prevent type II diabetes mellitus because cells become more sensitive to insulin for several hours after exercise. After several months of training, serum lipid profiles improve; cholesterol is less likely to be deposited into artery walls. This is one of the ways exercise helps to prevent heart disease. We've already discussed how weight-bearing exercise increases bone density.

Strength training causes some increase in muscle size, which is the result of an increase in the size and number of myofibrils. Following a bout of resistance training, there is an increase in the rate of amino acid entry into muscle fibers (cells). Regular training leads to more contractile protein per muscle fiber. Muscle hypertrophy is also due to increased vascularization and greater muscle glycogen stores.

It is likely that daily physical activity was a potent influence in the evolution of the body's systems for maintaining homeostasis. We were made to move, and always have until fairly recent history. The human body still works best when it works out.

MUSCULAR ATROPHY AND HYPERTROPHY

Muscular atrophy (A-trō-fē) refers to a wasting away of muscles. Individual muscle fibers decrease in size owing to a progressive loss of myofibrils. Muscles atrophy if they are not used (*disuse atrophy*). Bedridden individuals and people with casts may experience atrophy because the flow of impulses to the inactive muscle is greatly reduced. If the nerve supply to a muscle is cut, it will undergo complete atrophy (*denervation atrophy*). In about six months to two years, the muscle will be one-quarter its original size and the muscle fibers will be replaced by fibrous tissue. The transition to fibrous tissue, when complete, cannot be reversed. Battery-operated transcutaneous muscle stimulations (TMSs) are used to maintain the strength of muscles temporarily inactivated by trauma or stroke.

Muscular hypertrophy (hī-PER-trō-fē) is the reverse of atrophy. It is an increase in the diameters of muscle fibers owing to the production of more myofibrils, mitochondria, sarcoplasmic reticulum, nutrients, and energy-supplying molecules (ATP and phosphocreatine). Hypertrophic muscles result from very forceful muscular activity or repetitive muscular activity at moderate levels. It is believed that the number of muscle fibers does not increase after birth. During childhood, the increase in the *size* of muscle fibers appears to be at least partially under the control of human growth hormone (hGH), which is produced by the anterior pituitary gland. A further increase in the *size* of muscle fibers appears to be due to the hormone testosterone, produced by the testes. The influence of testosterone probably accounts for the generally larger muscles in males than females. More forceful muscular contractions, as in weight lifting, also contribute to larger muscles.

HOW SKELETAL MUSCLES PRODUCE MOVEMENT

ORIGIN AND INSERTION

Skeletal muscles produce movements by pulling on tendons, which in turn pull on bones. Most muscles cross at least one joint and are attached to the articulating bones that form the joint (Figure 8-7). When the muscle contracts, it draws one bone toward the other. The two bones do not move equally. One is held nearly in its original position; the attachment of a muscle tendon to the stationary bone is called the *origin*. The attachment of the other muscle tendon to the movable bone is the *insertion*. A good analogy is a spring on a door. The part of the spring attached to the door represents the insertion; the part attached to the frame is the origin. The fleshy portion of the muscle between the tendons of the origin and insertion is called the *belly (gaster)*.

LEVER SYSTEMS

In producing a body movement, bones act as levers and joints function as fulcrums of these levers. A *lever* may be defined as a rigid rod that moves about on some fixed point called a *fulcrum*. A fulcrum may be symbolized as △ . A lever is always acted on at two different points by two different forces, the *resistance* Ⓡ

and the *effort* (E). Whereas resistance is the force that opposes movement, effort is the force exerted to achieve an action. The resistance may be the weight of the body part that is to be moved. The effort is the muscular contraction, which must be applied to the bone at the insertion to produce motion. The biceps brachii flexing the forearm at the elbow as a weight is lifted is an example (Figure 8-7b). When the forearm is raised, the elbow is the fulcrum. The weight of the forearm plus the weight in the hand is the resistance. The shortening of the biceps brachii pulling the forearm up is the effort.

GROUP ACTIONS

Most movements require several skeletal muscles acting in groups rather than individually. Also, most skeletal muscles are arranged in opposing pairs at joints, that is, flexors–extensors, abductors–adductors, and so on. Bending the elbow is an example. A muscle that causes a desired action is referred to as the *prime mover (agonist)*. In this instance, the biceps brachii is the prime mover (see Figure 8-16). While the biceps brachii is contracting, another muscle, called the *antagonist*, is relaxing, here, the triceps brachii (see Figure 8-16). The antagonist has an effect opposite to that of the prime mover; that is, the antagonist relaxes and yields to the movement of the prime mover. Do not assume, however, that the biceps brachii is always the prime mover and the triceps brachii is always the antagonist. For example, when straightening the elbow, the triceps brachii serves as the prime mover and the biceps brachii functions as the antagonist. If the prime mover and antagonist contracted together with equal force, there would be no movement, as in an isometric contraction.

Most movements also involve muscles called *synergists*, which help the prime mover function more efficiently by reducing unnecessary movement.

Some muscles in a group also act as *fixators*, which stabilize the origin of the prime mover so that the prime mover can act more efficiently. Under different conditions and depending on the movement, many muscles act at various times as prime movers, antagonists, synergists, or fixators.

NAMING SKELETAL MUSCLES

The names of most of the nearly 700 skeletal muscles are based on specific characteristics.

1. Muscle names may indicate the *direction of the muscle fibers*. *Rectus* fibers run parallel to the midline of the body. *Transverse* fibers run perpendicular to the midline. *Oblique* fibers are diagonal to the midline. Examples include the rectus abdominis, transversus abdominis, and external oblique.
2. A muscle may be named according to *location*. The temporalis is near the temporal bone; the tibialis anterior is near the front of the tibia.
3. *Size* is another characteristic. *Maximus* means largest, *minimus* smallest, *longus* long, and *brevis* short. Examples include the gluteus maximus, gluteus minimus, adductor longus, and peroneus brevis.

FIGURE 8-7 Relationship of skeletal muscles to bones. (a) Skeletal muscles produce movements by pulling on bones. (b) Bones serve as levers, and joints act as fulcrums for the levers. Here the lever–fulcrum principle is illustrated by the movement of the forearm lifting a weight. Note where the resistance and effort are applied in this example.

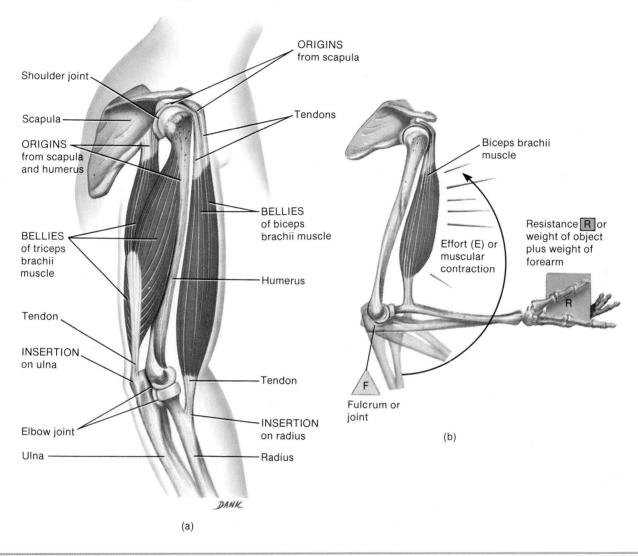

Which muscle produces the desired action? Which muscle performs the opposite action of the one producing the desired action? Which muscle prevents unwanted movements?

4. Some muscles are named for their *number of origins*. The biceps brachii has two origins, the triceps brachii three, and the quadriceps femoris four.
5. Other muscles are named on the basis of *shape*. Common examples include the deltoid (meaning triangular), trapezius (meaning trapezoid), serratus anterior (meaning sawtoothed), and rhomboideus major (meaning rhomboid or diamond shaped).
6. Muscles may be named after their *origin* and *insertion*. The sternocleidomastoid originates on the sternum and clavicle and inserts at the mastoid process of the temporal bone.
7. Still another characteristic of muscles used for naming is *action*. See Exhibit 8-1.

PRINCIPAL SKELETAL MUSCLES

Exhibits 8-2 through 8-14 list the principal muscles of the body with their origins, insertions, and actions. (By no means have all the muscles of the body been included.) For each exhibit, an *overview* section provides a general orientation to the muscles and their functions or unique characteristics.

The figures that accompany the exhibits contain superficial and deep, anterior and posterior, or medial and lateral views to show each muscle's position as clearly as possible. As you study groups of muscles, refer to Figure 8-8 to see how each group is related to all others.

FIGURE 8-8 Principal superficial skeletal muscles. (a) Anterior view.

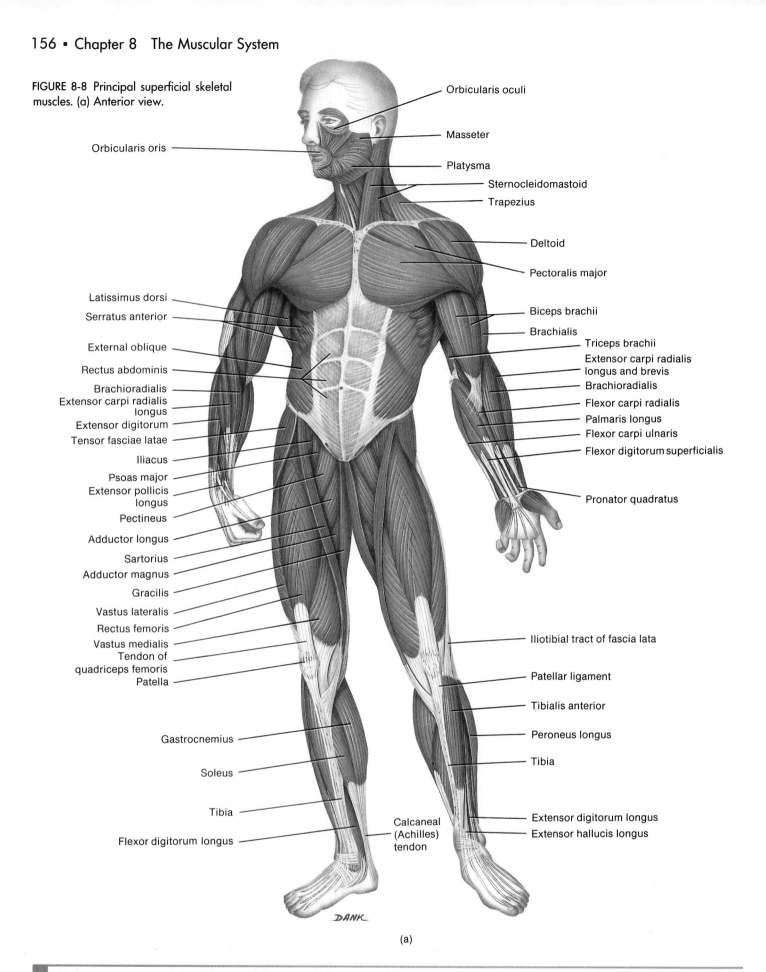

Orbicularis oculi

Masseter

Platysma

Sternocleidomastoid

Trapezius

Orbicularis oris

Deltoid

Pectoralis major

Latissimus dorsi

Serratus anterior

Biceps brachii

Brachialis

Triceps brachii

External oblique

Extensor carpi radialis longus and brevis

Rectus abdominis

Brachioradialis

Brachioradialis

Flexor carpi radialis

Extensor carpi radialis longus

Palmaris longus

Extensor digitorum

Flexor carpi ulnaris

Tensor fasciae latae

Flexor digitorum superficialis

Iliacus

Psoas major

Extensor pollicis longus

Pronator quadratus

Pectineus

Adductor longus

Sartorius

Adductor magnus

Gracilis

Vastus lateralis

Rectus femoris

Vastus medialis

Iliotibial tract of fascia lata

Tendon of quadriceps femoris

Patella

Patellar ligament

Tibialis anterior

Peroneus longus

Gastrocnemius

Tibia

Soleus

Tibia

Extensor digitorum longus

Flexor digitorum longus

Calcaneal (Achilles) tendon

Extensor hallucis longus

DANK

(a)

Circle a muscle named for direction of fibers. Underline a muscle named for shape.
Put an × next to a muscle named for action. Put a square around a muscle named for
location. Put a · next to a muscle named for size.

FIGURE 8-8 (*Continued*) (b) Posterior view.

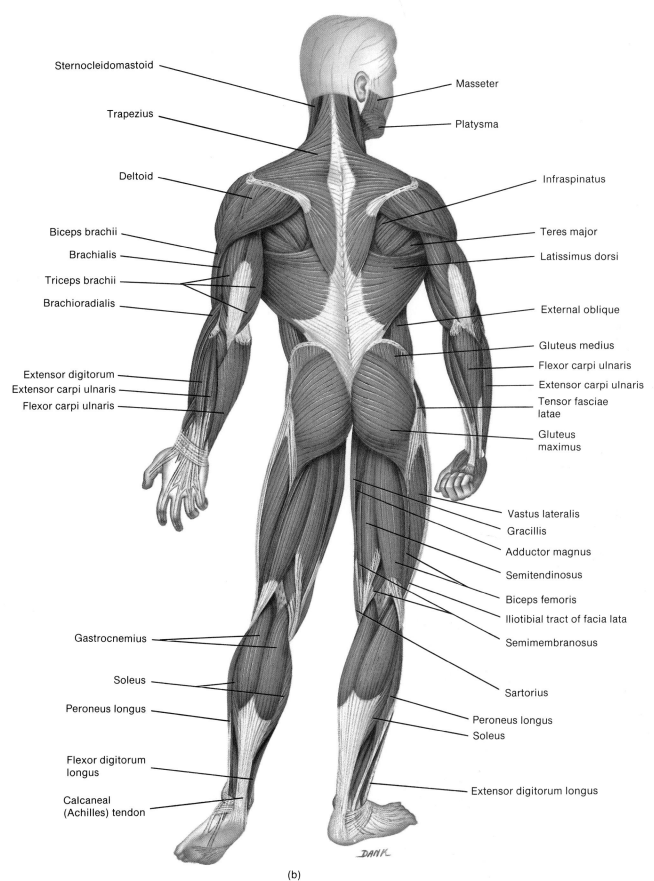

Sternocleidomastoid

Trapezius

Deltoid

Biceps brachii

Brachialis

Triceps brachii

Brachioradialis

Extensor digitorum
Extensor carpi ulnaris
Flexor carpi ulnaris

Gastrocnemius

Soleus

Peroneus longus

Flexor digitorum
longus

Calcaneal
(Achilles) tendon

Masseter

Platysma

Infraspinatus

Teres major

Latissimus dorsi

External oblique

Gluteus medius

Flexor carpi ulnaris

Extensor carpi ulnaris

Tensor fasciae
latae

Gluteus
maximus

Vastus lateralis

Gracillis

Adductor magnus

Semitendinosus

Biceps femoris

Iliotibial tract of facia lata

Semimembranosus

Sartorius

Peroneus longus

Soleus

Extensor digitorum longus

DANK

(b)

EXHIBIT 8-1
Principal Actions of Muscles

Action	Definition	Example
Flexor	Decreases the angle at a joint.	Flexor carpi radialis.
Extensor	Increases the angle at a joint.	Extensor carpi ulnaris.
Abductor	Moves a bone away from the midline.	Abductor hallucis.
Adductor	Moves a bone closer to the midline.	Adductor longus.
Levator	Produces an upward movement.	Levator scapulae.
Depressor	Produces a downward movement.	Depressor labii inferioris.
Supinator	Turns the palm upward or anteriorly.	Supinator.
Pronator	Turns the palm downward or posteriorly.	Pronator teres.
Sphincter	Decreases the size of an opening.	External anal sphincter.
Tensor	Makes a body part more rigid.	Tensor fasciae latae.
Rotator	Moves a bone around its longitudinal axis.	Obturator internus.

EXHIBIT 8-2
Muscles of Facial Expression (Figure 8-9)

Overview: The muscles in this group provide humans with the ability to express a wide variety of emotions, including frowning, surprise, fear, and happiness. The muscles themselves lie within the layers of superficial fascia. As a rule, they arise from the fascia or bones of the skull and insert into the skin. Because of their insertions, the muscles of facial expression move the skin rather than a joint when they contract.

Muscle	Origin	Insertion	Action
Epicranius (*epi* = over; *crani* = skull)	The epicranius muscle is divided into two portions: the frontalis over the frontal bone and the occipitalis over the occipital bone. The two muscles are united by a strong aponeurosis, the galea aponeurotica, which covers the superior and lateral surfaces of the skull.		
Frontalis (*front* = forehead)	Galea aponeurotica.	Skin superior to orbit.	Draws scalp forward, raises eyebrows, and wrinkles skin of forehead horizontally.
Occipitalis (*occipito* = base of skull)	Occipital and temporal bones.	Galea aponeurotica.	Draws scalp backward.
Orbicularis Oris (*orb* = circular; *or* = mouth)	Muscle fibers surrounding opening of mouth.	Skin at corner of mouth.	Closes lips, compresses lips against teeth, protrudes lips, and shapes lips during speech.
Zygomaticus Major (*zygomatic* = cheek bone; *major* = greater)	Zygomatic bone.	Skin at angle of mouth and orbicularis oris.	Draws angle of mouth upward and outward as in smiling or laughing.
Buccinator (*bucc* = cheek)	Maxilla and mandible.	Orbicularis oris.	Major cheek muscle; compresses cheek as in blowing air out of mouth and causes cheeks to cave in, producing the action of sucking.
Platysma (*platy* = flat, broad)	Fascia over deltoid and pectoralis major muscles.	Mandible, muscles around angle of mouth, and skin of lower face.	Draws outer part of lower lip downward and backward as in pouting; depresses mandible.
Oribcularis Oculi (*ocul* = eye)	Medial wall of orbit.	Circular path around orbit.	Closes eye.
Levator Palpebrae Superioris (*palpebrae* = eyelids) (see also Figure 8-11)	Roof of orbit.	Skin of upper eyelid.	Elevates upper eyelid.

FIGURE 8-9 Muscles of facial expression. (a) Anterior superficial view. (b) Anterior deep view.

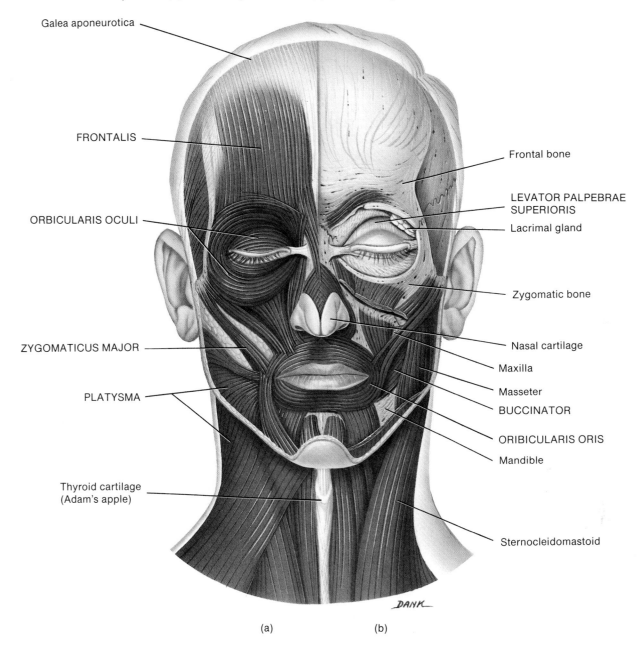

Galea aponeurotica

FRONTALIS

ORBICULARIS OCULI

ZYGOMATICUS MAJOR

PLATYSMA

Thyroid cartilage
(Adam's apple)

Frontal bone

LEVATOR PALPEBRAE
SUPERIORIS

Lacrimal gland

Zygomatic bone

Nasal cartilage

Maxilla

Masseter

BUCCINATOR

ORIBICULARIS ORIS

Mandible

Sternocleidomastoid

DANK

(a)   (b)

Which muscles cause frowning? Smiling? Pouting? Squinting?

EXHIBIT 8-3
Muscles That Move the Lower Jaw
(*Figure 8-10*)

Overview: Muscles that move the lower jaw are also known as muscles of mastication because they are involved in biting and chewing. These muscles also assist in speech.

Muscle	Origin	Insertion	Action
Masseter (*maseter* = chewer)	Maxilla and zygomatic arch.	Mandible.	Elevates mandible as in closing mouth, assists in side-to-side movement of mandible and protracts (protrudes) mandible.
Temporalis (*tempora* = temples)	Temporal bone.	Mandible.	Elevates and retracts mandible, and assists in side-to-side movement of mandible.

FIGURE 8-10 **Muscles that move the lower jaw seen in superficial right lateral view.**

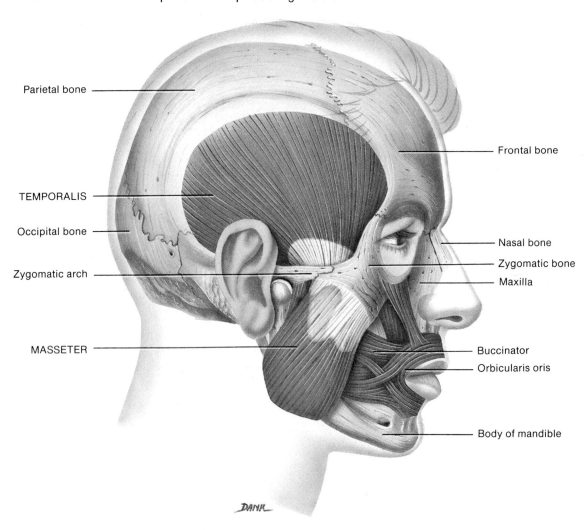

DANK

What is the attachment to the movable bone called? What is the attachment to the stationary bone called?

EXHIBIT 8-4
Muscles That Move the Eyeballs—Extrinsic Muscles[a] (Figure 8-11)

Overview: Two types of muscles are associated with the eyeball, extrinsic and intrinsic. *Extrinsic muscles* originate outside the eyeball and are inserted on its outer surface (sclera). *Intrinsic muscles* originate and insert entirely within the eyeball.

    Movements of the eyeballs are controlled by three pairs of extrinsic muscles. Two pairs of rectus muscles move the eyeball in the direction indicated by their respective names—superior, inferior, lateral, and medial. One pair of muscles, the oblique muscles—superior and inferior—rotate the eyeball on its axis. The extrinsic muscles of the eyeballs are among the fastest contracting and most precisely controlled skeletal muscles of the body.

Muscle	Origin	Insertion	Action
Superio Rectus (*superior* = above; *rectus* = in this case, muscle fibers running parallel to long axis of eyeball)	Tendinous ring attached to bony orbit around optic foramen.	Superior and central part of eyeball.	Rolls eyeball upward.
Inferior Rectus (*inferior* = below)	Same as above.	Inferior and central part of eyeball.	Rolls eyeball downward.
Lateral Rectus	Same as above.	Lateral side of eyeball.	Rolls eyeball laterally.
Medial Rectus	Same as above.	Medial side of eyeball.	Rolls eyeball medially.
Superior Oblique (*oblique* = in this case, muscle fibers running diagonally to long axis of eyeball)	Same as above.	Eyeball between superior and lateral recti.	Rotates eyeball on its axis; directs cornea downward and laterally; note that it moves through a ring of fibrocartilaginous tissue called the trochlea (*trochlea* = pulley).
Inferior Oblique	Maxilla	Eyeball between inferior and lateral recti.	Rotates eyeball on its axis; directs cornea upward and laterally.

[a] Muscles situated on the outside of the eyeball.

FIGURE 8-11 Extrinsic muscles of the eyeball seen in lateral view of the right eyeball.

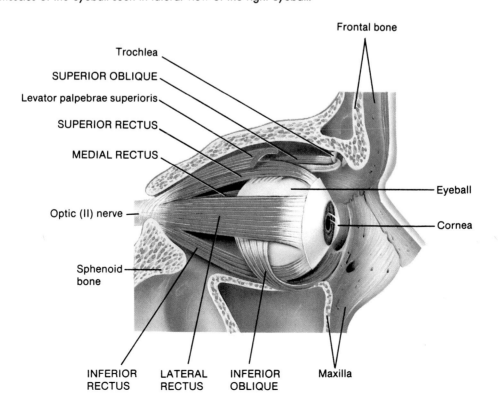

EXHIBIT 8-5
Muscles That Act on the Anterior
Abdominal Wall (Figure 8-12)

Overview: The anterior and lateral abdominal wall are composed of skin, fascia, and four pairs of flat, sheetlike muscles: rectus abdominis, external oblique, internal oblique, and transversus abdominis. The anterior surfaces of the rectus abdominis muscles are interrupted by three transverse fibrous bands of tissue called *tendinous intersections*. The aponeuroses of the external oblique, internal oblique, and transversus abdominis muscles meet at the midline to form the *linea alba* (white line), a tough fibrous band that extends from the xiphoid process of the sternum to the symphysis pubis. The inferior free border of the external oblique aponeurosis, plus some collagenous fibers, forms the *inguinal ligament*, which separates the thigh from the body wall.

Muscle	Origin	Insertion	Action
Rectus Abdominis (*rectus* = fibers parallel to midline; *abdomino* = abdomen)	Pubis and symphysis pubis.	Cartilage of fifth to seventh ribs and xiphoid process.	Compresses abdomen to aid in defecation, urination, forced expiration, and childbirth and flexes vertebral column.
External Oblique (*external* = closer to surface; *oblique* = fibers diagonal to midline)	Lower eight ribs.	Ilium and linea alba (midline aponeurosis).	Contraction of both compresses abdomen; contraction of one side alone bends vertebral column laterally.
Internal Oblique (*internal* = farther from surface)	Ilium, inguinal ligament, and thoracolumbar fascia.	Cartilage of last three or four ribs.	Compresses abdomen; contraction of one side alone bends vertebral column laterally.
Transversus Abdominis (*transverse* = fibers perpendicular to midline)	Ilium, inguinal ligament, lumbar fascia, and cartilages of last six ribs.	Sternum, linea alba, and pubis.	Compresses abdomen.

FIGURE 8-12 Muscles of the anterior abdominal wall. (a) Superficial view. (b) Deep view.

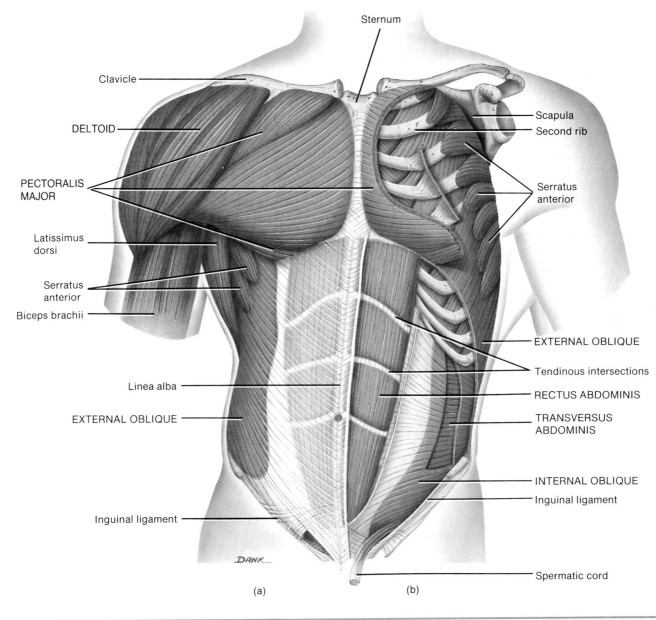

Sternum

Clavicle

Scapula
Second rib

DELTOID

PECTORALIS
MAJOR

Serratus
anterior

Latissimus
dorsi

Serratus
anterior

Biceps brachii

EXTERNAL OBLIQUE

Tendinous intersections

Linea alba

RECTUS ABDOMINIS

EXTERNAL OBLIQUE

TRANSVERSUS
ABDOMINIS

INTERNAL OBLIQUE

Inguinal ligament

Inguinal ligament

DANK

Spermatic cord

(a)   (b)

Which abdominal muscle aids in urination?

EXHIBIT 8-6
Muscles Used in Breathing (Figure 8-13)

Overview: The muscles described here are attached to the ribs and by their contraction and relaxation alter the size of the thoracic cavity during normal breathing. In forced breathing, other muscles are involved as well.

Muscle	Origin	Insertion	Action
Diaphragm (*dia* = across; *phragma* = wall)	Sternum, costal cartilages of last six ribs, and lumbar vertebrae.	Central tendon.	Forms floor of thoracic cavity, pulls central tendon downward during inspiration and thus increases vertical length of thorax.
External Intercostals (*external* = closer to surface; *inter* = between; *costa* = rib)	Inferior border of rib above.	Superior border of rib below.	May elevate ribs during inspiration and thus increase lateral, anterior, and posterior dimensions of thorax.
Internal Intercostals (*internal* = farther from surface)	Superior border of rib below.	Inferior border of rib above.	May draw adjacent ribs together during forced expiration and thus decrease lateral, anterior, and posterior dimensions of thorax.

FIGURE 8-13 Muscles used in breathing. (a) Anterior superficial view. (b) Anterior deep view.

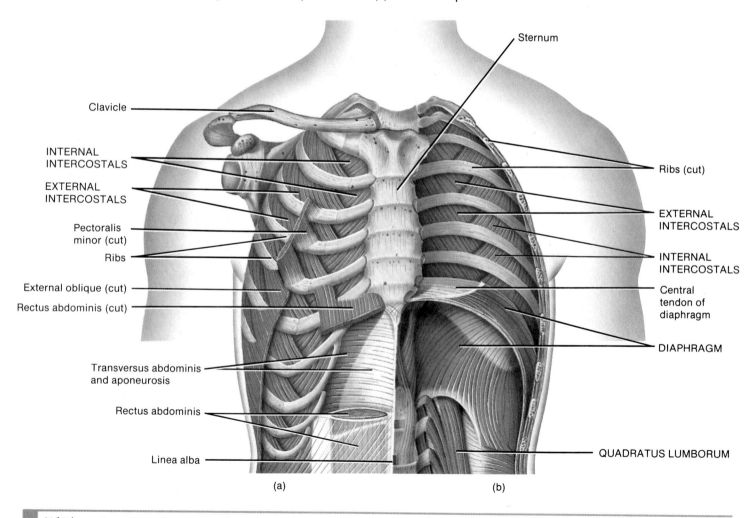

Clavicle

INTERNAL INTERCOSTALS

EXTERNAL INTERCOSTALS

Pectoralis minor (cut)

Ribs

External oblique (cut)

Rectus abdominis (cut)

Transversus abdominis and aponeurosis

Rectus abdominis

Linea alba

Sternum

Ribs (cut)

EXTERNAL INTERCOSTALS

INTERNAL INTERCOSTALS

Central tendon of diaphragm

DIAPHRAGM

QUADRATUS LUMBORUM

(a)   (b)

Which respiratory muscles are used in expiration? Inspiration?

EXHIBIT 8-7
Muscles That Move the Pectoral
(Shoulder) Girdle (Figure 8-14)

Overview: Muscles that move the pectoral (shoulder) girdle originate on the axial skeleton and insert on the clavicle or scapula. The muscles can be distinguished into *anterior* and *posterior* groups. The principal action of the muscles is to stabilize the scapula so that it can function as a stable point of origin for most of the muscles that move the humerus (arm).

Muscle	Origin	Insertion	Action
ANTERIOR			
Subclavius (*sub* = under; *clavius* = clavicle	First rib.	Clavicle.	Depresses clavicle.
Pectoralis Minor (*pectus* = breast, chest, thorax; *minor* = lesser)	Third through fifth ribs.	Scapula.	Depresses and moves scapula and elevates third through fifth ribs during forced inspiration when scapula is fixed.
Serratus Anterior (*serratus* = saw-toothed; *anterior* = front)	Upper eight or nine ribs.	Scapula.	Rotates scapula upward and laterally and elevates ribs when scapula is fixed.
POSTERIOR			
Trapezius (*trapezoides* = trapezoid-shaped)	Occipital bone and spines of seventh cervical and all thoracic vertebrae.	Clavicle and scapula.	Elevates clavicle, adducts scapula, rotates scapula upward, elevates or depresses scapula, and extends head.
Levator Scapulae (*levator* = raises; *scapulae* = scapula)	Upper four or five cervical vertebrae.	Scapula.	Elevates scapula and slightly rotates it downward.
Rhomboideus Major (*rhomboides* = rhomboid or diamond-shaped).	Spines of second to fifth thoracic vertebrae.	Scapula.	Adducts scapula and slightly rotates it downward.

FIGURE 8-14 Muscles that move the pectoral (shoulder) girdle. (a) Anterior deep view. (b) Anterior deeper view.

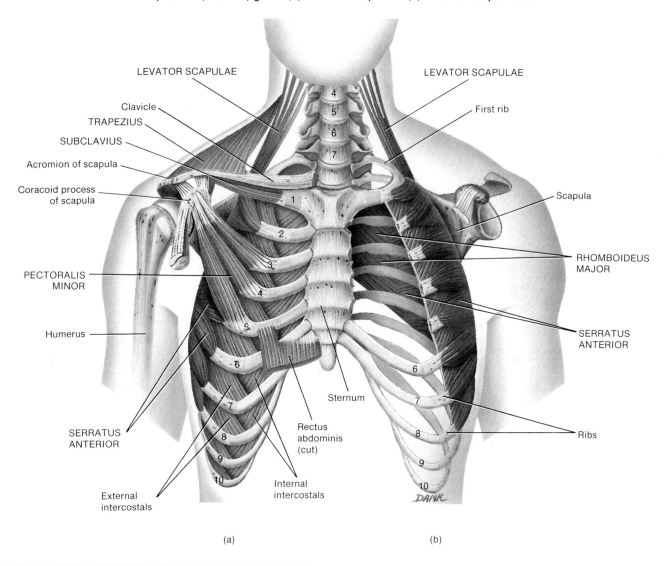

(a)   (b)

Which muscles originate on the ribs? The vertebrae?

FIGURE 8-14 (*Continued*) (c) Posterior superficial view. (d) Posterior deep view.

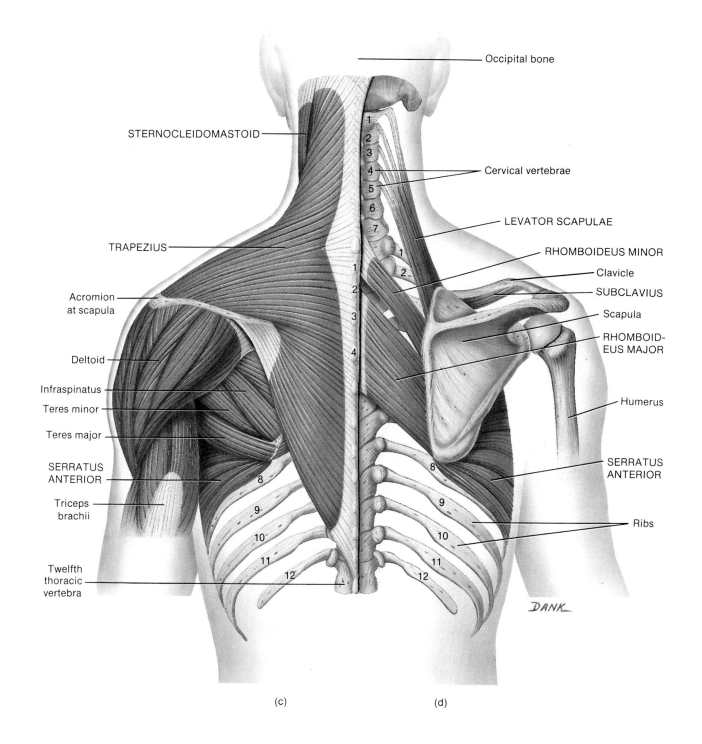

(c)   (d)

EXHIBIT 8-8
Muscles That Move the Arm (*Humerus*)
(*Figure 8-15*)

Overview: Of the muscles that cross the shoulder joint, only two of them (pectoralis major and latissimus dorsi) do not originate on the scapula. These two muscles are thus designated as *axial muscles*, since they originate on the axial skeleton. The remaining muscles, the *scapular muscles*, arise from the scapula.

The strength and stability of the shoulder joint are not provided by the shape of the articulating bones or its ligaments. Instead, four deep muscles of the shoulder and their tendons—subscapularis, supraspinatus, infraspinatus, and teres minor—strengthen and stabilize the shoulder joint. The muscles and their tendons are so arranged as to form a nearly complete circle around the joint. This arrangement is referred to as the *rotator* (*musculotendinous*) cuff and is a common site of injury to baseball pitchers, especially tearing of the supraspinatus muscle.

Muscle	Origin	Insertion	Action
AXIAL			
Pectoralis Major (see also Figure 8-12)	Clavicle, sternum, cartilages of second to sixth ribs.	Humerus.	Flexes, adducts, and rotates arm medially.
Latissimus Dorsi (*latissimus* = widest; *dorsum* = back)	Spines of lower six thoracic vertebrae, lumbar vertebrae, crests of sacrum and ilium, lower four ribs.	Humerus.	Extends, adducts, and rotates arm medially; draws arm downward and backward.
SCAPULAR			
Deltoid (*delta* = triangular) (see also Figure 8-12a)	Clavicle and scapula.	Humerus.	Abducts, flexes, extends, and rotates arm.
Subscapularis (*sub* = below; *scapularis* = scapula)	Scapula.	Humerus.	Rotates arm medially.
Supraspinatus (*supra* = above; *spinatus* = spine of scapula)	Scapula.	Humerus.	Assists deltoid muscle in abducting arm.
Infraspinatus (*infra* = below) (see also Figure 8-14c)	Scapula.	Humerus.	Rotates arm laterally; adducts arm.
Teres Major (*teres* = long and round) (see also Figure 8-14c)	Scapula.	Humerus.	Extends arm; assists in adduction and medial rotation of arm.

FIGURE 8-15 **Muscles that move the arm (humerus) as seen in anterior deep view.**

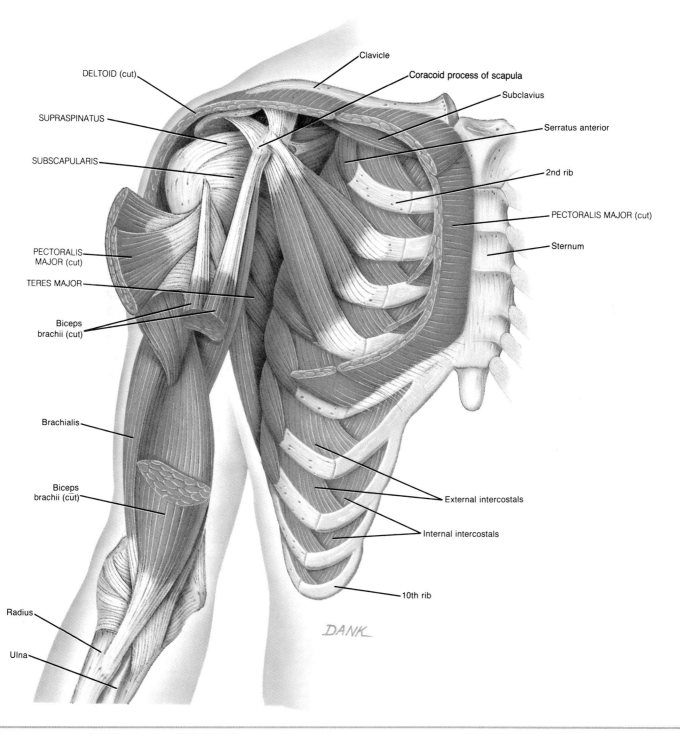

Circle the muscles that flex the arm. Underline those that extend the arm.

EXHIBIT 8-9
Muscles That Move the Forearm (*Radius and Ulna*) (*Figure 8-16*)

Overview: Most of the muscles that move the forearm (radius and ulna) are divided into *flexors* and *extensors*. Recall that the elbow joint is a hinge joint, capable only of flexion and extension. Whereas the biceps brachii, brachialis, and brachioradialis are flexors of the elbow joint, the triceps brachii is an extensor. Other muscles that move the forearm are concerned with supination and pronation.

Muscle	Origin	Insertion	Action
FLEXORS			
Biceps Brachii (*biceps* = two heads of origin; *brachion* = arm)	Scapula.	Radius.	Flexes and supinates forearm; flexes arm.
Brachialis	Humerus.	Ulna.	Flexes forearm.
Brachioradialis (*radialis* = radius) (see also Figure 8-17b)	Humerus.	Radius.	Flexes forearm; supinates and semipronates forearm.
EXTENSOR			
Triceps Brachii (*triceps* = three heads of origin)	Scapula and humerus.	Ulna.	Extends forearm; extends arm.
SUPINATOR			
Supinator[a] (*supination* = turning palm upward or anteriorly)	Humerus and ulna.	Radius.	Supinates forearm and hand.
PRONATOR			
Pronator Teres (*pronation* = turning palm downward or posteriorly) (see Figure 8-17a)	Humerus and ulna.	Radius.	Pronates forearm and hand.

[a] Not illustrated.

FIGURE 8-16 Muscles that move the forearm. (a) Anterior view. (b) Posterior view.

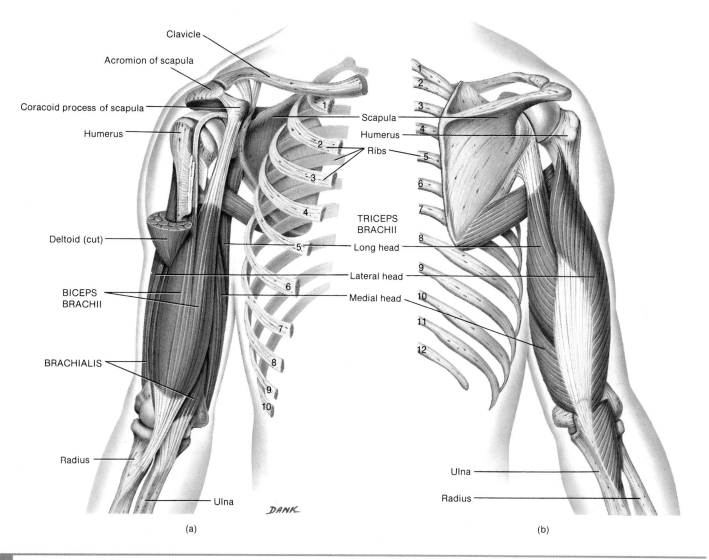

Clavicle

Acromion of scapula

Coracoid process of scapula

Humerus

Deltoid (cut)

BICEPS BRACHII

BRACHIALIS

Radius

Ulna

DANK

(a)

Scapula

Humerus

Ribs

TRICEPS BRACHII

Long head

Lateral head

Medial head

Ulna

Radius

(b)

Identify the three flexors of the forearm.

EXHIBIT 8-10
Muscles That Move the Wrist and Fingers (Figure 8-17)

Overview: Muscles that move the wrist, hand, and fingers are many and varied. However, as you will see, their names for the most part give some indication of their origin, insertion, or action. On the basis of location and function, the muscles are divided into two groups—anterior and posterior. The *anterior muscles* function as flexors. They originate on the humerus and typically insert on the carpals, metacarpals, and phalanges. The bellies of these muscles form the bulk of the proximal forearm. The *posterior muscles* function as extensors. These muscles arise on the humerus and insert on the metacarpals and phalanges. Each of the two principal groups is also divided into superficial and deep muscles.

The tendons of the muscles of the forearm that attach to the wrist or continue into the hand, along with blood vessels and nerves, are held close to bones by fascia. The tendons are also surrounded by tendon sheaths. At the wrist, the deep fascia is thickened into fibrous bands called *retinacula* (*retinere* = retain). The *flexor retinaculum* (*transverse carpal ligament*) is located over the palmar surface of the carpal bones. Through it pass the long flexor tendons of the digits and wrist and the median nerve. The *extensor retinaculum* (*dorsal carpal ligament*) is located over the dorsal surface of the carpal bones. Through it pass the extensor tendons of the wrist and digits.

Muscle	Origin	Insertion	Action
ANTERIOR			
Flexor Carpi Radialis (*flexor* = decreases angle at joint; *carpus* = wrist; *radialis* = radius)	Humerus.	Second and third metacarpals.	Flexes and abducts wrist.
Flexor Carpi Ulnaris (*ulnaris* = ulna)	Humerus and ulna.	Pisiform, hamate, and fifth metacarpal.	Flexes and adducts wrist.
Palmaris Longus (*palma* = palm)	Humerus.	Flexor retinaculum.	Flexes wrist.
Flexor Digitorum Profundus[a] (*digit* = finger or toe; *profundus* = deep)	Ulna.	Bases of distal phalanges.	Flexes distal phalanges of each finger.
Flexor Digitorum Superficialis (*superficialis* = closer to surface)	Humerus, ulna, and radius.	Middle phalanges.	Flexes middle phalanges of each finger.
POSTERIOR			
Extensor Carpi Radialis Longus (*extensor* = increases angle at joint; *longus* = long)	Humerus.	Second metacarpal.	Extends and abducts wrist.
Extensor Carpi Ulnaris	Humerus and ulna.	Fifth metacarpal.	Extends and adducts wrist.
Extensor Digitorum	Humerus.	Second through fifth phalanges.	Extends phalanges.

[a] Not illustrated.

FIGURE 8-17 Muscles that move the wrist, hand, and fingers. (a) Superficial anterior view. (b) Superficial posterior view.

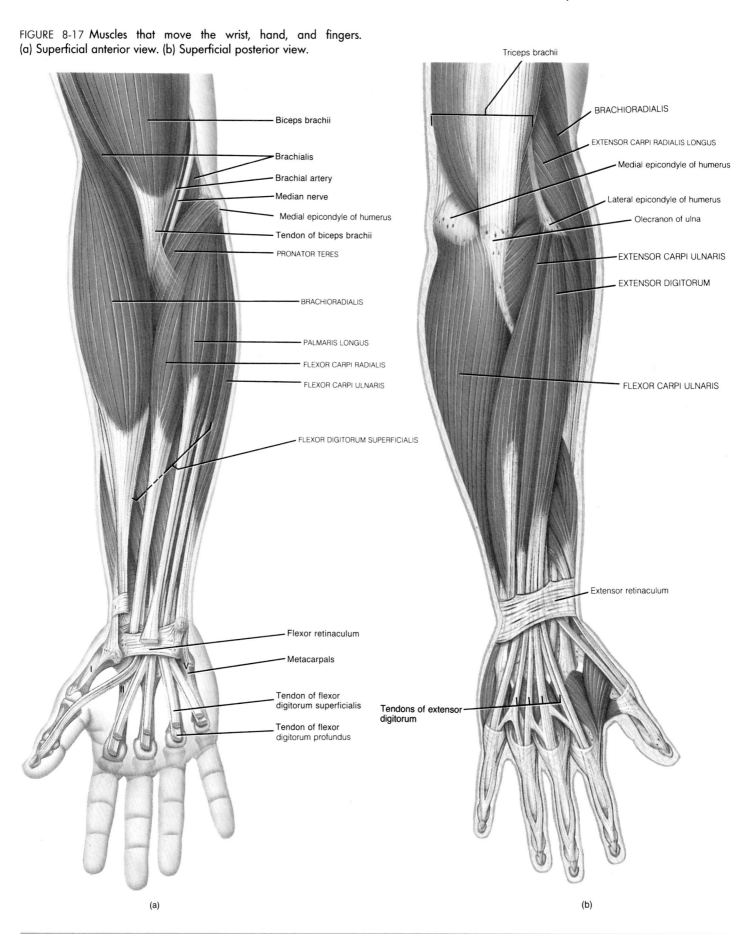

(a)

Biceps brachii

Brachialis

Brachial artery

Median nerve

Medial epicondyle of humerus

Tendon of biceps brachii

PRONATOR TERES

BRACHIORADIALIS

PALMARIS LONGUS

FLEXOR CARPI RADIALIS

FLEXOR CARPI ULNARIS

FLEXOR DIGITORUM SUPERFICIALIS

Flexor retinaculum

Metacarpals

Tendon of flexor digitorum superficialis

Tendon of flexor digitorum profundus

(b)

Triceps brachii

BRACHIORADIALIS

EXTENSOR CARPI RADIALIS LONGUS

Medial epicondyle of humerus

Lateral epicondyle of humerus

Olecranon of ulna

EXTENSOR CARPI ULNARIS

EXTENSOR DIGITORUM

FLEXOR CARPI ULNARIS

Extensor retinaculum

Tendons of extensor digitorum

Circle three muscles that flex the wrist. Underline three muscles that extend the wrist.

EXHIBIT 8-11
Muscles That Move the Vertebral Column
(*Figure 8-18*)

Overview: The muscles that move the vertebral column are quite complex because they have multiple origins and insertions and there is considerable overlapping among them.

Muscle	Origin	Insertion	Action
Sternocleidomastoid (*sternum* = breastbone; *cleido* = clavicle; *mastoid* = mastoid process of temporal bone) (see Figure 8-14c)	Sternum and clavicle.	Temporal bone.	Contractions of both muscles flex the cervical part of the vertebral column, draw the head forward, and elevate chin; contraction of one muscle rotates face toward side opposite contracting muscle.
Rectus Abdominis (see Figure 8-12)	Pubis and symphysis pubis.	Cartilages of fifth through seventh ribs and sternum.	Flexes vertebral column at lumbar spine and compresses abdomen.
Quadratus Lumborum (*quadratus* = squared, four-sided; *lumb* = lumbar region) (see Figure 8-19a)	Ilium.	Twelfth rib and upper four lumbar vertebrae.	Flexes vertebral column laterally.
Sacrospinalis (Erector Spinae)	This posterior muscle consists of three groupings: iliocostalis, longissimus, and spinalis. These groups, in turn, consist of a series of overlapping muscles. The iliocostalis group is laterally placed, the longissimus group is intermediate in placement, and the spinalis is medially placed. All three groups of muscles extend various parts of the vertebral column.		

FIGURE 8-18 Muscles that move the vertebral column as seen in posterior view.

LONGISSMUS
(intermediate)

SPINALIS
(medial)

ILIOCOSTALIS
(lateral)

1
2
3
4
5
6
7
8
9
10
11
12

DANK

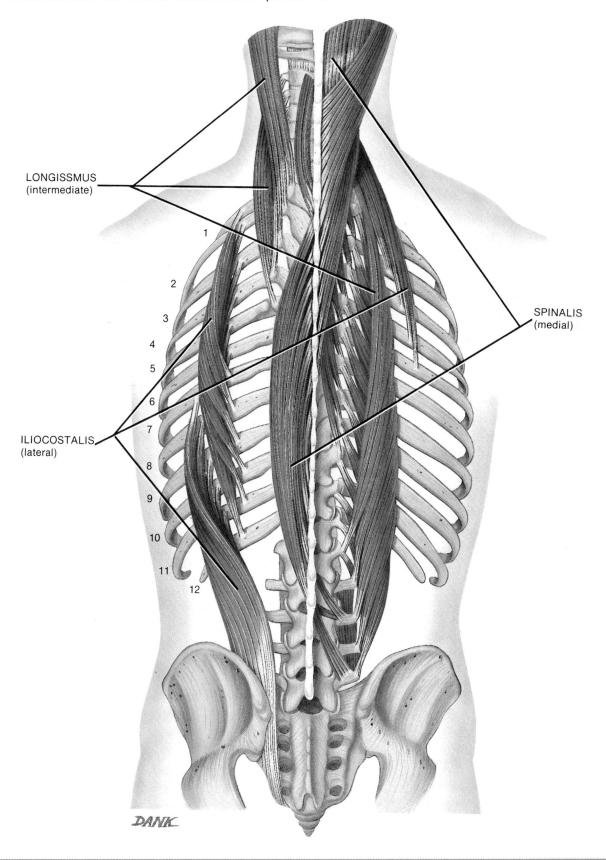

Which muscles extend the spinal column? Which muscles flex it?

EXHIBIT 8-12
Muscles That Move the Thigh (Femur)
(Figure 8-19)

Overview: Muscles of the lower extremities are larger and more powerful than those of the upper extremities since lower extremity muscles function in stability, locomotion, and maintenance of posture. Upper extremity muscles are characterized by versatility of movement. In addition, muscles of the lower extremities frequently cross two joints and act equally on both. The majority of muscles that act on the thigh (femur) originate on the pelvic (hip) girdle and insert on the 'femur. The anterior muscles are the psoas major and iliacus, together referred to as the iliopsoas muscle. The remaining muscles (except for the pectineus, adductors, and tensor fasciae latae) are posterior muscles. Technically, the pectineus and adductors are components of the medial compartment of the thigh, but they are included in this exhibit because they act on the thigh. The tensor fasciae latae muscle is laterally placed. The *fascia lata* is a deep fascia of the thigh that encircles the entire thigh. It is well developed laterally, where together with the tendons of the gluteus maximus and tensor fasciae latae muscles it forms a structure called the *iliotibial tract*. The tract inserts into the lateral condyle of the tibia.

Muscle	Origin	Insertion	Action
Psoas Major (psoa = muscle of loin)	Lumbar vertebrae.	Femur.	Flexes and rotates thigh laterally; flexes vertebral column.
Iliacus (*iliacus* = ilium)	Ilium.	Tendon of psoas major.	Flexes and rotates thigh laterally.
Gluteus Maximus (*glutos* = buttock; *maximus* = largest)	Ilium, sacrum, coccyx, and aponeurosis of sacrospinalis.	Iliotibial tract of fascia lata and femur.	Extends and rotates thigh laterally.
Gluteus Medius (*media* = middle)	Ilium.	Femur.	Abducts and rotates thigh medially.
Gluteus Minimus (*minimus* = smallest)	Ilium.	Femur.	Abducts and rotates thigh medially.
Tensor Fasciae Latae (*tensor* = makes tense; *fascia* = band; *latus* = wide)	Ilium.	Tibia by way of the iliotibial tract.	Flexes and abducts thigh.
Adductor Longus (*adductor* = moves part closer to midline; *longus* = long)	Pubis and symphysis pubis.	Femur.	Adducts, laterally rotates, and flexes thigh.
Adductor Magnus (*magnus* = large)	Pubis and ischium.	Femur.	Adducts, flexes, laterally rotates and extends thigh (anterior part flexes, posterior part extends).
Piriformis (*pirum* = pear; *forma* = shape)	Sacrum.	Femur.	Rotates thigh laterally and abducts it.
Pectineus (*pecten* = comb shaped)	Pubis.	Femur.	Flexes and adducts thigh.

FIGURE 8-19 **Muscles that move the thigh (femur). (a) Anterior superficial view.**

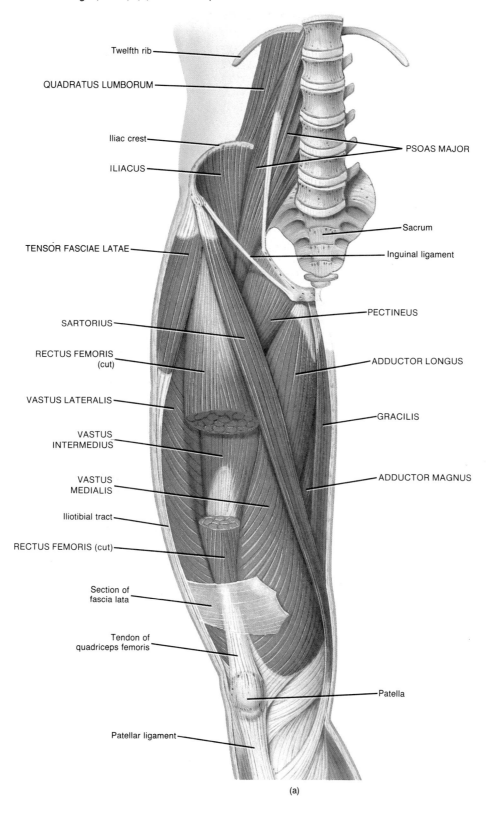

Twelfth rib

QUADRATUS LUMBORUM

Iliac crest

ILIACUS

PSOAS MAJOR

TENSOR FASCIAE LATAE

Sacrum

Inguinal ligament

SARTORIUS

PECTINEUS

RECTUS FEMORIS
(cut)

ADDUCTOR LONGUS

VASTUS LATERALIS

VASTUS
INTERMEDIUS

GRACILIS

VASTUS
MEDIALIS

ADDUCTOR MAGNUS

Iliotibial tract

RECTUS FEMORIS (cut)

Section of
fascia lata

Tendon of
quadriceps femoris

Patella

Patellar ligament

(a)

Put an × next to the muscles that are part of the quadriceps femoris.

FIGURE 8-19 (*Continued*) (b) Posterior deep view.

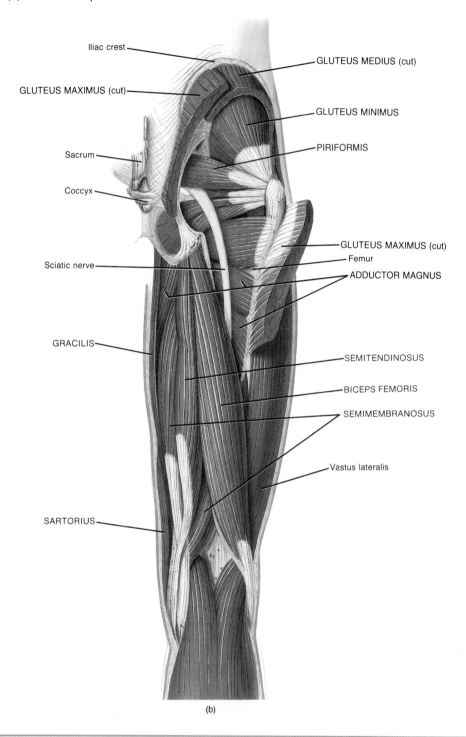

Iliac crest

GLUTEUS MEDIUS (cut)

GLUTEUS MAXIMUS (cut)

GLUTEUS MINIMUS

PIRIFORMIS

Sacrum

Coccyx

GLUTEUS MAXIMUS (cut)

Femur

ADDUCTOR MAGNUS

Sciatic nerve

GRACILIS

SEMITENDINOSUS

BICEPS FEMORIS

SEMIMEMBRANOSUS

Vastus lateralis

SARTORIUS

(b)

Put an × next to the hamstring muscles.

EXHIBIT 8-13
Muscles That Act on the Leg
(*Tibia and Fibula*) (*Figure 8-19*)

Overview: The muscles that act on the leg (tibia and fibula) originate in the hip and thigh and are separated into compartments by deep fascia. The *medial (adductor) compartment* is so named because its muscles adduct the thigh. The adductor magnus, adductor longus, and pectineus muscles, components of the medial compartment, are included in Exhibit 8-12 because they act on the femur. The gracilis, the other muscle in the medial compartment, not only adducts the thigh but also flexes the leg. For this reason, it is included in this exhibit.

The *anterior (extensor) compartment* is so designated because its muscles act to extend the leg, and some also flex the thigh. It is composed of the quadriceps femoris and sartorius muscles. The quadriceps femoris muscle is a composite muscle that includes four distinct parts, usually described as four separate muscles (rectus femoris, vastus lateralis, vastus intermedius, and vastus medialis). The common tendon for the four muscles is known as the *patellar ligament* and attaches to the tibial tuberosity. The rectus femoris and sartorius muscles are also flexors of the thigh.

The *posterior (flexor) compartment* is so named because its muscles flex the leg (but also extend the thigh). Included are the hamstrings (biceps femoris, semitendinosus, and semimembranosus). The hamstrings are so named because their tendons are long and stringlike in the popliteal area. The *popliteal fossa* is a diamond-shaped space on the posterior aspect of the knee bordered laterally by the tendons of the biceps femoris and medially by the semitendinosus and semimembranosus muscles.

Muscle	Origin	Insertion	Action
MEDIAL (ADDUCTOR) COMPARTMENT			
Adductor Magnus **Adductor Longus**	See Exhibit 8-12.		
Pectineus			
Gracilis (*gracilis* = slender)	Symphysis pubis and pubis.	Tibia.	Adducts thigh and flexes leg.
ANTERIOR (EXTENSOR) COMPARTMENT			
Quadriceps Femoris (*quadriceps* = four heads of origin; *femoris* = femur)			
Rectus Femoris (*rectus* = fibers parallel to midline)	Ilium.	Patella.	All four heads extend leg; rectus portion alone also flexes thigh.
Vastus Lateralis (*vastus* = large; *lateralis* = lateral)	Femur.		
Vastus Medialis (*medialis* = medial)	Ilium.	Patella and tibial tuberosity through patellar ligament (tendon of quadriceps).	
Vastus Intermedius (*intermedius* = middle)	Femur.		
Sartorius (*sartor* = tailor; refers to cross-legged position of tailors)	Ilium.	Tibia.	Flexes leg; flexes thigh and rotates it laterally, thus crossing leg.
POSTERIOR (FLEXOR) COMPARTMENT			
Hamstrings			
Biceps Femoris (*biceps* = two heads of origin)	Ischium and femur.	Fibula and tibia.	Flexes legs and extends thigh.
Semitendinosus (*semi* = half; *tendo* = tendon)	Ischium.	Tibia.	Flexes leg and extends thigh.
Semimembranosus (*membran* = membrane)	Ischium.	Tibia.	Flexes leg and extends thigh.

EXHIBIT 8-14
Muscles That Move the Foot and Toes
(Figure 8-20)

Overview: The musculature of the leg, like that of the thigh, is divided into three compartments by deep fascia. The *anterior compartment* consists of muscles that dorsiflex the foot. In a situation analogous to the wrist, the tendons of the muscles of the anterior compartment are held firmly to the ankle by thickenings of deep fascia called the *superior extensor retinaculum* (*transverse ligament of the ankle*) and *inferior extensor retinaculum* (*cruciate ligament of the ankle*).

The *lateral compartment* contains muscles that plantar flex and evert the foot. The *posterior compartment* consists of muscles that are divisible into superficial and deep groups. The superficial muscles share a common tendon of insertion, the calcaneal (Achilles) tendon that inserts into the calcaneus bone of the ankle. The superficial and most deep muscles plantar flex the foot.

Muscle	Origin	Insertion	Action
ANTERIOR COMPARTMENT			
Tibialis Anterior (*tibialis* = tibia; *anterior* = front)	Tibia.	First metatarsal and first cuneiform.	Dorsiflexes and inverts foot.
Extensor Digitorum Longus (*extensor* = increases angle at joint).	Tibia and fibula.	Middle and distal phalanges of four outer toes.	Dorsiflexes and everts foot and extends toes.
LATERAL COMPARTMENT			
Peroneus Longus (*perone* = fibula; *longus* = long)	Fibula and tibia.	First metatarsal and first cuneiform.	Plantar flexes and everts foot.
POSTERIOR COMPARTMENT			
Gastrocnemius (*gaster* = belly; *kneme* = long)	Femur.	Calcaneus by way of calcaneal (Achilles) tendon.	Plantar flexes foot and flexes leg.
Soleus (*soleus* = sole of foot).	Fibula and tibia.	Calcaneus by way of calcaneal (Achilles) tendon.	Plantar flexes foot.
Tibialis Posterior (*posterior* = back).	Tibia and fibula.	Second, third, and fourth metatarsals; navicular; all three cuneiforms; and cuboid.	Plantar flexes and inverts foot.
Flexor Digitorum Longus (*flexor* = decreases angle at joint; *digitorum* = finger or toe)	Tibia.	Distal phalanges of four outer toes.	Plantar flexes and inverts foot; flexes toes.

CARDIAC MUSCLE TISSUE

The heart wall is chiefly composed of *cardiac muscle tissue*. Although it is striated in appearance like skeletal muscle, it is involuntary (see Exhibit 4-3). Unlike skeletal muscle fibers, which contain several peripherally located nuclei, cardiac muscle cells have only a single centrally located nucleus. Cardiac muscle cells are also quadrangular, not elongated.

Cardiac muscle fibers form two separate networks. The muscular walls and partition of the upper chambers (atria) of the heart compose one network. The muscular walls and partition of the lower chambers (ventricles) of the heart compose the other network. Within each network, the cardiac muscle fibers branch and interconnect with one another (see Exhibit 4-3). At the same time, each fiber in a network is separated from the next fiber by an irregular transverse thickening of the sarcolemma called an ***intercalated*** (in-TER-ka-lāt-ed) ***disc***. These discs contain ***gap junctions***, which help conduct muscle action potentials from one muscle fiber to another. When a single fiber in a network is stimulated, all the fibers in the network become stimulated. Thus, each network contracts as a functional unit. As you will see in Chapter 15, when the fibers of the atria contract as a unit, blood moves into the ventricles. Then, when the ventricular fibers contract as a unit, blood is pumped into arteries.

Under normal resting conditions, cardiac muscle tissue contracts and relaxes rapidly, continuously, and rhythmically about 75 times a minute without stopping. This is a major physiological difference between cardiac and skeletal muscle tissue. Accordingly, cardiac muscle tissue requires a constant supply of oxygen. Energy generation occurs in numerous large mitochondria. Another difference is the source of stimulation. While skeletal muscle tissue ordinarily contracts only when stimulated by a nerve impulse, cardiac muscle tissue can contract without extrinsic (external) nerve stimulation. It is stimulated by a conducting tissue of specialized intrinsic (internal) muscle within the heart. Nerve stimulation merely causes the conducting tissue to increase or decrease its rate of discharge. This spontaneous, rhythmical self-excitation is called ***autorhythmicity***.

Cardiac muscle tissue also has an extra-long refractory period, lasting several tenths of a second, which allows time for the heart to relax between beats. The long refractory period permits heart rate to be increased significantly but prevents the heart itself from undergoing incomplete or complete tetanus. Tetanus of heart muscle would stop blood flow through the body and result in death.

SMOOTH MUSCLE TISSUE

Like cardiac muscle tissue, ***smooth muscle tissue*** is involuntary. Smooth muscle fibers are considerably smaller than skeletal muscle

FIGURE 8-20 Muscles that move the foot and toes. (a) Superficial posterior view. (b) Deep posterior view.

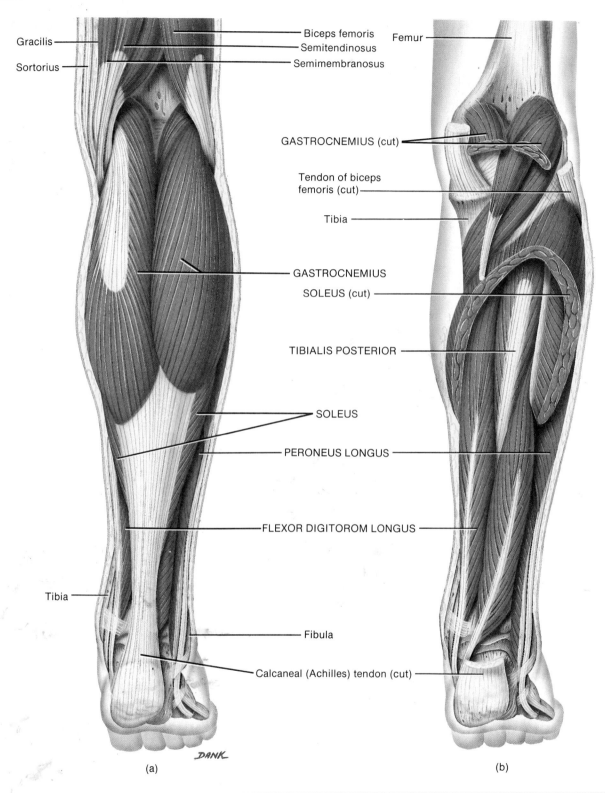

Gracilis

Sortorius

Biceps femoris

Semitendinosus

Semimembranosus

Femur

GASTROCNEMIUS (cut)

Tendon of biceps femoris (cut)

Tibia

GASTROCNEMIUS

SOLEUS (cut)

TIBIALIS POSTERIOR

SOLEUS

PERONEUS LONGUS

FLEXOR DIGITOROM LONGUS

Tibia

Fibula

Calcaneal (Achilles) tendon (cut)

DANK

(a)

(b)

Circle the muscles that dorsiflex the foot.

FIGURE 8-20 (*Continued*) (c) Superficial anterior view. (d) Superficial right lateral view.

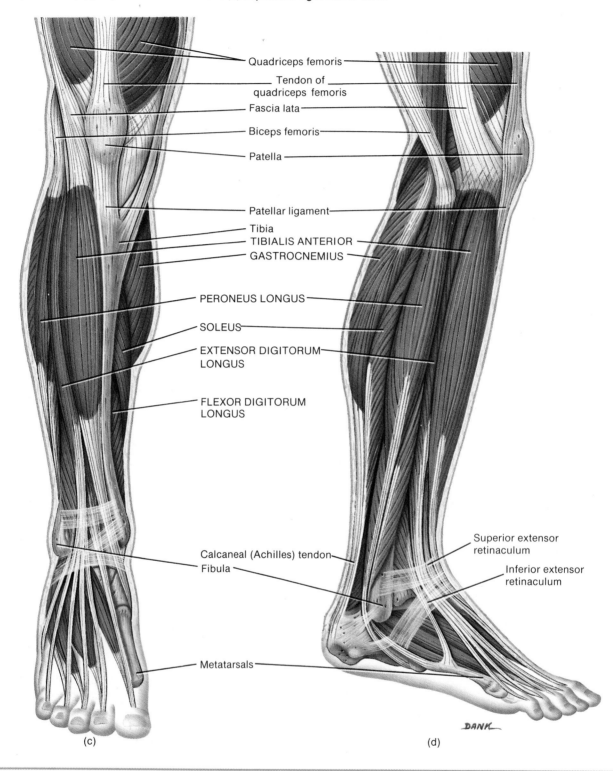

Quadriceps femoris
Tendon of quadriceps femoris
Fascia lata
Biceps femoris
Patella
Patellar ligament
Tibia
TIBIALIS ANTERIOR
GASTROCNEMIUS
PERONEUS LONGUS
SOLEUS
EXTENSOR DIGITORUM LONGUS
FLEXOR DIGITORUM LONGUS
Calcaneal (Achilles) tendon
Fibula
Metatarsals
Superior extensor retinaculum
Inferior extensor retinaculum
DANK
(c)   (d)

Circle the muscles that dorsiflex the foot.

fibers. The fibers are widest in their midportion and tapered at both ends, and within each fiber is a single, oval, centrally located nucleus (see Exhibit 4-3). In addition to thick and thin myofilaments, smooth muscle fibers also contain *intermediate filaments*. None of these various myofilaments has a regular pattern, and thus, it is nonstriated, or smooth.

The intermediate filaments stretch between structures called *dense bodies*, some of which are attached to the sarcolemma (others are dispersed in the sarcoplasm). In contraction, those attached to the sarcolemma cause a lengthwise shortening of the fiber. Note in Figure 8-21 how this causes a bubblelike appearance.

There are two kinds of smooth muscle tissue, visceral and multiunit. The more common type is *visceral (single-unit) muscle tissue.* It is found in wraparound sheets that form part of the walls of small arteries and veins and hollow viscera such as the stomach, intestines, uterus, and urinary bladder. The terms *smooth muscle tissue* and *visceral muscle tissue* are sometimes used interchangeably. The fibers in visceral muscle tissue are tightly bound together to form a continuous network. They contain gap junctions to facilitate muscle action potential conduction between fibers. When a neuron stimulates one fiber, the muscle action potential travels over the other fibers so that contraction occurs in a wave over many adjacent fibers. Whereas skeletal muscle fibers contract as individual units, visceral muscle cells contract in sequence as the action potential spreads from one cell to another.

The second kind of smooth muscle tissue, *multiunit smooth muscle tissue*, consists of individual fibers. Each has its own motor-nerve endings. Whereas stimulation of a single visceral muscle fiber causes contraction of many adjacent fibers, stimulation of a single multiunit fiber causes contraction of only that fiber, as with skeletal muscle. Multiunit smooth muscle tissue is found in the walls of large arteries, in large airways to the lungs, in the arrector pili muscles attached to hair follicles, and in the internal eye muscles.

Smooth muscle tissue exhibits several important physiological differences from striated muscle tissue. First, the duration of contraction of smooth muscle fibers is 5 to 500 times longer than in skeletal muscle fibers. Second, smooth muscle tissue can undergo sustained, long-term tone, which is important in the gastrointestinal tract (where the wall of the tract maintains a steady pressure on its contents), in the walls of the blood vessels called arterioles that maintain a steady pressure on blood, and in the wall of the urinary bladder that maintains a steady pressure on urine.

Smooth muscle is not under voluntary control. Some smooth muscle fibers contract in response to nerve impulses from the autonomic (involuntary) nervous system. Others contract in response

FIGURE 8-21 Smooth muscle fiber before contraction (left) and after contraction (right).

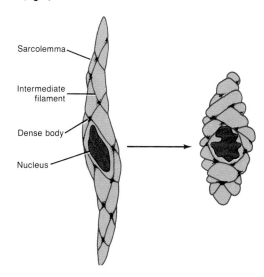

Sarcolemma

Intermediate filament

Dense body

Nucleus

Name four ways smooth muscle tissue is functionally different from striated muscle.

to hormones or local factors such as pH, oxygen and carbon dioxide levels, temperature, and ion concentrations.

Finally, unlike skeletal muscle fibers, smooth muscle fibers can stretch considerably without becoming tense or taut. This is important because it permits smooth muscle to accommodate great changes in size while still retaining the ability to contract. Smooth muscle in the wall of hollow organs such as the uterus, stomach, intestines, and urinary bladder can stretch as the viscera enlarge, while the pressure within them remains the same.

■ COMMON DISORDERS ■

Fibrosis

The formation of fibrous connective tissue in locations where it normally does not exist is called *fibrosis*.

Fibromyalgia

Fibromyalgia (*algia* = painful condition) refers to a group of common nonarticular rheumatic disorders characterized by pain, tenderness, and stiffness of muscles, tendons, and surrounding soft tissues.

Muscular Dystrophies

Muscular dystrophies (*dystrophy* = degeneration) are hereditary muscle-destroying diseases characterized by degeneration of individual muscle fibers (cells), which leads to a progressive atrophy of the skeletal muscle.

Myasthenia Gravis (MG)

Myasthenia (mī-as-THĒ-nē-a) *gravis* (**MG**) is a weakness of skeletal muscles caused by an abnormality at the neuromuscular junction that prevents muscle fibers from contracting. Myasthenia gravis is an autoimmune disorder caused by antibodies that bind to the receptors and hinder the attachment of ACh.

Abnormal Contractions

One kind of abnormal muscular contraction is a *spasm*, a sudden involuntary contraction of large groups of muscles. *Tremor* is a rhythmic, involuntary, purposeless contraction of opposing muscle groups. A *fasciculation* is an involuntary, brief twitch of a muscle visible under the skin. It occurs irregularly and is not associated with movement of the affected muscle. A *fibrillation* is similar to a fasciculation except that it is not visible under the skin. A *tic* is a spasmodic twitching made involuntarily by muscles that are ordinarily under voluntary control.

MEDICAL TERMINOLOGY AND CONDITIONS

Electromyography or *EMG* (e-lek'-trō-mī-OG-ra-fē; *electro* = electricity; *myo* = muscle; *graph* = to write) The recording and study of electrical changes that occur in muscle tissue.

Gangrene (GANG-rēn; *gangraena* = an eating sore) Death of a soft tissue, such as muscle, that results from interruption of its blood supply.

Myalgia (mī-AL-jē-a; *algia* = painful condition) Pain in or associated with muscles.

Myoma (mī-Ō-ma; *oma* = tumor) A tumor consisting of muscle tissue.

Myomalacia (mī'-ō-ma-LĀ-shē-a; *malaco* = soft) Softening of a muscle.

Myopathy (mī-OP-a-thē; *pathos* = disease) Any disease of muscle tissue.

Myosclerosis (mī'-ō-skle-RŌ-sis; *scler* = hard) Hardening of a muscle.

Myositis (mī'-ō-SĪ-tis; *itis* = inflammation of) Inflammation of muscle fibers (cells).

Myospasm (MĪ-o-spazm) Spasm of a muscle.

Myotonia (mī-ō-TO-nē-a; *tonia* = tension) Increased muscular excitability and contractility with decreased power of relaxation; tonic spasm of the muscle.

Paralysis (pa-RAL-a-sis; *para* = beyond; *lyein* = to loosen) Loss or impairment of motor (muscular) function resulting from a lesion of nervous or muscular origin.

Trichinosis (trik'-i-NŌ-sis) A myositis caused by the parasitic worm *Trichinella spiralis*, which may be found in the muscles of humans, rats, and pigs. People contract the disease by eating insufficiently cooked infected pork.

Volkmann's contracture (FŌLK-manz kon-TRAK-tur; *contra* = against) Permanent contraction of a muscle due to replacement of destroyed muscle fibers (cells) with fibrous tissue that lacks ability to stretch.

Wryneck or *torticollis* (RĪ-neck; *tortus* = twisted; *collum* = neck) Spasmodic contraction of several superficial and deep muscles of the neck that produces twisting of the neck and an unnatural position of the head.

STUDY OUTLINE

Characteristics (p. 142)

1. Excitability is the property of receiving and responding to stimuli.
2. Contractility is the ability to shorten and thicken (contract).
3. Extensibility is the ability to be stretched (extend).
4. Elasticity is the ability to return to original shape after contraction or extension.

Functions (p. 142)

1. Through contraction, muscle tissue performs three important functions.
2. These functions are motion, maintenance of posture, and heat production.

Types (p. 142)

1. Skeletal muscle tissue is attached to bones. It is striated and voluntary.
2. Cardiac muscle tissue forms most of the wall of the heart. It is striated and involuntary.
3. Visceral muscle tissue is located in viscera. It is nonstriated (smooth) and involuntary.

Skeletal Muscle Tissue (p. 142)

Connective Tissue Components (p. 143)

1. The term fascia is applied to a sheet or broad band of fibrous connective tissue underneath the skin or around muscles and organs of the body.
2. Other connective tissue components are epimysium, covering the entire muscle; perimysium, covering fasciculi; and endomysium, covering fibers. All are extensions of deep fascia.
3. Tendons and aponeuroses are extensions of connective tissue beyond muscle fibers that attach the muscle to bone or other muscle.

Nerve and Blood Supply (p. 143)

1. Nerves convey impulses (action potentials) for muscular contraction.
2. Blood provides nutrients and oxygen for contraction.

Histology (p. 144)

1. Skeletal muscle consists of fibers (cells) covered by a sarcolemma. The fibers contain sarcoplasm, nuclei, mitochondria, myoglobin, high-energy molecules, sarcoplasmic reticulum, and transverse tubules.
2. Each fiber also contains myofibrils that consist of thin and thick myofilaments. The myofilaments are compartmentalized into sarcomeres.
3. Thin myofilaments are composed of actin, tropomyosin, and troponin; thick myofilaments consist of myosin.
4. Projecting myosin heads are called cross bridges and contain actin- and ATP-binding sites.

Contraction (p. 144)

Sliding-Filament Mechanism (p. 144)
1. A muscle action potential travels over the sarcolemma and enters the transverse tubules and sarcoplasmic reticulum.
2. The action potential leads to the release of calcium ions from the sarcoplasmic reticulum.
3. Actual contraction is brought about when the thin myofilaments of a sarcomere slide toward each other as myosin cross bridges pull on actin myofilaments.

Neuromuscular Junction (p. 144)
1. A motor neuron transmits a nerve impulse (nerve action potential) to a skeletal muscle, which stimulates contraction.
2. A neuromuscular junction refers to an axon terminal of a motor neuron and the portion of the muscle fiber sarcolemma close to, but not touching, it (motor end plate).

Physiology of Contraction (p. 146)
1. When a nerve impulse (nerve action potential) reaches an axon terminal, the synaptic vesicles of the terminal release acetylcholine (ACh), which initiates a muscle action potential in the sarcolemma. The action potential then travels into the transverse tubules and sarcoplasmic reticulum.
2. The sarcoplasmic reticulum releases calcium ions that combine with troponin, causing it to pull on tropomyosin and thus expose myosin-binding sites on actin.
3. When ATP splits into ADP + P, the released energy activates (energizes) myosin cross bridges.
4. Activated cross bridges attach to actin and the cross bridges move toward the center of the sarcomere (power stroke); the movement of cross bridges results in the sliding of thin myofilaments.

Energy for Contraction (p. 148)
1. The immediate, direct source of energy for muscle contraction is ATP.
2. Muscle fibers generate ATP continuously. This involves phosphocreatine and the metabolism of nutrients such as glycogen.

Homeostasis (p. 149)
1. Oxygen debt is the amount of O_2 needed to convert accumulated lactic acid into CO_2 and H_2O. It occurs during strenuous exercise and is paid back by rapid breathing after exercising until homeostasis between muscular activity and oxygen requirements is restored.
2. Muscle fatigue results from diminished availability of oxygen, toxic effects of carbon dioxide, and lactic acid built up during exercise.
3. The heat given off during muscular contraction maintains the homeostasis of body temperature.

Motor Unit (p. 150)
1. A motor neuron and the muscle fibers it stimulates form a motor unit.
2. A single motor unit may innervate as few as 10 or as many as 2000 muscle fibers.

All-or-None Principle (p. 150)
1. The weakest stimulus capable of causing contraction is a threshold stimulus.
2. Muscle fibers of a motor unit contract to their fullest extent or not at all.

Kinds of Contractions (p. 150)
1. The various kinds of contractions are twitch, tetanus, treppe, isotonic, and isometric.
2. A record of a contraction is called a myogram. The refractory period is the time when a muscle has temporarily lost excitability. Skeletal muscles have a short refractory period. Cardiac muscle has a long refractory period.
3. Wave summation is the increased strength of a contraction resulting from the application of a second stimulus before the muscle has completely relaxed after a previous stimulus.

Muscle Tone (p. 151)
1. A sustained partial contraction of portions of a skeletal muscle results in muscle tone.
2. Tone is essential for maintaining posture.

Muscular Atrophy and Hypertrophy (p. 154)
1. Muscular atrophy refers to a state of wasting away of muscles.
2. Muscular hypertrophy refers to an increase in the diameter of muscle fibers.

How Skeletal Muscles Produce Movement (p. 154)

1. Skeletal muscles produce movement by pulling on bones.
2. The attachment to the stationary bone is the origin. The attachment to the movable bone is the insertion.
3. Bones serve as levers and joints as fulcrums. The lever is acted on by two different forces: resistance and effort.
4. The prime mover produces the desired action. The antagonist produces an opposite action. The synergist assists the prime mover by reducing unnecessary movement. The fixator stabilizes the origin of the prime mover so that it can act more efficiently.

Naming Skeletal Muscles (p. 154)

1. The names of most skeletal muscles indicate specific characteristics.
2. The major descriptive categories are direction of fibers, location, size, number of origins (or heads), shape, origin and insertion, and action.

Principal Skeletal Muscles (p. 155)

1. The principal skeletal muscles of the body are grouped according to region in Exhibits 8-2 through 8-14.
2. In studying groups, refer to Figure 8-8 to see how each group is related to all others.

Cardiac Muscle Tissue (p. 180)

1. This muscle tissue is found only in the heart. It is striated and involuntary.

2. The fibers are quadrangular and usually contain a single centrally placed nucleus.
3. The fibers branch freely and are connected via gap junctions.
4. Intercalated discs provide strength and aid action potential conduction.
5. Unlike skeletal muscle tissue, cardiac muscle tissue contracts and relaxes rapidly, continuously, and rhythmically. Energy is supplied by numerous, large mitochondria.
6. Cardiac muscle tissue can contract without extrinsic stimulation and can remain contracted longer than skeletal muscle tissue.
7. Cardiac muscle tissue has a long refractory period, which prevents tetanus.

Smooth Muscle Tissue (p. 180)

1. Smooth muscle tissue is nonstriated and involuntary.

2. Smooth muscle fibers contain intermediate filaments, in addition to thin and thick myofilaments, and dense bodies.
3. Visceral (single-unit) smooth muscle is found in the walls of viscera. The fibers are arranged in a network so that contraction occurs in a wave over many adjacent fibers.
4. Multiunit smooth muscle is found in blood vessels and the eye. The fibers operate singly rather than as a unit.
5. The duration of contraction and relaxation of smooth muscle is longer than in skeletal muscle.
6. Smooth muscle fibers contract in response to nerve impulses, hormones, and local factors.
7. Smooth muscle fibers can stretch considerably without developing tension.

REVIEW QUESTIONS

1. How is the muscular system related to the skeletal system? What are the three basic functions of the muscular system? (p. 142)
2. What are the four characteristics of muscle tissue? (p. 142)
3. How can the three types of muscle tissue be distinguished on the basis of location, appearance, and nervous control? (p. 142)
4. What is fascia? What are the functions of deep fascia? (p. 143)
5. Define epimysium, perimysium, endomysium, tendon, and aponeurosis. (p. 143)
6. Describe the nerve and blood supply to a skeletal muscle. (p. 143)
7. In considering the contraction of skeletal muscle tissue, describe the following: neuromuscular junction, motor unit, role of calcium, sources of energy, and sliding-filament mechanism. (p. 144)
8. Explain the energy sources for contraction. (p. 148)
9. Discuss each of the following as examples of muscle homeostasis: oxygen debt, fatigue, and heat production. (p. 150)
10. What is the all-or-none principle? Relate it to a threshold and subthreshold stimulus. (p. 150)
11. Define each of the following contractions and state the importance of each: twitch, tetanus, treppe, isotonic, and isometric. (p. 150)
12. What is a myogram? Describe the latent period, contraction period, and relaxation period of a twitch contraction. Construct a diagram to illustrate your answer. (p. 150)
13. Define the refractory period. What is wave summation? (p. 151)
14. What is muscle tone? Why is it important? Distinguish between atrophy and hypertrophy. (p. 151)

15. What is the relationship between shivering (uncontrolled muscular contractions) and body temperature? How does sweating (cooling of the skin) after strenuous exercise relate to the homeostasis of body temperature?
16. Using the terms origin and insertion in your discussion, describe how skeletal muscles produce body movements by pulling on bones. (p. 154)
17. What is a lever? Fulcrum? Apply these terms to the body, and indicate the nature of the forces that act on levers. (p. 154)
18. Define the role of the prime mover, antagonist, synergist, and fixator in producing body movements. (p. 154)
19. Select any of the muscles presented in Exhibits 8-2 through 8-14 and see if you can determine how each gets its name. In addition, refer to the prefixes, suffixes, roots, and definitions in each exhibit as a guide. Select as many muscles as you wish, so that you understand the concept involved. (p. 158–180)
20. Again using Exhibits 8-2 through 8-14 as references, select several muscles from each exhibit and see if you can determine their actions based on their origins and insertions. (p. 158–180)
21. Compare skeletal, cardiac, and smooth muscle with regard to differences in structure and function. (p. 180)
22. Define the following: fibrosis, fibromyalgia, muscular dystrophies, myasthenia gravis (MG), tremor, fasciculation, fibrillation, and tic. (p. 183)
23. Refer to the glossary of medical terminology and conditions associated with the muscular system. Be sure that you can define each term. (p. 184)

Nervous Tissue

STUDENT OBJECTIVES

1. Describe how the nervous system helps maintain homeostasis.
2. Classify the organs of the nervous system according to function.
3. Contrast the features and functions of neuroglia and neurons.
4. Describe how a nerve impulse is generated and conducted.
5. Define a synapse and list the factors involved in the conduction of a nerve impulse across a synapse.
6. List the necessary conditions for the regeneration of nervous tissue.

A LOOK AHEAD

ORGANIZATION
HISTOLOGY
  Neuroglia
  Neurons
    Structure
    Classification
FUNCTIONS
  Nerve Impulses
    Ion Channels in Plasma
      Membranes
    Membrane Potentials
    Excitability
    All-or-None Principle
    Saltatory Conduction
    Speed of Nerve Impulses
  Conduction Across Synapses
    Excitatory Transmission
    Inhibitory Transmission
    Integration at Synapses
    Alteration of Synaptic
      Conduction
  Graded Potentials
  Regeneration

The body has two control centers, one for rapid response—the nervous system—and one that makes slower but no less important adjustments for maintaining homeostasis—the endocrine system. The nervous system has three functions. First, it senses changes within the body and in the outside environment; this is its sensory function. Second, it interprets the changes; this is its integrative function. Third, it responds to the interpretation by initiating action in the form of muscular contractions or glandular secretions; this is its motor function.

The branch of medical science that deals with the normal functioning and disorders of the nervous system is called **neurology** (noo-ROL-ō-jē; *neuro* = nerve or nervous system; *logos* = study of).

ORGANIZATION

The nervous system has two principal divisions, the **central nervous system (CNS)** and the **peripheral** (pe-RIF-er-al) **nervous system (PNS)** (Figure 9-1).

The CNS consists of the brain and spinal cord and is the control center for the entire nervous system. Within the CNS, incoming sensory information is integrated and correlated, thoughts and emotions are generated, and muscles are stimulated to contract and glands to secrete via outgoing nerve impulses.

The PNS consists of nerves emerging from the brain (cranial nerves) and from the spinal cord (spinal nerves). These nerves carry impulses inward from receptors to the CNS and outward from the CNS to muscles and glands. The sensory (afferent) component of the PNS consists of nerve cells, called sensory (afferent) neurons, that conduct impulses from receptors in various parts of the body to the CNS. The motor (efferent) component consists of nerve cells, called motor (efferent) neurons, that conduct impulses from the CNS to muscles and glands.

Based on the origin of the incoming sensory information and the part of the body that responds, the PNS may be subdivided into a **somatic** (*soma* = body) **nervous system (SNS)** and **autonomic** (*auto* = self; *nomos* = law) **nervous system (ANS)**. The SNS consists of sensory neurons that convey information from cutaneous and special sense receptors primarily in the head, body wall, and extremities to the CNS (Chapter 22) and motor neurons from the CNS that conduct impulses to skeletal muscles only. Since the motor response of the SNS are under conscious control, it is voluntary.

The ANS consists of sensory neurons that convey homeostatic information from receptors primarily in the viscera to the CNS and motor neurons from the CNS that conduct impulses to smooth muscle, cardiac muscle, and glands. Since the motor responses of the ANS are under subconscious control, it is involuntary.

HISTOLOGY

Despite the complexity of the nervous system, it consists of only two kinds of cells: neurons and neuroglia. Neurons are specialized for nerve impulse conduction and for all special functions, such as thinking, controlling muscle activity, and regulating glands. Neuroglia serve to support and protect the neurons.

NEUROGLIA

The **neuroglia** (noo-ROG-lē-a; *neuro* = nerve; *glia* = glue) or **glial cells** serve numerous support and protective functions for the neurons. Neuroglia are generally smaller than neurons and outnumber them by 5 to 10 times. There are four types of neuroglia: astrocytes, oligodendrocytes, microglia, and ependymocytes. Exhibit 9-1 summarizes their functions. Neuroglia are a common source of gliomas (tumors), and it is estimated that gliomas account for 40 to 45 percent of brain tumors.

FIGURE 9-1 Organization of the nervous system.

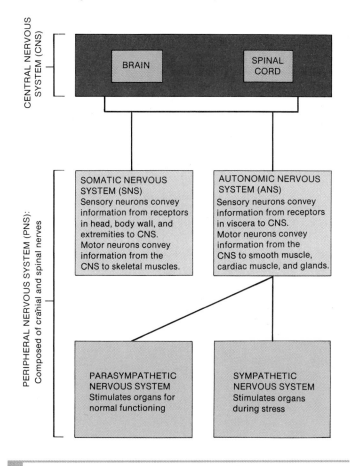

Which parts of the nervous system are voluntary?

NEURONS

Neurons conduct impulses from one part of the body to another. They are the basic information-processing units of the nervous system. The terms nerve cell and neuron are used interchangeably.

Structure

Neurons have three distinct parts: (1) cell body, (2) dendrites, and (3) axon (Figure 9-2a). The ***cell body*** contains a well-defined nucleus and nucleolus surrounded by a granular cytoplasm along with typical organelles. Also located in the cytoplasm are two structures found only in neurons: chromatophilic substance and neurofibrils. ***Chromatophilic substance*** (***Nissl bodies***) is a type of rough (granular) endoplasmic reticulum whose function is to make proteins. Newly produced proteins pass from the cell body into the cytoplasmic processes (to be explained shortly), mainly the axon. These proteins replace those lost during metabolism and are used for growth of neurons and regeneration of nerve fibers (also to be explained shortly). ***Neurofibrils*** are long, thin filaments that extend into nerve fibers and help support them; they also help transport nutrients. The inability of neurons to regenerate in adults is related to the absence of a mitotic apparatus.

Neurons have two kinds of cytoplasmic processes: dendrites and axons. ***Dendrites*** (*dendro* = tree) are short, thick, highly branched extensions of the cytoplasm. A neuron usually has several main dendrites. They function to receive impulses and conduct them toward the cell body.

The second type is the ***axon***, a single long, thin extension that sends impulses to another neuron or tissue. Axons vary in length from a few millimeters (1 mm = 0.04 inch) in the brain to a meter (3.28 ft) or more between the spinal cord and toes. Along the length of an axon, there may be side branches called ***axon collaterals***. The axon and its collaterals terminate by branching into many fine filaments called ***axon terminals***. The ends of axon terminals contain bulblike structures called ***synaptic end bulbs***, which contain sacs called ***synaptic vesicles*** that store chemicals called neurotransmitters. Neurotransmitters determine whether an impulse passes from one neuron to another or from a neuron to another tissue (Chapter 8).

Nerve fiber is a general term for any process projecting from the cell body, in other words, a dendrite or an axon. Most commonly, though, it refers to an axon and its sheaths. As you will see later, a nerve is a group of nerve fibers outside the CNS. Figure 9-2 shows a typical nerve fiber. Many axons, especially large ones in the PNS, are surrounded by a multilayered white,

EXHIBIT 9-1
Neurolgia of Central Nervous System

Type	Description	Function
Astrocytes (*astro* = star; *cyte* = cell)	Star-shaped cells with numerous processes.	Twine around neurons to form supporting network in brain and spinal cord; attach neurons to their blood vessels.
Oligodendrocytes (*oligo* = few; *dendro* = tree)	Resemble astrocytes in some ways, but processes are fewer and shorter.	Support neurons in CNS and produce a fatty (phospholipid) myelin sheath around axons of neurons of CNS.
Microglia (*micro* = small; *glia* = glue)	Small cells with few processes; normally stationary but may migrate to site of injury; also called *brain macrophages*.	Engulf and destroy microbes and cellular debris by phagocytosis.
Ependyma (**Ependymocytes**) (*ependyma* = upper garment)	Epithelial cells arranged in a single layer; many are ciliated.	Form a continuous lining for the ventricles of the brain (spaces that form cerebrospinal fluid) and the central canal of the spinal cord; probably assist in the circulation of cerebrospinal fluid.

FIGURE 9-2 Structure of a typical neuron as exemplified by an efferent (motor) neuron. (a) An efferent neuron. Arrows indicate the direction in which nerve impulses travel. The break indicates that the process is actually longer than shown. (b) Sectional planes through a myelinated fiber.

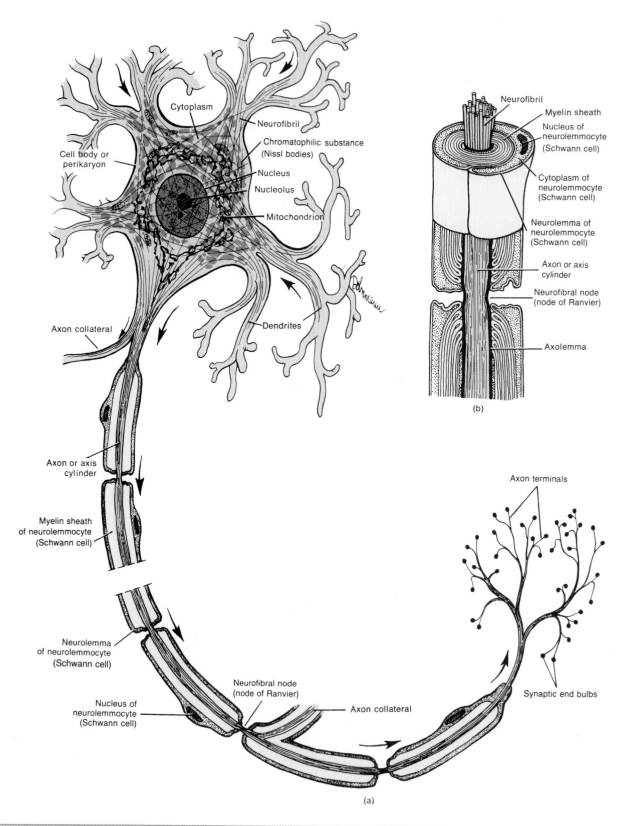

Which processes receive impulses? Which send impulses?

phospholipid, segmented covering called the **myelin sheath**. Axons with such a covering are said to be **myelinated**, while those without it are **unmyelinated**. The function of the myelin sheath is to increase the speed of nerve impulse conduction and to insulate and maintain the axon. Myelin is responsible for the color of the white matter in the nerves, brain, and spinal cord.

The myelin sheath of axons of the PNS is produced by **neurolemmocytes (Schwann cells)**. In forming a sheath, a neurolemmocyte wraps around the axon in a spiral, but in such a way that its cytoplasm and nucleus end up in the outside layer. The inner portion, of up to 20 to 30 layers, is the myelin sheath. The outer nucleated, cytoplasmic layer (the layer that encloses the sheath) is called the **neurolemma (sheath of Schwann)**.

The neurolemma is found only around fibers in the PNS. It functions in the regeneration of injured axons by forming a tube in which a regenerating axon grows. Between the segments of the myelin sheath are unmyelinated gaps called **neurofibral nodes** or **nodes of Ranvier** (ron-VĒ-ā). Unmyelinated fibers are also enclosed by neurolemmocytes, but they do not have multiple wrappings; they only contain a neurolemma.

Nerve fibers of the CNS may also be myelinated or unmyelinated. Myelination of CNS axons is accomplished by oligodendrocytes in somewhat the same manner that neurolemmocytes myelinate PNS axons but with one important difference. The oligodendrocytes merely deposit the sheath. There is no neurolemma, so fibers of the CNS cannot regenerate. Myelinated axons of the CNS also have neurofibril nodes, but they are not so numerous.

Myelin sheaths are first laid down during the later part of the fetal development and during the first year of life. The amount of myelin increases from birth to maturity and its presence greatly increases the rate of nerve impulse conduction. Since myelination is still in progress during infancy, an infant's responses to stimuli are not as rapid or coordinated as those of an older child or an adult.

Classification

The different neurons in the body are classified by structure and function.

The structural classification is based on the number of processes extending from the cell body. **Multipolar neurons** usually have several dendrites and one axon (Figure 9-2a). Most neurons in the brain and spinal cord are of this type. **Bipolar neurons** have one dendrite and one axon and are found in the retina of the eye, the inner ear, and the nose. **Unipolar neurons** have only one process extending from the cell body. The single process divides into a central branch, which functions as an axon, and a peripheral branch, which functions as a dendrite (Figure 9-3). Unipolar neurons originate in the embryo as bipolar neurons, and during development, the axon and dendrite fuse into a single process. Unipolar neurons are found in dorsal (sensory) root ganglia of spinal nerves (see Figure 10-2).

The functional classification of neurons is based on the direction in which they transmit impulses. **Sensory (afferent) neurons** transmit impulses from receptors in the skin, sense organs, muscles, joints, and viscera to the brain and spinal cord. **Motor (efferent) neurons** convey impulses from the brain and spinal cord to effectors, which may be either muscles or glands (see Figure 9-2a). Other

FIGURE 9-3 Structure of a typical afferent (sensory) neuron. Arrows indicate the direction in which the nerve impulse travels. The break indicates that the process is actually longer than shown.

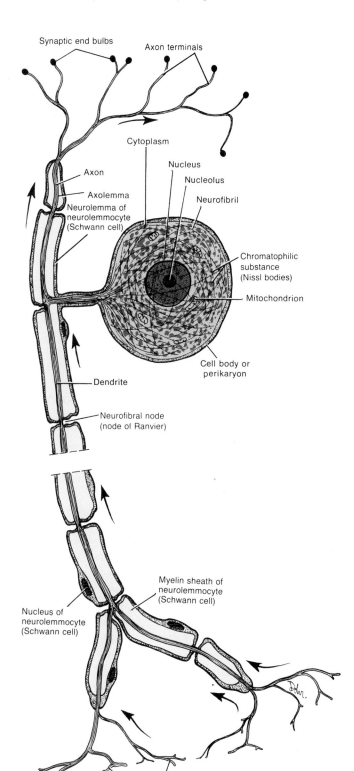

What are the three structural types of neurons? What are the three functional types of neurons?

Biofeedback: Expanding the Nervous System

The nervous system is often compared to an elaborate communication network in which huge volumes of information are continually being processed. Research scientists are exploring ways to enhance this communication system to help it function more effectively in both health and disease. One of these enhancements has been the development of biofeedback technology.

The use of instrumentation to measure biological functions is of course not new. In a sense, you use biofeedback when you take your temperature, pulse, or blood pressure. The difference is that biofeedback has self-regulation as its goal.

Biofeedback instruments give you information about what is happening in your body and enable you to use this information to gain control of the variables being monitored. For example, people who suffer from tension headaches can use biofeedback to learn how to relax tense muscles to decrease the activity of the efferent neurons that are overactivating head and neck muscles. Electrodes are attached to one of the offending muscles, typically the frontalis muscle of the forehead. The electrodes detect muscle electrical activity and send this information to the biofeedback instrument, which converts it into a signal such as a beep or flashing light that can be received by the patient. As the patient relaxes his/her forehead, the beeps or lights will slow down; if the muscle gets more tense, they will speed up. The instrument responds instantaneously to any change in muscle electrical activity, so the patient receives immediate feedback on how he/she is doing in his/her attempts to relax.

Afferent nerves already send information to the brain about muscle tension levels, so to some degree we can sense how tense a muscle is; but some muscles are more difficult to tune into, and chronic muscle tension can make muscle tightness feel normal. Biofeedback instruments amplify the signal sent by the muscles and direct it to our awareness. The machine informs us of even the smallest changes in activity that would normally be undetectable. Thus, biofeedback provides a sort of expansion of the nervous system. It amplifies and relays neural activity and allows parts of the body to communicate more effectively; so in a sense it helps the nervous system with its sensory, integrative, and motor functions.

Types of Biofeedback

If a body response can be monitored, biofeedback can be applied. Those most commonly monitored are the responses that are most accessible. Electromyography (EMG) senses the electrical activity of muscles, as in the example above. Electroencephalography (EEG) gives information about brain wave activity. Using EEG, subjects learn to produce the types of brain waves associated with relaxed or creative mental states. The galvanic skin response (GSR), also known as electrodermal response (EDR), is what lie detector tests use. Changes in the electrical conductivity of the skin reflect minute changes in sweat gland activity and skin cell membrane permeability, which occur in response to stress.

Biofeedback instruments can also sense changes in skin temperature, which reflect vasodilation (opening) of peripheral blood vessels. The arteries contain smooth muscle that controls blood flow. The muscle is in turn controlled by the autonomic nervous system as well as chemical changes in the blood. For example, you learned that during exercise, these muscles relax in the arteries near the skin in order to direct blood to the skin to get rid of excess heat. The more relaxed a person, the greater the peripheral vasodilation. (Perhaps this is why a person who is nervous about trying something is said to have "cold feet.") Blood pressure, heart rate and rhythm, and even stomach acid secretion can also be used for biofeedback.

How Does It Work?

No one knows exactly how autonomic functions are brought under voluntary control. Even somatic functions that we call voluntary are not always subject to conscious control. For example, a skeletal muscle spasm may not stop even though you will it to do so.

An exciting discovery of the early biofeedback research in the 1960s was the fact that functions previously believed to be outside an animal's or person's control could consciously be regulated. Rats learned to slow their heart rates to receive rewards, and people learned to lower blood pressure and increase hand temperature to hear fewer beeps.

People learning to use biofeedback usually work with a therapist who adjusts the instruments and gives suggestions. When biofeedback is used to decrease sympathetic arousal, patients usually practice a relaxation technique. If they are trying to warm their hands, they might imagine lying in the sun (with plenty of sunscreen) or putting their hands into very warm water. As the beeps slow down, they begin to associate certain images, thoughts, and sensations with relaxation. Important in this process is an attitude of passive attention, a "trying not to try." Willing the arteries in the hand to open only leads to more tension and less opening. Instead, the person must simply be aware, tune in, relax, and notice what seems to work.

It is interesting that what seems to work varies a great deal from person to person. Some people find imagining a soothing sound, like ocean waves, helps them achieve a relaxed state. Others find visual images helpful, such as the house they lived in as a child. Some people focus on dream images, meaningful words or prayers, or imagining physical sensations or heaviness and warmth.

Clinical Applications

Biofeedback is most widely used in the treatment of stress-related illnesses. The sympathetic division of the autonomic nervous system becomes overactive when people perceive themselves to be under stress. Biofeedback can help a person learn to decrease sympathetic activity and restore the resting homeostasis maintained by the parasympathetic division of the autonomic nervous system. You'll learn more about the relationship between stress and illness in Chapter 13, but you probably already know that stress is associated with many disorders such as ulcers, colitis, hypertension, tension and migraine headaches, and muscle tension problems such as low back pain. Biofeedback has been effective in treating all of these. It is also helpful for a disorder called Raynaud's disease, whose symptoms are cold extremities due to peripheral vasoconstriction. People with Raynaud's disease can learn to warm their hands and feet by dilating constricted arteries. Temporomandibular joint (TMJ) problems and insomnia are also sometimes treated with biofeedback.

Biofeedback is used for certain forms of muscle rehabilitation. It can help a person regain use of muscles following stroke or trauma. An EMG can sense very low levels of muscle activity and help the patient learn to increase the level to produce a muscle contraction. Biofeedback is also used to help reduce the firing of nerves that produce muscle spasms and spastic movement.

Psychological anxiety disorders often respond to biofeedback training. Mental relaxation is achieved as patients learn to induce a state of physical relaxation.

Take phobias for example. Phobias are excessive and irrational fears triggered by ordinary things, such as dogs or snakes, or experiences, such as taking exams or flying in airplanes. Students with exam phobia might begin their treatment sessions by learning some deep breathing and physical relaxation exercises. The first few sessions would be spent simply mastering these relaxation skills. When the students have become adept at sensing muscle tension and countering it with relaxation, they then imagine they are about to take an exam. Guess what happens! Right, the needle goes off the chart. Gradually, however, they learn to maintain physical relaxation in the face of the imagined exam. After several sessions they may move on to actually taking an exam, using the biofeedback equipment to monitor their state of physiological arousal.

An important part of biofeedback training is learning to transfer the skills learned during practice sessions to real-life situations. A person must be able to regulate blood pressure while driving in traffic, talking to friends, and performing a job, not just when hooked up to the biofeedback machine.

As biofeedback research continues to probe the workings of the nervous system, one of the most intriguing observations is that the more deeply into the nervous system you get, the harder it is to tell where the body ends and the mind begins. This illustration of the intimate connection between thoughts and emotions and autonomic and somatic functions supports the notion that a psychosomatic (literally "mind–body") illness is not "all in a person's mind."

neurons, called *association (connecting* or *interneuron) neurons*, carry impulses from sensory neurons to motor neurons and are located in the brain and spinal cord only.

FUNCTIONS

Two striking features of neurons are (1) their highly developed ability to produce and conduct electrical messages called nerve impulses and (2) their limited ability to regenerate.

NERVE IMPULSES

Nerve impulses are like tiny electric currents which pass along neurons. These impulses result from movement of ions (electrically charged particles) in and out through the plasma membranes of neurons. Before discussing changes in electrical charge, it is necessary to understand the kinds of channels in plasma membranes that permit movement of ions between the external and internal surfaces of the neuron's membrane.

Ion Channels in Plasma Membranes

Plasma membranes contain a variety of *ion channels*, that is, integral proteins (see Chapter 3) that span the membrane. These channels are usually highly selective with respect to which ions pass through. Some ionic channels, called *leakage* or *passive ion channels*, are always open. Most ion channels, however, are subject to regulation in which they spend some time open (conducting) and some time closed (nonconducting). These ion channels are known as *gated* or *active ion channels* (see Figure 9-5). In such channels, the passage of ions is controlled (gated) by protein molecules that form a gate which can change its shape to open or

close the channel in response to various signals. For example, *voltage-gated ion channels* open and close in response to changes in voltage of the plasma membrane. By contrast, *chemical-gated ion channels* open and close in response to chemicals, such as neurotransmitters and hormones, which bind to the channels.

Membrane Potentials

In a resting neuron (one that is not conducting an impulse), there is a difference in electrical charges on the outside and inside of the plasma membrane. The outside has a positive charge and the inside a negative charge. What causes this difference? There are two factors. One is the different numbers of potassium (K^+) and sodium (Na^+) ions on either side of the membrane. There are 30 times more K^+ ions inside the cell than outside. At the same time, there are about 15 times more Na^+ ions outside than inside. The other factor is the presence of large negatively charged ions (organic phosphates and proteins) trapped in the cell. Let us examine how these factors work to cause the difference in charge.

Even when a nerve cell is not conducting an impulse, it is *actively* transporting Na^+ ions out of the cell and K^+ ions in, at the same time, by means of the *sodium–potassium pump* (Figure 9-4a). (The mechanism of active transport may be reviewed in Figure 3-8.)

Since Na^+ ions are positive and are actively transported out, a positive charge develops outside the membrane. Even though K^+ ions are also positive and are actively transported into the cell, there are not enough K^+ ions to balance the even larger number of negative ions trapped in the cell.

In addition, the operation of the sodium–potassium pump creates a concentration and electrical gradient for Na^+ and K^+ ions, which means K^+ ions tend to diffuse (leak) out of the cell, and Na^+ ions tend to diffuse in. Moreover, membranes tend to be much

FIGURE 9-4 Development of the resting membrane potential. (a) Schematic representation of the sodium–potassium pump and distribution of ions. Note that the large number of Na^+ ions outside the membrane results in an external positive charge. Although there are more K^+ ions inside the cell membrane than outside, there are many more negative ions inside the membrane. This results in a net internal negative charge. (b) Simplified representation of a polarized membrane.

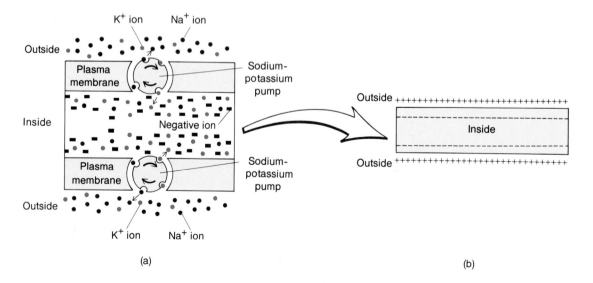

(a)

(b)

more permeable to K^+ ions, so K^+ ions diffuse out along their concentration gradient easily and quickly while Na^+ ions enter along their gradient much more slowly. The final result is a net positive charge outside and net negative charge inside. The difference in charge on either side of the membrane of a resting neuron is the *resting membrane potential*. Such a membrane is said to be *polarized* (Figure 9-4b). Said another way, a polarized membrane is one that is positive on the outside and negative on the inside. The electrical charge on the inside of a polarized membrane is -70 millivolts (mV). One millivolt equals one-thousandth of a volt; a 1.5-volt battery will power an ordinary flashlight. The membrane permeability, and thus the resting membrane potential, can be altered by neurotransmitters as well as other mechanisms.

Excitability

We will now look at how a nerve impulse, or nerve action potential, is generated. The ability of nerve cells to respond to stimuli and convert them into nerve impulses is called *excitability*. A *stimulus* is anything in the environment capable of altering the resting membrane potential.

If a stimulus of adequate strength (threshold stimulus) is applied to a polarized membrane, the membrane's permeability to Na^+ ions greatly increases at the point of stimulation (Figure 9-5), and voltage-gated sodium channels open to permit Na^+ ions to enter. The Na^+ ions also move in because they are attracted to the negative ions on the inside of the membrane. With more Na^+ ions entering than leaving at this point, the resting membrane potential begins to change. At first, the charge *inside* the membrane shifts from negative (-70 mV) toward 0 and then to positive ($+30$ mV). In relation to the inside of the membrane, the outside is now relatively negatively charged. This change is called *depolarization* and the membrane is said to be *depolarized* when it is positively charged on the inside and negative on the outside. Throughout depolarization, the Na^+ ions continue to rush inside until the membrane potential is *reversed*: the inside of the membrane becomes positive and the outside negative.

Once depolarization has taken place at a specific point on the membrane, that point immediately becomes *repolarized*, that is, its resting potential is restored. Just as depolarization results from changes in membrane permeability, so does repolarization. Repolarization, however, involves voltage-gated potassium channels. When the membrane is polarized (resting), the potassium channel gate is nearly closed and K^+ ions remain inside (Figure 9-5a). When the membrane becomes depolarized, the gate opens and K^+ ions rapidly diffuse out. At the same time, the voltage-gated sodium channels are closing, so the net effect is few Na^+ ions entering and many K^+ ions exiting. The loss of positive ions leaves the inner membrane negative again, that is, repolarized. Only a small fraction of ions move during a single depolarization or repolarization, so many action potentials may occur before the sodium–potassium pump must restore the concentration of ions to their original sites.

Once the events of depolarization and repolarization have taken place, we say that a *nerve impulse* (*nerve action potential*) has occurred. An *action potential* is a rapid change in membrane potential that involves a depolarization followed by a repolarization (restoration of the resting potential, that is, a return to negative inside, positive outside). The impulse self-propagates along the outside surface of the membrane of a neuron. Of all the cells of the body, only muscle fibers and nerve cells produce action potentials.

For the nerve impulse to communicate information to another part of the body, it must be propagated (transmitted) along the neuron. A nerve impulse generated at any one point on the membrane excites (depolarizes) adjacent portions of the membrane, causing depolarization of the adjacent areas, opening of new voltage-gated sodium channels, inward movement of Na^+ ions, and development of new nerve impulses at successive points along the membrane (Figure 9-5b–d).

Following depolarization, repolarization returns the cell to its resting membrane potential and the neuron is now prepared to receive another stimulus and conduct it in the same manner. In fact, until repolarization occurs, the neuron cannot conduct another nerve impulse. The period of time during which the neuron cannot generate another nerve action potential is called the *refractory period*.

The electrical changes associated with a nerve impulse are illustrated in Figure 9-5e.

Generally, a nerve impulse travels in only one direction along a neuron. A sensory neuron is stimulated at its dendrite by a receptor, a structure sensitive to changes in the environment. Association and motor neurons are stimulated at their dendrites or cell bodies by another neuron.

The initiation and conduction of a *muscle action potential* are basically similar to a nerve action potential (nerve impulse), although the duration of a muscle action potential is considerably longer, while the velocity of conduction of a nerve action potential is about 18 times faster.

All-or-None Principle

Any stimulus strong enough to initiate a nerve impulse is referred to as a *threshold stimulus*. A single nerve cell, just like a single muscle fiber, transmits an action potential according to the *all-or-none principle:* If a stimulus is strong enough to generate a nerve action potential, the impulse is conducted along the entire neuron at maximum strength, unless conduction is altered by conditions such as toxic materials in cells or fatigue.

An analogy helps in understanding this principle. If a long trail of gunpowder were spilled along the ground and ignited at one end, it would send a blazing signal down the entire length of the trail. It would not matter how tiny or great the triggering spark or flame or explosion was; the blazing signal moving along the trail of gunpowder would be just the same, a maximal one (or none at all if it never started).

Any stimulus weaker than a threshold stimulus is a *subthreshold stimulus*. If it occurs only once, it is incapable of initiating a nerve impulse. If, however, a second stimulus or a series of subthreshold stimuli is quickly applied to the neuron, the cumulative effect may be sufficient to initiate an impulse. This phenomenon is called summation and is discussed shortly.

Saltatory Conduction

Thus far we have been considering nerve impulse conduction for unmyelinated fibers. The step-by-step depolarization of each adja-

FIGURE 9-5 Initiation and propagation of a nerve impulse. (a–d) The stippled area containing the straight arrow represents the region of the membrane that is propagating the nerve impulse. The curved arrows represent local currents. (e) Record of potential changes of a nerve impulse.

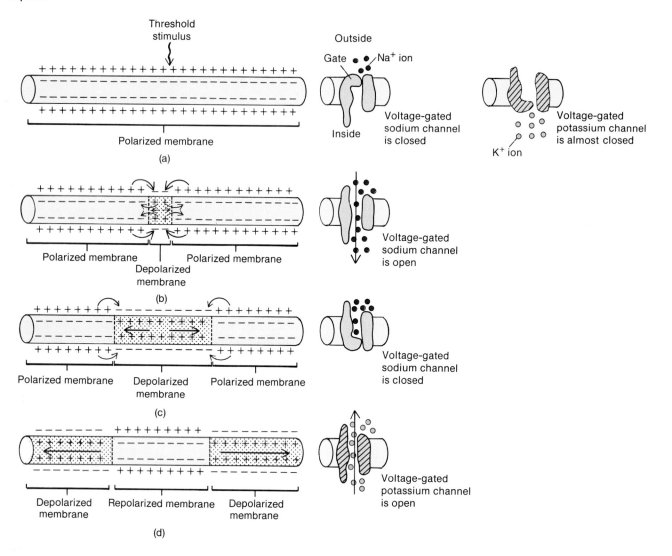

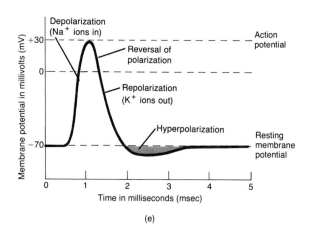

In a polarized membrane, what is the charge just inside the membrane?

cent area of the axon or dendrite membrane is called *continuous conduction*. In myelinated fibers, conduction is somewhat different. The myelin in a myelin sheath is a phospholipid and does not conduct electric current. It is a fatty insulating layer that virtually inhibits the movement of ions. However, as you recall, a myelin sheath is interrupted at various intervals called neurofibril nodes (nodes of Ranvier). Depolarization can occur at the nodes and nerve impulses can be generated and conducted. In a myelinated fiber, the nerve impulse jumps from node to node (Figure 9-6). This type of impulse conduction is called *saltatory conduction* (*saltare* = leaping).

Saltatory conduction is important in maintaining homeostasis. Since an impulse jumps long intervals as it moves from one node to the next, it travels much faster than in the step-by-step depolarization. This is especially important when split-second responses are necessary. Saltatory conduction is also more efficient. Because saltatory conduction does not require depolarization of large areas

FIGURE 9-6 Saltatory conduction. (a) The nerve impulse at the first node generates a local current that passes to the second node. (b) At the second node, the local current generates a nerve impulse. Then, the nerve impulse from the second node generates a local current that passes to the third node, and so on. After the nerve impulse jumps from node to node, each node becomes repolarized.

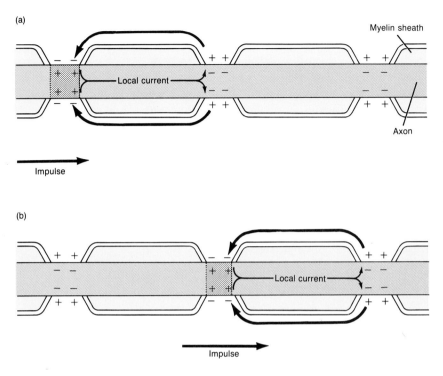

Do A, B, or C fibers conduct impulses fastest? What type of information does the nervous system transmit most quickly and why is this important?

of the plasma membrane, there is little leakage of Na^+ ions and K^+ ions each time a nerve impulse is transmitted. This results in a lower expenditure of energy by the sodium–potassium pump.

Speed of Nerve Impulses

The speed of a nerve impulse is independent of stimulus strength. Rather, if a neuron receives a threshold stimulation, the speed of the nerve impulse is determined by temperature, the diameter of the fiber, and the presence or absence of myelin.

When warmed, nerve fibers conduct impulses at higher speeds; when cooled, at lower speeds. Consequently, pain resulting from injured tissue can be reduced by the application of cold because the nerve fibers carrying the pain sensation are partially blocked.

Fibers with large diameters conduct impulses faster than those with small ones. Fibers with the largest diameter are called *A fibers* and are all myelinated and, therefore, capable of saltatory conduction. They transmit impulses at speeds up to 130 m/sec. A fibers are located in the axons of large sensory nerves that relay impulses associated with touch, pressure, position of joints, heat, and cold, and in all motor nerves that convey impulses to the skeletal muscles. Sensory A fibers generally connect the brain and spinal cord with sensors that detect danger in the outside

environment. Motor A fibers stimulate the muscles that can do something about the situation. If you touch a hot object, information about the heat passes over sensory A fibers to the spinal cord where it is relayed to motor A fibers that stimulate the hand muscles to withdraw instantaneously. A fibers are located where split-second reaction may mean survival or prevention of serious injury.

B fibers have a middle-sized diameter; they are also myelinated and therefore capable of saltatory conduction. They conduct impulses at speeds of about 10 m/sec. B fibers transmit impulses from the skin and viscera to the brain and spinal cord.

C fibers have the smallest diameter. They are unmyelinated, so their conduction is continuous, at the rate of about 0.5 m/sec. C fibers conduct impulses for pain from the skin and viscera. The motor functions of B and C fibers include constricting and dilating the pupils, increasing and decreasing the heart rate, and contracting and relaxing the urinary bladder—functions of the autonomic (involuntary) nervous system.

CONDUCTION ACROSS SYNAPSES

A nerve impulse is conducted not only along the length of a neuron but also from one neuron to another or to an effector such as a muscle or gland. Impulses from a neuron to a muscle fiber are conducted across a ***neuromuscular junction***, which was discussed in Chapter 8 (see Figure 8-4). Between a neuron and glandular cell, the impulse crosses a ***neuroglandular junction.***

Impulses are conducted from one neuron to another or from a neuron to another cell such as a muscle fiber or glandular cell across a ***synapse***—a junction between the cells. Within a synapse, the involved cells approach, but do not quite touch, one another. The synapse is essential for homeostasis because of its ability to transmit certain impulses and inhibit others. Most diseases of the brain and many psychiatric disorders result from a disruption of synaptic communication. Synapses are also the sites of action for most drugs that affect the brain, both therapeutic and addictive substances.

Figure 9-7b shows the three parts of a synapse between neurons. The ***presynaptic neuron*** is a neuron located before a synapse. The ***postsynaptic neuron*** is located after a synapse. The space between, filled with extracellular fluid, is the ***synaptic cleft***. Axon terminals of neurons end in bulblike structures called ***synaptic***

end bulbs. The synaptic end bulbs of a presynaptic neuron may synapse with the dendrites, cell body, or axon of a postsynaptic neuron.

There are two types of synapse, electrical and chemical. At ***electrical synapses*** an action potential passes from one neuron to another through small, tubular, protein structures called ***gap junctions*** (Figure 9-7a). Recall that gap junctions are also found in visceral smooth muscle and cardiac muscle fibers (cells) for conducting muscle action potentials between muscle fibers.

In a ***chemical synapse***, a neuron secretes a chemical substance called a ***neurotransmitter*** that acts on receptors of the next neuron, a muscle fiber at a neuromuscular junction, or a glandular cell at a neuroglandular junction (see Figure 8-4c). Almost all synapses in the CNS are chemical synapses. Neurotransmitters are made by the neuron, usually from amino acids, and are stored in the synaptic end bulbs in small membranous sacs called ***synaptic vesicles*** (Figure 9-7b). (The various kinds of neurotransmitters will be studied in Chapter 10.)

How are neurotransmitters released? When a nerve impulse ar-

FIGURE 9-7 **Types of synapses. (a) Electrical. (b) Chemical.**

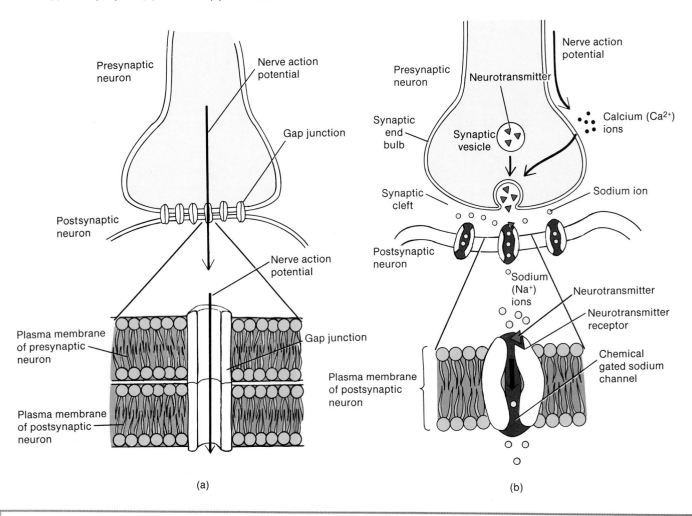

(a)

(b)

At chemical synapses, there is only one-way impulse conduction because only synaptic end bulbs of which neurons release neurotransmitter?

rives at the synaptic end bulb of a presynaptic neuron and depolarization occurs, voltage-gated calcium channels in the bulb open. Calcium ions flood in from interstitial fluid, attract synaptic vesicles to the plasma membrane, and help liberate the neurotransmitter molecules from the vesicles into the synaptic cleft (see Figure 8-4c). (In general, each neuron liberates only one type of neurotransmitter.) What happens once the neurotransmitter enters the synaptic cleft depends on the chemical nature of the neurotransmitter and how it interacts with the receptors on the postsynaptic plasma membrane. In general, though, a neurotransmitter will produce an excitatory transmission (create a new nerve impulse) or an inhibitory transmission (prevent further nerve impulses).

At a chemical synapse, there is only *one-way impulse conduction*—from a presynaptic axon to a postsynaptic cell. This is because only synaptic end bulbs of presynaptic neurons can release neurotransmitter. As a result, nerve impulses must move forward over their pathways. They cannot back up into another presynaptic neuron, a situation that would seriously disrupt homeostasis. One-way impulse conduction is crucial in preventing nerve impulse conduction along improper pathways.

Excitatory Transmission

In an *excitatory transmission,* the neurotransmitter lowers (makes less negative) the postsynaptic neuron's membrane potential so that a new nerve impulse can be generated across the synapse. If the potential is lowered enough, a nerve impulse is initiated, but generally, the release of a neurotransmitter by a single presynaptic end bulb is not sufficient to develop an action potential. However, the postsynaptic neuron does become more excitable to impulses from presynaptic neurons and is thus prepared for subsequent stimuli that can trigger an impulse because it is already partially depolarized. This near-threshold excitation is called *facilitation*.

If several presynaptic end bulbs release their neurotransmitter at about the same time, however, the combined effect may initiate a nerve impulse—an effect known as *summation*. The more presynaptic end bulbs that release their neurotransmitter simultaneously, in other words, the greater the summation, and therefore the depolarization, the greater the probability an impulse will be initiated.

Once an excitatory neurotransmitter attaches to a receptor site, it must be rapidly inactivated or it will stimulate (depolarize) the postsynaptic neuron, muscle, or gland indefinitely (see Figure 8-4c). This would prevent repolarization of the membrane, which must occur for normal neuron functioning.

Inhibitory Transmission

In *inhibitory transmission*, the neurotransmitter inhibits nerve impulse generation at a synapse. Whereas excitatory transmission makes the postsynaptic neuron's resting membrane potential *less* negative (causes depolarization), inhibitory transmission makes the postsynaptic neuron's resting membrane potential *more* negative. This is referred to as *hyperpolarization* (see Figure 9-5e). With the cell interior even more negative in comparison to the outside, it is more difficult for the neuron to generate an impulse because its membrane potential is even farther from threshold than it was in its resting state. The resting membrane potential is made more

negative as follows. When the neurotransmitter attaches to the receptor site, the membrane becomes more permeable to K^+ and/ or Cl^- ions. Permeability to Na^+ ions is not affected. When voltage-gated potassium channels are open, K^+ ions move to the exterior of the membrane. When voltage-gated chloride channels are open, Cl^- ions move to the interior of the membrane. As a result, there is an increase in internal negativity, or hyperpolarization.

Integration at Synapses

A single postsynaptic neuron receives signals from many presynaptic neurons. Some neurotransmitters produce excitation and some produce inhibition. The sum of all the effects, excitatory and inhibitory, determines the final effect on the postsynaptic neuron. Thus, the postsynaptic neuron is an *integrator*. It receives signals, integrates them, and then responds accordingly. The postsynaptic neuron may respond in the following ways:

1. If the excitatory effect is greater than the inhibitory effect, but less than the threshold level of stimulation, the result is *facilitation*, that is, near-threshold excitation so that subsequent stimuli can more easily generate a nerve impulse.
2. If the excitatory effect is greater than the inhibitory effect, but equal to or higher than the threshold level of stimulation, the result is *generation of one or many nerve impulses*, one after another.
3. If the inhibitory effect is greater than the excitatory effect, the membrane hyperpolarizes, and the result is inhibition of the postsynaptic neuron and thus an *inability to generate a nerve impulse*.

Alteration of Synaptic Conduction

There are many ways that *synaptic conduction can be altered* by disease, drugs, and pressure. In Chapter 8 it was noted that *myasthenia gravis* results from antibodies directed against acetylcholine receptors on skeletal muscle fiber membranes at neuromuscular junctions, causing dysfunctions in skeletal muscular contractions. *Alkalosis*, an increase in pH above 7.45, results in increased excitability of neurons that can cause lightheadedness, numbness around the mouth, tingling in the fingertips, nervousness, and muscle spasms and convulsions. *Acidosis*, a decrease in pH below 7.35, results in a progressive depression of neuronal activity that can produce apathy, weakness, and coma.

The plant derivative *curare* competes for acetylcholine receptor sites and can thus prevent muscular contractions. Curare-like drugs are often used in surgery to increase muscle relaxation. *Neostigmine* is an anticholinesterase agent that combines with acetylcholinesterase to inactivate it for several hours. Neostigmine is the antidote to curare used in surgery or in case of accidental curare poisoning and can be used to treat myasthenia gravis.

Diisopropyl fluorophosphate is a very powerful nerve gas found in many insecticides. It inactivates acetylcholinesterase for up to several weeks, making it a particularly lethal drug. It may cause nausea, diarrhea, sweating, bronchial constriction, excess respiratory mucus, generalized weakness, and fasciculation of skeletal muscles. The *botulism toxin* inhibits the release of acetylcholine, thus inhibiting muscle contraction. It is the substance involved in one type of food poisoning. *Hypnotics, tranquilizers,* and *anesthetics* depress synaptic conduction by increasing the threshold for excitation of neurons, whereas *caffeine, benzedrine,* and *nicotine* reduce the threshold for excitation of neurons and result in facilitation.

Crack, a potent form of cocaine that is smoked rather than sniffed, interferes with the normal functioning of neurotransmitters—dopamine (DA), norepinephrine (NE), and serotonin (5-HT)—that are involved in the regulation of mood and motor functions. Once a neurotransmitter has accomplished its function, it is inactivated and returned to the presynaptic neuron for resynthesis. Initially, crack inhibits the inactivation of the neurotransmitters. The resultant buildup of dopamine, in particular, has been linked to feelings of euphoria. Inhibition or inactivation of neurotransmitters also can cause convulsions, accelerated and abnormal heart rate, vasoconstriction and high blood pressure, weight loss, insomnia, and susceptibility to disease. Repeated use of crack may produce temporary shortage of neurotransmitters, resulting in depression, anxiety, and craving for more crack. Heavy, prolonged use of crack may eventually deplete neurotransmitters to the point where euphoria no longer occurs and depression is persistent.

Pressure has an effect on nerve impulse transmission also. If excessive or prolonged pressure is applied to a nerve, as when crossing one's legs, impulse transmission is interrupted, and part of the body may "go to sleep," producing a tingling sensation. This sensation is caused by an accumulation of waste products and a depressed circulation of blood.

GRADED POTENTIALS

As we have just seen, certain changes in membrane potential generate action potentials. Once generated, action potentials are propagated at maximum strength, according to the all-or-none principle. In addition to action potentials, neurons can also develop graded potentials, which differ considerably from action potentials. **Graded potentials** are localized changes in membrane potential (depolarizations or hyperpolarizations) that are very short-lived and decrease in intensity as they travel along a nerve fiber.

These potentials are called "graded" because their magnitude varies directly with the intensity of the stimulus—the greater the voltage, the farther it travels along the nerve fiber. Moreover, graded potentials do not have a refractory period. This means that if a second stimulus is applied to a neuron before a graded potential resulting from the first stimulus disappears, the second stimulus adds to the first to produce an even greater response. Thus, graded potentials are capable of summation. When graded potentials reach threshold, they can initiate action potentials. Graded potentials are discussed in Chapter 12.

REGENERATION

Unlike the cells of epithelial tissue, neurons have only limited powers of **regeneration**, that is, a natural ability to renew themselves. Around six months of age, nerve cells lose their mitotic apparatus (centrioles and mitotic spindles) and their ability to reproduce. Thus, when a neuron is damaged or destroyed, it cannot be replaced by other neurons. A neuron destroyed is permanently lost, and only some types of damage may be repaired.

In the peripheral nervous system (PNS), damage to some types of myelinated axons and dendrites can be repaired if the cell body remains intact and if the cell that performs the myelination remains active. If the myelinating cell is a neurolemmocyte (Schwann cell), it helps regeneration. These cells proliferate following axonal damage, and their neurolemmas form a tube that assists in regeneration. Axons in the central nervous system (CNS) are myelinated by oligodendrocytes, which do not form neurolemmas to assist in regeneration and do not survive following axonal damage. An added complication in the CNS is that following axonal damage, astrocytes appear to stop axons from regenerating by activating what is called a physiological stop-pathway. In addition, following axonal damage, astroglial proliferation causes rapid scar tissue formation, and scar tissue is an actual physical barrier to regeneration. Thus, an injury to the brain or spinal cord has permanent effects. An injury to a nerve in the PNS may repair itself before scar tissue forms. As a result, some nerve function may be restored.

STUDY OUTLINE

Organization (p. 188)

1. The nervous system helps control and integrate all body activities by sensing changes (sensory), interpreting them (integrative), and responding to them (motor).
2. The nervous system has two principal divisions: central nervous system (CNS) and peripheral nervous system (PNS).
3. The CNS consists of the brain and spinal cord. Within the CNS, incoming sensory information is integrated and correlated, thoughts and emotions are generated, and muscles are stimulated

to contract and glands to secrete via outgoing nerve impulses.
4. The PNS consists of nerves emerging from the brain (cranial nerves) and from the spinal cord (spinal nerves). The nerves carry impulses inward from receptors to the CNS and outward from the CNS to muscles and glands.
5. Based on the origin of incoming sensory information and the part of the body that responds, the PNS is subdivided into a somatic nervous system (SNS) and autonomic nervous system (ANS).
6. The SNS consists of sensory neurons that convey information

from cutaneous and special sense receptors primarily in the head, body wall, and extremities to the CNS and motor neurons that conduct impulses to skeletal muscle only. The SNS is voluntary.

7. The ANS consists of sensory neurons that convey homeostatic information from receptors primarily in the viscera to the CNS and motor neurons from the CNS that conduct impulses to smooth muscle, cardiac muscle, and glands. The ANS is involuntary.

Histology (p. 188)

Neuroglia (p. 188)

1. Neuroglia are specialized tissue cells that support neurons, attach neurons to blood vessels, produce the myelin sheath around axons of the CNS, and carry out phagocytosis.

2. Neuroglial cells include astrocytes, oligodendrocytes, microglia, and ependyma.

Neurons (p. 189)

1. Neurons (nerve cells) consist of a cell body, dendrites that receive stimuli, and a single axon that sends impulses to another neuron or to an effector, a muscle or gland.

2. On the basis of structure, neurons are multipolar, bipolar, and unipolar.

3. On the basis of function, sensory (afferent) neurons conduct impulses to the CNS; association neurons conduct impulses to other neurons, including motor neurons; and motor (efferent) neurons conduct impulses to effectors.

Functions (p. 194)

Nerve Impulses (p. 194)

1. The nerve impulse (nerve action potential) is the body's quickest way of controlling and maintaining homeostasis.

2. Plasma membranes contain voltage-gated and chemical-gated ion channels. A channel is an integral protein.

3. The membrane of a nonconducting neuron is positive outside and negative inside, owing to the differing numbers of K^+ and Na^+ ions, large, negatively charged proteins, and the operation of the sodium–potassium pump. Such a membrane is said to be polarized.

4. When a stimulus causes the inside of the cell membrane to become positive and the outside negative, the membrane is said to have an action potential, which travels along the membrane. The traveling action potential is a nerve impulse. The ability of a neuron to respond to a stimulus and convert it into a nerve impulse is called excitability.

5. Restoration of the resting membrane potential is called repolarization. The period of time during which the membrane recovers and cannot initiate another action potential is called the refractory period.

6. According to the all-or-none principle, if a stimulus is strong enough to generate an action potential, the impulse travels at a constant and maximum strength. A stronger stimulus will not cause a larger impulse.

7. Nerve impulse conduction that occurs as a step-by-step process is called continuous conduction.

8. Conduction in which the impulse jumps from node to node is called saltatory conduction.

9. Fibers with larger diameters conduct impulses faster than those with smaller diameters; myelinated fibers conduct impulses faster than unmyelinated.

Conduction Across Synapses (p. 197)

1. Nerve impulse conduction can occur from one neuron to another or from a neuron to an effector.

2. The junction between neurons, neurons and muscle fibers, or neurons and glandular cells is called a synapse. Synapses may be electrical or chemical.

3. At a chemical synapse, there is only one-way nerve impulse conduction from a presynaptic axon to a postsynaptic dendrite, cell body, or axon.

4. In an excitatory transmission, the neurotransmitter depolarizes or lowers (makes less negative) the postsynaptic neuron's membrane potential, so that a new impulse can be generated across the synapse.

5. Facilitation refers to a state of near-threshold excitation, so that subsequent stimuli can generate an impulse more easily.

6. If several presynaptic end bulbs release their neurotransmitter at about the same time, the combined effect may generate a nerve impulse, resulting in the phenomenon referred to as summation.

7. In an inhibitory transmission, the neurotransmitter raises (makes more negative or hyperpolarizes) the postsynaptic neuron's membrane potential and thus inhibits an impulse at a synapse.

8. The postsynaptic neuron is an integrator. It receives signals, integrates them, and then responds accordingly.

9. Synaptic conduction may be altered by disease, drugs, and pressure.

Graded Potentials (p. 200)

1. Graded potentials are localized changes in membrane potential.

2. They are short-lived and decrease in intensity as they travel along a nerve fiber.

Regeneration (p. 200)

1. At about six months of age, the neuron loses its mitotic apparatus and is no longer able to divide.

2. Nerve fibers in the PNS have a neurolemma and are thus capable of regeneration.

3. Axons and dendrites in the CNS do not have a neurolemma, which means injury to the brain or spinal cord has permanent effects.

REVIEW QUESTIONS

1. Describe the three basic functions of the nervous system in maintaining homeostasis. (p. 188)
2. Distinguish between the central and peripheral nervous systems and describe the components and functions of each subdivision. (p. 188)
3. What are neuroglia? List the principal types and their functions. Why are they important clinically? (p. 188)
4. Define a neuron. Diagram and label a neuron. Next to each part list its function. (p. 189)
5. What is a myelin sheath? Why is it important? (p. 191)
6. Discuss the structural and functional classification of neurons. Give an example of each. (p. 191)
7. Define excitability. (p. 195)
8. Outline the principal steps in the origin and conduction of a nerve impulse (action potential). (p. 195)
9. Define the following: resting membrane potential, polarized membrane, nerve impulse (nerve action potential), depolarized membrane, repolarized membrane, and refractory period. (p. 195)
10. What is the all-or-none principle? Relate it to threshold stimulus, subthreshold stimulus, and summation. (p. 195)
11. What is continuous conduction? What is saltatory conduction and why is it important? (p. 196)
12. What factors determine the speed of nerve impulses? Give several examples. (p. 197)
13. Define a synapse. Distinguish between electrical and chemical synapses. (p. 198)
14. What events are involved in the conduction of a nerve impulse across a synapse. (p. 198)
15. Why does one-way impulse conduction occur at chemical synapses? (p. 199)
16. Describe some of the factors that might alter synaptic conduction. (p. 199)
17. Describe how neurons regenerate. (p. 200)

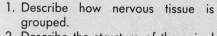

Central and Somatic Nervous Systems

STUDENT OBJECTIVES

1. Describe how nervous tissue is grouped.
2. Describe the structure of the spinal cord.
3. Explain the functions of the spinal cord as a conduction pathway and a reflex center.
4. Describe how a reflex arc works and its relationship to homeostasis.
5. Name the 31 pairs of spinal nerves and describe the composition, coverings, and branches of a spinal nerve.
6. Describe how the brain is protected.
7. Name the principal parts of the brain and explain the functions of each.
8. Compare the functions of the motor, sensory, and association areas of the cerebrum.
9. Describe an electroencephalogram (EEG) and explain its significance in the diagnosis of certain disorders.
10. Discuss the various classes of neurotransmitters.
11. Identify the 12 pairs of cranial nerves by name, number, type, location, and function.
12. Define common disorders and medical terminology and conditions associated with the central nervous system.

A LOOK AHEAD

GROUPING OF NEURAL TISSUE
SPINAL CORD
    Protection and Coverings
        Vertebral Canal
        Meninges
    General Features
    Structure in Cross Section
    Functions
        Impulse Conduction
        Reflex Center
        Reflex Arc and Homeostasis
SPINAL NERVES
    Names
    Composition and Coverings
    Distribution
        Branches
        Plexuses
        Intercostal (Thoracic) Nerves
BRAIN
    Principal Parts
    Protection and Coverings
    Cerebrospinal Fluid (CSF)
    Blood Supply
    Brain Stem
        Medulla Oblongata
        Pons
        Midbrain
    Diencephalon
        Thalamus
        Hypothalamus
    Reticular Activating System (RAS),
        Consciousness, and Sleep
    Cerebrum
        Lobes
        Brain Lateralization (Split Brain
          Concept)
        White Matter
        Basal Ganglia (Cerebral Nuclei)
        Limbic System
        Functional Areas of
          Cerebral Cortex
        Electroencephalogram (EEG)
    Cerebellum
NEUROTRANSMITTERS
CRANIAL NERVES
COMMON DISORDERS
MEDICAL TERMINOLOGY AND
    CONDITIONS

I n this chapter we will examine the structure and function of the central nervous system—the brain and spinal cord. We will also take a look at spinal nerves and cranial nerves, which are part of the somatic nervous system. But first, let us look at how neural tissue is grouped in the body.

GROUPING OF NEURAL TISSUE

A *nerve* is a bundle of fibers (axons and/or dendrites) outside the central nervous system (CNS). Cell bodies of neurons that lie outside the CNS are generally found in groups called *ganglia* (GANG-lē-a; *ganglion* = knot).

A *tract* is a bundle of fibers in the CNS. Tracts may run long distances up or down the spinal cord. Tracts also exist in the brain and connect parts of the brain with each other and with the spinal cord. Spinal tracts that conduct impulses up the cord and carry sensory impulses are called *ascending tracts*. Spinal tracts that carry impulses down the cord carry motor impulses and are called *descending tracts*. Tracts consist of myelinated fibers, which make up the white matter of the CNS.

The term *white matter* refers to groups of myelinated axons from many neurons. Myelin is a phospholipid and has a whitish color that gives white matter its name. The *gray matter* of the nervous system contains either neuron cell bodies and dendrites or bundles of unmyelinated axons. The absence of myelin in these areas accounts for the gray color.

In the brain, gray matter is found covering its outer surface and in deeper regions called nuclei. A *nucleus* is similar to a ganglion, but it contains unmyelinated dendrites as well as cell bodies of neurons. In the spinal cord, gray matter is located internally in regions called *horns*, surrounded by white matter called *columns* (see Figure 10-2).

We will now examine the structure and functions of the spinal cord in detail.

SPINAL CORD

PROTECTION AND COVERINGS

Vertebral Canal

The spinal cord is located in the vertebral canal of the vertebral column. Since the wall of the vertebral canal is essentially a ring of bone, the cord is well protected. Additional protection is provided by the meninges, cerebrospinal fluid, and vertebral ligaments.

Meninges

The *meninges* (me-NIN-jēz) are connective tissue coverings that run continuously around the spinal cord and brain. They are called, respectively, the *spinal meninges* and the *cranial meninges* (see Figure 10-6). The outermost of the three layers of the meninges is called the *dura mater* (DYOO-ra MĀ-ter). This means "tough mother," since its tough white fibrous connective tissue helps to protect the delicate structures of the CNS. It provides additional support to the brain by folding down between the lobes. Although it attaches the brain to the inner surface of the cranial bones, it is not attached to the vertebrae. The spinal cord is also protected by a cushion of fat and connective tissue located in the *epidural space*, a space between the dura mater and vertebral canal. The spinal dura mater continues as the cranial dura mater of the brain. The tube of spinal dura mater ends just below the spinal cord around the second lumbar vertebra.

The middle layer is called the *arachnoid* (a-RAK- noyd), or spider layer, because of its delicate weblike appearance. It is also continuous with the arachnoid of the brain.

The inner layer is the *pia mater* (PĪ-a MĀ-ter), or delicate mother, a transparent fibrous membrane that adheres to the surface of the spinal cord and brain. It contains numerous blood vessels. Between the arachnoid and the pia mater is the *subarachnoid space*, where cerebrospinal fluid circulates.

Cerebrospinal fluid is removed from the subarachnoid space between the third and fourth or fourth and fifth lumbar vertebrae by a *spinal tap*. Inflammation of the meninges is known as *meningitis*.

GENERAL FEATURES

The length of the adult *spinal cord* ranges from 42 to 45 cm (16 to 18 inches). It extends from the foramen magnum of the occipital bone to the second lumbar vertebra (Figure 10-1). It does not run the entire length of the vertebral column. Consequently, nerves arising from the lowest portion of the cord angle down the vertebral canal like wisps of flowing hair. They are appropriately named the *cauda equina* (KAW-da ē-KWĪ-na), meaning horse's tail.

The cord is not a straight, narrow structure. It has two conspicu-

FIGURE 10-1 **Spinal cord and spinal nerves seen in posterior view.**

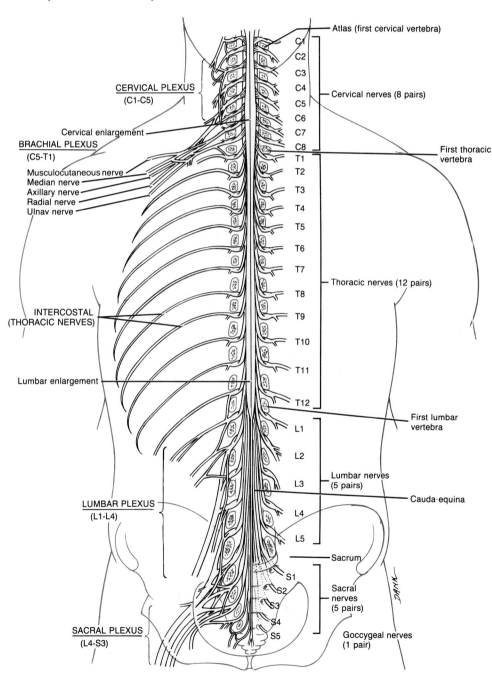

The spinal nerves are part of which division of the nervous system? What is the total number of spinal nerves? What is the cauda equina?

ous enlargements. The *cervical enlargement* contains nerves that supply the upper extremities; the *lumbar enlargement* contains nerves supplying the lower extremities.

The cord consists of 31 *spinal segments,* each giving rise to a pair of spinal nerves (see Figure 10-2).

STRUCTURE IN CROSS SECTION

The cord is divided into right and left halves by two grooves, the deep *anterior median fissure* and the shallower *posterior median sulcus* (see Figure 10-3). As previously described, the spinal cord contains a centrally located H-shaped mass of gray matter surrounded by white matter (Figure 10-2). In the center of the gray matter is the *central canal,* which runs the length of the cord and contains cerebrospinal fluid. The sides of the H are divided into regions called *horns,* named relative to their location: anterior, lateral, and posterior. The gray matter consists mainly of association and motor neurons that serve as relay stations for impulses (more on this later). The white matter is also organized into regions called anterior, lateral, and posterior *columns*. The columns consist of myelinated axons organized into sensory (ascending) and motor (descending) tracts that convey impulses between the brain and spinal cord.

FUNCTIONS

A major function of the spinal cord is to convey sensory impulses from the periphery to the brain and to conduct motor impulses from the brain to the periphery. A second principal function is related to reflexes. Both functions are essential to maintaining homeostasis.

Impulse Conduction

The vital function of conveying sensory and motor information to and from the brain is carried out by the ascending and descending tracts of the cord. The names of the tracts indicate the white column in which the tract travels, where the tract originates, and where it terminates. Since the origin and termination are specified, the direction of impulse conduction is also indicated by the name. For example, the anterior spinothalamic tract is located in the *anterior* white column, it originates in the *spinal* cord, and it terminates in the *thalamus* (a region of the brain). It is, therefore, an ascending (sensory) tract. The principal ascending and descending tracts are shown in Figure 10-3.

Sensory information transmitted from receptors up the spinal cord to the brain is conducted along two general pathways: posterior column and spinothalamic. The *posterior column pathway* (fasciculus gracilis and fasciculus cuneatus) carries impulses related to proprioception (awareness of the activities of muscles, tendons, and joints and equilibrium), discriminative touch (ability to recognize exactly what part of the body is touched), two-point discrimination (ability to distinguish that two points on the skin are touched even though they are close together), and vibrations. The *spinothalamic pathway* conveys impulses for pain and temperature (lateral spinothalamic) and light touch and pressure (anterior spinothalamic).

The sensory systems keep the central nervous system aware of the external and internal environments. Responses to this information are brought about by the motor systems, which enable us to move about and change our relationship to the world around us. As sensory information is conveyed to the central nervous system, it becomes part of a large pool of sensory input. We do not respond

FIGURE 10-2 **Spinal cord.** The organization of gray and white matter in the spinal cord as seen in cross section. The front of the figure has been sectioned at a lower level than the back so that you can see what is inside the posterior root ganglion, posterior root of the spinal nerve, anterior root of the spinal nerve, and the spinal nerve.

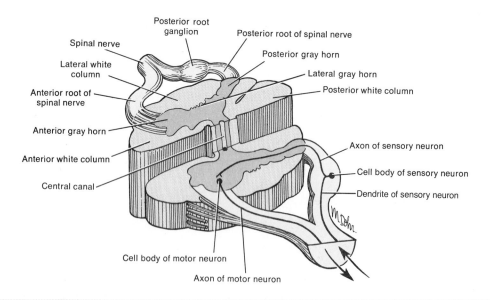

Posterior root ganglion
Posterior root of spinal nerve
Spinal nerve
Posterior gray horn
Lateral white column
Lateral gray horn
Anterior root of spinal nerve
Posterior white column
Anterior gray horn
Axon of sensory neuron
Anterior white column
Cell body of sensory neuron
Central canal
Dendrite of sensory neuron
Cell body of motor neuron
Axon of motor neuron

The posterior root contains which type of nerve fibers? The anterior root contains which type of nerve fibers?

FIGURE 10-3 **Selected tracts of the spinal cord. Ascending (sensory) tracts are indicated in pink; descending (motor) tracts are shown in blue.**

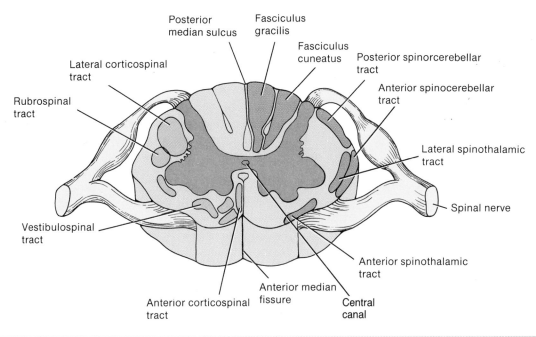

From where to where do ascending tracts carry impulses? From where to where do descending tracts carry impulses?

actively to every bit of input the central nervous system receives. Rather, each piece of incoming information is integrated with all the other information arriving from activated sensory receptors. The integration process occurs not just once but at many stations along the pathways of the central nervous system with the help of association neurons. It occurs within the spinal cord, brain stem, cerebellum, and cerebral motor cortex. As a result, a motor response to make a muscle contract or a gland secrete can be initiated at any of these levels. The cerebral motor cortex assumes the major role for controlling precise, discrete, muscular movements. The basal ganglia largely integrate semivoluntary movements like walking, swimming, and laughing. The cerebellum, although not a control center, assists the motor cortex and basal ganglia by making body movements smooth and coordinated.

When the input reaches the highest center, sensory–motor integration occurs. This involves not only using information contained within that center but also information coming to it from other centers in the central nervous system. After integration occurs, the output of the center is sent down the spinal cord in two major descending motor pathways: the pyramidal pathways and extrapyramidal pathways.

Pyramidal pathways (lateral corticospinal, anterior corticospinal, corticobulbar) convey impulses to skeletal muscles that result in precise movements. *Extrapyramidal pathways* (rubrospinal, tectospinal, vestibulospinal) help coordinate head movements with visual stimuli, maintain skeletal muscle tone and posture, and play a major role in equilibrium by regulating muscle tone in response to movements of the head.

Reflex Center

The second principal function of the spinal cord is to serve as a center for *reflexes*, fast responses to changes in the environment that help us to maintain homeostasis. Spinal nerves are the paths of communication between the spinal cord tracts and the periphery. Each pair of spinal nerves is connected to the cord at two points called roots (see Figure 10-2). The *posterior* or *dorsal (sensory) root* contains sensory nerve fibers only and conducts impulses from the periphery to the spinal cord, specifically, the posterior (dorsal) gray horn. Each posterior root also has a swelling, the *posterior* or *dorsal (sensory) root ganglion*, which contains the cell bodies of the sensory neurons from the periphery. The other point of attachment of a spinal nerve to the cord is the *anterior* or *ventral (motor) root*. It contains motor nerve axons only and conducts impulses from the spinal cord to the periphery.

The cell bodies of the motor neurons are located in the gray matter of the cord. If the motor impulse supplies a skeletal muscle, the cell bodies are located in the anterior gray horn. If, however, the impulse supplies smooth muscle, cardiac muscle, or a gland through the autonomic nervous system, the cell bodies are located in the lateral gray horn.

Reflex Arc and Homeostasis

The path an impulse follows from its origin in the dendrites or cell body of a neuron in one part of the body to its termination elsewhere in the body is called a *conduction pathway*. All conduction pathways consist of several connected neurons. One kind of pathway is known as a ***reflex arc.*** A reflex arc consists of two or more neurons over which impulses are conducted from a receptor to the brain or spinal cord and then to an effector. The basic components of a reflex arc are as follows (Figure 10-4):

1. **Receptor.** Portion of a sensory neuron or specialized cell that generates a nerve impulse in a neuron. A receptor responds to a change (stimulus) in the internal or external environment to produce a nerve impulse.
2. **Sensory neuron.** Passes the impulse from the receptor to its axonal termination in the CNS.
3. **Center.** A region in the CNS where an incoming sensory impulse may be transmitted or inhibited. In the simplest reflex arcs, the sensory impulse is transmitted directly to a motor neuron. In more complex reflex arcs, the sensory impulse is relayed to an association neuron, which may relay the impulse to other association neurons as well as to a motor neuron.
4. **Motor neuron.** Transmits the impulse generated by the sensory or association neuron in the center to the organ that will respond.
5. **Effector.** The organ of the body (a muscle or a gland) that responds to the motor impulse. This response is called a reflex action.

When we were discussing skeletal muscle physiology, we described skeletal muscle as voluntary. However, if we had to think about every action we do, we might suffer severe injuries by the time we decided on the right action and we would be so preoccupied with our body movements we would have little time left for other more serious considerations. As noted earlier, a reflex is a rapid response to changes in the external or internal environment to help the body maintain homeostasis. Reflexes that result in the contraction of skeletal muscles are ***somatic reflexes***. One example

is the ***patellar reflex*** (knee jerk), a two-neuron reflex arc that helps us to remain standing erect despite the effects of gravity. Another is the three-neuron reflex arc involved in the ***withdrawal reflex*** that protects us from serious cuts or burns by causing immediate withdrawal from a source of injury, usually before we are even aware of any pain. During a physical examination when a physician tests somatic reflexes, the responses are a reflection of the health of the nervous system.

The functions of smooth and cardiac muscles and many glands are regulated by reflexes also. Later in the book we will examine the reflex nature of swallowing, coughing, sneezing, digestion, urination, and defecation. Reflexes that cause contraction of smooth or cardiac muscle or secretion by glands are called ***visceral (autonomic) reflexes***. Like somatic reflexes these also have a protective function in that they help to maintain homeostasis by regulating such actions as heart rate, blood pressure, and the rate and depth of respirations.

SPINAL NERVES

NAMES

Spinal nerves, which connect the CNS to sensory receptors, muscles, and glands, are components of the somatic nervous system (SNS) of the peripheral nervous system (PNS). The 31 pairs of spinal nerves are named and numbered according to the region and level of the spinal cord from which they emerge (see Figure 10-1). The first cervical pair emerges between the atlas and the occipital bone. All other spinal nerves leave the vertebral column from the intervertebral foramina (holes) between vertebrae. There are 8 pairs of cervical nerves, 12 pairs of thoracic, 5 pairs of lumbar, 5 pairs of sacral, and 1 pair of coccygeal nerves.

FIGURE 10-4 Components of a generalized reflex arc. The arrows show the direction of nerve impulse conduction.

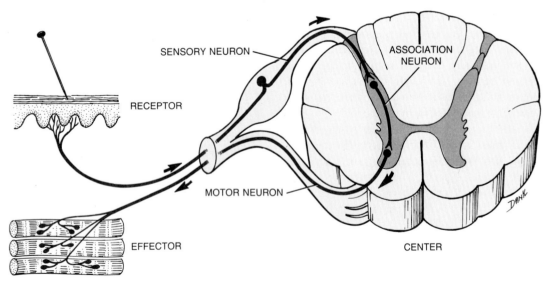

SENSORY NEURON

ASSOCIATION NEURON

RECEPTOR

MOTOR NEURON

EFFECTOR

CENTER

What is an effector? What initiates the nerve impulse? Part of the axon of a motor neuron is located in which root of the spinal nerve? The center is located in which division of the nervous system?

COMPOSITION AND COVERINGS

A *spinal nerve* has two points of attachment to the cord, a posterior (dorsal) root and an anterior (ventral) root (see Figure 10-2). These roots unite to form a spinal nerve at the intervertebral foramen. Since the posterior root contains sensory fibers and the anterior root contains motor fibers, a spinal nerve is composed of both and, therefore, is a *mixed nerve*. Each spinal nerve consists of bundles of fibers wrapped in connective tissue and supplied with blood vessels, very similar to the structure of skeletal muscles but, of course, even longer and thinner.

DISTRIBUTION

Branches

After a spinal nerve leaves its intervertebral foramen, it divides into several branches, known as *rami* (RĀ-mī). The *dorsal ramus* (RĀ-mus) innervates (supplies) the deep muscles and skin of the back. The *ventral ramus* of a spinal nerve innervates the superficial back muscles, all the structures of the extremities, and the lateral and ventral trunk. In addition, spinal nerves give off a *meningeal branch*. This branch reenters the spinal canal through the intervertebral foramen and supplies the vertebrae, vertebral ligaments, blood vessels of the spinal cord, and the meninges. Other branches of a spinal nerve are the *rami communicantes* (ko-myoo-nī-KAN-tēz), components of the autonomic nervous system which are discussed in the next chapter.

Plexuses

The ventral rami of spinal nerves, except for thoracic nerves T2–T11, do not go directly to the body structures they supply. Instead, they form networks on either side of the body by joining with adjacent nerves. Such a network is called a *plexus* (*plexus* = braid). The principal plexuses are the cervical plexus, brachial plexus, lumbar plexus, and sacral plexus (see Figure 10-1). Emerging from the plexuses are nerves bearing names that are often descriptive of the general regions they supply or the course they take. Each of the nerves, in turn, may have several branches named for the specific structures they innervate.

The *cervical plexus* supplies the skin and muscles of the head, neck, and upper part of the shoulders, connects with some cranial nerves, and supplies the diaphragm. The *brachial plexus* constitutes the nerve supply for the upper extremities and a number of neck and shoulder muscles. The *lumbar plexus* supplies the abdominal wall, external genitals, and part of the lower extremities. The *sacral plexus* supplies the buttocks, perineum, and lower extremities. The sciatic nerve, the longest nerve in the body, arises from the sacral plexus.

Intercostal (Thoracic) Nerves

Spinal nerves T2–T11 do not enter into the formation of plexuses. They are known as *intercostal* (*thoracic*) *nerves* and go directly to the structures they supply, muscles between ribs, abdominal muscles, and skin of the chest and back (see Figure 10-1).

Now we will consider the principal parts of the brain, how the brain is protected, and how it is related to the spinal cord and cranial nerves.

BRAIN

PRINCIPAL PARTS

The *brain* is one of the largest organs of the body, weighing about 1300 g (3 lb). It is mushroom shaped and divided into four principal parts: brain stem, diencephalon, cerebrum, and cerebellum (Figure 10-5). The *brain stem*, the stalk of the mushroom, consists of the medulla oblongata, pons, and midbrain. Above the brain stem is the *diencephalon* (dī-en-SEF-a-lon), consisting of the thalamus and hypothalamus. The *cerebrum* spreads over the diencephalon. It has two sides called hemispheres and occupies most of the cranium. Below the cerebrum and behind the brain stem is the *cerebellum*.

PROTECTION AND COVERINGS

The brain is protected by the cranium and meninges. The *cranial meninges*, as extensions of the spinal meninges, have the same names: the outermost *dura mater*, middle *arachnoid*, and innermost *pia mater* (Figure 10-6a).

CEREBROSPINAL FLUID (CSF)

The brain, as well as the rest of the CNS, is further protected against injury by *cerebrospinal fluid* (*CSF*). CSF circulates through the subarachnoid space, around the brain and spinal cord, and through the ventricles of the brain.

The *ventricles* (VEN-tri-kuls) are cavities in the brain that communicate with each other, with the central canal of the spinal cord, and with the subarachnoid space (Figure 10-6a). There are four ventricles, two *lateral ventricles*, one *third ventricle*, and one *fourth ventricle*. The third and fourth communicate with each other through the *cerebral aqueduct*.

The entire CNS contains between 80 and 150 ml (3 to 5 oz) of cerebrospinal fluid. It is a clear, colorless liquid. It contains proteins, glucose, urea, salts, and lymphocytes. CSF has two principal functions related to homeostasis: protection and circulation. It serves as a shock-absorbing medium to protect the brain and spinal cord from jolts that would otherwise be traumatic. The fluid also buoys the brain so that it "floats" in the cranial cavity. In its circulatory function, CSF delivers nutritive substances filtered from blood and removes wastes and toxic substances produced by brain and spinal cord cells.

Cerebrospinal fluid is formed by filtration and secretion from *choroid* (KŌ-royd; *chorion* = delicate) *plexuses*, specialized concentrations of networks of capillaries in the ventricles (Figure 10-6a). Various components of the choroid plexuses form a *blood–cerebrospinal fluid barrier* that permits certain substances to enter the fluid but prohibits others that might be harmful.

FIGURE 10-5 **Brain.** Principal parts of the brain seen in sagittal section. The infundibulum and pituitary gland are discussed in conjunction with the endocrine system in Chapter 13.

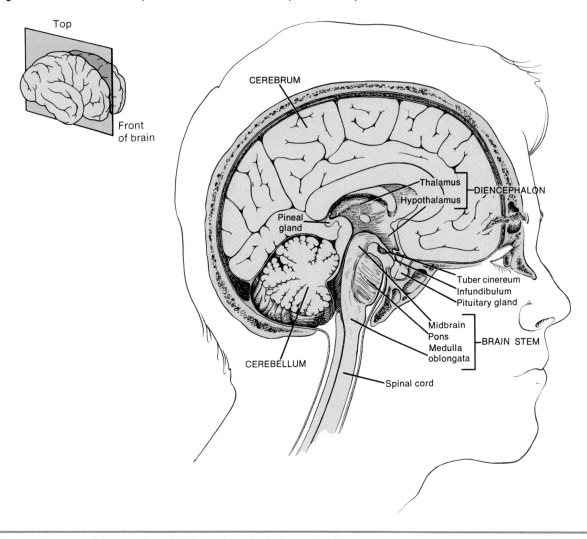

What is the continuation of the spinal cord? Where does the limbic system lie?

CSF circulates continually. From the ventricles, it flows into the subarachnoid space around the back of the brain and downward around the posterior surface of the spinal cord, up the anterior surface of the spinal cord, and around the anterior part of the brain. From there it is gradually reabsorbed into veins, mostly into a vein called the superior sagittal sinus. The absorption actually occurs through **arachnoid villi**, which are fingerlike projections of the arachnoid that push into the superior sagittal sinus (Figure 10-6b). Normally, cerebrospinal fluid is absorbed as rapidly as it is formed.

If an obstruction (tumor or a congenital blockage) or inflammation interferes with the drainage of fluid from the ventricles into the subarachnoid space and fluid accumulates in the ventricles, fluid pressure inside the brain increases and, if the fontanels have not yet closed, the head bulges to relieve the pressure. This is **internal hydrocephalus** (*hydro* = water; *enkephalos* = brain). If an obstruction causes CSF to accumulate in the subarachnoid space, the condition is termed **external hydrocephalus**. Hydrocephalus

is treated by inserting a shunt in the ventricles to drain off the excess fluid into a vein in the neck.

BLOOD SUPPLY

The brain is well supplied with oxygen and nutrients from a special circulatory route at the base of the brain called the **cerebral arterial circle** (**circle of Willis**) (Chapter 16). Although the brain comprises only about 2 percent of total body weight, it requires about 20 percent of the body's oxygen supply. The brain is one of the most metabolically active organs of the body, and the amount of oxygen it uses varies with the degree of mental activity.

If the blood flow to the brain is interrupted even briefly, unconsciousness may result. A 1- or 2-minute interruption may weaken the brain cells by starving them of oxygen. If the cells are totally deprived of oxygen for 4 minutes, many are permanently injured because lysosomes of brain cells are extremely sensitive to decreased

FIGURE 10-6 **Meninges and ventricles of the brain.** (a) Brain, ventricles, spinal cord, and meninges seen in sagittal section. Arrows indicate the direction of flow of cerebrospinal fluid.

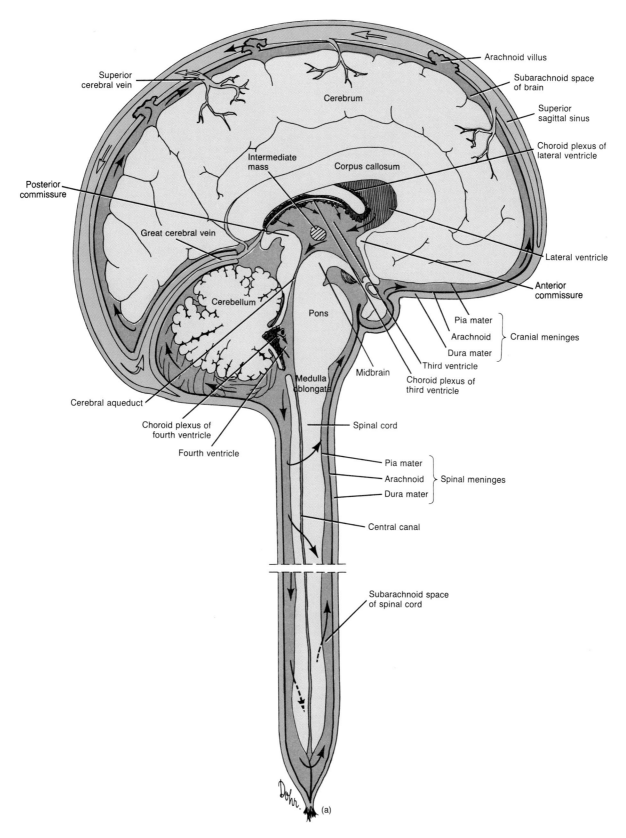

Superior cerebral vein

Cerebrum

Arachnoid villus

Subarachnoid space of brain

Superior sagittal sinus

Choroid plexus of lateral ventricle

Intermediate mass

Corpus callosum

Posterior commissure

Great cerebral vein

Lateral ventricle

Anterior commissure

Cerebellum

Pons

Pia mater

Arachnoid

Dura mater

Cranial meninges

Third ventricle

Midbrain

Choroid plexus of third ventricle

Medulla oblongata

Cerebral aqueduct

Spinal cord

Choroid plexus of fourth ventricle

Pia mater

Arachnoid

Dura mater

Spinal meninges

Fourth ventricle

Central canal

Subarachnoid space of spinal cord

(a)

CSF is formed by which structures in the ventricles? Where is CSF reabsorbed?

FIGURE 10-6 (*Continued*) (b) Frontal section through the superior portion of the brain showing the relationship of the superior sagittal sinus to the arachnoid villi.

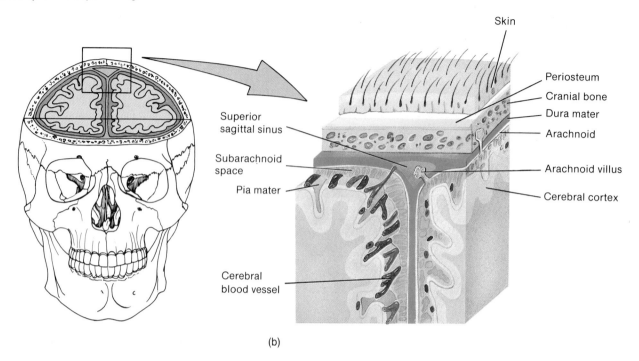

(b)

oxygen. If the condition persists, they break open and release enzymes that bring about self-destruction of brain cells.

Blood supplying the brain also contains glucose, the principal source of energy for brain cells. Because carbohydrate storage in the brain is limited, the supply of glucose must be continuous. If blood entering the brain has a low glucose level, mental confusion, dizziness, convulsions, and loss of consciousness may occur.

Glucose, oxygen, and certain ions pass rapidly from the circulating blood into brain cells. Other substances enter slowly or not at all. The different rates of passage of certain materials from the blood into most parts of the brain are due to the **blood–brain barrier (BBB)**. Brain capillary walls have more tightly connected cells with a thicker basement membrane, plus they are surrounded by astrocytes, thus forming a barrier to all but the smallest molecules or those selectively admitted through active transport. Brain cells are protected from harmful substances this way. Unfortunately, most antibiotics cannot enter either. Trauma, inflammation, and toxins can cause a breakdown of the blood–brain barrier.

BRAIN STEM

Medulla Oblongata

The **medulla oblongata** (me-DULL-la ob′-long-GA-ta) is a continuation of the spinal cord. It forms the inferior part of the brain stem (Figures 10-7 and 10-5).

The medulla contains all ascending and descending tracts running between the spinal cord and other parts of the brain. These tracts constitute the white matter of the medulla. Some tracts cross as they pass through the medulla. Let us see how this crossing occurs and what it means.

In the medulla are two roughly triangular structures called **pyramids** (Figures 10-7 and 10-8). The pyramids contain the largest motor tracts that pass from the outer region of the cerebrum (cerebral cortex) to the spinal cord. Most of the fibers in the left pyramid cross to the right side, and most of the fibers in the right pyramid cross to the left. This crossing is called the **decussation** (dē′-ku-SĀ-shun) **of pyramids**. Decussation explains why one side of the cerebral cortex controls the opposite side of the body. Motor fibers that originate in the left cerebral cortex activate muscles on the right side of the body, and vice versa.

Similarly, most sensory fibers also cross over in the medulla so that nearly all sensory impulses received on one side of the body are perceived in the opposite side of the cerebral cortex.

Throughout the brain stem—medulla, pons, and midbrain—is a group of widely scattered neurons that give a netlike (reticular) appearance and is referred to as the **reticular formation**. Neurons of the reticular formation receive and integrate input from the cerebral cortex, hypothalamus, thalamus, cerebellum, and spinal cord. The neurons also send impulses to all levels of the CNS. Thus, the reticular formation is ideally suited to govern the activity of the nervous system.

A severe blow to the mandible twists and distorts the brain stem and overwhelms the reticular formation by sending a sudden volley of nerve impulses to the brain, resulting in unconsciousness.

Three vital reflex centers are in the medulla. The **cardiac center** regulates heartbeat and force of contraction; a portion of the **respiratory center** adjusts the basic rhythm of breathing; and the **vasomotor center** regulates the diameter of blood vessels. Centers in the me-

FIGURE 10-7 Brain stem. Ventral surface of the brain, showing the structure of the brain stem in relation to the cranial nerves and associated structures.

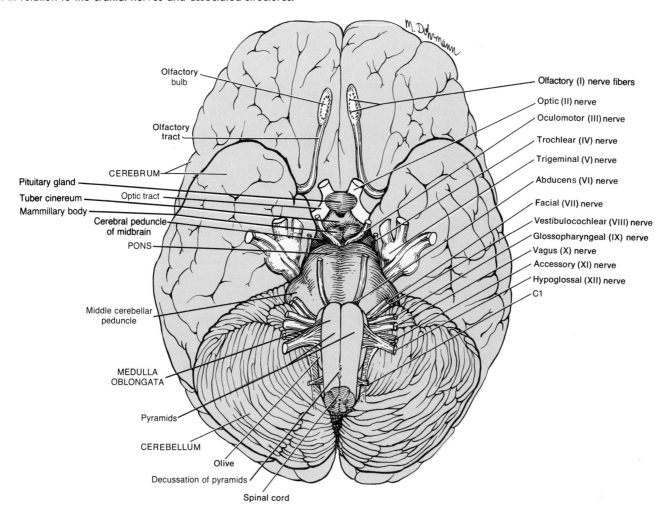

Olfactory bulb
Olfactory tract
CEREBRUM
Pituitary gland
Tuber cinereum
Mammillary body
Cerebral peduncle of midbrain
PONS
Middle cerebellar peduncle
MEDULLA OBLONGATA
Pyramids
CEREBELLUM
Olive
Decussation of pyramids
Spinal cord
Optic tract

Olfactory (I) nerve fibers
Optic (II) nerve
Oculomotor (III) nerve
Trochlear (IV) nerve
Trigeminal (V) nerve
Abducens (VI) nerve
Facial (VII) nerve
Vestibulocochlear (VIII) nerve
Glossopharyngeal (IX) nerve
Vagus (X) nerve
Accessory (XI) nerve
Hypoglossal (XII) nerve
C1

What part of the brain contains pyramids? Contains cerebral peduncles? Literally means bridge? Controls the autonomic nervous system? Relays all sensory impulses, except smell?

dulla considered nonvital include swallowing, vomiting, coughing, sneezing, and hiccuping. Finally, the following cranial nerves originate in the medulla: vestibulochochlear, accessory, vagus, and hypoglossal nerves (Figures 10-7 and 10-8 and Exhibit 10-1).

In view of the many vital activities controlled by the medulla, it is not surprising that a hard blow to the base of the skull can be fatal.

Pons

The *pons*, which means bridge, lies directly above the medulla and in front of the cerebellum (Figures 10-5 and 10-7). Like the medulla, the pons consists of nuclei and scattered white fibers. The pons connects the spinal cord with the brain and parts of the brain with each other. The transverse fibers in the pons connect it with the cerebellum. Longitudinal fibers connect the spinal cord and medulla with the higher brain centers.

Cranial nerves originating in the pons include the trigeminal, abducens, facial, and the vestibular branch of the vestibulocochlear nerves (Figure 10-7 and Exhibit 10-1). The pons also contains another part of the *respiratory center* which helps regulate breathing (see Chapter 18).

Midbrain

The *midbrain* extends from the pons to the lower portion of the diencephalon (Figures 10-5 and 10-7). It contains a pair of fiber bundles referred to as *cerebral peduncles* (pe-DUNG-kulz). The cerebral peduncles contain motor fibers that connect the cerebral cortex to the pons and spinal cord, and sensory fibers that connect the spinal cord to the thalamus. The cerebral peduncles constitute the main connection for tracts between upper and lower parts of the brain and spinal cord. The midbrain also contains structures called *colliculi* (ko-LIK-yoo-lī), which serve as reflex centers for

FIGURE 10-8 **Details of the medulla showing the decussation of pyramids.**

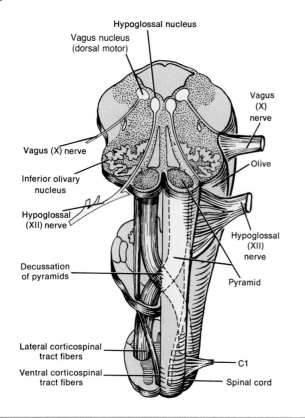

What does decussation mean? Why is it important?

movements of the eyeballs, head, and trunk in response to visual and auditory stimuli.

Cranial nerves originating in the midbrain include the oculomotor and trochlear nerves (see Figure 10-7 and Exhibit 10-1).

DIENCEPHALON

The ***diencephalon*** (*dia* = through; *enkephalos* = brain) consists principally of the thalamus and hypothalamus. See Figure 10-5.

Thalamus

The ***thalamus*** (THAL-a-mus; *thalamos* = inner chamber) is an oval structure above the midbrain that consists of mostly gray matter organized into nuclei (Figure 10-9). Some nuclei in the thalamus serve as relay stations for all sensory impulses, except smell, from the spinal cord, brain stem, cerebellum, and parts of the cerebrum to the cerebral cortex. Other nuclei interpret certain sensory impulses, such as pain, temperature, light touch, and pressure.

Hypothalamus

The ***hypothalamus*** (*hypo* = under) is the small portion of the diencephalon that lies between the thalamus and pituitary gland, the "master gland" of the body (Figure 10-9).

Despite its small size, nuclei in the hypothalamus control many body activities, most of them related to homeostasis.

1. The hypothalamus controls and integrates the autonomic nervous system through which it regulates heart rate, movement of food

FIGURE 10-9 **Frontal section of the diencephalon showing the thalamus and hypothalamus. Structures of the basal ganglia (cerebral nuclei) are also identified.**

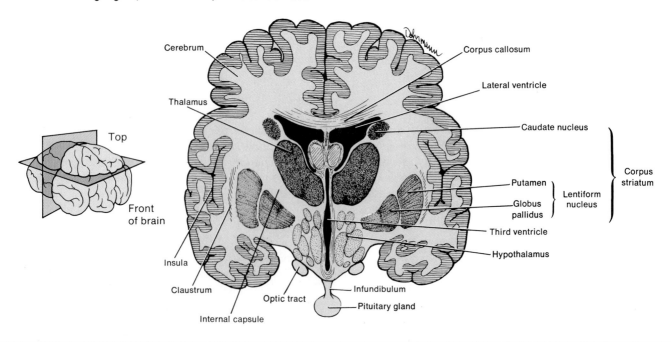

Which is the larger of the two diencephalon structures? Which is the smaller? Where are the basal ganglia found in the cerebrum? Basal ganglia are masses of what kind of matter?

through the gastrointestinal tract, and contraction of the urinary bladder.

2. It receives and integrates sensory impulses from the viscera.

3. It is the principal connection between the nervous system and the endocrine system—the two major control systems of the body. The hypothalamus controls the endocrine system by producing substances called regulating hormones (or factors) that stimulate or inhibit the release of hormones by the pituitary gland (see Chapter 13). The hypothalamus also produces two hormones, antidiuretic hormone (ADH) and oxytocin (OT), which are transported to and stored in the pituitary gland, to be released when needed. ADH decreases urine volume in order to regulate blood volume and electrolyte balance and under certain conditions, such as severe blood loss, can constrict arterioles to raise blood pressure. OT brings about uterine contractions during labor and assists in milk ejection by the mammary glands. (See Chapter 13.)

4. It controls normal body temperature.

5. It is the center for mind-over-body phenomena.

6. It is associated with feelings of rage and aggression.

7. It contains a *hunger center* and a *satiety center*.

8. It contains a *thirst center*.

9. It is one of the centers that maintains consciousness and sleep patterns.

RETICULAR ACTIVATING SYSTEM (RAS), CONSCIOUSNESS, AND SLEEP

Humans awaken and sleep in a fairly constant 24-hour rhythm called *circadian* (ser-KĀ-dē-an) *rhythm*. When the brain is aroused or awake, it is in a state of readiness and able to react consciously to various stimuli. The aroused state depends largely on the reticular formation. Because stimulating a portion of the reticular formation increases cortical activity, this area is also known as the *reticular activating system (RAS)*. When the pons and midbrain part of the RAS is stimulated, impulses pass upward to widespread areas of the cerebral cortex. When the other part of the RAS, the thalamic part, is stimulated, signals from specific parts of the thalamus activate specific parts of the cerebral cortex. Apparently the pons and midbrain part of the RAS produces consciousness and the thalamic part causes *arousal*, that is, awakening from deep sleep.

Following arousal, the RAS and cerebral cortex continue to activate each other through a feedback system consisting of many circuits. The result is a state of wakefulness called *consciousness*. The RAS is the physical basis of consciousness. It continuously sifts and selects, forwarding only the essential, unusual, or dangerous to the conscious mind. Since humans experience different levels of consciousness (alertness, attentiveness, relaxation, inattentiveness), it is assumed that the level of consciousness depends on the number of feedback circuits operating at the time.

Consciousness may be altered by cocaine and amphetamines (which produce extreme alertness), meditation (which causes relaxed, focused consciousness), and alcohol and anesthetics (which cause various levels of unconsciousness called anesthesia). Damage or disease can produce a lack of consciousness called *coma*.

Inactivation of the RAS produces *sleep*, a state of partial unconsciousness from which an individual can be aroused. Just as there are different levels of consciousness, there are different levels of

sleep. Normal sleep consists of non–rapid eye movement (NREM) sleep and rapid eye movement (REM) sleep.

NREM sleep or *slow wave sleep* consists of four stages, each of which gradually merges into the next.

Stage 1. A transition stage between waking and sleep that normally lasts from 1 to 7 minutes. Respirations are regular, pulse is even. If awakened, the person will frequently say she or he had not been sleeping.

Stage 2. The first stage of true sleep, though light. Fragments of dreams may be experienced.

Stage 3. Moderately deep sleep that occurs about 20 minutes after falling asleep. Body temperature begins to fall and blood pressure decreases. It is difficult to awaken the person.

Stage 4. Deep sleep. The person responds slowly if awakened. If bed-wetting and sleepwalking occur, it is during this stage.

In a typical 7- or 8-hour sleep period, a person goes from stages 1 to 4 of NREM sleep, then ascends to stages 3 and 2, and then to REM sleep within the first 50 to 90 minutes.

In *REM sleep*, there are significant physiological changes. Muscle tone is depressed (except for rapid movements of the eyes), and respirations and pulse rate increase and are irregular. Blood pressure fluctuates considerably. Most dreaming occurs during REM sleep. Following REM sleep, the person descends again to stages 3 and 4 of NREM sleep.

REM and NREM sleep alternate throughout the night. The REM periods start out lasting from 5 to 10 minutes and gradually lengthen until the final one lasts about 50 minutes. As much as 50 percent of an infant's sleep is REM as contrasted with 20 percent for adults. Most sedatives significantly decrease REM sleep. As a person ages, the average time spent sleeping and the percentage of REM sleep decrease.

CEREBRUM

Supported on the brain stem and forming the bulk of the brain is the *cerebrum* (see Figure 10-5). The surface is composed of gray matter and is called the *cerebral cortex* (*cortex* = rind or bark). The cortex consists of six layers of nerve cell bodies, beneath which lies the cerebral white matter.

During embryonic development, when there is a rapid increase in brain size, the gray matter of the cortex enlarges much faster than the underlying white matter. As a result, the cortical region rolls and folds upon itself. The folds are called *gyri* (JĪ-ri) or *convolutions* (Figure 10-10).

The deep grooves between folds are *fissures*; the shallow grooves are *sulci* (SUL-sī). The *longitudinal fissure* separates the cerebrum into right and left halves, or *hemispheres*. Internally, however, the hemispheres are connected by a large bundle of transverse fibers composed of white matter called the *corpus callosum* (kal-LŌ-sum; *corpus* = body; *callosus* = hard).

Lobes

Each cerebral hemisphere is subdivided into four lobes by sulci or fissures. The lobes are named frontal, parietal, temporal, and occipital (Figure 10-10). The *central sulcus* separates the frontal and parietal lobes. A major gyrus, the *precentral gyrus*, is located immediately anterior to the central sulcus. The gyrus is a landmark for the primary motor area of the cerebral cortex. Another major gyrus, the *postcentral gyrus*, is located immediately posterior to the central sulcus. This gyrus is a landmark for the general sensory area of the cerebral cortex. The *lateral cerebral sulcus* separates the frontal and temporal lobes. The *parietooccipital sulcus* separates the parietal and occipital lobes. A fifth part of the cerebrum, the *insula*, lies deep within the cerebrum and cannot be seen in an external view.

As you will see later, the olfactory and optic nerves are associated with specific lobes of the cerebrum.

Brain Lateralization (Split-Brain Concept)

On gross examination, the brain appears the same on both sides. However, detailed examination reveals anatomical differences between the two hemispheres. For example, in left-handed people the parietal and occipital lobes of the right hemisphere and the frontal lobe of the left hemisphere are typically narrower.

In addition to structural differences, there are also functional differences. The left hemisphere is more important for right-handed control, spoken and written language, numerical and scientific skills, and reasoning in most people. Conversely, the right hemisphere is more important for left-handed control, musical and artistic awareness, space and pattern perception, insight, imagination, and

FIGURE 10-10 Lobes and fissures of the cerebrum. (a) Right lateral view. Since the insula cannot be seen externally, it has been projected to the surface. (b) Superior view. The insert to the left indicates the relative differences among a gyrus, sulcus, and fissure.

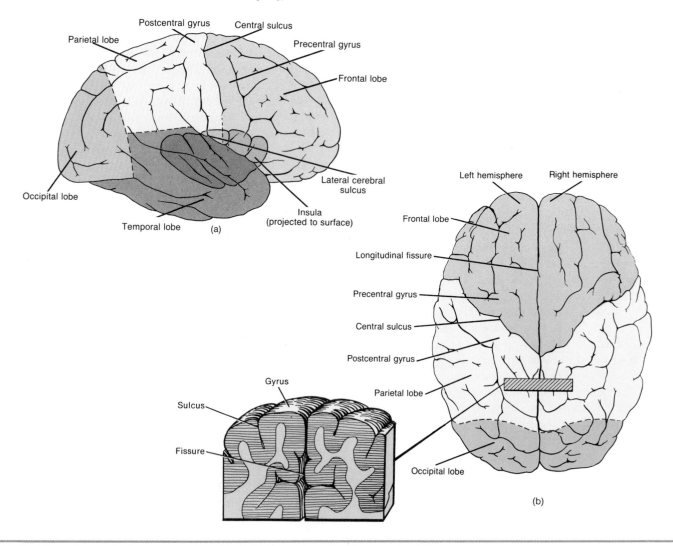

Explain how the cerebral cortex comes to be folded in on itself. What are the folds called? What are the deep grooves called?

generating mental images of sight, sound, touch, taste, and smell in order to compare relationships.

White Matter

The white matter underlying the cortex consists of myelinated axons running in three principal directions.

1. **Association fibers** transmit impulses between gyri in the same hemisphere.
2. **Commissural fibers** transmit impulses from the gyri in one cerebral hemisphere to the corresponding gyri in the opposite cerebral hemisphere. One example is the corpus callosum.
3. **Projection fibers** transmit impulses from the cerebrum to other parts of the brain and spinal cord.

Basal Ganglia (Cerebral Nuclei)

The *basal ganglia (cerebral nuclei)* are paired masses of gray matter embedded in the white matter of each hemisphere (see Figure 10-9).

The largest basal ganglion of each hemisphere is the *corpus striatum* (strī-Ā-tum; *corpus* = body; *striatus* = striped). It consists of the *caudate* (*cauda* = tail) *nucleus* and the *lentiform* (*lenticula* = shaped like a lentil or lens) *nucleus*. The lentiform nucleus, in turn, is subdivided into the *putamen* (pu-TĀ-men; *putamen* = shell) and the *globus pallidus* (*globus* = ball; *pallid* = pale).

The caudate nucleus and the putamen control large subconscious movements of skeletal muscles, such as swinging the arms while walking (such gross movements are also consciously controlled by the cerebral cortex). The globus pallidus regulates muscle tone required for specific body movements.

Damage to the basal ganglia results in uncontrollable shaking (tremor) or involuntary muscle movements, such as in Parkinson's disease. Destruction of a substantial portion of the caudate nucleus, as in a stroke, results in total *paralysis* of the side of the body opposite to the damage.

Limbic System

Certain parts of the cerebral hemispheres and diencephalon constitute the *limbic* (*limbus* = border) *system*. It is a wishbone-shaped group of structures that encircles the brain stem and assumes a primary function in emotions such as pain, pleasure, anger, rage, fear, sorrow, sexual feelings, docility, and affection. It is sometimes, therefore, called the ''visceral'' or ''emotional'' brain. Although behavior is a function of the entire nervous system, the limbic system controls most of its involuntary aspects, the aspects related to survival, and animal experiments suggest that it has a major role in controlling the overall pattern of behavior. Together with portions of the cerebrum, the limbic system also functions in memory; memory impairment results from lesions in the limbic system.

Functional Areas of Cerebral Cortex

The functions of the cerebrum are numerous and complex. In a general way, the cerebral cortex is divided into three areas. The *sensory areas* interpret sensory impulses, the *motor areas* control muscular movement, and the *association areas* are concerned with emotional and intellectual processes.

■ **Sensory Areas** The *primary somesthetic* (sō-mes-THET-ik; *soma* = body; *aisthesis* = perception) *area* or *general sensory area* receives sensations from cutaneous, muscular, and visceral receptors in various parts of the body. Each point of the area receives sensations from specific parts of the body. Some parts of the body are represented by large areas in the general sensory area. These include the lips, face, and thumb. Other body parts, such as the trunk and lower extremities, are represented by relatively small areas. The space allotment in the general sensory area is directly proportional to the number of specialized sensory receptors in each respective part of the body. Thus, there are numerous receptors in the skin of the lips, but relatively few in the skin of the trunk. Essentially, the amount of space given a particular part of the body is determined by the functional importance of the part and its need for sensitivity. The major function of the primary somesthetic area is to localize exactly the points of the body where the sensations originate. In Figure 10-11 the general sensory area is designated by the areas numbered 1, 2, and 3.*

Posterior to the primary somesthetic area is the *somesthetic association area* (numbered 5 and 7 and Figure 10-11), which receives input from the thalamus or lower portions of the brain and the primary somesthetic area. It integrates and interprets sensations. This area permits you to determine the exact shape and texture of an object without looking at it, to determine the orientation of one object to another as they are felt, and to sense the relationship of one body part to another. The somesthetic association area also stores memories of past sensory experiences. Thus, you can compare sensations with previous experiences.

Other sensory areas of the cerebral cortex include:

1. **Primary visual area** (area 17). Located in the occipital lobe, it receives sensory impulses from the eyes and interprets shape, color, and movement.
2. **Visual association area** (areas 18 and 19). Located in the occipital lobe, it receives sensory signals from the primary visual area and the thalamus. It relates present to past visual experiences in order to recognize and evaluate what is seen.
3. **Primary auditory area** (areas 41 and 42). Located in the temporal lobe, it interprets the basic characteristics of sound such as pitch and rhythm.
4. **Auditory association area** (area 22). Located in the temporal lobe, it determines if a sound is speech, music, or noise. It also interprets the meaning of speech by translating words into thoughts.
5. **Primary gustatory area** (area 43). Located at the base of the postcentral gyrus, it interprets sensations related to taste.
6. **Primary olfactory area.** Located in the temporal lobe, it interprets sensations related to smell.
7. **Gnostic** (NOS-tik; *gnosis* = knowledge) **area** (areas 5, 7, 39, and 40). Located among the somesthetic, visual, and auditory

* These numbers, as well as most of the others shown, are based on K. Brodmann's cytoarchitectural map of the cerebral cortex. His map, first published in 1909, attempts to correlate structure and function.

In Search of Morpheus

While those who get it usually take it for granted, people who don't would do almost anything for it. Although it's extremely valuable, you can't buy it. You can't borrow it, steal it, or give it away. And the harder you try to get it, the less likely you are to succeed. What is it? A good night's sleep.

Almost everyone has trouble sleeping occasionally and experiences nights when they toss and turn, look at the clock, get upset, and toss and turn some more. At any one time, about 15 to 20 percent of adults in North America complain of problems sleeping. Fortunately, there are no serious consequences to missing a good night's sleep once in a while, although even a few sleepless nights can dampen one's good humor and interfere with one's mental alertness (students who pull all-nighters take note!).

Sleep problems often last for a relatively short period of time and go away on their own even without treatment. Some people, however, suffer from chronic insomnia, or difficulty sleeping, that may persist for months or even years. Chronic sleep deprivation can seriously impair one's mental and physical health and ability to function effectively.

Insomnia

Insomnia may include any or all of the following problems:

1. Taking a long time to fall asleep.
2. Awakening frequently during the night.
3. Awakening too early in the morning.
4. Feeling tired and dissatisfied with one's sleep upon awakening.

The first problem is usually the complaint of young people, while those middle-aged and older are more likely to experience the second and third problems. Insomnia may originate during times of stress, anxiety, or depression, but then continue even though the person may no longer feel stressed, anxious, or depressed.

Most people need about seven to eight hours of sleep each night, but there is no standard amount of sleep that guarantees well-being. Enough sleep is whatever you need to awaken feeling alert and refreshed. Sleep requirements vary among individuals and decrease with age. Sometimes a person sleeps "enough hours" but does not sleep well. This results in a deficiency of stage 4 (deep) sleep, and so he or she does not feel rested. Stage 4 sleep is the most restorative, and people who are deprived of sleep will spend more time in stage 4 sleep on subsequent nights.

Insomnia that continues for more than a few weeks is a problem requiring medical attention, because occasionally it may signal a more serious health problem. One of these is sleep apnea, a potentially dangerous condition that occurs when the sleeper stops breathing for about 30 seconds and awakens, snoring loudly. Sleep disorder clinics treat insomnia and problems like sleep apnea.

Sleeplessness is sometimes a medication side effect. Overuse of caffeine, alcohol, sleeping pills, and other drugs can also cause insomnia.

Why Not Take a Sleeping Pill?

Sleeping pills are one of the most commonly prescribed drugs in North America. An occasional dose of sleep medication can be helpful when taken as directed, but, in general, medication only worsens the problem. Sleeping pills disturb the sleep cycle, so sleep is less satisfying even though you may get more of it. They often leave the user with a "hangover" that results in daytime fatigue. The user quickly builds up a tolerance to the medication so that it becomes less effective within a week. One of the biggest dangers of sleeping pills is addiction, which can be very difficult to overcome.

Some sleeping pills can even cause insomnia by suppressing the brain's production of dopamine, a neurotransmitter that helps you go to sleep. Sleeping pills pose a danger of overdose, especially if taken with alcohol or other drugs. Many people believe alcohol will help them relax and go to sleep, but like sleeping pills, alcohol disrupts the sleep cycle. While it may help you fall asleep, it usually produces light, restless sleep, and the sleeper often awakens suddenly during the night, unable to go back to sleep.

The biggest problem with sleeping pills is that they do not address the real causes of the sleeping problem: bad habits and stress.

Sleep Therapy

Sleep experts believe that insomnia often begins during times of stress, such as going through a divorce, moving to a new town, or losing a job. The insomnia should disappear once the situation is resolved. When it continues it is often because the person has developed a poor sleep environment and/or poor sleep habits.

Sometimes the sleep environment is the problem. Look at the example of Goldilocks: Is the bed too hard or too soft? The sleep environment should be comfortable, restful, and associated with relaxation and sleep. Sleep and work areas should be separate. Changing simple physical characteristics of the sleeping area often improves sleep quality. Most people sleep best when the room temperature is about 60 to 65°F. Noise level can be a more difficult problem. Tapes can provide a soothing sound that helps cover traffic and other noise. Some people even resort to earplugs. Shades that block light can help darken rooms with windows near street lights or keep out the early summer sun.

Regular use of stimulants can interfere with the ability to fall and stay asleep. A decrease in the consumption of caffeine (found in coffee, tea, and cola) often improves sleep quality. People who are very sensitive to caffeine may find that even one cup of coffee after dinner will keep them up for several hours. The nicotine in cigarettes is a stimulant and should be avoided. Heavy smokers experience less REM and less stage 4 sleep.

A large meal before bed can inhibit sleep. Heavy foods and/or large quantities of food are associated with poor sleep quality. A light snack, however, can help you sleep better.

Exercise helps decrease muscle tension and improve sleep quality. EMG studies have found lower resting muscle activity following exercise. Exercise can also improve the ability to manage stress, to feel less worried and more in control. But beware: Exercise too close to bedtime can wind you up instead of down. Sleep experts generally recommend exercising in the afternoon. If an exercise program is making sleep worse, you are probably overdoing it. Too much exercise increases sympathetic arousal (you'll read more about the stress response in Chapter 13), which gears you up to fight or flee, not slow down and sleep.

Sleep comes more easily to those who go to bed with a relaxed mental attitude. It is helpful to relax for at least an hour before bed. Read, listen to music, knit a sweater. Avoid activities that wind you up. A pre-bed routine helps get you ready for sleep.

If sleep is still elusive, a person can try a more rigorous approach called "stimulus control," which consists of behaviors designed to help a person associate going to bed with sleep and sleepiness. Examples of stimulus control instructions include the following. Lie down to sleep only when sleepy. Don't read, watch TV, study, or eat in bed. If you haven't fallen asleep after 10 minutes, get up and do something else until you get sleepy again. Troubled sleepers are also advised to get up at the same time every day and not to nap during the day.

People who consistently have trouble sleeping might try this: Decrease the time you spend in bed by an hour or more until 90 percent of that time is spent sleeping. When you are finally sleeping well, start adding 15 minutes per day to your time in bed.

Stress Management for Insomniacs

An important step to overcoming insomnia is to figure out whether psychological stress might be one of the possible causes. If so, getting help for the stresses, tensions, and anxieties that are contributing to the insomnia can be beneficial. Resolving conflicts and dealing with problem situations are better than worrying. And a sense of control is the best sleeping aid.

Some research has identified people with insomnia as high-strung. They take things too seriously, and have difficulty resolving problems. Chronic worriers must address those thought patterns that discourage good sleep. They can learn to identify worrisome thoughts and replace them with a more useful mental outlook, often with the help of a therapist. Active problem solving during the night is not conducive to deep sleep.

Stress management techniques can help people with sleep problems in two ways. First of all, some techniques, such as time management and communication skills, can help people address their problems and worries more effectively. Second, many stress management techniques increase a person's ability to simply let go of tension and relax.

Many stress management techniques have been used successfully for the treatment of insomnia, including biofeedback (discussed in the Wellness Box in Chapter 9), self-hypnosis, and various other methods of relaxation training. These techniques often focus on inducing the physical sensations associated with deep relaxation, such as feelings of warmth and heaviness in the arms and legs, and on suggestions to let go, to enjoy feeling tired and drowsy. Sometimes one of the barriers to successful sleep is the insomniac's dread of going to bed, and his or her fears of another miserable, restless night. These stress management techniques help insomniacs "change the channel" of their current thought patterns while trying to fall asleep to ones that are more conducive to a good night's sleep.

FIGURE 10-11 Functional areas of the cerebrum. This right lateral view indicates the sensory and motor areas of the right hemisphere. Although Broca's area is in the left hemisphere in most people, it is shown here to indicate its location.

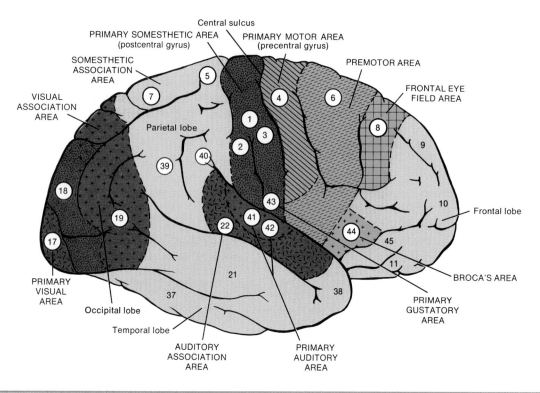

Which area of the cerebrum integrates sensory interpretations? Translates thoughts into speech? Controls skilled muscular movements? Interprets sensations related to taste? Interprets pitch and rhythm? Interprets shape, color, and movement of objects? Controls voluntary scanning movements of eyes?

association areas, the gnostic area receives impulses from these areas, as well as from the taste and smell areas, the thalamus, and lower portions of the brain stem. It integrates sensory interpretations from the association areas and impulses from other areas so that a common thought can be formed from the various sensory inputs. It then transmits signals to other parts of the brain to cause the appropriate motor response.

■ **Motor Areas** Just as the general sensory area of the cortex has been mapped to reflect the amounts of sensory information coming from different body parts, the *motor cortex* has been mapped to indicate which specific areas control particular muscle groups. There is a motor cortex in both the left and right cerebral hemispheres. Just as in the general sensory area, different muscle groups are not represented equally in the motor cortex. The degree of representation is proportional to the precision of movement required of a particular body part. For example, the thumb, fingers, lips, tongue, and vocal cords have large representations. The trunk has a relatively small representation. Sensory and motor representations are not identical for the same part of the body.

The *primary motor area* (area 4) is located in the precentral gyrus of the frontal lobe (Figure 10-11). Like the primary somes-

thetic area, the primary motor area consists of regions that control specific muscles or groups of muscles. Stimulation of a specific point of the primary motor area results in a muscular contraction, usually on the opposite side of the body.

The *premotor area* (area 6) is anterior to the primary motor area. It is concerned with skilled movements, learned motor activities of a complex and sequential nature. It generates impulses that cause a specific group of muscles to contract in a specific sequence, for example, writing.

The *frontal eye field area* (area 8) in the frontal lobe is sometimes included in the premotor area. This area controls voluntary scanning movements of the eyes—searching for a word in a dictionary, for instance.

The *language areas* are significant parts of the motor cortex. The translation of speech or written words into thought involves sensory areas—primary auditory, auditory association, primary visual, visual association, and gnostic—as we just described. The translation of thoughts into speech involves the *motor speech area* (area 44) or *Broca's* (BRŌ-kaz) *area*, located in the frontal lobe just superior to the lateral cerebral sulcus. This area coordinates contractions of speech (larynx, pharynx, and mouth) and breathing muscles to enable you to translate your thoughts into speech.

Broca's area and other language areas are located in the left cerebral hemisphere of most individuals regardless of whether they are left-handed or right handed. Injury to the sensory or motor speech areas results in *aphasia* (a-FĀ-zē-a; *a* = without; *phasis* = speech), an inability to speak; *agraphia* (*a* = without; *graph* = write), an inability to write; *word deafness*, an inability to understand spoken words; or *word blindness*, an inability to understand written words.

▪ **Association Areas** The *association areas* of the cerebrum are made up of association tracts that connect motor and sensory areas and occupy the greater portion of the lateral surfaces of the occipital, parietal, and temporal lobes and the frontal lobes anterior to the motor areas. They are concerned with memory, emotions, reasoning, will, judgment, personality traits, and intelligence—all integrative functions of the cerebrum.

At this point, we will take a closer look at memory. Although memory has been studied by scientists for decades, there is still no satisfactory explanation for how we remember. *Memory* is the ability to recall thoughts. For an experience to become part of memory, it must produce changes in the central nervous system that represent the experience. Such a memory trace in the brain is called an *engram*.

Memory is classified into two categories, short-term and long-term. *Short-term memory* lasts only seconds or hours and is the ability to recall bits of information. One example is looking up a telephone number and remembering it long enough to dial. *Long-term memory*, on the other hand, lasts from days to years. If you frequently use a telephone number, it becomes part of long-term memory and can be retrieved for use for quite a long period. Such reinforcement use is called *memory consolidation*.

Our brain receives many stimuli but we are conscious of only a few of them. Of all the information that comes to our consciousness, only about 1 percent goes into long-term memory, and most of that is forgotten. It is a feature of memory that we can remember short lists easier than long ones. This is to say the obvious, but human memory does not record everything like an endless magnetic tape. It cannot remember or record long lists of details. Another feature of memory is that even when details are lost, the concept or main idea is retained. Then, we can often explain the idea—not like replaying a tape—but with our own selection of words.

Electroencephalogram (EEG)

Brain cells generate electrical activity as a result of literally millions of action potentials (nerve impulses) of individual neurons. The electrical potentials are called *brain waves* and indicate activity of the cerebral cortex. Brain waves pass easily through the skull and can be detected by sensors called electrodes. A record of such waves is called an *electroencephalogram* (*EEG*). The EEG is used clinically in the diagnosis of epilepsy, infectious disease, tumors, trauma, and hematomas. In cases of doubt, a flat EEG is increasingly being taken as one criterion of *brain death*.

CEREBELLUM

The *cerebellum* is the second-largest portion of the brain. It is behind the medulla and pons and below the occipital lobes of the cerebrum (see Figure 10-5).

The surface of the cerebellum, called the *cortex*, consists of gray matter. Beneath are *white matter tracts* that resemble branches of a tree. Deep within the white matter are masses of gray matter, the *cerebellar nuclei*. The cerebellum is attached to the brain stem by three paired bundles of fibers called *cerebellar peduncles* (see Figure 10-7).

The cerebellum is a motor area of the brain. It is concerned with subconscious skeletal movements required for coordination, balance, and maintaining posture.

Damage to the cerebellum through trauma or disease creates symptoms involving skeletal muscles. The effects are on the same side of the body as the damaged side of the cerebellum because of a double crossing of tracts within the cerebellum. There may be lack of muscle coordination, called *ataxia* (*a* = without; *taxis* = order). Blindfolded people with ataxia cannot touch the tip of their nose with a finger because they cannot coordinate movement with their sense of where a body part is located. Another sign of ataxia is a changed speech pattern due to uncoordinated speech muscles. Cerebellar damage may also result in *disturbances of gait* (staggering or abnormal walking movements) and *severe dizziness*.

NEUROTRANSMITTERS

Numerous substances are either known or suspected *neurotransmitters*. These substances establish lines of communication between nerve cells; they facilitate, excite, or inhibit postsynaptic neurons in the CNS.

Based on their chemistry, neurotransmitters fall into the following classes:

1. **Acetylcholine (ACh).** This is in a class by itself.
2. **Biogenic amines.** Contain an amine group that includes dopamine (DA), norepinephrine (NE), serotonin (5-HT), and histamine.
3. **Amino acids.** Included are gamma aminobutyric acid (GABA), an amino acid derivative; glycine; and glutamate.
4. **Neuropeptides.** Consist of chains of 2 to 40 amino acids; include enkephalins, endorphins, substance P, dynorphin, angiotensin II, cholecystokinin (CCK), and regulating hormones (or factors).

Exhibit 10-1 summarizes these neurotransmitters.

CRANIAL NERVES

Cranial nerves, like spinal nerves, are part of the somatic nervous system (SNS) of the peripheral nervous system. Of the 12 pairs of *cranial nerves*, 10 originate from the brain stem, but all leave the skull through foramina (holes) of the skull (see Figure 10-7). The cranial nerves are designated with Roman numerals and with names. The Roman numerals indicate the order in which the nerves arise from the brain (front to back). The names indicate the distribution or function.

Some cranial nerves contain only sensory fibers and thus are called *sensory nerves*. The remainder contain both sensory and motor fibers and are referred to as *mixed nerves*. At one time it was believed that the oculomotor, trochlear, abducens, accessory,

EXHIBIT 10-1
Summary of Representative Neurotransmitters

Substance	Comment
ACETYLCHOLINE (ACh)	Found in many parts of cerebral cortex, all skeletal neuromuscular junctions, and some portions of autonomic nervous system; usually excitatory; inactivated by acetylcholinesterase (AChE); blockage of ACh receptors in skeletal muscles leads to myasthenia gravis.
BIOGENIC AMINES	
Dopamine (DA)	Concentrated in substantia nigra of midbrain, hypothalamus, and limbic system; generally excitatory; involved in emotional responses and subconscious movements of skeletal muscles; decreased levels associated with Parkinson's disease; an excess of DA might be involved in schizophrenia.
Norepinehprine (NE)	Released at some neuromuscular and neuroglandular junctions; concentrated in the brain stem; also found in cerebral cortex, hypothalamus, cerebellum, and spinal cord; usually excitatory; may be related to arousal, dreaming, and regulation of mood; inactivated by catechol-*O*-methyltransferase monoamine oxidase (MAO).
Serotonin (5-HT)	Found in brain stem, limbic system, hypothalamus, cerebellum, and spinal cord; generally inhibitory; may be involved in inducing sleep, sensory perception, temperature regulation, and control of mood.
Histamine	Found in hypothalamus; also produced by mast cells during inflammation.
AMINO ACIDS	
Gamma Aminobutyric Acid (GABA)	Concentrated in midbrain, thalamus, hypothalamus, cerebellum, and occipital lobes of cerebrum; inhibitory; probably a target for antianxiety drugs.
Glycine	Found in spinal cord and retina; inhibitory in spinal cord.
Glutamate	Found in brain and spinal cord; excitatory in brain.
NEUROPEPTIDES	
Substance P	Found in sensory nerves, spinal cord pathways, and parts of brain associated with pain; stimulates perception of pain; endorphins exert their pain-inhibiting properties by suppressing release of substance P.
Enkephalins	Concentrated in thalamus, hypothalamus, limbic system, and spinal cord pathways that relay pain impulses; inhibit pain impulses by suppressing substance P.
Endorphins	Concentrated in pituitary gland, hypothalamus, thalamus, and brain stem; inhibit pain by inhibiting substance P; may have a role in memory and learning, sexual activity, control of body temperature; have been linked to depression and schizophrenia.
Dynorphin	Found in posterior pituitary gland, hypothalamus, and small intestine; 50 times more powerful than beta endorphin; may be related to controlling pain and registering emotions.
Angiotensin II	Produced from renin released by kidneys; may regulate blood pressure in brain.
Cholecystokinin (CCK)	Found in cerebral cortex and small intestine; may be related to the regulation of feeding.
Regulating Hormones (or factors)	Chemicals produced by hypothalamus that regulate the release of hormones by the anterior pituitary gland.

and hypoglossal nerves were entirely motor; however, it is now known that they contain some sensory fibers from proprioceptors (receptors for body position and movement) in muscles they supply. They are, however, primarily motor in function, serving to stimulate

skeletal muscle contraction. The cell bodies of sensory fibers are found outside the brain, whereas the cell bodies of motor fibers lie in nuclei within the brain.

Exhibit 10-2 summarizes cranial nerves and clinical applications.

EXHIBIT 10-2
Summary of Cranial Nerves[a]

Nerve (Type)	Location	Function and Clinical Application
Olfactory (I) (Sensory)	Arises in lining of nose, and terminates in primary olfactory areas of cerebral cortex.	**Function:** Smell. **Clinical application:** Loss of the sense of smell, called *anosmia*, may result from head injuries in which the cribriform plate of the ethmoid bone is fractured and from lesions along the olfactory pathway.
Optic (II) (Sensory)	Arises in retina of the eye, and terminates in visual areas of cerebral cortex.	**Function:** Sight. **Clinical application:** Fractures in the orbit, lesions along the visual pathway, and diseases of the nervous system may result in visual field defects and loss of visual acuity. A defect of vision is called *anopsia*.
Oculomotor (III) (Mixed, primarily motor)	**Motor portion:** Originates in midbrain and distributes to upper eyelid and four extrinsic eyeball muscles (superior rectus, medial rectus, inferior rectus, and inferior oblique); parasympathetic innervation to ciliary muscle of eyeball and sphincter muscle of iris. **Sensory portion:** Consists of afferent fibers from proprioceptors in eyeball muscles and terminates in midbrain.	**Motor function:** Movement of eyelid and eyeball, accommodation of lens for near vision, and constriction of pupil. **Sensory function:** Muscle sense (proprioception). **Clinical application:** A lesion in the nerve causes *strabismus* (weakness of external eye muscles), *ptosis* (drooping) of the upper eyelid, pupil dilation, the movement of the eyeball downward and outward on the damaged side, a loss of accommodation for near vision, and double vision (*diplopia*).
Trochlear (IV) (Mixed, primarily motor)	**Motor portion:** Originates in midbrain and is distributed to superior oblique muscle, an extrinsic eyeball muscle. **Sensory portion:** Consists of afferent fibers from proprioceptors in superior oblique muscle and terminates in midbrain.	**Motor function:** Movement of eyeball. **Sensory function:** Muscle sense (proprioception). **Clinical application:** In trochlear nerve paralysis, the head is tilted to the affected side and diplopia and strabismus occur.
Trigeminal (V) (Mixed)	**Motor portion:** Originates in pons and terminates in muscles used in chewing. **Sensory portion:** Consists of three branches: *ophthalmic*—contains sensory fibers from skin over upper eyelid, eyeball, lacrimal glands, nasal cavity, side of nose, forehead, and anterior half of scalp; *maxillary*—contains sensory fibers from mucosa of nose, palate, parts of pharynx, upper teeth, upper lip, cheek, and lower eyelid; *mandibular*—contains sensory fibers from anterior two-thirds of tongue, lower teeth, skin over mandible, and side of head in front of ear. The three branches terminate in pons. Sensory portion also consists of afferent fibers from proprioceptors in muscles of mastication.	**Motor function:** Chewing. **Sensory function:** Conveys sensations for touch, pain, and temperature from structures supplied; muscle sense (proprioception). **Clinical application:** Injury results in paralysis of the muscles of mastication and a loss of sensation of touch and temperature. *Neuralgia* (pain) of one or more branches of trigeminal nerve is called trigeminal neuralgia (*tic douloureux*).
Abducens (VI) (Mixed, primarily motor)	**Motor portion:** Originates in pons and is distributed to lateral rectus muscle, an extrinsic eyeball muscle. **Sensory portion:** Consists of afferent fibers from proprioceptors in lateral rectus muscle and terminates in pons.	**Motor function:** Movement of eyeball. **Sensory function:** Muscle sense (proprioception). **Clinical application:** With damage to this nerve, the affected eyeball cannot move laterally beyond the midpoint and the eye is usually directed medially.

continued

[a] A mnemonic device used to remember the names of the nerves is: "Oh, Oh, Oh, to touch and feel very green vegetables—AH!" The initial letter of each word corresponds to the initial letter of each pair of cranial nerves.

EXHIBIT 10-2 (*Continued*)

Nerve (Type)	Location	Function and Clinical Application
Facial (VII) **(Mixed)**	***Motor portion:*** Originates in pons and is distributed to facial, scalp, and neck muscles, parasympathetic distribution to lacrimal, sublingual, submandibular, nasal, and palatine glands. ***Sensory portion:*** Arises from taste buds on anterior two-thirds of tongue and terminates in gustatory areas of cerebral cortex. Also consists of afferent fibers from proprioceptors in muscles of face and scalp.	***Motor function:*** Facial expression and secretion of saliva and tears. ***Sensory function:*** Muscle sense (proprioception). ***Clinical application:*** Injury produces paralysis of the facial muscles, called *Bell's palsy*, loss of taste, and the eyes remain open, even during sleep.
Vestibulocochlear **(VIII)** **(Sensory)** **(formerly called** **the auditory** **nerve)**	***Cochlear branch:*** Arises in spiral organ (organ of Corti) and terminates in thalamus. ***Vestibular branch:*** Arises in semicircular canals, saccule, and utricle and terminates in thalamus.	***Cochlear branch function:*** Conveys impulses associated with hearing. ***Vestibular branch function***: Conveys impulses associated with equilibrium. ***Clinical application***: Injury to the cochlear branch may cause *tinnitus* (ringing) or deafness. Injury to the vestibular branch may cause *vertigo* (dizziness), *ataxia*, and *nystagmus* (involuntary rapid movement of the eyeball).
Glossopharyngeal **(IX)** **(Mixed)**	***Motor portion:*** Originates in medulla and is distributed to swallowing muscles of pharynx; parasympathetic distribution to parotid gland. ***Sensory portion:*** Arises from taste buds on posterior one-third of tongue and terminates in thalamus. Also consists of afferent fibers from proprioceptors in swallowing muscles supplied.	***Motor function:*** Secretion of saliva. ***Sensory function:*** Taste and regulation of blood pressure; muscle sense (proprioception). ***Clinical application:*** Injury results in pain during swallowing, reduced secretion of saliva, loss of sensation in the throat, and loss of taste.
Vagus (X) **(Mixed)**	***Motor portion***: Originates in medulla and terminates in muscles of pharynx, larynx, respiratory passageways, lungs, esophagus, heart, stomach, small intestine, most of large intestine, and gallbladder; parasympathetic fibers innervate involuntary muscles and glands of the gastrointestinal (GI) tract. ***Sensory portion***: Arises from essentially same structures supplied by motor fibers and terminates in medulla and pons. Also consists of afferent fibers from proprioceptors in muscles supplied.	***Motor function:*** Visceral muscle movement. ***Sensory function:*** Sensations from organs supplied; muscle sense (proprioception). ***Clinical application:*** Severing of both nerves in the upper body interferes with swallowing, paralyzes vocal cords, and interrupts sensations from many organs. Injury to both nerves in the abdominal area has little effect, since the abdominal organs are also supplied by autonomic fibers from the spinal cord.
Accessory (XI) **(Mixed, primarily** **motor)**	***Motor portion:*** Consists of a bulbar portion and a spinal portion. Bulbar portion originates from medulla and supplies voluntary muscles of pharynx, larynx, and soft palate. Spinal portion originates from cervical spinal cord and supplies sternocleidomastoid and trapezius muscles. ***Sensory portion:*** Consists of afferent fibers from proprioceptors in muscles supplied.	***Motor function:*** Cranial portion mediates swallowing movements; spinal portion mediates movement of head. ***Sensory function***: Muscle sense (proprioception). ***Clinical application***: If damaged, the sternocleidomastoid and trapezius muscles become paralyzed, with resulting inability to turn the head or raise the shoulders.
Hypoglossal (XII) **(Mixed, primarily** **motor)**	***Motor portion:*** Originates in medulla and supplies muscles of tongue. ***Sensory portion:*** Consists of fibers from proprioceptors in tongue muscles that terminate in medulla.	***Motor function:*** Movement of tongue during speech and swallowing. ***Sensory function:*** Muscle sense (proprioception). ***Clinical application:*** Injury results in difficulty in chewing, speaking, and swallowing. The tongue, when protruded, curls toward the affected side and that side becomes atrophied, shrunken, and deeply furrowed.

■ COMMON DISORDERS ■

Cerebrovascular Accident (CVA)

The most common brain disorder is a ***cerebrovascular accident (CVA)***, also called a ***stroke*** or ***cerebral apoplexy***. A CVA is characterized by a relatively abrupt onset of persisting neurological symptoms due to the destruction of brain tissue (infarction) resulting from disorders in the blood vessels that supply the brain. CVAs may be classified as (1) *ischemic*, the most common type, due to a decreased blood supply, and (2) *hemorrhagic*, due to a blood vessel that bursts.

Transient Ischemic Attack (TIA)

A *transient ischemic attack* (*TIA*) is an episode of temporary cerebral dysfunction caused by an interference of the blood supply to the brain.

Spinal Cord Injury

The spinal cord may be damaged in various ways. Depending on the location and extent of the injury, paralysis may occur. *Paralysis* refers to total loss of motor function resulting from damage to nervous tissue or a muscle. Paralysis may be classified as follows: *monoplegia* (*mono* = one; *plege* = stroke), paralysis of one extremity only; *diplegia* (*di* = two), paralysis of both upper extremities or both lower extremities; *paraplegia* (*para* = beyond), paralysis of both lower extremities; *hemiplegia* (*hemi* = half), paralysis of the upper extremity, trunk, and lower extremity on one side of the body; and *quadriplegia* (*quad* = four), paralysis of the two upper and two lower extremities.

Complete transection of the spinal cord means that the cord is severed, thus cutting all ascending and descending tracts. It results in a loss of all sensations and voluntary movement below the level of the transection.

Hemisection is a partial transection. It is characterized, below the hemisection, by a loss of proprioception, tactile discrimination, and feeling of vibration on the same side as the injury; paralysis on the same side; and loss of feelings of pain and temperature on the opposite side.

Following transection, there is an initial period of *spinal shock* that lasts from a few days to several weeks, during which time all reflex activity is abolished.

Neuritis

Neuritis is inflammation of a single nerve, two or more nerves in separate areas, or many nerves simultaneously.

Sciatica

Sciatica (sī-AT-i-ka) is a type of neuritis characterized by severe pain along the path of the sciatic nerve or its branches.

Shingles

Shingles is an acute infection of the peripheral nervous system. It is caused by *herpes zoster* (HER-pēz ZOS-ter), the chickenpox virus.

Brain Tumors

A *brain tumor* refers to any benign or malignant growth within the cranium.

Poliomyelitis

Poliomyelitis (*infantile paralysis*), or simply *polio*, is caused by a virus called poliovirus. The virus destroys motor nerve cell bodies, specifically in the anterior horns of the spinal cord and in the nuclei of the cranial nerves and produces paralysis and atrophy of skeletal muscle.

Cerebral Palsy (CP)

The term *cerebral palsy* (*CP*) refers to a group of motor disorders resulting in loss of muscle control. It is caused by damage to the motor areas of the brain.

Parkinson's Disease (PD)

Parkinson's disease (*PD*) is a progressive disorder of the central nervous system that typically affects its victims around age 60. The cause is unknown. In PD, there is a degeneration of DA-producing neurons in the substantia nigra, and the severe reduction of DA in the basal ganglia brings about most of the symptoms: tremor, impaired motor performance, rigid facial muscles, impaired walking, poor posture, autonomic dysfunctions, and sensory complaints.

Multiple Sclerosis (MS)

Multiple sclerosis (*MS*) is the progressive destruction of the myelin sheaths of neurons in the central nervous system that interferes with the transmission of nerve impulses from one neuron to another. Although the cause of MS is unclear, there is some evidence that it results from a virus that triggers an autoimmune response.

Epilepsy

Epilepsy is characterized by short, recurrent, periodic attacks of motor, sensory, or psychological malfunction. The attacks, called *epileptic seizures*, are initiated by abnormal and irregular discharges of electricity from millions of neurons in the brain.

Dyslexia

In *dyslexia* (dis-LEK-sē-a; *dys* = difficulty; *lexis* = words), the brain's ability to translate images received from the eyes or ears into understandable language is impaired. The condition is unrelated to basic intellectual capacity. Some peculiarity in the brain's organizational pattern distorts the ability to read, write, and count.

Tay–Sachs Disease

Tay–Sachs disease is an inherited disease in which neurons of the brain degenerate because of excessive amounts of a lipid called ganglioside. There is no known cure, and children who are born with the disorder usually die before age 5.

Headache

One of the most common human afflictions is *headache* or *cephalgia* (*enkephalos* = brain; *algia* = painful condition). Based on origin, two general types are distinguished: intracranial (within the brain) and extracranial (outside the brain). Serious headaches of intracranial origin are caused by brain tumors, blood vessel abnormalities, inflammation of the brain or meninges, decrease in oxygen supply to the brain, or damage to brain cells. Extracranial headaches are related to infections of the eyes, ears, nose, and sinuses and are commonly felt as headaches because of the location of these structures. Tension headaches are extracranial headaches associated with stress, fatigue, and anxiety and usually occur in the occipital and temporal muscles.

Trigeminal Neuralgia (Tic Douloureux)

Irritation of the trigeminal (V) nerve causes *trigeminal neuralgia* or *tic douloureux* (doo-loo-ROO), characterized by brief but extreme pain in the face and forehead on the affected side.

Reye's Syndrome (RS)

Reye's (RĪZ) *syndrome* (*RS*) seems to occur following a viral infection, particularly chickenpox or influenza. Aspirin at normal doses is believed to be a risk factor and most often, children and teenagers are affected. The disease is characterized by vomit-

ing and brain dysfunction (disorientation, lethargy, and personality changes) and may progress to coma and death.

Alzheimer's Disease (AD)

Alzheimer's (ALTZ-hī-merz) *disease,* or *AD,* is a disabling neurological disorder that afflicts about 5 percent (about 1 million people) of the population over age 65. Its causes are unknown, its effects are irreversible, and it has no cure. It is the fourth leading cause of death among the elderly following heart disease, cancer, and stroke.

Delirium

Delirium (de-LIR-ē-um; *deliria* = off the tract), is a temporary disorder of abnormal cognition (perception, thinking, and memory) accompanied by fever, disturbances of the sleep–wake cycle and speech, and hyperactive or hypoactive movements.

MEDICAL TERMINOLOGY AND CONDITIONS

Agnosia (ag-NŌ-zē-a; *a* = without; *gnosis* = knowledge) Inability to recognize the significance of sensory stimuli.

Analgesia (an-al-JĒ-zē-a; *an* = without; *algia* = painful condition) Pain relief.

Anesthesia (an'-es-THĒ-zē-a; *esthesia* = feeling) Loss of feeling.

Apraxia (a-PRAK-sē-a; *pratto* = to do) Inability to carry out purposeful movements in the absence of paralysis.

Dementia (de-MEN-shē-a; *de* = away from; *mens* = mind) An organic mental disorder that results in permanent or progressive loss of intellectual abilities (memory, judgment, abstract thinking) and changes in personality.

Electroconvulsive therapy (ECT) (e-lek-trō-con-VUL-siv THER-a-pē) A form of shock therapy in which convulsions are induced by passing a brief electric current through the brain.

Huntington's chorea (HUNT-ing-tunz kō-RĒ-a; *choreia* = dance) A rare hereditary disease characterized by involuntary jerky movements and mental deterioration that terminates in dementia.

Nerve block Loss of sensation in a region, such as in local dental anesthesia, from injection of a local anesthetic.

Neuralgia (noo-RAL-jē-a; *neur* = nerve) Attacks of pain along the entire course or branch of a peripheral sensory nerve.

Paresis (pa-RĒ-sis) Slight or partial paralysis.

Spastic (SPAS-tik; *spas* = draw or pull) An increase in muscle tone (stiffness) associated with an increase in tendon reflexes and abnormal reflexes (Babinski sign).

Viral encephalitis (VĪ-ral en'-sef-a-LĪ-tis) An acute inflammation of the brain caused directly by various viruses or by an allergic reaction to one of the many viruses that are normally harmless to the central nervous system. If the virus affects the spinal cord as well, it is called *encephalomyelitis.*

STUDY OUTLINE

Grouping of Neural Tissue (p. 204)

1. A nerve is a bundle of nerve fibers outside the central nervous system.
2. A ganglion is a collection of neuron cell bodies outside the central nervous system.
3. A tract is a bundle of fibers of similar function in the central nervous system.
4. White matter is a group of myelinated axons.
5. Gray matter is a collection of neuron cell bodies and dendrites or unmyelinated axons.

6. A nucleus is a mass of nerve cell bodies and unmyelinated dendrites in the gray matter of the brain.
7. A horn is an area of gray matter in the spinal cord.

Spinal Cord (p. 204)

Protection and Coverings (p. 204)

1. The spinal cord is protected by the vertebral canal, meninges (dura mater, arachnoid, pia mater), cerebrospinal fluid, and vertebral ligaments.

2. The meninges are three connective tissue coverings that run continuously around the spinal cord and brain: dura mater, arachnoid, and pia mater.

3. Removal of cerebrospinal fluid from the subarachnoid space is called a spinal puncture. The procedure is used to diagnose pathologies and to introduce antibiotics or contrast media.

General Features (p. 205)

1. The spinal cord begins as a continuation of the medulla oblongata and terminates at about the second lumbar vertebra.

2. It contains cervical and lumbar enlargements that serve as points of origin for nerves to the extremities.

3. Nerves arising from the lowest portion of the cord are called the cauda equina.

4. The gray matter in the spinal cord is divided into horns and the white matter into columns.

5. In the center of the spinal cord is the central canal, which runs the length of the spinal cord and contains cerebrospinal fluid.

6. There are ascending (sensory) tracts and descending (motor) tracts.

Structure in Cross Section (p. 206)

1. Parts of the spinal cord observed in cross section are the central canal; anterior, posterior, and lateral gray horns; anterior, posterior, and lateral white columns; and ascending and descending tracts.

2. The spinal cord conveys sensory and motor information by way of the ascending and descending tracts, respectively.

Functions (p. 206)

1. One major function of the spinal cord is to convey sensory impulses from the periphery to the brain and to conduct motor impulses from the brain to the periphery.

2. The other major function is to serve as a reflex center. The posterior root, posterior root ganglion, and anterior root are involved in conveying an impulse.

3. A reflex arc is the shortest route that can be taken by an impulse from a receptor to an effector. Its basic components are a receptor, a sensory neuron, a center, a motor neuron, and an effector.

4. A reflex is a quick, involuntary response to a stimulus that passes along a reflex arc. Reflexes represent the body's principal mechanisms for responding to changes (stimuli) in the internal and external environment.

Spinal Nerves (p. 208)

Names (p. 208)

1. The 31 pairs of spinal nerves are named and numbered according to the region and level of the spinal cord from which they emerge.

2. There are 8 pairs of cervical, 12 pairs of thoracic, 5 pairs of lumbar, 5 pairs of sacral, and 1 pair of coccygeal nerves.

Composition and Coverings (p. 209)

1. Spinal nerves are attached to the spinal cord by means of a posterior root and an anterior root. All spinal nerves are mixed.

2. Individual fibers, bundles of nerves, and the entire nerve are wrapped in connective tissue.

Distribution (p. 209)

1. Branches of a spinal nerve include the dorsal ramus, ventral ramus, meningeal branch, and rami communicantes.

2. The ventral rami of spinal nerves, except for T2–T11, form networks of nerves called plexuses.

3. The principal plexuses are called the cervical, brachial, lumbar, and sacral plexuses. The plexuses branch and subbranch.

4. Nerves T2–T11 do not form plexuses and are called intercostal (thoracic) nerves. They are distributed directly to the structures they supply in intercostal spaces.

Brain (p. 209)

Principal Parts (p. 209)

1. The principal parts of the brain are the brain stem, diencephalon, cerebrum, and cerebellum.

2. The brain stem consists of the medulla oblongata, pons, and midbrain. The diencephalon consists of the thalamus and hypothalamus.

Protection and Coverings (p. 209)

1. The brain is protected by cranial bones, meninges, and cerebrospinal fluid.

2. The cranial meninges are continuous with the spinal meninges and are named dura mater, arachnoid, and pia mater.

Cerebrospinal Fluid (CSF) (p. 209)

1. Cerebrospinal fluid is formed in the choroid plexuses and circulates continually through the subarachnoid space, ventricles, and central canal.

2. Cerebrospinal fluid protects by serving as a shock absorber. It also delivers nutritive substances from the blood and removes wastes.

Blood Supply (p. 210)

1. Any interruption of the oxygen supply to the brain can weaken, permanently damage, or kill brain cells.

2. Glucose deficiency may produce dizziness, convulsions, and unconsciousness.

3. The blood–brain barrier (BBB) explains the differential rates of passage of certain material from the blood into the brain.

Brain Stem (p. 212)

1. The medulla oblongata is continuous with the upper part of the spinal cord. It contains nuclei that are reflex centers for regulating heart rate, respiratory rate, vasoconstriction, swallowing, coughing, vomiting, sneezing, and hiccuping. The vestibulocochlear, accessory, vagus, and hypoglossal nerves also originate there.

2. The pons is superior to the medulla. It connects the spinal cord with the brain and links parts of the brain with one another. It relays impulses related to voluntary skeletal movements from the cerebral cortex to the cerebellum. The trigeminal, abducens, facial, and vestibular branch of the vestibulocochlear nerves also originate there. The reticular formation of the pons contains the pneumotaxic and apneustic centers for respiration.

3. The midbrain connects the pons and diencephalon. It conveys motor impulses from the cerebrum to the cerebellum and cord, sensory impulses from cord to thalamus, and regulates auditory and visual reflexes. The oculomotor and trochlear nerves also originate there.

Diencephalon (p. 214)

1. The diencephalon consists of the thalamus and hypothalamus.
2. The thalamus is superior to the midbrain and contains nuclei that serve as relay stations for all sensory impulses, except smell, to the cerebral cortex. It also registers conscious recognition of pain and temperature and some awareness of light touch and pressure.
3. The hypothalamus is inferior to the thalamus. It controls and integrates the autonomic nervous system, receives sensory impulses from viscera, connects the nervous and endocrine systems, coordinates mind-over-body phenomena, functions in rage and aggression, controls body temperature, regulates food and fluid intake, and maintains the waking state and sleep patterns.

Reticular Activating System (RAS), Consciousness, and Sleep (p. 215)

1. The RAS functions in arousal (awakening from deep sleep) and consciousness (wakefulness).
2. Sleep is a state of partial unconsciousness from which an individual can be aroused.
3. Sleep consists of nonrapid eye movement sleep (NREM) and rapid eye movement sleep (REM).

Cerebrum (p. 215)

1. The cerebrum is the largest part of the brain. Its cortex contains convolutions, fissures, and sulci.
2. The cerebral lobes are named the frontal, parietal, temporal, and occipital.
3. The left hemisphere is more important for right-handed control, spoken and written language, numerical and scientific skills, and reasoning; the right hemisphere is more important for left-handed control, musical and artistic awareness, space and pattern perception, insight, imagination, and generating mental images of sight, sound, touch, taste, and smell.
4. The white matter is under the cortex and consists of myelinated axons running in three principal directions.

5. The basal ganglia are paired masses of gray matter in the cerebral hemispheres. They help to control muscular movements.
6. The limbic system is found in the cerebral hemispheres and diencephalon. It functions in emotional aspects of behavior and memory.
7. The sensory areas of the cerebral cortex interpret sensory impulses. The motor areas govern muscular movement. The association areas are concerned with emotional and intellectual processes.
8. Memory is the ability to recall thoughts and is generally classified into two kinds: short-term and long-term memory.
9. Brain waves generated by the cerebral cortex are recorded as an electroencephalogram (EEG). They may be used to diagnose epilepsy, infections, and tumors.

Cerebellum (p. 221)

1. The cerebellum occupies the inferior and posterior aspects of the cranial cavity. It consists of two hemispheres with a cortex of gray matter and interior of white matter tracts.
2. It is attached to the brain stem by three pairs of cerebellar peduncles.
3. The cerebellum coordinates skeletal muscles and maintains normal muscle tone and body equilibrium.

Neurotransmitters (p. 221)

1. Numerous substances are either known or suspected neurotransmitters that function to facilitate, excite, or inhibit postsynaptic neurons.
2. Neurotransmitters are classified as follows: acetylcholine (ACh); biogenic amines (dopamine, norepinephrine, serotonin, and histamine); amino acids (gamma aminobutyric acid, glycine, and glutamate); and neuropeptides (enkephalin, endorphin, substance P, and dynorphin). See Exhibit 10-1 for a summary of neurotransmitters.

Cranial Nerves (p. 221)

1. Twelve pairs of cranial nerves originate from the brain. Like spinal nerves, they are part of the PNS. See Exhibit 10-2 for a summary of cranial nerves.

REVIEW QUESTIONS

1. Define the following groupings of neural tissue: nerve, ganglion, tract, white matter, gray matter, nucleus, and horn (column). (p. 204)
2. Explain the location and composition of the spinal and cranial meninges. Describe the location and significance of the subarachnoid space. Define meningitis. (p. 204)
3. Describe the general features of the spinal cord and define cauda equina. What is a spinal segment? (p. 205)

4. Describe the function of the spinal cord as a conduction pathway and reflex center. (p. 206)
5. What is a reflex arc? List and define the components of a reflex arc. (p. 208)
6. Define a reflex. How are reflexes related to the maintenance of homeostasis? (p. 207)
7. Define a spinal nerve. Why are all spinal nerves classified as mixed nerves? (p. 209)

8. Explain how a spinal nerve is attached to the spinal cord and wrapped in its connective tissue coverings? (p. 209)

9. How are spinal nerves named and numbered? (p. 209)

10. Describe the branches of a typical spinal nerve. (p. 209)

11. What is a plexus? Describe the principal plexuses and the regions they supply. (p. 209)

12. Identify the four principal parts of the brain and the components of each, where applicable. (p. 209)

13. Where is cerebrospinal fluid (CSF) formed? Describe its circulation. (p. 209)

14. What is the blood–brain barrier (BBB)? What are its advantages and disadvantages to brain function? (p. 212)

15. Describe the location and structure of the medulla. Define decussation of pyramids. Why is it important? List the principal functions of the medulla. (p. 212)

16. Describe the location and structure of the pons. What are its functions? (p. 213)

17. Describe the location and structure of the midbrain. What are its functions? (p. 213)

18. What is the reticular formation? (p. 212)

19. Describe the location and structure of the thalamus. List its functions. (p. 214)

20. Where is the hypothalamus located? Explain some of its major functions. (p. 214)

21. What is the reticular activating system (RAS)? How does it function in consciousness and sleep? (p. 215)

22. What are the four stages of non–rapid eye movement (NREM) sleep? How is NREM sleep different from rapid eye movement (REM) sleep? (p. 215)

23. Where is the cerebrum located? Describe the cortex, convolutions, fissures, and sulci of the cerebrum. (p. 215)

24. Describe brain lateralization (split-brain concept). (p. 216)

25. Describe the organization of cerebral white matter. Indicate the function of each group of fibers. (p. 219)

26. What are basal ganglia? Name the important basal ganglia and list the function of each. (p. 217)

27. Define the limbic system. Explain several of its functions. (p. 217)

28. What is meant by a sensory area of the cerebral cortex? List, locate, and give the function of each sensory area. (p. 217)

29. What is meant by a motor area of the cerebral cortex? List, locate, and give the function of each motor area. (p. 220)

30. What is an association area of the cerebral cortex? What are its functions? (p. 220)

31. What is memory? What are the two kinds? (p. 221)

32. Define an electroencephalogram (EEG). What is the diagnostic value of an EEG? (p. 221)

33. Describe the location and function of the cerebellum. (p. 221)

34. Define a neurotransmitter and classify them according to chemical structure. (p. 221)

35. Define a cranial nerve. How are cranial nerves named and numbered? Distinguish between a mixed and a sensory cranial nerve. (p. 221)

36. For each of the 12 pairs of cranial nerves, list (a) its name, number, and type; (b) its location; and (c) its function. In addition, list the effects of damage, where applicable. (p. 223)

37. Define the following: cerebrovascular accident (CVA), transient ischemic attack (TIA), spinal cord injury, neuritis, sciatica, shingles, brain tumor, poliomyelitis, cerebral palsy (CP), Parkinson's disease (PD), multiple sclerosis (MS), epilepsy, dyslexia, Tay–Sachs disease, headache, trigeminal neuralgia, Reye's syndrome (RS), Alzheimer's disease (AD), and delirium. (p. 224)

38. Refer to the glossary of medical terminology and conditions associated with the nervous system. Be sure that you can define each term. (p. 226)

Autonomic Nervous System

1. Describe the main characteristics of the autonomic nervous system.
2. Compare the somatic and autonomic nervous systems.
3. Describe the functions of the autonomic nervous system.
4. Explain how the autonomic nervous system is controlled.

A LOOK AHEAD

COMPARISON OF SOMATIC AND
    AUTONOMIC NERVOUS
    SYSTEMS
STRUCTURE OF THE AUTONOMIC
    NERVOUS SYSTEM
  Visceral Efferent Pathways
    Preganglionic Neurons
    Autonomic Ganglia
    Postganglionic Neurons
  Sympathetic Division
  Parasympathetic Division
FUNCTIONS OF THE AUTONOMIC
    NERVOUS SYSTEM
  Neurotransmitters
  Activities

The part of the nervous system that regulates smooth muscle, cardiac muscle, and glands is the *autonomic nervous system* (*ANS*). (Recall that together the ANS and somatic nervous system comprise the peripheral nervous system.) The ANS consists of visceral efferent neurons organized into nerves, ganglia, and plexuses. Functionally, it usually operates without conscious control. The system was originally named *autonomic* because physiologists thought it was autonomous or self-governing, that is, that it functioned with no control from the central nervous system. It is now recognized that many parts of the CNS are connected to the ANS and exert considerable influence over it. Autonomic centers in the cerebral cortex are connected to autonomic centers of the thalamus, which in turn connect with the hypothalamus. It is at the level of the hypothalamus that the major control and integration of the ANS occur. (This connection between the CNS and ANS is the reason why anxiety, which can result from either conscious or subconscious stimulation in the cerebral cortex, results in sweaty palms and palpitations and why a shocking or horrifying sight can result in lowered blood pressure and fainting.) The hypothalamus receives input from areas of the nervous system concerned with emotions and receptors associated with visceral functions, olfaction (smell), gustation (taste), and changes in temperature and electrolyte composition of the blood.

COMPARISON OF SOMATIC AND AUTONOMIC NERVOUS SYSTEMS

Whereas the somatic nervous system consists of sensory neurons that convey information from receptors primarily in the head, body wall, and extremities and motor neurons that produce conscious movement in skeletal muscles, the autonomic nervous system (visceral efferent nervous system) regulates visceral activities, and it generally does so involuntarily and automatically. For example, the ANS regulates the size of the pupils, accommodation for near vision, dilation of blood vessels, rate and force of the heartbeat, movements of the gastrointestinal tract, and secretion by most glands.

The ANS is generally considered to be entirely motor. All its axons are efferent fibers, which transmit impulses from the central nervous system to visceral effectors. Autonomic fibers are thus called *visceral efferent (motor) fibers*. *Visceral effectors* include cardiac muscle, smooth muscle, and glandular epithelium. This does not mean there are no afferent (sensory) impulses from visceral effectors. Impulses that arise from receptors in viscera travel along afferent neurons of spinal nerves to the spinal cord or cranial nerves to the lower portions of the brain. Impulses from afferent neurons are delivered to various autonomic centers and the returning efferent (motor) impulse usually produces an adjustment in a visceral effector without conscious recognition. Some visceral sensations, however, such as hunger, nausea, fullness of the urinary bladder, and pain from viscera, produce conscious recognition.

The autonomic nervous system consists of two divisions: *sympathetic* and *parasympathetic*. Many organs innervated by the ANS receive visceral efferent neurons from both divisions. In general, impulses from one division stimulate the organ to start or increase activity, whereas impulses from the other decrease the organ's activity. Organs that receive impulses from both sympathetic and parasympathetic fibers are said to have *dual innervation*. Thus, autonomic innervation may be excitatory or inhibitory. In the somatic nervous system, stimulation of a skeletal muscle is always excitatory. When the neuron ceases to stimulate a muscle, contraction stops altogether.

In the somatic nervous system, all motor pathways involve only one efferent neuron from the CNS which synapses directly on a skeletal muscle and secretes only acetylcholine (ACh). In the ANS, there are two efferent neurons and a ganglion between them. The first neuron runs from the CNS to a ganglion, where it synapses with the second efferent neuron. This neuron ultimately synapses on a visceral effector. The second neuron may release either ACh or norepinephrine (NE).

Exhibit 11-1 presents a summary of the principal differences between the somatic and autonomic nervous systems.

STRUCTURE OF THE AUTONOMIC NERVOUS SYSTEM

VISCERAL EFFERENT PATHWAYS

Autonomic motor pathways involve two efferent (motor) neurons. One extends from the CNS to a ganglion; the other from the ganglion to the effector (muscle or gland).

The first efferent neuron in an autonomic pathway is called a *preganglionic neuron* (Figure 11-1). Its cell body is in the brain or spinal cord. Its myelinated axon, called a *preganglionic fiber*, passes out of the CNS as part of a cranial or spinal nerve. At

some point, the fiber separates from the nerve and travels to an autonomic ganglion, where it synapses with the second neuron in the efferent pathway, called a postganglionic neuron.

The *postganglionic neuron* lies entirely outside the CNS. Its cell body and dendrites (if it has dendrites) are located in the autonomic ganglion, where the synapse with one or more preganglionic fibers occurs. The axon of a postganglionic neuron, called a *postganglionic fiber*, is unmyelinated and terminates in a visceral effector.

Thus, preganglionic neurons convey efferent impulses from the CNS to autonomic ganglia. Postganglionic neurons relay the impulses from autonomic ganglia to visceral effectors.

Preganglionic Neurons

In the sympathetic division, the preganglionic neurons have their cell bodies in the lateral gray horns of the twelve thoracic segments and first two or three lumbar segments of the spinal cord (Figure 11-2). The cell bodies of the preganglionic neurons of the parasympathetic division are located in the nuclei of cranial nerves III,

---

EXHIBIT 11-1
Comparison of Somatic and Autonomic Nervous Systems

	Somatic	**Autonomic**
Effectors	Skeletal muscles.	Cardiac muscle, smooth muscle, glandular epithelium.
Type of Control	Voluntary.	Involuntary.
Neural Pathway	One efferent neuron extends from CNS and synapses directly on a skeletal muscle.	One efferent neuron extends from the CNS and synapses with another efferent neuron in a ganglion; the second neuron synapses on a visceral effector.
Action on Effector	Always excitatory.	May be excitatory or inhibitory.
Neurotransmitters	Acetylcholine (ACh).	Acetylcholine (ACh) or norepinephrine (NE).

FIGURE 11-1 Relationship between preganglionic and postganglionic (sympathetic) neurons.

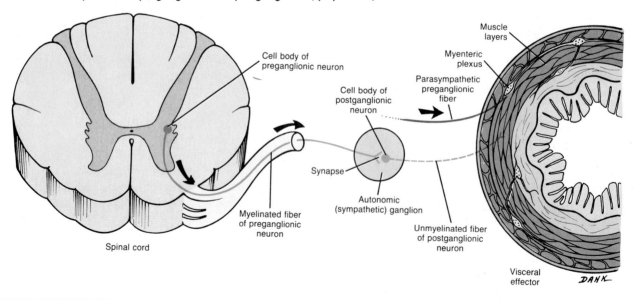

What are the four ways in which the somatic and autonomic nervous systems differ?

FIGURE 11-2 Structure of the autonomic nervous system. Although the sympathetic division is shown only on the left side of the figure and the parasympathetic division is shown only on the right side, keep in mind that each division is actually on both sides of the body (bilateral symmetry).

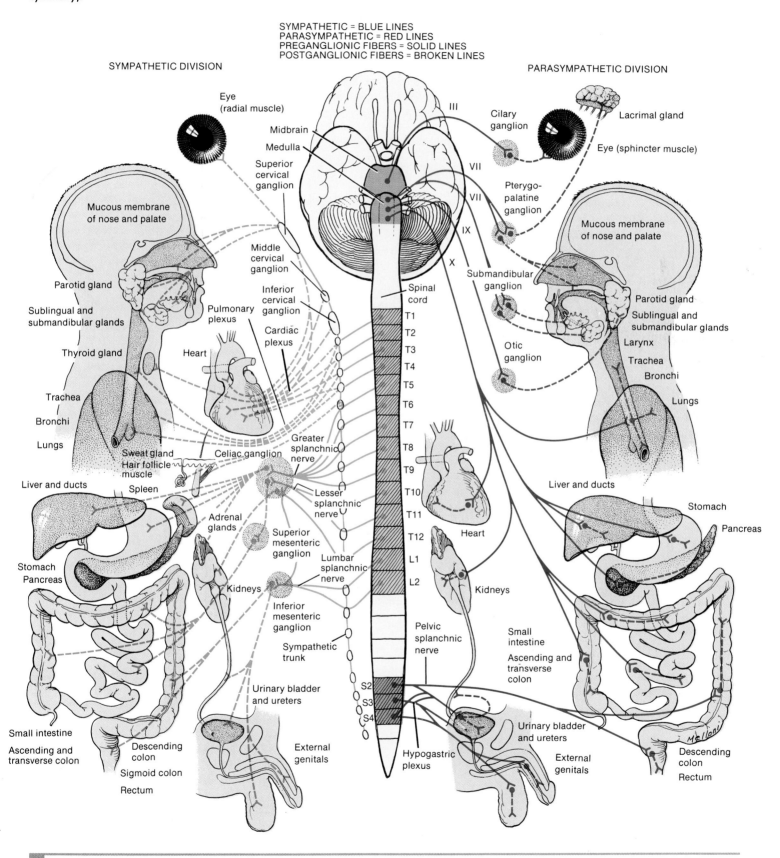

SYMPATHETIC = BLUE LINES
PARASYMPATHETIC = RED LINES
PREGANGLIONIC FIBERS = SOLID LINES
POSTGANGLIONIC FIBERS = BROKEN LINES

Compare the sympathetic and parasympathetic nervous systems with regard to types of ganglia, location of ganglia, preganglionic fiber synapses, and distribution.

VII, IX, and X in the brain stem and in the lateral gray horns of the second through fourth sacral segments of the spinal cord.

Autonomic Ganglia

Autonomic pathways always include *autonomic ganglia*, where synapses between visceral efferent neurons occur.

Autonomic ganglia may be divided into three groups. A *sympathetic trunk ganglion* lies on either side of the vertebral column, from the base of the skull to the coccyx (Figure 11-3). It receives preganglionic fibers only from the sympathetic division (see Figure 11-2). Because of this, sympathetic preganglionic fibers tend to be short.

The second kind of autonomic ganglion is called a *prevertebral ganglion* (Figure 11-3). It lies anterior to the spinal column and close to the large abdominal arteries from which the names of prevertebral ganglia are derived, the celiac ganglion, the superior mesenteric ganglion, and the inferior mesenteric ganglion (Figure 11-2). Prevertebral ganglia receive preganglionic fibers from the sympathetic division.

The third kind of autonomic ganglion belongs to the parasympathetic division and is called a *terminal ganglion*. The ganglia of this group are located at the end of a visceral efferent pathway very close to visceral effectors or actually within the walls of visceral effectors. Terminal ganglia receive preganglionic fibers

FIGURE 11-3 Ganglia of the sympathetic division of the autonomic nervous system.

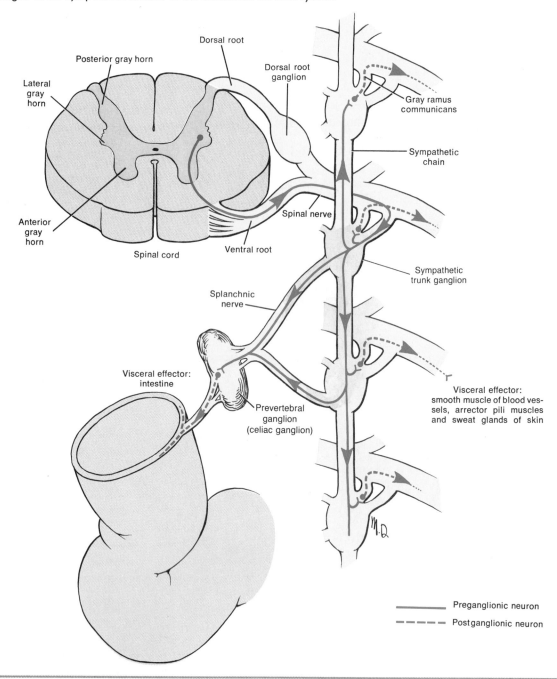

Trace a preganglionic fiber leaving the spinal cord through the anterior root of a spinal nerve to its terminating point. What is that termination point? What happens there?

from the parasympathetic division. The preganglionic fibers do not pass through sympathetic trunk ganglia (Figure 11-2). Because of this, parasympathetic preganglionic fibers tend to be long.

Postganglionic Neurons

Axons from preganglionic neurons of the sympathetic division pass to ganglia of the sympathetic trunk. They can either synapse with postganglionic neurons in the ganglia of the *sympathetic chain* (preganglionic fibers on which the ganglia are strung), or they can continue, without synapsing, through the chain ganglia to a prevertebral ganglion and synapse with the postganglionic neurons there. Each sympathetic preganglionic fiber synapses with several postganglionic fibers in the ganglion, and the postganglionic fibers pass to several visceral effectors. After exiting their ganglia, the postsynaptic fibers innervate their visceral effectors.

Axons from preganglionic neurons of the parasympathetic division pass to terminal ganglia near or within a visceral effector. In the ganglion, the presynaptic neuron usually synapses with four or five postsynaptic neurons to a single visceral effector. After exiting their ganglia, the postsynaptic fibers supply their visceral effectors.

With this background in mind, we can now examine some specific structural features of the sympathetic and parasympathetic divisions.

SYMPATHETIC DIVISION

Preganglionic fibers are myelinated and leave the spinal cord through the anterior root of a spinal nerve. After exiting, they pass to the nearest sympathetic trunk ganglion on the same side. A preganglionic fiber may terminate (synapse) in several ways, but in most cases, the synapse will be with a large number, usually 20 or more, of postganglionic cell bodies in a ganglion. Often the postganglionic fibers then terminate in widely separated organs of the body. Thus, an impulse that starts in a single preganglionic neuron may reach several visceral effectors. For this reason, most sympathetic responses have widespread effects on the body.

PARASYMPATHETIC DIVISION

Preganglionic cell bodies of the parasympathetic division are found in nuclei in the brain stem and the lateral gray horn of the second through fourth sacral segments of the spinal cord (Figure 11-2). Their fibers emerge as part of a cranial or spinal nerve. Preganglionic fibers of both the cranial and sacral outflows end in terminal ganglia, where they synapse with postganglionic neurons.

The sympathetic and parasympathetic divisions are compared in Exhibit 11-2.

FUNCTIONS OF THE AUTONOMIC NERVOUS SYSTEM

NEUROTRANSMITTERS

Autonomic fibers, like other axons of the nervous system, release neurotransmitters at synapses as well as at points of contact with visceral effectors (smooth and cardiac muscle and glands). These points are called *neuroeffector junctions*. Neuroeffector junctions may be either neuromuscular or neuroglandular junctions. On the basis of the neurotransmitter produced, autonomic fibers may be classified as either cholinergic or adrenergic.

Cholinergic (kō'-lin-ER-jik) *fibers* release *acetylcholine (ACh)* and include the following: (1) all sympathetic and parasympathetic preganglionic axons, (2) all parasympathetic postganglionic axons, and (3) a few sympathetic postganglionic axons, such as those to sweat glands and blood vessels in skeletal muscles. Since acetylcholine is quickly inactivated by the enzyme *acetylcholinesterase (AChE),* the effects of cholinergic fibers are short-lived and local.

EXHIBIT 11-2 *Structural Features of Sympathetic and Parasympathetic Divisions*	
Sympathetic	**Parasympathetic**
Contains sympathetic trunk and prevertebral ganglia.	Contains terminal ganglia.
Ganglia are close to the CNS and distant from visceral effectors.	Ganglia are near or within visceral effectors.
Each preganglionic fiber synapses with many postganglionic neurons that pass to many visceral effectors.	Each preganglionic fiber usually synapses with four or five postganglionic neurons that pass to a single visceral effector.
Distributed throughout the body, including the skin.	Distribution limited primarily to head and viscera of thorax, abdomen, and pelvis.

Adrenergic (ad'-ren-ER-jik) *fibers* produce *norepinephrine (NE)*. Most sympathetic postganglionic axons are adrenergic. Since norepinephrine is inactivated much more slowly than acetylcholine and since norepinephrine may enter the bloodstream, the effects of sympathetic stimulation are longer lasting and more widespread than parasympathetic stimulation.

ACTIVITIES

As noted earlier, impulses from one division stimulate the organ's activities, whereas impulses from the other inhibit the organ's activities. The stimulating division may be either the sympathetic or the parasympathetic, depending on the organ. For example, sympathetic impulses increase heart activity, whereas parasympathetic impulses decrease it. On the other hand, parasympathetic impulses increase digestive activities, whereas sympathetic impulses inhibit them. The actions of the two systems are integrated to help maintain homeostasis. A summary of ANS functions is presented in Exhibit 11-3.

The parasympathetic division is primarily concerned with activities that conserve and restore body energy. For instance, parasympathetic impulses to the digestive glands and smooth muscle of the gastrointestinal tract normally dominate over sympathetic impulses. Thus, energy-supplying food can be digested and absorbed by the body.

Stress Management: When You Can't Fight or Flee

The fight-or-flight arousal of the sympathetic nervous system is very helpful when you encounter a snarling dog or need to escape from a burning building. But how often in real life do we encounter a source of stress that gives us a chance to fight or run away? Traffic jams, problematic professors, family conflict, financial problems: instead of fight-or-flight, it is sit-and-stew.

A stressor is a source of stress, a situation that requires some form of adaptation or response. Heat, cold, crowds, noise, exercise, emotional excitement, psychological distress, and pain are stressors. Stressors may be positive, such as holidays, vacation travel, and getting married. These require just as much adjustment as negative stressors, such as divorce, losing a job, or getting a speeding ticket. Any change in life-style is associated with a certain amount of stress since change requires adaptation.

It is not only big events that have an impact on our stress levels. Have you ever lost your wallet? Run out of gas? Had trouble changing the ribbon on your printer? Some psychologists believe that these daily "hassles" have a greater impact on our health than major stressors, such as moving to a new town or getting a new job.

All stressors, large and small, elicit the fight-or-flight response. In a healthy person, the fight-or-flight response lasts a short time and is followed by recovery, sometimes called the relaxation response. The actions of the sympathetic and parasympathetic nervous systems are balanced. Blood pressure goes up, but it quickly returns to normal. Distress and disease result when the fight-or-flight response dominates homeostasis, when it becomes the physiological norm. When our stress response begins to harm us and/or the people important to us, it's time to make some changes in life-style and outlook. It's time to practice stress management.

Coping with Stress

Stress management consists of improving the way you cope or deal with stress. It means changing either the source of stress, one's stress response, or both.

When you rearrange your day to meet a new deadline or go to your professor to ask for help with a paper, you are using *direct* coping strategies. In these situations you are trying to confront and deal directly with the source of stress. If your coping attempts help alleviate the impact of the stressor, they are *adaptive.* (This doesn't mean you always get what you want, but at least you feel better for trying.) If they make the situation worse, they're considered *maladaptive.* Procrastinating on an important project or getting into a heated argument and calling your professor nasty names won't help you meet your deadline or get your paper written.

Direct coping is always the first stress-management step to take. But sometimes there is nothing we can do to improve a situation, or we still feel stressed even though we've done all we can. When our coping behavior does not directly address the situation but makes us feel better in some other way, it is called *palliative.* Adaptive palliative responses include talking to a friend, going for a walk, and listening to music. These responses help to decrease sympathetic arousal and make us feel more relaxed. Maladaptive palliative responses include drug and alcohol abuse, driving too fast, and taking your anger out on others.

Changing Your Stress Response

It is helpful to think of the stress response in terms of behavioral, psychological, and physical components. Various stress-management techniques are designed to change each of these components. It's important to note that these stress response components can function together in a positive feedback cycle, increasing the sympathetic response. Let's say you are anxious about an upcoming exam. Your psychological response—anxiety—causes sympathetic arousal. As you feel your heart pound and your muscles tense, you become more anxious than ever. If you procrastinate studying and instead watch television, you start to get more worried about the work you should be doing, and the sympathetic arousal continues.

Coping strategies are considered part of your behavioral response. Stress-management programs usually teach participants ways to improve both direct and palliative coping behaviors. Study skills, time management, problem-solving techniques, effective communication skills, and assertiveness training can improve one's ability to confront sources of stress. Simply getting organized can help relieve stress. Have you ever spent 10 or 20 minutes looking for a copy of an assignment due tomorrow? Then you know the frustrations of disorganization. Developing a systematic system for organizing your school (and other) materials, and an environment conducive to productive studying can help you become a more efficient student. When you can accomplish more in a given amount of study time, you have more time left over for other activities, and you feel less stressed.

Positive health behaviors are adaptive, and can help decrease the negative effects of stress. These include getting enough sleep, eating a healthful diet, avoiding too much caffeine, and getting regular exercise.

Stress-management programs teach participants to cultivate adaptive behaviors such as enjoying hobbies and recreational activities, spending time with friends and pets, and taking "mini-vacations" throughout the day to unwind. These can be something as simple as enjoying a cup of tea or watching the birds at the feeder.

Psychological intervention consists of changing the way we perceive stressors and the way we talk to ourselves. People vary considerably in their assessment of what is stressful. Some of us find waiting for a restaurant meal intolerable, while others would enjoy talking with friends and scarcely notice the wait. This suggests that there is something more to stress than a simple response to a given stimulus: our perception and interpretation of events have a strong effect on how we respond both physically and psychologically. People who perceive waiting for a meal as valuable time wasted become anxious and tense, and the sympathetic nervous system is in high gear. Alternatively, others may see the wait simply as a harmless situation over which they have no control and decide they may as well use the opportunity to relax.

Irrational beliefs can be a source of stress, especially if they lead to negative emotions, depression, anxiety, and anger. An example of an irrational belief is feeling that you must always be perfect at everything you do. This is obviously impossible, but many people let this belief keep them from trying new things, and from being satisfied with their best efforts. It also leads to procrastination: If you do a project at the last minute, you have an excuse for not doing it perfectly.

One's self-esteem plays a crucial role in the psychological appraisal of stressors. If you are confident of your ability to complete an upcoming assignment, you feel less stress than a classmate who is sure the project is beyond his capability. Developing a realistic understanding of our abilities encourages an appropriate response to demands. A positive attitude and philosophy of life and a good sense of humor are important stress-management tools.

Some stress-management techniques consist of more formal exercises designed to reduce sympathetic arousal. Relaxation exercises teach participants to become aware of excess muscle tension and to "let go." Meditation, in which the person focuses on an image, word, or feeling, calms both the body and mind. In some techniques, participants imagine their bodies feeling warm and heavy, feelings associated with relaxation. Slow, deep breathing is often used to reduce tension.

Stress Resistance

Why is it that some people have a high tolerance for stress, while others practically fall apart under similar conditions? Why do some people develop stress-related illnesses, while other people seem immune to the negative effects of stress? Researchers have studied the characteristics of people with high levels of stress resistance and observed the following qualities:

1. *Sense of control.* Control doesn't mean you can control every aspect of a situation, but it does mean you have some control in how you respond. People in control have a sense of options. A sense of control reduces feelings of helplessness and hopelessness, which are associated with stress-related illnesses.
2. *Positive outlook.* Stress-resistant people see demands as challenges to be met rather than obstacles to be overcome. They feel a commitment to their families, friends, and jobs. Their social support systems provide a buffer from stress, and they feel that life has purpose and meaning. They see the glass as half-full, rather than half-empty.
3. *Commitment to regular exercise.* Exercise is a natural tranquilizer, and people who exercise regularly report lower levels of stress. Exercise can reduce muscle tension, improve self-concept, and increase a person's sense of control.

The sympathetic division, in contrast, is primarily concerned with processes involving the expenditure of energy. When the body is in homeostasis, the main function of the sympathetic division is to counteract the parasympathetic effects just enough to carry out normal processes requiring energy. During extreme stress, however, the sympathetic dominates the parasympathetic. Confronted with stress, the body becomes alert and capable of unusual feats. Fear stimulates the sympathetic division as do a variety of other emotions and physical activities.

Activation of the sympathetic division sets into operation a series of physiological responses collectively called the *fight-or-flight response.* It produces the following effects:

1. The pupils of the eyes dilate.
2. Heart rate, force of contraction, and blood pressure increase.
3. The blood vessels of the skin and viscera constrict.
4. Blood vessels of organs involved in fighting off danger—skeletal muscles, cardiac muscles, brain, and lungs—dilate to allow faster flow of blood.
5. Bronchioles dilate to provide more oxygen to muscles for energy production.
6. Blood sugar level rises as liver glycogen converts to glucose for extra energy.
7. The medullae of the adrenal glands produce epinephrine and norepinephrine to intensify and prolong the sympathetic effects just described.
8. Processes not essential for meeting the stress are inhibited; for example, muscular movements of the gastrointestinal tract and digestive secretions slow down or stop.

EXHIBIT 11-3
Activities of Autonomic Nervous System

Visceral Effector	Effect of Sympathetic Stimulation	Effect of Parasympathetic Stimulation
GLANDS		
Sweat	Stimulates local secretion.	Stimulates generalized secretion.
Lacrimal (Tear)	No known functional innervation.	Stimulates secretion.
Adrenal Medulla	Promotes epinephrine and norepinephrine secretion.	No known functional innervation.
Liver	Promotes glycogenolysis and gluconeogenesis; decreases bile secretion.	Promotes glycogen synthesis; increases bile secretion.
Kidney	Secretion of renin.	No effect.
Pancreas	Inhibits secretion of enzymes and insulin; promotes secretion of glucagon.	Promotes secretion of enzymes and insulin.
SMOOTH MUSCLE		
Radial Muscle of Iris	Contraction that results in dilation of pupil.	No known functional innervation.
Sphincter Muscle of Iris	No known functional innervation.	Contraction that results in constriction of pupil.
Ciliary Muscle of Eye	Relaxation that results in far vision.	Contraction that results in near vision.
Salivary Glands	Vasoconstriction, which decreases secretion.	Stimulates vasodilation.
Gastric Glands	Vasoconstriction, which inhibits secretion.	Stimulates secretion.
Intestinal Glands	Vasoconstriction, which inhibits secretion.	Stimulates secretion.
Lungs (Smooth Muscle of Bronchi)	Dilation.	Constriction.
Heart (Smooth Muscle of Coronary Vessels)	Dilation.	Constriction.
Skin and Mucosa Arterioles	Constriction.	No known functional innervation for most.
Skeletal Muscle Arterioles	Constriction or dilation.	No known functional innervation.
Abdominal Viscera Arterioles	Constriction.	No known functional innervation for most.
Cerebral Arterioles	Slight constriction.	No known functional innervation.
Systemic Veins	Constriction and dilation.	No known functional innervation.
Gallbladder and Ducts	Relaxation.	Contraction.
Stomach	Decreases motility and tone; contracts sphincters.	Increases motility and tone; relaxes sphincters.
Intestines	Decreases motility and tone; contracts sphincters.	Increases motility and tone; relaxes sphincters.
Kidney	Constriction of blood vessels that results in decreased urine volume.	No effect.
Ureter	Increases motility.	Decreases motility.

Spleen	Contraction and discharge of stored blood into general circulation.	No known functional innervation.
Urinary Bladder	Relaxation of muscular wall; contraction of internal sphincter.	Contraction of muscular wall; relaxation of internal sphincter.
Arrector Pili of Hair Follicles	Contraction that results in erection of hairs.	No known functional innervation.
Uterus	Inhibits contraction if nonpregnant; stimulates contraction if pregnant.	Minimal effect.
Sex Organs	In male, contraction of smooth muscle of ductus (vas) deferens, seminal vesicle, prostate; results in ejaculation. In female, reverse uterine peristalsis.	Vasodilation and erection in both sexes.
CARDIAC MUSCLE		
Heart	Increases rate and strength of contraction.	Decreases rate and strength of contraction.

STUDY OUTLINE

Comparison of Somatic and Autonomic Nervous Systems (p. 231)

1. The somatic nervous system produces conscious movement in skeletal muscles.
2. The autonomic nervous system regulates visceral activities, that is, activities of smooth muscle, cardiac muscle, and glands, and it usually operates without conscious control.
3. It is regulated by centers in the brain, in particular, by the cerebral cortex and hypothalamus.
4. A single somatic efferent neuron synapses on skeletal muscles; in the ANS, there are two efferent neurons—one from the CNS to a ganglion and one from a ganglion to a visceral effector.
5. Somatic efferent neurons release acetylcholine (ACh) and autonomic efferent neurons release either acetylcholine (ACh) or norepinephrine (NE).

Structure of the Autonomic Nervous System (p. 232)

1. The ANS consists of visceral efferent neurons organized into nerves, ganglia, and plexuses.
2. It is entirely motor. All autonomic axons are efferent fibers.

3. Efferent neurons are preganglionic (with myelinated axons) and postganglionic (with unmyelinated axons).
4. The autonomic system consists of two principal divisions: sympathetic and parasympathetic.
5. Autonomic ganglia are classified as sympathetic trunk ganglia (on sides of spinal column), prevertebral ganglia (anterior to spinal column), and terminal ganglia (near or inside visceral effectors).

Functions of the Autonomic Nervous System (p. 235)

1. Autonomic fibers release neurotransmitters at synapses. On the basis of the neurotransmitter produced, these fibers may be classified as cholinergic or adrenergic.
2. Cholinergic fibers release acetylcholine (ACh). Adrenergic fibers produce norepinephrine (NE).
3. Acetylcholine (ACh) interacts with nicotinic receptors on postganglionic neurons and muscarinic receptors on certain visceral effectors.
4. Sympathetic responses are widespread and, in general, concerned with energy expenditure. Parasympathetic responses are restricted and are typically concerned with energy restoration and conservation.

REVIEW QUESTIONS

1. What are the principal differences between the somatic nervous system and the autonomic nervous system? (p. 231)
2. Describe how the hypothalamus controls and integrates the autonomic nervous system. (p. 231)
3. Relate the role of visceral efferent fibers and visceral effectors to the autonomic nervous system. (p. 231)
4. Distinguish between preganglionic neurons and postganglionic neurons with respect to location and function. (p. 232)
5. What is an autonomic ganglion? Describe the location and function of the three types of autonomic ganglia. (p. 234)
6. On what basis are the sympathetic and parasympathetic divisions of the autonomic nervous system differentiated structurally and functionally? (p. 235)
7. Discuss the distinction between cholinergic and adrenergic fibers of the autonomic nervous system. (p. 235)
8. Give examples of the antagonistic effects of the sympathetic and parasympathetic divisions of the autonomic nervous system. (p. 238)

12

Sensations

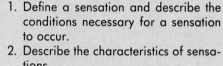
1. Define a sensation and describe the conditions necessary for a sensation to occur.
2. Describe the characteristics of sensations.
3. Compare the different types of receptors in the body.
4. Describe the importance of the general senses such as touch, pressure, vibration, heat, cold, pain, and proprioception.
5. Describe the receptors and pathways to the brain for the special senses of smell, taste, vision, hearing, and equilibrium.
6. Describe how smell, taste, vision, hearing, and equilibrium occur.
7. Define common disorders and medical terminology and conditions associated with the sense organs.

A LOOK AHEAD

SENSATIONS
  Definition
  Characteristics
  Classification of Receptors
    Location
    Stimulus Detected
    Simplicity or Complexity
GENERAL SENSES
  Cutaneous Sensations
    Tactile Sensations
    Thermoreceptive Sensations
    Pain Sensations
  Proprioceptive Sensations
    Receptors
SPECIAL SENSES
OLFACTORY SENSATIONS
  Structure of Receptors
  Stimulation of Receptors
  Olfactory Pathway
GUSTATORY SENSATIONS
  Structure of Receptors
  Stimulation of Receptors
  Gustatory Pathway
VISUAL SENSATIONS
  Accessory Structures of Eye
  Structure of Eyeball
    Fibrous Tunic
    Vascular Tunic
    Retina (Nervous Tunic)
    Lens
    Interior
  Physiology of Vision
    Retinal Image Formation
    Stimulation of Photoreceptors
    Visual Pathway
AUDITORY SENSATIONS AND
    EQUILIBRIUM
  External Ear
  Middle Ear
  Internal Ear
  Sound Waves
  Physiology of Hearing
  Physiology of Equilibrium
    Static Equilibrium
    Dynamic Equilibrium
COMMON DISORDERS
MEDICAL TERMINOLOGY AND
    CONDITIONS

Having examined the structure of the nervous system and its activities, we will now see how its different parts cooperate in receiving sensory information and transmitting motor impulses that result in movement or secretion.

SENSATIONS

Consider what would happen if you could not feel the pain of a hot pot handle or an inflamed appendix, or if you could not see, hear, smell or taste, or maintain your balance. In short, if you could not "sense" your environment and make the necessary homeostatic adjustments, you could not survive very well on your own.

DEFINITION

Sensation refers to a state of awareness of external or internal conditions of the body. For a sensation to occur, four conditions must be satisfied.

1. A *stimulus*, or change in the environment, capable of initiating a nerve impulse (nerve action potential) must be present.
2. A *receptor* or *sense organ* must pick up the stimulus and convert it to a nerve impulse by way of a generator potential (described shortly). A receptor or sense organ is specialized nervous tissue that is extremely sensitive to internal or external stimuli.
3. The impulse must be *conducted* along a neural pathway from the receptor or sense organ to the brain.
4. A region of the brain must *translate* the nerve impulse into a sensation.

A stimulus received by a receptor may be light, heat, pressure, mechanical energy, or chemical energy. When an adequate stimulus is applied to a receptor, it responds by altering its membrane's permeability to small ions. This results in a change in the resting membrane potential called a *generator potential*.

When the generator potential reaches the threshold level, it initiates a nerve impulse that is transmitted along the nerve fiber. The function of a generator potential is to convert a stimulus into a nerve impulse.

Receptors vary in their complexity. The simplest are free dendrite endings in the skin (for example, pain receptors). Others are housed in complex sense organs such as the eye. Regardless of complexity, all sense receptors contain the dendrites of sensory neurons, either along or in close association with specialized cells of other tissues.

A receptor converts a stimulus into a nerve impulse, and only after that impulse has been conducted to a region of the spinal cord or brain can it be translated into a sensation. The nature of the sensation and the type of reaction generated vary with the level of the central nervous system at which the sensation is translated.

Sensory fibers terminating in the spinal cord generate spinal reflexes without action by the brain. Sensory fibers terminating in the lower brain stem bring about more complex motor reactions. At the lower brain stem, they cause subconscious motor reactions. Sensory impulses that reach the thalamus are localized crudely in the body and sorted by specific sensations such as touch, pressure, pain, position, hearing, or taste. When sensory information reaches the cerebral cortex, we experience precise localization. It is at this level that memories of previous sensory information are stored and the perception of sensation occurs on the basis of past experience.

CHARACTERISTICS

Conscious sensations and perceptions occur in the cerebral cortical regions. In other words, you see, hear, and feel in the brain. You seem to see with your

241

eyes and feel pain in an injured part of your body but that is because the cortex interprets the sensation as coming from the stimulated sense receptor. ***Projection*** is the name of the process by which the brain refers sensations to their point of stimulation.

A second characteristic of many sensations is ***adaptation***, that is, a decrease in sensitivity to continued stimuli. In fact, the perception of a sensation may actually disappear, even though the stimulus is still being applied. For example, when you first get into a tub of hot water, you feel a burning sensation. But soon the sensation becomes one of comfortable warmth, even though the stimulus (hot water) is still present, and in time, even the sensation of warmth disappears.

Sensations are also characterized by ***afterimages***, that is, some sensations persist even though the stimulus has been removed. This is the reverse of adaptation. When you look at a bright light, then look away or close your eyes and still see the light for several seconds afterward, you are experiencing afterimage.

Although all nerve impulses are the same, one sensation can be distinguished from another, such as sights from sounds. ***Modality*** refers to that specific characteristic of each sensation which allows it to be distinguished from other types. Modality refers to the fact that nerve impulses generated in the eyes are interpreted by the occipital lobe as sight, whereas those from the ears are interpreted by the temporal lobes as hearing.

CLASSIFICATION OF RECEPTORS

Receptors are classified on the basis of location, stimulus detected, and simplicity or complexity.

Location

Terms that identify receptor location include the following:

Exteroceptors (eks'-ter-ō-SEP-tors) provide information about the external environment. They transmit sensations of hearing, sight, smell, taste, touch, pressure, temperature, and pain. Exteroceptors are located near the surface of the body.

Enteroceptors (en-ter-ō-SEP-tors) provide information about the internal environment. These sensations arise from within the body such as pain, pressure, fatigue, hunger, thirst, and nausea. Enteroceptors are located in blood vessels and viscera.

Proprioceptors (prō'-prē-ō-SEP-tors) provide information about body position and movement. Such sensations give us information about muscle tension, the position and tension of our joints, and equilibrium. Proprioceptors are located in muscles, tendons, joints, and the internal ear.

Stimulus Detected

The types of stimulus detected are designated by specific terms. ***Mechanoreceptors*** detect mechanical deformation of the receptor or adjacent cells. Stimuli so detected are related to touch, pressure, vibration, proprioception, hearing, equilibrium, and blood pressure. ***Thermoreceptors*** detect changes in temperature. ***Nociceptors*** detect pain, usually as a result of physical or chemical damage to tissues. ***Photoreceptors*** in the retina of the eye detect light. ***Chemoreceptors*** detect taste in the mouth, smell in the nose, and chemicals, such as oxygen, carbon dioxide, water, and glucose in body fluids.

Simplicity or Complexity

Finally, receptors may be classified according to the simplicity or complexity of their structure and the neural pathway involved. ***Simple receptors*** and neural pathways are associated with the **general senses**—touch, pressure, vibration, heat, cold, and pain. ***Complex receptors*** and neural pathways are associated with the ***special senses***—smell, taste, sight, hearing, and equilibrium.

GENERAL SENSES

CUTANEOUS SENSATIONS

Cutaneous (*cuta* = skin) ***sensations*** include tactile sensations (touch, pressure, vibration), thermoreceptive sensations (cold and heat), and pain. Receptors for these sensations are in the skin, connective tissue, and the ends of the gastrointestinal tract.

The cutaneous receptors are distributed over the body surface in such a way that some areas are densely populated with receptors and are consequently very sensitive, while other areas contain only a few and are insensitive.

Cutaneous receptors have simple structures. They consist of the dendrites of sensory neurons that may or may not be enclosed in a capsule of epithelial or connective tissue. Recall that cutaneous sensations are interpreted by the parietal lobes.

Tactile Sensations

Even though the ***tactile*** (*tact* = touch) ***sensations*** are divided into touch, pressure, and vibration, they are all detected by mechanoreceptors.

▪ **Touch** ***Touch sensations*** come from stimulation of tactile receptors in the skin or tissues immediately beneath the skin. ***Light touch*** refers to the perception that something has touched the skin, though its exact location, shape, size, or texture cannot be determined. ***Discriminative touch*** refers to the ability to recognize exactly what point of the body is touched.

Tactile receptors for touch include hair root plexuses, free nerve endings, tactile discs, corpuscles of touch, and type II cutaneous mechanoreceptors (Figure 12-1). ***Hair root plexuses*** are dendrites arranged in networks around the roots of hairs. If a hair shaft is moved, the dendrites are stimulated. Hair root plexuses detect movements mainly on the surface of the body when hairs are disturbed.

The receptors called ***free nerve endings*** are the branching ends of the dendrites of certain sensory neurons. Free nerve endings are found everywhere in the skin and many other tissues. They are important in pain reception, but they also respond to objects in continuous contact with the skin, such as clothing.

Tactile, or ***Merkel's*** (MER-kelz), ***discs*** are modified epidermal cells found in hairless skin. Their basal ends are in contact with dendrites of sensory neurons. ***Corpuscles of touch,*** or ***Meissner's*** (MĪS-nerz) ***corpuscles,*** are egg-shaped receptors containing a mass of dendrites. Both tactile discs and corpuscles of touch are located in the dermal papillae of the skin, especially in the fingertips, palms, and soles. They are also abundant in the eyelids, tip of

FIGURE 12-1 Structure and location of cutaneous receptors.

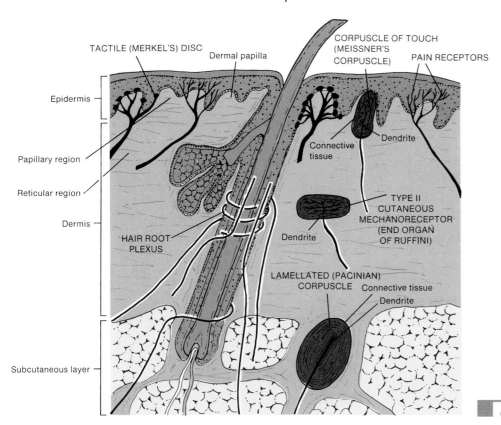

Circle the receptors associated with touch.

the tongue, lips, nipples, clitoris, and tip of the penis. Both receptors function in discriminative touch.

Type II cutaneous mechanoreceptors, or *end organs of Ruffini*, are embedded deep in the dermis and in deeper tissues of the body. They detect heavy and continuous touch.

■ **Pressure** *Pressure sensations* generally occur in deeper tissues. In comparison with touch, pressure lasts longer, varies less in intensity, and is felt over a larger area. Pressure is really sustained touch.

Pressure receptors are free nerve endings, type II cutaneous mechanoreceptors, and lamellated corpuscles. *Lamellated*, or *Pacinian* (pa-SIN-ē-an), *corpuscles* (Figure 12-1) consist of connective tissue, layered like an onion and enclosing dendrites. They are located in subcutaneous tissues, around joints, tendons, and muscles, and in the mammary glands, external genitalia, and certain viscera.

■ **Vibration** *Vibration sensations* result from rapidly receptive sensory signals from tactile receptors.

Receptors for vibration are corpuscles of touch which detect low-frequency vibration, and lamellated corpuscles, for higher-frequency vibration.

Thermoreceptive Sensations

The *thermoreceptive* (*therm* = heat) *sensations* are heat and cold. The exact nature of *thermoreceptive receptors* is not known, but they might be free nerve endings. Thermoreceptive receptors can detect temperatures from as low as 10°C (50°F) to as high as 45°C (113°F). Below 10°C, nerve impulses cannot be generated; for this reason cold is an excellent anesthetic. Also, below 10°C, pain receptors rather than cold receptors are stimulated, which produces the sensation of freezing. Likewise, above 45°C, pain receptors rather than thermoreceptors are stimulated, producing the sensation of burning. Since freezing and burning feel almost alike, very cold stimuli may be reported as hot and vice versa. Thermoreceptors tend to adapt to continuous stimulation, which can easily lead to tissue injuries due to burns or frostbite.

Pain Sensations

Pain is indispensable for a normal life, providing us with information about tissue-damaging stimuli and thus often enabling us to protect ourselves from greater damage. Pain initiates our search for medical assistance, and our description and indication of the location of the pain help pinpoint the underlying cause of disease.

The receptors for *pain*, called *nociceptors* (nō-sē-SEP-tors; *noci* = harmful), are free nerve endings. Pain receptors are found in practically every tissue of the body and they respond to any type of stimulus. When stimuli for other sensations reach a certain threshold, they stimulate pain receptors as well. Excessive stimulation of a sense organ causes pain. Other stimuli include excessive stretching of a structure, prolonged muscular contractions, inadequate blood flow to an organ, or the presence of certain chemical substances.

During tissue irritation or injury, chemicals such as prostaglandins are released, which stimulate nociceptors. This explains why

Flexibility: Taking It to the Limit

Unless you were in physical education class a long time ago, your instructors probably warned "Don't bounce!" whenever you assumed a stretching position. But ask your parents about physical education classes and sports programs, and how they warmed up for activity. They might remember doing ballistic stretches, where muscles were forcefully pulled then released. An example of ballistic stretching is bending over quickly with straight knees to touch your toes, bouncing back up to a standing position, and repeating this eight or ten times. If you pull harder on a muscle, it will stretch farther, right?

Only if someone takes away your muscle spindles. When a muscle is forcefully stretched, these proprioceptive receptors sense that something is wrong. In order to correct the situation, they tell the muscle to contract. It's fruitless or even counterproductive to try to stretch a muscle while it's contracting. At best, ones ballistic efforts to lengthen the muscle are done in vain, and if too forceful they can even cause tightness and injury. As scientists have come to understand the physiology of the proprioceptors, more effective methods for increasing flexibility have been devised.

Flexibility: A Question of Degree

Flexibility refers to a joint's range of motion (ROM). Range of motion is the maximum ability to move the bones of a joint through an arc. For example, if an injured knee can go from a 50° angle fully flexed to a 120° angle fully extended, the knee's ROM is 70°. Physical therapists measure improved joint mobility by an increase in its ROM.

When we think of flexibility we often think of athletes, especially those involved in sports such as dance, gymnastics, figure skating, and diving, in which a great degree of flexibility is required for success. Flexibility varies from person to person, and many athletes are born with flexible bodies. Nevertheless, athletes must still incorporate stretching exercises into their daily workouts in order to increase and/or maintain this flexibility.

Athletes in other sports not noted for their flexibility requirements often include some sort of stretching exercises into their workouts to increase or maintain flexibility as well. Flexibility improves performance and prevents injury. Runners, for example, use stretching exercises to prevent muscle tightness in the lower back and hamstring muscles, and to improve stride length and overall running form.

A minimum amount of flexibility is necessary for everyone, simply for the performance of daily tasks. Limitation in your shoulder's ROM can make it difficult to pull a shirt on over your head, lift a dish from a shelf, or even brush your hair. As we age, muscles tend to shorten, especially if we do not do stretching exercises regularly. Shorter muscles decrease a joint's ROM. This may be part of the reason age is associated with stiff joints. It's important to remember, however, that this stiffness is due more to lack of exercise than to getting older *per se*.

Emotional tension, poor posture, and long periods of sitting at a desk can also shorten muscles and lead to tension and fatigue. Tightness in the low back muscles and hamstring muscles is associated with low back pain. Flexibility also prevents injury due to lack of joint mobility. Muscle injury can occur from something as simple as lunging to catch a falling stack of papers if that lunge is beyond the muscle's ability to stretch. Some degree of flexibility is important for its own sake: it's more fun to live in a limber body that likes to move. Thus, adequate flexibility is considered to be one of the components of health-related physical fitness and wellness. Regular stretching is important for everyone.

Mind and Muscle

The first step to take in increasing flexibility is to relax. Sounds simple, right? But if you ever visit an exercise class, you'll notice some people who are all tense, rigid, and hunched up, because the stretching position is uncomfortable and the muscles tighten up in protest. These people figure they'd better push a little harder, and tense up even more. Even though they are not bouncing, this rigid stretch is not accomplishing much.

When the muscle is stretched too fast or too hard, the muscle spindle senses the stretch and sends a message via a sensory neuron to the central nervous system, which in turn activates motor neurons to contract the muscle. This should sound familiar: it's the reflex arc from Chapter 10 (Figure 10-4). This particular action is called the stretch reflex. So to stretch effectively, you've got to stretch in a way that doesn't alarm the muscle spindles. One of these ways is called static stretching.

A good static stretch is slow and gentle. You get into a stretching position that is not uncomfortable and relax. Then, continuing to relax and breathe deeply, you reach just a little farther . . . and a little farther, holding the stretch for at least 30 seconds. If you have difficulty relaxing, you know you have stretched too far and ease up until you feel a stretch but no strain.

Often after holding a stretch for 10 seconds or more, you can feel the muscle relax even more, which results in a greater increase in flexibility. This response is called the inverse stretch reflex. Tendon organs respond to increases in tension, produced by either contraction or stretching, by telling the muscle to relax in order to prevent injury. Its action is to inhibit muscle contraction. This is why people rarely injure themselves from contracting a muscle too much. An exception is arm wrestlers, who learn to override this reflex (disinhibition) and can actually rupture muscles and tendons, or break bones when the muscle contraction and the tension exerted by the opponent become greater than the strength of the connective tissue. It's smarter to give up!

Physical therapists have designed some special stretching procedures that go beyond static stretching and take advantage of the muscle spindle and tendon organ functions. These techniques are called proprioceptive neuromuscular facilitation, or PNF for short.

One PNF technique is the contract–relax method, whose goal is to activate the inverse stretch reflex by first contracting the muscle to be stretched. The stretch is slow and gentle, so the muscle spindles keep quiet.

Try this: first stretch your neck by moving your head to your left shoulder. Keep your shoulders relaxed. Note how far you can go. Hold that position but put your left hand on the right side of your head to make an isometric contraction (muscle tension but no movement). Hold for 5 to 10 seconds. Now with the same hand gently pull your head a little closer to your shoulder. Be sure not to pull too hard. Did your stretch increase? This is an exercise physical therapists often teach to patients with neck and upper back tension.

The contract–relax with agonist-contraction (CRAC) method starts out the same as the contract–relax method, but then you also apply resistance to the muscle group opposite the one being stretched. In general, when one muscle group contracts, the group opposite it relaxes; this is called reciprocal inhibition. Try it: after you've done the contract–relax sequence above, move your left hand to the left side of your head and push, without moving your head, creating tension on the left side of the neck. Hold for 5 to 10 seconds. Now relax and pull very gently on the right side again, moving the head a little farther toward the left shoulder. Compare the flexibility on the right side of your neck with that on the left. Any difference?

Versions of PNF stretches using partners are sometimes taught, but it is extremely important that the partner not force the stretch too far. The best partner is yourself, a trained instructor, or a physical therapist.

Stretching It

When a relaxed muscle is physically stretched, its ability to elongate is limited by connective tissue structures, such as epimysium, perimysium, endomysium, and tendons. Collagenous tissue stretches best when slow, gentle force is applied at elevated tissue temperatures. An external source of heat such as hot packs or ultrasound can be used. But aerobic exercise is the best way to raise muscle temperature, if possible. That's where the name "warm-up" comes from. It's important to warm up *before* stretching, not vice versa. Stretching cold muscles does not increase flexibility and may even cause injury.

The best workout sequence is (1) warm up with repetitive movements of large muscle groups until you feel warm (about 5 to 10 minutes), (2) perform the aerobic and muscular strength/endurance portions of your workout, decreasing intensity at the end of the workout to gradually decrease heart rate, and then (3) stretch all muscle groups for 10 to 15 minutes.

pain persists even after the initial trauma occurs since nociceptors adapt only slightly or not at all to the presence of these substances, which are only slowly removed from the tissues following an injury. If there were adaptation to pain, it would cease to be sensed and irreparable damage could result.

Recognition of the kind and intensity of most pain occurs in the cerebral cortex. In most instances of somatic pain, the cortex projects the pain back to the stimulated area. If you burn your finger, you feel the pain in your finger. In most instances of visceral pain, however, the sensation is not projected back to the point of stimulation. Rather, the pain is felt in the skin overlying the stimulated organ or in a surface area far from the stimulated organ. This phenomenon is called *referred pain*. It occurs because the area to which the pain is referred and the visceral organ involved are innervated by the same segment of the spinal cord. For example, afferent fibers from the heart as well as from the skin over the heart and left upper extremity enter spinal cord segments T1–T4. Thus, the pain of a heart attack is typically felt in the skin over the heart and along the left arm.

A kind of pain frequently experienced by patients who have had a limb amputated is called *phantom pain*. They still experience sensations such as itching, pressure, tingling, or pain in the extremity as if the limb were still there. An explanation for this phenomenon is that the remaining proximal portions of the sensory nerves that previously received impulses from the limb are being stimulated by the trauma of the amputation. Stimuli from these nerves are interpreted by the brain as coming from the nonexistent (phantom) limb.

Pain may be controlled by interfering with nerve impulse transmission to the cerebrum. This may be done by a variety of analgesic or anesthetic drugs, surgery, acupuncture, shiatsu, hypnosis, relaxation, massage, biofeedback, and electrical stimulation directly to the affected nerves or via the skin over those nerve pathways.

PROPRIOCEPTIVE SENSATIONS

An awareness of the activities of muscles, tendons, and joints and of equilibrium is provided by the *proprioceptive* (*proprio* = one's own), or *kinesthetic* (kin′-es-THET-ik), *sense*. It informs us of the degree to which muscles are contracted, the amount of tension in tendons, the change of position of joints, and the position of the head relative to the ground and in response to movements (equilibrium). Proprioception tells us the location and rate of movement of one body part in relation to others, so we can walk, type, or dress without using our eyes. It also allows us to estimate weight and determine the muscular work necessary to perform a task.

Most proprioceptors adapt only slightly. This feature is advantageous since the brain must be apprised of the status of different parts of the body at all times so that adjustments can be made to ensure coordination.

The afferent pathway for muscle sense consists of impulses generated by proprioceptors via cranial and spinal nerves to the central nervous system. Impulses for conscious proprioception pass along ascending tracts in the cord, where they are relayed to the thalamus and cerebral cortex. The sensation is registered in the general sensory area in the parietal lobe of the cerebral cortex

posterior to the central sulcus. Proprioceptive impulses that have resulted in reflex action pass to the cerebellum along spinocerebellar tracts and contribute to subconscious proprioception.

Receptors

Proprioceptive receptors are located in skeletal muscles, tendons in and around synovial joints, and the internal ear.

▪ **Muscle Spindles** *Muscle spindles* are delicate proprioceptive receptors between skeletal muscle fibers (cells). When a muscle is stretched, the spindle is stretched and it sends an impulse to the CNS, indicating how much and how fast the muscle is changing its length. Within the CNS, the information is integrated to coordinate muscle activity.

▪ **Tendon Organs** *Tendon organs (Golgi tendon organs)* protect tendons and their associated muscles from damage resulting from excessive tension. A tendon organ is a thin capsule of connective tissue that encloses a few bundles of tendons. Sensory neurons penetrate the capsule and their terminal branches twine around the tendons. When tension is applied to a tendon, tendon organs relay the information to the central nervous system.

▪ **Joint Kinesthetic Receptors** There are several types of *joint kinesthetic receptors* in and around synovial joints. Encapsulated receptors respond to pressure. Small lamellated (Pacinian) corpuscles respond to acceleration and deceleration. Articular ligaments contain receptors that respond to excessive strain on a joint.

▪ **Maculae and Cristae** The proprioceptors in the internal ear are the macula of the saccule and utricle and the cristae in the semicircular ducts. Their function in equilibrium is discussed later in the chapter.

SPECIAL SENSES

The special senses—smell, taste, sight, hearing, and equilibrium—have receptor organs that are structurally more complex than receptors for general sensations. The sense of smell is the least specialized, and the sense of sight, the most. Like the general senses, the special senses allow us to detect changes in our environment.

OLFACTORY SENSATIONS

STRUCTURE OF RECEPTORS

The receptors for the *olfactory* (ol-FAK-tō-rē; *olfact* = smell) *sense* or sense of smell are located in the superior and middle nasal conchae of the ethmoid bone and upper portion of the nasal septum (Figure 12-2). They consist of three kinds of cells: supporting, olfactory, and basal. *Supporting cells* are columnar epithelial cells of the mucous membrane lining the nose. The *olfactory cells* are neurons whose distal end contains a dendrite that terminates in six to eight cilia called *olfactory hairs*. These hairs react to odors in the air and then stimulate the olfactory cells. The opposite

FIGURE 12-2 Olfactory receptors. (a) Location of receptors in nasal cavity. (b) Enlarged aspect of olfactory receptors.

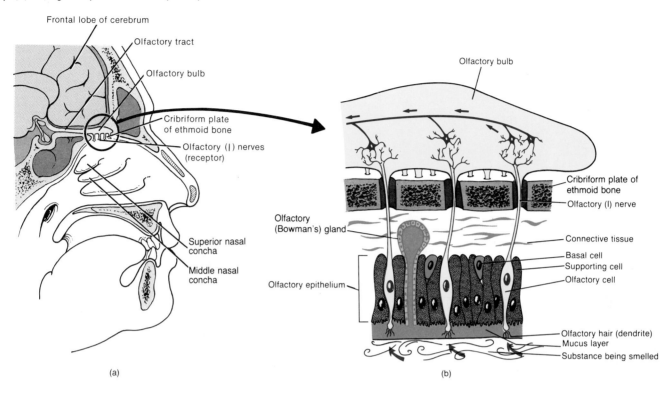

(a)

(b)

List, in order, the structures involved in the olfactory pathway.

end of each olfactory cell contains an axon. **Basal cells** are believed to produce new supporting cells. Within the connective tissue beneath the olfactory epithelium are **olfactory (Bowman's) glands** that produce mucus. It moistens the surface of the olfactory epithelium and serves as a solvent for odoriferous substances. The continuous secretion of mucus also serves to freshen the surface film of fluid and prevents continuous stimulation of olfactory hairs by the same odor.

STIMULATION OF RECEPTORS

For a substance to be smelled, it must be capable of becoming a gas so that the gaseous particles can enter the nostrils. Also, the substance must be water-soluble so that it can dissolve in the nasal mucus to make contact with olfactory cells. Finally, the substance must be lipid-soluble to pass through the plasma membranes of olfactory hairs and initiate an impulse.

Many attempts have been made to classify sensations of smell. One scheme includes seven classes of primary sensations: camphoraceous, musky, floral, pepperminty, ethereal, pungent, and putrid. However, there may be as many as 50 or more primary sensations.

It is believed that olfactory cells react to olfactory stimuli in the same way that most sensory receptors react to their specific stimuli: first a generator potential is developed followed by initiation of a nerve impulse. The **chemical theory** says different receptors in the olfactory hairs react with particular olfactory substances

(stimuli). The interaction between the chemical receptor and the substance alters the permeability of the plasma membrane so that a generator potential is developed.

OLFACTORY PATHWAY

The axons of the olfactory cells unite to form the **olfactory (I) nerves,** which pass through holes in the cribriform plate of the ethmoid bone (Figure 12-2). The olfactory (I) nerves terminate in two masses of gray matter called the **olfactory bulbs,** which lie beneath the frontal lobes of the cerebrum. Axons of the olfactory (I) nerves synapse with dendrites of neurons inside the olfactory bulbs. Axons of these olfactory neurons form the **olfactory tract,** which then conveys impulses to the primary olfactory area of the cerebral cortex. There, the impulses are interpreted as odor and give rise to the sensation of smell.

GUSTATORY SENSATIONS

STRUCTURE OF RECEPTORS

The receptors for **gustatory** (GUS-ta-tō′-rē; *gust* = taste) **sensations,** or sensations of taste, are located in the taste buds of the tongue, soft palate, and throat (Figure 12-3). The **taste buds** consist of three kinds of cells: supporting, gustatory, and basal. The **sup-**

FIGURE 12-3 Gustatory receptors. (a) Structure of a taste bud. (b) Locations of four taste zones.

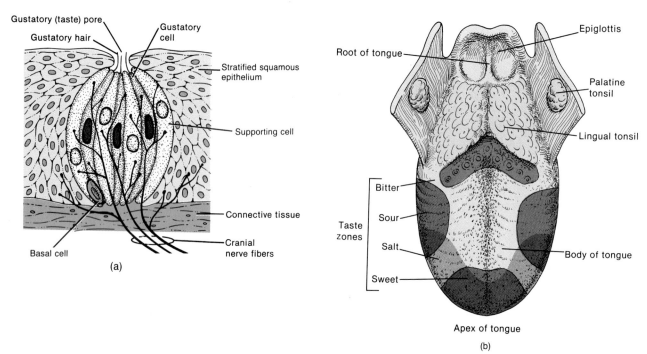

Trace, in order, the structures involved in the gustatory pathway.

porting cells are a specialized epithelium that forms a capsule inside of which are *gustatory cells*. Each gustatory cell contains a hairlike process (*gustatory hair*) that projects to the external surface through an opening in the taste bud called the *taste pore*. Gustatory cells make contact with taste stimuli through the taste pore. *Basal cells* are found at the edge and bottom of the taste bud. These cells produce supporting and gustatory cells whose average life span is about 10 days.

STIMULATION OF RECEPTORS

For gustatory cells to be stimulated, substances must be in solution in saliva so they can enter taste pores. Once the taste substance makes contact with plasma membranes of the gustatory hairs, a generator potential is developed. This is presumed to occur as a result of a membrane receptor–taste substance interaction, similar to the way olfactory cells interact with olfactory stimuli. Then the generator potential initiates a nerve impulse.

There are four primary taste sensations: sour, salt, bitter, and sweet. All other "tastes," such as chocolate, pepper, and coffee, are combinations of these four that are modified by accompanying olfactory sensations.

Persons with colds or allergies sometimes complain that they cannot taste their food. Actually, their taste sensations are probably operating normally, but their olfactory sensations are not. Much of what we think of as taste is actually smell because odors from foods pass upward to stimulate the olfactory system. In fact, a given concentration of a substance stimulates the olfactory system thousands of times more than it stimulates the gustatory system.

Certain regions of the tongue react more strongly to particular primary tastes than others. The tip of the tongue reacts to all four primary taste sensations but is highly sensitive to sweet and salty substances. The posterior portion is highly sensitive to bitter substances, and the lateral edges are sensitive to sour substances (Figure 12-3b).

GUSTATORY PATHWAY

Taste impulses are conveyed from the gustatory cells in taste buds along cranial nerves V, VII, IX, and X to the medulla and then to the thalamus. They terminate in the primary gustatory area in the parietal lobe of the cerebral cortex.

VISUAL SENSATIONS

The study of the structure, function, and diseases of the eye is known as *ophthalmology* (of'-thal-MOL-ō-jē; *ophthalmo* = eye; *logos* = study of). A physician who specializes in the diagnosis and treatment of eye disorders with drugs, surgery, and corrective lenses is known as an *ophthalmologist,* whereas an *optometrist* has a doctorate in optometry and is licensed to test the eyes and treat visual defects by prescribing corrective lenses. An *optician* is a technician who fits, adjusts, and dispenses corrective lenses on prescription of an ophthalmologist or optometrist.

Vision involves the eyeball, the optic (II) nerve, the occipital lobe of the cerebrum, and a number of accessory structures.

ACCESSORY STRUCTURES OF EYE

The *accessory structures* are the eyebrow, eyelids, eyelashes, and the lacrimal apparatus (Figure 12-4). The *eyebrows* help protect the eyeballs from foreign objects, perspiration, and direct rays of the sun. The upper and lower *eyelids* shade the eyes during sleep, protect the eyes from excessive light and foreign objects, and spread lubricating secretions over the eyeballs (by blinking).

The *lacrimal* (*lacrima* = tear) *apparatus* refers to the glands, ducts, canals, and sacs that manufacture and drain tears. A *lacrimal gland* produces tears. Each is about the size and shape of an almond and has ducts that empty tears onto the surface of the upper lid. Tears are a watery solution containing salts, some mucus, and a bactericidal enzyme called *lysozyme*. Tears clean, lubricate, and moisten the surface of the eyeball exposed to air to prevent drying out and loss of the optical qualities of the cornea.

Normally, tears are cleared away by evaporation or by passing into the nasal cavities as fast as they are produced. If, however, an irritating substance makes contact with the eye, the lacrimal glands are stimulated to oversecrete and tears accumulate. This is a protective mechanism since the tears dilute and wash away the irritant. Humans are unique in that they have the ability to cry to express certain emotions. In response to parasympathetic stimulation, the lacrimal glands produce excessive tears that may spill over the edges of the eyelids and even fill the nasal cavity with fluid.

FIGURE 12-4 Accessory structures of the eye seen in anterior view.

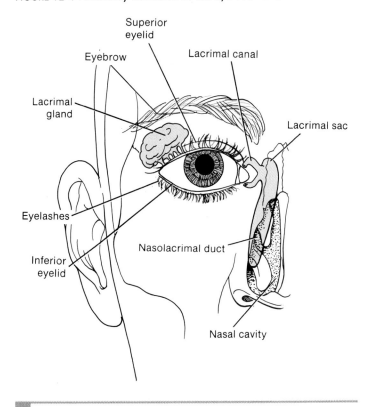

Superior eyelid
Eyebrow
Lacrimal canal
Lacrimal gland
Lacrimal sac
Eyelashes
Nasolacrimal duct
Inferior eyelid
Nasal cavity

▪ What are tears and what is their function?

STRUCTURE OF EYEBALL

The adult *eyeball* measures about 2.5 cm (1 inch) in diameter and is divided into three layers: fibrous tunic, vascular tunic, and retina or nervous tunic (Figure 12-5).

Fibrous Tunic

The *fibrous tunic* is the outer coat of the eyeball consisting of the posterior sclera and the anterior cornea. The *sclera* (SKLE-ra; *skleros* = hard), the "white of the eye," is a white coat of dense fibrous tissue that covers all the eyeball except the anterior colored portion (iris). The sclera gives shape to the eyeball, makes it more rigid, and protects its inner parts. The *cornea* (KOR-nē-a) is a nonvascular, transparent fibrous coat that covers the iris. The cornea's outer surface is covered by an epithelial layer called the *conjunctiva*. At the junction of the sclera and cornea is an opening known as the *scleral venous sinus* (*canal of Schlemm*).

The cornea bends light rays entering the eyeball in order to produce a clear image. If the cornea is not curved properly, the image is not focused on the area of sharpest vision on the retina and blurred vision occurs. A defective cornea can be removed and replaced with a donor cornea of similar diameter, a procedure called a *corneal transplant*. Corneal transplants are considered the most successful type of transplantation because corneas do not contain blood vessels and the body is less likely to reject them.

Vascular Tunic

The *vascular tunic* is the middle layer of the eyeball and is composed of the choroid, ciliary body, and iris. The *choroid* (KŌ-royd) is a thin, dark brown membrane that lines most of the internal surface of the sclera. It contains numerous blood vessels and a large amount of pigment. The choroid absorbs light rays so they are not reflected within the eyeball and its blood supply nourishes the retina.

At the front of the eye, the choroid becomes the *ciliary* (SIL-ē-ar'-ē) *body*. The ciliary body consists of the *ciliary processes*, folds that secrete aqueous humor, and the *ciliary muscle*, a smooth muscle that alters the shape of the lens for near or far vision.

The *iris* (*irid* = colored circle) is the doughnut-shaped colored portion of the eyeball. It consists of circular and radial smooth muscle fibers. The black hole in the center of the iris is the *pupil*, through which light enters the eyeball. The iris regulates the amount of light entering the eyeball. When the eye is stimulated by bright light, the circular muscles of the iris contract and decrease the size of the pupil, called constriction. When the eye must adjust to dim light, the radial muscles contract and the pupil dilates or becomes larger. These muscles are controlled by the autonomic nervous system (see Exhibit 11-3).

Retina (Nervous Tunic)

The third and inner coat of the eye, the *retina* (*nervous tunic*), lies only in the posterior portion of the eye. Its primary function is image formation. The retina is one of the few places in the body where blood vessels can be seen directly. With a special reflecting light called an *ophthalmoscope*, it is possible to examine the retina and detect vascular changes associated with hypertension,

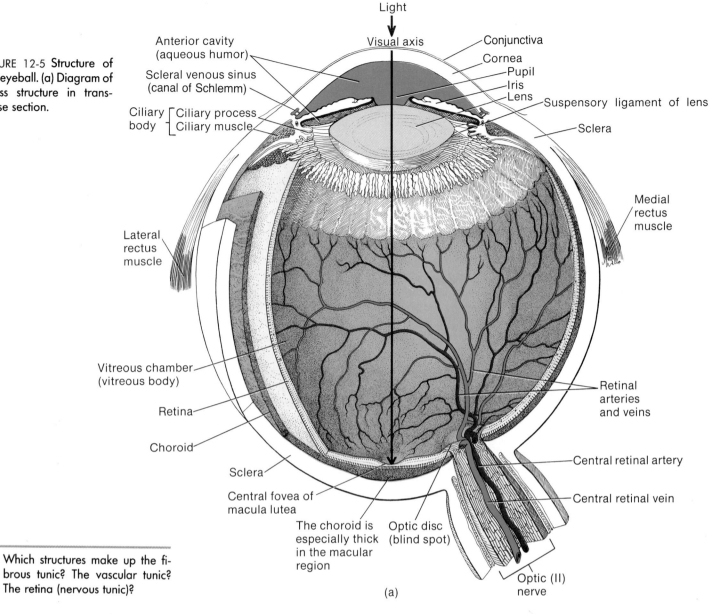

FIGURE 12-5 **Structure of the eyeball. (a)** Diagram of gross structure in transverse section.

Light

Visual axis

Anterior cavity (aqueous humor)

Scleral venous sinus (canal of Schlemm)

Ciliary body [Ciliary process / Ciliary muscle]

Conjunctiva

Cornea

Pupil

Iris

Lens

Suspensory ligament of lens

Sclera

Medial rectus muscle

Lateral rectus muscle

Vitreous chamber (vitreous body)

Retina

Choroid

Sclera

Central fovea of macula lutea

The choroid is especially thick in the macular region

Optic disc (blind spot)

Retinal arteries and veins

Central retinal artery

Central retinal vein

Optic (II) nerve

(a)

Which structures make up the fibrous tunic? The vascular tunic? The retina (nervous tunic)?

FIGURE 12-5 (*Continued*) **(b)** Microscopic structure of the retina exaggerated for emphasis. The arrows pointing upward indicate the direction of the signals passing through the nervous layer of the retina that ultimately result in a nerve impulse that passes into the optic (II) nerve.

Which photoreceptor is specialized for vision in dim light and allows us to see shapes and movement? Which photoreceptor is specialized for color and visual acuity?

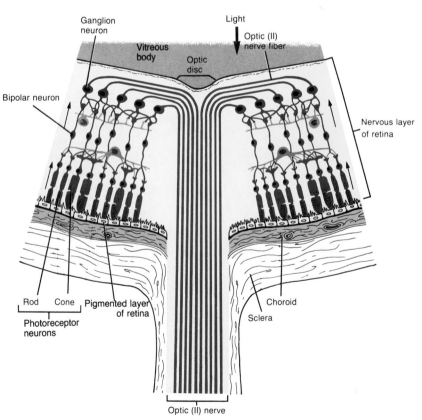

Ganglion neuron

Light

Optic (II) nerve fiber

Vitreous body

Optic disc

Bipolar neuron

Nervous layer of retina

Rod Cone Pigmented layer of retina

Photoreceptor neurons

Choroid

Sclera

Optic (II) nerve

(b)

atherosclerosis, and diabetes. The retina consists of an inner nervous tissue layer (visual portion) and an outer pigmented layer (nonvisual portion).

The nervous layer of the retina contains three zones of neurons, *photoreceptor neurons, bipolar neurons,* and *ganglion neurons.* The dendrites of the photoreceptor neurons are called rods and cones because of their shapes. They are visual receptors highly specialized for stimulation by light rays. Functionally, rods and cones develop generator potentials. *Rods* are specialized for vision in dim light. They also allow us to discriminate between different shades of dark and light and to see shapes and movement. *Cones* are specialized for color vision and sharpness of vision (*visual acuity*). They are stimulated only by bright light, which is why we cannot see color by moonlight.

There are about 3 million cones and 100 million rods. Cones are most densely concentrated in the *central fovea,* a small depression in the center of the macula lutea. The *macula lutea* (MAK-yoo-la LOO-tē-a), or yellow spot, is in the exact center of the retina, right on the visual axis of the eye. The fovea is the area of sharpest vision because of its high concentration of cones. Rods are absent from the fovea and macula and increase in density toward the periphery of the retina. It is for this reason that you can see better at night when not looking directly at an object.

When information has passed through the photoreceptor neurons, it is conducted to the bipolar neurons, then to the ganglion neurons. The axons of the ganglion neurons extend posteriorly to a small area of the retina called the *optic disc (blind spot)* where they all exit as the optic (II) nerve. The optic disc is called the blind spot because it contains no photoreceptors. The optic nerve transmits visual impulses to the occipital lobe of the cerebral cortex for interpretation as sight (Chapter 10).

A frequently encountered problem related to the retina is a *detached retina,* which may occur in trauma, such as a blow to the head. The actual detachment occurs between the sensory part of the retina and the underlying pigmented layer. Fluid accumulates between these layers, resulting in distorted vision and blindness. Often, the retina may be surgically reattached.

Lens

The *lens* is a biconvex transparent structure that normally focuses light rays onto the retina. It is constructed of numerous layers of protein fibers encased in a connective tissue capsule. *Suspensory ligaments* attached to the capsule hold the lens in position behind the pupil. A loss of transparency of the lens is known as a *cataract.*

Interior

The interior of the eyeball is a large space divided into two cavities by the lens, the anterior cavity and the vitreous chamber. The *anterior cavity* lies in front of the lens and is filled with a watery fluid, similar to cerebrospinal fluid, called the *aqueous* (*aqua* = water) *humor.* The fluid is secreted into the anterior cavity by choroid plexuses of the ciliary bodies. It then is drained off into the scleral venous sinus (canal of Schlemm) and reenters the blood.

The pressure in the eye, called *intraocular pressure (IOP),* is produced mainly by the aqueous humor. The intraocular pressure, along with the vitreous body, maintains the shape of the eyeball

and keeps the retina smoothly applied to the choroid so the retina will be well nourished and form clear images. Normal intraocular pressure (about 16 mm Hg) is maintained by drainage of the aqueous humor as described above. Besides maintaining intraocular pressure, the aqueous humor also helps nourish the lens and cornea since neither has blood vessels.

The second, and larger, cavity of the eyeball is the *vitreous chamber.* It lies behind the lens and contains a clear jellylike substance called the *vitreous body.* This substance contributes to intraocular pressure, helps prevent the eyeball from collapsing, and holds the retina flush against the eyeball. The vitreous body, unlike the aqueous humor, does not undergo constant replacement. It is formed during embryonic life and is not replaced thereafter.

A summary of structures associated with the eyeball is presented in Exhibit 12-1.

EXHIBIT 12-1
Summary of Structures Associated with the Eyeball

Structure	Function
Fibrous tunic	
Sclera	Provides shape and protects inner parts.
Cornea	Admits and refracts light.
Vascular Tunic	
Choroid	Provides blood supply and absorbs light.
Ciliary Body	Secretes aqueous humor and alters shape of lens for near or far vision (accommodation).
Iris	Regulates amount of light that enters eyeball.
Retina (Nervous Tunic)	Receives light, converts light into generator potentials and nerve impulses, and transmits impulses to the optic (II) nerve.
Lens	Refracts light.
Anterior Cavity	Contains aqueous humor that helps maintain shape of eyeball, and refracts light.
Vitreous Chamber	Contains vitreous body that helps maintain shape of eyeball, keeps retina applied to choroid, and refracts light.

PHYSIOLOGY OF VISION

Before light can reach the rods and cones of the retina to result in image formation, it must pass through the cornea, aqueous humor, pupil, lens, and vitreous body. For vision to occur, light reaching the rods and cones must form an image on the retina.

Retinal Image Formation

The formation of an image on the retina requires four basic processes, all concerned with focusing light rays: (1) refraction of light rays, (2) accommodation of the lens, (3) constriction of the pupil, and (4) convergence of the eyes. Accommodation and pupil size are functions of the ciliary muscle and the muscles of the iris. They are termed *intrinsic eye muscles,* since they are inside the eyeball. Convergence is a function of the voluntary muscles attached to the outside of the eyeball called the *extrinsic eye muscles* (see Figure 8-11).

▪ **Refraction of Light Rays** When light rays traveling through a transparent medium (such as air) pass into a second transparent medium with a different density (such as water), they bend at the surface of the two media. This is *refraction* (Figure 12-6a). The eye has four such media of refraction: cornea, aqueous humor, lens, and vitreous body. Most refraction occurs at the cornea.

The degree of refraction that takes place at each surface in the eye is very precise. In general, though, the nearer the object the more that light rays from it must be refracted to fall exactly on the central fovea in order to produce a clear image. This change in refraction is brought about by the lens in a process called accommodation.

▪ **Accommodation of the Lens** If the surface of a lens curves outward, as in a convex lens, the lens will refract incoming rays toward each other so they eventually intersect. The greater the curve, the more acutely it bends the rays toward each other. The lens of the eye is biconvex. Furthermore, it can change focusing power by becoming moderately curved at one moment and greatly curved the next. This increase in the curvature of the lens is called *accommodation* (Figure 12-7).

In near vision, the ciliary muscle contracts, pulling the ciliary process and choroid forward toward the lens, releasing the tension on the lens and suspensory ligament. Because of its elasticity, the lens shortens, thickens, and bulges, becoming more convex and refracting more acutely. In far vision, the ciliary muscle is relaxed and the lens becomes flatter. With aging, the lens loses elasticity and, therefore, its ability to accommodate.

The normal eye, known as an *emmetropic* (em'-e-TROP-ik) *eye*, can sufficiently refract light rays from an object to focus a clear image on the retina. Many individuals, however, do not have normal vision because of improper refraction. They may have *myopia* (mi-Ō-pē-a) or nearsightedness, *hypermetropia* (hī'-per-mē-TRO-pē-a) or farsightedness, or *astigmatism* (a-STIG-ma-tizm), irregularities in the surface of the lens or cornea. Myopia

FIGURE 12-6 Normal and abnormal refraction in the eyeball. (a) Refraction of light rays passing from air into water. (b) In the normal (emmetropic) eye, light rays from an object are bent sufficiently by the refracting media and converged on the central fovea. A clear image is formed. (c) In the nearsighted (myopic) eye, the image is focused in front of the retina. The condition may result from an elongated eyeball or thickened lens. (d) Correction is by use of a concave lens that diverges entering light rays so that they have to travel further through the eyeball and are focused directly on the retina. (e) In the farsighted (hypermetropic) eye, the image is focused behind the retina. The condition results from a shortened eyeball or a thin lens. (f) Correction is by a convex lens that converges entering light rays so that they focus directly on the retina. An astigmatism is an irregular curvature of the cornea or lens. As a result, horizontal and vertical rays are focused at two different points on the retina, and the image is not focused on the area of sharpest vision of the retina. This results in blurred or distorted vision. Suitable glasses correct the refraction of an astigmatic eye.

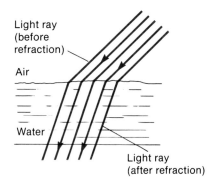

(a) Refraction of light rays

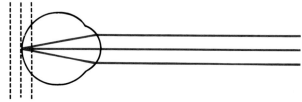

(b) Normal (emmetropic) eye

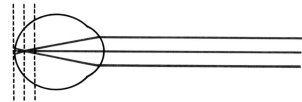

(c) Nearsighted (myopic) eye, uncorrected

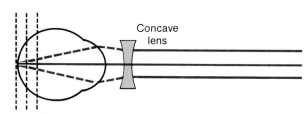

(d) Nearsighted (myopic) eye, corrected

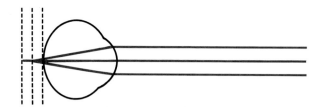

(e) Farsighted (hypermetropic) eye, uncorrected

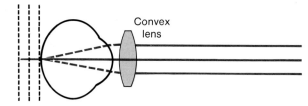

(f) Farsighted (hypermetropic) eye, corrected

What are the four refracting media?

FIGURE 12-7 Accommodation. (a) For objects 6 m (20 ft) or more away. (b) For objects closer than 6 m.

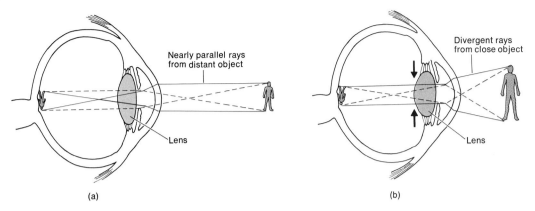

(a)

(b)

What is accommodation?

and hypermetropia are illustrated and explained in Figure 12-6c–f. The inability to focus on nearby objects due to loss of elasticity of the lens with age is called *presbyopia* (prez-bē-OP-ē-a).

▪ **Constriction of Pupil** Part of the accommodation mechanism involves the circular muscle fibers of the iris, which constrict the pupil. *Constriction of the pupil* means narrowing the diameter of the hole through which light enters the eye. This action occurs simultaneously with accommodation so that light rays do not enter the eye at the periphery of the lens. Rays entering at the periphery would not be brought to focus on the retina and would result in blurred vision.

▪ **Convergence** Because of the way their eyes are positioned in the skull, many animals see a set of objects off to the left with one eye and an entirely different set off to the right with the other. This characteristic doubles their field of vision and allows them to detect predators. In humans, both eyes normally focus on only one set of objects—a characteristic called *single binocular vision*.

With single binocular vision, when we stare straight ahead at a distant object, the incoming light rays are aimed directly at both pupils and are refracted to identical spots on the retinas of both eyes. But as we move closer to the object, our eyes must move toward the nose for the light rays to hit the same points on both retinas. The term *convergence* refers to this movement of the two eyeballs toward the nose so they are both directed to the object being viewed. The nearer the object, the greater the convergence required. Convergence is brought about by the automatic action of the extrinsic eye muscles. Convergence and binocular vision are also necessary for *depth perception* and for perceiving a single object that has a three-dimensional appearance (versus two two-dimensional views of the same object or scene).

▪ **Inverted Image** Images are focused upside down on the retina. They also undergo mirror reversal; that is, light reflected from the right side of an object hits the left side of the retina and vice versa. Note in Figure 12-7 how the reflected light from the top of the object hits below the central fovea and vice versa. We do not see an inverted world, however, because the brain learns early in life to coordinate visual images with the exact location of objects. The brain stores memories of reaching and touching objects and automatically turns visual images right-side-up and right-side-around.

Stimulation of Photoreceptors

▪ **Excitation of rods** After an image is formed on the retina by refraction, accommodation, constriction of the pupil, and convergence, light impulses must be converted into nerve impulses. The initial step is the development of generator potentials by rods and cones. To understand how this occurs, we will first examine the role of photopigments.

A *photopigment* is a substance that can absorb light and undergo a change in structure to produce a generator potential. The photopigment in rods is called *rhodopsin* and is composed of a protein called *scotopsin* and a derivative of vitamin A called *retinal*.

Rhodopsin is a highly unstable compound in the presence of even very small amounts of light. Any amount of light in a darkened room will trigger a complex series of chemical reactions that ultimately cause the depolarization of rods, thereby initiating visual impulses. In low-light situations, scotopsin and retinal will quickly recombine back into rhodopsin and its production is able to keep pace with its rate of breakdown, thereby providing night vision. Rods usually are nonfunctional in daylight, since rhodopsin is broken down faster than it can be re-formed. After going from bright sunlight into a dark room, it will take several minutes before the rods will function again. The period of adjustment is the time needed for the completely dissociated rhodopsin to reform.

Night blindness, which is also referred to as *nyctalopia* (nik′-ta-LŌ-pē-a), is the lack of normal night vision following the adjustment period. It is most often caused by vitamin A deficiency.

▪ **Excitation of Cones** Cones are the receptors for bright light and color. As in rods, photopigment breakdown produces the generator potential. The photopigment in cones also contains retinal, but the protein is different. Unlike rhodopsin, the photopigment of the cones requires bright light for breakdown and it re-forms quickly. There are three types of cones, each containing a different combination of retinal and protein. Each of the three types has a different maximum absorption of light of a different wavelength, and thus each responds best to light of a given color. One type of cone responds best to red light, the second to green, and the third to blue. Just as an artist can obtain almost any color by mixing colors, the cones can perceive any color by differential stimulation. When looking at an object, if all three types of cones are stimulated, the object will be perceived to be white in color; if none is stimulated, the object looks black.

If one of the types of cones is missing from the retina, an individual cannot distinguish some colors from others and is said to be *color-blind*. The most common type is *red–green color blindness* in which usually one photopigment is missing. Color blindness is an inherited condition that affects males far more frequently than females. The inheritance of the condition is discussed in Chapter 24 and illustrated in Figure 24-12.

Visual Pathway

Figure 12-5b shows that the retina (nervous tunic) is composed of three zones of neurons: photoreceptor (rods and cones), bipolar, and ganglion. For light to reach rods and cones and stimulate them to produce generator potentials, it must first pass through the ganglion and bipolar zones.

Once generator potentials are developed by rods and cones, the potentials produce signals in bipolar neurons. In response, bipolar neurons become excited and then transmit the excitatory visual signal on to ganglion cells. The ganglion cells become depolarized and initiate nerve impulses. The cell bodies of the ganglion cells lie in the retina, and their axons leave the eyeball as the *optic (II) nerve* (Figure 12-8) and pass through the *optic chiasma* (kī-AZ-ma), a point where some fibers cross to the opposite side and others remain uncrossed. On passing through the optic chiasma, the fibers, now part of the *optic tract*, terminate in the thalamus. Here the fibers synapse with neurons whose axons pass to the visual areas in the occipital lobes of the cerebral cortex. Because of the crossing at the optic chiasma, the right primary visual area of the cortex interprets visual sensations from the left side of an object and the left primary visual area interprets visual sensations from the right side of an object.

AUDITORY SENSATIONS AND EQUILIBRIUM

In addition to containing receptors for sound waves, the ear also contains receptors for equilibrium. The ear is divided into three principal regions: external ear, middle ear, and internal ear.

EXTERNAL EAR

The *external ear* is designed to collect sound waves and pass them inward (Figure 12-9a). It consists of a pinna, external auditory canal, and eardrum.

The *pinna* is a flap of elastic cartilage shaped like the flared end of a trumpet and attached to the head by ligaments and muscles.

The *external auditory* (*audire* = hearing) *canal* is a curved tube that extends from the pinna to the eardrum. The canal contains a few hairs and specialized glands called *ceruminous* (se-ROO-mi-nus) *glands*, which secrete *cerumen* (earwax). The hairs and cerumen (se-ROO-min) help prevent foreign objects from entering the ear.

The *eardrum (tympanic membrane)* is a thin, semitransparent partition of fibrous connective tissue between the external auditory canal and the middle ear.

MIDDLE EAR

The *middle ear* is a small, epithelial-lined, air-filled cavity between the eardrum of the external ear and a thin bony partition that contains two small openings, the oval window and the round window, of the inner ear (Figure 12-9a–c).

The posterior wall of the middle ear communicates with the mastoid air cells of the temporal bone through a chamber called the *tympanic antrum*. This anatomical fact explains why a middle ear infection may spread to the temporal bone, causing mastoiditis, or even to the brain.

An opening in the wall of the middle ear leads directly into the *auditory (Eustachian) tube*, which connects the middle ear with the upper part of the throat. The auditory tube equalizes air pressure on both sides of the eardrum. Abrupt changes in external or internal air pressure might otherwise cause the eardrum to rupture. During swallowing and yawning, the tube opens to relieve pressure, which explains why the sudden pressure change in an airplane may be equalized by deliberately swallowing or pinching the nose closed, closing the mouth, and gently forcing air up from the lungs.

Extending across the middle ear and attached to it by means of ligaments are three exceedingly small bones called *auditory ossicles* (OS-si-kuls) and named for their shapes—the malleus, incus, and stapes, commonly called the hammer, anvil, and stirrup (Figure 12-9). Tiny skeletal muscles control the amount of movement of these bones to prevent damage to the inner ear by excessively loud noises. The stapes fits into a small opening in the thin bony partition between the middle and internal ear called the *oval window*. Directly below the oval window is another opening, the *round window*.

INTERNAL EAR

The *internal ear* is divided into the outer bony labyrinth and inner membranous labyrinth, which lies within the bony labyrinth (Figure 12-10). The *bony labyrinth* is a series of cavities in the temporal bone. It can be divided into three areas named on the basis of shape—vestibule, cochlea, and semicircular canals. The bony labyrinth contains a fluid called *perilymph*. This fluid, which is similar to cerebrospinal fluid, surrounds the inner *membranous labyrinth*, a series of sacs and tubes having the same general form as the bony labyrinth. The membranous labyrinth contains a fluid called *endolymph*, which is chemically similar to intracellular fluid.

The *vestibule* is the middle part of the bony labyrinth. The membranous labyrinth in the vestibule consists of two sacs called the *utricle* (YOO-tri-kul = little bag) and *saccule* (SAK-yool). Each gives off a branch to form the *endolymphatic duct*, which ultimately enlarges to form the *endolymphatic sac*.

FIGURE 12-8 Afferent pathway for visual impulses. The dark circle in the center of the visual fields is the macula lutea. The center of the macula lutea is the central fovea, the area of sharpest vision.

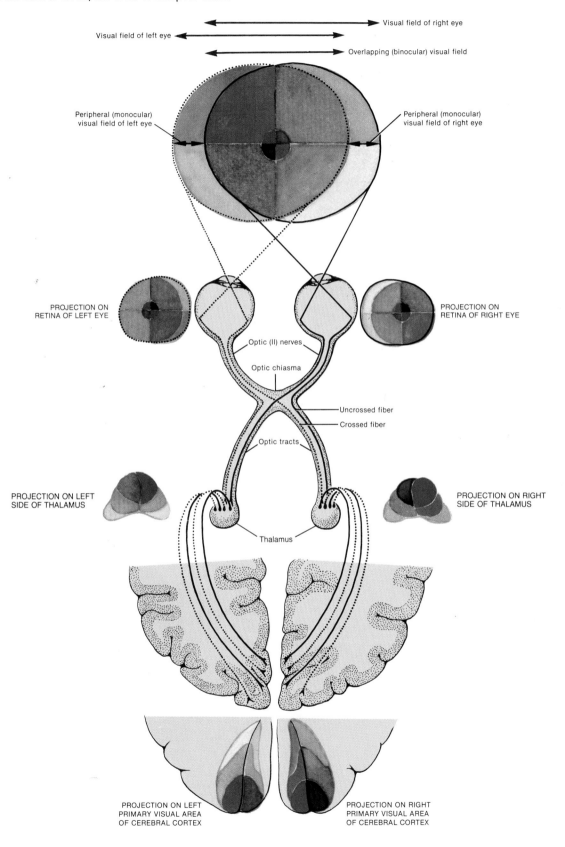

What is the optic chiasma and why is it significant?

FIGURE 12-9 **Structure of the auditory apparatus.** (a) Divisions of the right ear into external, middle, and internal portions seen in a frontal section through the right side of the skull. (b) Details of the middle ear and bony labyrinth of the internal ear. (c) Auditory ossicles of the middle ear.

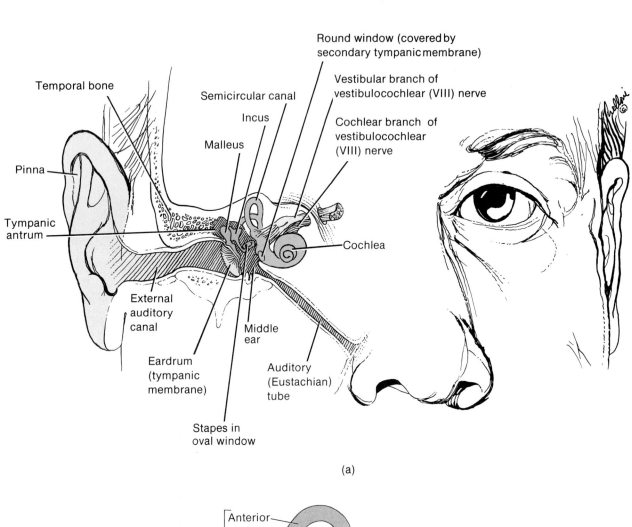

(a)

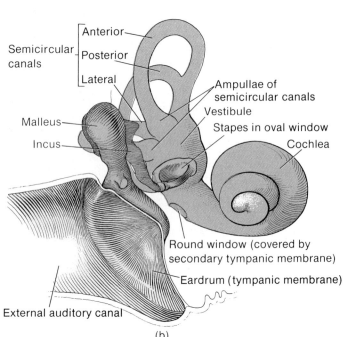

(b)

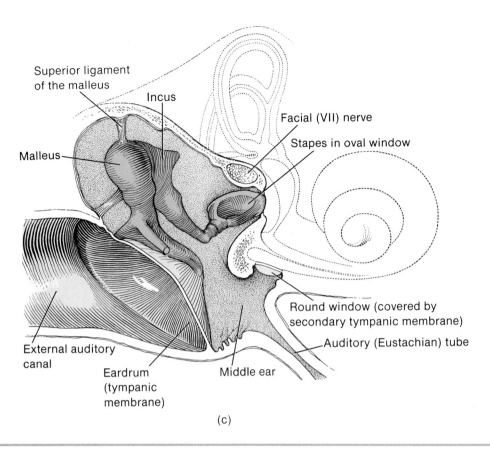

Superior ligament
of the malleus

Incus

Facial (VII) nerve

Stapes in oval window

Malleus

Round window (covered by
secondary tympanic membrane)

Auditory (Eustachian) tube

External auditory
canal

Eardrum
(tympanic
membrane)

Middle ear

(c)

Where are the receptors for hearing? For equilibrium?

FIGURE 12-10 **Details of the internal
ear.** (a) The outer, blue-colored area
belongs to the bony labyrinth. The
inner, pink-colored area belongs to
the membranous labyrinth.

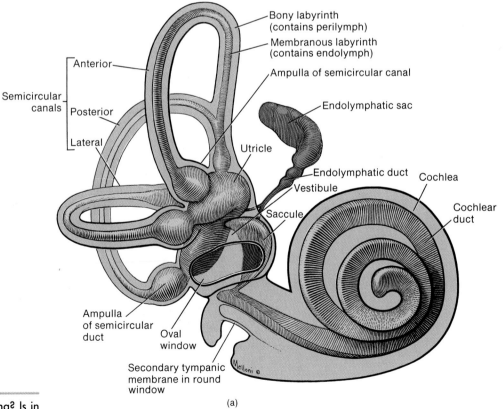

Bony labyrinth
(contains perilymph)

Membranous labyrinth
(contains endolymph)

Ampulla of semicircular canal

Endolymphatic sac

Anterior

Semicircular
canals

Posterior

Lateral

Utricle

Endolymphatic duct

Cochlea

Vestibule

Saccule

Cochlear
duct

Ampulla
of semicircular
duct

Oval
window

Secondary tympanic
membrane in round
window

(a)

Which structure is the organ of hearing? Is in
contact with hair cells of the spiral organ? Contains the utricle and saccule?

FIGURE 12-10 (*Continued*) (b) Relationship of the scala tympani, cochlear duct, and scala vesitbuli as seen in sections through the cochlea. The arrows indicate the transmission of sound waves, which are discussed shortly. (c) The origins of the vestibular and cochlear branches of the vestibulocochlear (VIII) nerve. (d) Detail of a section through one turn of the cochlea.

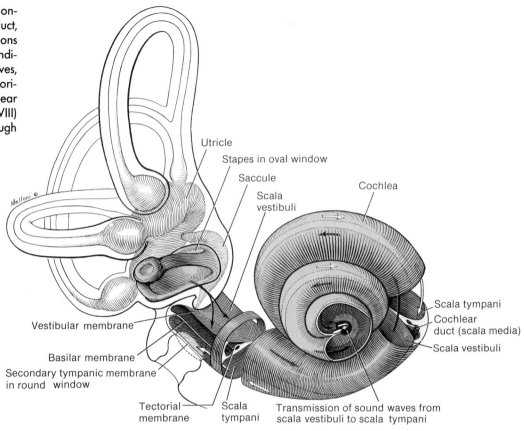

(b)

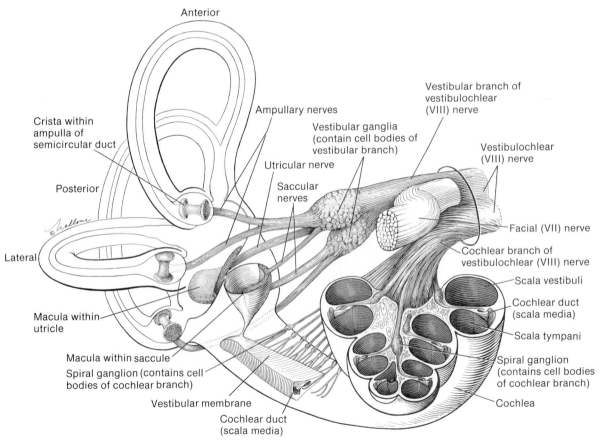

(c)

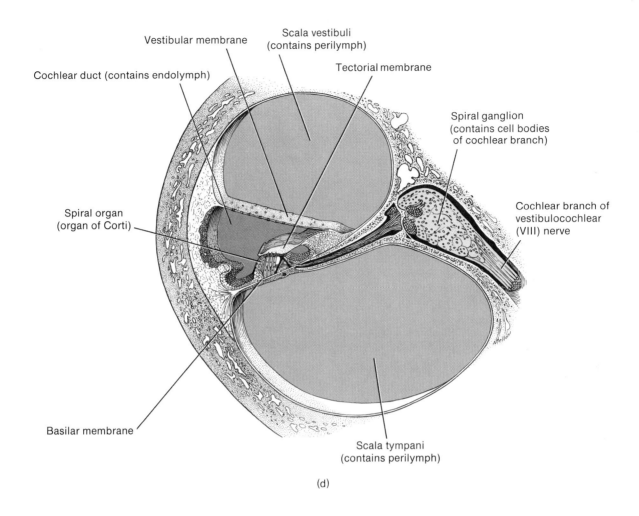

Vestibular membrane

Scala vestibuli
(contains perilymph)

Cochlear duct (contains endolymph)

Tectorial membrane

Spiral ganglion
(contains cell bodies
of cochlear branch)

Spiral organ
(organ of Corti)

Cochlear branch of
vestibulocochlear
(VIII) nerve

Basilar membrane

Scala tympani
(contains perilymph)

(d)

FIGURE 12-10 (*Continued*) (e) Enlargement of the spiral organ (organ of Corti).

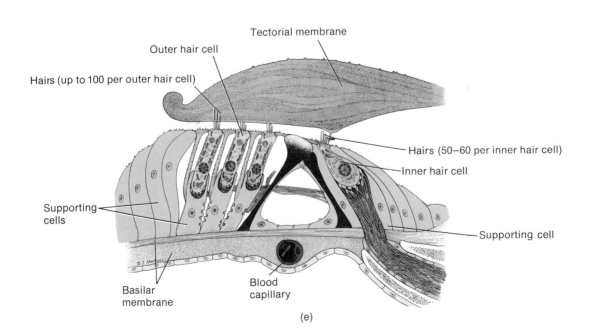

Tectorial membrane

Outer hair cell

Hairs (up to 100 per outer hair cell)

Hairs (50–60 per inner hair cell)

Inner hair cell

Supporting cells

Supporting cell

Basilar membrane

Blood capillary

(e)

Behind the vestibule are the three bony *semicircular canals*. The anterior and posterior semicircular canals are vertical, whereas the lateral one is horizontal. One end of each canal enlarges into a swelling called the *ampulla* (am-POOL-la; *ampulla* = little jar). Inside the bony semicircular canals are the *semicircular ducts,* which communicate with the utricle of the vestibule.

In front of the vestibule is the *cochlea* (KOK-lē-a), a bony spiral canal that resembles a snail's shell. A cross section through the cochlea shows that it is divided into three channels, the *scala vestibuli* which ends at the oval window, the *scala tympani* which ends at the round window, and the *cochlear duct (scala media).* Between the scala media and the scala vestibuli is the *vestibular membrane*. Between the scala media and scala tympani is the *basilar membrane*.

Resting on the basilar membrane is the *spiral organ (organ of Corti),* the organ of hearing. The spiral organ consists of supporting cells and hair cells, which are the receptors for auditory sensations. The hair cells have long hairlike processes at their free ends that extend into the endolymph of the cochlear duct. The basal ends of the hair cells are in contact with fibers of the cochlear branch of the vestibulocochlear (VIII) nerve. Over the hair cells and in contact with them is the *tectorial (tectum = cover) membrane*, a delicate and flexible gelatinous membrane.

Hair cells of the spiral organ are easily damaged by exposure to high-intensity noises such as those produced by jet planes and loud music. They rearrange into disorganized patterns or they and their supporting cells degenerate.

SOUND WAVES

Sound waves are produced from the alternate compression and decompression of air molecules. They originate from a vibrat-

ing object, much the same way that waves travel over the surface of water. The sounds heard most acutely by human ears are from sources that vibrate at frequencies between 1000 and 4000 cycles per second (Hz). The entire range extends from 20 to 20,000 Hz.

The frequency of vibration is its pitch. The greater the vibration, the higher the pitch. Also, the greater the force of the vibration, the louder the sound. Intensity or loudness is measured in *decibels (dB).* The point at which a person can just detect sound from silence is 0 dB. Between 115 and 120 dB is the pain threshold. In the range between 0 and 120 dB, rustling leaves have a decibel rating of 15, normal conversation 45, crowd noise 60, a vacuum cleaner 75, and a pneumatic drill 90. Hearing loss may result from prolonged exposure to sounds over 90 dB.

PHYSIOLOGY OF HEARING

The events involved in the physiology of hearing sound waves are as follows (Figure 12-11):

1. Sound waves that reach the ear are directed by the pinna into the external auditory canal.
2. When the waves strike the eardrum, the alternate compression and decompression of the air causes the eardrum to vibrate. The distance the eardrum moves is always very small and is relative to the force and velocity of the sound waves. It vibrates slowly in response to low-frequency sounds and rapidly in response to high-frequency sounds.
3. The central area of the eardrum is connected to the malleus, which also starts to vibrate. The vibration is then picked up by the incus, which transmits the vibration to the stapes.

FIGURE 12-11 Physiology of hearing. The numbers correspond to the events listed in the text. The cochlea has been uncoiled in order to illustrate the transmission of sound waves and their subsequent distortion of the vestibular or basilar membranes of the cochlear duct.

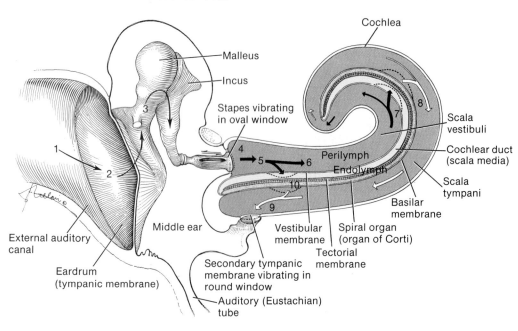

4. As the stapes moves back and forth, it pushes the oval window in and out.

5. The movement of the oval window sets up waves in the perilymph of the cochlea.

6. As the oval window bulges inward, it pushes the perilymph of the scala vestibuli and pressure waves are propagated through the scala vestibuli. If the inward movement of the stapes at the oval window is slow, the pressure in the perilymph pushes the perilymph from the scala vestibuli to the scala tympani and eventually to the round window, causing it to bulge outward into the middle ear. (See number 9 in Figure 12-11.)

7. As the pressure moves through the perilymph of the scala vestibuli, it pushes the vestibular membrane inward and increases the pressure of the endolymph inside the cochlear duct.

8. As a result, the basilar membrane moves slightly and bulges into the scala tympani.

9. As the pressure moves through the scala tympani, perilymph moves toward the round window, causing it to bulge outward into the middle ear. If the stapes vibrates rapidly, pressure in the perilymph does not have time to pass from the scala vestibuli to the scala tympani to the round window. Instead, pressure in the perilymph in the scala vestibuli is transmitted through the basilar membrane and eventually to the round window. As a result, the area of the basilar membrane near the oval and round windows vibrates.

10. When the basilar membrane vibrates, the hair cells of the spiral organ move against the tectorial membrane. The movement of the hairs develops generator potentials that ultimately lead to the generation of nerve impulses.

The function of hair cells is to convert a mechanical force (stimulus) into an electrical signal (nerve impulse). It is believed to occur this way: When the hairs at the top of the cell are moved, the hair cell membrane depolarizes, producing the generator potential. Depolarization spreads through the cell and causes the release of a neurotransmitter from the hair cell, which excites a sensory nerve fiber at the base of the hair cell.

The impulses are then passed on to the cochlear branch of the vestibulocochlear (VIII) nerve (Figure 12-11) and cochlear nuclei in the medulla. Here most impulses cross to the opposite side and then travel to the midbrain, thalamus, and finally to the auditory area of the temporal lobe of the cerebral cortex.

Differences in pitch result when sound waves of various frequencies cause different regions of the basilar membrane to vibrate more intensely than others. Loudness is determined by the intensity of sound waves. High-intensity waves cause greater vibration of the basilar membrane. Thus, more hair cells are stimulated and more impulses reach the brain.

PHYSIOLOGY OF EQUILIBRIUM

There are two kinds of *equilibrium*. *Static equilibrium* refers to the position of the body (mainly the head) relative to the ground (gravity). *Dynamic equilibrium* is the maintenance of body position (mainly the head) in response to sudden movements such as rotation, acceleration, and deceleration. The receptor organs for equilibrium are in the internal ear, in the saccule, the utricle, and the semicircular ducts.

Static Equilibrium

The walls of the utricle and saccule contain a small, flat region called a *macula* (see Figure 12-10c). The maculae are the receptors for static equilibrium. They provide sensory information regarding the position of the head in space and are essential for maintaining posture.

The maculae resemble the spiral organ (organ of Corti). They consist of two kinds of cells: *hair (receptor) cells* and *supporting cells* (Figure 12-12). Hair cells contain long extensions of the cell membrane consisting of many *stereocilia* and one extremely long *kinocilium*. Floating over the hair cells is a thick, jellylike substance called the *gelatinous otolithic membrane*. A layer of calcium carbonate crystals, called *otoliths* (*oto* = ear; *lithos* =

FIGURE 12-12 **Structure of the macula. (a) Overall structure of a section of the macula.**

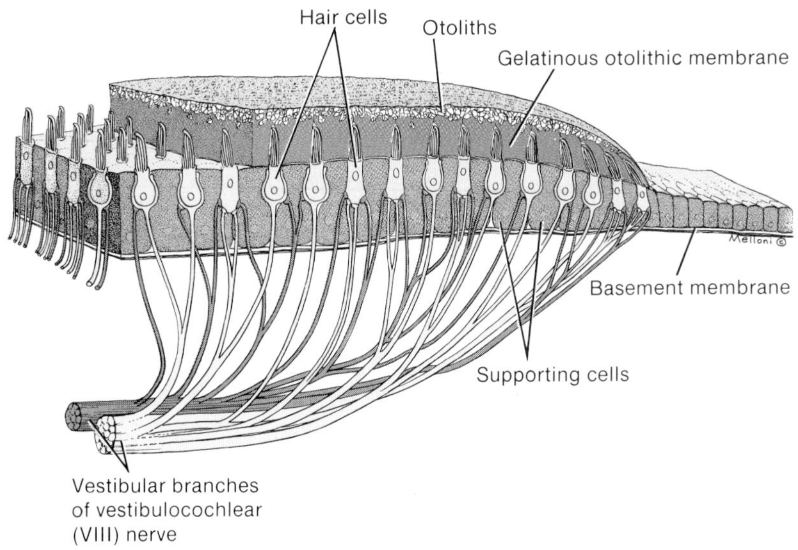

(a)

FIGURE 12-12 (*Continued*) (b) Details of several hair cells.

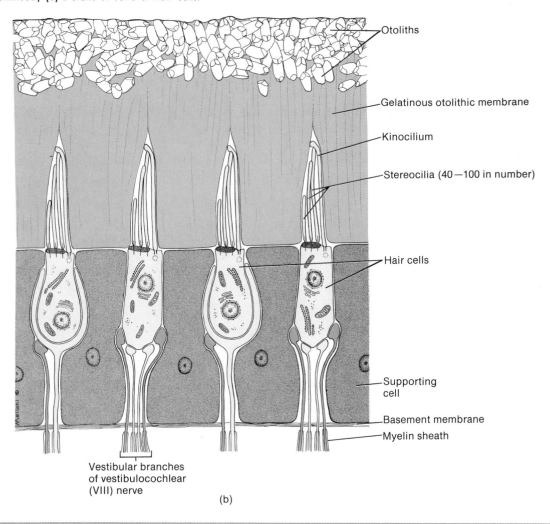

Otoliths

Gelatinous otolithic membrane

Kinocilium

Stereocilia (40—100 in number)

Hair cells

Supporting cell

Basement membrane

Myelin sheath

Vestibular branches of vestibulocochlear (VIII) nerve

(b)

What are the maculae and how do they function?

stone), extends over the surface of the gelatinous otolithic membrane.

The gelatinous otolithic membrane sits on top of the macula like a discus on a greased cookie sheet. If you tilt your head forward, the membrane (the discus in our analogy) slides downhill over the hair cells in the direction of the tilt. Similarly, if you are sitting in a car that suddenly jerks forward, the membrane, due to its inertia, slides backward and stimulates the hair cells. As the otoliths move, they pull on the gelatinous otolithic membrane, which pulls on the stereocilia and makes them bend. The movement of the stereocilia initiates a nerve impulse that is then transmitted to the vestibular branch of the vestibulocochlear (VIII) nerve (see Figure 12-10a).

Most of the vestibular branch fibers enter the brain stem and terminate in the medulla. The remaining fibers enter the cerebellum. Fibers from the medulla form a tract that extends from the brain stem into the spinal cord. This tract sends impulses to the cranial nerves that control eye movements [oculomotor (III), trochlear (IV), and abducens (VI)] and to the accessory (XI) nerve that helps control head and neck movements. In addition, fibers from

the medulla form the vestibulospinal tract that conveys impulses to skeletal muscles that regulate body tone in response to head movements. Various pathways between the medulla, cerebellum, and cerebrum enable the cerebellum to assume a key role in helping the body to maintain static equilibrium. The cerebellum continuously receives updated sensory information from the utricle and saccule concerning static equilibrium. Using this information, the cerebellum sends continuous impulses to the motor areas of the cerebrum, in response to input from the utricle and saccule, causing the motor system to increase or decrease its impulses to specific skeletal muscles in order to maintain static equilibrium.

Dynamic Equilibrium

The three semicircular ducts maintain dynamic equilibrium (see Figure 12-10). The ducts are positioned at right angles to one another in three planes: the two vertical ones are the anterior and posterior semicircular ducts; the horizontal one is the lateral duct. This positioning permits detection of an imbalance in three planes. In the ampulla, the dilated portion of each duct, there is a small elevation called the *crista* (Figure 12-13). Each crista is composed of a group of *hair cells* and *supporting cells* covered by a jellylike material called the *cupula*. When the head moves, the endolymph in the semicircular ducts flows over the hairs and bends them. The movement of the hairs stimulates sensory neurons, and the impulses pass over the vestibular branch of the vestibulocochlear (VIII) nerve. The impulses follow the same pathways as those for static equilibrium and are eventually sent to the muscles that must contract to maintain body balance in the new position.

A summary of the structures of the ear related to hearing and equilibrium is presented in Exhibit 12-2.

FIGURE 12-13 Semicircular ducts and dynamic equilibrium. (a) Position of a cristae with the head upright (above) and when the head moves (below). (b) Enlarged aspect of a crista. The ampullary nerves are branches of the vestibular division of the vestibulocochlear (VIII) nerve.

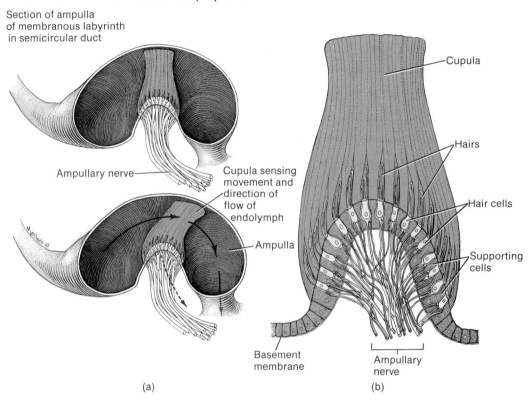

Section of ampulla of membranous labyrinth in semicircular duct

Ampullary nerve

Cupula sensing movement and direction of flow of endolymph

Ampulla

Cupula

Hairs

Hair cells

Supporting cells

Basement membrane

Ampullary nerve

(a)

(b)

What is dynamic equilibrium? Static equilibrium?

EXHIBIT 12-2
Summary of Structures of the Ear Related to Hearing and Equilibrium

Structure	Function
External Ear	
Pinna	Collects sound waves.
External Auditory Canal	Directs sound waves to eardrum.
Eardrum	Sound waves cause it to vibrate, which, in turn, causes the malleus to vibrate.
Middle Ear	
Auditory (Eustachian) Tube	Equalizes pressure on both sides of the eardrum.
Auditory Ossicles	Transmit sound waves from eardrum to oval window.
Internal Ear	
Utricle	Contains macula, receptor for static equilibrium.
Saccule	Contains macula, receptor for static equilibrium.
Semicircular Ducts	Contain cristae, receptors for dynamic equilibrium.
Cochlea	Contains a series of fluids, channels, and membranes that transmit sound waves to the spiral organ (organ of Corti), the spiral organ of hearing; the spiral organ generates nerve impulses and transmits them to the cochlear branch of the vestibulocochlear (VIII) nerve.

▪ COMMON DISORDERS ▪

Cataract

A *cataract*, meaning "waterfall," is a clouding of the lens or its capsule so that it becomes opaque or milk white.

Glaucoma

Glaucoma is a group of disorders characterized by an abnormally high intraocular pressure (IOP), owing to a buildup of aqueous humor inside the eyeball. Glaucoma can progress from mild visual impairment to a point where neurons of the retina are destroyed, resulting in degeneration of the optic disc, visual field defects, and blindness.

Conjunctivitis (Pinkeye)

Conjunctivitis (*pinkeye*) is an inflammation of the conjunctiva, which is the membrane that lines the insides of the eyelids and covers the cornea.

Trachoma

Trachoma (tra-KŌ-ma) is a serious form of chronic contagious conjunctivitis, which is caused by a bacterium called *Chlamydia trachomatis*.

Deafness

Deafness is significant or total hearing loss. *Sensorineural deafness* is caused by an impaired cochlea or cochlear branch of the vestibulocochlear (VIII) nerve. *Conduction deafness* is caused by impaired external and middle ear mechanisms for transmitting sounds to the cochlea. Among the factors that contribute to deafness are atherosclerosis, which reduces blood supply to the ears; repeated exposure to loud noise, which destroys hair cells of the spiral organ (organ of Corti); certain drugs, such as streptomycin; disease; impacted cerumen; injury to the eardrum; and aging, which results in thickening of the eardrum, stiffened auditory ossicle joints, and decreased numbers of hair cells due to diminished cell division.

Labyrinthine Disease

Labyrinthine (lab'-i-RIN-thēn) *disease* refers to a malfunction of the internal ear that is characterized by deafness, tinnitus (ringing in the ears), vertigo (hallucination of movement), nausea, and vomiting. There may also be blurred vision, nystagmus (rapid, involuntary movement of the eyeballs), and a tendency to fall in a certain direction.

Ménière's Syndrome

Ménière's (men-YAIRZ) *syndrome* is characterized by an increased amount of endolymph that enlarges the internal ear. Among the symptoms are fluctuating hearing loss, attacks of vertigo, and roaring tinnitus (ringing in the ears). The cause of Ménière's syndrome is unknown.

Vertigo

Vertigo (*vertex* = whorl) is a sensation of spinning or movement in which the world is revolving or the person is revolving in space.

Otitis Media

Otitis media is an acute bacterial infection of the middle ear. It is characterized by pain, malaise, fever, and a reddening and outward bulging of the eardrum, which may rupture without prompt treatment.

Motion Sickness

Motion sickness is a functional disorder brought on by repetitive angular, linear, or vertical motion and characterized by various symptoms, primarily nausea and vomiting.

MEDICAL TERMINOLOGY AND CONDITIONS

Achromatopsia (a-krō'-ma-TOP-sē-a; *a* = without; *chrom* = color) Complete color blindness.

Ametropia (am'-e-TRŌ-pē-a; *ametro* = disproportionate; *ops* = eye) Refractive defect of the eye resulting in an inability to focus images properly on the retina.

Anopsia (an-OP-sē-a; *opsia* = vision) A defect of vision.

Astereognosis (a-ster-ē-og-NŌ-sis; *stereos* = solid; *gnosis* = knowledge) Loss of ability to recognize objects or to appreciate their form by touching them.

Audiometer (aw-dē-OM-e-ter; *audire* = to hear; *metron* = to measure) An instrument used to measure hearing by producing acoustic stimuli of known frequency and intensity.

Blepharitis (blef-a-RĪ-tis; *blepharo* = eyelid; *itis* = inflammation of) An inflammation of the eyelid.

Dyskinesia (dis'-ki-NĒ-zē-a; *dys* = difficult; *kinesis* = movement) Abnormality of motor function characterized by involuntary, purposeless movements.

Epiphora (e-PIF-ō-ra; *epi* = above) Abnormal overflow of tears.

Eustachitis (yoo'-stā-KĪ-tis) An inflammation or infection of the auditory (Eustachian) tube.

Exotropia (ek'-sō-TRŌ-pē-a; *ex* = out; *tropia* = turning) Turning outward of the eyes.

Keratitis (ker'-a-TĪ-tis; *kerato* = cornea) An inflammation or infection of the cones.

Kinesthesis (kin'-es-THĒ-sis; *aisthesis* = sensation) The sense of perception of movement.

Labyrinthitis (lab'-i-rin-THĪ-tis) An inflammation of the internal ear.

Mydriasis (mi-DRĒ-a-sis) Dilated pupil.

Myringitis (mir'-in-JĪ-tis; *myringa* = eardrum) An inflammation of the eardrum; also called *tympanitis.*

Nystagmus (nis-TAG-mus; *nystazein* = to nod) A rapid involuntary movement of the eyeballs.

Otalgia (o-TAL-jē-a; *oto* = ear; *algia* = pain) Earache.

Otosclerosis (ō'-tō-skle-RŌ-sis; *oto* = ear; *sclerosis* = hardening) Pathological process that may be hereditary in which new bone is deposited around the oval window. The result may be immobilization of the stapes, leading to deafness.

Photophobia (fō'-tō-FŌ-bē-a; *photo* = light; *phobia* = fear) Abnormal visual intolerance to light.

Presbyopia (pres'-bē-Ō-pē-a; *presby* = old) Inability to focus on nearby objects owing to loss of elasticity of the crystalline lens. The loss is usually caused by aging.

Ptosis (TŌ-sis; *ptosis* = fall) Falling or drooping of the eyelid. (This term is also used for the slipping of any organ below its normal position.)

Retinoblastoma (ret'-i-nō-blas-TŌ-ma; *blast* = bud; *oma* = tumor) A tumor arising from immature retinal cells and accounting for 2 percent of childhood malignancies.

Scotoma (skō-TŌ-ma; *scotoma* = darkness) An area of reduced or lost vision in the visual field. Also called a *blind spot* (other than the normal blind spot or optic disc).

Strabismus (stra-BIZ-mus) An imbalance in the extrinsic eye muscles that a person cannot overcome. In *convergent strabismus* (*cross-eye*), the visual axes converge. In *divergent strabismus* (*walleye*), the visual axes diverge. *Amblyopia* is the term used to describe the loss of vision in an otherwise normal eye that, because of muscle imbalance, cannot focus in sync with the other eye.

Tinnitus (ti-NĪ-tus) A ringing, roaring, or clicking in the ears.

STUDY OUTLINE

Sensations (p. 241)

Definition (p. 241)
1. Sensation is a state of awareness of external and internal conditions of the body.
2. The prerequisites for a sensation to occur are reception of a stimulus, conversion of the stimulus into a nerve impulse by a receptor, conduction of the impulse to the brain, and translation of the impulse into a sensation by a region of the brain.
3. Each stimulus is capable of causing the membrane of a receptor to depolarize. This is called the generator potential.

Characteristics (p. 241)
1. Projection occurs when the brain refers a sensation to the point of stimulation.
2. Adaptation is the loss of sensation even though the stimulus is still applied.
3. An afterimage is the persistence of the sensation even though the stimulus is removed.
4. Modality is the property by which one sensation is distinguished from another.

Classification of Receptors (p. 242)
1. According to location, receptors are classified as exteroceptors, enteroceptors, and proprioceptors.
2. On the basis of type of stimulus detected, receptors are classified as mechanoreceptors, thermoreceptors, nociceptors, electromagnetic receptors, and chemoreceptors.
3. In terms of simplicity or complexity, simple receptors are associated with general senses and complex receptors are associated with special senses.

General Senses (p. 242)

Cutaneous Sensations (p. 242)
1. Cutaneous sensations include tactile sensations (touch, pressure, vibration), thermoreceptive sensations (heat and cold), and pain. Receptors for these sensations are located in the skin, connective tissues, and the ends of the gastrointestinal tract.
2. Receptors for touch are hair root plexuses, free nerve endings, tactile (Merkel's) discs, corpuscles of touch (Meissner's corpuscles), and type II cutaneous mechanoreceptors (end organs of Ruffini). Receptors for pressure are free nerve endings, type II cutaneous mechanoreceptors, and lamellated (Pacinian) corpuscles. Receptors for vibration are corpuscles of touch and lamellated corpuscles.
3. Pain receptors (nociceptors) are located in nearly every body tissue.
4. Referred pain is felt in the skin near or away from the organ sending pain impulses.
5. Phantom pain is the sensation of pain in a limb that has been amputated.

Proprioceptive Sensations (p. 246)
1. Receptors located in skeletal muscles, tendons, in and around joints, and the internal ear convey impulses related to muscle tone, movement of body parts, and body position.
2. The receptors include muscle spindles, tendon organs (Golgi tendon organs), joint kinesthetic receptors, and the maculae and cristae.

Olfactory Sensations (p. 246)
1. The receptors for olfaction, the olfactory cells, are in the nasal epithelium.
2. Substances to be smelled must be gaseous, water-soluble, and lipid-soluble.
3. Olfactory cells convey impulses to olfactory (I) nerves, olfactory bulbs, olfactory tracts, and the cerebral cortex.

Gustatory Sensations (p. 247)
1. The receptors for gustation, the gustatory cells, are located in taste buds.
2. Substances to be tasted must be in solution in saliva.
3. The four primary tastes are salt, sweet, sour, and bitter.
4. Gustatory cells convey impulses to cranial nerves V, VII, IX, and X, the medulla, thalamus, and cerebral cortex.

Visual Sensations (p. 248)
1. Accessory structures of the eyes include the eyebrows, eyelids, eyelashes, and the lacrimal apparatus.
2. The eye is constructed of three coats: (a) fibrous tunic (sclera and cornea), (b) vascular tunic (choroid, ciliary body, and iris), and (c) retina (nervous tunic), which contains rods and cones.
3. The anterior cavity contains aqueous humor; the vitreous chamber contains the vitreous body.
4. The refractive media of the eye are the cornea, aqueous humor, lens, and vitreous body.
5. Retinal image formation involves refraction of light, accommodation of the lens, constriction of the pupil, convergence, and inverted image formation.
6. Improper refraction may result from myopia (nearsightedness), hypermetropia (farsightedness), and astigmatism (corneal or lens abnormalities).
7. Rods and cones develop generator potentials and ganglion cells initiate nerve impulses.
8. Impulses from ganglion cells are conveyed through the retina to the optic (II) nerve, the optic chiasma, the optic tract, the thalamus, and the cortex.

Auditory Sensations and Equilibrium (p. 254)
1. The ear consists of three anatomical subdivisions: (a) the external ear (pinna, external auditory canal, and eardrum), (b) the middle

ear (auditory or Eustachian tube, ossicles, oval window, and round window), and (c) the internal ear (bony labyrinth and membranous labyrinth). The internal ear contains the spiral organ (organ of Corti), the organ of hearing.

2. Sound waves enter the external auditory canal, strike the eardrum, pass through the ossicles, strike the oval window, set up waves in the perilymph, strike the vestibular membrane and scala tympani, increase pressure in the endolymph, strike the basilar membrane, and stimulate hairs on the spiral organ. A sound impulse is then initiated.

3. Static equilibrium is the orientation of the body relative to the pull of gravity. The maculae of the utricle and saccule are the sense organs of static equilibrium.

4. Dynamic equilibrium is the maintenance of body position in response to movement. The cristae in the semicircular ducts are the sense organs of dynamic equilibrium.

REVIEW QUESTIONS

1. Define a sensation and a sense receptor. What prerequisites are necessary for the perception of a sensation? (p. 241)
2. Describe the following characteristics of a sensation: projection, adaptation, afterimage, modality. (p. 241)
3. Classify receptors on the basis of location, stimulus detected, and simplicity or complexity. (p. 242)
4. Distinguish between a general sense and a special sense. (p. 242)
5. What is a cutaneous sensation? Distinguish tactile, thermoreceptive, and pain sensations. (p. 242)
6. For each of the following cutaneous sensations, describe the receptor involved in terms of structure, function, and location: touch, pressure, vibration, thermoreceptive, and pain. (p. 242)
7. Why are pain receptors important? What is referred pain? Phantom pain? (p. 243)
8. What is the proprioceptive sense? Where are the receptors for this sense located? (p. 246)
9. Describe muscle spindles, tendon organs (Golgi tendon organs), and joint kinesthetic receptors. (p. 246)
10. What are the necessary conditions for substances to be smelled? (p. 247)
11. Discuss the origin and path of an impulse that results in smelling. (p. 247)
12. How are gustatory receptors stimulated? (p. 248)
13. Discuss how an impulse for taste travels from a taste bud to the brain. (p. 248)
14. Describe the importance of the following accessory structures of the eye: eyelids, eyelashes, and eyebrows. (p. 249)
15. What is the function of the lacrimal apparatus? Explain how it operates. (p. 249)
16. By means of a labeled diagram, indicate the principal parts of the eye. (p. 250)

17. Describe the location and contents of the chambers of the eye. What is intraocular pressure (IOP)? How is the scleral venous sinus (canal of Schlemm) related to this pressure? (p. 251)
18. Explain how each of the following events is related to vision: (a) refraction of light, (b) accommodation of the lens, (c) constriction of the pupil, and (d) convergence. (p. 251)
19. Explain inverted image formation. (p. 253)
20. Distinguish emmetropia, myopia, hypermetropia, and astigmatism by means of a diagram. (p. 252)
21. How are rods and cones excited? Relate your discussion to the rhodopsin cycle by means of a diagram. (p. 253)
22. Describe the path of a visual impulse from the optic (II) nerve to the brain. (p. 254)
23. Diagram the principal parts of the external, middle, and internal ear. Describe the function of each part labeled. (p. 256)
24. Explain the events involved in the transmission of sound from the pinna to the spiral organ (organ of Corti). (p. 260)
25. What is the afferent pathway for sound impulses from the cochlear branch of the vestibulocochlear (VIII) nerve to the brain? (p. 261)
26. Compare the function of the maculae in the saccule and utricle in maintaining static equilibrium with the role of the cristae in the semicircular ducts in maintaining dynamic equilibrium. (p. 261)
27. Describe the path of an impulse that results in static and dynamic equilibrium. (p. 262)
28. Define the following: cataract, glaucoma, conjunctivitis, trachoma, deafness, labyrinthine disease, Ménière's syndrome, vertigo, otitis media, and motion sickness. (p. 264)
29. Refer to the medical terminology and conditions associated with the sense organs. Be sure that you can define each term listed. (p. 265)

13

The Endocrine System

A LOOK AHEAD

ENDOCRINE GLANDS
CHEMISTRY OF HORMONES
MECHANISM OF HORMONAL ACTION
    Overview
    Receptors
    Interaction with Plasma Membrane
        Receptors
    Interaction with Intracellular
        Receptors
    Prostaglandins (PGs) and Hormones
CONTROL OF HORMONAL
        SECRETIONS: FEEDBACK
        CONTROL
PITUITARY
    Anterior Pituitary
        Human Growth Hormone (hGH)
        Thyroid-Stimulating Hormone
         (TSH)
        Adrenocorticotropic Hormone
         (ACTH)
        Follicle-Stimulating Hormone
         (FSH)
        Luteinizing Hormone (LH)
        Prolactin (PRL)
        Melanocyte-Stimulating Hormone
         (MSH)
    Posterior Pituitary
        Oxytocin (OT)
        Antidiuretic Hormone (ADH)
THYROID
    Function and Control of
        Thyroid Hormones
    Calcitonin (CT)
PARATHYROIDS
    Parathyroid Hormone (PTH)
ADRENALS
    Adrenal Cortex
        Mineralocorticoids
        Glucocorticoids
        Gonadocorticoids
    Adrenal Medulla
        Epinephrine and
         Norepinephrine (NE)
PANCREAS
    Glucagon
    Insulin
OVARIES AND TESTES
PINEAL
THYMUS
OTHER ENDOCRINE TISSUES
STRESS AND THE GENERAL
        ADAPTATION SYNDROME
        (GAS)
    Stressors
    Alarm Reaction
    Resistance Reaction
    Exhaustion
    Stress and Disease
COMMON DISORDERS
MEDICAL TERMINOLOGY AND
        CONDITIONS

The nervous and endocrine systems working together coordinate the functioning of all of the body systems. The nervous system controls homeostasis through electrical impulses delivered over neurons. The endocrine system releases chemical messengers called **hormones** (*hormone* = set in motion) into the bloodstream. Whereas the nervous system sends messages to a specific set of cells (muscle fibers, gland cells, or other neurons), the endocrine system as a whole sends messages to cells in virtually any part of the body. The nervous system causes muscles to contract and glands to secrete, and the endocrine system brings about changes in the metabolic activities of body tissues. Neurons act within milliseconds; hormones can take up to several hours or more to bring about their responses. Finally, the effects of nervous system stimulation are generally brief compared to the effects of endocrine stimulation.

A comparison between the nervous and endocrine regulation of homeostasis is presented in Exhibit 13-1.

Obviously, the body could not function if these systems were to pull in opposite directions. The nervous and endocrine systems coordinate their activities like an interlocking supersystem. Certain parts of the nervous system stimulate or inhibit the release of hormones, and hormones, in turn, may stimulate or inhibit the flow of nerve impulses.

Although the ***effects of hormones*** are many and varied, their actions can be categorized into five broad areas:

1. Hormones regulate the chemical composition and volume of the internal environment.
2. Hormones help regulate organic metabolism and energy balance.
3. Hormones help the body cope with emergency environmental demands such as infection, trauma, emotional stress, dehydration, starvation, hemorrhage, and temperature extremes.
4. Hormones assume a role in the coordinated, sequential integration of growth and development.
5. Hormones contribute to the basic processes of reproduction, including gamete (egg and sperm) production, fertilization, nourishment of the embryo and fetus, delivery, and nourishment of the newborn.

The science concerned with the structure and functions of the endocrine glands and the diagnosis and treatment of disorders of the endocrine system is called ***endocrinology*** (en'-dō-kri-NOL-ō-jē; *endo* = within; *crin* = to secrete; *logos* = study of).

EXHIBIT 13-1
Summary of Comparison between Nervous and Endocrine Regulation of Homeostasis

	Nervous System	**Endocrine System**
Mechanism of Control	Electrical impulses.	Hormones.
Cells Affected	Muscle fibers, gland cells, other neurons.	Virtually all body cells.
Type of Action That Results	Muscular contraction or glandular secretion.	Changes in metabolic activities.
Onset of Action	Within milliseconds.	Up to several hours or more.
Duration of Action	Generally brief.	Generally longer.

ENDOCRINE GLANDS

The endocrine glands make up the *endocrine system*. The body contains two kinds of glands, exocrine and endocrine. *Exocrine glands* secrete their products onto a free surface or into ducts. The ducts carry the secretions into body cavities, into the lumina of various organs, or to the body's surface. Exocrine glands include sudoriferous (sweat), sebaceous (oil), mucous, and digestive glands. *Endocrine glands*, by contrast, secrete their products (hormones) into the extracellular space around the secretory cells, rather than into ducts. The secretion then passes into capillaries to be transported in the blood. The endocrine glands include the pituitary, thyroid, parathyroids, adrenals, pineal, and thymus. In addition, several organs of the body contain endocrine tissue but are not exclusively endocrine glands. These include the hypothalamus, pancreas, ovaries, testes, kidneys, stomach, small intestine, and placenta. The locations of many organs of the endocrine system and endocrine-containing organs are illustrated in Figure 13-1.

CHEMISTRY OF HORMONES

Chemically, hormones are grouped into three classes: (1) amines, (2) proteins and peptides, and (3) steroids.

1. **Amines**. These are the simplest hormone molecules. They are produced from an amino acid and are water-soluble.
2. **Proteins and peptides**. These hormones consist of chains of amino acids. Protein and peptide hormones are also water-soluble.
3. **Steroids**. These hormones are derived from cholesterol and are lipid-soluble.

Exhibit 13-2 contains a summary of the classes of hormones, examples of each, and sites of production.

The one function all hormones have in common is maintaining homeostasis by changing the physiological activities of cells.

MECHANISM OF HORMONAL ACTION

OVERVIEW

The amount of hormone released by an endocrine gland or tissue is determined by the body's *need* for the hormone at any given time (Figure 13-2). This is the basis on which the endocrine system operates. Hormone-producing cells are sent information from sensing and signaling systems that permit them to regulate the amount and duration of hormone release. Various characteristics of the internal environment, for example, the blood levels of glucose, Na^+, K^+, and O_2, are detected by these sensors and the information is received by the endocrine cells responsible for regulating that specific characteristic or substance. Feedback systems (described later) attempt to regulate endocrine cell activity to maintain normal hormone production so that there is no overproduction or underproduction of a hormone.

Once a hormone is released by a secretory cell, it is absorbed into the blood. Some hormones, such as steroids, may be carried

EXHIBIT 13-2
Classes of Hormones, Examples, and Sites of Production

Class	Example	Where Produced
Amines	Thyroxine (T_4) and triiodothyronine (T_3).	Thyroid gland.
	Epinephrine and norepinephrine (NE) (catecholamines)	Medulla (inner zone) of adrenal gland.
Proteins and Pepties	Oxytocin (OT).	Hypothalamus.
	Insulin.	Pancreas.
	Anterior pituitary hormones (such as human growth hormone and thyroid-stimulating hormone)	Anterior pituitary.
	Calcitonin (CT).	Thyroid gland.
	Parathyroid hormone (PTH).	Parathyroid glands.
Steroids	Aldosterone, cortisol, and androgens.	Cortex (outer zone) of adrenal gland.
	Testosterone.	Testes.
	Estrogens and progesterone.	Ovaries.

in the blood attached to plasma proteins. Other hormones, such as catecholamines, proteins, and peptide hormones, may be carried in free form.

Although a given hormone travels throughout the body in the blood, it will only affect some specific cells called *target cells*. Most body cells have *receptors* that bind to one or more hormones; that is, they are target cells for only those hormones. Once a hormone binds to a target cell, it causes that cell to respond in a manner that will maintain some substance or characteristic of the internal environment in homeostasis, after which the hormone is metabolized by the target cell and excreted by the liver or kidneys.

RECEPTORS

The body's cells are constantly exposed to equal concentrations of approximately 50 different hormones and yet they respond to only selected ones. The reasons for their selective *response* are due to differences in the solubility of hormones (as described above) and whether the hormone binds with a plasma membrane receptor or an intracellular receptor (described shortly). Although there are thousands of receptors per target cell, only specific receptors will recognize a given hormone and allow it to attach to that receptor. Therefore, each hormone will influence only its target cells but not other cells in the body.

Although a given hormone will only bind with specific cells, different types of cells can possess receptors for the same hormone and their responses will be very different from one another. Once a hormone binds to a cell's receptors, the combination starts a chain of events within the target cell to produce the effects desired by that hormone, for example, increasing blood level of calcium (see Figure 6-4). If blood calcium is getting too low, parathyroid

FIGURE 13-1 Location of many endocrine glands and organs containing endocrine tissue.

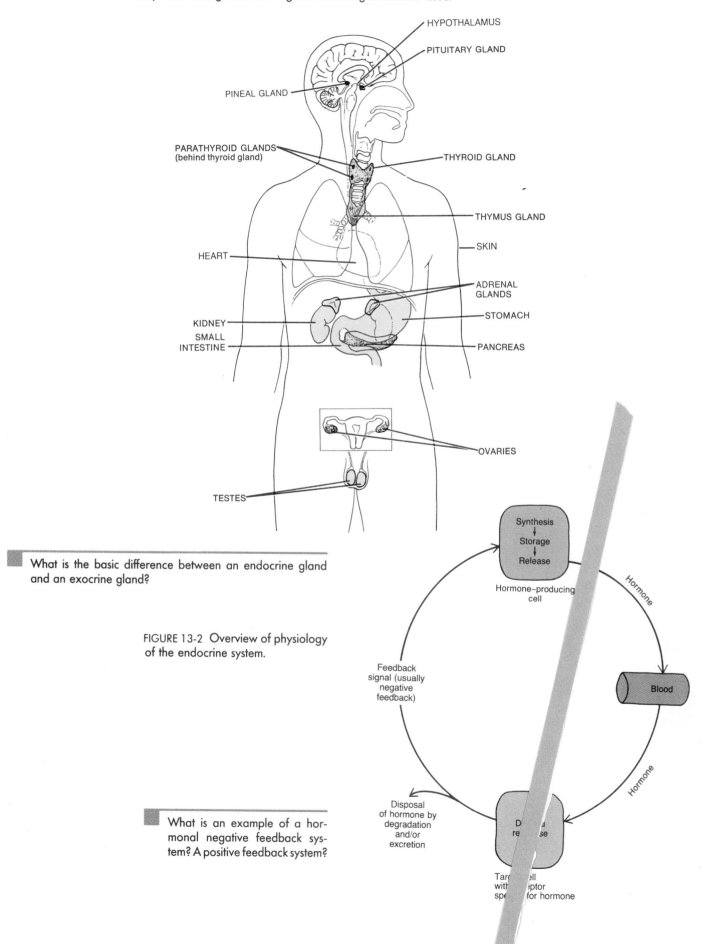

HYPOTHALAMUS

PITUITARY GLAND

PINEAL GLAND

PARATHYROID GLANDS
(behind thyroid gland)

THYROID GLAND

THYMUS GLAND

SKIN

HEART

ADRENAL
GLANDS

KIDNEY

STOMACH

SMALL
INTESTINE

PANCREAS

OVARIES

TESTES

Synthesis
↓
Storage
↓
Release

Hormone–producing
cell

Hormone

Blood

Feedback
signal (usually
negative
feedback)

Hormone

Disposal
of hormone by
degradation
and/or
excretion

Target cell
with receptor
specific for hormone

■ What is the basic difference between an endocrine gland and an exocrine gland?

FIGURE 13-2 Overview of physiology of the endocrine system.

■ What is an example of a hormonal negative feedback system? A positive feedback system?

hormone (PTH) will cause bone cells to release calcium from bones, the cells of the gastrointestinal tract to absorb more calcium, and the cells of the kidney tubules to reabsorb calcium from fluid that will become urine. Thus, one hormone may interact with various types of cells, which will respond in their own particular manner to achieve an overall homeostatic goal.

Generally, hormones cause target cells to alter their *rate* of function. The specific way in which hormones produce this effect depends on whether they interact with (1) plasma membrane receptors or (2) intracellular receptors.

INTERACTION WITH PLASMA MEMBRANE RECEPTORS

Most amine, protein, and peptide hormones use plasma membrane receptors (integral proteins) because they are not lipid-soluble; they are water-soluble and, consequently, cannot penetrate the phospholipid bilayer of the plasma membrane. When water-soluble hormones are released from an endocrine gland, they circulate in the blood, reach a target cell, and bring a specific message to that cell. The hormone is called the *first messenger*. However, since they can only deliver their messages up to the plasma membrane, they must rely on molecules called *second messengers* to take the message inside the cell where hormonal responses take place.

The best known of the second messengers is *cyclic AMP* (Figure 13-3). When a hormone attaches to a plasma membrane receptor, the synthesis of cyclic AMP increases. Cyclic AMP is synthesized from ATP, the main energy-storing chemical in cells, in a process that requires an enzyme, *adenylate cyclase*, on the inner surface of the plasma membrane. When the first messenger (hormone) attaches to its receptor, the adenylate cyclase is activated in the plasma membrane. The adenylate cyclase then converts ATP into cyclic AMP in the cytoplasm of the cell (Figure 13-4a) so cyclic AMP can act as the second messenger and, ultimately, cell function will be altered according to the hormone's message.

FIGURE 13-3 Structure of cyclic AMP.

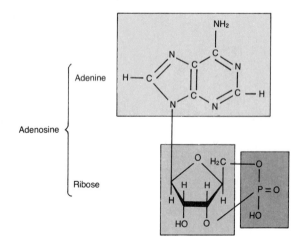

Why is cyclic AMP called a "second messenger"?

Cyclic AMP does not *directly* produce the given physiological response indicated by the hormone. Instead, cyclic AMP activates one or more enzymes called *protein kinases*. They can add a phosphate group from ATP to a protein. As a result, the protein is altered and it catalyzes the physiological response. Such responses include regulating enzymes, inducing secretion, activating protein synthesis, and altering plasma membrane permeability. Cyclic AMP is rapidly degraded by an enzyme called *phosphodiesterase*, which means, for certain hormones, to exert their effects, increased synthesis is required.

Other substances that serve as second messengers are calcium (Ca^{2+}) ions; cyclic guanosine monophosphate (cyclic GMP), a nucleotide that functions like cyclic AMP but activates different enzymes, and probably prostaglandins.

INTERACTION WITH INTRACELLULAR RECEPTORS

Steroid hormones and thyroid hormones alter cell function by activating genes. After separating from their carrier proteins, these hormones pass through the plasma membrane (since they are lipid-soluble) and through the cytoplasm of the cell to bind with intracellular receptors, primarily receptors in the nucleus. After the binding, the receptor undergoes a transformation and activates certain genes of the nuclear DNA to form proteins, usually enzymes, that, in turn, catalyze the effect indicated by the hormone (Figure 13-4b).

PROSTAGLANDINS (PGs) AND HORMONES

Prostaglandins (pros'-ta-GLAN-dins) or *PGs* are lipids that mimic hormones. They are secreted into blood in minute quantities and are potent in their action. Prostaglandins are also called *local hormones* because their site of action is the immediate area in which they are produced. This differentiates them from *circulating hormones*, which act on distant targets. In addition, prostaglandins are synthesized not by specialized endocrine tissues, as are circulating hormones, but by nearly every body cell. Chemical and mechanical stimuli lead to prostaglandin release.

Prostaglandins are believed to be the regulators of cell metabolism by increasing or decreasing cyclic AMP formation. In this way, prostaglandins can alter the responses of cells to a hormone whose action involves cyclic AMP. Prostaglandins are rapidly inactivated, especially in the lungs, liver, and kidneys.

Prostaglandins are important in the normal physiology of smooth muscle (such as uterine contraction), secretion, blood flow, reproduction, platelet function, respiration, nerve impulse transmission, fat metabolism, and the immune response. Prostaglandins are also involved in certain pathologies; they help induce inflammation, promote fever, and intensify pain. Drugs such as aspirin and acetaminophen (Tylenol) inhibit prostaglandin synthesis and thus reduce fever and decrease pain.

CONTROL OF HORMONAL SECRETIONS: FEEDBACK CONTROL

The amount of hormone released by an endocrine gland or tissue is determined by the body's need for the hormone at any given

FIGURE 13-4 **Proposed mechanisms of hormonal action. (a) Interaction with plasma membrane receptors in which there is an increase in the synthesis of cyclic AMP. (b) Activation of genes by a steroid or thyroid hormone.**

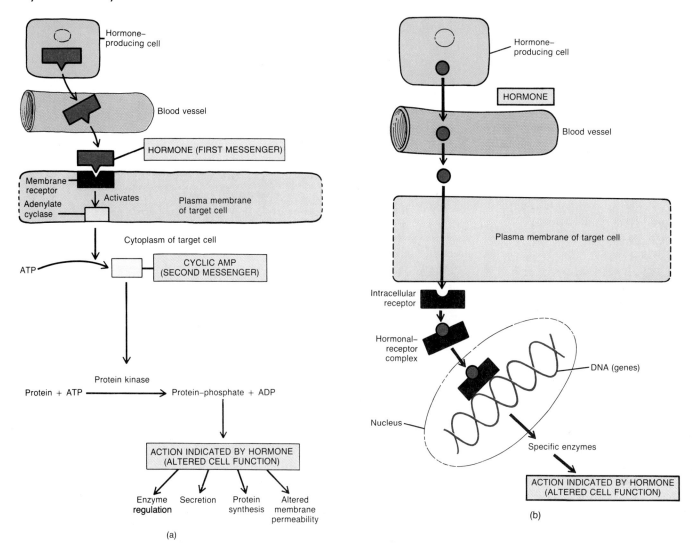

(a)

(b)

Which mechanism is used by water-soluble hormones? Lipid-soluble hormones? Why?

time. Most hormones are released in short bursts unless repeatedly stimulated; in which case, blood levels of the hormone increase. On the other hand, if there is no stimulation, bursts are minimal or inhibited, and blood levels of the hormone decrease. Secretion is normally regulated so that there is no overproduction or underproduction of a particular hormone. This regulation is one of the very important ways the body attempts to maintain homeostasis. If the regulating mechanism does not operate properly and hormonal levels are excessive or deficient, disorders result, several of which are discussed at the end of the chapter.

Hormonal secretions are most often regulated by *negative feedback control* (see Figures 13-2 and 1-10). Information regarding the hormone level or its effect is fed back to the gland, which then responds accordingly. Here we will describe the three negative feedback systems for controlling hormonal secretion.

In the first type of negative feedback system, control of the

hormone is triggered by levels of certain substances in blood. For example, blood calcium level is controlled by parathyroid hormone (PTH), produced by the parathyroid glands, and calcitonin (CT), produced by the thyroid gland. If blood calcium level is low, this serves as a stimulus for the parathyroids to release more PTH (see Figure 13-13). PTH then exerts its effects in various parts of the body until the blood calcium level is raised to normal. A high blood calcium level serves as a stimulus for the parathyroids to cease their production of PTH. However, the thyroid gland increases its production of CT to lower blood calcium level to normal. Note that in negative feedback control the body's response (increased or decreased calcium level) is opposite (negative) to the stimulus (low or high calcium level).

Other hormones regulated according to blood levels of certain chemicals include insulin, produced by the pancreas and controlled by blood levels of glucose; and aldosterone, produced by the adrenals and controlled by blood volume and blood levels of potassium.

In the second type of negative feedback system, the hormone is released as a direct result of nerve impulses that stimulate the endocrine gland. Epinephrine and norepinephrine (NE) are released from the adrenals in response to sympathetic nerve impulses during stress. Antidiuretic hormone (ADH) is released from the posterior pituitary in response to nerve impulses from the hypothalamus (see Figure 13-10).

In the third type of negative feedback system, the hormone is controlled through chemical secretions from the hypothalamus called *regulating hormones* (or *factors*). If the structure of the secretion is known, it is called a *regulating hormone;* if its structure is unknown, it is referred to as a *regulating factor* (see Figure 13-7). These hypothalamic secretions that stimulate the release of the hormone into the blood are called *releasing hormones* (or *factors*). Those that prevent the release of the hormone are called *inhibiting hormones* (or *factors*).

One of the few exceptions to the rule of negative feedback control is oxytocin (OT). The regulating system for the release of OT from the pituitary gland is a positive feedback cycle; that is, the output intensifies the input (see Figure 13-9). Another exception is the luteinizing hormone (LH) surge that results in ovulation (see Chapter 23).

As we discuss the effects of various hormones in this chapter, we will also describe how the secretions are controlled. At that time you will be able to see which type of negative feedback system is operating.

PITUITARY

The hormones of the *pituitary gland* regulate so many body activities that the pituitary has been nicknamed the "master gland." It is a small, round structure that is attached to the hypothalamus of the brain by a stalklike structure, the *infundibulum* (see Figure 13-5). It is divided into an anterior lobe and a posterior lobe. Both are connected to the hypothalamus. The *anterior lobe* constitutes about 75 percent of the total weight of the gland and forms the glandular part of the pituitary. Blood vessels connect the anterior lobe with the hypothalamus. The *posterior lobe* contains axon terminations of neurons whose cell bodies are in the hypothalamus. Nerve fibers connect the posterior lobe directly with the hypothalamus.

ANTERIOR PITUITARY

The anterior lobe of the pituitary releases hormones that regulate a whole range of body activities from growth to reproduction. The release of these hormones is controlled by *regulating hormones* (or *factors*) from the hypothalamus.

These hormones (or factors) are delivered to the anterior lobe directly from the hypothalamus through a system of blood vessels without first circulating through the heart (Figure 13-5). The short route allows the regulating hormones (or factors) to act quickly on the anterior lobe and prevents their dilution or destruction.

The anterior pituitary secretes seven hormones (Figure 13-6):

1. **Human growth hormone (hGH),** which controls general body growth

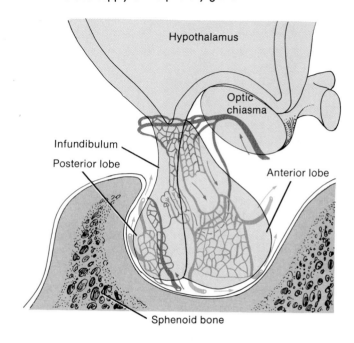

FIGURE 13-5 **Blood supply of the pituitary gland.**

Hypothalamus

Optic chiasma

Infundibulum

Posterior lobe

Anterior lobe

Sphenoid bone

How do the anterior and posterior lobes differ with respect to how they release hormones?

2. **Prolactin (PRL),** which initiates milk production by the mammary glands
3. **Adrenocorticotropic hormone (ACTH),** which stimulates the adrenal cortex to secrete its hormones
4. **Melanocyte-stimulating hormone (MSH),** which is related to skin pigmentation
5. **Thyroid-stimulating hormone (TSH),** which controls the thyroid gland
6. **Follicle-stimulating hormone (FSH),** which stimulates the production of eggs and sperm in the ovaries and testes, respectively
7. **Luteinizing hormone (LH),** which stimulates other sexual and reproductive activities

Human Growth Hormone (hGH)

Human growth hormone (hGH or GH) causes body cells to grow. It acts on the skeleton and skeletal muscles, in particular, to increase their rate of growth and maintain their size once growth is attained. hGH causes cells to grow and multiply by increasing the rate at which amino acids enter cells and are built up into proteins. hGH also promotes fat catabolism and plays a role in the efficient use and conversion of glucose. hGH does not act directly. Instead, it stimulates the liver to synthesize and secrete small proteins called *somatomedins* (sō′-ma-tō-MĒ-dins) which mediate most of its effects.

The release of hGH from the anterior pituitary is apparently controlled by at least two regulating hormones, *growth hormone releasing hormone (GHRH)* and *growth hormone inhibiting hormone (GHIH)*. GHRH is released by the hypothalamus into the bloodstream; it circulates to the anterior pituitary and stimulates the release of hGH. On the other hand, GHIH inhibits the release of hGH.

FIGURE 13-6 **Hormones produced by the anterior pituitary gland and their general functions.**

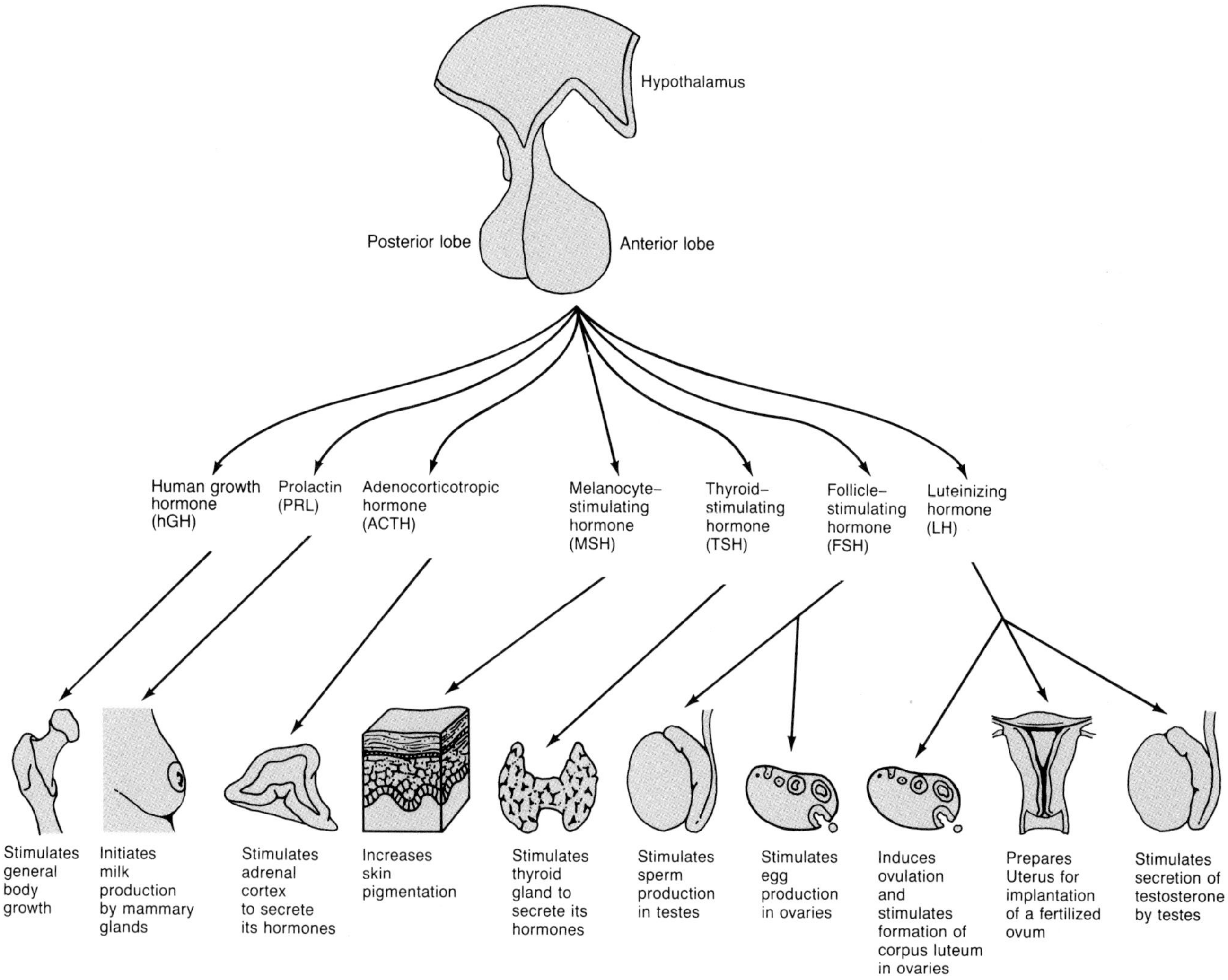

Hypothalamus

Posterior lobe   Anterior lobe

| Human growth hormone (hGH) | Prolactin (PRL) | Adenocorticotropic hormone (ACTH) | Melanocyte-stimulating hormone (MSH) | Thyroid-stimulating hormone (TSH) | Follicle-stimulating hormone (FSH) | Luteinizing hormone (LH) |

| Stimulates general body growth | Initiates milk production by mammary glands | Stimulates adrenal cortex to secrete its hormones | Increases skin pigmentation | Stimulates thyroid gland to secrete its hormones | Stimulates sperm production in testes | Stimulates egg production in ovaries | Induces ovulation and stimulates formation of corpus luteum in ovaries | Prepares Uterus for implantation of a fertilized ovum | Stimulates secretion of testosterone by testes |

Why is the anterior pituitary called the "master gland" of the body?

Among the stimuli that promote hGH secretion is *hypoglycemia*, that is, low blood sugar level. When blood sugar level is low, the hypothalamus is stimulated to secrete GHRH, which causes the release of hGH. Somatomedins, under the influence of hGH, raise blood sugar level by converting glycogen into glucose and releasing it into the blood. As soon as blood sugar level returns to normal, GHRH secretion shuts off (Figure 13-7). This is a negative feedback system.

The secretion of hGH peaks at the end of the adolescent growth spurt. Genetic differences control when this will occur and thus account for differences in height.

Thyroid-Stimulating Hormone (TSH)

Thyroid-stimulating hormone (TSH) stimulates the production and secretion of hormones from the thyroid gland. Secretion is controlled by the hypothalamic regulating hormone, *thyrotropin releasing hormone (TRH).* Release of TRH depends on blood levels of

thyroxine and the body's metabolic rate and operates according to a negative feedback system.

Adrenocorticotropic Hormone (ACTH)

Adrenocorticotropic hormone (ACTH) controls the production and secretion of certain adrenal cortex hormones. Secretion is controlled by the hypothalamic regulating hormone, *corticotropin releasing hormone (CRH).* Release of CRH depends on a number of stimuli and hormones and operates as a negative feedback system.

Follicle-Stimulating Hormone (FSH)

In the female, *follicle-stimulating hormone (FSH)* is transported from the anterior pituitary by the blood to the ovaries, where it initiates the development of ova each month. FSH also stimulates cells in the ovaries to secrete estrogens, or female sex hormones. In the male, FSH stimulates the testes to produce testosterone, a

FIGURE 13-7 Regulation of the secretion of human growth hormone (hGH). Like other hormones of the adenohypophysis, the secretion of hGH is controlled by regulating hormones. Like most hormones of the body, hGH secretion and inhibition involve negative feedback systems.

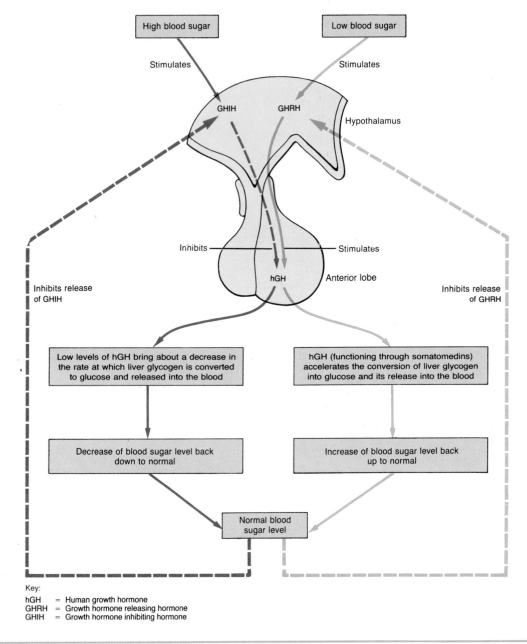

Key:
hGH   = Human growth hormone
GHRH = Growth hormone releasing hormone
GHIH = Growth hormone inhibiting hormone

Explain why regulation of human growth hormone (hGH) represents a negative feedback cycle.

male sex hormone. Secretion of FSH is controlled by the hypothalamic regulating hormone called **gonadotropin releasing hormone (GnRH).** High levels of estrogens and testosterone inhibit the release of GnRH, and thus FSH, by a negative feedback system.

Luteinizing Hormone (LH)

In the female, **luteinizing** (LOO-tē-in′-īz-ing) **hormone (LH)**, to-

gether with FSH, stimulates ovulation, the release of a secondary oocyte (future ovum) by the ovary. LH also stimulates formation of the corpus luteum in the ovary, which secretes progesterone (another female sex hormone). Estrogens and progesterone prepare the uterus for implantation of a fertilized ovum and prepare the mammary glands for milk secretion. In the male, LH stimulates the testes to develop and secrete large amounts of testosterone. Secretion of LH, like that of FSH, is controlled by GnRH.

Prolactin (PRL)

Prolactin (PRL), together with other female hormones, initiates milk secretion by the mammary glands. The development of the mammary glands during puberty and pregnancy is under the control of estrogens and progesterone. The secretion of these glands is directly affected by PRL levels. The mammary glands must be primed by female sex hormones.

PRL has both an inhibitory and an excitatory negative control system. During menstrual cycles, *prolactin inhibiting factor (PIF)*, a regulating factor from the hypothalamus, inhibits the release of PRL. As levels of estrogens and progesterone fall during the late phase of the menstrual cycle, secretion of PIF diminishes and the blood level of PRL rises. However, its rising level does not last long enough to have much effect on the breasts, which may be tender because of the presence of PRL just before menstruation. As the menstrual cycle starts up again and the level of estrogens again rises, PIF is again secreted and the PRL level drops.

PRL levels rise during pregnancy. Apparently a regulating factor from the hypothalamus, called *prolactin releasing factor (PRF)*, stimulates PRL secretion after long periods of inhibition. PRL levels fall after delivery and rise again during breast-feeding. A nursing infant causes a reduction in the hypothalamic secretion of PIF.

Melanocyte-Stimulating Hormone (MSH)

The exact role of *melanocyte-stimulating hormone (MSH)* in humans is unknown, but administration of MSH for several days produces a darkening of the skin. In the absence of the hormone, the skin may be pallid. Secretion of MSH is stimulated by a hypothalamic regulating factor called *melanocyte-stimulating-hormone releasing factor (MRF)*. It is inhibited by a *melanocyte-stimulating-hormone inhibiting factor (MIF)*.

POSTERIOR PITUITARY

In a strict sense, the posterior pituitary is not an endocrine gland since it does not *make* hormones. Instead, it *stores* hormones. The posterior lobe contains axon terminals of secretory neurons of the hypothalamus called *neurosecretory cells* (Figure 13-8). The cell bodies, in the hypothalamus, produce two hormones, *oxytocin (OT)* and *antidiuretic hormone (ADH)*, which are transported through the axons to the posterior lobe for storage.

Oxytocin (OT)

Oxytocin (ok′-sē-TŌ-sin) or *OT* stimulates the contraction of the smooth muscle cells in the pregnant uterus and the contractile cells of the mammary glands. It is released in large quantities just prior to giving birth (Figure 13-9). When labor begins, the cervix of the uterus is distended. This distention signals the neurosecretory cells in the hypothalamus that stimulate the synthesis of OT. The OT is transported to the posterior lobe and from there, released into the blood and carried to the uterus to reinforce uterine contractions. As the contractions become more forceful, more OT is synthesized. As the baby's head passes through the uterine cervix, further distention of the cervix triggers the release of still more

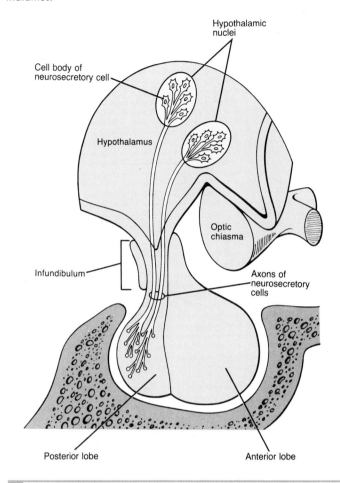

FIGURE 13-8 **The posterior pituitary lobe's connection to the hypothalamus.**

Which hormones are produced in the hypothalamus and stored in the posterior lobe?

OT. Thus, a positive feedback cycle is established. The cycle is broken by the birth of the infant. OT is used clinically to induce labor (the trade name is Syntocinon or Pitocin). In addition, Pitocin is used in the immediate postpartum period to increase uterine tone and control hemorrhage.

OT affects milk ejection. Milk formed by the glandular cells of the breasts is stored until the baby begins active sucking. This initiates a mechanism similar to that involved in forming and releasing OT for uterine muscle contractions. OT is transported from the posterior lobe via the blood to the mammary glands where it stimulates smooth muscle cells around the glandular cells and ducts to contract and eject milk. This response is called the *milk letdown reflex*.

Antidiuretic Hormone (ADH)

An *antidiuretic* is any chemical substance that prevents excessive urine production. The principal physiological activity of *antidiuretic hormone (ADH)* is its effect on urine volume. ADH causes the kidneys to remove water from fluid that will become urine and

FIGURE 13-9 **Regulation of the secretion of oxytocin (OT) during labor.**

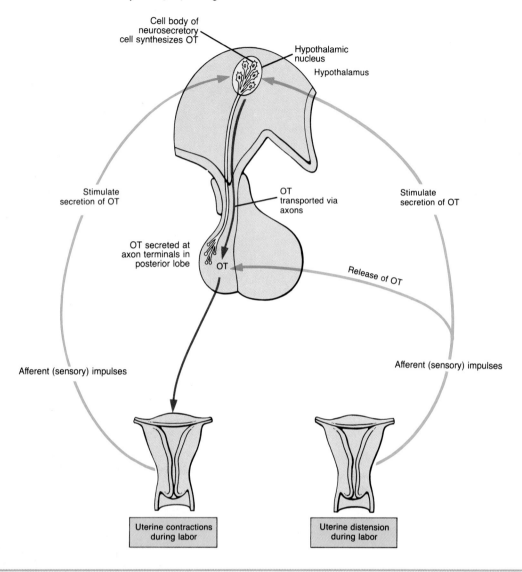

Cell body of
neurosecretory
cell synthesizes OT

Hypothalamic
nucleus

Hypothalamus

Stimulate
secretion of OT

OT
transported via
axons

Stimulate
secretion of OT

OT secreted at
axon terminals in
posterior lobe

OT

Release of OT

Afferent (sensory) impulses

Afferent (sensory) impulses

Uterine contractions
during labor

Uterine distension
during labor

Explain why the regulation of oxytocin (OT) represents a positive feedback cycle.

return it to the bloodstream, thus decreasing urine volume (antidiuresis). In the absence of ADH, urine output may be increased 10-fold.

ADH can also raise blood pressure by bringing about constriction of arterioles. For this reason, ADH is also referred to as *vasopressin*. If there is a severe loss of blood volume due to hemorrhage, ADH output increases.

The amount of ADH normally secreted varies with the body's needs (Figure 13-10). When the body is dehydrated, receptors in the hypothalamus called *osmoreceptors* detect the low water concentration in the blood and stimulate the neurosecretory cells in the hypothalamus to synthesize ADH, which is then transported to the posterior lobe, released into the bloodstream, and transported to the kidneys. The kidneys respond by reabsorbing water back into the blood and decreasing urine output. ADH also decreases the rate at which perspiration is produced during dehydration. By contrast, if the blood contains a higher than normal water concentra-

tion, the reverse process takes place and the concentration of water is returned to normal.

Secretion of ADH can also be altered by a number of other conditions. Pain, stress, trauma, anxiety, acetylcholine (ACh), nicotine, high blood levels of sodium (Na^+) ions, and drugs such as morphine, tranquilizers, and some anesthetics stimulate secretion of the hormone. Alcohol inhibits secretion and thereby increases urine output. This may be why thirst is one symptom of a hangover.

THYROID

The *thyroid gland* is located just below the larynx and in front of the trachea. It consists of two lobes connected by a mass of tissue called an *isthmus* (IS-mus) (Figure 13-11). It has a rich blood supply and, thus, can deliver high levels of hormone in a short period of time if necessary.

FIGURE 13-10 **Regulation of the secretion of antidiuretic hormone (ADH).**

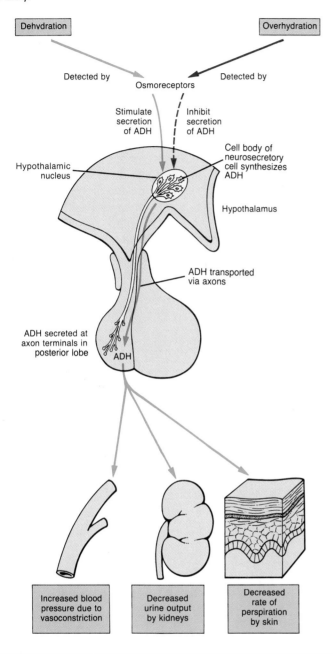

List several other stimuli that might promote secretion of antidiuretic hormone (ADH).

FIGURE 13-11 **Thyroid gland. (a) Location of the thyroid gland in anterior view. (b) Histology of the thyroid gland showing a single thyroid follicle.**

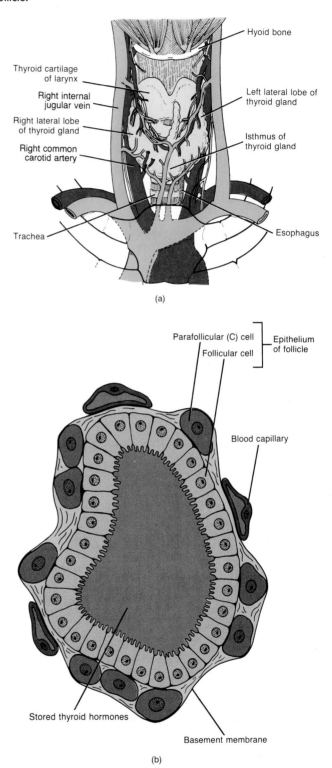

(a)

(b)

What hormones are produced by parafollicular cells? By follicular cells?

The thyroid gland is filled with microscopic spherical sacs called *thyroid follicles,* which consist of *follicular cells* and *parafollicular (C) cells* (Figure 13-11b). The follicular cells manufacture *thyroxine* (thi-ROK-sēn) or T_4, since it contains four atoms of iodine, and *triiodothyronine* (trī-ī′-ōd-ō-THĪ-rō-nēn) or T_3, since it contains three atoms of iodine. Together, these hormones are referred to as the *thyroid hormones*. They both contain iodine and both have the same effect. The parafollicular cells produce *calcitonin* (kal-si-TŌ-nin) or *CT*.

FUNCTION AND CONTROL OF THYROID HORMONES

The thyroid hormones function to (1) regulate metabolism, (2) regulate growth and development, and (3) regulate the activity of the nervous system. Thyroid hormones regulate metabolism by stimulating virtually all aspects of carbohydrate and lipid breakdown in most cells of the body and increasing the rate of protein synthesis.

Thyroid hormones help regulate tissue growth and development, especially in children. With human growth hormone (hGH), they accelerate body growth, particularly the growth of nervous tissue. Deficiency of thyroid hormones during fetal development results in fewer and smaller neurons, defective myelination of axons, and mental retardation. During the early years of life, thyroid hormone deficiency results in small stature and poor development of certain organs.

Finally, thyroid hormones increase the reactivity of the nervous system. This results in increased blood flow, increased and more forceful heartbeats, increased blood pressure, increased motility of the gastrointestinal tract, and increased nervousness.

The secretion of thyroid hormones is stimulated by several factors (Figure 13-12). If thyroid hormone levels in the blood fall below normal or the metabolic rate decreases, chemical sensors in the hypothalamus detect the change in blood chemistry and stimulate the hypothalamus to secrete a regulating hormone called thyrotropin releasing hormone (TRH). TRH stimulates the anterior pituitary to secrete thyroid-stimulating hormone (TSH). Then TSH stimulates the thyroid to release thyroid hormones until the metabolic rate returns to normal. Conditions that increase the body's need for energy—a cold environment, high altitude, pregnancy— also trigger this negative feedback system and increase the secretions of thyroid hormones.

Thyroid activity can be inhibited by a number of other factors. When large amounts of certain sex hormones (estrogens and androgens) are circulating in the blood, for example, TSH secretion diminishes. Aging slows down the activities of most glands, and thyroid production may decrease. This is one of the factors that often accounts for weight gain as people age.

FIGURE 13-12 **Regulation of the secretion of thyroid hormones.**

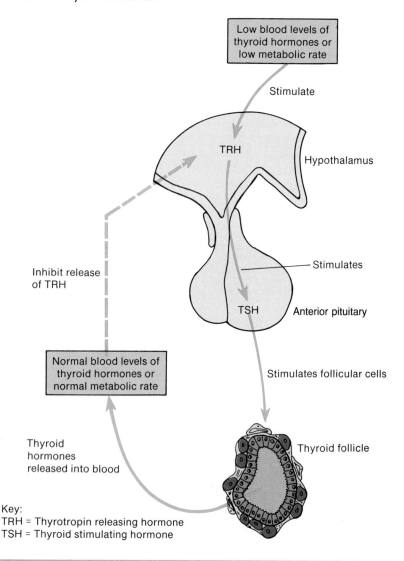

Low blood levels of thyroid hormones or low metabolic rate

Stimulate

TRH

Hypothalamus

Inhibit release of TRH

Stimulates

TSH   Anterior pituitary

Normal blood levels of thyroid hormones or normal metabolic rate

Stimulates follicular cells

Thyroid hormones released into blood

Thyroid follicle

Key:
TRH = Thyrotropin releasing hormone
TSH = Thyroid stimulating hormone

What are the three main functions of thyroid hormones?

CALCITONIN (CT)

The hormone produced by the parafollicular cells of the thyroid gland is *calcitonin* (kal-si-TŌ-nin) or *CT*. It is involved in the homeostasis of blood calcium (Ca^{2+}) and phosphate (HPO_4^{2-}) levels. CT lowers the amount of calcium and phosphate in the blood by (1) inhibiting bone breakdown (specifically, by inhibiting osteoclasts—bone-destroying cells), (2) accelerating the uptake of calcium and phosphate by the bones, (3) inhibiting parathyroid hormone (PTH) (to be explained shortly), and (4) increasing urinary excretion of calcium and phosphate. CT is used to treat postmenopausal osteoporosis, in conjunction with adequate calcium and vitamin D intake.

Blood calcium level directly controls the secretion of CT by a negative feedback system (Figure 13-13).

PARATHYROIDS

Attached to the posterior surfaces of the thyroid gland are small, round masses of tissue called the *parathyroid glands*. Usually, two parathyroids, superior and inferior, are attached to each thyroid lobe (Figure 13-14a). There may be as few as two or as many as six parathyroid glands.

Microscopically, the parathyroids contain two kinds of epithelial cells (Figure 13-14b). The more numerous cells, called *principal (chief) cells,* are believed to be the major producer of *parathyroid hormone (PTH).* The other kind of cell, called an *oxyphil cell,* makes reserve hormone.

PARATHYROID HORMONE (PTH)

Parathyroid hormone (PTH) helps to control the homeostasis of calcium (Ca^{2+}) and phosphate (HPO_4^{2-}) ions in the blood, but in a different way than calcitonin. When adequate vitamin D is present, PTH helps activate vitamin D and increases the rate of calcium, phosphate, and some magnesium absorption from the gastrointestinal tract into the blood. PTH also increases the number and activity of osteoclasts (bone-destroying cells), which causes bone tissue to break down and additional calcium and phosphate to be released into the blood. Finally, PTH increases the rate at which the kidneys remove calcium and magnesium from urine that is being formed and returns it to the blood, and inhibits the transportation of phosphate from urine into blood so that phosphate is excreted in urine. More phosphate is lost through the urine than is gained from the bones.

The overall effect of PTH with respect to ions, then, is to decrease blood phosphate level and increase blood calcium and magnesium levels. As far as blood calcium level is concerned, PTH and CT have opposite functions.

When the calcium level of the blood falls, more PTH is released (see Figure 13-13). Conversely, when the calcium level of the blood rises, less PTH (and more CT) is secreted. This is another example of a negative feedback control system.

ADRENALS

The body has two *adrenal glands,* one of which is located superior to each kidney (Figure 13-15). Each adrenal gland is divided into two regions: the outer *adrenal cortex,* which makes up the bulk of the gland, and the inner *adrenal medulla.* Each region produces different hormones. Like the thyroid, the adrenals are very vascular organs.

ADRENAL CORTEX

The adrenal cortex is subdivided into three zones, and each zone secretes different groups of steroid hormones. The outer zone se-

FIGURE 13-13 **Regulation of the secretion of the parathyroid hormone (PTH) and calcitonin (CT). The figure on the left is a posterior view; the one on the right is an anterior view.**

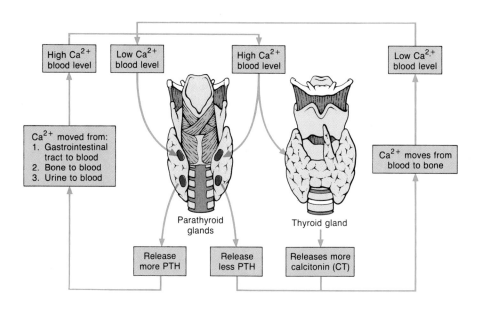

Explain how PTH and CT have opposite effects on blood calcium level.

FIGURE 13-14 Parathyroid glands. (a) Location of the parathyroid glands in posterior view. (b) Histology of the parathyroid glands.

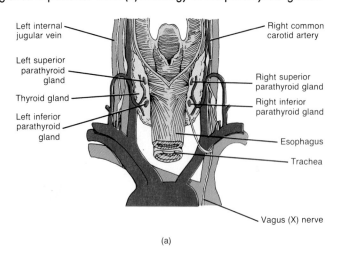

(a)

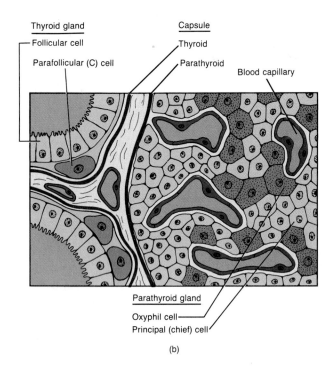

(b)

The parathyroids are attached to which other endocrine gland?

FIGURE 13-15 Location of the adrenal glands.

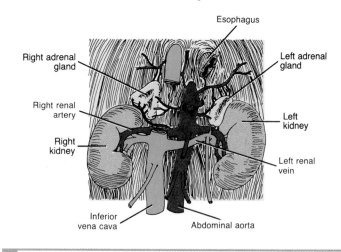

Into which two regions is the adrenal cortex divided?

cretes mineralocorticoids (min′-er-al-ō-KOR-ti-koyds). The middle zone secretes glucocorticoids (gloo′-kō-KOR-ti-koyds). The inner zone synthesizes mainly the sex hormones called gonadocorticoids (gō-na-dō-KOR-ti-koyds) or adrenal sex hormones and, of these, the primary hormones are the male hormones called androgens.

Mineralocorticoids

Mineralocorticoids help control the homeostasis of water, sodium (Na^+), and potassium (K^+). Although the adrenal cortex secretes several mineralocorticoids, the one responsible for about 95 percent

of mineralocorticoid activity is aldosterone (al-do-STĒR-ōn). Aldosterone acts on certain cells in the kidneys to increase their reabsorption of sodium (Na^+) ions from the urine, return them to the blood, and thus prevent rapid depletion of sodium from the body. At the same time, aldosterone stimulates excretion of potassium (K^+) ions, so that large amounts of potassium are lost in the urine.

These two basic functions—conservation of sodium and elimination of potassium—have important secondary effects. First, most sodium reabsorption occurs through an exchange reaction whereby hydrogen (H^+) ions, which are acidic, pass from the blood into the urine to replace sodium ions. This makes the blood less acidic and prevents acidosis, a blood pH lower than 7.35. Also, the increase in sodium in the blood causes water to move by osmosis from the fluid that eventually becomes urine into the blood, which increases blood volume and, thus, raises blood pressure. When ADH is present, even more water is reabsorbed.

The control of aldosterone secretion involves several mechanisms. One of these is the *renin–angiotensin* (an′-jē-ō-TEN-sin) *pathway* (Figure 13-16). A decrease in blood pressure caused by a drop in blood volume due to dehydration, sodium (Na^+) ion deficiency, or hemorrhage stimulates certain kidney cells, called juxtaglomerular cells, to secrete into the blood an enzyme called *renin* (RĒ-nin) (see Figure 21-7). Renin converts *angiotensinogen*, a protein produced by the liver, into *angiotensin I,* which is then converted into *angiotensin II* by an enzyme in the lungs. Angiotensin II stimulates the adrenal cortex to produce more aldosterone. Aldosterone brings about increased Na^+ reabsorption and water follows, thus increasing fluid volume and restoring blood pressure to normal. Angiotensin II is also a powerful vasoconstrictor , which helps to further elevate blood pressure.

A second mechanism for the control of aldosterone is potassium (K^+) ion concentration. An increased K^+ concentration in extracellular fluid directly stimulates aldosterone secretion and causes the kidneys to eliminate excess K^+. A decreased K^+ concentration in extracellular fluid reverses the process.

FIGURE 13-16 **Proposed mechanism for the regulation of the secretion of aldosterone by the renin—angiotensin pathway.**

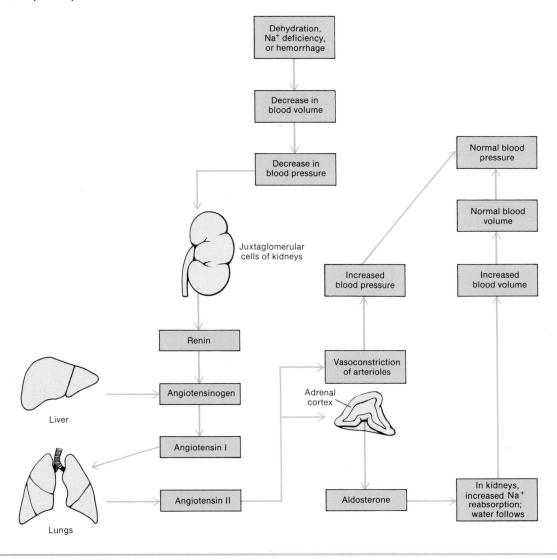

What are the two basic functions of aldosterone?

Glucocorticoids

Glucocorticoids are concerned with metabolism and resistance to stress. Three glucocorticoids are *cortisol (hydrocortisone), corticosterone,* and *cortisone.* Cortisol is the most abundant and is responsible for most glucocorticoid activity. The glucocorticoids have the following effects on the body:

1. Glucocorticoids work with other hormones to promote normal metabolism. Their role is to make sure enough energy is available. They increase the rate at which proteins are catabolized and amino acids are removed from cells and transported to the liver to be synthesized into new proteins, such as the enzymes needed for metabolic reactions. Or, the liver may convert the amino acids to glucose if the body's reserves of glycogen and fat are low. This conversion of a substance other than carbohy-

drate into glucose is called ***gluconeogenesis*** (gloo′-kō-nē′-ō-JEN-e-sis). Glucocorticoids also release fatty acids from adipose tissue as an additional energy source.

2. Glucocorticoids work in many ways to provide resistance to stress. A sudden increase in available glucose by way of gluconeogenesis makes the body more alert and supplies energy for combating a range of stresses from fright and temperature extremes to high altitude and surgery, for example. Glucocorticoids also make the blood vessels more sensitive to vessel-constricting chemicals, thereby raising blood pressure, which is especially advantageous if the stress happens to be blood loss, which causes a drop in blood pressure.

3. Glucocorticoids are antiinflammatory compounds; that is, they reduce the number of cells and inhibit the release of chemicals that cause inflammation. Unfortunately, they also retard connective tissue regeneration and thereby slow wound healing. High

doses of glucocorticoid drugs cause atrophy of the thymus gland, spleen, and lymph nodes, thus depressing the body's ability to fight disease. However, high doses may be useful in the treatment of chronic inflammation.

The control of glucocorticoid secretion is a typical negative feedback mechanism (Figure 13-17). The two principal stimuli are stress and low blood level of glucocorticoids. Either condition stimulates the hypothalamus to secrete a regulating hormone called *corticotropin releasing hormone* (*CRH*)*,* which initiates the release of ACTH from the anterior lobe of the pituitary, which then stimulates glucocorticoid secretion. Low levels of glucocorticoids also influence the anterior lobe directly to release adrenocorticotrophic hormone (ACTH).

Gonadocorticoids

The adrenal cortex secretes both male and female *gonadocorticoids* (*adrenal sex hormones*). These are estrogens and androgens.

The concentration of sex hormones secreted by adult male adrenals is so low as to be insignificant. In females, androgens contribute to sex drive (libido). Androgens help in the prepubertal growth spurt and early development of axillary and pubic hair in boys and girls.

ADRENAL MEDULLA

The adrenal medulla consists of sympathetic ANS postganglionic cells that are specialized to secrete hormones. Hormone secretion is directly controlled by the ANS and the gland responds rapidly to a stimulus.

Epinephrine and Norepinephrine (NE)

The two principal hormones synthesized by the adrenal medulla are *epinephrine* and *norepinephrine* (*NE*), also called adrenaline and noradrenaline, respectively. Epinephrine constitutes about 80 percent of the gland's total secretion and is more potent than norepi-

FIGURE 13-17 Regulation of the secretion of glucocorticoids.

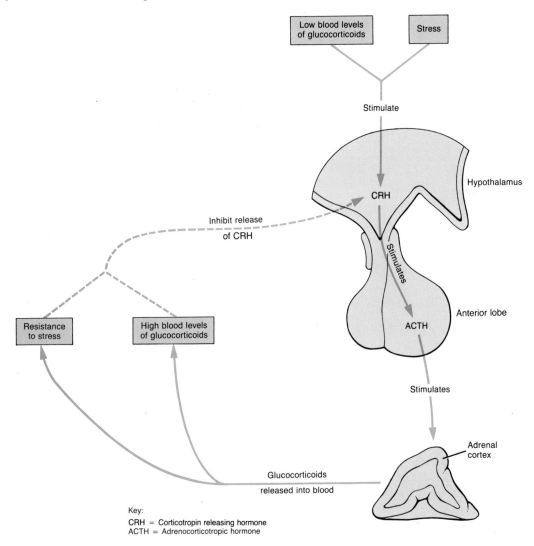

Key:
CRH = Corticotropin releasing hormone
ACTH = Adrenocorticotropic hormone

Name the three glucocorticoids. Which is responsible for most glucocorticoid activity?

nephrine. Both hormones are *sympathomimetic* (sim'-pa-thō-mi-MET-ik); that is, they produce effects that mimic those brought about by the sympathetic division of the autonomic nervous system. To a large extent, they are responsible for the fight-or-flight response. Like the glucocorticoids of the adrenal cortices, they help the body resist stress, but unlike the cortical hormones, they are not essential for life.

Under stress, impulses received by the hypothalamus are conveyed to the adrenal medulla, which increases the output of epinephrine and norepinephrine. Both hormones increase blood pressure by increasing heart rate and constricting blood vessels; accelerate the rate of respiration; dilate respiratory passageways; decrease the rate of digestion; increase the efficiency of muscular contractions; increase blood sugar level; and stimulate cellular metabolism. Hypoglycemia (low blood sugar) may also stimulate secretion of epinephrine and norepinephrine.

PANCREAS

The *pancreas* can be classified as both an endocrine and an exocrine gland. We will treat its endocrine functions here and discuss its exocrine functions with the digestive system (see Chapter 19). The pancreas is a flattened organ located behind and slightly below the stomach (Figure 13-18a).

The endocrine portion of the pancreas consists of clusters of cells called *pancreatic islets* or *islets of Langerhans* (LAHNG-er-hanz) (Figure 13-18b). At least three major kinds of cells are found in these clusters: (1) *alpha cells* that secrete the hormone glucagon; (2) *beta cells* that secrete the hormone insulin, and (3) *delta cells* that secrete growth hormone inhibiting hormone (GHIH) which inhibits secretion of insulin and glucagon. The islets are infiltrated by blood capillaries and surrounded by cells that form the exocrine part of the gland. Glucagon and insulin are the chief regulators of blood sugar level.

GLUCAGON

Glucagon (GLOO-ka-gon) increases the blood glucose level (Figure 13-19). It does this by accelerating the conversion of glycogen in the liver into glucose (glycogenolysis) and the conversion in the liver of other nutrients, such as amino acids, glycerol, and lactic acid, into glucose (gluconeogenesis). The liver then releases the glucose into the blood, and the blood sugar level rises.

Secretion of glucagon is directly controlled by the level of blood sugar by a negative feedback system. When the blood sugar level falls below normal, chemical sensors in the alpha cells of the islets stimulate the cells to secrete glucagon. When blood sugar rises, the cells are no longer stimulated and production slackens.

FIGURE 13-18 Pancreas. (a) Location of the pancreas. (b) Histology of the pancreas. Shown is a single pancreatic islet (islet of Langerhans).

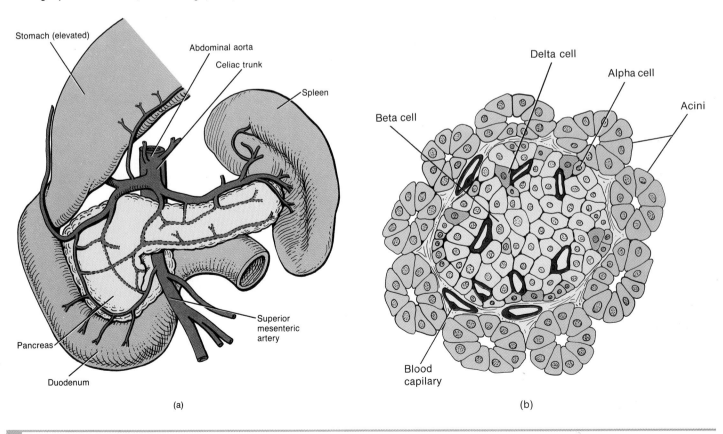

(a)

(b)

What hormone is secreted by alpha cells? Beta cells?

FIGURE 13-19 Regulation of the secretion of glucagon and insulin.

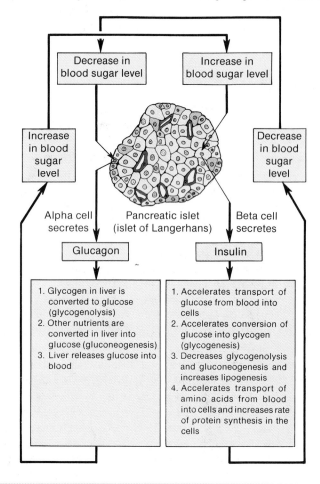

How are the actions of glucagon and insulin opposite?

If for some reason the self-regulating device fails and the alpha cells secrete glucagon continuously, hyperglycemia (high blood sugar level) may result. Exercise and largely or entirely protein meals that raise the amino acid level of the blood also cause an increase in glucagon secretion. Glucagon secretion is inhibited by GHIH and insulin.

INSULIN

Insulin's action is opposite that of glucagon. It decreases blood sugar level. It does this in several ways (Figure 13-19). It accelerates the transport of glucose from blood into cells, especially skeletal muscle fibers; glucose entry into cells depends on the presence of insulin receptors. It accelerates the conversion of glucose into glycogen (glycogenesis) and decreases glycogenolysis and gluconeogenesis. Finally, insulin also stimulates the conversion of glucose or other nutrients into fatty acids (lipogenesis), and helps stimulate protein synthesis.

The regulation of insulin secretion, like that of glucagon secretion, is directly determined by the level of sugar in the blood and is based on a negative feedback system. Insulin secretion is stimulated by several other hormones including ACTH, human growth hormone, and gastrointestinal hormones (to be discussed

in Chapter 19). High blood levels of amino acids also stimulate insulin release, but GHIH and glucagon inhibit its secretion.

OVARIES AND TESTES

The female gonads, called the *ovaries*, are paired oval bodies located in the pelvic cavity. They produce *estrogens* and *progesterone (PROG)*, which are responsible for the development and maintenance of the female sexual characteristics. Along with FSH and LH of the pituitary gland, the sex hormones regulate the menstrual cycle, maintain pregnancy, and prepare the mammary glands for lactation. The ovaries (and placenta) also produce a hormone called *relaxin (RLX)*, which relaxes the symphysis pubis and helps dilate the uterine cervix toward the end of pregnancy. The ovaries also produce *inhibin*, a hormone that inhibits secretion of FSH and that might be important toward the end of the menstrual cycle.

The male has two oval glands, called *testes*, that lie in the scrotum. The testes produce *testosterone*, the primary male sex hormone, which stimulates the development of male sexual characteristics. The testes also produce inhibin that inhibits secretion of FSH. The detailed structure of the ovaries and testes and the specific roles of gonadotropic hormones and sex hormones will be discussed in Chapter 23.

PINEAL

The *pineal* (PĪN-ē-al) *gland* is attached to the roof of the third ventricle, near the thalamus in the brain (see Figure 13-1). The pineal gland starts to accumulate calcium at about the time of puberty; the deposits are referred to as *brain sand*. Contrary to a once widely held belief, there is no evidence that the pineal atrophies with age and that the presence of brain sand indicates atrophy. In fact, the presence of brain sand may even indicate increased secretory activity. Although the anatomy of the pineal gland is well known, its physiology is still obscure. One hormone secreted by the pineal gland is *melatonin*, which appears to inhibit reproductive activities by inhibiting gonadotropic hormones. Melatonin is produced during darkness; its formation is interrupted in light. Light entering the eyes stimulates the retina to transmit impulses to the pineal gland that inhibit melatonin secretion.

THYMUS

The *thymus gland* plays two roles in the body's immune system. The larger role, producing antibodies, will be discussed in Chapter 17. The other role is hormonal and has to do with the maturation of T lymphocytes, a type of white blood cell essential for the development of immunity. The thymic hormones are called *thymosin, thymic humoral factor (THF), thymic factor (TF),* and *thymopoietin*. There is some evidence that thymic hormones may retard aging.

OTHER ENDOCRINE TISSUES

There are body tissues other than those normally classified as endocrine glands that contain endocrine tissue and thus secrete hormones.

The gastrointestinal tract synthesizes several hormones that regulate digestion in the stomach and small intestine, including *stomach gastrin, enteric gastrin, secretin, cholecystokinin (CCK), enterocrinin,* and *gastric inhibitory peptide (GIP).* The actions of these hormones and their control of secretion are summarized in Exhibit 19-1.

The placenta produces *human chorionic gonadotropin (hCG), estrogens, progesterone (PROG), relaxin,* and *human chorionic somatomammotropin (hCS),* all of which are related to pregnancy (see Chapter 24). At its peak of activity, the placenta takes over some of the functions of the pituitary gland.

If the kidneys are not receiving adequate oxygen, they release an enzyme called *renal erythropoietic factor* into the blood where it brings about the production of a hormone called *erythropoietin* (ē-rith′-rō-POY-ē-tin) that stimulates red blood cell production (see Chapter 14).

Cardiac muscle fibers of the heart produce a hormone called *atrial natriuretic factor (ANF)* when they are stretched. The overall function of ANF is to lower blood pressure by inhibiting the actions of renin, aldosterone, and ADH and thereby decreasing blood volume.

Recall from Chapter 5 that the skin helps to synthesize the hormone vitamin D.

STRESS AND THE GENERAL ADAPTATION SYNDROME (GAS)

Homeostatic mechanisms attempt to counteract the everyday stresses of living. If they are successful, the internal environment maintains normal physiological limits of chemistry, temperature, and pressure. If a stress is extreme, unusual, or long-lasting, however, the normal mechanisms may not be sufficient. In this case, the stress triggers a wide-ranging set of bodily changes called the *general adaptation syndrome (GAS)*. Hans Selye, a world authority on stress, introduced the concept of GAS. Unlike the homeostatic mechanisms, the general adaptation syndrome does not maintain a normal environment. In fact, it does just the opposite in order to prepare the body to meet an emergency. For instance, blood pressure and blood sugar level are raised above normal.

It should be pointed out that it is impossible to remove all stress from our everyday lives. In fact, some stress is productive, but other stress is harmful. The term *eustress* refers to productive stress, and *distress* refers to harmful stress.

STRESSORS

The hypothalamus is the body's watchdog. It has sensors that detect changes in the chemistry, temperature, and pressure of the blood and it is informed of emotions through tracts that connect it with the emotional centers of the cerebral cortex. When the hypothalamus senses stress, it initiates the general adaptation syndrome.

A *stressor* may be almost any disturbance—heat or cold, environmental poisons, surgery, or a strong emotional reaction. Stressors vary among different people and even in the same person at different times.

When a stressor appears, it stimulates the hypothalamus to initiate the syndrome through two pathways. The first pathway produces an immediate set of responses called the alarm reaction. The second pathway, called the resistance reaction, is slower to start, but its effects last longer.

ALARM REACTION

The *alarm reaction*, or *fight-or-flight response*, is actually a complex of reactions initiated by hypothalamic stimulation of the sympathetic nervous system and the adrenal medulla (Figure 13-20a). The reactions are immediate and short-lived, mobilizing the body's resources for immediate physical activity. In essence, the alarm reaction brings tremendous amounts of glucose and oxygen to the organs that are most active in warding off danger. These are the brain, which must become highly alert; the skeletal muscles, which may have to fight off an attacker; and the heart, which must work furiously to pump enough materials to the brain and muscles. See Chapter 11 on activities of the autonomic nervous system to review the specific fight-or-flight responses.

Sympathetic impulses to the adrenal medulla increase its secretion of epinephrine and norepinephrine (NE). These hormones supplement and prolong many sympathetic responses—increasing heart rate and strength, increasing blood pressure, constricting blood vessels, accelerating the rate of breathing, widening respiratory passageways, increasing the rate of glycogen catabolism to increase the blood sugar level, and decreasing the rate of digestion.

If you group the stress responses of the alarm stage by function, you will note that they are designed to increase circulation rapidly, promote catabolism for energy production, and decrease nonessential activities. If the stress is great enough, the body mechanisms may not be able to cope and death can result. During the alarm reaction, digestive, urinary, and reproductive activities are inhibited.

RESISTANCE REACTION

The second stage of the GAS is the *resistance reaction* (Figure 13-20b). Unlike the short-lived alarm reaction that is initiated by nerve impulses from the hypothalamus, the resistance reaction is initiated by regulating hormones secreted by the hypothalamus and is a long-term reaction. The regulating hormones are corticotropin releasing hormone (CRH), growth hormone releasing hormone (GHRH), and thyrotropin releasing hormone (TRH).

CRH stimulates the anterior pituitary to increase secretion of ACTH. ACTH stimulates the release of mineralocorticoids and glucocorticoids.

The mineralocorticoids conserve sodium ions by the body and eliminate hydrogen ions that could make the blood more acidic. Thus, during stress, a lowering of body pH is prevented. Sodium retention also leads to water retention, thus maintaining the high blood pressure that is typical of the alarm reaction. This would make up for fluid lost through severe bleeding.

The glucocorticoids bring about a number of stress reactions; these were detailed earlier in the chapter.

Two other regulating hormones are secreted by the hypothalamus in response to stress: TRH and GHRH. TRH causes the anterior pituitary to secrete thyroid-stimulating hormone (TSH); GHRH

Diabetes and the Wellness Life-Style

Now that you understand some of the ways in which the endocrine system's hormonal team helps to regulate blood glucose level, you can appreciate the difficulty that arises when insulin, one of the leading players, is missing from the game. You can also see why the isolation of insulin about 70 years ago was one of the greatest achievements of modern medicine. Before insulin was available, people with type I diabetes seldom survived for more than a few months after the appearance of disease symptoms. But although insulin injections mean the difference between life and death, they are no cure, and insulin administration can still not be given according to biological demand as is done by the pancreas. People with diabetes must learn how to control the effects of the disease.

About one million people in the United States have type I diabetes. Continuous and informed self-care are more critical for diabetes than for any other disease. If blood glucose homeostasis is not maintained, the result is either hypoglycemia or hyperglycemia, either of which, if uncontrolled, can lead to loss of consciousness and death. The risk of long-term complications is also greater for diabetics who are unable to maintain glycemic (blood glucose) control. For the diabetic, wellness and quality of life mean achieving good blood glucose regulation by controlling those factors with greatest impact on blood glucose levels: insulin, diet, and physical activity.

Insulin

Insulin is a complex protein, and if taken orally, it would be broken down by the digestive system into its constituent amino acids. It is therefore given by injection into subcutaneous tissue in a variety of areas, such as the thigh, from which it is gradually absorbed into the bloodstream. There are several types of insulin available, and they vary in strength and in the length of time they take to reach peak concentration in the bloodstream. Patient needs vary greatly, and the physician–patient team must work very closely together to decide which types of insulin are most suitable.

The patient must understand when his insulin dosage peaks and time his meals and activity accordingly, so that the insulin is there when it is needed. People with diabetes need to develop a regular schedule so that insulin, meals, and activity are balanced with one another.

Diet

The most important ingredients in the diabetic diet are balance and consistency. It used to be that diabetics were instructed to avoid carbohydrates to decrease blood sugar and the need for insulin. Unfortunately, diets low in carbohydrates are high in fats and protein, which accelerate artery and kidney disease, for which diabetics are already at increased risk. Today, diabetics are taught how to combine carbohydrates, fats, and proteins nutritiously in order to reduce heart disease risk, control weight, and consume all the nutrients necessary for good health.

Most diabetics use the Exchange List Diet, originally developed by dieticians for diabetics but now also used for many diet prescriptions. Canada's Food Guide, Weight Watchers, and other diet groups use this system to promote a balanced diet and control weight. Foods are divided into groups according to their nutritional composition: bread, vegetables, milk, meat, fruits, and fats. A person is allowed so many "exchanges," or items, from each group per day. An exchange is a specific amount of food from a given group. For example, one slice of bread or half a cup of cooked pasta each equal one bread exchange. Vegetarians can count certain plant proteins such as peanut butter and lentils for meat group exchanges and substitute other calcium sources for milk exchanges. The result is a flexible, well-balanced diet.

The diet takes some time to learn, but it is relatively adaptable to individual food preferences and health needs, and it teaches people how to plan healthful meals.

Diabetics should avoid alcohol, since it blocks the enzymes in the liver responsible for glycolysis (conversion of glucose into pyruvic acid) and gluconeogenesis (production of glucose from noncarbohydrate sources).

Physical Activity

People with diabetes used to be discouraged from participation in sports and exercise programs, since it was thought that exercise posed an excessive metabolic challenge and that glycemic control was easier to attain without it. But physicians now encourage regular physical activity for the majority of diabetics, since the health benefits of exercise appear to make it worth the work of adjusting insulin administration and meal schedules. Diabetes doubles a person's risk of heart disease and stroke. Regular aerobic exercise can help decrease this risk by helping to control weight and blood pressure and improve cholesterol metabolism. Both weight loss and exercise can lead to an increased number of insulin receptors, and exercise may also improve the ability of membrane receptors to bind to insulin, so that insulin is used more effectively.

Perhaps the most important health benefit of exercise is the boost in self-esteem and sense of control it can provide. The ability to participate in physical activity and sports is especially crucial to the emotional well-being of children and adolescents with diabetes, who long to do everything their peers can do. Living with diabetes can be very difficult, and denial is a common response. Involvement in sports and other physical activity can provide a positive motivation for children and adolescents to master the self-care skills required for glycemic control.

People with diabetes must get clearance from their physicians to change activity level, since exercise can be harmful if blood glucose is not well regulated, or if certain complications, such as nerve damage, have already developed. The patient will need to work closely with the physician and dietician as an exercise routine is developed, since insulin dosage and meals will need to be adjusted.

Diabetics need to understand the effects of exercise on metabolism. Exercise causes a decrease in blood glucose in diabetics with mild to moderately elevated blood sugar levels. This may be due to improved insulin absorption from the injection sites, increased sensitivity to glucose of the insulin receptors, and/or an inability of the liver to produce enough glucose to make up for that used by the active skeletal muscles. If blood glucose falls too much, hypoglycemia results, and may necessitate a lowering of the insulin dose and/or an increase in caloric intake. The effectiveness of these strategies on long-term blood sugar control has been questioned by some experts, since individual response is quite variable. Thus, individualization of diabetic self-care measures is vital to optimal glycemic control.

Exercise can lead to either hypo- or hyperglycemia, depending on preexercise blood glucose and insulin concentrations. Imagine what would happen if the diabetic began exercising with low blood glucose and elevated insulin levels. Insulin would inhibit liver glycolysis and gluconeogenesis at a time when energy is desperately needed. The result: hypoglycemia.

Now imagine what would happen if the person began exercising with low insulin levels but elevated blood glucose. During exercise, glucagon and the catecholamines increase liver glucose production, but without sufficient insulin the glucose cannot enter the muscles. The result this time is hyperglycemia and ketosis.

If the person begins exercising with normal glucose and insulin levels, the amount of glucose being used by the muscles is equal to that being produced by the liver and no problems arise. This is not as easy as it sounds and may require weeks of trial and error combined with blood glucose monitoring and adjustment of insulin and meals. General recommendations for these adjustments are the following: For light exercise, such as a mile of walking, or a half hour of bicycling, the diabetic is advised to increase carbohydrate intake, and to decrease insulin only if it is injected close to the exercise period. For moderate exercise, such as endurance activity that lasts from 30 minutes to 2 hours, the diabetic will probably need to decrease insulin and increase carbohydrate intake. For all-day activities such as hiking and cross-country skiing, substantial decreases in insulin (50 percent or more) may be required.

Physical activity is best tolerated when it is consistent and predictable. This means a similar number of calories should be expended at the same time each day, if possible.

The diabetic must also be aware of factors that alter absorption of insulin from the injection site. Cold weather can slow insulin absorption, whereas warm weather can accelerate it, because of variation in blood flow. Insulin should not be injected just before exercise, since the vasodilation that occurs to expend heat will hasten absorption.

FIGURE 13-20 Responses to stressors during the general adaptation syndrome (GAS). (a) During the alarm stage. (b) During the resistance stage. Colored arrows indicate immediate reactions. Black arrows indicate long-term reactions.

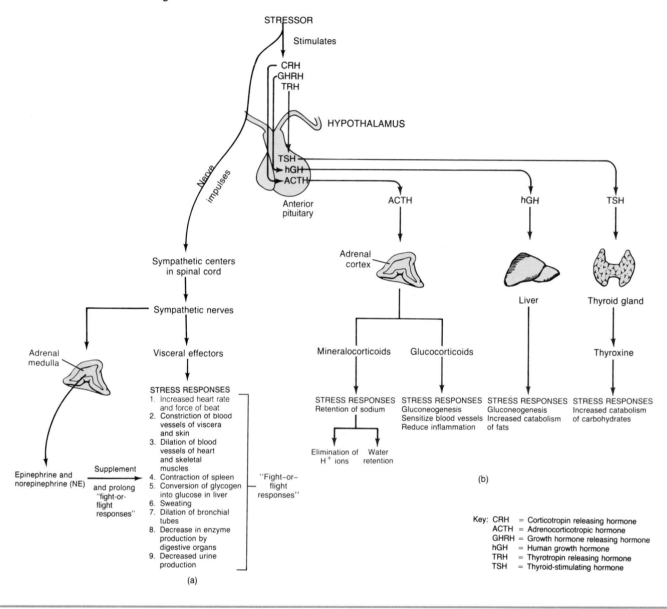

How does the general adaptation syndrome (GAS) differ from homeostasis?

causes it to secrete human growth hormone (hGH). TSH stimulates the thyroid to secrete thyroxine and triiodothyronine, which increase the catabolism of carbohydrates. hGH stimulates the catabolism of fats and the conversion of glycogen to glucose. The combined actions of TSH and hGH increase catabolism and thereby supply additional energy for the body.

The resistance stage of the GAS allows the body to continue fighting a stressor long after the alarm reaction dissipates. It provides the energy, functional proteins, and circulatory changes required for meeting emotional crises, performing strenuous tasks, fighting infection, or resisting the threat of bleeding to death. However, cells use glucose at the same rate it enters the bloodstream, which means blood sugar level returns to normal and blood pH is brought

under control by the kidneys as they excrete more hydrogen ions. Blood pressure remains abnormally high, though, because the retention of water increases the volume of blood.

Generally, the resistance stage is successful in seeing us through a stressful situation, and our bodies then return to normal. Occasionally, the resistance stage fails to combat the stressor. In this case, the general adaptation syndrome moves into the stage of exhaustion.

EXHAUSTION

A major cause of exhaustion is loss of potassium ions. When the mineralocorticoids stimulate the kidneys to retain sodium ions, potassium and hydrogen ions are traded off for sodium ions and secreted in the fluid that will become urine. As the chief positive ion in cells, potassium is partly responsible for controlling the water concentration of the cytoplasm. As the cells lose more and more potassium, they function less and less effectively. Finally, they start to die. This condition is called the *stage of exhaustion*. Unless it is rapidly reversed, vital organs cease functioning and the person dies.

Another cause of exhaustion is depletion of the adrenal glucocorticoids. In this case, blood glucose level suddenly falls, and the cells do not receive enough nutrients. A final cause of exhaustion is weakening organs. A long-term or strong resistance reaction puts heavy demands on the body, particularly on the heart, blood vessels, and adrenal cortex. They may not be up to handling the demands, or they may fail under the strain. In this respect, ability to handle stressors is determined to a large degree by general health.

STRESS AND DISEASE

Although the exact role of stress in human diseases is not known, it is becoming quite clear that stress can lead to certain diseases. Stress-related conditions include gastritis, ulcerative colitis, irritable bowel syndrome, peptic ulcers, hypertension, asthma, rheumatoid arthritis (RA), migraine headaches, anxiety, and depression. Individuals under stress are also at greater risk of developing chronic disease or dying prematurely.

■ COMMON DISORDERS ■

Disorders of the endocrine system, in general, involve underproduction or overproduction of hormones.

Pituitary Gland Disorders

Several disorders of the anterior pituitary involve human growth hormone (hGH). If hGH is undersecreted during the growth years, bone growth is slow and the epiphyseal plates close before normal height is reached. This condition is called *pituitary dwarfism*. Other organs of the body also fail to grow, and the individual is childlike physically in many respects.

Oversecretion of hGH during childhood results in *giantism* (*gigantism*), an abnormal increase in the length of long bones. The person grows to be very large, but body proportions are about normal.

Oversecretion of hGH during adulthood is called *acromegaly* (ak'-rō-MEG-a-lē), which is shown in Figure 13-21a. The hormone cannot produce further lengthening of the long bones because the epiphyseal plates are already closed. Instead, the bones of the hands, feet, cheeks, and jaws thicken. Other tissues also grow; the eyelids, lips, tongue, and nose enlarge, and the skin thickens and furrows, especially on the forehead and soles of the feet.

The principal abnormality associated with dysfunction of the posterior lobe is *diabetes insipidus* (in-SIP-i-dus). Diabetes insipidus is the result of undersecretion of ADH, usually caused by damage to the posterior lobe or hypothalamus.

Thyroid Gland Disorders

Undersecretion of thyroid hormones during fetal life or infancy results in *cretinism* (KRĒ-tin-izm), shown in Figure 13-21b. Two outstanding clinical symptoms of the cretin are dwarfism and mental retardation.

Hypothyroidism (*hypo* = under) during the adult years produces *myxedema* (mix-e-DĒ-ma) characterized by an edema that causes the facial tissues to swell and look puffy. Myxedema is more common in females.

Oversecretion of thyroid hormones may be due to an autoimmune disease called *exophthalmic* (ek'-sof-THAL-mik) *goiter* (GOY-ter). This disease, like myxedema, is also more frequent in females. One of its primary symptoms is an enlarged thyroid, called a *goiter*. Two other symptoms are an edema behind the eye, which causes the eye to protrude *(exophthalmos)*, which is shown in Figure 13-21c, and an abnormally high metabolic rate.

Goiter is a symptom of other thyroid disorders as well. In North America, it is typically caused by ingestion of excess iodine. It may also be genetically caused. This condition is called *simple goiter* (Figure 13-21d). Simple goiter may also be caused by a lower-than-average amount of iodine in the diet.

Parathyroid Gland Disorders

A deficiency of calcium owing to *hypoparathyroidism* causes neurons to depolarize without the usual stimulus. As a result, nervous impulses increase and cause muscle twitches, spasms, and convulsions. This condition is called *tetany*. Hypoparathyroidism results from surgical removal of the parathyroids or from damage caused by parathyroid disease, infection, hemorrhage, or mechanical injury.

Adrenal Gland Disorders

Oversecretion of the mineralocorticoid aldosterone results in *aldosteronism*, characterized by an increase in sodium and decrease in potassium in the blood. If potassium depletion is great, neurons cannot depolarize and muscular paralysis results. Excessive retention of sodium, and therefore of water, increases the volume of the blood and also causes hypertension.

Undersecretion of glucocorticoids (and aldosterone) results in *Addison's disease*. Increased potassium and decreased sodium lead to low blood pressure, dehydration, hypoglycemia, excessive skin pigmentation, arrhythmias, and potential cardiac arrest.

Cushing's syndrome is an oversecretion of glucocorticoids,

FIGURE 13-21 Photograph of various endocrine disorders. (a) Acromegaly. (b) Cretinism. (c) Exophthalmos. (d) Simple goiter. (e) Cushing's syndrome. (f) Virilism.

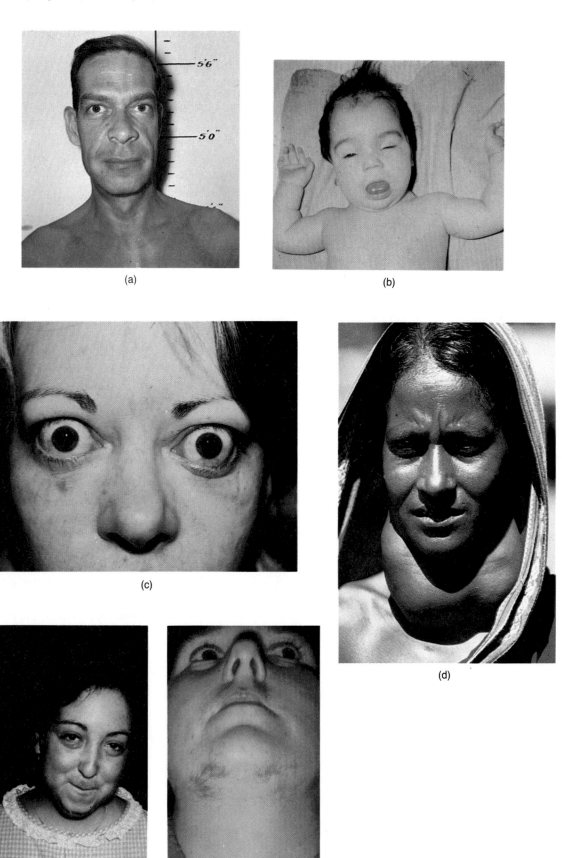

(a)

(b)

(c)

(d)

(e)

(f)

Courtesy of Lester Bergman and Associates.

especially cortisol and cortisone (Figure 13-21e). It is characterized by a redistribution of fat resulting in spindly legs accompanied by a ''moon face,'' ''buffalo hump'' on the back, and pendulous abdomen.

The *adrenogenital syndrome* usually refers to a group of enzyme deficiencies that block the synthesis of glucocorticoids. In an attempt to compensate, the anterior pituitary secretes more ACTH. As a result, excess androgens are produced, causing *virilism*, or masculinization.

Pancreatic Endocrine Disorders

Diabetes mellitus (MEL-i-tus) refers to a group of hereditary diseases, all of which ultimately lead to elevated blood glucose (hyperglycemia) and excretion of glucose in the urine as hyperglycemia increases. Diabetes mellitus is also characterized by the three ''polys'': an inability to reabsorb water, resulting in increased urine production (*polyuria*), excessive thirst (*polydipsia*), and excessive eating (*polyphagia*).

Two major types of diabetes mellitus have been distinguished, type I and type II. *Type I diabetes*, which occurs abruptly, is characterized by a deficiency of insulin due to a decline in the number of insulin-producing beta cells. This is called *insulin-dependent diabetes* because periodic administration of insulin is required to treat it. It is also called *juvenile-onset diabetes* because it most commonly develops in people younger than age 20, though it persists throughout life. People who develop type I diabetes appear to have certain genes that make them more susceptible, but some triggering factor, usually a viral infection, is required. Insulin deficiency accelerates the breakdown of the body's fat reserves, which produces organic acids called ketones. Excess ketones cause a form of acidosis called *ketosis*, which lowers the pH of the blood and can result in death.

Type II diabetes is much more common than type I, representing more than 90 percent of all cases. Type II most often occurs in people who are over 40 and overweight. It is also called *maturity-onset diabetes*. Clinical symptoms are mild, and high glucose levels in the blood can usually be controlled by diet, exercise, and/or antidiabetic drugs such as glyburide (DiaBeta). Many type II diabetics have a sufficient amount or even a surplus of insulin in the blood. For these individuals, diabetes arises not from a shortage of insulin but probably because cells lose insulin receptors and become less sensitive to insulin. Type II diabetes is therefore also called *noninsulin-dependent diabetes*.

Oversecretion results in *hypoglycemia* and is much rarer than undersecretion and is generally the result of a malignant tumor in an islet. The principal symptom is a decreased blood glucose level, which stimulates the secretion of epinephrine, glucagon, and human growth hormone. As a consequence, anxiety, sweating, tremor, increased heart rate, and weakness occur. Moreover, brain cells do not have enough glucose to function efficiently. This condition leads to mental disorientation, convulsions, unconsciousness, and shock.

MEDICAL TERMINOLOGY AND CONDITIONS

Hyperplasia (hī′-per-PLĀ-zē-a; *hyper* = over; *plas* = grow) Increase in the number of cells owing to an increase in the frequency of cell division.

Hypertrophy (hī-PER-trō-fē) Increase in the size of cells without cell division.

Hypoplasia (hī′-pō-PLĀ-zē-a; *hypo* = under) Defective development of tissue.

Neuroblastoma (noo′-rō-blas-TŌ-ma; *neuro* = nerve) Malignant tumor arising from the adrenal medulla associated with metastases to bones.

Thyroid (THĪ-roid) *storm* An aggravation of all symptoms of hyperthyroidism characterized by unregulated hypermetabolism with fever and rapid heart rate; results from trauma, surgery, and unusual emotional stress or labor.

STUDY OUTLINE

Endocrine Glands (p. 270)

1. Both the endocrine and nervous systems help maintain homeostasis.
2. Hormones help regulate the internal environment, respond to stress, help regulate growth and development, and contribute to reproductive processes.
3. Exocrine glands (sweat, sebaceous, digestive) secrete their products through ducts into body cavities or onto body surfaces.

4. Endocrine glands secrete hormones into the blood.

Chemistry of Hormones (p. 270)

1. Chemically, hormones are classified as amines, proteins and peptides, and steroids.
2. Amines are produced from an amino acid (e.g., thyroid hormones); protein and peptide hormones consist of chains of amino

acids (e.g., oxytocin); steroid hormones are derived from cholesterol (e.g., aldosterone).

Mechanism of Hormonal Action (p. 270)

1. The amount of hormone released is determined by the body's need for the hormone.
2. Cells that respond to the effects of hormones are called target cells.
3. Receptors are found in the plasma membrane and inside the cell, primarily in the nucleus of target cells.
4. The combination of hormone and receptor activates a chain of events in a target cell in which the physiological effects of the hormone are expressed.
5. One mechanism of hormonal action involves interaction of hormone with plasma membrane receptors; the second messenger may be cyclic AMP or calcium ions.
6. Another mechanism of hormonal action involves interaction of a hormone with intracellular receptors usually in the nucleus.
7. Prostaglandins (PGs) are lipids that can increase or decrease cyclic AMP formation and thus regulate hormone responses that use cyclic AMP.

Control of Hormonal Secretions: Feedback Control (p. 272)

1. A negative feedback control mechanism prevents overproduction or underproduction of a hormone.
2. Hormone secretions are controlled by levels of circulating hormone itself, nerve impulses, and regulating hormones (or factors).

Pituitary (p. 274)

1. The pituitary is attached to the hypothalamus and consists of an anterior lobe and a posterior lobe.
2. Hormones of the pituitary are released or inhibited by regulating hormones (or factors) produced by the hypothalamus.
3. The blood vessels from the hypothalamus to the anterior pituitary transport hypothalamic regulating hormones (or factors).
4. The anterior pituitary consists of cells that produce human growth hormone (hGH), prolactin (PRL), thyroid-stimulating hormone (TSH), follicle-stimulating hormone (FSH), luteinizing hormone (LH), adrenocorticotropin hormone (ACTH), and melanocyte-stimulating hormone (MSH).
5. hGH stimulates body growth through small proteins called somatomedins and insulinlike growth factors (IGF) and is controlled by GHIF (growth hormone inhibiting factor) and GHRF (growth hormone releasing factor).
6. TSH regulates thyroid gland activities and is controlled by TRH (thyrotropin releasing hormone).
7. ACTH regulates the activities of the adrenal cortex and is controlled by CRH (corticotropin releasing hormone).
8. FSH regulates the activities of the ovaries and testes and is controlled by GnRH (gonadotropin releasing hormone).
9. LH regulates female and male reproductive activities and is controlled by GnRH.
10. PRL helps initiate milk secretion and is controlled by PIF (prolactin inhibiting factor) and PRF (prolactin releasing factor).

11. MSH increases skin pigmentation and is controlled by MRF (melanocyte-stimulating-hormone releasing factor) and MIF (melanocyte-stimulating-hormone inhibiting factor).
12. The posterior lobe contains axon terminals of secretory neurons whose cell bodies reside in the hypothalamus.
13. Hormones made by the hypothalamus and stored in the posterior lobe are oxytocin or OT (stimulates contraction of uterus and ejection of milk) and antidiuretic hormone or ADH (stimulates water reabsorption by the kidneys and arteriole constriction).
14. OT secretion is controlled by uterine distention and sucking during nursing; ADH is controlled primarily by water concentration.

Thyroid (p. 278)

1. The thyroid gland is located below the larynx.
2. The thyroid consists of thyroid follicles composed of follicular cells, which secrete the thyroid hormones thyroxine (T_4) and triiodothyronine (T_3), and parafollicular cells, which secrete calcitonin (CT).
3. Thyroid hormones regulate the rate of metabolism, growth and development, and the reactivity of the nervous system. Secretion is controlled by TRH.
4. Calcitonin (CT) lowers the blood level of calcium. Secretion is controlled by its own level in blood.

Parathyroids (p. 281)

1. The parathyroids are embedded on the posterior surfaces of the thyroid.
2. The parathyroids consist of principal and oxyphil cells.
3. Parathyroid hormone (PTH) regulates the homeostasis of calcium and phosphate by increasing blood calcium level and decreasing blood phosphate level. Secretion is controlled by its own level in blood.

Adrenals (p. 281)

1. The adrenal glands are located above the kidneys. They consist of an outer cortex and inner medulla.
2. The cortex is divided into three zones and each zone secretes different groups of hormones.
3. The outer zone secretes mineralocorticoids; the middle zone secretes glucocorticoids; and the inner zone secretes gonadocorticoids.
4. Mineralocorticoids (e.g., aldosterone) increase sodium and water reabsorption and decrease potassium reabsorption. Secretion is controlled by the renin–angiotensin pathway and blood level of potassium.
5. Glucocorticoids (e.g., cortisol) promote normal metabolism, help resist stress, and serve as antiinflammatories. Secretion is controlled by CRH.
6. Gonadocorticoids (adrenal sex hormones) secreted by the adrenal cortex have minimal effects.
7. Medullary secretions are epinephrine and norepinephrine (NE), which produce effects similar to sympathetic responses. They are released under stress.

Pancreas (p. 285)

1. The pancreas is behind and slightly below the stomach. It is both an endocrine and an exocrine gland.
2. The endocrine portion consists of pancreatic islets or islets of Langerhans, made up of three types of cell—alpha, beta, and delta.
3. Alpha cells secrete glucagon, beta cells secrete insulin, and delta cells secrete growth hormone inhibiting hormone (GHIH).
4. Glucagon increases blood sugar level. Secretion is controlled by its own level in the blood.
5. Insulin decreases blood sugar level. Secretion is controlled by its own level in the blood.

Ovaries and Testes (p. 286)

1. Ovaries are located in the pelvic cavity and produce sex hormones related to the development of female sexual characteristics, menstrual cycle, pregnancy, and lactation.
2. Testes lie inside the scrotum and produce sex hormones related to the development of male sexual characteristics.

Pineal (p. 286)

1. The pineal is attached to the roof of the third ventricle near the thalamus.
2. It has calcium deposits referred to as brain sand.
3. It secretes melatonin, which possibly regulates reproductive activities by inhibiting gonadotropic hormones.

Thymus (p. 286)

1. The thymus gland secretes several hormones related to immunity.
2. Thymosin, thymic humoral factor (THF), thymic factor (TF), and thymopoietin promote the maturation of T cells.

Other Endocrine Tissues (p. 286)

1. The gastrointestinal tract synthesizes stomach and intestinal gastrin, secretin, cholecystokinin (CCK), enterocrinin, and gastric inhibitory peptide (GIP).
2. The placenta produces human chorionic gonadotropin (hCG), estrogens, progesterone, relaxin, and human chorionic somato-mammotropin (hCS).
3. The kidneys release an enzyme that produces erythropoietin.
4. The heart secretes atrial natriuretic factor (ANF).

5. The skin when exposed to sunlight helps synthesize vitamin D.

Stress and the General Adaptation Syndrome (GAS) (p. 287)

1. If stress is extreme or unusual, it triggers the general adaptation syndrome (GAS).
2. Unlike the homeostatic mechanisms, this syndrome does not maintain a constant internal environment. In fact, it does the opposite, preparing the body to meet an emergency.

Stressors (p. 287)

1. The stimuli that produce the general adaptation syndrome are called stressors.
2. A stressor can be almost any disturbance, from surgical operations and poisons to strong emotional responses.

Alarm Reaction (p. 287)

1. The alarm reaction is initiated by nerve impulses from the hypothalamus to the sympathetic division of the autonomic nervous system and adrenal medulla.
2. Responses are the immediate and short-lived, fight-or-flight responses that increase circulation, promote catabolism for energy production, and decrease nonessential activities.

Resistance Reaction (p. 287)

1. The resistance reaction is initiated by regulating factors secreted by the hypothalamus.
2. The regulating factors are CRH, GHRH, and TRH.
3. CRH stimulates the anterior pituitary to increase its secretion of ACTH, which in turn stimulates the adrenal cortex to secrete its mineralocorticoids and glucocorticoids.
4. Glucocorticoids are produced in especially high concentrations during stress, with many distinct physiological effects.
5. Resistance reactions are long term and accelerate catabolism to provide energy to counteract stress.

Exhaustion (p. 290)

1. The stage of exhaustion results from dramatic changes during alarm and resistance reactions.
2. Exhaustion is caused mainly by loss of potassium, depletion of adrenal glucocorticoids, and weakened organs. If stress is too great, it may lead to death.

Stress and Disease (p. 291)

1. It appears that stress can lead to certain diseases.
2. Stress-related conditions include gastritis, peptic ulcers, hypertension, and migraine headaches.

REVIEW QUESTIONS

1. Contrast endocrine and nervous system control of homeostasis. (p. 269)
2. Distinguish between an endocrine gland and an exocrine gland. What are the five principal actions of hormones? (p. 270)
3. What is a hormone? (p. 269)
4. Describe the chemical classification of hormones. Give an example of each. (p. 270)
5. Explain how receptors are related to hormones. (p. 270)

6. Describe the mechanism of hormonal action involving (a) interaction with plasma membrane receptors and (b) interaction with intracellular receptors. (p. 272)

7. Define a prostaglandin (PG). Give several functions of prostaglandins. How are prostaglandins related to hormonal action? (p. 272)

8. How are negative feedback systems related to hormonal control? Discuss three models of operation. (p. 273)

9. In what respect is the pituitary gland actually two glands? (p. 274)

10. What hormones are produced by the anterior lobe of the pituitary gland? What are their functions? How are they controlled? (p. 274)

11. Discuss the function and regulation of hormones of the posterior lobe of the pituitary gland. (p. 277)

12. Describe the location of the thyroid gland. Discuss the physiological effects of the thyroid hormones. How are these hormones regulated? (p. 278)

13. Describe the function and control of calcitonin (CT). (p. 280)

14. Where are the parathyroids located? What are the functions of parathyroid hormone (PTH)? (p. 281)

15. Describe the hormones produced by the adrenal cortex in terms of type, normal function, and control. (p. 282)

16. What relationship does the adrenal medulla have to the autonomic nervous system? What is the action of adrenal medullary hormones? (p. 284)

17. Describe the location of the pancreas. What are the actions of glucagon and insulin? How are the hormones controlled? (p. 285)

18. Why are the ovaries and testes considered to be endocrine glands? (p. 286)

19. Where is the pineal gland located? What are its assumed functions? (p. 286)

20. How are hormones of the thymus gland related to immunity? (p. 286)

21. List the hormones secreted by the gastrointestinal tract, placenta, kidneys, and heart. (p. 287)

22. Define the general adaptation syndrome (GAS). What is a stressor? (p. 287)

23. Outline the reactions of the body during the alarm stage, resistance stage, and stage of exhaustion. What is the central role of the hypothalamus during stress? (p. 287)

24. How is stress related to disease? (p. 291)

25. Define the following: pituitary dwarfism, giantism, acromegaly, diabetes insipidus, cretinism, myxedema, exophthalmic goiter, simple goiter, tetany, aldosteronism, Addison's disease, Cushing's syndrome, adrenogenital syndrome, gynecomastia, type I diabetes, and type II diabetes. (p. 291)

26. Refer to the glossary of medical terminology and conditions associated with the endocrine system. Be sure that you can define each term. (p. 293)

The Cardiovascular System: Blood

STUDENT OBJECTIVES

1. Describe how blood is related to homeostasis.
2. List the characteristics and functions of blood.
3. Explain where and how blood is produced.
4. Describe the functions of red blood cells, white blood cells, and platelets.
5. List the components and importance of blood plasma.
6. Describe how blood clots and why clotting is important.
7. Describe the principal blood types and their relationship to transfusions.
8. Define common disorders and medical terminology and conditions associated with blood.

A LOOK AHEAD

FUNCTIONS OF BLOOD
PHYSICAL CHARACTERISTICS OF
    BLOOD
COMPONENTS OF BLOOD
    Formed Elements
        Origin
        Erythrocytes (Red Blood Cells)
        Leucocytes (White Blood Cells)
        Thrombocytes (Platelets)
    Plasma
HEMOSTASIS OF BLOOD
    Vascular Spasm
    Platelet Plug Formation
    Coagulation
        Extrinsic Pathway
        Intrinsic Pathway
        Retraction and
            Fibrinolysis
        Hemostatic Control
            Mechanisms
        Intravascular Clotting
GROUPING (TYPING) OF BLOOD
    ABO
    Rh
COMMON DISORDERS
MEDICAL TERMINOLOGY AND
    CONDITIONS

The blood, heart, and blood vessels together make up the ***cardiovascular system.*** In this chapter, we will consider blood. The heart and blood vessels are discussed in the next two chapters. A closely related system, the ***lymphatic system,*** is discussed in Chapter 17.

The cardiovascular system is designed to transport blood to all cells of the body. Blood contains all the nutrients and oxygen required by cells; therefore, it must reach virtually every cell in the body. This is accomplished by the heart, which serves as a pump, and blood vessels, which take blood from the heart to body cells.

The substance that bathes cells is called ***interstitial fluid.*** Interstitial fluid, in turn, is serviced by blood and lymph. Blood picks up oxygen from the lungs, nutrients from the gastrointestinal tract, hormones from endocrine glands, and enzymes from still other parts of the body. It transports these substances to all the tissues, where they diffuse from microscopic blood vessels into interstitial fluid. From the interstitial fluid, these substances enter the cells and wastes from the cells enter the blood.

The blood carries carbon dioxide and metabolic waste to the lungs, kidneys, and sweat glands for elimination from the body. Some wastes must be processed by the liver before they can be excreted (Chapter 19).

Sometimes disease-causing organisms (pathogens) are able to invade the blood and interstitial fluid. The way in which the lymphatic system filters body fluids to prevent the spread of pathogens throughout the body is discussed in Chapter 17.

Blood inside blood vessels, interstitial fluid around body cells, and lymph inside lymph vessels constitute the body's ***internal environment.*** Because body cells are too specialized to adjust to more than very limited changes in their environment, the internal environment must be kept within normal physiological limits. This condition we have called homeostasis. In preceding chapters, we have discussed how the internal environment is kept in homeostasis. Now we will look at that environment itself, beginning with ***blood.***

The branch of science concerned with the study of blood and blood-forming tissues and the disorders associated with them is called ***hematology*** (hēm'-a-TOL-ō-jē; *hem* = blood; *logos* = study of).

FUNCTIONS OF BLOOD

Blood is a liquid connective tissue that performs a number of critical functions.

1. **It transports:** oxygen from the lungs to the cells of the body; carbon dioxide from the cells to the lungs; nutrients from the digestive organs to the cells; waste products from the cells to the kidneys, lungs, and sweat glands; hormones from endocrine glands to the cells; heat from various cells.
2. **It regulates:** pH through buffers; normal body temperature through the heat-absorbing and coolant properties of its water content; the water content of cells, principally through dissolved sodium ions (Na^+).
3. **It protects against:** blood loss through the clotting mechanism; foreign microbes and toxins through phagocytic white blood cells or specialized plasma proteins such as antibodies.

PHYSICAL CHARACTERISTICS OF BLOOD

Blood is a viscous fluid: it is heavier, thicker, and more viscous than water. The viscosity (adhesiveness or stickiness) of blood, may be felt by touching it. Its temperature is about 38°C (100.4°F), its pH range is 7.35 to 7.45 (slightly alkaline), and its salt (NaCl) concentration is 0.90 percent. The blood volume of an average-sized male is 5 to 6 liters (5 to 6 qt); an average-sized female has 4 to 5 liters. Blood constitutes about 8 percent of the total body weight.

COMPONENTS OF BLOOD

Blood is composed of two portions: 45 percent of its volume is composed of formed elements (cells and cell fragments) and 55 percent is plasma (liquid containing dissolved substances) (Figure 14-1).

FORMED ELEMENTS

The *formed elements* of the blood are (Figure 14-2):

I. **Erythrocytes** (red blood cells)
II. **Leucocytes** (white blood cells)
   A. **Granular leucocytes** (granulocytes)
     1. Neutrophils
     2. Eosinophils
     3. Basophils
   B. **Agranular leucocytes** (agranulocytes)
     1. Lymphocytes
     2. Monocytes
   C. **Thrombocytes** (platelets)

FIGURE 14-1 Components of blood in a normal adult.

Origin

The process by which blood cells are formed is called ***hemopoiesis*** (hē-mō-poy-Ē-sis). In the adult, hemopoiesis takes place in red bone marrow of the humerus and femur, sternum, ribs, vertebrae, pelvis, and lymphoid tissue. Red blood cells, granular leucocytes, and platelets are produced in red bone marrow. Agranular leucocytes arise from red bone marrow and lymphoid tissue in the spleen, tonsils, and lymph nodes.

All blood cells originate from ***hemocytoblasts*** (hē′-mō-SĪ-tō-blasts), immature cells that undergo differentiation into five types of cells from which the major types of blood cells develop (Figure 14-2).

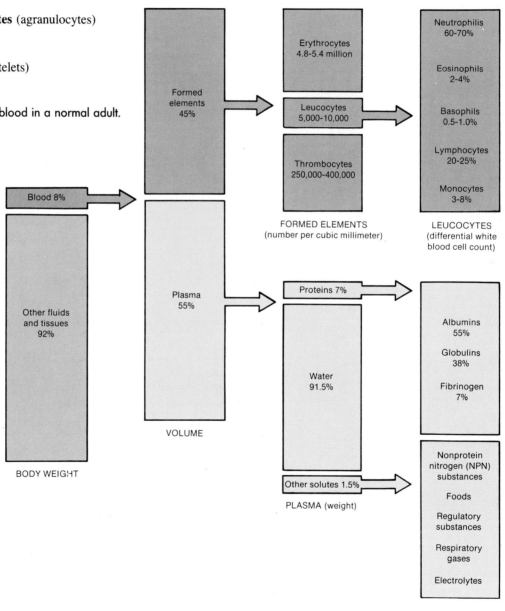

What are the normal values for the following physical characteristics of blood: pH? Salt concentration? Percent body weight? Volume?

FIGURE 14-2 Origin, development, and structure of blood cells.

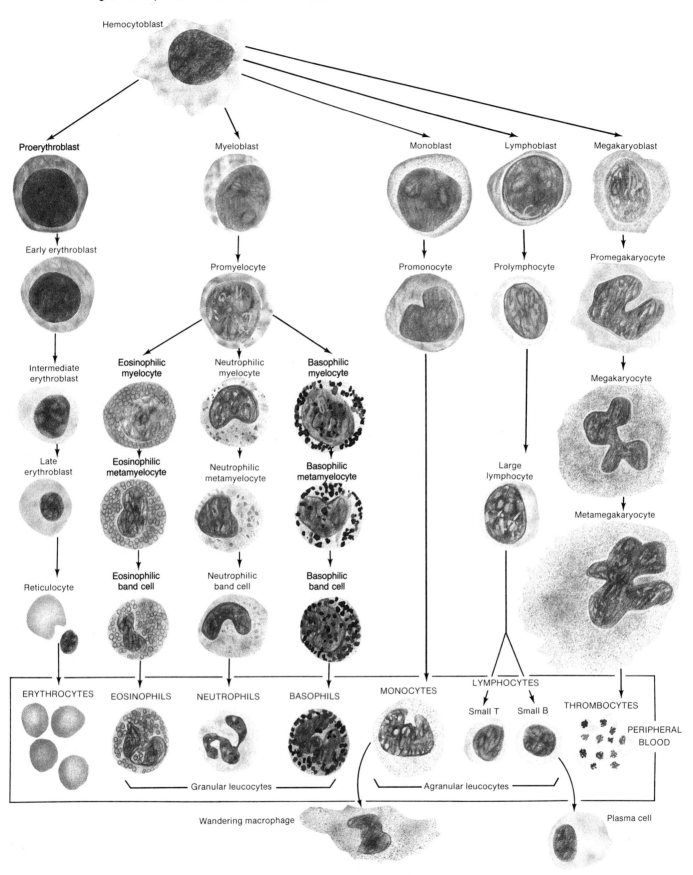

Hemocytoblast

Proerythroblast   Myeloblast   Monoblast   Lymphoblast   Megakaryoblast

Early erythroblast

Promyelocyte   Promonocyte   Prolymphocyte   Promegakaryocyte

Intermediate
erythroblast

Eosinophilic   Neutrophilic   Basophilic
myelocyte   myelocyte   myelocyte

Late
erythroblast

Eosinophilic   Neutrophilic   Basophilic
metamyelocyte   metamyelocyte   metamyelocyte

Megakaryocyte

Large
lymphocyte

Reticulocyte

Eosinophilic   Neutrophilic   Basophilic
band cell   band cell   band cell

Metamegakaryocyte

ERYTHROCYTES   EOSINOPHILS   NEUTROPHILS   BASOPHILS   MONOCYTES   LYMPHOCYTES

Small T   Small B   THROMBOCYTES

PERIPHERAL
BLOOD

⎣———— Granular leucocytes ————⎦   ⎣———— Agranular leucocytes ————⎦

Wandering macrophage   Plasma cell

■ List the functions of blood.

Erythrocytes (Red Blood Cells)

■ **Structure** *Erythrocytes* (e-RITH-rō-sīts) or *red blood cells* (**RBCs**) are biconcave discs averaging about 8 μm in diameter* (Figure 14-3a). Mature red blood cells are quite simple in structure. They have no nucleus or other organelles and cannot divide or carry on extensive metabolic activities. Essentially, they consist of a selectively permeable plasma membrane, cytoplasm, and a red pigment called *hemoglobin.* Hemoglobin carries oxygen to body cells and is responsible for the red color of blood. Normal values for hemoglobin are 14 to 20 g/100 ml of blood in infants, 12 to 15 g/100 ml in adult females, and 14 to 16.5 g/100 ml in adult males. As you will see later, certain proteins (antigens) on the surfaces of red blood cells are responsible for the various blood groups such as ABO and Rh groups.

■ **Functions** The hemoglobin in erythrocytes combines with oxygen and with carbon dioxide in order to then transport them through blood vessels. The hemoglobin molecule consists of four identical proteins called *globin* and four nonprotein pigments called *hemes,* each of which is attached to a protein and contains iron (Figure 14-3b). As the erythrocytes pass through the lungs, each of the four iron atoms in the hemoglobin molecules combines with a molecule of oxygen to form oxyhemoglobin. The oxygen is transported as oxyhemoglobin to other tissues of the body. In the tissues, the iron–oxygen reaction reverses, and the oxygen is released to diffuse into the interstitial fluid and cells. On the return trip, the globin portion combines with carbon dioxide from the interstitial fluid to form carbaminohemoglobin, which is transported to the lungs, where the carbon dioxide is released and then exhaled. Although some carbon dioxide is transported by hemoglobin, the greater portion is transported in blood plasma (see Chapter 18). Since erythrocytes lack a nucleus, their capacity for carrying oxygen is increased greatly. Moreover, since they lack mitochondria and generate ATP anaerobically, they do not consume any of the oxygen that they transport.

■ **Life Span and Number** Red blood cells live only about 120 days because of wear and tear on their fragile plasma membranes as they squeeze through blood capillaries. Worn-out red blood cells are phagocytized by macrophages in the spleen, liver, and bone marrow. The red blood cell's hemoglobin is subsequently recycled (Figure 14-4). The globin is split from the heme portions and broken down into amino acids that may be reused by other cells for protein synthesis. The heme is broken down into iron and biliverdin. The iron is stored in the liver in two forms: ferritin and hemosiderin. *Ferritin* consists of iron bound to a protein in the liver called apoferritin. *Hemosiderin* is an extremely insoluble form of iron. Iron is transported in the blood by combining with a protein called *transferrin.* Transferrin combines with iron that has been absorbed from the gastrointestinal tract or iron released from storage and transports it to bone marrow to be reused for hemoglobin synthesis. Transferrin also transports iron to the liver for storage. The noniron portion of heme is converted into *biliverdin,* a greenish pigment, and then into *bilirubin.* Bilirubin is released

* 1 μm = 1/25,000 of an inch or 1/1000 of a millimeter (mm).

FIGURE 14-3 Erythrocytes. (a) Shape of an erythrocyte in surface view (top) and side view (bottom). (b) Diagram of a hemoglobin molecule. The four globin (protein) portions of the molecule are indicated in green and blue and the four heme (iron-containing) portions are in the center of each globin molecule.

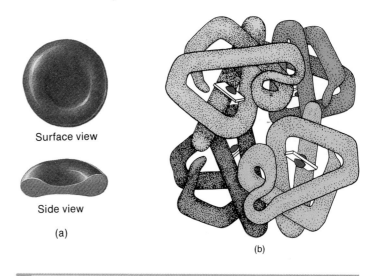

Surface view

Side view

(a)

(b)

What happens to hemoglobin after red blood cells die?

into plasma and is transported to the liver to be excreted in bile, which passes from the liver to the small intestine. In the large intestine, bilirubin is converted by bacteria into *urobilinogen,* most of which is eliminated in feces in the form of a brown pigment (*stercobilin*), which gives feces its characteristic color.

■ **Production** The process by which erythrocytes are formed is called *erythropoiesis* (e-rith´-rō-poy-Ē-sis). It takes place in red bone marrow (see Figure 14-2). Normally, erythropoiesis and red blood cell destruction proceed at the same pace. A healthy male has about 5.4 million red blood cells per cubic millimeter (mm^3) of blood, and a healthy female about 4.8 million. The higher value in the male is caused by higher levels of testosterone, which stimulate the production of red blood cells. To maintain normal quantities of erythrocytes, the body must produce new mature cells at the astonishing rate of 2 million per second. If the body suddenly needs more erythrocytes or if erythropoiesis is not keeping up with red blood cell destruction, a homeostatic mechanism steps up production (Figure 14-5). The mechanism is triggered by the reduced supply of oxygen for body cells, called *hypoxia* (hī-POKS-ē-a). If certain kidney (or liver) cells become oxygen-deficient, they release an enzyme called *renal erythropoietic factor* (**REF**) that converts a plasma protein into the hormone *erythropoietin* (*poiem* = to make). This hormone travels to the red bone marrow where it stimulates production of red blood cells. One medical use of erythropoietin is to increase the amount of blood that can be collected by individuals who choose to donate their own blood before surgery. Erythropoiesis increases at high altitudes where the air contains less oxygen and as a result of disease conditions that produce hypoxia, such as pneumonia or the anemias (discussed later).

FIGURE 14-4 **Recycling of red blood cells.**

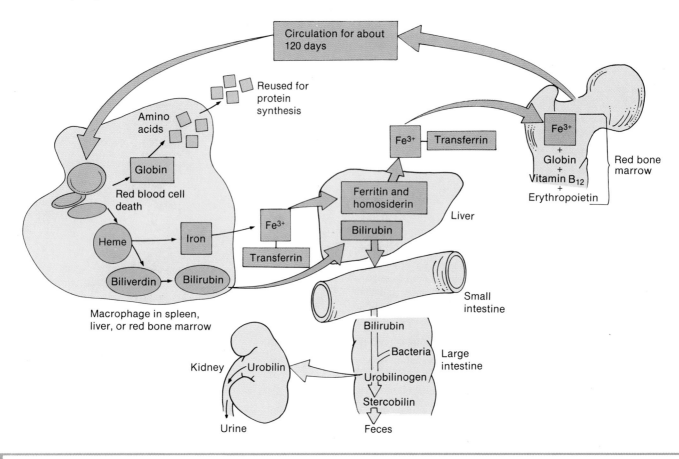

Trace how red blood cell components are recycled.

In order for the bone marrow to produce adequate numbers of healthy red blood cells, individuals must consume adequate amounts of iron, protein, folic acid, and vitamin B_{12}, and produce enough *intrinsic factor (IF)* in the stomach to absorb the vitamin B_{12}. Regular aerobic exercise also increases erythropoiesis.

The rate of erythropoiesis is measured by a procedure called a *reticulocyte* (re-TIK-yoo-lō-sīt) *count.* Reticulocytes are an intermediate cell in the development of a mature red blood cell (see Figure 14-2). Some reticulocytes are normally released in the bloodstream before they become mature red blood cells. The percentage of reticulocytes in a blood sample should range from 0.5 to 1.5. A *hematocrit (Hct)* is the percentage of blood made up of the RBCs. In adult males the average is 40 to 54 percent; in females it is 38 to 46 percent (athletes may have somewhat higher values). The results of a hematocrit test help a doctor to determine the rate of hemopoiesis.

Leucocytes (White Blood Cells)

▪ **Structure and Types** Unlike red blood cells, *leucocytes* (LOO-kō-sīts) or *white blood cells (WBCs)* have nuclei and do not contain hemoglobin (see Figure 14-2). Leucocytes fall into two major groups. The first is *granular leucocytes.* They have large characteristic granules in their cytoplasm and possess lobed nuclei. The three kinds of granular leucocytes are identified on the basis of their specific granules and are called *neutrophils* (10 to 12 μm in diameter), *eosinophils* (10 to 12 μm in diameter), and *basophils* (8 to 10 μm in diameter).

The second group is called *agranular leucocytes* because their cytoplasmic granules do not stain easily and cannot be seen with an ordinary light microscope. *Lymphocytes* (7 to 15 μm in diameter) and *monocytes* (14 to 19 μm in diameter) are the two types in this group.

Just as red blood cells have surface proteins, so do white blood cells and all other nucleated cells in the body. These proteins, called **HLA (human leucocyte associated) antigens,** are unique for each person (except for identical twins) and can be used to identify a tissue. If an incompatible tissue is transplanted, it is rejected by the recipient as foreign, due, in part, to differences in donor and recipient HLA antigens. The HLA antigens are used to type tissues to help prevent rejection.

▪ **Functions** The skin and mucous membranes of the body are continuously exposed to microbes and their toxins. Some of these microbes are capable of invading deeper tissues to cause disease. Once they enter the body, it is the function of leucocytes to combat them by phagocytosis or antibody production. Neutrophils and monocytes are actively *phagocytotic*—they can ingest bacteria and dispose of dead cells (see Figure 3-6a,b). Neutrophils (NOO-trō-

FIGURE 14-5 Production of red blood cells.

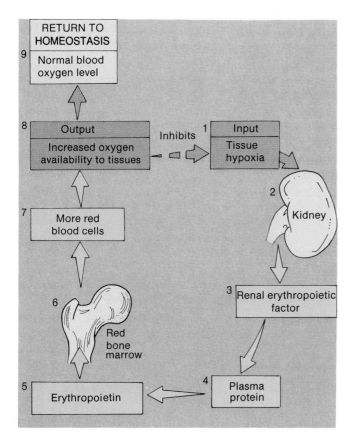

What are the three causes of hypoxia?

the production of antibodies and is capable of reacting specifically with the antibody. Most antigens are proteins, and most are not synthesized by the body. Rather, antigens make up the cell structures and enzymes of bacteria. They also make up the toxins released by bacteria. When antigens enter the body, they stimulate certain lymphocytes, called B cells, to become *plasma cells* (Figure 14-6). The plasma cells then produce antibodies, globulin-type proteins that attach to antigens and inactivate them. This is called the *antigen–antibody response.* Eosinophils destroy the antigen–antibody complexes. The antigen–antibody response is discussed in more detail in Chapter 17.

Other lymphocytes are called *T cells.* One group of T cells, called *cytotoxic (killer) T cells,* that is activated by certain antigens react by destroying them directly or indirectly by recruiting other lymphocytes and macrophages. T cells are especially effective against bacteria, viruses, fungi, transplanted cells, and cancer cells. Other types of T cells and their functions are discussed in Chapter 17.

The antigen–antibody response helps us combat infection and gives us immunity to some diseases. It is also responsible for

FIGURE 14-6 Antigen–antibody response. An antigen entering the body stimulates a B cell to develop into an antibody-producing plasma cell. The antibodies attach to the antigen, cover it, and render it harmless.

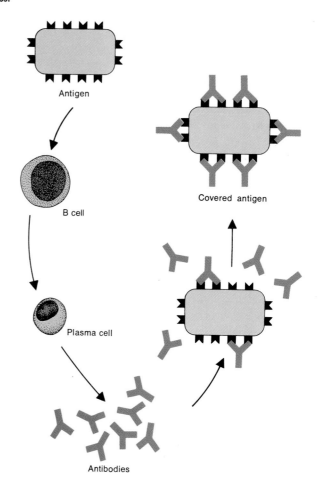

What are the five results of the antigen–antibody response?

fils) respond first to bacterial invasion, carrying on phagocytosis and releasing the enzyme lysozyme, which destroys certain bacteria. Monocytes (MON-ō-sīts) take longer to reach the site of infection than do neutrophils, but once they arrive, they do so in larger numbers and destroy more microbes. Monocytes that have migrated to infected tissues are called *wandering (tissue) macrophages.* They clean up cellular debris following an infection. Most leucocytes possess, to some degree, the ability to crawl through minute spaces between the cells that form the walls of capillaries and through connective and epithelial tissue. This movement, like that of amoebas, is called *diapedesis* (dī′-a-pe-DĒ-sis).

Eosinophils (ē′-ō-SIN-ō-fils) are believed to release enzymes, such as histaminase, that combat the effects of inflammation, allergic reactions, and certain parasitic worms. Thus, a high eosinophil count frequently indicates an allergic condition or a parasitic infection. Eosinophils also phagocytize antigen–antibody complexes (to be explained shortly).

Basophils (BĀ-sō-fils) are also believed to be involved in allergic reactions. Once they leave the capillaries and enter the tissues, they are known as mast cells and function to liberate heparin, histamine, and serotonin, substances that intensify the inflammatory reaction (see Chapter 17).

Some lymphocytes (LIM-fō-sīts) develop into cells that produce antibodies. *Antibodies* (AN-ti-bod′-ēz) are proteins that inactivate antigens. An *antigen* (AN-ti-jen) is any substance that stimulates blood types, allergies, and the body's rejection of organs transplanted from an individual with a different genetic makeup.

An increase in the number of white blood cells present in the blood usually indicates an inflammation or infection. Because each type of white cell plays a different role, determining the *percentage* of each type in the blood assists in diagnosing the condition. A *differential white blood cell count* is the number of each kind of white cell in 100 white blood cells. A normal differential blood count falls within the following percentages:

Neutrophils	60–70%
Eosinophils	2–4%
Basophils	0.5–1%
Lymphocytes	20–25%
Monocytes	3–8%
	100%

Particular attention is paid to the neutrophils in a differential white blood cell count. A high neutrophil count usually indicates a response to invading bacteria. A high monocyte count generally indicates a chronic infection. Eosinophils and basophils are elevated during allergic reactions. High lymphocyte counts indicate antigen–antibody reactions.

■ **Life Span and Number** Bacteria exist everywhere in the environment and have continuous access to the body through the mouth, nose, and pores of the skin. Furthermore, many cells, especially those of epithelial tissue, age and die daily, and their remains must be removed by leucocytes that actively ingest bacteria and cellular debris. However, a leucocyte can phagocytize only a certain number of substances before they interfere with its own metabolic activities and bring on its death. Consequently, the life span of most leucocytes is only a few days. During a period of infection they may live only a few hours.

Leucocytes are far less numerous than red blood cells, averaging from 5000 to 10,000 cells per cubic millimeter (mm^3) of blood. Red blood cells, therefore, outnumber white blood cells about 700 to 1. The term *leucocytosis* (loo′-kō-sī-TŌ-sis) refers to an increase in the number of white blood cells. If the increase exceeds 10,000/mm^3, a pathological condition is usually indicated. An abnormally low level of white blood cells (below 5000/mm^3) is termed *leucopenia* (loo-kō-PĒ-nē-a).

■ **Production** Granular leucocytes are produced in red bone marrow. Agranular leucocytes are produced in both red bone marrow and lymphoid tissue. The developmental sequences for the five types of leucocytes are shown in Figure 14-2. White blood cells develop under the influence of substances called *colony-stimulating factors* (*CSFs*). In addition to stimulating the development of various white blood cells, CSFs also enhance their functions in fighting infection and inflammation. CSFs also play a role in immunity (Chapter 17) and are now being tested to evaluate their effectiveness in treating AIDS, cancer, and bone marrow suppression.

Thrombocytes (Platelets)

■ **Structure** One of the cell types that hemocytoblasts differentiate into is called a megakaryoblast (see Figure 14-2). Megakaryoblasts ultimately transform into megakaryocytes, large cells that shed

fragments of cytoplasm. Each fragment becomes enclosed by a piece of the cell membrane and is called a *thrombocyte* (THROM-bō-sīt) or *platelet*. Platelets are disc-shaped cells without a nucleus but contain many granules that play a role in blood clotting. They average from 2 to 4 μm in diameter.

■ **Function** Platelets prevent fluid loss by initiating a chain of reactions that results in blood clotting. This mechanism is described shortly.

■ **Life Span and Number** Platelets have a short life span, only 5 to 9 days. Between 250,000 and 400,000 platelets appear in each cubic millimeter (mm^3) of blood.

■ **Production** Platelets are produced in red bone marrow according to the developmental sequence shown in Figure 14-2.

A summary of the formed elements in blood is presented in Exhibit 14-1.

The most commonly ordered hematology test is the *complete blood count* (*CBC*). It examines blood for the numbers and shapes of red blood cells, white blood cells, and platelets; differential white blood cell count; hemoglobin; and hematocrit.

PLASMA

When the formed elements are removed from blood, a straw-colored liquid called *plasma* is left. Plasma consists of 91.5 percent water and 8.5 percent solutes. The solutes consist of plasma proteins, nutrients, gases, electrolytes, waste products of metabolism, enzymes, and hormones. Some of the proteins in plasma are found elsewhere in the body, but those confined to blood are called *plasma proteins. Albumins,* which are synthesized by the liver and constitute 55 percent of plasma proteins, are largely responsible for blood's osmotic pressure, which is essential in maintaining blood volume. This is discussed in detail in Chapter 16. *Globulins* comprise 38 percent of the plasma proteins. Alpha and beta globulins are produced by the liver and assist in the transport of fats and fat-soluble vitamins in the blood. Gamma globulins refer to the proteins produced by plasma cells during the process of developing immunity. These are discussed in more detail in Chapter 17. *Fibrinogen* makes up about 7 percent of plasma proteins and functions in the blood-clotting mechanism along with platelets. It is also produced by the liver.

HEMOSTASIS OF BLOOD

Hemostasis refers to the stoppage of bleeding. When blood vessels are damaged, three basic mechanisms prevent blood loss: (1) vascular spasm, (2) platelet plug formation, and (3) blood coagulation (clotting). These mechanisms are useful for preventing hemorrhage in smaller blood vessels, but extensive hemorrhage requires medical treatment.

VASCULAR SPASM

When a blood vessel is damaged, the smooth muscle in its wall contracts immediately. Such a *vascular spasm* reduces blood loss

EXHIBIT 14-1
Summary of the Formed Elements in Blood

Formed Element	Number	Diameter (μm)	Life Span	Function
Erythrocyte (red blood cell)	4.8 million/mm^3 in females; 5.4 million/mm^3 in males.	8	120 days.	Transports oxygen and carbon dioxide.
Leucocyte (white blood cell)	5000–10,000/mm^3.		Few hours to a few days.[a]	
Granular				
Neutrophil	60–70% of total.	8–10		Phagocytosis.
Eosinophil	2–4% of total.	10–12		Combats the effects of histamine in allergic reactions, phagocytizes antigen–antibody complexes, and destroys certain parasitic worms.
Basophil	0.5–1% of total.	8–10		Liberates heparin, histamine, and serotonin in allergic reactions that intensify the inflammatory response.
Agranular				
Lymphocyte	20–25% of total.	7–15		Immunity (antigen–antibody reactions).
Monocyte	3–8% of total.	14–19		Phagocytosis (after transforming into wandering macrophages).
Thrombocyte (platelet)	250,000–400,000/mm^3.	2–4	5–9 days.	Blood clotting.

[a] Some lymphocytes, called T and B memory cells, can live throughout life once they are established. Most white blood cells, however, have life spans ranging from a few hours to a few days.

for several minutes to several hours, during which time the other hemostatic mechanisms go into operation. The spasm is probably caused by damage to the vessel wall and from reflexes initiated by pain receptors.

PLATELET PLUG FORMATION

When platelets come into contact with parts of a damaged blood vessel, their characteristics change drastically. They begin to enlarge and their shapes become even more irregular. They also become sticky and begin to adhere to collagen fibers in the wound. They produce substances that activate more platelets, causing them to stick to the original platelets. The accumulation and attachment of large numbers of platelets form a mass called a ***platelet plug.*** Initially, it is loose, but it becomes quite tight when reinforced by fibrin threads formed during coagulation. A platelet plug is very effective in a small vessel and can stop blood loss completely if the hole is small.

COAGULATION

Normally, blood maintains its liquid state as long as it remains in the vessels. If it is drawn from the body, however, it thickens and forms a gel. Eventually, the gel separates from the liquid. The straw-colored liquid, called ***serum,*** is simply plasma minus its clotting proteins. The gel is called a ***clot (thrombus)*** and consists of a network of insoluble fibers in which the blood cells are trapped.

The process of clotting is called ***coagulation.*** If the blood clots too easily, the result can be ***thrombosis***—clotting in an unbroken blood vessel. If the blood takes too long to clot, bleeding can result.

Clotting involves various chemicals known as ***coagulation factors.*** In plasma, these factors are called ***plasma coagulation factors.*** A few ***platelet coagulation factors*** are released by platelets. Tissue factor is a protein released by damaged tissues which also acts as a coagulation factor.

Clotting is a complex process in which coagulation factors activate each other. That is, the first coagulation factor activates the second, the second activates the third, and so on. We will describe clotting in three basic stages:

Stage 1. Formation of ***prothrombin activator.***
Stage 2. ***Conversion of prothrombin*** (a plasma protein formed by the liver) into the enzyme ***thrombin,*** by prothrombin activator.
Stage 3. ***Conversion of fibrinogen*** (another plasma protein formed by the liver) ***into insoluble fibrin*** by thrombin. Fibrin

Cholesterol, Lipoproteins, and Life-Style

Cholesterol. Who would have guessed this white, waxy substance would generate so many articles in the popular press, so much discussion among scientists, so many new food product labels, and so much public controversy?

Cholesterol is one of the many substances in the body that use the cardiovascular system to get to where they need to go. During the past several years, population and clinical studies have repeatedly shown a strong association between the amount of cholesterol in the blood (serum cholesterol) and risk of artery disease (atherosclerosis). The higher the cholesterol level, the greater one's risk of artery disease. When cholesterol levels are lowered through life-style change or drugs, the risk of heart disease declines as well. In fact, along with hypertension and smoking, a high level of serum cholesterol (hypercholesterolemia) is considered one of the major modifiable risk factors for heart disease.

Despite this negative association, cholesterol is biologically very useful. This lipid is an important component of cell membranes and is synthesized into steroid hormones, including testosterone and estrogens. The liver uses cholesterol to produce bile salts, which are sent to the small intestine to assist lipid digestion and absorption. It is also used to make vitamin D.

Some of the confusion surrounding the cholesterol–health connection arises because cholesterol is found in two places: in our bodies and in the foods we consume. Most of the cholesterol in our bodies is produced by the liver, although some is produced by other body cells and some does come from food. Because our bodies can manufacture cholesterol, it is not considered an essential nutrient; you don't need to consume cholesterol to be healthy. Dietary cholesterol is found mostly in animal foods, such as meat, eggs, and dairy products. Yet vegans, who consume none of these things, still manufacture sufficient cholesterol for good health.

Lipoproteins

Since lipids are not soluble in the blood, an aqueous environment, they are transported by special carriers, one type of which is called lipoproteins. Lipoproteins are manufactured by the liver and are composed of triglycerides, cholesterol, phospholipids, and protein. The protein coat allows lipoproteins to dissolve in the blood. These lipoproteins are categorized by density, as measured in a centrifuge. The denser the lipoprotein, the less fat it contains. (Remember from Chapter 4 that fat is less dense than protein.)

High-density lipoproteins (HDLs) are the densest and are 8 percent triglycerides and 30 percent cholesterol. They are commonly called the "good guy" cholesterol, because the higher a person's HDL level, the *lower* his or her heart disease risk.

Low-density lipoproteins (LDLs) are less dense than HDLs and are composed of about 10 percent triglycerides and 47 percent cholesterol. As serum LDL levels rise, so does a person's risk of heart disease.

The life of a lipoprotein goes something like this. The liver takes fats that have come from the gastrointestinal tract (chylomicrons) and packages them into very-low-density lipoproteins (VLDLs), which carry triglycerides to the liver, muscles, and fat cells to be used for energy or stored as fat. VLDLs do not stay in the blood for very long. When the triglyceride has been delivered, VLDLs break up into smaller LDLs, which then transport the cholesterol throughout the body to wherever it is needed, and eventually return it to the liver. But despite their useful delivery function, LDLs appear to leave cholesterol behind in the artery walls.

HDLs are sent into the bloodstream from the liver, where they seem to pick up not only circulating cholesterol but also cholesterol that has been deposited in artery walls. HDLs bring this cholesterol back to the liver, where it is made into something useful or excreted as bile salts.

Cholesterol Screening

While not so long ago most people couldn't even pronounce cholesterol, many can now tell you their serum cholesterol levels, their HDL and LDL cholesterol levels, and recite dietary recommendations for cholesterol control in four-part harmony. This is because of a massive public health campaign that evolved from the epidemiological studies finding a strong association between serum cholesterol level and atherosclerosis. Both the medical community and the public were targeted by this campaign, the result of which is a heightened awareness of the importance of blood cholesterol testing and control. Like high blood pressure, hypercholesterolemia is a silent disease; people with this condition often have no apparent symptoms until atherosclerosis has progressed to a dangerous degree, and they suffer a heart attack or stroke.

When attempting to prevent a progressive disease such as atherosclerosis, the earlier the intervention, the better the results. People in their twenties who discover that their serum cholesterol is high can significantly delay the progression of their atherosclerosis through various life-style measures, and drug treatment if necessary. That's why the National Cholesterol Education Program recommends cholesterol testing for North Americans over age 20, with check-up tests every five years. The American Academy of Pediatrics also recommends cholesterol tests for children over age 2 who have a family history of early heart disease.

Initial cholesterol screening usually consists of a measure of all blood lipoproteins, called total cholesterol. A desirable level is less than 200 mg/dl. People whose total cholesterol is between 200 and 239 mg/dl are advised to follow the advice on life-style change given in the next section, and get tested again in a year. However, people in this category with two or more risk factors for heart disease are advised to have their blood analyzed for specific lipoproteins, as are people whose total cholesterol is 240 mg/dl and over. People with a high total cholesterol who also have a high LDL level (160 mg/dl or more) will receive aggressive treatment. Aggressive treatment usually consists of life-style change to begin with, and drug therapy if life-style change has not had a sufficient effect after six months.

An HDL level under 35 mg/dl is also considered a risk factor, even if total cholesterol is low! In fact, some experts believe that the ratio of total cholesterol to HDLs (a ratio of 4.5 or higher is considered too high) is the best predictor of heart disease.

Life-Style

Life-style change is recommended for both the prevention and treatment of hypercholesterolemia. To understand the general relationship of life-style to levels of LDL and HDL cholesterol, we might think of our bodies as being in either a fat-accumulation or a fat-utilization mode. In fat-accumulation mode, a high intake of dietary fat and calories combined with a low level of physical activity means that more fat is going into storage than is being taken out. The fat-storage process, encouraged by insulin, calls for more VLDL carriers to transport the fat to the adipose tissue. These become LDLs, which may deposit cholesterol on arterial walls.

On the other hand, a person in fat-utilization mode, getting plenty of exercise and eating a well-balanced diet, regularly uses fat for fuel. This person maintains a more healthful lipoprotein homeostasis. Fat storage is balanced with fat utilization, and there is no excessive need for LDLs. The disease process of atherosclerosis does not occur, or at least occurs at a much slower rate.

How do you get into a fat-utilization mode? Follow a balanced diet low in dietary fat, especially saturated fat, and cholesterol, lose excess weight, and exercise at a moderate intensity. These are the three recommendations most likely to help increase HDL cholesterol while lowering LDL and total cholesterol, especially important if your cholesterol level is above 200 mg/dl.

Dietary fat, especially saturated fat (found in foods such as meats, whole milk and cream, cheese, coconut, and palm oil) has a greater effect on serum cholesterol than does dietary cholesterol alone, although dietary cholesterol is still important. Since cholesterol and saturated fats are often found in the same foods, such as egg yolks and butter, by cutting down on these foods people will often automatically decrease their consumption of dietary cholesterol as well.

Replacing high-fat foods with those high in soluble fiber (found in oats, beans, fruits, and vegetables) and eating more fish also help to reduce serum cholesterol. Studies show that most people, especially those with hypercholesterolemia, can decrease their cholesterol levels by about 15 to 25 percent using dietary methods alone. Each 1-percent decrease in cholesterol is associated with approximately a 2-percent reduction in heart disease risk. More guidelines for a "heart-healthy" diet are presented in the Wellness Box in Chapter 15.

Physical activity is associated with an increase in HDL levels. It is not known exactly how this occurs, but many studies have shown that the more a person exercises, especially if the activity is aerobic, the greater the increase in HDLs. Both exercise amount and intensity seem to be important.

Exercise and diet can both help a person lose excess weight. Excess body fat not only raises serum cholesterol, but is itself also a risk factor for heart disease.

forms the threads of the clot. (Cigarette smoke contains at least two substances that interfere with fibrin formation.)

Prothrombin activator is formed along both the extrinsic and intrinsic pathways of blood clotting (Figure 14-7).

Extrinsic Pathway

The *extrinsic pathway* of blood clotting has fewer steps than the intrinsic pathway and occurs rapidly, within seconds if trauma is severe. It is so named because the formation of prothrombin activator is initiated by *tissue factor (TF)*, also called *thromboplastin,* which is found on the surfaces of cells *outside* the cardiovascular system. Damaged tissues release tissue factor (Figure 14-7a). Following several additional reactions that require calcium (Ca^{2+}) ions, tissue factor eventually converts into prothrombin activator. This completes the extrinsic pathway and stage 1 of clotting. In stage 2, prothrombin activator and Ca^{2+} ions convert prothrombin into thrombin. In stage 3, thrombin, in the presence of Ca^{2+} ions, converts fibrinogen, which is soluble, to fibrin, which is insoluble.

Thrombin has a positive feedback effect in accelerating the formation of prothrombin activator. Once clotting begins and thrombin is formed, a positive feedback cycle is set up in which thrombin accelerates production of prothrombin activator, which in turn accelerates the production of more thrombin, and so on. If unchecked, a clot would continue to get larger and larger as a result of the positive feedback cycle. Blood flow partly prevents this. In addition, there are substances in blood that inhibit or destroy coagulating factors. Blood clot control is discussed in greater detail shortly.

Intrinsic Pathway

The *intrinsic pathway* of blood clotting is more complex than the extrinsic pathway and it operates more slowly, usually requiring several minutes. The intrinsic pathway is so named because the formation of prothrombin activator is initiated by a tissue factor found on the surfaces of endothelial cells that line blood vessels and monocytes, cells *within* the cardiovascular system. The intrinsic pathway is triggered when blood comes into contact with the tissue factor of damaged endothelial cells or monocytes (Figure 14-7b). This damages the platelets, causing them to release phospholipids. Then, following several additional reactions that require the presence of Ca^{2+} ions, prothrombin activator is formed. This completes the intrinsic pathway and stage 1 of clotting. From this point on, the reactions for stages 2 and 3 are similar to those of the extrinsic pathway. Once thrombin is formed, it causes more platelets to adhere to each other, resulting in the release of more platelet phospholipids. This is another example of a positive feedback cycle.

Once the clot is formed, it plugs the ruptured area of the blood vessel and thus prevents bleeding. Permanent repair of the blood vessel can then take place. In time, fibroblasts form connective tissue in the ruptured area and new endothelial cells repair the lining of the blood vessel.

Hemophilia (*hemo* = blood; *philein* = to love) refers to several different hereditary deficiencies of coagulation. The effects of all the forms of the disorder are so similar that they are hardly distin-

guishable from one another, but each is a deficiency of a different blood-clotting factor.

Hemophilia is characterized by spontaneous or traumatic subcutaneous and intramuscular hemorrhaging, nosebleeds, blood in the urine, and joint pain and damage due to joint hemorrhaging. Treatment includes applying pressure to accessible bleeding sites and transfusing fresh plasma or the appropriate deficient clotting factor to relieve the bleeding tendency. The inheritance of hemophilia is discussed in Chapter 24.

Retraction and Fibrinolysis

Normal coagulation involves two additional events after clot formation: clot retraction and fibrinolysis. *Clot retraction* is the tightening of the fibrin clot. The fibrin threads attached to the damaged surfaces of the blood vessel gradually contract because of platelets pulling on them. As the clot retracts, it pulls the edges of the damaged vessel closer together. Thus, the risk of hemorrhage is further decreased. During retraction, some serum escapes between the fibrin threads, but the formed elements in blood remain trapped in the fibrin threads. Normal clot retraction depends on an adequate number of platelets. Platelets in the clot bind various fibrin threads together and release a coagulation factor that strengthens and stabilizes the clot and other factors that help compress the clot.

The second event following clot formation is *fibrinolysis* (fibrin-OL-i-sis), dissolution of the blood clot. When a clot is formed, an inactive plasma enzyme called *plasminogen* is incorporated into the clot. Plasminogen is activated into *plasmin,* an active plasma enzyme, by thrombin, among other substances. Once plasmin is formed, it can dissolve the clot by digesting fibrin threads and inactivating the fibrinogen, prothrombin, and coagulation factors. In addition to dissolving large clots in tissues, plasmin also removes very small clots in intact blood vessels before the clot can grow and impair blood flow to tissues.

Clot formation is a vital mechanism that prevents excessive loss of blood from the body. To form clots, the body needs calcium and vitamin K. Vitamin K is not involved in actual clot formation but is required for the synthesis of prothrombin and certain coagulation factors. The vitamin is normally produced by bacteria that live in the large intestine. Applying a thrombin or fibrin spray on a rough surface such as gauze may also promote clotting of bleeding wounds.

Hemostatic Control Mechanisms

Even though thrombin has a positive feedback effect on producing a blood clot, clot formation occurs locally at the site of damage; it does not extend beyond the wound site into general circulation. One reason for this is that some of the coagulation factors are carried away by blood flow so that their concentrations are not high enough to bring about widespread clotting. In addition, fibrin itself has the ability to absorb and inactivate up to nearly 90 percent of thrombin formed from prothrombin. This helps stop the spread of thrombin into the blood and thus inhibits clotting except at the wound.

A number of substances that inhibit coagulation are present in blood. Such substances are called *anticoagulants. Heparin* is an

FIGURE 14-7 Blood clotting. (a) Extrinsic pathway. (b) Intrinsic pathway. (c) Scanning electron micrograph of fibrin threads and red blood cells at a magnification of about 1500 to 2000×. (Courtesy of Fisher Scientific Company and S.T.E.M. Laboratories, Inc. Copyright © 1975.)

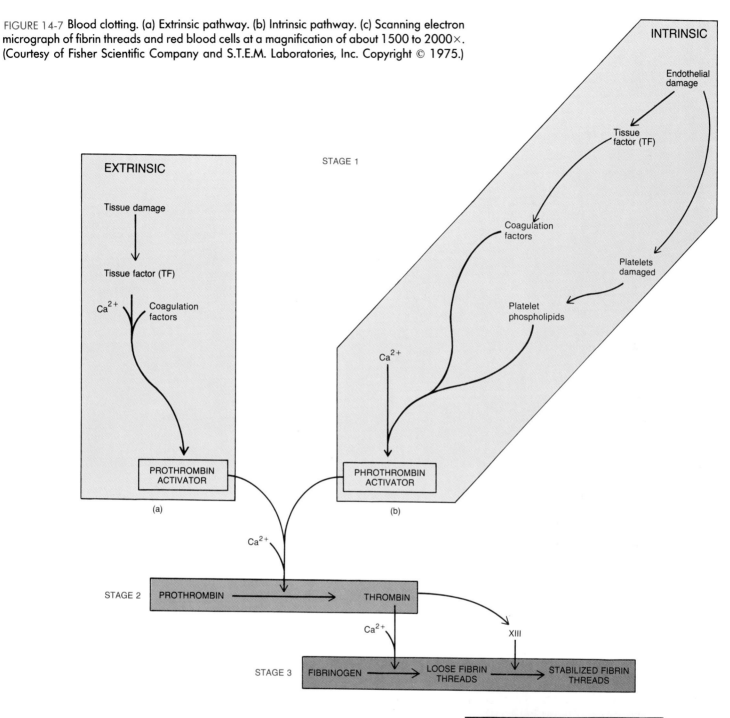

INTRINSIC

Endothelial damage

Tissue factor (TF)

Coagulation factors

Platelets damaged

Platelet phospholipids

Ca²⁺

PHROTHROMBIN ACTIVATOR

(b)

EXTRINSIC

STAGE 1

Tissue damage

Tissue factor (TF)

Ca²⁺ Coagulation factors

PROTHROMBIN ACTIVATOR

(a)

Ca²⁺

STAGE 2 PROTHROMBIN ⟶ THROMBIN

Ca²⁺   XIII

STAGE 3 FIBRINOGEN ⟶ LOOSE FIBRIN THREADS ⟶ STABILIZED FIBRIN THREADS

What three mechanisms stop bleeding?

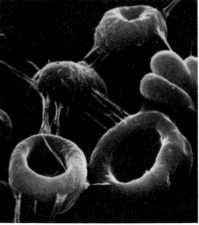

(c)

anticoagulant produced by mast cells and basophils located in endothelial cells lining blood vessels. It inhibits the conversion of prothrombin to thrombin, thereby preventing thrombus formation.

Intravascular Clotting

Even though the body has anticoagulating mechanisms, blood clots sometimes form in intact vessels. Such clots may be initiated by roughened endothelial surfaces of a blood vessel as a result of atherosclerosis, trauma, or infection. These conditions induce adhesion of platelets. Intravascular clots may also form when blood flows too slowly, allowing coagulation factors in local areas to increase in concentration and initiate coagulation. Clotting in an unbroken blood vessel (usually a vein) is called *thrombosis*. A thrombus may dissolve spontaneously, but if not, there is the possibility it will become dislodged and carried in the blood. If the clot occurs in an artery, the clot may block the circulation to a vital organ. A blood clot, bubble of air, fat from broken bones, or a piece of debris transported by the bloodstream is called an *embolus* (*em* = in; *bolus* = a mass). When an embolus becomes lodged in the lungs, the condition is called *pulmonary embolism*.

GROUPING (TYPING) OF BLOOD

The surfaces of red blood cells contain genetically determined antigens called *agglutinogens* (ag'-loo-TIN-ō-jens). There are at least 300 blood group systems based on agglutinogens, but the two major ones are ABO and Rh.

ABO

ABO blood grouping is based on two agglutinogens called *A* and *B* (Figure 14-8). Individuals whose red blood cells manufacture only agglutinogen *A* are said to have blood type A. Those who manufacture only agglutinogen *B* are type B. Individuals who manufacture both *A* and *B* are type AB. Those who manufacture neither are type O.

These four blood types are not equally distributed. They occur as follows in the white population in North America: type A, 41 percent; type B, 10 percent; type AB, 4 percent; type O, 45 percent. Among blacks, the frequencies are: type A, 27 percent; type B, 20 percent; type AB, 7 percent; and type O, 46 percent.

The blood plasma of people who are type A, B, or O contains antibodies referred to as *agglutinins* (a-GLOO-ti-nins). These are agglutinin *a* (anti-A), which attacks agglutinogen A, and agglutinin *b* (anti-B), which attacks B. Agglutinins are produced spontaneously about two months after birth and reach peak levels at about 8 to 10 years of age. They develop without known exposure to agglutinogens that would induce their formation. It is possible that small amounts of *A* or *B* agglutinogens enter the body in food, bacteria, or other ways and stimulate the formation of these agglutinins.

The agglutinins formed by each of the four blood types are shown in Figure 14-8. You do not produce agglutinins that attack the agglutinogens of your own red blood cells, but you do have an agglutinin against any agglutinogen you do not have. In an incompatible blood transfusion, the donated red blood cells are attacked by the recipient's agglutinins, causing the blood cells to *agglutinate* (clump). Agglutinated cells become lodged in small capillaries throughout the body and, over a period of hours, the

FIGURE 14-8 Agglutinogens (antigens) and agglutinins (antibodies) involved in the ABO blood grouping system.

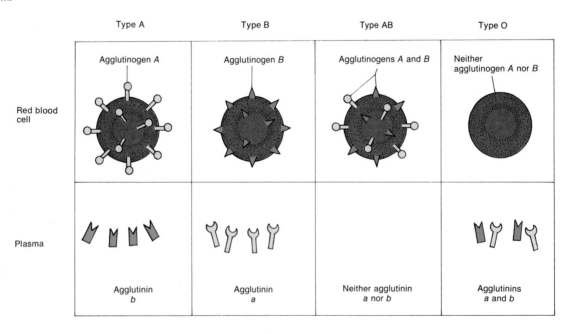

What is the other common blood grouping system besides ABO? Explain its basis for classification.

cells swell, rupture, and release hemoglobin into the blood. Such a reaction is called **hemolysis** (*lysis* = dissolve). The degree of agglutination depends on the amount of agglutinin in the blood. This reaction is another example of an antigen–antibody response (Figure 14-9).

Blood transfusions are most frequently given when blood volume is low, for example, during shock. When blood is transfused, care must be taken to avoid incompatibilities that can result in agglutination, for agglutinated cells can block blood vessels and lead to kidney or brain damage and death, and the liberated hemoglobin may cause kidney damage.

As an example of an incompatible blood transfusion, consider what happens if an individual with type A blood receives a transfusion of type B blood. The recipient's blood (type A) contains *A*

FIGURE 14-9 Compatibility of blood types as viewed through a microscope. (a) Since the blood type shown has no agglutinins to attack the agglutinogens, no agglutination (clumping) occurs. (b) But if a blood type has incompatible agglutinins, they will attack the agglutinogens and agglutination (clumping) occurs. (Courtesy of Lester Bergman and Associates)

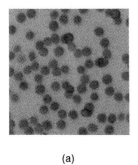   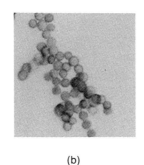

(a)   (b)

What happens when cells agglutinate? Why is this dangerous?

agglutinogens and *b* agglutinins. The donor's blood (type B) contains *B* agglutinogens and *a* agglutinins. Given this situation, two things can happen. First, the *b* agglutinins in the recipient's plasma will attack the *B* agglutinogens on the donor's red blood cells, causing hemolysis of the red blood cells. Second, the *a* agglutinins in the donor's plasma will attack the *A* agglutinogens on the recipient's red blood cells, causing their hemolysis. However, the latter reaction is usually not serious because the donor's *a* agglutinins become so diluted in the recipient's plasma that they do not cause any significant hemolysis of the recipient's red blood cells. Thus, a person with type A blood may not receive type B or AB blood, but may receive type A or type O. The interaction of the four blood types of the ABO system are summarized in Exhibit 14-2.

Individuals with type AB blood do not have any agglutinins in their plasma and are therefore called universal recipient's because they can *theoretically* receive blood from donors of all four blood types. They have no agglutinins to attack donated red blood cells (Figure 14-10). Individuals with type O blood have no agglutinogens on their red blood cells and are therefore referred to as universal donors because they can *theoretically* donate blood to recipients of all four blood types. Type O persons requiring blood may receive

FIGURE 14-10 Since type AB can *theoretically* receive from all other blood types, it is called the universal recipient. Type O is called the universal donor since it can *theoretically* give to all other blood types.

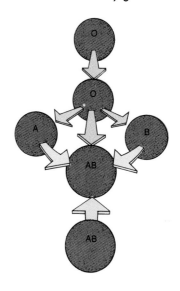

What blood types are compatible with type A blood? Incompatible?

only type O blood, though in practice, use of the terms universal recipient and universal donor is misleading and dangerous because there are other agglutinogens and agglutinins in blood besides those associated with the ABO system that can cause transfusion problems. Thus, blood should be carefully matched before transfusion, except in extreme emergency.

Knowledge of blood types is also used in paternity suits, linking suspects to crimes, and as part of studies to establish a relationship among ethnic groups. In about 80 percent of the population, soluble antigens of the ABO type appear in saliva and other bodily fluids. In criminal investigations it has been possible to type such fluids from saliva residues on a cigarette or from semen in cases of rape.

Rh

The **Rh system** of blood classification is so named because it was first worked out in the blood of the **Rhesus** monkey. Like the ABO grouping, the Rh system is based on agglutinogens that lie on the surfaces of red blood cells. Individuals whose red blood cells have the Rh agglutinogens are designated **Rh⁺**. Those who lack Rh agglutinogens are designated **Rh⁻**. It is estimated that 85 percent of whites and 88 percent of blacks in North America are Rh^+, whereas 15 percent of whites and 12 percent of blacks are Rh^-.

Under normal circumstances, human plasma does not contain anti-Rh agglutinins. Unlike the ABO system in which agglutinins develop spontaneously in blood, in the Rh system, agglutinins develop only upon greater exposure to Rh agglutinogens. For example, if an Rh^- person receives Rh^+ blood, the body starts to make anti-Rh agglutinins that will remain in the blood. If a second transfusion of Rh^+ blood is given later, the previously formed anti-Rh

EXHIBIT 14-2 *Summary of ABO System Interactions*				
Blood Type	Agglutinogen (antigen)	Agglutinin (antibody)	Compatible Donor Blood Types	Incompatible Donor Blood Types
A	*A*	*b*	A, O	B, AB
B	*B*	*a*	B, O	A, AB
AB	*A, B*	Neither *a* nor *b*	A, B, AB, O	—
O	Neither *A* nor *B*	*a, b*	O	A, B, AB

agglutinins will react against the donated blood and a severe reaction may occur.

The most common problem with Rh incompatibility may arise during pregnancy. During pregnancy, a small amount of the fetus's blood may leak from the placenta into the mother's bloodstream, with the greatest possibility of transfer occurring at delivery. If the fetus is Rh$^+$ and the mother is Rh$^-$, she, upon exposure to the Rh$^+$ fetal cells or cellular fragments, will make anti-Rh agglutinins. If she becomes pregnant again, her anti-Rh agglutinins will cross the placenta and make their way into the bloodstream of the baby. If the fetus is Rh$^-$, no problem will occur, since Rh$^-$ blood does not have the Rh agglutinogen. If the fetus is Rh$^+$, hemolysis may occur in the fetal blood. The hemolysis brought on by fetal–maternal incompatibility is called **hemolytic disease of the newborn** (**HDN**) or **erythroblastosis fetalis.**

When a baby is born with this condition, blood is slowly removed and replaced a little at a time, with Rh$^-$ blood. It is even possible to transfuse blood into the unborn child if the disease is diagnosed before birth. More important, though, the disorder can be prevented with an injection of anti-Rh gamma$_2$-globulin preparation, administered to Rh$^-$ mothers at 26 to 28 weeks of gestation and right after delivery, miscarriage, or abortion. These agglutinins tie up the fetal agglutinogens, if present, so the mother cannot respond to the foreign agglutinogens by producing agglutinins. Thus, the fetus of the next pregnancy is protected. In the case of an Rh$^+$ mother, there are no complications, since she cannot make anti-Rh agglutinins. In cases where Rh incompatibility is not implicated in hemolytic disease of the newborn, incompatibility of other blood groups such as the ABO group is frequently involved.

■ COMMON DISORDERS ■

Anemia

Anemia is a condition in which the oxygen-carrying capacity of the blood is impaired. There are several types of anemia (described next) and all lead to fatigue and intolerance to cold, both of which are related to lack of oxygen needed for energy and heat production, and to paleness, which is due to low hemoglobin content.

Nutritional anemia arises from inadequate diet, one without sufficient amounts of iron, the necessary amino acids, or vitamin B_{12}.

Pernicious anemia is the insufficient production of erythrocytes because of lack of vitamin B_{12}, resulting from an inability of the stomach to produce intrinsic factor, which allows absorption of vitamin B_{12}.

An excessive loss of erythrocytes through bleeding is called **hemorrhagic anemia.**

Hemolytic anemia is characterized by distortion in the shape of erythrocytes that are progressing toward hemolysis. The premature destruction of red blood cells may result from inherent defects, parasites, toxins, and antibodies from incompatible blood (Rh$^-$ mother and Rh$^+$ fetus, for instance).

Destruction or inhibition of red bone marrow results in *aplas-

tic anemia.* Toxins, radiation, and certain medications are causes.

The erythrocytes of a person with **sickle-cell anemia** (**SCA**) manufacture an abnormal kind of hemoglobin. When an erythrocyte gives up its oxygen to the interstitial fluid, its hemoglobin tends to form long, stiff, rodlike structures that bend the erythrocyte into a sickle shape. The sickled cells rupture easily. Prolonged oxygen reduction may eventually cause extensive tissue damage. Furthermore, they tend to get stuck in blood vessels and can cut off blood supply to an organ altogether.

Polycythemia

The term *polycythemia* (pol′-ē-sī-THĒ-mē-a) refers to an abnormal increase in the number of red blood cells.

Infectious Mononucleosis (IM)

Infectious mononucleosis (*IM*) is a contagious disease primarily affecting lymphoid tissue throughout the body but also affecting the blood. It is caused by the Epstein–Barr virus (EBV). It occurs mainly in children and young adults. The virus most commonly enters the body through intimate oral contact.

Chronic Fatigue Syndrome

Chronic fatigue syndrome typically occurs among young adults, primarily females. It is characterized by extreme fatigue that impairs normal activities for at least six months and the absence of known diseases that might produce similar symptoms.

Leukemia

Clinically, *leukemia* is classified on the basis of the duration and character of the disease, that is, acute or chronic. Acute leukemia is a malignant disease of blood-forming tissues charac-terized by uncontrolled production and accumulation of immature leucocytes. In chronic leukemia, there is an accumulation of mature leucocytes in the bloodstream because they do not die at the end of their normal life span. The *human T-cell leukemia-lymphoma virus-1* (*HTLV-1*) is strongly associated with some types of leukemia.

In acute leukemia, anemia and bleeding problems result from the crowding out of normal bone marrow cells by the overproduc-tion of immature cells, preventing normal production of red blood cells and platelets. The abnormal accumulation of imma-ture leucocytes may be reduced by using x-rays and antileukemic drugs.

MEDICAL TERMINOLOGY AND CONDITIONS

Apheresis (a-FER-e-sis; *aphairesis* = removal) Procedure in which blood is withdrawn from the body, its components are selectively separated, the undesirable component causing disease is removed, and the remainder is returned to the body.

Autologous intraoperative transfusion (*AIT*) Procedure in which blood lost during surgery is suctioned from the patient, treated with an anticoagulant, filtered of debris, and centri-fuged (spun at high speed) to recover the red blood cells. Then the red blood cells are washed in saline solution and returned to the patient.

Autologous (aw-TOL-o-gus; *auto* = self) *transfusion* (trans-FYOO-zhun) Donating one's own blood for up to 6 weeks before elective surgery to ensure an abundant supply and also reduce transfusion complications such as those that may be associated with AIDS and hepatitis. Also called *predona-tion.*

Blood bank A stored supply of blood for future use by the donor or other individuals. Since blood banks have now assumed additional and diverse functions (immunohematol-ogy reference work, continuing medical education, bone and tissue storage, and clinical consultation), they are more appro-priately referred to as *centers of transfusion medicine.*

Citrated (SIT-rāt-ed) *whole blood* Whole blood protected from coagulation by CPD (citrate phosphate dextrose) or a similar compound.

Cyanosis (sī'-a-NŌ-sis; *cyano* = blue) A reduced oxyhemoglo-bin in blood that causes a slightly bluish, dark purple discol-oration most easily seen in the nail beds and mucous mem-branes.

Direct (*immediate*) *transfusion* (*trans* = through) Transfer of blood directly from one person to another without exposing the blood to a storage container.

Exchange transfusion Removing blood from the recipient while alternately replacing it with donor blood. This method is used for treating hemolytic disease of the newborn (HDN) and poisoning. This may be done intrauterine.

Gamma globulin (GLOB-yoo-lin) Solution of globulins from nonhuman blood consisting of antibodies that react with spe-cific pathogens, such as measles, epidemic hepatitis, tetanus, and possibly poliomyelitis viruses.

Hemochromatosis (hē-mō-krō'-ma-TŌ-sis; *heme* = iron; *chroma* = color) Disorder of iron metabolism characterized by excess deposits of iron in tissues, especially the liver and pancreas, that result in bronze coloration of the skin, cirrhosis, diabetes mellitus, and bone and joint abnormalities.

Hemorrhage (HEM-or-ij; *rrhage* = bursting forth) Bleeding, either internal (from blood vessels into tissues) or external (from blood vessels directly to the surface of the body).

Indirect (*mediate*) *transfusion* Transfer of blood from a donor to a container and then to the recipient, permitting blood to be stored for an emergency.

Multiple myeloma (mī'-e-LŌ-ma) Malignant disorder of plasma cells in bone marrow.

Platelet (PLĀT-let) *concentrates* A preparation of platelets ob-tained from freshly drawn whole blood and used for transfu-sions in platelet-deficiency disorders such as hemophilia.

Reciprocal (re-CIP-rō-cal) *transfusion* Transfer of blood from a person who has recovered from a contagious infection into a patient suffering with the same infection. An equal amount of blood is returned from the patient to the well person.

Septicemia (sep'-ti-SĒ-mē-a; *sep* = decay; *emia* = condition of blood) Toxins or disease-causing bacteria in the blood. Also called "blood poisoning."

Thrombocytopenia (throm'-bō-sī'-tō-PĒ-nē-a; *thrombo* = clot; *penia* = poverty) Very low platelet count that results in a tendency to bleed from capillaries.

Transfusion (trans-FYOO-zhun) Transfer of whole blood, blood components (red blood cells only or plasma only), or bone marrow directly into the bloodstream.

Venesection (vēn'-e-SEK-shun; *veno* = vein) Opening of a vein for withdrawal of blood.

Whole blood Blood containing all formed elements, plasma, and plasma solutes in natural concentration.

STUDY OUTLINE

Functions of Blood (p. 298)

1. Blood transports oxygen, carbon dioxide, nutrients, wastes, and hormones.
2. It helps to regulate pH, body temperature, and water content of cells.
3. It prevents blood loss through clotting and combats toxins and microbes through special combat-unit cells.

Physical Characteristics of Blood (p. 298)

1. The cardiovascular system consists of blood, the heart, and blood vessels.
2. Blood inside blood vessels, interstitial fluid around body cells, and lymph inside lymph vessels constitute the body's internal environment.
3. Physical characteristics of blood include temperature, 38°C (100.4°F); pH, 7.35 to 7.45; and salinity, 0.85 to 0.90 NaCl. Blood constitutes about 8 percent of body weight.

Components of Blood (p. 299)

1. The formed elements in blood include erythrocytes (red blood cells), leucocytes (white blood cells), and thrombocytes (platelets).
2. Blood cells are formed by a process called hemopoiesis.
3. Red bone marrow is responsible for producing red blood cells, granular leucocytes, and platelets; lymphoid tissue and red bone marrow tissue produce agranular leucocytes.

Erythrocytes (Red Blood Cells) (p. 301)

1. Erythrocytes are biconcave discs without nuclei and containing hemoglobin.
2. The function of red blood cells is to transport oxygen and carbon dioxide.
3. Red blood cells live about 120 days. A healthy male has about 5.4 million/mm^3 of blood; a healthy female about 4.8 million/mm^3.
4. Erythrocyte formation, called erythropoiesis, occurs in adult red marrow of certain bones.
5. A reticulocyte count is a diagnostic test that indicates the rate of erythropoiesis.
6. A hematocrit (Hct) measures the percentage of red blood cells in whole blood.

Leucocytes (White Blood Cells) (p. 302)

1. Leucocytes are nucleated cells. The two principal types are granular (neutrophils, eosinophils, basophils) and agranular (lymphocytes and monocytes).
2. The general function of leucocytes is to combat inflammation and infection. Neutrophils and monocytes (wandering macrophages) do so through phagocytosis.
3. Eosinophils combat the effects of histamine in allergic reactions, phagocytize antigen–antibody complexes, and combat parasitic worms.
4. Basophils liberate heparin, histamine, and serotonin in allergic reactions, which intensify the inflammatory response.
5. Lymphocytes, in response to the presence of foreign substances called antigens, differentiate into tissue plasma cells that produce antibodies. Antibodies attach to the antigens and render them harmless. This antigen–antibody response combats infection and provides immunity.
6. A differential white blood cell count is a diagnostic test in enumerating the percentage of each type of white blood cell in a sample.
7. White blood cells usually live for only a few hours or a few days. Normal blood contains 5000 to 10,000/mm^3.

Thrombocytes (Platelets) (p. 304)

1. Thrombocytes are disc-shaped structures without nuclei.
2. They are formed from megakaryocytes and are involved in clotting.
3. Normal blood contains 250,000 to 400,000/mm^3.

Plasma (p. 304)

1. The liquid portion of blood, called plasma, consists of 91.5 percent water and 8.5 percent solutes.
2. Principal solutes include proteins, albumins, globulins, fibrinogens, nutrients, gases, electrolytes, wastes, enzymes, and hormones.

Hemostasis of Blood (p. 304)

1. Hemostasis refers to the prevention of blood loss.
2. The three hemostatic mechanisms are vascular spasm, platelet plug formation, and blood coagulation.
3. In vascular spasm, the smooth muscle of a blood vessel wall contracts to stop bleeding.
4. Platelet plug formation is the clumping of platelets to stop bleeding.
5. A clot is a network of insoluble protein (fibrin) in which formed elements of blood are trapped.
6. The chemicals involved in clotting are known as coagulation factors. There are two kinds, plasma and platelet coagulation factors.
7. Blood clotting involves two pathways, intrinsic and extrinsic.
8. Normal coagulation also involves clot retraction (tightening of the clot) and fibrinolysis (dissolution of the clot).
9. Clotting in an unbroken blood vessel is called thrombosis. A thrombus that moves from its site of origin is called an embolus.
10. Anticoagulants, such as heparin, prevent clotting.

Grouping (Typing) of Blood (p. 310)

1. ABO and Rh systems are based on antigen–antibody responses.

2. In the ABO system, agglutinogens (antigens) *A* and *B* determine blood type. Plasma contains agglutinins (antibodies), designated as *a* and *b*, that clump agglutinogens foreign to the individual.

3. In the Rh system, individuals whose erythrocytes have Rh agglu-tinogens are classified as Rh^+. Those who lack the antigen are Rh^-.

4. A disorder due to Rh incompatibility between mother and fetus is hemolytic disease of the newborn or HDN (erythroblastosis fetalis).

REVIEW QUESTIONS

1. List the functions and physical characteristics of blood. (p. 298)

2. Describe the origin of blood cells. (p. 299)

3. What is the function of erythrocytes? Define reticulocyte count and hematocrit (Hct). (p. 301)

4. Describe the classification of leucocytes. What are their functions? (p. 302)

5. What is the importance of diapedesis and phagocytosis in fighting bacterial invasion? (p. 302)

6. What is a differential white blood cell count? What is its significance? Distinguish between leucocytosis and leucopenia. (p. 304)

7. Describe the function of thrombocytes. (p. 304)

8. Compare erythrocytes, leucocytes, and thrombocytes with respect to size, number per mm^3, and life span. (p. 305)

9. What are the major components in plasma? What do they do? (p. 304)

10. Define hemostasis. Explain the mechanism involved in vascular spasm and platelet plug formation. (p. 304)

11. Describe the process of clot formation. What is fibrinolysis? Why does blood usually not remain clotted in vessels? (p. 305)

12. What is the basis for ABO blood grouping? What are agglutinogens and agglutinins? (p. 310)

13. What is the basis for the Rh system? How does hemolytic disease of the newborn or HDN (erythroblastosis fetalis) occur? How may it be prevented? (p. 311)

14. Define the following: anemia, polycythemia, infectious mononucleosis (IM), chronic fatigue syndrome, and leukemia. (p. 312)

15. Refer to the glossary of medical terminology and conditions associated with blood. Be sure that you can define each term. (p. 313)

The Cardiovascular System: Heart

■ STUDENT OBJECTIVES

1. Describe the structure of the heart and its functions as the center of the cardiovascular system.
2. Explain how problems with the blood vessels of the heart may lead to serious disorders.
3. Describe how nerve impulses travel through the heart to maintain heartbeat.
4. Explain the importance of an electrocardiogram (ECG).
5. Describe the phases of a heartbeat.
6. Explain the factors that affect heart rate.
7. List and explain the risk factors involved in heart disease.
8. Define common disorders and medical terminology and conditions associated with the heart.

A LOOK AHEAD

LOCATION OF HEART
PERICARDIUM
HEART WALL
CHAMBERS OF THE HEART
GREAT VESSELS OF THE HEART
VALVES OF THE HEART
    Atrioventricular (AV) Valves (Cuspid
        Valves)
    Semilunar Valves
BLOOD SUPPLY
CONDUCTION SYSTEM
ELECTROCARDIOGRAM (ECG OR EKG)
BLOOD FLOW THROUGH THE HEART
CARDIAC CYCLE
    Phases
    Timing
    Sounds
CARDIAC OUTPUT (CO)
HEART RATE
    Autonomic Control
    Chemicals
    Temperature
    Emotions
    Sex and Age
RISK FACTORS IN HEART
        DISEASE
COMMON DISORDERS
MEDICAL TERMINOLOGY AND
        CONDITIONS

The *heart* is the center of the cardiovascular system. It is a hollow, muscular organ that weighs between 250 and 350 grams (9 to 12 oz) and beats over 100,000 times a day to pump 7000 liters (1835 gallons) of blood per day through over 60,000 miles of blood vessels. The blood vessels form a network of tubes that carry blood from the heart to the tissues of the body and then return it to the heart.

The study of the heart and diseases associated with it is known as *cardiology* (kar-dē-OL-ō-jē; *cardio* = heart).

LOCATION OF HEART

The heart is situated between the lungs, in the mediastinum (see Figure 1-7). About two-thirds of the heart lies to the left of the body's midline (Figure 15-1). The heart is shaped like a blunt cone about the size of your closed fist. Its pointed end, the *apex,* is formed by the tip of the left ventricle, a lower chamber of the heart, and rests on the diaphragm. The major blood vessels are attached to the *base* of the heart formed by the atria (upper chamber of the heart), mostly the left atrium.

PERICARDIUM

The heart is enclosed and held in place by the *pericardium.* It consists of two portions, the fibrous pericardium and the serous pericardium (Figure 15-2). The outer *fibrous pericardium* is very heavy fibrous connective tissue and prevents overdistention of the heart, provides a tough protective membrane around the heart, and anchors the heart in the mediastinum. The inner *serous pericardium* is a thinner, more delicate membrane that forms a double layer around the heart. The outer *parietal layer* of the serous pericardium is directly beneath the fibrous pericardium. The inner *visceral layer* of the serous pericardium, also called the *epicardium,* is beneath the parietal layer, attached to the myocardium (muscle) of the heart. Between the parietal and visceral layers of the serous pericardium is a thin film of serous fluid that holds the two layers together much like a thin film of water binds two microscopic slides. The fluid, known as *pericardial fluid,* prevents friction between the membranes as the heart moves. The space occupied by the pericardial fluid is a potential space (not an actual space) called the *pericardial cavity.*

An inflammation of the pericardium is known as *pericarditis.* Pericarditis with a buildup of pericardial fluid or extensive bleeding into the pericardium, if untreated, is a life-threatening condition. Since the pericardium cannot stretch to accommodate the excessive fluid or blood buildup, the heart is subjected to compression. This compression is known as *cardiac tamponade* (tam'-pon-ĀD) and can result in cardiac failure.

HEART WALL

The wall of the heart (Figure 15-2) is divided into three layers: epicardium (external layer), myocardium (middle layer), and endocardium (inner layer). The *epicardium* (which is also the visceral layer of serous pericardium) is the thin, transparent outer layer of the wall. It is composed of serous tissue and epithelium.

The *myocardium* consists of cardiac muscle tissue, which constitutes the bulk of the heart. Cardiac muscle fibers (cells) are involuntary, striated, and branched, and the tissue is arranged in interlacing bundles of fibers. The myocardium is responsible for the pumping action of the heart.

The *endocardium* is a thin layer of simple squamous epithelium that lines the inside of the myocardium and covers the valves of the heart and the tendons that hold them open. It is continuous with the epithelial lining of the large blood vessels.

317

FIGURE 15-1 Position of the heart and associated blood vessels in the thoracic cavity. In this and subsequent illustrations, vessels that carry oxygenated blood are shown in red; vessels that carry deoxygenated blood are shown in blue.

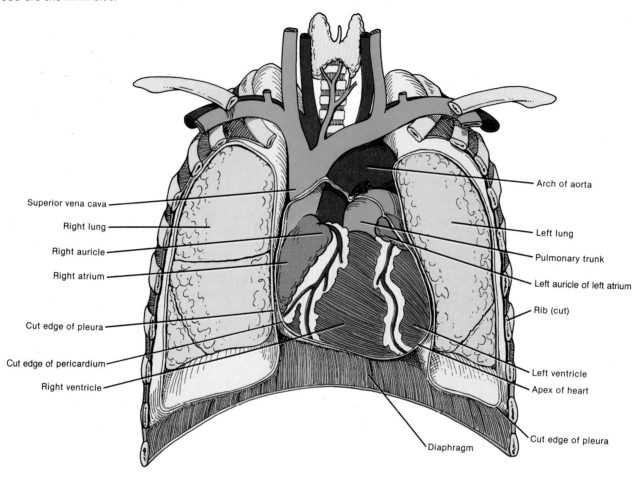

Superior vena cava

Right lung

Right auricle

Right atrium

Cut edge of pleura

Cut edge of pericardium

Right ventricle

Arch of aorta

Left lung

Pulmonary trunk

Left auricle of left atrium

Rib (cut)

Left ventricle

Apex of heart

Cut edge of pleura

Diaphragm

What is the mediastinum? What is the apex of the heart? The base?

FIGURE 15-2 Structure of the pericardium and heart wall.

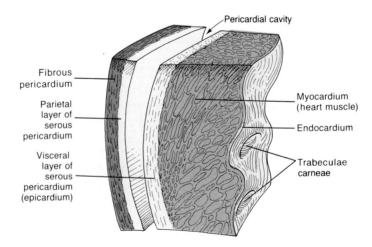

Pericardial cavity

Fibrous pericardium

Parietal layer of serous pericardium

Visceral layer of serous pericardium (epicardium)

Myocardium (heart muscle)

Endocardium

Trabeculae carneae

What is the composition of the epicardium? The myocardium? The endocardium?

Inflammations of the epicardium, myocardium, and endocardium are referred to as *epicarditis*, *myocarditis*, and *endocarditis*.

CHAMBERS OF THE HEART

The interior of the heart is divided into four cavities called *chambers* that receive and pump blood (Figure 15-3). The two upper chambers are called the right and left *atria* (*atrium* = court or hall). Each atrium has an appendage called an *auricle* (OR-i-kul; *auris* = ear), so named because its shape resembles a dog's ear. The auricle increases the atrium's surface area. The atria are separated by a partition called the *interatrial septum*. The two lower chambers are the right and left *ventricles*. They are separated from each other by an *interventricular septum*. Externally, a groove known as the *coronary sulcus* (SUL-kus) separates the atria from the ventricles. It encircles the heart and contains fat and coronary blood vessels.

The thickness of the four chambers varies according to their function (Figure 15-3c). The atria are thin-walled because they need only enough cardiac muscle tissue to pump the blood to the ventricles below. Both ventricles pump blood into arteries, blood vessels that carry blood away from the heart. The right ventricle has a thicker wall because it must send blood to the lungs and back to the left atrium. The left ventricle has the thickest wall, since it must pump blood at high pressure through literally thousands of miles of vessels in the head, trunk, and extremities.

GREAT VESSELS OF THE HEART

The right atrium receives deoxygenated blood (blood that has given up its oxygen to cells) from all parts of the body except the lungs. It receives the blood through three veins. The *superior vena cava* (*SVC*) brings blood mainly from parts of the body above the heart; the *inferior vena cava* (*IVC*) brings blood mostly from parts of the body below the heart; and the *coronary sinus* drains blood from most of the vessels supplying the wall of the heart (Figure 15-3b, c). The right atrium then delivers the blood into the right ventricle, which pumps it into the *pulmonary trunk*. The pulmonary trunk divides into a *right* and *left pulmonary artery*, each of which carries blood to the corresponding lung. In the lungs, the blood releases its carbon dioxide and takes on oxygen. Oxygenated blood (blood that has not given up its oxygen to cells) is transported to the left atrium via four *pulmonary veins*. The blood then passes into the left ventricle, which pumps the blood into the *ascending aorta*. From here the blood is passed into the *coronary arteries, arch of the aorta, thoracic aorta,* and *abdominal aorta*. These blood vessels and their branches transport the blood to all body parts.

VALVES OF THE HEART

As each chamber of the heart contracts, it pushes a portion of blood into a ventricle or out of the heart through an artery. To keep the blood from flowing backward, the heart has four *valves* composed of dense connective tissue.

ATRIOVENTRICULAR (AV) VALVES (CUSPID VALVES)

Atrioventricular (*AV*) *valves* or *cuspid valves* lie between the atria and ventricles (Figure 15-3c). The atrioventricular valve between the right atrium and right ventricle is also called the *tricuspid valve* because it consists of three cusps (flaps). These cusps are fibrous tissues that grow out of the walls of the heart and are covered with endocardium. The pointed ends of the cusps project into the ventricle. Tendonlike cords called *chordae tendineae* (KOR-dē TEN-di-nē) connect the pointed ends to small modified cardiac muscle projections located on the inner surface of the ventricles called *papillary muscles*. The chordae tendineae are like the spokes of an umbrella which keep the valves from pushing up into the atria when the ventricles contract, much the same way that an umbrella might be blown inside out by a very strong wind. This action prevents backflow of blood to conserve the workload of the heart.

The atrioventricular valve between the left atrium and left ventricle is called the *bicuspid* (*mitral*) *valve*. It has two cusps that work in the same way as the cusps of the tricuspid valve. For blood to pass from an atrium to a ventricle, an atrioventricular (AV) valve must open. As you will see later, the opening and closing of the valves is due to pressure differences across the valves. When blood moves from an atrium to a ventricle, the valve is pushed open, the papillary muscles relax, and the chordae tendineae slacken (Figure 15-4a). When a ventricle contracts, the pressure of the ventricular blood drives the cusps upward until their edges meet and close the opening (Figure 15-4b). At the same time, contraction of the papillary muscles and tightening of the chordae tendineae help prevent the valve from swinging upward into the atrium.

SEMILUNAR VALVES

Both arteries that leave the heart have a valve that is designed to prevent blood from flowing back into the heart. These valves are referred to as *semilunar valves* (see Figure 15-3c). The *pulmonary semilunar valve* lies in the opening where the pulmonary trunk leaves the right ventricle. The *aortic semilunar valve* is situated at the opening between the left ventricle and the aorta.

Both valves consist of three semilunar (half-moon or crescent-shaped) cusps. Each cusp is attached to the artery wall. The free borders of the cusps curve outward and project into the opening inside the blood vessel. Like the atrioventricular valves, the semilunar valves permit blood to flow in only one direction—in this case, from the ventricles into the arteries.

Both sets of valves open and close in response to the pressure of the blood against them, as will be discussed in more detail later.

The heart valves are attached to dense rings of fibrous tissue around the proximal ends of the pulmonary trunk and aorta. These rings help to prevent dilation of the atrial and ventricular outlets during myocardial contractions. This supportive tissue also helps to prevent backflow of blood. These rings and other masses of fibrous tissue between the atria and ventricles and in the walls of the septa are referred to as the *cardiac skeleton*.

FIGURE 15-3 Structure of the heart. (a) Anterior external view. (b) Posterior external view. (c) Anterior internal view. (d) Path of blood through the heart.

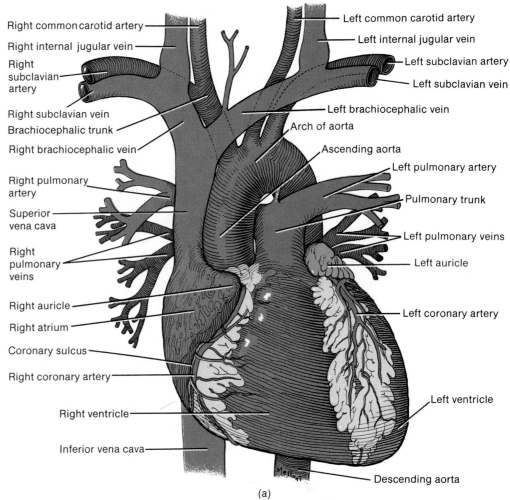

Right common carotid artery
Right internal jugular vein
Right subclavian artery
Right subclavian vein
Brachiocephalic trunk
Right brachiocephalic vein
Right pulmonary artery
Superior vena cava
Right pulmonary veins
Right auricle
Right atrium
Coronary sulcus
Right coronary artery
Right ventricle
Inferior vena cava

Left common carotid artery
Left internal jugular vein
Left subclavian artery
Left subclavian vein
Left brachiocephalic vein
Arch of aorta
Ascending aorta
Left pulmonary artery
Pulmonary trunk
Left pulmonary veins
Left auricle
Left coronary artery
Left ventricle
Descending aorta

(a)

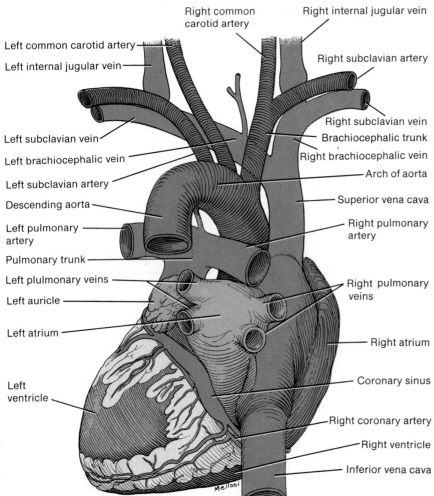

Right common carotid artery
Right internal jugular vein
Left common carotid artery
Left internal jugular vein
Left subclavian vein
Left brachiocephalic vein
Left subclavian artery
Descending aorta
Left pulmonary artery
Pulmonary trunk
Left plulmonary veins
Left auricle
Left atrium
Left ventricle

Right subclavian artery
Right subclavian vein
Brachiocephalic trunk
Right brachiocephalic vein
Arch of aorta
Superior vena cava
Right pulmonary artery
Right pulmonary veins
Right atrium
Coronary sinus
Right coronary artery
Right ventricle
Inferior vena cava

(b)

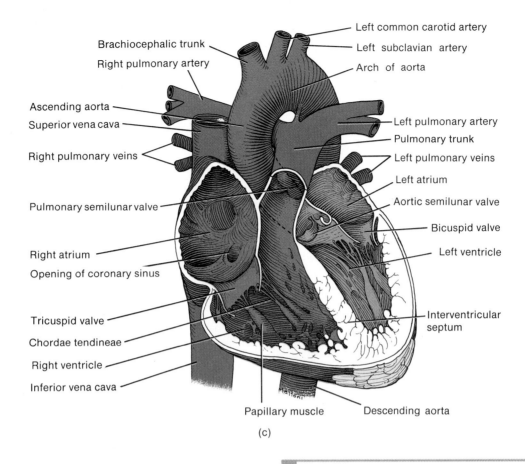

Left common carotid artery
Left subclavian artery
Arch of aorta
Brachiocephalic trunk
Right pulmonary artery
Ascending aorta
Superior vena cava
Right pulmonary veins
Left pulmonary artery
Pulmonary trunk
Left pulmonary veins
Left atrium
Aortic semilunar valve
Pulmonary semilunar valve
Bicuspid valve
Right atrium
Left ventricle
Opening of coronary sinus
Tricuspid valve
Interventricular septum
Chordae tendineae
Right ventricle
Inferior vena cava
Papillary muscle
Descending aorta

(c)

Trace the path of blood through the heart, starting in the right atrium and ending in the aorta.

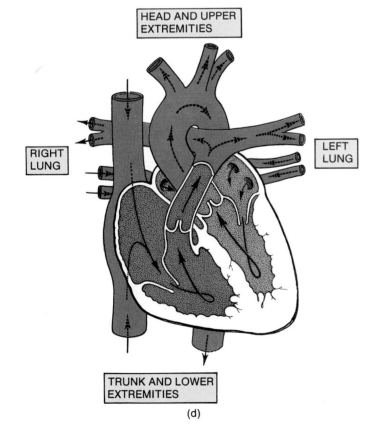

HEAD AND UPPER EXTREMITIES

RIGHT LUNG

LEFT LUNG

TRUNK AND LOWER EXTREMITIES

(d)

FIGURE 15-4 Atrioventricular (AV) valves. (a) Bicuspid valve open. (b) Bicuspid valve closed. The tricuspid valve operates in a similar manner.

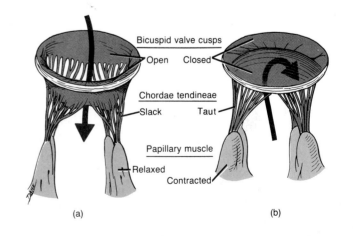

Bicuspid valve cusps
Open   Closed
Chordae tendineae
Slack   Taut
Papillary muscle
Relaxed
Contracted

(a)   (b)

What is the function of the heart valves?

BLOOD SUPPLY

The wall of the heart, like any other tissue, has its own blood vessels. Nutrients could not possibly diffuse through all three layers of the heart wall. The flow of blood through the numerous vessels in the myocardium is called *coronary circulation.*

The principal coronary vessels are the *left* and *right coronary arteries,* which originate as branches of the ascending aorta (see Figure 15-3a). Each artery branches and subbranches to deliver oxygen and nutrients and collect carbon dioxide and wastes. Most of the deoxygenated blood, which carries the carbon dioxide and wastes, is collected by a large vein on the back of the heart, the *coronary sinus* (see Figure 15-3b), which empties into the right atrium.

Most heart problems result from faulty coronary circulation. If a reduced oxygen supply weakens cells but does not actually kill them, the condition is called *ischemia* (is-KĒ-mē-a). *Angina pectoris* (an-JĪ-na, or AN-ji-na, PEK-to-ris), meaning "chest pain," results from ischemia of the myocardium. Common causes include stress, strenuous exertion after a heavy meal, atherosclerosis, coronary artery spasm, hypertension, fever, anemia, hyperthyroidism, and aortic stenosis. Symptoms include chest pain, accompanied by tightness or pressure, labored breathing, and a sensation of foreboding. Sometimes weakness, dizziness, and perspiration occur.

A much more serious problem is *myocardial infarction* (in-FARK-shun), or *MI,* commonly called a heart attack. *Infarction* means death of an area of tissue because of an interrupted blood supply. Myocardial infarction may result from a thrombus or embolus in one of the coronary arteries. Tissue dies and is replaced by noncontractile scar tissue, which causes the heart muscle to lose some of its strength. The aftereffects depend partly on the size and location of the infarcted, or dead, area. In addition to killing normal heart tissue, an infarction may disturb the heart's conducting system. These are the heart attacks that cause sudden death (ventricular fibrillation) which may, nevertheless, be reversed by timely cardiopulmonary resuscitation (CPR). Individuals who survive an MI may develop new blood vessels (collaterals) that carry blood through secondary channels (collateral circulation) following obstruction of the principal channel.

CONDUCTION SYSTEM

Cardiac muscle fibers form two separate networks—one atrial and one ventricular. Each fiber is in physical contact with other fibers in the networks by transverse thickenings of the sarcolemma called *intercalated discs* (see Exhibit 4-3). Within the discs are *gap junctions* that aid in the conduction of muscle action potentials between cardiac muscle fibers. The gap junctions provide bridges for the spread of excitation (muscle action potentials) from one fiber to another. Thus, the spread of excitation is exceedingly rapid, and the atria contract as one unit and the ventricles as another. The intercalated discs also link cardiac muscle fibers to one another so they do not pull apart. Each network contracts as a functional unit.

The heart is innervated by the autonomic nervous system (Chapter 11), but the autonomic nerves only increase or decrease the time it takes to complete a cardiac cycle (heartbeat); that is, they do not initiate contraction. The chamber walls can go on contracting and relaxing without any direct stimulus from the nervous system. This is possible because the heart has an intrinsic regulating system called the *conduction system.* The conduction system is composed of specialized muscle tissue that generates and distributes the action potentials that stimulate the cardiac muscle fibers (cells) to contract. These tissues are found in the sinoatrial (SA) node, the atrioventricular (AV) node, the atrioventricular (AV) bundle (bundle of His), the bundle branches, and the conduction myofibers (Purkinje fibers). All cardiac muscle is capable of *self-excitation;* that is, it spontaneously and rhythmically generates action potentials that result in contraction of the muscle. The normal resting rate of self-excitation of the sinoatrial node is about 75 times per minute in adults. Since this rate occurs faster than that of other cardiac muscle fibers, the SA node is called the "pacemaker."

The *sinoatrial (SA) node* is located in the right atrial wall just below the opening to the superior vena cava (Figure 15-5a). The SA node initiates each heartbeat and thereby sets the basic pace for the heart rate. Since the SA node spontaneously generates action potentials faster than other components of the conduction system, nerve impulses from the SA node spread to the other areas and stimulate them so frequently they are not able to generate action potentials at their own inherent rates. Thus, the faster SA node sets the rhythm for the rest of the heart. The rate set by the SA node may be altered by nerve impulses from the autonomic nervous system or by bloodborne chemicals such as thyroid hormones and epinephrine.

When an action potential is initiated by the SA node, it spreads out over both atria, causing them to contract and at the same time depolarizing the slowly conducting *atrioventricular (AV) node.* The AV node is one of the last portions of the atria to be depolarized, which allows time for the atria to empty their blood into the ventricles before the ventricles begin their contraction; the atria finish their contraction before the ventricles begin theirs.

From the AV node, a tract of conducting fibers called the *atrioventricular (AV) bundle (bundle of His)* runs through the cardiac skeleton and toward the heart's apex as the *right* and *left bundle branches.* The AV bundle distributes the action potential over the ventricles. Actual contraction of the ventricles is stimulated by the *conduction myofibers (Purkinje fibers)* that emerge from the bundle branches and distribute the action potential to all of the ventricular myocardial cells at about the same time.

ELECTROCARDIOGRAM (ECG OR EKG)

Transmission of action potentials through the conduction system generates electric currents that can be detected on the body's surface. A recording of the electrical changes that accompany the heartbeat is called an *electrocardiogram (ECG* or *EKG).*

Each portion of a heartbeat produces a different action potential. These action potentials are graphed as a series of up-and-down waves during an ECG. Three clearly recognizable waves normally accompany each cardiac cycle (Figure 15-5b). The first, called the *P wave,* is a small upward wave. It indicates atrial depolarization—the spread of an action potential from the SA node through the two atria. A fraction of a second after the P wave begins, the

FIGURE 15-5 Conduction system of the heart. (a) Location of the nodes and bundles of the conduction system. The arrows indicate the flow of action potentials through the atria. (b) Normal electrocardiogram of a single heartbeat, enlarged for emphasis. Recall from Chapter 9 that a millivolt (mV) is equal to one-thousandth of a volt; a 1.5-volt battery will power an ordinary flashlight.

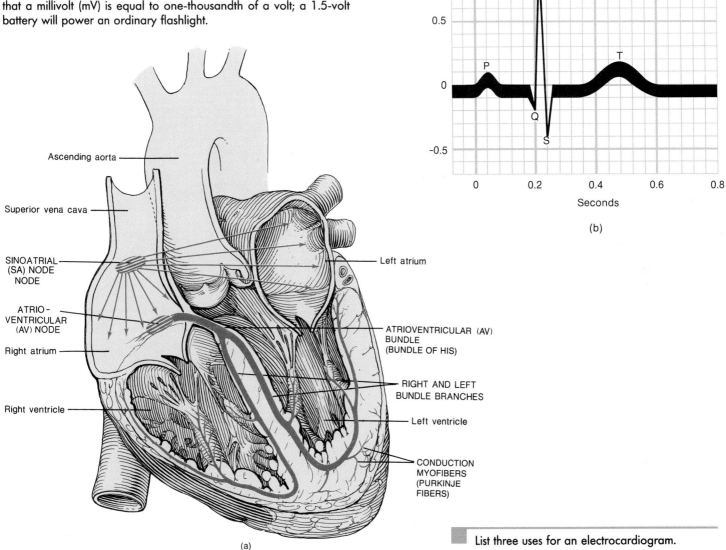

(a)

(b)

List three uses for an electrocardiogram.

atria contract. The second wave, called the **QRS wave,** begins as a downward deflection, continues as a large, upright, triangular wave, and ends as a downward wave at its base. This wave represents ventricular depolarization, that is, the spread of the action potential through the ventricles. The third wave is a dome-shaped **T wave.** This wave indicates ventricular repolarization. There is no wave to show atrial repolarization because the stronger QRS wave masks this event.

The ECG is useful in diagnosing abnormal cardiac rhythms and conduction patterns and in following the course of recovery from a heart attack. It can also detect the presence of fetal life.

In major disruptions of the conduction system, an irregular heart rhythm may occur. In one type of rhythm disturbance, the ventricles fail to receive atrial action potential, causing the ventricles and atria to beat independently of each other. Normal heart rhythm can be restored and maintained with an **artificial pacemaker,** a device that sends out small electrical charges to stimulate the heart.

BLOOD FLOW THROUGH THE HEART

The movement of blood through the heart is directly related to changes in blood pressure, which is caused by changes in the size of the chambers, and which brings about the opening and closing of the valves. When the walls of the atria are stimulated to contract by the SA node, the size of the atrial chambers is decreased, which thereby increases blood pressure within them. This increased blood pressure forces the AV valves open and atrial blood flows into the ventricles. As previously stated, the ventricle walls stay relaxed until after the atria are finished contracting. When a chamber's wall is relaxed, the blood pressure in that chamber is decreased. Blood flows through the heart from areas of higher blood pressure to areas of lower blood pressure.

When the ventricle walls contract, ventricular blood pressure increases to a higher level than that in the arteries so ventricular blood pushes the semilunar valves open and blood flows into the arteries. Also, at the same time, the shape of the AV valve cusps causes them to be pushed shut, to prevent backflow of ventricular blood into the atria where the pressure is now lower since the walls there have relaxed.

CARDIAC CYCLE

In a normal heartbeat, the two atria contract while the two ventricles relax. Then, when the two ventricles contract, the two atria relax. The term *systole* (SIS-tō-lē) refers to the phase of contraction; *diastole* (dī'-AS-tō-lē) is the phase of relaxation. One *cardiac cycle,* or complete heartbeat, consists of atrial systole and ventricular diastole occurring simultaneously followed by ventricular systole and atrial diastole occurring simultaneously.

PHASES

For the purposes of our discussion, we will divide the cardiac cycle into the following phases (as you read the description, refer to Figure 15-6):

1. **Atrial systole (contraction).** Between heartbeats, all four chambers of the heart are in diastole (relaxation). During this time, pressure in the heart is low and most of the blood flows passively into the atria and then through the open atrioventricular valves. The semilunar valves are closed. When the SA node stimulates the atria, the atria contract and force the remaining blood into the ventricles. This contraction produces the P wave in an ECG.
2. **Ventricular diastole (relaxation).** During this time, pressure in the ventricles is low since the ventricular walls are relaxed. This permits the ventricles to fill with blood from the atria. The semilunar valves are closed.
3. **Ventricular systole (contraction).** Near the end of atrial contraction, the action potential passes through the AV node and into the ventricles, causing their walls to contract. This produces the QRS wave in an ECG. Shortly after ventricular contraction begins, pressure in the ventricles increases rapidly, thus exceeding atrial pressure. The high pressure causes the atrioventricular valves to close. It also causes the semilunar valves to open, which forces blood into the aorta and pulmonary trunk.
4. **Atrial diastole (relaxation).** At this time, the walls of the atria are relaxed and the atria fill with blood in preparation for the next cardiac cycle. Under normal conditions, blood flows continuously into the atria. About 70 percent of atrial blood flows into the ventricles by gravity. The remaining 30 percent is pushed into the ventricles by atrial contractions.

TIMING

If the average heart rate (HR) is 75 times per minute, then each cardiac cycle requires about 0.8 sec. In a complete cycle, the atria are in systole 0.1 sec and in diastole 0.7 sec. The ventricles are in systole 0.3 sec and in diastole 0.5 sec. The last 0.4 sec of the cycle is the *relaxation period* and all chambers are in diastole. When the heart beats faster than normal, the relaxation period is shortened accordingly.

FIGURE 15-6 Cardiac cycle. (a) Atrial systole and ventricular diastole. (b) Atrial diastole and ventricular systole.

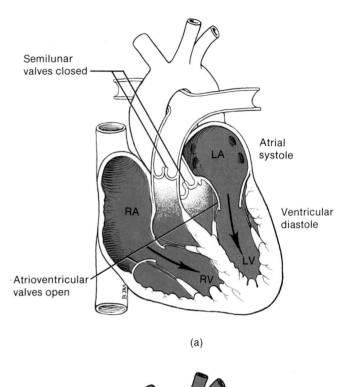

(a)

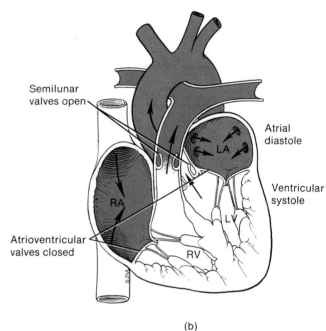

(b)

SOUNDS

The sound of the heartbeat comes primarily from turbulence in blood flow created by the closure of the valves, not from the contraction of the heart muscle. The first sound, *lubb,* is a long, booming sound from the AV valves closing after ventricular systole begins. The second sound, a short, sharp sound, *dupp,* is from the semilunar valves closing at the end of ventricular systole. There is a pause between cycles. Thus, the cardiac cycle is heard as lubb, dupp, pause; lubb, dupp, pause; lubb, dupp, pause.

CARDIAC OUTPUT (CO)

Although the heart can beat independently, it is regulated by events occurring in the rest of the body. All body cells must receive a certain amount of oxygenated blood each minute to maintain health and life. When cells are very active, as during exercise, they need even more blood. During rest periods, cellular need is reduced, and the heart cuts back on its output.

The amount of blood ejected from the left ventricle into the aorta (or right ventricle into the pulmonary trunk) per minute is called the *cardiac output (CO)*. Cardiac output is determined by (1) the amount of blood pumped by the left (or right) ventricle during each beat and (2) the number of heartbeats per minute. The amount of blood ejected by a ventricle during each systole is called the *stroke volume (SV)*. In a resting adult, stroke volume averages 70 ml and heart rate is about 75 beats per minute. The average cardiac output, then, in a resting adult is

$$\text{Cardiac output} = \text{stroke volume} \times \text{beats per minute}$$
$$= 70 \text{ ml} \times 75/\text{min}$$
$$= 5250 \text{ ml/min or } 5.25 \text{ liters/min}$$

Factors that increase stroke volume or heart rate tend to increase cardiac output. Factors that decrease stroke volume or heart rate tend to decrease cardiac output.

One factor that determines the force of ventricular contraction is the length of cardiac muscle fibers (cells). The more cardiac fibers are stretched by the filling of a chamber with blood, the stronger the walls will contract to eject the blood. This relationship is referred to as *Starling's law of the heart*. The situation is somewhat like stretching a rubber band; the more you stretch it, the harder it contracts. The operation of Starling's law of the heart is important in maintaining equal blood output from both ventricles.

HEART RATE

Cardiac output depends on heart rate as well as stroke volume. In fact, changing heart rate is the body's principal mechanism of short-term control over cardiac output and blood pressure. The sinoatrial (SA) node that initiates contraction would, if left to itself, set an unvarying heart rate. However, the body's need for blood supply varies under different conditions, so several regulatory mechanisms exist. They are stimulated by such factors as chemicals present in the body, temperature, emotional state, and age.

In certain pathological states, stroke volume may fall dangerously low. If the ventricular myocardium is weak or damaged by an infarction, it cannot contract strongly. Or blood volume may be reduced by excessive bleeding, causing stroke volume to fall because the cardiac fibers are not sufficiently stretched. In these cases, the body attempts to maintain a safe cardiac output by increasing the rate and strength of contraction.

The heart rate is regulated by several factors, but the most important control of rate and strength of contraction is the autonomic nervous system.

AUTONOMIC CONTROL

Within the medulla of the brain is a group of neurons called the *cardiac center.* One part is known as the *cardioacceleratory center (CAC).* Arising from this center are sympathetic fibers that travel down a tract in the spinal cord and then pass outward to *cardiac nerves* that innervate the conduction system, atria, and ventricles (Figure 15-7). When the cardioacceleratory center is stimulated, nerve impulses travel along the sympathetic fibers. This causes them to release norepinephrine (NE), which increases the rate of heartbeat and the strength of contraction.

The cardiac center also contains a group of neurons that form the *cardioinhibitory center (CIC).* Arising from this center are parasympathetic fibers that reach the heart via the *vagus (X) nerve.* When this center is stimulated, nerve impulses are transmitted along the parasympathetic fibers to the conduction system and atria that cause the release of acetylcholine (ACh). This decreases the rate of heartbeat and strength of contraction by slowing the SA and AV nodes.

The autonomic control of the heart is therefore the result of opposing sympathetic (stimulatory) and parasympathetic (inhibitory) influences. Receptors in the cardiovascular system inform the cardiac center so that a balance between stimulation and inhibition is maintained. For example, *baroreceptors* respond to changes in blood pressure.

Baroreceptors are strategically located in certain arteries and veins. If there is an increase in blood pressure, the baroreceptors send nerve impulses that stimulate the cardioinhibitory center and inhibit the cardioacceleratory center (Figure 15-8). As a result, heart rate and force of contraction decrease, cardiac output decreases, and blood pressure decreases. If, on the other hand, blood pressure falls, baroreceptors do not stimulate the cardioinhibitory center and the cardioacceleratory center is free to dominate. As a result, heart rate and force of contraction increase, cardiac output increases, and blood pressure increases.

CHEMICALS

Certain chemicals in the body have an effect on heart rate. For example, epinephrine, produced by the adrenal medulla in response to sympathetic stimulation, increases the excitability of the SA node, which, in turn, increases the rate and strength of contraction. Elevated levels of potassium (K^+) or sodium (Na^+) decrease the heart rate and strength of contraction. Excess potassium apparently interferes with the generation of nerve impulses and excess sodium interferes with calcium (Ca^{2+}) participation in muscular contraction. An excess of calcium increases heart rate and strength of contraction.

When oxygen demands of the body are low, heart rate is decreased. In response to increased oxygen demands, heart rate increases.

TEMPERATURE

Increased body temperature, from fever or strenuous exercise, for example, causes the AV node to discharge impulses faster and thereby increases heart rate. Decreased body temperature, from exposure to cold or deliberately cooling the body prior to surgery, decreases heart rate and strength of contraction.

Diet and Heart Disease: Redefining the Good Life

A person's diet has a strong influence on four risk factors associated with heart disease: high blood cholesterol level, high blood pressure, obesity, and diabetes mellitus. So if a person is watching his cholesterol, blood pressure, weight, or blood sugar, he should also be watching his diet.

Researchers attribute our high incidence of these risk factors, as well as our high rates of coronary artery disease, partly to an overconsumption of high-fat, low-fiber foods. The incidence of cardiovascular disease is much lower in countries where less fat and fewer calories are consumed and in groups within North America who eat low-fat diets. The greatest nutritional problem in the developed nations of the world is not a vitamin deficiency. It's not too little calcium, iron, or protein. Instead, our problem might be called malnourishment due to overnourishment: too many calories and too much fat, sugar, salt, and alcohol.

More than a century ago, Sir Richard Burton (1821–1890) recognized the relationship between overnourishment and health. He said: "Gluttony is the source of all our infirmities and the fountain of all our diseases. As a lamp is choked by a superabundance of oil, and a fire extinguished by excess of fuel, so is the natural health of the body destroyed by intemperate diet."

Public health officials have issued a number of dietary guidelines and recommendations to help people eat a "heart-healthy" diet, a diet that can help prevent obesity, high blood pressure, high blood cholesterol, and type II diabetes and decrease one's risk of developing coronary artery disease, the leading cause of death in North America.

Eat Less Fat, Saturated Fat, and Cholesterol

We need only a very small amount of fat in the diet to maintain good health; extra fat is extra calories, which contribute to obesity. Recent research shows that a given number of calories consumed as fat results in more fat storage than the same number of calories consumed as carbohydrate or protein. In other words, it is metabolically efficient to store fat in adipose tissue but not as efficient to turn carbohydrate or protein into fat.

In addition, 1 gram of fat contains over twice as many calories as 1 gram of carbohydrate or protein. We say that foods high in fats are "calorically dense." A pastry and a piece of fruit may be about the same size, but guess which one has more than twice as many calories as the other.

A diet high in fat, particularly saturated fat, also raises blood cholesterol level. Polyunsaturated fats seem to decrease blood cholesterol, but at the expense of HDL levels, which decline as well. It is advisable not to replace saturated fat with some other type of fat, but to let a decrease in saturated fat intake decrease total fat intake as well.

Foods high in saturated fat include butter, cream, whole milk, cheese, some shortenings and margarines, and palm and coconut oils. Some food labels list how much and what types of fat have been added to food products. Lean cuts of meat, poultry, and fish have less saturated fat than fatty cuts such as prime rib. Poultry fat is found under the skin, so by trimming the skin you remove most of the fat. Cooking methods such as broiling and baking are preferable to frying and breading.

Cholesterol intake should be limited, too, by restricting consumption of egg yolks, meats, and organ meats. Cholesterol is found in muscle tissue itself, so simply trimming the visible fat from meats does little to decrease its cholesterol content.

Eat More Fish

Research has suggested that fish consumption is associated with a decreased risk of heart disease. Several studies have shown that fish oil may prevent heart disease by decreasing platelet stickiness and perhaps by lowering total serum cholesterol as well. Cold-water fish such as salmon, mackerel, trout, bluefish, and herring contain a special type of polyunsaturated fat called omega-3 fatty acids.

People who consume diets high in fish oils produce less thromboxane, which promotes platelet clumping and is a powerful vasoconstrictor. Fewer clots and more open arteries mean a lower risk of blockage. Although fish oil contains substantial amounts of cholesterol, some studies have shown that people who consumed fish oil reduced their serum cholesterol levels.

Scientists are reluctant to recommend fish-oil supplements at this point, because their safety for long-term consumption has not been proved. But all agree that eating more fish is a good idea. In fact, it appears that fish may contain beneficial components in addition to oils. A study from the Netherlands found that people who eat only one or two fish meals a week, regardless of the type of fish consumed, had a lower incidence of heart disease than people who ate less fish.

Eat More Foods High in Complex Carbohydrates and Fiber

Carbohydrate foods do not increase total serum cholesterol and are less likely to cause obesity than fatty foods, although too much of any food means extra calories. Complex carbohydrates include grains such as wheat, rice, corn, oats, and their products such as cereals, breads, and pasta; peas and beans, such as split peas, lentils, kidney beans, and chick-peas; and starchy vegetables such as potatoes, yams, and winter squashes.

Complex carbohydrates are "nutritionally dense," as opposed to the calorically dense fatty foods mentioned above. Nutritionally dense foods have a lot of nutrition per calorie. These foods are rich in vitamins and minerals. They are also high in the type of fiber (water-soluble) that helps decrease serum cholesterol.

Some research has suggested that people consuming a diet high in complex carbohydrates can maintain or even lose weight while eating as much as they want of these foods. They don't count calories or portions, and eat until they are full. Researchers theorize that people consuming a diet high in these foods feel full and quit eating before they have consumed too many calories. Just the opposite can happen with calorically dense foods. Since a lot of calories are contained in a small volume, you can eat hundreds of calories in a very short period of time and feel full.

It is unfortunate that dieters have shunned complex carbohydrates for years. Many people would still choose a steak rather than a potato when trying to lose weight. Yet even a small (3.5 oz) steak is high in fat and may

have over 400 calories, while a baked potato has almost no fat and only about 90 calories. It's important to prepare these complex carbohydrate foods without high-fat sauces or other added fats, however. The 90-calorie baked potato becomes a high-fat, high-calorie dish as soon as you add a tablespoon of sour cream.

Complex carbohydrate foods cause a slower rise in blood sugar than simple sugars such as white and brown sugar, honey, and corn syrup, and thus lead to a more moderate insulin response. This is helpful for people with type II diabetes, who have poor blood sugar control. It is also helpful for people trying to lose weight, since insulin encourages energy (including fat) storage.

Eat Less Sugar and Salt

Simple sugars are found naturally in fruits, vegetables, and dairy products and in a more concentrated form in dextrose and other sugar products. Concentrated sugar sources contribute to obesity because they provide empty calories; that is, they have no other beneficial nutrients in them. They are often consumed with fats in pies, pastries, cakes, cookies, and candies.

Hypertension is more prevalent in population groups who consume high levels of salty foods. A taste for salt developed early in life can lead to hypertension in later years. Some research suggests that it may take years for a high salt intake to cause hypertension, so most people do not realize that their salt intake is harming their health.

Salt is found in most prepared foods, such as soups, sauces, canned fish and meat; condiments such as soy sauce and steak sauce; pickled and cured foods; and salty foods such as potato chips and pretzels. High sodium levels are also present in some vegetables, such as celery and mushrooms.

Drink Less Alcohol

Too much alcohol is associated with increased risk of hypertension. No one can say exactly how much is too much, but most authorities agree one or two drinks a day are probably safe. Alcoholic drinks are empty calories and can thus contribute to obesity.

People are often looking for a magic food that will lower cholesterol or melt fat away, something that will make them sexy, slim, and healthy. There's no one food that can prevent heart disease or even a short-term eating plan that will reverse atherosclerosis. A heart-healthy diet consists of making the right daily food choices that add up to a lifetime of good dietary habits.

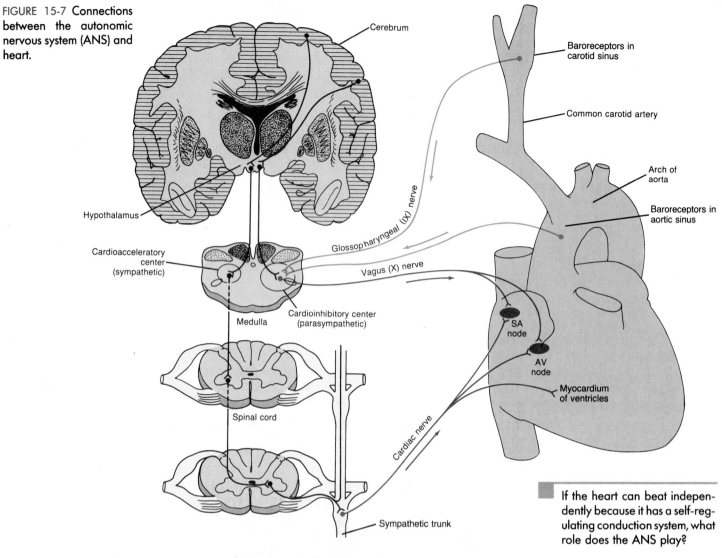

FIGURE 15-7 Connections between the autonomic nervous system (ANS) and heart.

Cerebrum

Baroreceptors in carotid sinus

Common carotid artery

Arch of aorta

Baroreceptors in aortic sinus

Hypothalamus

Glossopharyngeal (IX) nerve

Cardioacceleratory center (sympathetic)

Vagus (X) nerve

Medulla

Cardioinhibitory center (parasympathetic)

SA node

AV node

Myocardium of ventricles

Spinal cord

Cardiac nerve

Sympathetic trunk

If the heart can beat independently because it has a self-regulating conduction system, what role does the ANS play?

FIGURE 15-8 Control of heart rate and force of contraction by baroreceptors and aortic reflex.

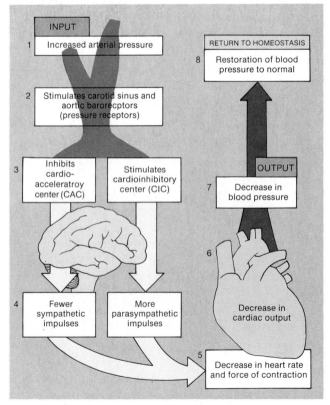

INPUT

1 Increased arterial pressure

RETURN TO HOMEOSTASIS

8 Restoration of blood pressure to normal

2 Stimulates carotid sinus and aortic barorecptors (pressure receptors)

3 Inhibits cardioacceleratroy center (CAC)

Stimulates cardioinhibitory center (CIC)

OUTPUT

7 Decrease in blood pressure

4 Fewer sympathetic impulses

More parasympathetic impulses

6 Decrease in cardiac output

5 Decrease in heart rate and force of contraction

List four factors that affect heart rate.

328

EMOTIONS

Strong emotions such as fear, anger, and anxiety, along with a multitude of physiological stressors, increase heart rate through the general adaptation syndrome (see Figure 13-20). Mental states such as depression and grief tend to stimulate the cardioinhibitory center and decrease heart rate.

SEX AND AGE

Sex is another factor: the heartbeat is somewhat faster in normal females than normal males. Age is yet another factor: the heartbeat is fastest at birth, moderately fast in youth, average in adulthood, and below average in old age.

RISK FACTORS IN HEART DISEASE

It is estimated that one in every five persons who reaches age 60 will have a myocardial infarction (heart attack). One in every four persons between 30 and 60 has the potential to be stricken. Heart disease is epidemic in North America, despite the fact that some of the causes can be foreseen and prevented. Research indicates that people with combinations of certain risk factors are likely to have heart attacks. *Risk factors* are characteristics, symptoms, or signs present in a person free of disease that are statistically associated with an excessive rate of development of a disease. Among the major risk factors in heart disease are

1. High blood cholesterol level
2. High blood pressure
3. Cigarette smoking
4. Obesity
5. Lack of regular exercise
6. Diabetes mellitus
7. Genetic predisposition (family history of heart disease at an early age)
8. Sex
9. Age
10. Fibrinogen level
11. Left ventricular hypertrophy (enlarged left ventricle)

The first five risk factors all contribute to increasing the heart's work load. High blood cholesterol is discussed shortly, and hypertension is discussed in the next chapter. Cigarette smoking, through the effects of nicotine, stimulates the adrenal gland to oversecrete aldosterone, epinephrine, and norepinephrine (NE)—the latter two being powerful vasoconstrictors. Overweight people develop miles of extra capillaries to nourish fat tissue. The heart has to work harder to pump the blood through more vessels.

Without exercise, venous return gets less help from contracting skeletal muscles. In addition, regular exercise strengthens the smooth muscle of blood vessels and enables them to assist general circulation. Exercise also increases cardiac efficiency and output. In diabetes mellitus, fat metabolism dominates glucose metabolism. As a result, cholesterol levels get progressively higher and result in plaque formation, a situation that may lead to high blood pressure. High blood pressure drives fat into the vessel wall, encouraging atherosclerosis.

Up to age 50, there is a 10- to 15-year lag in the extent of heart disease in females compared with males; after that, the rates of disease in both sexes are similar. Regardless of sex, the incidence of heart disease increases with age. The higher the level of fibrinogen, the higher the risk of heart disease (more so in males than females). Fibrinogen enhances blood clot formation. Both hypertension and obesity contribute to left ventricular hypertrophy (enlargement).

■ COMMON DISORDERS ■

Coronary Artery Disease (CAD)

In *coronary artery disease (CAD)*, the heart muscle does not receive an adequate amount of blood because of an interruption of its blood supply. Two of the principal causes are atherosclerosis and coronary artery spasm. Another is a thrombus or an embolus in a coronary artery.

Atherosclerosis
Atherosclerosis (ath′-er-ō-skle-RŌ-sis) is a process in which fatty substances, especially cholesterol (CH) and triglycerides (ingested fats), are deposited in the walls of medium-sized and large arteries in response to certain stimuli. It is believed that the first event in atherosclerosis is damage to the endothelial lining of the artery. Contributing factors include high blood pressure (hypertension), carbon monoxide in cigarettes, diabetes mellitus, and high cholesterol level.

Coronary Artery Spasm
Atherosclerosis results in a fixed obstruction to blood flow. Obstruction can also be caused by *coronary artery spasm,* in which the smooth muscle of a coronary artery undergoes a sudden contraction, resulting in vasoconstriction. Coronary artery spasm typically occurs in individuals with atherosclerosis and may result in chest pain, heart attacks, and sudden death. Factors related to coronary artery spasm include smoking, stress, and a vasoconstrictor chemical released by platelets.

Congenital Defects

A defect that exists at birth, and usually before, is called a *congenital defect.*

In *coarctation* (kō′-ark-TĀ-shun) *of the aorta,* a segment of the aorta is too narrow, reducing the flow of oxygenated blood to the body.

In *patent ductus arteriosus,* the ductus arteriosus (temporary blood vessel) between the aorta and the pulmonary trunk, which normally closes shortly after birth, remains open.

Interatrial septal defect is failure of the fetal foramen ovale (described in Chapter 16) between the two atria to close after birth.

Interventricular septal defect is caused by an incomplete closure of the interventricular septum.

Valvular stenosis is a narrowing of one of the valves regulating blood flow in the heart.

Tetralogy of Fallot (tet-RAL-ō-jē; fal-ō) is a combination of four defects: an interventricular septal defect, an aorta that emerges from both ventricles instead of from the left ventricle only, a stenosed pulmonary semilunar valve, and an enlarged right ventricle.

Arrhythmia (a-RITH-me-a) or *dysrhythmia* is a general term referring to an irregularity in heart rhythm. It results when there is a disturbance in the conduction system of the heart, either due to faulty production or conduction of electrical impulses. Arrhythmias are caused by factors such as caffeine, nicotine, alcohol, anxiety, certain drugs, hyperthyroidism, potassium deficiency, and certain heart diseases. One serious arrhythmia is called a **heart block.** The most common blockage is in the atrioventricular (AV) node, which conducts impulses from the atria to the ventricles. This disturbance is called *atrioventricular (AV) block.*

In *atrial flutter,* the atrial rhythm averages between 240 and 360 beats per minute. The condition is essentially rapid atrial contractions accompanied by AV block. *Atrial fibrillation* is asynchronous contraction of the atrial muscles that causes the atria to contract irregularly and still faster. When the muscle fibrillates, the muscle fibers of the atrium quiver individually instead of contracting together. The quivering cancels out the pumping of the atrium. *Ventricular fibrillation* (*VF*) is characterized by asynchronous, haphazard, ventricular muscle contractions. Ventricular contraction becomes ineffective and circulatory failure and death occur.

Congestive Heart Failure (CHF)

Congestive heart failure (*CHF*) is a chronic or acute state that results when the heart is not capable of supplying the oxygen demands of the body.

Cor Pulmonale (CP)

Cor pulmonale (kor pul-mōn-ALE; *cor* = heart; *pulmon* = lung), or *CP,* refers to enlargement of the right ventricle from disorders that bring about hypertension (high blood pressure) in the circulation of blood in the lungs.

MEDICAL TERMINOLOGY AND CONDITIONS

Angiocardiography (an′-jē-ō-kar′-dē-OG-ra-fē; *angio* = vessel; *cardio* = heart; *graph* = writing) X-ray examination of the heart and great blood vessels after injection of a radiopaque dye into the bloodstream.

Aortic insufficiency (ā-OR-tik in′-su-FISH-en-sē) An improper closure of the aortic semilunar valve that permits a backflow of blood.

Aortography (ā′-or-TOG-ra-fē) X-ray examination of the aorta and its main branches after injection of a radiopaque dye.

Cardiac arrest (KAR-dē-ak a-REST) A clinical term meaning cessation of an effective heartbeat. The heart quivers ineffectively (ventricular fibrillation).

Cardiomegaly (kar′-dē-ō-MEG-a-lē; *mega* = large) Heart enlargement.

Constrictive pericarditis (kon-STRIK-tiv per′-i-kar-DĪ-tis) A shrinking and thickening of the pericardium that prevents heart muscle from expanding and contracting normally.

Incompetent valve (in-KOM-pe-tent VALV) Any valve that does not close properly, thus permitting a backflow of blood; also called *valvular insufficiency.*

Palpitation (pal′-pi-TĀ-shun) A fluttering of the heart or abnormal rate or rhythm of the heart.

Pancarditis (pan′-kar-DĪ-tis; *pan* = all) Inflammation of the whole heart including the inner layer (endocardium), heart muscle (myocardium), and outer sac (pericardium).

Paroxysmal tachycardia (par′-ok-SIZ-mal tak′-e-KAR-dē-ā) A period of rapid heartbeats that begins and ends suddenly.

Stokes–Adams syndrome Sudden attacks of unconsciousness, sometimes with convulsions, that may accompany heart block.

Sudden cardiac death The unexpected cessation of circulation and breathing due to an underlying heart disease such as ischemia, myocardial infarction, or a disturbance in cardiac rhythm.

STUDY OUTLINE

Location of Heart (p. 317)

1. The heart is situated between the lungs in the mediastinum.
2. About two-thirds of its mass is to the left of the midline.

Pericardium (p. 317)

1. The pericardium consists of an outer fibrous layer and an inner serous pericardium.
2. The serous pericardium is composed of a parietal and visceral layer.
3. Between the parietal and visceral layers of the serous pericardium is the pericardial cavity, a space filled with pericardial fluid that prevents friction between the two membranes.

Wall; Chambers; Vessels; and Valves (pp. 317)

1. The wall of the heart has three layers: epicardium, myocardium, and endocardium.
2. The chambers include two upper atria and two lower ventricles.
3. The blood flows through the heart from the superior and inferior venae cavae and the coronary sinus to the right atrium, through the tricuspid valve to the right ventricle, through the pulmonary trunk to the lungs, through the pulmonary veins into the left atrium, through the bicuspid valve to the left ventricle, and out through the aorta.
4. Four valves prevent backflow of blood in the heart.
5. Atrioventricular (AV) valves or cuspid valves, between the atria and their ventricles, are the tricuspid valve on the right side of the heart and the bicuspid valve on the left.
6. The chordae tendineae and their papillary muscles stop blood from backflowing into the atria.
7. The two arteries that leave the heart both have a semilunar valve.

Blood Supply (p. 322)

1. Coronary circulation delivers oxygenated blood to the myocardium and removes carbon dioxide from it.
2. Deoxygenated blood returns to the right atrium via the coronary sinus.
3. Complications of this system are angina pectoris and myocardial infarction (MI).

Conduction System (p. 322)

1. The conduction system consists of tissue specialized for action potential conduction.
2. Components of this system are the sinoatrial (SA) node (pacemaker), atrioventricular (AV) node, atrioventricular (AV) bundle (bundle of His), bundle branches, and conduction myofibers (Purkinje fibers).

Electrocardiogram (ECG or EKG) (p. 322)

1. The record of electrical changes during each cardiac cycle is referred to as an electrocardiogram (ECG).
2. A normal ECG consists of a P wave (spread of action potential from SA node over atria), QRS wave (spread of action potential through ventricles), and T wave (ventricular repolarization).
3. The ECG is used to diagnose abnormal cardiac rhythms and conduction patterns, detect the presence of fetal life, and follow the course of recovery from a heart attack.

Blood Flow Through the Heart (p. 323)

1. Blood flows through the heart from areas of higher to lower pressure.
2. The pressure developed is related to the size and volume of a chamber.
3. The movement of blood through the heart is controlled by the opening and closing of the valves and the contraction and relaxation of the myocardium.

Cardiac Cycle (p. 324)

1. A cardiac cycle consists of the systole (contraction) and diastole (relaxation) of both atria plus the systole and diastole of both ventricles followed by a short pause.
2. The phases of the cardiac cycle are (a) atrial systole, (b) ventricular diastole, (c) ventricular systole, and (d) atrial diastole.
3. With an average heartbeat of 75/min, a complete cardiac cycle requires 0.8 sec.
4. The first heart sound (lubb) represents the closing of the atrioventricular valves. The second sound (dupp) represents the closing of semilunar valves.

Cardiac Output (CO) (p. 325)

1. Cardiac output (CO) is the amount of blood ejected by the left ventricle into the aorta per minute. It is calculated as follows: CO = stroke volume × beats per minute.
2. Stroke volume (SV) is the amount of blood ejected by a ventricle during each systole.

Heart Rate (p. 325)

1. Heart rate and strength of contraction may be increased by sympathetic stimulation from the cardioacceleratory center (CAC) in the medulla and decreased by parasympathetic stimulation from the cardioinhibitory center (CIC) in the medulla.
2. Baroreceptors are nerve cell receptors that respond to changes in blood pressure. They act on the cardiac centers in the medulla.
3. Other influences on heart rate include chemicals (epinephrine, sodium, potassium), temperature, emotion, sex, and age.

Risk Factors in Heart Disease (p. 329)

1. Risk factors in heart disease include high blood cholesterol, high blood pressure, cigarette smoking, obesity, lack of regular exercise, diabetes mellitus, genetic predisposition, sex, age, fibrinogen level, and left ventricular hypertrophy.

REVIEW QUESTIONS

1. Describe the location of the heart. (p. 317)
2. Distinguish the subdivisions of the pericardium. What is the purpose of this structure? (p. 317)
3. Compare the three layers of the heart wall according to composition, location, and function. (p. 317)
4. Define atria and ventricles. What vessels enter or exit the atria and ventricles? (p. 319)
5. Describe the principal valves in the heart and how they operate. (p. 319)
6. Distinguish between angina pectoris and myocardial infarction (MI). (p. 322)
7. Describe the structure and function of the heart's conducting system. (p. 322)
8. Define and label the waves of a normal electrocardiogram (ECG). Explain why the ECG is an important diagnostic tool. (p. 322)
9. Define the cardiac cycle. (p. 324)
10. List the principal events of atrial systole, ventricular diastole, ventricular systole, and atrial diastole. (p. 324)
11. Describe heart sounds. (p. 324)
12. What is cardiac output (CO)? How is it calculated? (p. 325)
13. Distinguish between the cardioacceleratory center (CAC) and cardioinhibitory center (CIC) with respect to the regulation of heart rate. (p. 325)
14. What is a baroreceptor? (p. 325)
15. Explain how each of the following affects heart rate: chemicals, temperature, emotions, sex, and age. (p. 325)
16. Describe the risk factors involved in heart disease. (p. 329)
17. Define the following: coronary artery disease (CAD), atherosclerosis, coronary artery spasm, coarctation of the aorta, patent ductus arteriosus, septal defect, valvular stenosis, tetralogy of Fallot, arrhythmia, heart block, atrial flutter, atrial fibrillation, ventricular fibrillation (VF), congestive heart failure (CHF), and cor pulmonale (CP). (p. 329)
18. Refer to the glossary of medical terminology and conditions associated with the heart. Be sure that you can define each term. (p. 330)

16

The Cardiovascular System: Blood Vessels

STUDENT OBJECTIVES

1. Describe the functions of the various types of blood vessels.
2. Explain why blood flows through blood vessels.
3. Explain what causes blood pressure and specifically what factors are involved.
4. Discuss how materials are exchanged between blood and body cells.
5. Describe how blood returns to the heart.
6. Define pulse and blood pressure and describe how they are measured.
7. Compare the various major routes that blood takes through different regions of the body.
8. Explain the benefits of exercise on the cardiovascular system.
9. Define common disorders and medical terminology and conditions associated with blood vessels.

A LOOK AHEAD

ARTERIES
ARTERIOLES
CAPILLARIES
VENULES
VEINS
BLOOD RESERVOIRS
PHYSIOLOGY OF CIRCULATION
  Blood Flow
    Blood Pressure
    Resistance
  Factors That Affect Arterial Blood
    Pressure
    Cardiac Output (CO)
    Blood Volume
    Peripheral Resistance
  Homeostasis of Blood
    Pressure Regulation
    Vasomotor Center
    Baroreceptors
    Chemoreceptors
    Regulation by Higher Brain
      Centers
    Chemicals
    Autoregulation
  Capillary Exchange
  Factors That Aid Venous
    Return
    Pumping Action of the Heart
    Velocity of Blood Flow
    Skeletal Muscle
      Contractions and Valves
    Breathing
CHECKING CIRCULATION
  Pulse
  Measurement of Blood
    Pressure (BP)
SHOCK AND HOMEOSTASIS
CIRCULATORY ROUTES
  Systemic Circulation
  Pulmonary Circulation
  Cerebral Circulation
  Hepatic Portal Circulation
  Fetal Circulation
EXERCISE AND THE
  CARDIOVASCULAR SYSTEM
COMMON DISORDERS
MEDICAL TERMINOLOGY AND
  CONDITIONS

Blood vessels form a network of tubes that carry blood away from the heart to the tissues of the body and then return it to the heart. *Arteries* are the vessels that carry blood away from the heart to the tissues. Large arteries leave the heart and divide into medium-sized arteries that branch out into the various regions of the body. These medium-sized arteries then divide into small arteries, which, in turn, divide into still smaller arteries called *arterioles*. Arterioles within a tissue or organ branch into countless microscopic vessels called *capillaries*. Through the walls of the capillaries, substances are exchanged between the blood and body tissues. Before leaving the tissue, groups of capillaries reunite to form small veins called *venules*. These, in turn, merge to form progressively larger tubes called veins. *Veins* are blood vessels that convey blood from the tissues back to the heart.

ARTERIES

Arteries have walls constructed of three layers of tissue and a hollow core, called a *lumen*, through which the blood flows (Figure 16-1a). The inner layer is composed of simple squamous epithelium called *endothelium* and elastic tissue. The middle layer consists of smooth muscle and elastic connective tissue. The outer layer is composed principally of elastic and collagenous fibers.

As a result of the structure of the middle coat especially, arteries have two major properties: elasticity and contractility. When the ventricles of the heart contract and eject blood into the large arteries, the arteries expand to accommodate the extra blood. Then, as the ventricles relax, the elastic recoil of the arteries forces the blood onward.

The contractility of an artery comes from its smooth muscle, which is supplied by the sympathetic branch of the autonomic nervous system. When sympathetic stimulation increases, the smooth muscle contractions increase, thereby narrowing or constricting the lumen, a process called *vasoconstriction*. *Vasodilation*, or an increase in lumen size, results from a decrease in sympathetic stimulation and consequent relaxation of smooth muscle. Some vasoconstriction is necessary to maintain circulation (more detail on this later).

The smooth muscle layer of blood vessels, epecially arteries, also helps limit bleeding from wounds. When an artery is cut, the smooth muscle contracts, producing vascular spasm of the vessel (see Chapter 14). However, there is a limit to how much vasoconstriction can prevent hemorrhaging since the heart's pumping action causes blood to flow through arteries under great pressure.

ARTERIOLES

An *arteriole* is a small artery that delivers blood to capillaries. Arterioles closer to the arteries from which they branch have walls similar to arteries. But as they get smaller in size, they eventually consist of little more than a layer of endothelium covered by a few smooth muscle fibers.

Arterioles play a key role in regulating blood flow from arteries into capillaries. The smooth muscle of arterioles, like that of arteries, is subject to vasoconstriction and vasodilation. During vasoconstriction, blood flow to capillaries is restricted; during vasodilation, the flow is significantly increased. The relationship of arterioles to blood flow will be considered in detail later in the chapter.

CAPILLARIES

Capillaries are microscopic vessels that connect arterioles and venules (Figure 16-1c). They are found near almost every cell in the body, but their distribution varies with the activity of the tissue. Body tissues with high metabolic activity

FIGURE 16-1 Comparative structure of (a) an artery, (b) a vein, and (c) a capillary. The relative size of the capillary is enlarged for emphasis. Note the valve in the vein.

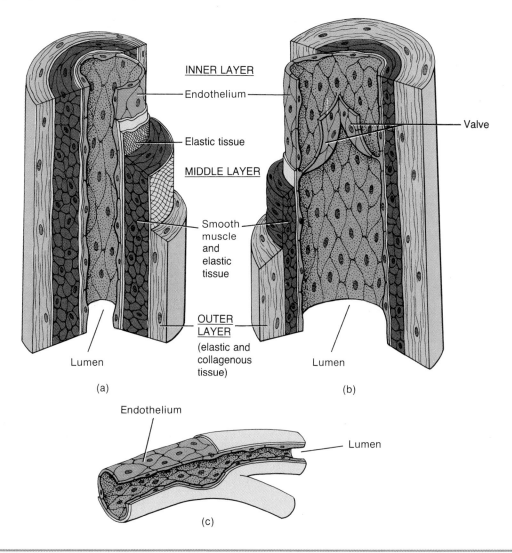

INNER LAYER

Endothelium

Elastic tissue

MIDDLE LAYER

Smooth muscle and elastic tissue

OUTER LAYER
(elastic and collagenous tissue)

Lumen

(a)

Valve

Lumen

(b)

Endothelium

Lumen

(c)

List four differences between arteries and veins.

(nervous, muscular, liver, kidneys) require more oxygen and nutrients so they need more capillaries to supply blood. Tissues with low activity, such as tendons, need fewer capillaries. The epidermis, cornea and lens of the eye, and cartilage do not contain capillaries.

The primary function of capillaries is to permit the exchange of nutrients and wastes between the blood and tissue cells. The structure of the capillaries is admirably suited to this purpose. Since capillary walls are composed of a single layer of endothelial cells, substances in the blood pass easily through them to reach tissue cells and vice versa. Depending on how tightly the endothelial cells are joined, different types of capillaries permit varying degrees of permeability. The structure of capillaries is vital for the tissues' homeostasis since the walls of all other vessels are too thick to permit the exchange of substances between blood and tissue cells.

In some regions, capillaries pass directly from arterioles to venules; in other places, they form extensive branching networks between the two vessels. These networks increase the surface area for diffusion and thereby allow a rapid exchange of large quantities

of materials. Blood normally flows through only a small portion of the capillary network when metabolic needs are low. But when a tissue becomes active, the entire capillary network fills with blood. The flow of blood in capillaries called metarterioles is controlled by smooth muscle fibers (cells) scattered throughout the proximal parts of the vessels that contract and relax (Figure 16-2). Other capillaries, called true capillaries, contain *precapillary sphincters*, rings of smooth muscle at their origin that open or shut to regulate blood flow through them.

VENULES

When several capillaries unite, they form small veins called *venules*. Venules collect blood from capillaries and drain it into veins. Venules are similar in structure to arterioles; their walls are thinner near the capillaries and thicker as they progress toward the heart. Large venules have the same structure as veins.

FIGURE 16-2 Details of a capillary network.

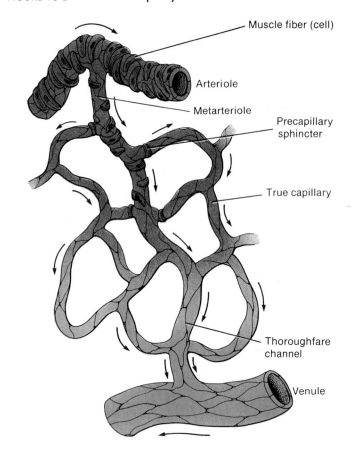

Muscle fiber (cell)

Arteriole

Metarteriole

Precapillary
sphincter

True capillary

Thoroughfare
channel

Venule

What is the function of extensive branching networks of capillaries?

VEINS

Veins are structurally similar to arteries but their outer layer is thicker, their middle layer is thinner, and the inner layer of endothelium may fold inward to form valves (see Figure 16-1b). Nevertheless, veins are still flexible enough to accommodate variations in the volume and pressure of blood passing through them.

By the time blood leaves the capillaries and moves into veins, it has lost a great deal of pressure. This can be observed in the blood leaving a cut vessel. Blood from a vein flows slowly and evenly. Blood from an artery gushes in rapid spurts. The structural differences between arteries and veins reflect this pressure difference. For example, the walls of veins are not as strong as those of arteries. The low pressure in veins, however, has its disadvantages. When you stand, the pressure pushing blood up the veins in your lower extremities is barely enough to balance the force of gravity pushing it back down. For this reason, many veins, especially those in the limbs, have valves that prevent backflow (see Figure 16-6).

In people with weak venous valves, gravity forces blood back down into the vein. This pressure overloads the vein and pushes its wall outward. After repeated overloading, the walls lose their

elasticity and become stretched and flabby. Such dilated and tortuous veins caused by incompetent valves are called **varicose veins (VVs).** They may be due to heredity, mechanical factors (prolonged standing and pregnancy), or aging. Veins close to the surface of the legs are highly susceptible to varicosities. Veins that lie deeper are not as vulnerable because surrounding skeletal muscles prevent their walls from overstretching.

BLOOD RESERVOIRS

The volume of blood in various parts of the cardiovascular system varies considerably. Veins, venules, and venous sinuses contain about 59 percent of the blood in the system, arteries about 13 percent, pulmonary vessels about 12 percent, the heart about 9 percent, and arterioles and capillaries about 7 percent. Since veins contain so much of the blood, they are referred to as **blood reservoirs**. They serve as storage depots for blood, which can be moved quickly to other parts of the body if the need arises, for example, to skeletal muscles when there is increased muscular activity.

A similar mechanism operates in cases of hemorrhage, when blood volume and pressure decrease. Vasoconstriction in venous reservoirs helps compensate for the blood loss. The principal blood reservoirs are the veins of the abdominal organs (especially the liver and spleen) and the skin.

PHYSIOLOGY OF CIRCULATION

BLOOD FLOW

Blood flow refers to the amount of blood that passes through a blood vessel in a given period of time. Blood flows or circulates through two major sets of blood vessels, to the lungs where blood gets rid of carbon dioxide and picks up oxygen (pulmonary circulation) and to the rest of the body where blood delivers oxygen and removes carbon dioxide (systemic circulation). In the discussion that follows, we will concentrate on systemic circulation.

Blood flow is determined by two factors: (1) blood pressure and (2) resistance (opposition), the force of friction as blood travels through blood vessels.

Blood Pressure

Blood pressure (BP) is the pressure exerted by blood on the wall of a blood vessel. In clinical use, the term refers to pressure in arteries. As you will see shortly, BP is influenced by cardiac output, blood volume, and resistance. Blood flows through its system of closed vessels because of different blood pressures in various parts of the cardiovascular system. Blood flow is directly proportional to blood pressure; that is, as pressure increases, flow increases. Blood always flows from regions of higher blood pressure to regions of lower blood pressure. The average pressure in the aorta is about 100 millimeters of mercury (mm Hg). Since the heart pumps in a pulsating manner, the systemic arterial pressure in a resting young adult fluctuates between 120 mm Hg (systolic) and 80 mm Hg (diastolic). As blood leaves the left ventricle and flows through systemic circulation, its pressure falls progressively to 0 mm Hg by the time it reaches the right atrium (Figure 16-3).

FIGURE 16-3 Blood pressures in various portions of systemic circulation of the cardiovascular system. The dashed line represents average arterial pressure.

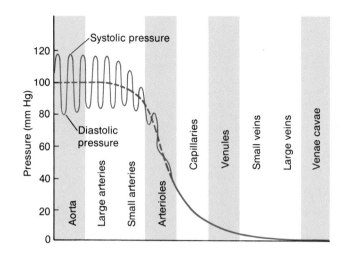

What is the importance of the differences in pressure in various parts of the systemic circuit?

Resistance

Resistance refers to the opposition to blood flow that results from friction between blood and blood vessel walls. Resistance is related to (1) blood viscosity, (2) blood vessel length, and (3) blood vessel radius.

The viscosity (adhesiveness or stickiness) of blood is a function of the ratio of red blood cells and solutes, especially plasma proteins, to fluid. Any condition that increases the viscosity of blood, such as dehydration, an unusually high number of red blood cells (polycythemia), or severe burns, increases blood pressure. A depletion of plasma proteins or red blood cells, as a result of anemia or hemorrhage, decreases blood viscosity and blood pressure. The longer a blood vessel, the greater the resistance as blood flows through it. The smaller the radius of the blood vessel, the greater resistance it offers to blood flow.

FACTORS THAT AFFECT ARTERIAL BLOOD PRESSURE

Three factors influence arterial blood pressure: (1) cardiac output, (2) blood volume, and (3) peripheral resistance.

Cardiac Output (CO)

Cardiac output (*CO*), the amount of blood ejected by the left ventricle into the aorta each minute, is the principal determinant of blood pressure. As noted in Chapter 15, CO is calculated by multiplying stroke volume (the amount ejected by a ventricle during each contraction) by heart rate. In a normal, resting adult, it is about 5.25 liters/min (70 ml × 75 beats/min). Also recall from Chapter 15 that CO is partially regulated by the cardioacceleratory center (CAC) and the cardioinhibitory center (CIC) and that certain chemicals (epinephrine, potassium, sodium, and calcium), temperature, emotions, sex, and age affect heart rate. By their effect on CO, these factors also affect blood pressure. Blood pressure varies directly with CO. Any increase in CO increases blood pressure. Conversely, any decrease lowers blood pressure.

Blood Volume

Blood pressure is directly proportional to the **volume of blood** in the cardiovascular system. The normal volume of blood in a human body is about 5 liters (5 qt). Any decrease in this volume, as from hemorrhage, decreases the amount of blood that is circulated through the arteries each minute. As a result, blood pressure drops. Conversely, anything that increases blood volume, such as high salt intake and therefore water retention, increases blood pressure.

Peripheral Resistance

Peripheral resistance refers to the opposition to blood flow in peripheral circulation, that is, away from the heart where most friction occurs. A major function of arterioles is to control peripheral resistance—and, therefore, blood pressure and flow—by changing their diameters. This regulation is governed by the vasomotor center in the medulla.

HOMEOSTASIS OF BLOOD PRESSURE REGULATION

In order to maintain homeostasis, blood pressure must be kept within a normal range. Whereas high blood pressure can do a great deal of damage to the heart, brain, and kidneys, low blood pressure can result in the delivery of inadequate amounts of oxygen and nutrients to body cells to meet their metabolic needs. In order to maintain a normal homeostatic range, there is a regulating center in the brain that receives input from receptors throughout the body, higher brain centers, and chemicals.

Vasomotor Center

In the medulla is a cluster of sympathetic neurons referred to as the **vasomotor** (*vas* = vessel; *motor* = movement) **center**. This center controls the diameter of blood vessels, especially arterioles of the skin and abdominal viscera. It is the integrating center for blood pressure control. It continually sends impulses to the smooth muscle in arteriole walls that result in a moderate state of vasoconstriction at all times, which helps maintain peripheral resistance and blood pressure. By increasing the number of sympathetic impulses, the vasomotor center brings about vasoconstriction and raises blood pressure. By decreasing the number of sympathetic impulses, it causes vasodilation and lowers blood pressure. In other words, the sympathetic division of the autonomic nervous system can bring about either vasoconstriction or vasodilation by varying the frequency of impulses.

The vasomotor center is modified by any number of inputs from baroreceptors, chemoreceptors, higher brain centers, and various chemicals, all of which influence blood pressure.

Baroreceptors

Baroreceptors are neurons sensitive to blood pressure that are located in the aorta, internal carotid arteries, and other large arteries in the neck and chest. They send impulses to the cardiac center to increase or decrease cardiac output and thus help regulate blood pressure (see Figure 15-7). This reflex acts not only on the heart but also on the arterioles. For example, if there is an increase in blood pressure, the baroreceptors stimulate the cardioinhibitory center (CIC) and inhibit the cardioacceleratory center (CAC), causing a decrease in cardiac output and consequent decrease in blood pressure. Or, if there is a decrease in blood pressure, the baroreceptors inhibit the CIC and stimulate the CAC to increase cardiac output and thus increase blood pressure. The baroreceptors also send impulses to the vasomotor center. In response, the vasomotor center either decreases sympathetic stimulation to arterioles and veins, resulting in vasodilation and a decrease in blood pressure, or increases sympathetic stimulation, resulting in vasoconstriction and an increase in blood pressure (Figure 16-4).

FIGURE 16-4 Vasomotor center control of blood pressure by stimulation of baroreceptors.

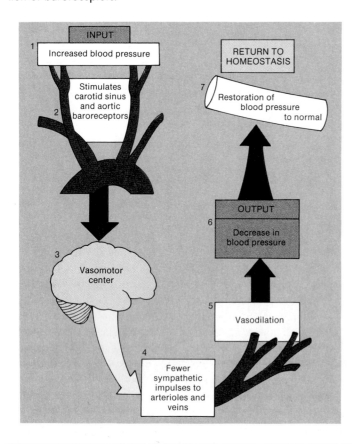

Why is this cycle a negative feedback system?

Chemoreceptors

Neurons that are sensitive to chemicals in the blood are called *chemoreceptors*. They are located in the two *carotid bodies*, small masses of tissue located where the common carotid arteries branch into external and internal carotid arteries, and in several *aortic bodies*, small masses of tissue near the aorta. Chemoreceptors are sensitive to lower than normal levels of oxygen (O_2) and even more so to higher than normal levels of carbon dioxide (CO_2) and hydrogen (H^+) ions and send impulses to the vasomotor center when these conditions exist. In response, the vasomotor center increases sympathetic stimulation to arterioles to bring about vasoconstriction and an increase in blood pressure.

Regulation by Higher Brain Centers

Higher brain centers, such as the cerebral cortex, influence blood pressure in response to strong emotions. During periods of intense anger or sexual excitement, for example, the cerebral cortex relays impulses to the hypothalamus, which travel on to the vasomotor center. From here, impulses to arterioles cause vasoconstriction and an increase in blood pressure. Also, sympathetic impulses to the adrenal medulla cause the release of epinephrine and norepinephrine, which prolong many sympathetic responses, including vasoconstriction and higher blood pressure. When a person is depressed or grieving, impulses from higher brain centers decrease vasomotor center stimulation, producing vasodilation and a decrease in blood pressure. A frequent result is fainting because blood flow to the brain is diminished.

Chemicals

Several *chemicals* affect blood pressure by causing vasoconstriction. Epinephrine and norepinephrine (NE), produced by the adrenal medulla, increase the rate and force of heart contractions and bring about vasoconstriction of arterioles in the skin and abdomen (see Figure 13-20). They also dilate cardiac and skeletal muscle arterioles. Antidiuretic hormone (ADH), produced by the hypothalamus, causes vasoconstriction during hemorrhage (see Figure 13-10). When arterial blood pressure decreases, certain kidney cells secrete renin, which raises blood pressure directly by causing vasoconstriction and indirectly by stimulating secretion of aldosterone (increases sodium ion concentration and water reabsorption) (see Figure 13-16). Histamine, produced by mast cells, and kinins, found in plasma, are vasodilators in the inflammatory response (see Chapter 17). Alcohol inhibits release of ADH, depresses the vasomotor center, and brings about vasodilation, which lowers blood pressure.

Autoregulation

Autoregulation is a local, automatic adjustment of blood flow in a given region of the body in response to the particular needs of that tissue. In most body tissues, oxygen is the principal, though not direct, stimulus for autoregulation. Autoregulation works in this way: In response to low oxygen supplies, the cells in the immediate area produce *vasodilator substances*, combinations of potassium (K^+) ions, hydrogen (H^+) ions, carbon dioxide (CO_2), lactic acid, and adenosine. These vasodilator substances cause local arterioles to dilate and precapillary sphincters to relax. The result is an increased flow of blood to the tissue, which restores oxygen levels to normal. Autoregulation is important for meeting the nutritional demands of active tissues, such as muscle tissue.

CAPILLARY EXCHANGE

For reasons to be discussed shortly, you will see that although blood pressure decreases consistently from the aorta to the venae cavae, the velocity of blood decreases as it flows through the aorta, arterioles, and capillaries and then increases as it passes into venules and veins. The velocity of blood flow in capillaries is the slowest in the cardiovascular system, which is important because this is what allows for the exchange of materials between blood and body tissues.

Blood usually does not flow steadily through capillary networks. Rather, it flows intermittently because of contraction and relaxation of the smooth muscle fibers of metarterioles and the precapillary sphincters of true capillaries. This intermittent contraction and relaxation may occur 5 to 10 times per minute and is called *vasomotion*. The most important factor in controlling vasomotion is the oxygen concentration in tissues, a mechanism similar to autoregulation in arterioles.

The movement of water and dissolved substances, except proteins, through capillary walls depends on opposing forces or pressures. At the arterial end of a capillary some forces push fluid out of capillaries into the surrounding interstitial (tissue) spaces, a process called *filtration*. To prevent fluid from moving in one direction only and accumulating in interstitial spaces, opposing forces push fluid from interstitial spaces into the venous end of blood capillaries. This process is called *reabsorption*.

Not all the fluid filtered at one end of the capillary is reabsorbed at the other end. When we discuss fluid dynamics in more detail in Chapter 22, you will see that some of the filtered fluid and any proteins that escape from blood into interstitial fluid are returned by the lymphatic system to the cardiovascular system. Thus, ultimately, there is a state of near equilibrium at the arterial and venous ends of a capillary.

FACTORS THAT AID VENOUS RETURN

A number of factors help the blood to return through the veins: (1) pumping action of the heart, (2) velocity of blood flow, (3) skeletal muscle contractions, (4) valves in veins, and (5) breathing.

Pumping Action of the Heart

Cardiac pumping creates pressure in veins to the heart. This is the major factor in blood return. In addition, during ventricular contraction in the cardiac cycle, the atrioventricular valves are pulled downward, thus increasing the size of the atria. This action sucks blood into the atria from the large veins and contributes to venous return, especially when heartbeat is fast.

Velocity of Blood Flow

Velocity of blood flow depends on the cross-sectional area of the blood vessel. Blood flows most rapidly where the cross-sectional area is least (Figure 16-5). Each time an artery branches, the total cross-sectional area of all the branches is greater than that of the original vessel. Thus, the velocity of blood decreases as it flows from the aorta to arteries to arterioles to capillaries. Velocity in the capillaries is the slowest in the cardiovascular system. Decreased velocity allows adequate exchange (diffusion) time between capillaries and tissues. As blood vessels leave capillaries and approach the heart, their cross-sectional area decreases. Therefore, the velocity of blood increases as it flows from capillaries to venules to veins to the heart.

Skeletal Muscle Contractions and Valves

Skeletal muscle contractions and valves in veins work in combination to aid venous blood return. Many veins, especially in the extremities, contain valves. When skeletal muscles contract, they tighten around the veins running through them and the valves open. This pressure drives the blood toward the heart—the action is called *milking* (Figure 16-6). When the muscles relax, the valves close to prevent backflow. Isometric contractions of leg muscles help venous return and also aid in preventing varicose veins in individuals whose jobs require them to stand for long periods of time. When people are immobilized through injury or disease, they don't have the advantage of these contractions, which means venous return is slower and the heart has to work harder. Periodic leg massage helps.

Breathing

Breathing is important in venous circulation because during inspiration, the diaphragm moves downward. This causes a decrease in pressure in the thoracic (chest) cavity and an increase in pressure in the abdominal cavity. As a result, blood moves from the abdominal veins into the thoracic veins. When the pressures reverse during expiration, blood in the veins is prevented from backflowing by the valves.

CHECKING CIRCULATION

PULSE

The alternate expansion and elastic recoil of an artery with each systole of the left ventricle is called *pulse*. Pulse is strongest in the arteries closest to the heart. It becomes weaker as it passes over the arterial system, and it disappears altogether in the capillaries. The radial artery at the wrist is most commonly used to feel the pulse. Others include the brachial artery along the medial side of the biceps brachii muscle; the common carotid artery, next to the voice box, which is frequently used in cardiopulmonary resuscitation; the popliteal artery behind the knee; and the dorsalis pedis artery above the instep of the foot.

The pulse rate is the same as the heart rate and averages between 70 and 80 beats per minute in the resting state. The term *tachycardia* (tak′-i-KAR-dē-a; *tachy* = fast) refers to a rapid heart or pulse rate (over 100/min). *Bradycardia* (brăd′-i-KAR-dē-a; *brady* = slow) indicates a slow heart or pulse rate (under 50/min).

FIGURE 16-5 Relationship between velocity of blood flow and total cross-sectional area in various blood vessels of the cardiovascular system. (a) Graph. (b) Exhibit.

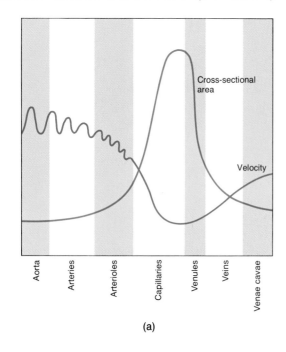

(a)

Relationship between Blood Velocity and Cross-Sectional Area

Vessel	Cross-Sectional Area (cm²)	Velocity (cm/sec)
Aorta	2.5	40
Arteries	20	10–40
Arterioles	40	0.1
Capillaries	2500	less than 0.1
Venules	250	0.3
Veins	80	0.3–5.0
Venae cavae	8	5–20

(b)

Does blood flow most rapidly when cross-sectional area is greatest or least?

MEASUREMENT OF BLOOD PRESSURE (BP)

In clinical use, the term **blood pressure (BP)** refers to the pressure in arteries exerted by the left ventricle when it undergoes systole (contraction) and the pressure remaining in the arteries when the ventricle is in diastole (relaxation). Blood pressure is usually taken in the left brachial artery, and it is measured by a **sphygmomanometer** (sfig′-mō-ma-NOM-e-ter; *sphygmo* = pulse). Following inflation of the pressure cuff, it is slowly deflated. When the first sound is heard through the stethoscope, a reading on the mercury column is made. This sound corresponds to **systolic blood pressure (SBP)**—the force with which blood is pushing against arterial walls during ventricular contraction. The pressure recorded on the mercury column when the sounds suddenly become faint is called **diastolic blood pressure (DBP)**. It measures the force of blood in arteries during ventricular relaxation. Whereas systolic pressure indicates the force of the left ventricular contraction, diastolic pressure provides information about the resistance of blood vessels.

The average blood pressure of a young adult male is 120 mm Hg systolic and 80 mm Hg diastolic, expressed as 120/80. In young adult females, the pressures are 8 to 10 mm Hg less. The difference between systolic and diastolic pressure is called **pulse pressure**. This pressure, which averages 40 mm Hg, provides information about the condition of the arteries. For example, conditions such as atherosclerosis and patent (open) ductus arteriosus greatly increase pulse pressure. The normal ratio of systolic pressure to diastolic pressure to pulse pressure is about 3:2:1.

SHOCK AND HOMEOSTASIS

Shock occurs when the cardiovascular system cannot deliver sufficient oxygen and nutrients to meet the needs of body cells. The underlying cause is inadequate cardiac output. As a result, cellular membranes dysfunction, cellular metabolism is abnormal, and cellular death may occur. Shock may be due to hemmorhage, dehydration, burns, and excessive vomiting, diarrhea, or sweating.

Symptoms of shock vary with the severity of the condition. Among the characteristic ones are the following:

1. Low blood pressure in which the systolic blood pressure is lower than 90 mm Hg as a result of generalized vasodilation and decreased cardiac output
2. Clammy, cool, pale skin due to vasoconstriction of skin blood vessels
3. Sweating due to sympathetic stimulation and increased levels of epinephrine
4. Reduced urine formation due to low blood pressure and increased levels of aldosterone and antidiuretic hormone (ADH)
5. Altered mental status due to reduced oxygen supply to the brain
6. Tachycardia (rapid heart rate and pulse) due to sympathetic stimulation and increased levels of epinephrine
7. Weak, rapid pulse due to generalized vasodilation and reduced cardiac output
8. Thirst due to loss of extracellular fluid

FIGURE 16-6 Role of skeletal muscle contractions and venous valves in returning blood to the heart. (a) When skeletal muscles contract, the valves open, and blood is forced toward the heart. (b) When skeletal muscles relax, the valves close to prevent the backflowing of blood from the heart.

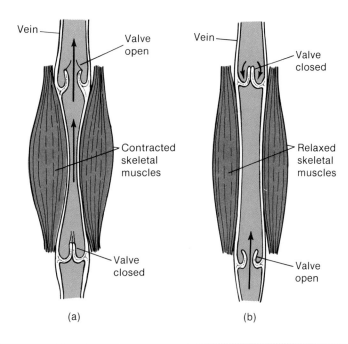

(a)   (b)

List three factors, other than valves, that aid venous return.

CIRCULATORY ROUTES

The arteries, arterioles, capillaries, venules, and veins are organized into routes that circulate blood throughout the body.

Figure 16-7 shows the basic *circulatory routes*. As noted earlier, the two basic circulatory routes are systemic and pulmonary. *Systemic circulation* includes all the oxygenated blood that leaves the left ventricle and the deoxygenated blood that returns to the right atrium after traveling to all the organs, including the nutrient arteries to the lungs. Three of the many subdivisions of systemic circulation are *coronary (cardiac) circulation,* which supplies the myocardium of the heart (see Chapter 15); *cerebral circulation*, which supplies the brain; and *hepatic portal circulation*, which runs from the gastrointestinal tract to the liver. Blood leaving the aorta and traveling through the systemic arteries is a bright red color. As it moves through the capillaries, it loses its oxygen and takes on carbon dioxide, so that the blood in the systemic veins is a dark red color.

When blood returns to the heart from the systemic route, it is pumped out of the right ventricle through the *pulmonary circulation* to the lungs. In the lungs, it loses its carbon dioxide and takes on oxygen. It is now bright red again. It returns to the left atrium of the heart and reenters systemic circulation.

Another major route—*fetal circulation*—exists only in the fetus and contains special structures that allow the developing fetus to exchange materials with its mother (see Figure 24-7).

SYSTEMIC CIRCULATION

The flow of blood from the left ventricle to all parts of the body and back to the right atrium is called *systemic circulation.* Its function is to carry oxygen and nutrients to body tissues and to remove carbon dioxide and other wastes. All systemic arteries branch from the *aorta*, which arises from the left ventricle of the heart.

As the aorta emerges from the left ventricle, it passes upward and deep to the pulmonary trunk. At this point, it is called the *ascending aorta*. The ascending aorta gives off two coronary branches to the heart muscle. Then it turns to the left, forming the *arch of the aorta*, before descending to the level of the fourth thoracic vertebra as the *descending aorta*. The descending aorta lies close to the vertebral bodies, passes through the diaphragm, and divides at the level of the fourth lumbar vertebra into two *common iliac arteries*, which carry blood to the lower extremities. The section of the descending aorta between the arch of the aorta and the diaphragm is referred to as the *thoracic aorta*. The section between the diaphragm and the common iliac arteries is termed the *abdominal aorta*. Each section of the aorta gives off arteries that continue to branch into distributing arteries leading to organs and finally into the arterioles and capillaries that service all the tissues of the body, except the air sacs of the lungs.

Blood is returned to the heart through the systemic veins. All the veins of the systemic circulation flow into either the *superior*

FIGURE 16-7 Circulatory routes. Systemic circulation is indicated by heavy black arrows, pulmonary circulation by thin black arrows in the pulmonary blood vessels, and hepatic portal circulation by thin colored arrows.

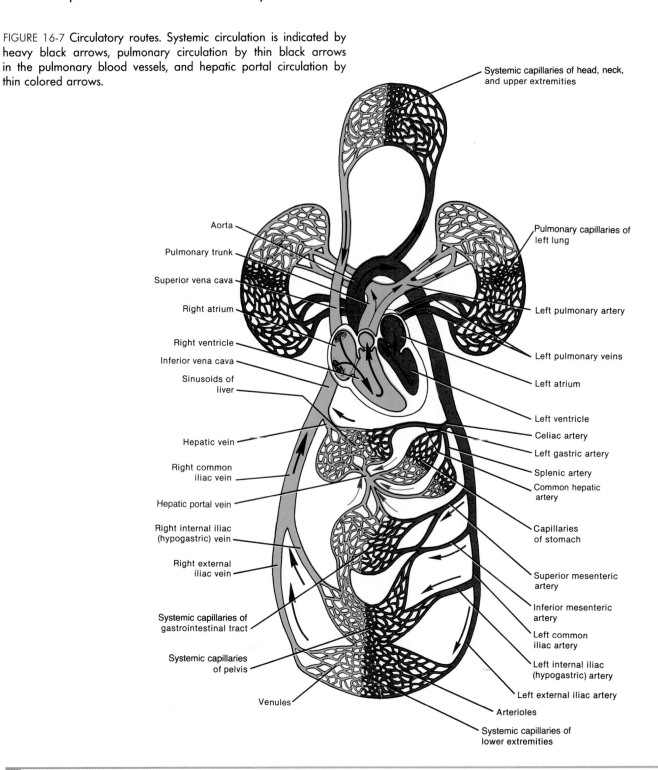

Systemic circulation includes all the oxygenated blood that leaves which ventricle?

or *inferior vena cava* or the *coronary sinus*. They, in turn, empty into the right atrium. The principal blood vessels of systemic circulation are shown in Figure 16-8.

PULMONARY CIRCULATION

The flow of deoxygenated blood from the right ventricle to the air sacs of the lungs and the return of oxygenated blood from the

air sacs to the left atrium is called *pulmonary circulation* (Figure 16-9). The *pulmonary trunk* emerges from the right ventricle and passes upward, backward, and to the left. It then divides into two branches. The *right pulmonary artery* runs to the right lung; the *left pulmonary artery* goes to the left lung. The pulmonary arteries are the only postnatal (after birth) arteries that carry deoxygenated blood. On entering the lungs, the branches divide and subdivide until ultimately they form capillaries around the air sacs

FIGURE 16-8 Major blood vessels of systemic circulation. (a) Arteries.

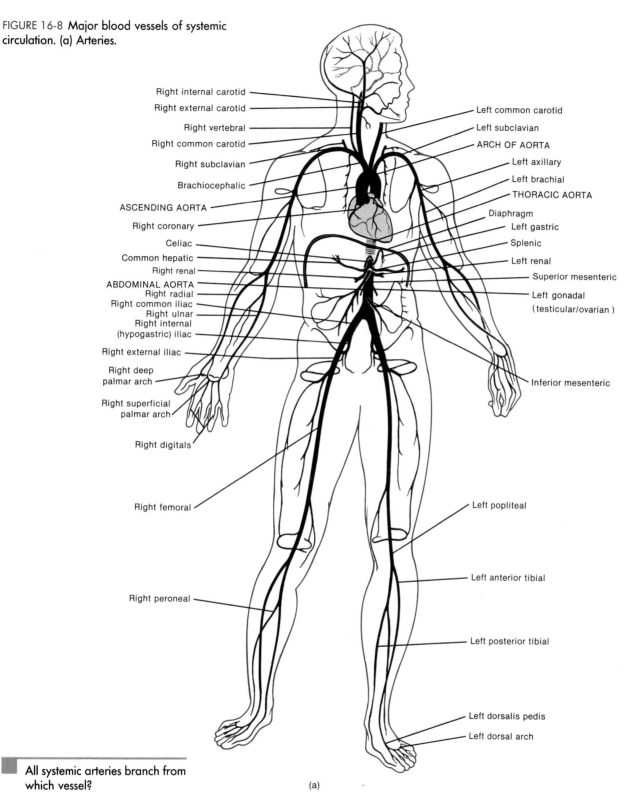

Right internal carotid

Right external carotid

Right vertebral

Right common carotid

Right subclavian

Brachiocephalic

ASCENDING AORTA

Right coronary

Celiac

Common hepatic

Right renal

ABDOMINAL AORTA

Right radial

Right common iliac

Right ulnar

Right internal (hypogastric) iliac

Right external iliac

Right deep palmar arch

Right superficial palmar arch

Right digitals

Right femoral

Right peroneal

Left common carotid

Left subclavian

ARCH OF AORTA

Left axillary

Left brachial

THORACIC AORTA

Diaphragm

Left gastric

Splenic

Left renal

Superior mesenteric

Left gonadal (testicular/ovarian)

Inferior mesenteric

Left popliteal

Left anterior tibial

Left posterior tibial

Left dorsalis pedis

Left dorsal arch

All systemic arteries branch from which vessel?

(a)

in the lungs. Carbon dioxide is passed from the blood into the air sacs to be breathed out of the lungs. Oxygen breathed in is passed from the air sacs into the blood. The capillaries unite, venules and veins are formed, and, eventually, two **pulmonary veins** exit from each lung to transport the oxygenated blood to the left atrium. The pulmonary veins are the only postnatal veins that carry oxygenated blood. Contractions of the left ventricle then send the blood into systemic circulation.

CEREBRAL CIRCULATION

Inside the cranium is a somewhat hexagonal arrangement of blood vessels at the base of the brain called the **cerebral arterial circle (circle of Willis)** (Figure 16-10). From this circle arises arteries supplying most of the brain. The circle is formed by the union of the internal carotid arteries and its branches, the **anterior cerebral arteries**, and the basilar artery and its branches, the **posterior**

FIGURE 16-8 (*Continued*) (b) Veins.

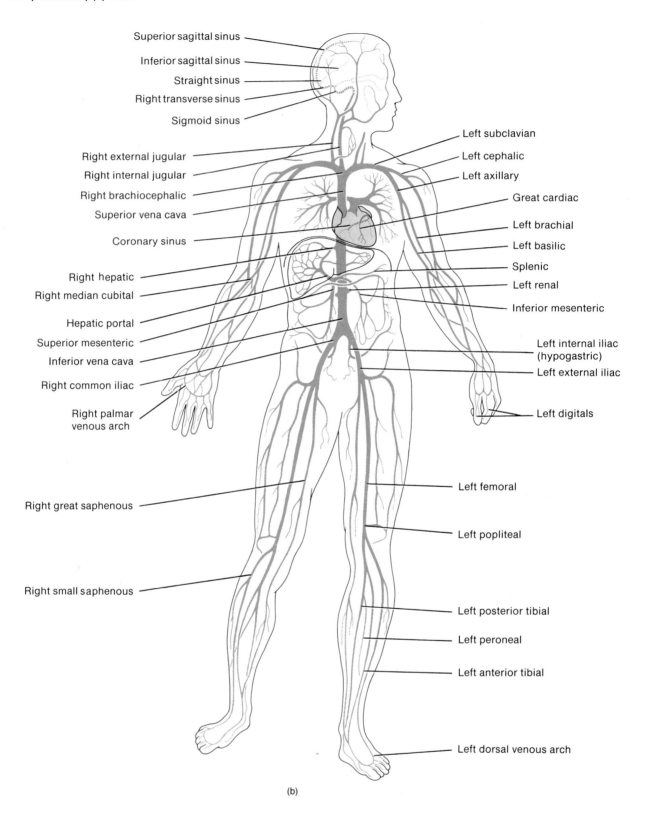

Superior sagittal sinus

Inferior sagittal sinus

Straight sinus

Right transverse sinus

Sigmoid sinus

Right external jugular

Right internal jugular

Right brachiocephalic

Superior vena cava

Coronary sinus

Right hepatic

Right median cubital

Hepatic portal

Superior mesenteric

Inferior vena cava

Right common iliac

Right palmar
venous arch

Right great saphenous

Right small saphenous

Left subclavian

Left cephalic

Left axillary

Great cardiac

Left brachial

Left basilic

Splenic

Left renal

Inferior mesenteric

Left internal iliac
(hypogastric)

Left external iliac

Left digitals

Left femoral

Left popliteal

Left posterior tibial

Left peroneal

Left anterior tibial

Left dorsal venous arch

(b)

All veins of systemic circulation flow into the venae cavae, which in turn empty into which atrium?

FIGURE 16-9 Pulmonary circulation.

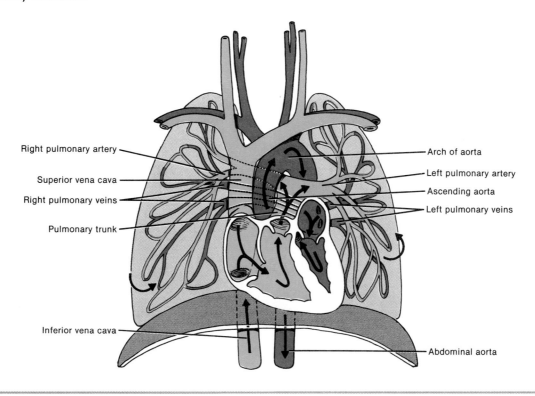

Right pulmonary artery

Superior vena cava

Right pulmonary veins

Pulmonary trunk

Inferior vena cava

Arch of aorta

Left pulmonary artery

Ascending aorta

Left pulmonary veins

Abdominal aorta

The pulmonary trunk emerges from which ventricle of the heart? Pulmonary veins return oxygenated blood to which artium of the heart?

FIGURE 16-10 Cerebral circulation as seen at the base of the brain.

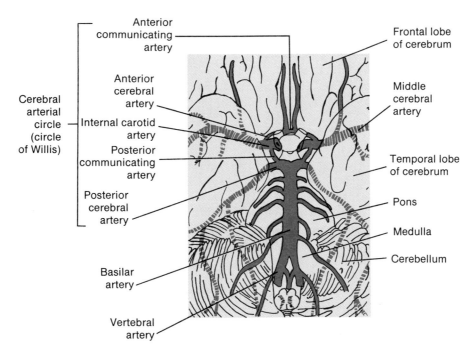

Anterior communicating artery

Anterior cerebral artery

Cerebral arterial circle (circle of Willis)

Internal carotid artery

Posterior communicating artery

Posterior cerebral artery

Basilar artery

Vertebral artery

Frontal lobe of cerebrum

Middle cerebral artery

Temporal lobe of cerebrum

Pons

Medulla

Cerebellum

What are the two functions of the cerebral arterial circle (circle of Willis)?

Hypertension Prevention

The public health campaigns to control high blood pressure have educated millions of people in North America about the significance of this disease. They are responsible for getting large numbers of people to seek regular blood pressure screening and to initiate medical therapy if their blood pressure is found to be high. Medical treatment for hypertension has been shown to effectively lower blood pressure and, consequently, an individual's risk for coronary heart disease and stroke. Improved blood pressure control has received partial credit for the decline in mortality and morbidity due to cardiovascular disease that has occurred over the past two decades.

Despite the enormous scope of these public health campaigns, there has been no decline in the number of Americans diagnosed with hypertension, currently about 58 million in the United States, about 23 percent of the population. More than 10 million are taking medication for this condition, at a cost of about $2.5 billion a year, more than is spent on medication for any other disease. Its prevalence and treatment make hypertension one of the costliest medical conditions for patients, businesses, and others who pick up the health-care tab. The importance of efforts to prevent this disease has never been greater.

While hypertension medication can be lifesaving, it does not correct the underlying cause of the disorder. Indeed, for most cases of hypertension, the cause is not known. At one time, it was thought that a rise in blood pressure was a normal and necessary part of the aging process, which is why hypertension not caused by some other underlying disease was called essential hypertension. Increased blood pressure was thought to be essential to maintain adequate circulation in older adults.

But blood pressure does not increase with age in many other cultures, especially in nonindustrialized countries. Yet 50 percent of all North Americans have hypertension by age 74. What is it about the American way of life that makes us so prone to hypertension?

Risk Factors

Hypertension appears to have both genetic and environmental causes, like coronary artery disease. And like heart disease, risk factors for the development of hypertension have been identified. Unmodifiable risk factors include family history of hypertension (people whose parents or siblings developed hypertension are at increased risk), age (risk increases with age), ethnicity (black Americans have greater risk), and gender (men have more hypertension between the ages of 18 and 54, and women have more hypertension after age 65).

People who fall into a high-risk group should not worry. (Worrying will only increase your blood pressure.) A wellness life-style can significantly reduce a person's risk of developing hypertension and decrease its severity should it develop. Life-style change is particularly effective in helping to lower blood pressure for those with borderline or mild hypertension, 80 percent of hypertensives.

Diet

As described in Chapter 15, a healthful diet can help to prevent hypertension. For example, decreasing sodium intake may cause a decrease in blood pressure as well. However, not everyone with hypertension experiences this effect; therefore, some argue that since not everyone is sodium sensitive, perhaps it is not important that everyone try to limit sodium intake. The counterargument to this is that sodium sensitivity often develops with age, so restricting sodium intake should begin early in life. Although we are born with a preference for sweet foods, a taste for salt appears to be acquired. Anyone at risk for hypertension, and indeed anyone planning to grow old, may benefit from not acquiring this taste.

Animal studies suggest that salty diets may damage arteries even without causing an increase in blood pressure. In one study, rats fed high-salt diets experienced only a slight rise in blood pressure, but all died much earlier than littermates on a low-salt diet. Autopsies revealed a great deal of arterial damage, including atherosclerosis, and associated damage to brain cells supplied by the damaged arteries. Epidemiological studies support the dangers of a high-salt diet as well. For example, people who live in northern Japan consume high-salt diets and also have a very high incidence of stroke.

A few studies have found a small association between calcium intake and blood pressure. Some people with hypertension do appear to respond to an increased intake of dietary calcium. An adequate intake of magnesium may also help regulate blood pressure. In animal studies, magnesium deficiency increased vascular muscle tone, thus increasing peripheral resistance to blood flow. Magnesium is found in legumes, nuts, whole grains, and meat.

The role of dietary fat in blood pressure regulation is not well understood. Decreasing fat intake will not automatically lower blood pressure or prevent its development in most people. However, fat intake is associated with other heart disease risk factors, most notably high blood cholesterol and obesity, so people following dietary measures to reduce their risk of hypertension will probably want to limit fat intake as well.

Alcohol consumption has been associated with hypertension: the greater the alcohol intake, the greater one's risk of hypertension. Many people find that their blood pressure improves when they decrease their alcohol consumption.

Obesity

People who are more than 20 percent above their desirable weight have twice as much hypertension. Even a small weight loss can significantly decrease blood pressure, even if the person does not lose all the extra weight. Extra fat in the abdominal region is more strongly associated with hypertension than excess fat in the hips and thighs.

Obesity is associated with insulin resistance. People with insulin resistance produce enough insulin, but the insulin receptors located in the cell membrane are not sensitive enough to respond correctly. Insulin can stimulate sympathetic nervous system activity, similar to the stress response that causes blood pressure to rise. Insulin may also lead to fluid retention, which increases blood pressure. Weight loss may help to restore insulin sensitivity.

Physical Activity

Many studies have found that regular aerobic exercise can normalize blood pressure, especially in people with borderline hypertension. Epidemiological studies have also found that exercise can improve the health of people with this condition. In fact, one study showed that hypertensives who exercised regularly had almost the same mortality rates as people with normal blood pressure.

Exercise may exert this protective effect partly by contributing to a person's weight control efforts. This is one of the reasons it has been difficult to measure the effect of exercise on high blood pressure. Subjects in exercise studies often lose extra weight, so we can't tell whether the reduction in blood pressure is due to the exercise or the weight loss. A few well-controlled studies have kept subjects from losing weight, and their results suggest that exercise exerts a beneficial effect in addition to its effects on body composition.

Hormonal changes that occur with exercise may lead to decreased peripheral resistance to blood flow. We know that peripheral resistance decreases when a person begins to exercise, in order to increase blood flow to the working muscles. In a healthy person, systolic blood pressure rises as cardiac output increases, but diastolic blood pressure actually drops. Researchers have hypothesized that exercising somehow helps increase peripheral resistance at rest as well, although the mechanisms for such a decrease are not understood.

Insulin sensitivity improves for many hours after exercise. As mentioned above, this could prevent the sympathetic nervous system activity and fluid retention that could contribute to high blood pressure.

Many people find that exercise helps them relax and decreases their stress reactivity. Things don't upset them as easily. Several studies have demonstrated an association between stress and hypertension, so exercise may be helpful because it reduces an individual's stress response.

Stress Management

Since blood pressure is so easy to measure, and since hypertension is such a prevalent problem, the effect of relaxation training on hypertension has been well-studied. As you learned in Chapter 13, blood pressure rises during the alarm phase of the stress response. In most individuals, blood pressure returns to normal once the alarm phase is over. It has been hypothesized that individuals under chronic stress may develop a chronic elevation in sympathetic nervous system activity, called sympathetic tone. This includes an elevation in the hormones associated with the stress response, including epinephrine and norepinephrine, which might cause chronic high blood pressure. Stress management techniques can help a person learn physical relaxation skills, decrease sympathetic tone, and thus reduce blood pressure.

cerebral arteries. Communicating arteries connect the internal carotids with the various branches. The circle equalizes blood pressure to the brain and provides alternate routes should any arteries become damaged.

HEPATIC PORTAL CIRCULATION

In *hepatic portal circulation*, venous blood from the gastrointestinal organs and spleen is delivered to the liver before returning to the heart. This blood is rich with substances absorbed from the gastrointestinal tract. The liver monitors these substances before they pass into the general circulation. It may remove surplus glucose and store it. It modifies certain digested substances so they may be used by cells. It also detoxifies harmful substances absorbed by the gastrointestinal tract and destroys bacteria by phagocytosis.

The hepatic portal system includes veins that drain blood from the pancreas, spleen, stomach, intestines, and gallbladder and converge to form the hepatic portal vein that enters the liver. At the same time the liver receives and acts on deoxygenated blood, it receives oxygenated blood from the systemic circulation via the hepatic artery to nourish its tissues and remove its wastes. Ultimately, all blood leaves the liver through the hepatic veins, which enter the inferior vena cava.

FETAL CIRCULATION

The circulatory system of a fetus, called *fetal circulation*, differs from an adult's because the lungs, kidneys, and digestive organs of a fetus are nonfunctional. The fetus derives its oxygen and nutrients from the maternal blood and eliminates its carbon dioxide and wastes into the maternal blood. The details of fetal circulation are discussed in Chapter 24.

EXERCISE AND THE CARDIOVASCULAR SYSTEM

No matter what a person's level of fitness, it can be improved at any age with regular exercise. However, of the various kinds of exercise, some are more effective than others for improving the cardiovascular system because large body muscles are involved. *Aerobics*, or any activity that works large muscles for at least 20 minutes, increases the flow of blood to the heart and accelerates metabolic rate for at least 20 minutes. Three to five such sessions a week are usually recommended for improving the health of the cardiovascular system. Brisk walking, running, bicycling, cross-country skiing, and swimming are examples of aerobic exercises.

Sustained exercise increases the oxygen demand of the muscles. Whether the demand is met depends primarily on the adequacy of cardiac output and proper function of the respiratory system. After several weeks of training, the healthy individual increases cardiac output and thereby increases the rate of oxygen delivery to the tissues.

Physical conditioning also affects blood pressure. After a conditioning period, hypertensive (high blood pressure) individuals show an average reduction in systolic pressure of 13 mm Hg. Not only does lower pressure itself help reduce myocardial oxygen requirements, but blood pressure response to a given work load is also less after training. Conditioning is useful in the management of hypertension.

Additional benefits to be gained from physical conditioning are an increase in high-density lipoprotein (HDL), a substance that seems to counteract cholesterol and heart disease, a decrease in triglyceride levels, and improved lung function. Exercise also helps reduce anxiety and depression, control weight, and increase the body's ability to dissolve blood clots by increasing fibrinolytic activity. Intense exercise increases blood levels of endorphins, the body's natural painkillers. This may explain the psychological "high" that runners experience with strenuous training and the "low" they feel when they miss regular workouts. Exercise also helps make bones stronger and thus may inhibit osteoporosis. Some research indicates that exercise may even offer some protection against cancer and diabetes.

A well-trained athlete can achieve a cardiac output of up to six times that of an untrained individual. This is because training develops and enlarges the heart, which is a muscle. However, even though an athlete's heart is larger, *resting* cardiac output is about the same as a healthy untrained person's. This is because stroke volume is increased while heart rate is decreased; the heart rate of a trained athlete is about 40 to 60 beats per minute.

▪ COMMON DISORDERS ▪

Hypertension

Hypertension, or high blood pressure, is the most common disease of the heart and blood vessels. Statistically, hypertension afflicts one out of every five American adults. Although there is some disagreement as to what defines hypertension, the consensus is that a blood pressure of 120/80 is normal in a healthy adult. Bordering high blood pressure is therefore defined as diastolic pressure between 85 and 89. Mild high blood pressure is diastolic pressure between 90 and 104. Moderate high blood pressure is diastolic pressure between 105 and 114. Severe high blood pressure is diastolic pressure of 115 or higher. Isolated systolic hypertension is systolic pressure greater than 160 in those whose diastolic pressure is less than 90. The lower the blood pressure, the less the risk of coronary artery disease (CAD). High blood pressure is of considerable concern because of the harm it can do to the heart, brain, and kidneys if it remains uncontrolled, but it can be controlled with diet, exercise, and/or medication.

Aneurysm

An *aneurysm* (AN-yoo-rizm) is a thin, weakened section of the wall of an artery or a vein that bulges outward, forming a balloonlike sac of the blood vessel. Common causes include atherosclerosis, syphilis, congenital blood vessel defects, and trauma.

Deep-Venous Thrombosis (DVT)

Venous thrombosis, the presence of a thrombus in a vein, typically occurs in deep veins of the lower extremities and is referred to as *deep-venous thrombosis* (*DVT*). A serious compliction is *pulmonary embolism,* in which the thrombus dislodges and finds its way into the pulmonary arterial blood flow.

MEDICAL TERMINOLOGY AND CONDITIONS

Aortography (ā′-or-TOG-ra-fē) X-ray examination of the aorta and its main branches after injection of radiopaque dye.

Arteritis (ar′-te-RĪ-tis; *itis* = inflammation of) Inflammation of an artery, probably due to an autoimmune response.

Cardiac arrest (KAR-dē-ak a-REST) Complete stoppage of the heartbeat.

Carotid endarterectomy (ka-ROT-id end′-ar-ter-EK-tō-mē) The removal of atherosclerotic plaque from the carotid artery to restore greater blood flow to the brain.

Claudication (klaw′-di-KĀ-shun) Pain and lameness caused by defective circulation of the blood in vessels of the limbs.

Compensation (kom′-pen-SĀ-shun) A change in the circulatory system made to compensate for some abnormality; an adjustment of the size of the heart or rate of heartbeat made to counterbalance a defect in structure or function; often used specifically to describe the maintenance of adequate circulation in spite of the presence of heart disease.

Hypercholesterolemia (hī-per-kō-les′-ter-ol-Ē-mē-a; *hyper* = over; *heme* = blood) An excess of cholesterol in the blood.

Hypotension (hī′-pō-TEN-shun; *hypo* = below; *tension* = pressure) Low blood pressure; most commonly used to describe an acute drop in blood pressure, as occurs in shock.

Normotensive (nor′-mō-TEN-siv) Characterized by normal blood pressure.

Occlusion (o-KLOO-shun) The closure or obstruction of the lumen of a structure such as a blood vessel.

Orthostatic (or′-thō-STAT-ik) *hypotension* (*ortho* = straight; *statikos* = causing to stand) An excessive lowering of systemic blood pressure on assuming an erect or semierect posture; it is usually a sign of disease. May be caused by excessive fluid loss, certain drugs (antihypertensives), and cardiovascular or neurogenic factors. Also called *postural hypotension.*

Phlebitis (fle-BĪ-tis; *phleb* = vein) Inflammation of a vein, often in a leg.

Raynaud's (rā-NOZ) *disease* A vascular disorder, primarily of females, characterized by bilateral attacks of ischemia, usually of the fingers and toes, in which the skin becomes pale and exhibits burning and pain; it is brought on by cold or emotional stimuli.

Shunt A passage between two blood vessels or between the two sides of the heart.

Syncope (SIN-kō-pē) A temporary cessation of consciousness; a faint. One cause might be insufficient blood supply to the brain.

Thrombectomy (throm-BEK-tō-mē; *thrombo* = clot) An operation to remove a blood clot from a blood vessel.

Thrombophlebitis (throm′-bō-fle-BĪ-tis) Inflammation of a vein with clot formation. Superficial thrombophlebitis occurs in veins under the skin, especially the calf.

White coat (*office*) *hypertension* A syndrome found in patients who have elevated blood pressures while being examined by health-care personnel but are otherwise normotensive.

STUDY OUTLINE

Arteries (p. 334)

1. Arteries carry blood away from the heart. Their walls consist of three coats.
2. The structure of the middle coat gives arteries their two major properties, elasticity and contractility.

Arterioles (p. 334)

1. Arterioles are small arteries that deliver blood to capillaries.

2. Through constriction and dilation, they assume a key role in regulating blood flow from arteries into capillaries and thus arterial blood pressure.

Capillaries (p. 334)

1. Capillaries are microscopic blood vessels through which materials are exchanged between blood and tissue cells.
2. Capillaries branch to form an extensive network throughout a tissue that increases the surface area and thus allows a rapid

exchange of large quantities of material.

3. Precapillary sphincters regulate blood flow through true capillaries.

Venules (p. 335)

1. Venules are small vessels that continue from capillaries and merge to form veins.
2. They drain blood from capillaries into veins.

Veins (p. 336)

1. Veins consist of the same three coats as arteries but have less elastic tissue and smooth muscle.
2. They contain valves to prevent backflow of blood.
3. Weak valves can lead to varicose veins (VVs) or hemorrhoids.

Blood Reservoirs (p. 336)

1. Systemic veins are collectively called blood reservoirs.
2. They store blood that, through vasoconstriction, can move to other parts of the body if the need arises.
3. The principal reservoirs are the veins of the abdominal organs (liver and spleen) and skin.

Physiology of Circulation (p. 336)

1. Blood flow is determined by blood pressure and resistance.
2. Blood flows from regions of higher to lower pressure. The established pressure gradient for systemic circulation is from aorta to large arteries to small arteries to arterioles to capillaries to venules to veins to venae cavae to right atrium.
3. Resistance refers to the opposition to blood flow caused by friction between blood and the walls of blood vessels.
4. Resistance is determined by blood viscosity, blood vessel length, and blood vessel radius.
5. Factors that affect arterial blood pressure include cardiac output (CO), blood volume, and peripheral resistance.
6. Any factor that increases cardiac output (CO) increases blood pressure.
7. As blood volume increases, blood pressure increases.
8. Peripheral resistance refers to the opposition to blood flow in peripheral circulation (away from the heart).
9. In order to maintain a normal homeostatic range of blood pressure, there is a regulating center in the brain that receives input from receptors throughout the body (baroreceptors, chemoreceptors), higher brain centers, and chemicals.
10. The vasomotor center in the medulla controls the diameter of arterioles to regulate blood pressure.
11. Baroreceptors, neurons sensitive to pressure, send impulses to the cardiac center and vasomotor center to regulate blood pressure.
12. Chemoreceptors, neurons sensitive to concentrations of oxygen, carbon dioxide, and hydrogen ions, send impulses to the vasomotor center to regulate blood pressure.
13. The cerebral cortex can influence the vasomotor center and this affects blood pressure.

14. Autoregulation refers to local, autonomic adjustment of blood flow in response to tissue needs.
15. The movement of water and dissolved substances (except proteins) through capillaries depends on pressure differences between the capillaries and interstitial spaces.
16. There is a state of near equilibrium at the arterial and venous ends of a capillary by which fluids exit and enter.
17. Blood return to the heart is maintained by the pumping action of the heart, velocity of blood flow, skeletal muscle contractions, valves in veins (especially in the extremities), and breathing.

Checking Circulation (p. 339)

Pulse (p. 339)

1. Pulse is the alternate expansion and elastic recoil of an artery with each heartbeat. It may be felt in any artery that lies near the surface or over a hard tissue.
2. A normal rate is between 70 and 80 beats per minute.

Measurement of Blood Pressure (p. 340)

1. Blood pressure is the pressure exerted by blood on the wall of an artery when the left ventricle undergoes systole and then diastole. It is measured by a sphygmomanometer.
2. Systolic blood pressure (SBP) is the force of blood recorded during ventricular contraction. Diastolic blood pressure (DBP) is the force of blood recorded during ventricular relaxation. The average blood pressure is 120/80 mm Hg.
3. Pulse pressure is the difference between systolic and diastolic pressure. It averages 40 mm Hg and provides information about the condition of arteries.

Shock and Homeostasis (p. 340)

1. Shock is a failure of the cardiovascular system to deliver adequate amounts of oxygen and nutrients to meet the metabolic needs of cells.
2. Symptoms include low blood pressure; clammy, cool, pale skin; sweating; decreased urinary output; altered mental state; acidosis; tachycardia; weak, rapid pulse; and thirst.

Circulatory Routes (p. 341)

1. The two major circulatory routes are systemic circulation and pulmonary circulation.
2. Three main subdivisions of systemic circulation are coronary circulation, hepatic portal circulation, and cerebral circulation.

Systemic Circulation (p. 341)

1. Systemic circulation takes oxygenated blood from the left ventricle through the aorta to all parts of the body, including lung tissue (but not the air sacs) and returns deoxygenated blood to the right atrium.
2. The aorta is divided into the ascending aorta, the arch of the aorta, and the descending aorta. Each section gives off arteries that branch to supply the whole body.
3. Deoxygenated blood is returned to the heart through the systemic veins. All the veins of systemic circulation flow into either

the superior or inferior vena cava or the coronary sinus. They in turn empty into the right atrium.

Pulmonary Circulation (p. 342)
1. Pulmonary circulation takes deoxygenated blood from the right ventricle to the air sacs of the lungs and returns oxygenated blood from the air sacs to the left atrium.
2. It allows blood to be oxygenated from systemic circulation.

Cerebral Circulation (p. 343)
1. Cerebral circulation arises from a somewhat hexagonal arrangement of arteries at the base of the brain called the cerebral arterial circle (circle of Willis).
2. The circle equalizes blood pressure to the brain and provides alternate routes for blood.

Hepatic Portal Circulation (p. 348)
1. Hepatic portal circulation collects deoxygenated blood from the veins of the pancreas, spleen, stomach, intestines, and gallbladder and directs it into the hepatic portal vein of the liver.
2. This routing allows the liver to extract and modify nutrients and detoxify harmful substances in the blood.
3. The liver also receives oxygenated blood from the hepatic artery.

Exercise and the Cardiovascular System (p. 348)
1. Aerobic exercises greatly benefit the cardiovascular system.
2. Benefits include increased cardiac output, increased delivery of oxygen to tissues, lower systolic blood pressure, increased high-density lipoprotein (HDL), weight control, and increased ability to dissolve blood clots.

REVIEW QUESTIONS

1. Describe the structural and functional differences among arteries, arterioles, capillaries, venules, and veins. (p. 334)
2. Discuss the importance of the elasticity and contractility of arteries. (p. 334)
3. What are blood reservoirs? Why are they important? (p. 336)
4. Explain how blood pressure and resistance relate to blood flow. (p. 336)
5. Explain the factors that contribute to resistance. (p. 337)
6. Describe how each of the following affects blood pressure: cardiac output (CO), blood volume, and peripheral resistance. (p. 337)
7. Discuss how the vasomotor center operates with baroreceptors, chemoreceptors, and higher brain centers to control blood pressure. (p. 337)
8. Explain the effects of different chemicals, such as epinephrine, norepinephrine (NE), antidiuretic hormone (ADH), histamine, kinins, and alcohol on blood pressure. (p. 338)
9. What is meant by autoregulation and how does it work? (p. 338)
10. Why is blood flow slowest in capillaries? (p. 339)
11. Why do veins contain valves, but arteries do not? (p. 339)
12. Define pulse. Where may pulse be felt? (p. 339)
13. Contrast tachycardia, bradycardia, and irregular pulse. (p. 339)
14. What is blood pressure (BP)? Compare the clinical significance of systolic and diastolic pressure. How are these pressures written? (p. 340)
15. Define pulse pressure. What does this pressure indicate? (p. 340)
16. Define shock. What are its symptoms and why do they occur? (p. 340)
17. What is meant by a circulatory route? Define systemic circulation. (p. 341)
18. Define pulmonary circulation. Diagram its route. (p. 342)
19. Describe cerebral circulation. Why is it important? (p. 343)
20. Why is hepatic portal circulation significant? (p. 348)
21. How does exercise benefit the cardiovascular system? (p. 348)
22. Define the following: hypertension, aneurysm, and deep-venous thrombosis (DVT). (p. 348)
23. Refer to the glossary of medical terminology and conditions associated with blood vessels. Be sure that you can define each term. (p. 349)

17

The Lymphatic System and Immunity

STUDENT OBJECTIVES

1. Describe the parts of the lymphatic system and their functions in maintaining homeostasis.
2. Explain how lymphatic tissue is organized and distributed throughout the body.
3. Describe the flow of lymph through the body.
4. Describe the various nonspecific mechanisms for resisting disease.
5. Explain how antigens (Ags) and antibodies (Abs) contribute to specific resistance to disease.
6. Compare the two principal types of immunity and their functions in protecting the body against disease.
7. Describe the relationship of immunology to cancer.
8. Define common disorders and medical terminology and conditions related to the lymphatic system and immunity.

A LOOK AHEAD

FUNCTIONS
LYMPH AND INTERSTITIAL FLUID
LYMPHATIC VESSELS
LYMPHATIC TISSUE
  Lymph Nodes
  Tonsils
  Spleen
  Thymus Gland
LYMPH CIRCULATION
  Route
  Maintenance
NONSPECIFIC RESISTANCE TO
  DISEASE
  Skin and Mucous
    Membranes
      Mechanical Factors
      Chemical Factors
  Antimicrobial Substances
    Interferon (IFN)
    Complement
    Properdin
  Phagocytosis
    Kinds of Phagocytes
    Mechanism
  Inflammation
    Symptoms
    Stages
  Fever
IMMUNITY (SPECIFIC
  RESISTANCE TO DISEASE)
  Antigens (Ags)
  Antibodies (Abs)
  Cellular and Humoral
    Immunity
      Formation of T Cells and B Cells
      T Cells and Cellular
        Immunity
      B Cells and Humoral
        Immunity
      Actions of Antibodies
  The Skin and Immunity
  Immunology and Cancer
COMMON DISORDERS
MEDICAL TERMINOLOGY AND
  CONDITIONS

The *lymphatic* (lim-FAT-ik) *system* consists of a fluid called lymph, vessels called lymphatic vessels (lymphatics) that transport lymph, and a number of structures and organs, all of which contain lymphatic (lymphoid) tissue (see Figure 17-3a). Lymphatic tissue is a specialized form of reticular connective tissue that contains large numbers of lymphocytes. The stroma (framework) of most lymphatic tissue is a meshwork of reticular fibers and reticular cells (fibroblasts and fixed macrophages).

Lymphatic tissue is found in various forms in the body. Lymphatic tissue not enclosed by a capsule is referred to as **diffuse lymphatic tissue.** It is found in mucous membranes of the gastrointestinal tract, respiratory passageways, urinary tract, and reproductive tract and in small amounts in almost every organ of the body. **Lymphatic nodules** are oval-shaped masses of lymphatic tissue that consist of an inner region of large lymphocytes (*germinal center*) and an outer region of small lymphocytes (*cortex*). Lymphatic nodules are not encapsulated either. Some lymphatic nodules are found singly in mucous membranes of the gastrointestinal tract, respiratory passageways, urinary tract, and reproductive tract. Other lymphatic nodules occur in clusters in specific parts of the body, such as the tonsils in the throat, aggregated lymphatic follicles (Peyer's patches) in the small intestine, and in the appendix. The **lymphatic organs** of the body—the lymphatic nodes, spleen, and thymus gland—all contain encapsulated lymphatic tissue. Since bone marrow produces lymphocytes, it is also part of the lymphatic system.

FUNCTIONS

The lymphatic system has several functions.

1. Lymphatic vessels drain tissue spaces of the protein-containing fluid (interstitial fluid) that escapes from blood capillaries. Then, the lymphatic vessels return the proteins, which cannot be directly reabsorbed by blood vessels, along with any fluid not directly reabsorbed, to the cardiovascular system.
2. Lymphatic vessels transport fats from the gastrointestinal tract to the blood (Chapter 19).
3. Lymphatic tissue functions in surveillance and defense. That is, lymphocytes, with the aid of macrophages, protect the body from foreign cells, microbes, and cancer cells.

LYMPH AND INTERSTITIAL FLUID

Whole blood does not flow into the tissue spaces; it remains in closed vessels of the cardiovascular system. However, certain constituents of the plasma move through the capillary walls, and once they move out of the blood, they are called interstitial fluid. Interstitial fluid and lymph are basically the same. The major difference between the two is location. When the fluid bathes the cells, it is called **interstitial fluid** or **intercellular fluid**. When it flows through the lymphatic vessels, it is called **lymph** (*lympha* = clear water).

Both fluids are similar in composition to plasma. The principal chemical difference is that they contain less protein than plasma because large plasma protein molecules cannot filter through the capillary wall.

Each day, about 3 liters of fluid seeps from blood into tissue spaces. This fluid, as well as the plasma proteins it contains, must be returned to the cardiovascular system to maintain normal blood volume and functions. We noted in Chapter 16 that most of the protein-containing fluid filtered from the arterial end of a blood capillary is reabsorbed at the venous end of the capillary. This fluid is returned to the blood directly. A small amount of fluid, however, is returned to the blood indirectly. It first passes into the lymphatic system and then into blood. Let us see how this occurs.

LYMPHATIC VESSELS

Lymphatic vessels originate as *lymph capillaries*, microscopic vessels in spaces between cells (Figure 17-1). Lymph capillaries may occur singly or in networks. They are found throughout the body, with the exception of avascular tissue, the central nervous system, splenic pulp, and bone marrow. They are slightly larger than blood capillaries and have a unique structure that permits interstitial fluid to flow into them, but not out. The endothelial cells that make up the wall of a lymph capillary are not attached end to end, but rather, the ends overlap. In addition, fine filaments anchor the cells to surrounding tissues. When pressure is greater in the interstitial fluid than lymph, the cells separate slightly, like a one-way swinging door, and fluid enters the lymph capillary. When pressure is greater inside the lymph capillary, the cells adhere more closely and fluid cannot escape back into interstitial fluid. Whereas blood capillaries link two larger blood vessels that form part of a circuit, lymph capillaries begin in the tissues and carry the lymph that forms there toward a larger lymphatic vessel.

Just as blood capillaries converge to form venules and veins, lymph capillaries unite to form larger and larger lymph vessels called *lymphatic vessels (lymphatics)* (Figures 17-2 and 17-3). Lymphatic vessels resemble veins in structure but have thinner walls, more valves, and contain lymph nodes at various intervals. Lymphatic vessels of the skin travel in loose subcutaneous tissue

FIGURE 17-1 Lymph capillaries. Relationship of lymph capillaries to tissue cells and blood capillaries.

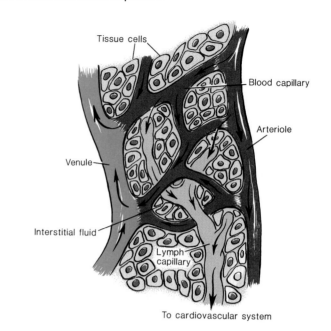

List three functions of the lymphatic system.

FIGURE 17-2 Schematic representation of the relationship of the lymphatic system to the cardiovascular system.

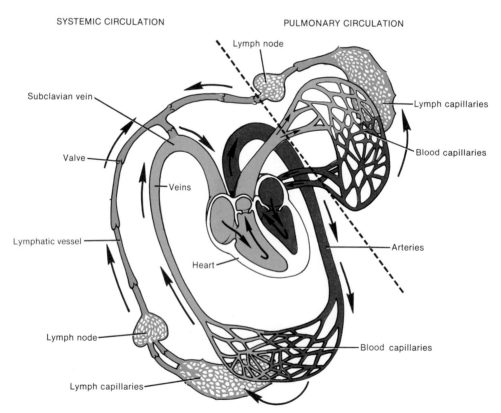

How do lymph capillaries differ from blood capillaries? How do lymphatic vessels differ from veins?

FIGURE 17-3 Lymphatic system. (a) Location of the principal components of the lymphatic system. (b) Details of the thoracic and right lymphatic ducts. (c) In this front view, the light gold area indicates those portions of the body drained by the right lymphatic duct. All other areas of the body are drained by the thoracic duct.

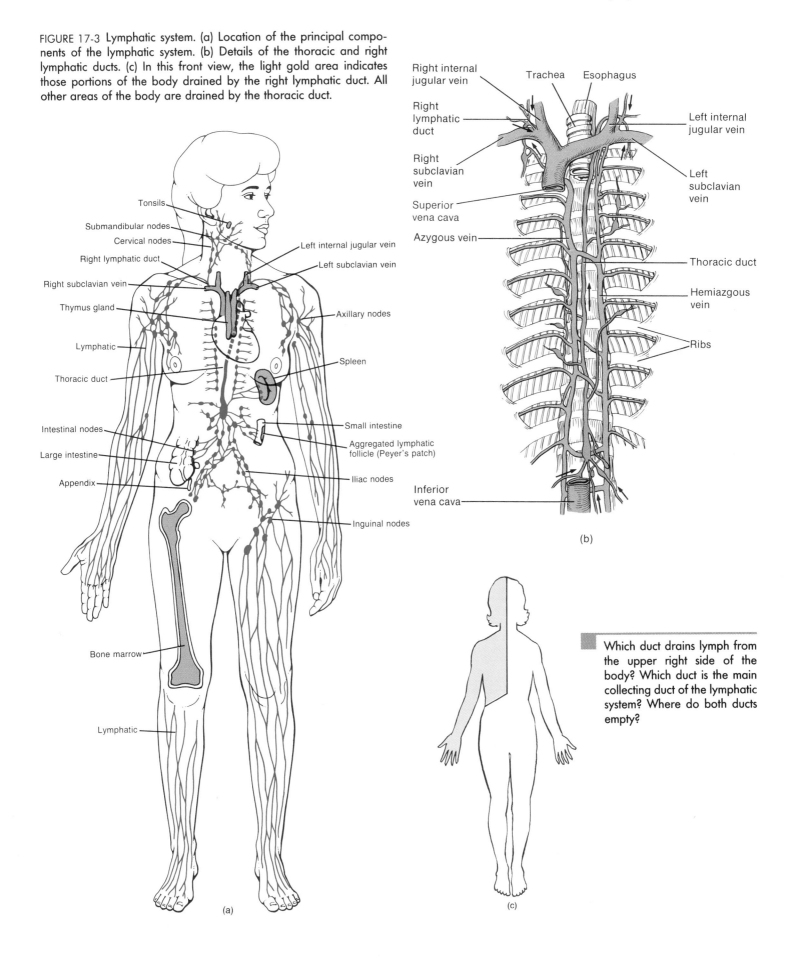

(a)

Tonsils

Submandibular nodes

Cervical nodes

Right lymphatic duct

Right subclavian vein

Thymus gland

Lymphatic

Thoracic duct

Intestinal nodes

Large intestine

Appendix

Bone marrow

Lymphatic

Left internal jugular vein

Left subclavian vein

Axillary nodes

Spleen

Small intestine

Aggregated lymphatic follicle (Peyer's patch)

Iliac nodes

Inguinal nodes

(b)

Right internal jugular vein

Right lymphatic duct

Right subclavian vein

Superior vena cava

Azygous vein

Inferior vena cava

Trachea   Esophagus

Left internal jugular vein

Left subclavian vein

Thoracic duct

Hemiazgous vein

Ribs

(c)

Which duct drains lymph from the upper right side of the body? Which duct is the main collecting duct of the lymphatic system? Where do both ducts empty?

and generally follow veins. Lymphatic vessels of the viscera generally follow arteries, forming networks around them. Ultimately, lymphatic vessels deliver lymph into two main channels: the thoracic duct and the right lymphatic duct (Figure 17-3). These are described shortly.

LYMPHATIC TISSUE

LYMPH NODES

Lymph nodes are oval or bean-shaped organs, about 0.1 to 2.5 cm (1/25 to 1 inch) in length, located along lymphatic vessels (Figure 17-4). Lymph enters a node through several *afferent lymphatic vessels* located on the convex side of the node. These vessels have valves that open toward the node so that the lymph is directed *inward*. Lymph leaves a node through one or two *efferent lymphatic vessels*, which contain valves that open away from the node to convey lymph *outward*. Efferent lymphatic vessels are found at the *hilus* (HĪ-lus), a depression on the concave side. Blood vessels are also found there.

Each node is covered by a *capsule* of dense connective tissue. Internally, nodes are divided into *nodules*, masses of reticular fibers, reticular cells, macrophages, and lymphocytes, which surround lymph channels called *lymph sinuses*. The nodular tissue filters lymph as it passes through the sinuses.

Lymph nodes are scattered throughout the body, usually in groups (see Figure 17-3a). Typically, these groups are arranged in two sets: *superficial* and *deep*.

Lymph nodes filter the lymph passing from tissue spaces through lymphatic vessels on its way back to the cardiovascular system. Reticular fibers trap foreign substances, which are then destroyed through one or more means: macrophages by phagocytosis, T cells (lymphocytes) by releasing various antimicrobial products, and plasma cells, descendants of B cells (lymphocytes) that produce antibodies that destroy them. Lymph nodes also produce lymphocytes, some of which can circulate to other parts of the body.

FIGURE 17-4 Structure of a lymph node. Note the path taken by circulating lymph as indicated by the arrows.

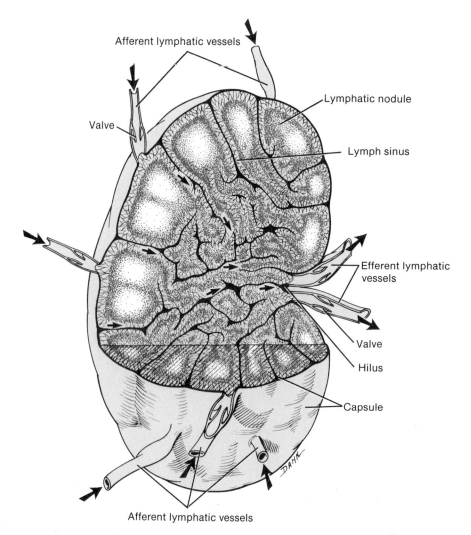

Afferent lymphatic vessels

Valve

Lymphatic nodule

Lymph sinus

Efferent lymphatic vessels

Valve

Hilus

Capsule

Afferent lymphatic vessels

List three functions of lymph nodes.

The location of lymph nodes and direction of lymph flow are important in the diagnosis and prognosis of the spread of cancer by *metastasis*. Cancer cells may be spread by way of the lymphatic system, producing aggregates of tumor cells where they lodge. Such secondary tumor sites are predictable by the direction of lymph flow from the organ primarily involved.

TONSILS

Tonsils are groups of large lymphatic nodules embedded in a mucous membrane and arranged in a ring at the junction of the oral cavity and pharynx. The single *pharyngeal* (fa-RIN-jē-al) *tonsil* or *adenoid* is embedded in the posterior wall of the nasopharynx (see Figure 18-2). The paired *palatine* (PAL-a-tīn) *tonsils* are situated between the pharyngopalatine and glossopalatine arches (see Figure 19-3). These are the ones commonly removed by a tonsillectomy. The paired *lingual* (LIN-gwal) *tonsils* are located at the base of the tongue and may also have to be removed by a tonsillectomy (see Figure 12-3b).

The tonsils produce lymphocytes and antibodies and are situated strategically to protect against invasion of foreign substances that are inhaled or ingested.

SPLEEN

The oval *spleen* is the largest mass of lymphatic tissue in the body. It lies between the stomach and diaphragm, in the upper left region of the abdominal cavity. The spleen is covered by a capsule of dense connective tissue and smooth muscle fibers (cells) that is then covered by a serous membrane, the peritoneum. The spleen contains reticular fibers and cells, lymphatic tissue, red blood cells, plasma cells, and granular leucocytes. It also has a hilus where the splenic artery enters and splenic vein exits. Since the spleen has no afferent lymphatic vessels or lymph sinuses, it does not filter lymph. It does, however, contain vascular sinuses which store blood, one of the spleen's main functions. During severe blood loss, sympathetic impulses cause the smooth muscle of the capsule to contract and release stored blood to maintain blood volume and pressure.

The spleen functions in immunity by producing B cells (lymphocytes), which develop into antibody-producing plasma cells. Also, cells within the spleen phagocytize bacteria and worn-out and damaged red blood cells and platelets. Finally, during early fetal development, the spleen participates in blood cell formation.

The spleen is the most frequently damaged organ in cases of abdominal trauma, particularly those involving severe blows to the lower left chest or upper abdomen that fracture the protecting ribs. Such a crushing injury may *rupture the spleen,* which causes severe hemorrhage and shock. Prompt removal of the spleen, called a *splenectomy*, is needed to prevent the patient from bleeding to death. The functions of the spleen are then assumed by other structures, particularly bone marrow.

THYMUS GLAND

The *thymus gland* is behind the sternum and between the lungs (Figure 17-5). The two *thymic lobes* are held together by connective

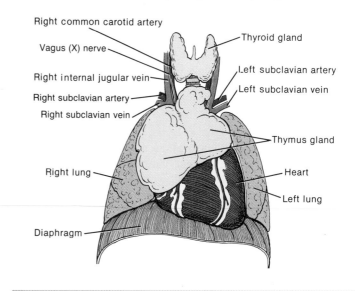

FIGURE 17-5 Location of the thymus gland in a young child.

Right common carotid artery
Vagus (X) nerve
Right internal jugular vein
Right subclavian artery
Right subclavian vein
Right lung
Diaphragm
Thyroid gland
Left subclavian artery
Left subclavian vein
Thymus gland
Heart
Left lung

What is the role of the thymus gland in immunity?

tissue. Each lobe is covered by a connective tissue *capsule*. Extensions of the capsule divide each lobe into smaller units called *lobules*. Each lobule consists of lymphocytes held in place by reticular tissue fibers.

The thymus gland is conspicuous in the infant but does not reach its maximum size until puberty. After puberty, much of the thymic tissue is replaced by fat and connective tissue but the gland continues to be functional.

Its role in immunity is to help produce and distribute to other lymphoid organs T cells (lymphocytes) that destroy invading microbes directly or indirectly by producing various substances. Recall from Chapter 13 that hormones produced by the thymus gland, such as thymosin and thymopoietin, promote the proliferation and maturation of T cells. This is discussed in detail shortly.

LYMPH CIRCULATION

ROUTE

Lymph from lymph capillaries is passed to lymphatic vessels that run toward lymph nodes. At the nodes, afferent vessels enter the capsules at numerous points and the lymph passes through the nodes. Efferent vessels from the nodes either run with afferent vessels into another node of the same group or pass on to another group of nodes. The efferent vessels from the last nodes in a chain unite to form *lymph trunks*.

The principal trunks pass their lymph into two main channels: the thoracic duct and the right lymphatic duct. The *thoracic duct* is the main collecting duct of the lymphatic system and receives lymph from the left side of the head, neck, and chest, the left upper extremity, and the entire body below the ribs. The *right lymphatic duct* drains lymph from the upper right side of the body (see Figure 17-3).

Ultimately, the thoracic duct empties all its lymph into the junction of the left internal jugular vein and left subclavian vein, and the right lymphatic duct empties all its lymph into the junction of the right internal jugular vein and right subclavian vein (see Figure 17-3). Thus, lymph is drained back into the blood, and the cycle repeats itself continuously.

MAINTENANCE

The flow of lymph from tissue spaces to the large lymphatic ducts to the subclavian veins is maintained primarily by the milking action of skeletal muscles. Skeletal muscle contractions compress lymphatic vessels and force lymph toward the subclavian veins. Lymphatic vessels, like veins, contain valves, and the valves ensure the movement of lymph toward the subclavian veins (see Figures 17-2 and 17-3).

Lymph flow is also maintained by respiratory movements, which create a pressure gradient between the two ends of the lymphatic system. Lymph flows from the abdominal region, where the pressure is higher, toward the thoracic region, where it is lower.

Edema, an excessive accumulation of interstitial fluid in tissue spaces, may be caused by an obstruction, such as an infected node, or a blockage of lymphatic vessels. Another cause is excessive lymph formation and increased permeability of blood capillary walls. Edema may also result from a rise in capillary blood pressure, in which interstitial fluid is formed faster than it is passed into lymphatic vessels.

NONSPECIFIC RESISTANCE TO DISEASE

The human body continually attempts to maintain homeostasis by counteracting the activities of disease-producing organisms, called *pathogens* (PATH-ō-jens), or their toxins. The ability to ward off disease is called *resistance*. Vulnerability or lack of resistance is called *susceptibility*. Defenses against disease may be grouped into two broad areas: nonspecific resistance and specific resistance. *Nonspecific resistance* is inherited and represents a wide variety of body reactions that provide a general response against invasion by a wide range of pathogens. *Specific resistance* or *immunity* is the production of a specific antibody against a specific pathogen or its toxin. We will first consider mechanisms of nonspecific resistance.

SKIN AND MUCOUS MEMBRANES

The skin and mucous membranes of the body are the first line of defense against disease-causing microorganisms. They possess certain mechanical and chemical factors that combat the initial attempt of a microbe to cause disease.

Mechanical Factors

The *intact skin*, as noted in Chapter 5, consists of two distinct portions called the epidermis and dermis. The closely packed cells of the epidermis, its many layers, and the presence of keratin, provide a formidable physical barrier to the entrance of microbes. In addition, periodic shedding of epidermal cells helps to remove microbes on the skin surface. The intact surface of healthy epidermis seems to be rarely penetrated by bacteria. But when the epithelial surface is broken, from cuts, burns, and so on, a subcutaneous infection often develops. The bacteria most likely to cause such an infection are staphylococci, which normally inhabit the hair follicles and sweat glands of the skin. Moreover, when the skin is moist, as in hot humid climates, dermal infections, especially fungus infections, such as athlete's foot, are common.

Mucous membranes, like the skin, also consist of two layers: an epithelial layer and an underlying connective tissue layer. Mucous membranes line body cavities that open to the exterior. Mucous membranes line the entire gastrointestinal, respiratory, urinary, and reproductive tracts. The epithelial layer of a mucous membrane secretes a fluid called *mucus*, which prevents the cavities from drying out. Because mucus is slightly viscous, it traps many microbes that enter the respiratory and gastrointestinal tracts. The mucous membrane of the nose has mucus-coated *hairs* that trap and filter air containing microbes, dust, and pollutants. The mucous membrane of the upper respiratory tract contains *cilia*, microscopic hairlike projections of the epithelial cells (see Figure 18-4), that move in such a manner that they pass inhaled dust and microbes that have become trapped in mucus toward the throat. This so-called "ciliary escalator" keeps the "mucus blanket" moving toward the throat at a rate of 1 to 3 cm per hour. Coughing and sneezing speed up the "escalator."

Microbes are also prevented from entering the lower respiratory tract by a small lid of cartilage called the *epiglottis* that covers the voice box during swallowing (see Figure 18-2).

The cleansing of the urethra by the *flow of urine* is another mechanical factor that prevents microbial colonization in the urinary system.

Defecation and *vomiting* might also be considered mechanical processes that expel microbes. For example, in response to microbes, such as those that might irritate the lining of the gastrointestinal tract, the muscle of the tract contracts vigorously to attempt to expel the microbes (diarrhea).

Some pathogens, however, thrive on the moist secretions of a mucous membrane, so they are able to penetrate the membrane if present in sufficient numbers. This penetration may be allowed by toxic products produced by the microbes, prior injury to the membrane by viral infections, or mucosal irritations. Although mucous membranes do inhibit the entrance of many microbes, they are less effective than the skin.

A mechanism that protects the eyes is the *lacrimal* (LAK-ri-mal) *apparatus* (see Figure 12-4), a group of structures that manufactures and drains away tears, which are spread over the surface of the eyeball by blinking. Normally, the tears are carried away by evaporation or pass into the nose as fast as they are produced. The continual washing action of tears helps to keep microbes from settling on the surface of the eye. If an irritating substance or large number of microbes make contact with the eye, the lacrimal glands oversecrete. Tears then accumulate more rapidly than they can be carried away, a protective mechanism that dilutes and washes away the irritating substance or microbes.

Saliva, produced by the salivary glands, washes microbes from the surface of the teeth and the mucous membrane of the mouth, much as tears wash the eyes. This helps to prevent colonization by microbes.

Chemical Factors

Certain chemical factors also account for the high degree of resistance of skin and mucous membranes to microbial invasion.

Oil glands of the skin secrete an oily substance called **sebum** that forms a protective film over the surface of the skin. The skin's acidity, with a pH of 3 to 5, probably discourages the growth of many microorganisms. Sweat glands of the skin produce **perspiration**, which flushes some microorganisms from the skin surface. Perspiration also contains **lysozyme**, an enzyme capable of breaking down cell walls of certain bacteria under certain conditions. Lysozyme is also found in tears, saliva, nasal secretions, and tissue fluids, where it exhibits its antimicrobial activity.

Gastric juice is produced by the glands of the stomach. It is a mixture of hydrochloric acid, enzymes, and mucus. The very high acidity of gastric juice (pH 1.2 to 3.0) destroys bacteria and most bacterial toxins.

ANTIMICROBIAL SUBSTANCES

In addition to the mechanical and chemical barriers of the skin and mucous membranes, the body also produces certain antimicrobial substances, including interferon, complement, and properdin. They provide a second line of defense should microbes penetrate the skin and mucous membranes.

Interferon (IFN)

Body cells infected with viruses produce a protein called **interferon** (in'-ter-FĒR-on) or **IFN**. IFN diffuses to uninfected neighboring cells and binds to surface receptors. This somehow enables uninfected cells to make another antiviral protein that inhibits viral replication and since viruses can cause disease only if they are able to replicate in cells, interferon appears to protect cells from invasion against many different viruses. A form of interferon has been incorporated into a nasal spray that may prevent the spread of colds among family members.

The suggested relationship between viruses and cancer has prompted scientists to study the effects of interferon on cancer patients. Clinical trials indicate that large doses of pure interferon have limited effects against some types of tumors, including melanoma and renal cell cancer, and no effect against others. Moreover, interferon produces side effects ranging from mild to life threatening.

Complement

Complement is actually a group of at least 20 proteins found in normal blood serum. The system is called complement because it "complements" certain immune and allergic reactions involving antibodies (these are discussed shortly). Complement proteins destroy microbes in various ways.

Some initiate reactions that create holes in the plasma membrane of the microbe, causing the contents of the microbe to leak out, a process called **cytolysis**. Some attract phagocytes to the site of inflammation, called **chemotaxis**, and increase blood flow to the area by causing vasodilation of arterioles. Others bind to the surface of a microbe and promote phagocytosis. This is called **opsonization** (op-sō-ni-ZĀ-shun).

Properdin

Properdin (prō-PER-din; *pro* = in behalf of; *perdere* = to destroy) is a complex of three proteins and, like complement, is also found in serum. The proteins become involved by interacting with polysaccharides on the surface of certain bacteria and fungi. Properdin, acting together with complement, destroys several types of bacteria by cytolysis, the enhancement of phagocytosis, and triggering of inflammatory responses.

PHAGOCYTOSIS

When microbes penetrate the skin and mucous membranes or bypass the antimicrobial substances in blood, the next nonspecific resistance of the body is phagocytosis. **Phagocytosis** (*phagein* = to eat; *cyto* = cell) is the ingestion and destruction of microbes or any foreign particles by cells called phagocytes, which are certain types of white blood cells and cells derived from them.

Kinds of Phagocytes

The phagocytes that participate in phagocytosis fall into two broad categories: granulocytes (microphages) and macrophages. The **granulocytes** of blood (see Figure 14-2) vary in the degree to which they function as phagocytes. Neutrophils are the most phagocytic. Eosinophils have some phagocytic capability, and the role of basophils is debatable.

When an infection occurs, both granulocytes (especially neutrophils) and monocytes migrate to the infected area. During this migration, the monocytes enlarge and develop into active phagocytic cells called **macrophages** (MAK-rō-fā-jez) (see Figure 14-2). Since these cells leave the blood and migrate to infected areas, they are called **wandering macrophages**. Some macrophages, called **fixed macrophages**, remain in certain tissues and organs of the body. Fixed macrophages are found in the skin and subcutaneous layer (histiocytes), liver (stellate reticuloendothelial cells), lungs (alveolar macrophages), brain (microglia), spleen, lymph nodes, and bone marrow (tissue macrophages).

Mechanism

For convenience of study, phagocytosis is divided into four phases: chemotaxis, adherence, ingestion, and digestion (Figure 17-6).

▪ **Chemotaxis** *Chemotaxis* (kē'-mō-TAK-sis; *chemo* = chemicals; *taxis* = arrangement) is the chemical attraction of phagocytes to microorganisms. Chemicals that attract phagocytes might come from the microbe themselves, damaged tissue cells, or complement.

▪ **Adherence** In *adherence*, the phagocyte attaches itself to the microorganism or other foreign material. Sometimes, adherence occurs easily and the microorganism is readily phagocytized (see Figure 3-9a,b). If the two surfaces do not readily adhere, the phagocyte traps the particle to be ingested against a rough surface, like a blood vessel, blood clot, or connective tissue fibers, where it cannot slide away.

Also, microorganisms can more readily be phagocytized if they are first coated with certain plasma proteins (complement) that promote attachment (opsonization).

FIGURE 17-6 **Phases of phagocytosis.**

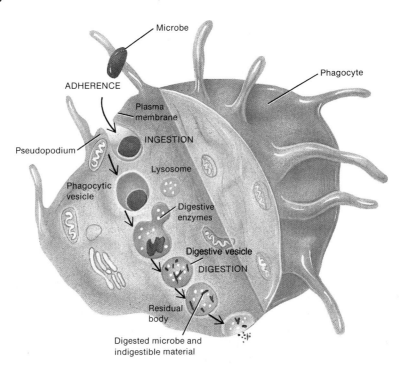

Which cells of the body function as phagocytes?

▪ **Ingestion** Following adherence, *ingestion* occurs. In the process, the cell membrane of the phagocyte extends projections, called pseudopods, that engulf the microorganism. When the pseudopods meet, they fuse, surrounding the microorganism with a sac called a *phagocytic vesicle*.

▪ **Digestion** In this phase, the phagocytic vesicle pinches off from the membrane and enters the cytoplasm, where there are lysosomes that contain digestive enzymes and bactericidal substances. Upon contact, the phagocytic vesicle and lysosome membranes fuse to form a single, larger structure called a *digestive vesicle*. Some bacteria are killed in only 10 to 30 minutes. Indigestible materials remain in structures called *residual bodies*, which the cell disposes of by exocytosis, a process in which the residual body migrates to the plasma membrane, fuses to it, ruptures, and releases its contents.

INFLAMMATION

When cells are damaged by microbes, physical agents, chemical agents, or even a clean surgical cut, the injury sets off an inflammation. The injury may be viewed as a form of stress.

Symptoms

Inflammation is a defensive response of the body to stress due to tissue damage that is characterized by four symptoms: *redness, pain, heat,* and *swelling*. A fifth symptom can be *loss of function*, depending on the site and extent of the injury. Inflammation is an attempt to dispose of microbes, toxins, or foreign material at the site of injury, to prevent their spread to other organs, and to prepare the site for tissue repair. Thus, the inflammatory response is an attempt to restore tissue homeostasis.

Stages

There are four basic stages of the inflammatory response: vasodilation and increased permeability of blood vessels, fibrin formation, phagocyte migration, and pus formation.

▪ **Vasodilation and Increased Permeability of Blood Vessels** Immediately following tissue damage, there is vasodilation and increased permeability of blood vessels in the area of the injury. *Vasodilation* is an increase in diameter of the blood vessels. *Increased permeability* means that substances normally retained in blood are permitted to pass from the blood vessels. Vasodilation allows more blood to go to the damaged area, and increased permeability permits defensive substances in the blood such as antibodies, phagocytes, and clot-forming chemicals to enter the injured area. The increased blood supply also removes toxic products released by the invading microorganisms and dead cells, preventing them from complicating the injury.

One of the principal substances that contribute to vasodilation, increased permeability, and other aspects of the inflammatory response is *histamine* (Figure 17-7). It is found in many body cells, especially mast cells in connective tissue, basophils, and blood

FIGURE 17-7 Inflammatory response. Histamine and other substances stimulate vasodilation, increased permeability of blood vessels, chemotaxis (attracting phagocytes to the area), diapedesis, and phagocytosis.

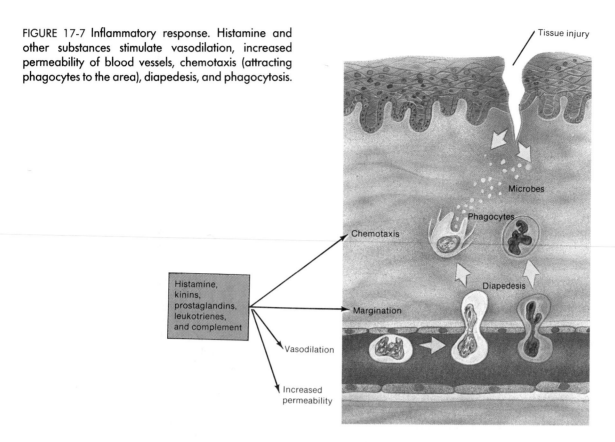

What causes each of the following symptoms of inflammation— redness, pain, heat, and swelling?

platelets, which release it when tissues are invaded or injured. Other inflammatory substances include kinins, prostaglandins (PGs), leukotrienes, and complement.

It is the dilation of arterioles and increased permeability of capillaries that produce heat, redness, and swelling. Heat results from the large amount of warm blood that accumulates in the area and, to a certain extent, from the heat energy produced by the metabolic reactions. The large amounts of blood are also responsible for the redness.

Pain results from injury to nerve fibers or from toxic chemicals released by microorganisms. Some of the substances that promote vasodilation and increased dilation, epecially kinins and prostaglandins, also affect nerve endings and cause pain.

■ **Fibrin Formation** Blood contains a soluble protein called *fibrinogen* (fī-BRIN-ō-jen). The increased permeability of capillaries causes leakage of fibrinogen into tissues. Fibrinogen is then converted to an insoluble, thick network called *fibrin* (FĪ-brin), which traps the invading organisms, preventing their spread. This network eventually forms a clot that isolates the invading microbes or their toxins.

■ **Phagocyte Migration** Generally, within an hour after the inflammatory process is initiated, phagocytes appear on the scene. As the flow of blood decreases, neutrophils begin to stick to the inner surface of the endothelium (lining) of blood vessels. This

is called *margination*. Then the neutrophils begin to squeeze through the wall of the blood vessel to reach the damaged area. This migration, which resembles amoeboid movement, is called *diapedesis* (dī'-a-pe-DĒ-sis) and it depends on chemotaxis. Neutrophils are attracted by microbes, kinins, complement, and other neutrophils. Neutrophils attempt to destroy the invading microbes by phagocytosis.

As the inflammatory response continues, monocytes follow the neutrophils into the infected area. Once in the tissue, monocytes become transformed into wandering macrophages that augment the phagocytic activity of fixed macrophages. In addition, fixed macrophages mobilize as wandering macrophages and migrate to the infected area. The neutrophils predominate in the early stages of infection but tend to die off rapidly. Macrophages then enter the picture. Macrophages are several times more phagocytic than neutrophils and large enough to engulf tissue that has been destroyed, neutrophils that have been destroyed, and invading microbes.

■ **Pus Formation** In all but very mild inflammations, pus is produced. *Pus* is a thick fluid that contains living, as well as nonliving, white blood cells and debris from other dead tissue. Pus formation continues until the infection subsides. At times, the pus pushes to the surface of the body or into an internal cavity for dispersal. Occasionally, the pus remains after the infection is terminated, in which case, it is gradually destroyed and absorbed by the body.

If the pus cannot drain out of the body, an abscess develops. An *abscess* is simply an excessive accumulation of pus in a confined space. Common examples are pimples and boils. When inflamed tissue is shed many times, it produces an open sore, called an *ulcer*, on the surface of an organ or tissue. Ulcers may result from a prolonged inflammatory response to continuously injured tissue. For instance, overproduction of hydrochloric acid in the stomach may steadily erode the epithelium lining the stomach.

FEVER

The most frequent cause of *fever*, an abnormally high body temperature, is infection from bacteria (and their toxins) and viruses. The high body temperature inhibits some microbial growth and speeds up body reactions that aid repair. (Fever is discussed in more detail in Chapter 20.)

The factors that contribute to nonspecific resistance are summarized in Exhibit 17-1.

IMMUNITY (SPECIFIC RESISTANCE TO DISEASE)

Despite the various kinds of nonspecific resistance, they all have one thing in common. They are designed to protect the body from *any* kind of pathogen. They are not specifically directed against a particular microbe. Specific resistance to disease, called *immunity*, involves the production of a specific type of cell or specific molecule (antibody) to destroy a particular antigen. If antigen 1 invades the body, antibody 1 is produced against it. If antigen 2 invades the body, antibody 2 is produced against it, and so on. The branch of science that deals with the responses of the body when challenged by antigens is referred to as *immunology* (im'-yoo-NOL-ō-jē; *immunis* = free).

Acquired immunity refers to immunity gained as a result of contact with an antigen. It may be obtained either actively or passively, or by natural or artificial means. Following are the types of acquired immunity:

1. **Naturally acquired active immunity** is obtained when a person's immune system comes into contact with microbes in the course of daily living and the body responds by producing antibodies and/or T cells. Examples include immunity to measles and chickenpox. This type of immunity may be lifelong or last only a few years.
2. **Naturally acquired passive immunity** involves the natural transfer of antibodies from an immunized donor to a nonimmunized recipient. For example, an expectant mother can pass some of her antibodies to the fetus across the placenta to provide immunity to diseases such as diphtheria, rubella, or polio. Certain antibodies are also passed from the mother to her nursing infant in breast milk. Naturally acquired passive immunity generally lasts only a few weeks or months.
3. **Artificially acquired active immunity** results from vaccination. The individual is given an antigen that is antigenic (stimulates antibody production) but nonpathogenic (causes few symptoms of the disease). The preparations used are called *vaccines*, which consist of killed or weakened living microbes, and *toxoids*,

EXHIBIT 17-1
Summary of Nonspecific Resistance

Component	Functions
SKIN AND MUCOUS MEMBRANES	
Mechanical Factors	
Intact Skin	Forms a physical barrier to the entrance of microbes.
Mucous Membranes	Inhibit the entrance of many microbes, but not as effective as intact skin.
Mucus	Traps microbes in respiratory and gastrointestinal tracts.
Hairs	Filter microbes and dust in nose.
Cilia	Together with mucus, trap and remove microbes and dust from upper respiratory tract.
Epiglottis	Prevents microbes and dust from entering trachea.
Flow of Urine	Washes microbes from urethra.
Defecation and Vomiting	Expel microbes.
Lacrimal Apparatus	Tears dilute and wash away irritating substances and microbes.
Saliva	Washes microbes from surfaces of teeth and mucous membranes of mouth.
Chemical Factors	
Sebum	Forms protective oily film over skin.
Acid pH of Skin	Discourages growth of many microbes.
Perspiration	Flushes microbes from skin.
Lysozyme	Antimicrobial substance in perspiration, tears, saliva, nasal secretions, and tissue fluids.
Gastric Juice	Destroys bacteria and most toxins in stomach.
ANTIMICROBIAL SUBSTANCES	
Interferon (IFN)	Protects uninfected host cells from viral infection.
Complement	Causes cytolysis of microbes, promotes phagocytosis, and contributes to inflammation.
Properdin	Works with complement to bring about same responses as complement.
PHAGOCYTOSIS	Ingestion and destruction of foreign particulate matter by granulocytes and macrophages.
INFLAMMATION	Confines and destroys microbes and repairs tissues.
FEVER	Inhibits microbial growth and speeds up body reactions that aid repair.

which contain inactivated bacterial toxins. Vaccines are used for vaccination against polio and mumps while toxoids are used against tetanus and diphtheria. Some vaccines, such as those for polio and measles, provide lifelong immunity; others, such as pneumococcal pneumonia and meningococcal meningitis, last only a few years.

4. **Artificially acquired passive immunity** is gained by an injection of antibodies from an outside source. For example, a person might be given gamma globulin (antibody-rich serum). Human antibodies produced by one person can also be transferred to another person.

We will examine the mechanism of immunity by first discussing the nature of antigens and antibodies.

ANTIGENS (Ags)

An *antigen (immunogen)*, or *Ag*, is any chemical substance that, when introduced into the body, causes the body to produce specific antibodies and/or specific cells called T cells that react with the antigen. Antigens thus have two important characteristics. The first is *immunogenicity* (im′-yoo-nō-jen-IS-it-ē), the ability to stimulate the formation of specific antibodies. The second is *reactivity*, the ability of the antigen to react specifically with the produced antibodies. Chemically, most antigens are proteins, nucleoproteins, lipoproteins, glycoproteins, or certain large polysaccharides.

The entire microbe, such as a bacterium or virus, or just parts of microbes may act as an antigen, for example, bacterial structures such as flagella, capsules, and cell walls; bacterial toxins are highly antigenic. Nonmicrobial examples of antigens include foods, drugs, pollen, and organs, tissue, cells, or serum of other humans, animals, or insects.

As a rule, antigens are foreign substances. They are not usually part of the chemistry of the body. The body's own substances, recognized as "self," do not act as antigens; substances identified as "nonself," however, stimulate antibody production. There are certain conditions, though, in which the distinction between self and nonself breaks down, and antibodies that attack the body are produced. These autoimmune diseases are described later.

Now let us look at some of the characteristics of antibodies.

ANTIBODIES (Abs)

An *antibody (Ab)* is a protein produced by the body in response to the presence of an antigen and is capable of combining specifically with the antigen. As you will see shortly, lymphocytes called B cells develop into plasma cells that produce the antibodies.

Antibodies do not form against the whole antigen. At specific regions on the surface of the antigen, called *antigenic determinant sites* (Figure 17-8), specific chemical groups of the antigen combine with the antibody. This combination depends on the size and shape of the determinant site and whether or not it corresponds to the chemical structure of the antibody. The combination is very much like the fit between a lock and key.

Most antibodies have two antigenic determinant sites; most antigens have two or more. An antigen with only one site, called a *hapten* (*haptein* = to grasp), can stimulate an immune response

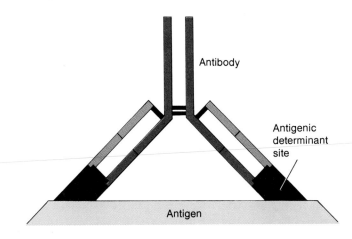

FIGURE 17-8 Relationship of an antigen to an antibody. Most antigens have more than one antigenic determinant site. Most antibodies have two reaction sites that match only specific antigenic determinant sites. Each antibody has reaction sites for specific antigenic determinant sites only.

What are antigenic determinant sites and why are they important?

only if it attaches to a larger carrier molecule so the combined molecule has two determinant sites.

Antibodies belong to a group of plasma proteins called globulins, and for this reason an antibody is also known as an *immunoglobulin* (im′-yoo-nō-GLOB-yoo-lin), or *Ig*. There are five different classes of immunoglobulins, designated as IgG, IgA, IgM, IgD, and IgE. Each has a distinct chemical structure and a specific biological role.

IgG antibodies are the most abundant and are found in blood, lymph, and the intestines. They protect against bacteria and viruses by enhancing phagocytosis, neutralizing toxins, and triggering the complement system.

IgA antibodies are found in tears, saliva, mucus, milk, gastrointestinal secretions, blood, and lymph. During stress, IgA levels decrease and this decrease can lower resistance to infection. They provide localized protection on mucous membranes.

IgM antibodies are the first antibodies to appear after an initial exposure to an antigen. They are found in blood, lymph, and on the surfaces of B cells and cause agglutination and lysis of microbes. ABO agglutinins are IgM antibodies.

IgD antibodies are found in blood, lymph, and on the surfaces of B cells. They may be involved in stimulating antibody-producing cells to manufacture antibodies.

IgE antibodies are located on mast and basophil cells and are involved in allergic reactions.

CELLULAR AND HUMORAL IMMUNITY

The body defends itself against invading agents by means of its immune system, which consists of two closely allied components. One component involves the formation of special lymphocytes called T cells that attack foreign agents and destroy them. This is *cellular (cell-mediated) immunity* and is particularly effective against fungi, parasites, intracellular viral infections, cancer cells, and foreign tissue transplants. ***Humoral (antibody-mediated) immunity*** refers to antibodies that are dissolved in blood plasma and lymph, the body's fluids or "humors." In the humoral system, specialized lymphocytes called B cells develop into plasma cells that produce antibodies. This type of immunity is particularly effective against bacterial and viral infections.

Both types of immunity originate in lymphoid tissue located in the lymph nodes, spleen, gastrointestinal tract, and bone marrow. Lymphoid tissue is strategically located to intercept an invading agent before it can spread too extensively into general circulation.

Formation of T Cells and B Cells

Both T cells and B cells develop in the embryo from lymphocytic stem cells in bone marrow, which they leave for two different areas of the body (Figure 17-9). About half of them first migrate to the thymus gland, where they are processed to become T cells. The name T cell is derived from the processing that occurs in the thymus gland. T cells then leave the thymus gland and become embedded in lymphoid tissue where thymic hormones continue to stimulate them and their descendants.

The remaining stem cells are processed elsewhere, possibly bone marrow, or fetal liver and spleen, or gut-associated lymphoid tissue. They become the B cells. They are given this name because

FIGURE 17-9 Origin and differentiation of T cells and B cells.

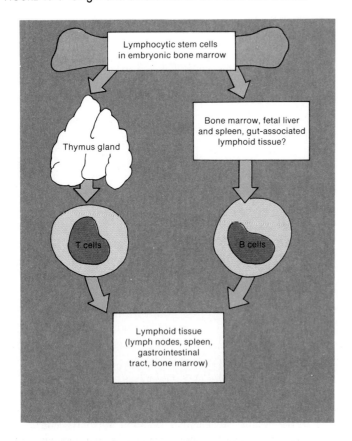

On what basis are T cells named? B cells?

in birds they are processed in the bursa of Fabricius, a small pouch of lymphoid tissue attached to the intestine. Once processed, B cells migrate to lymphoid tissue. Although both T cells and B cells occupy the same lymphoid tissues, they localize in separate areas.

T Cells and Cellular Immunity

Before T cells can attack foreign invaders (antigens), they must become ***sensitized***, that is, able to recognize them. Sensitization occurs as follows. Macrophages must first process and present the antigens to T cells (Figure 17-10). In the processing phase, the macrophage phagocytizes the antigen; in the presentation phase, the partially digested antigen is displayed on the surface of the macrophage, along with other antigens found on the surface of white blood cells called human leucocyte associated (HLA) antigens. Macrophages also participate in the immune response by secreting ***lymphokines***, powerful protein growth factors, such as interleukin-1 (IL-1) and interferons (IFNs), that stimulate T cell growth. In addition, lymphokines stimulate the reproduction of sensitized T cells to give rise to a very large population of sensitized T cells.

There are literally thousands of different kinds of T cells, since each must be able to recognize and attack a different specific antigen. Most of the time, the T cells are inactive. But once sensitized, they increase in size and differentiate into various cell types, all of which play an important role in the immune response.

1. **Cytotoxic (killer) T cells** migrate from lymphoid tissue to the site of invasion where they destroy antigens directly or indirectly. They attach to invading cells and secrete various lymphokines. One particular lymphokine, called ***perforin***, destroys antigens directly by making holes in the plasma membrane of the antigen, resulting in its lysis. Other lymphokines destroy antigens indirectly by recruiting additional lymphocytes, attracting macrophages, intensifying phagocytosis by macrophages, preventing macrophages from migrating away from the site of infection, and inducing rapid division of nonsensitized lymphocytes in the area. Cytotoxic T cells also secrete ***interferons***, which inhibit viral replication and enhance the killing action of cytotoxic T cells themselves. Cytotoxic T cells are especially effective against cancer cells, slowly developing bacterial diseases (such as tuberculosis and brucellosis), some viruses, fungi, and transplanted cells.

   Certain lymphocytes in the body resemble cytotoxic T cells. These ***natural killer (NK) cells*** are nonspecific, that is, they react against a wide variety of pathogens, and they can kill certain cells spontaneously, without interacting with lymphocytes or antigens. Like cytotoxic T cells, NK cells lyse target cells by releasing perforin. NK cells may be a separate line of T cells, but they are not cytotoxic T cells. NK cells provide an immediate first line of defense against cancer cells and perhaps other cells infected by agents other than viruses. NK cells produce interferons that inhibit viral replication and enhance their own killing action by lysis. It has been observed that the number of NK cells is decreased in cancer patients and that the decrease is related to the severity of the disease.

2. **Helper T cells** cooperate with B cells to amplify antibody

FIGURE 17-10 **Role of T cells in cellular immunity.**

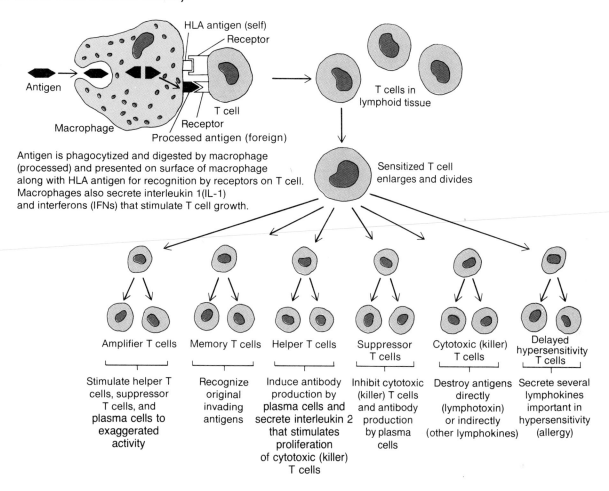

Antigen is phagocytized and digested by macrophage
(processed) and presented on surface of macrophage
along with HLA antigen for recognition by receptors on T cell.
Macrophages also secrete interleukin 1(IL-1)
and interferons (IFNs) that stimulate T cell growth.

Amplifier T cells	Memory T cells	Helper T cells	Suppressor T cells	Cytotoxic (killer) T cells	Delayed hypersensitivity T cells
Stimulate helper T cells, suppressor T cells, and plasma cells to exaggerated activity	Recognize original invading antigens	Induce antibody production by plasma cells and secrete interleukin 2 that stimulates proliferation of cytotoxic (killer) T cells	Inhibit cytotoxic (killer) T cells and antibody production by plasma cells	Destroy antigens directly (lymphotoxin) or indirectly (other lymphokines)	Secrete several lymphokines important in hypersensitivity (allergy)

production. Antigens must first interact with helper T cells before plasma cells (descendants of B cells) can produce antibodies. Helper T cells also secrete *interleukin-2 (IL-2),* a lymphokine that stimulates proliferation of cytotoxic T cells and proteins that amplify the inflammatory response and phagocytosis by macrophages.

3. **Amplifier T cells** somehow stimulate helper T cells, suppressor T cells, and plasma cells to higher levels of activity.

4. **Memory T cells** are descendants of other T cells that are programmed to recognize the original invading antigen. Should the pathogen invade the body at a later date, the memory cells initiate a far swifter reaction than during the first invasion. In fact, the second response is so swift that the pathogens are usually destroyed before any signs or symptoms of the disease occur. Memory cells permit sensitized persons to retain immunity for years.

5. **Delayed hypersensitivity T cells** secrete several lymphokines that play important roles in the hypersensitivity (allergy) response and rejection of transplanted tissue.

6. **Suppressor T cells** depress parts of the immune response by shutting down certain activities several weeks after infection activates them. They inhibit production of secretions of cytotoxic T cells and production of antibodies by plasma cells. The normal ratio of helper T cells to suppressor T cells is 2:1. Suppressor T cells are important in preventing autoimmune responses, where the immune system begins to attack the body's own tissues, as occurs, for example, in rheumatoid arthritis.

Cellular immunity augments our natural resistance and plays a crucial role in the initiation of humoral immunity.

B Cells and Humoral Immunity

The body contains not only thousands of different T cells but also thousands of different B cells, each capable of responding to a specific antigen. Whereas killer T cells leave lymphoid tissue to confront a foreign antigen, B cells do not. Instead, a macrophage prepares and presents an antigen to B cells in the lymph nodes, spleen, or lymphoid tissue of the gastrointestinal tract similar to the process previously described for T cells. B cells specific for the antigen are activated. Some of them enlarge and divide and develop into a population of *plasma cells* under the influence of thymic hormones (Figure 17-11). Plasma cells secrete antibodies. The division and development of B cells into plasma cells are also influenced by interleukin-1, secreted by macrophages, and lymphokines, secreted by helper T cells.

Giving Viruses a Cold Reception

The immune system learns to recognize specific disease agents through exposure to them. Antigen exposure can occur through vaccination or by natural means. We acquire immunity to chickenpox, measles, mumps, tetanus, cholera, smallpox, and many other life-threatening diseases. So, one might wonder, if the immune system can do such amazing things, why can't it defend us from the common cold?

We are susceptible to at least 200 different cold-causing viruses. The most common type, rhinoviruses (literally "nose viruses"), cause about 30 to 50 percent of all colds in adults. As soon as the immune system learns to recognize and defend us from one, another comes along, and then another. This antigenic diversity creates quite a challenge for the immune system, so much so that most people succumb to one to six colds per year.

Nonspecific Resistance

Cold symptoms are not produced directly by the cold virus, but by the body's nonspecific immune response as it fights the virus. When viruses invade the cells lining the nasal passages, your body responds with inflammation and the production of extra mucus. This causes nasal congestion, a "stuffy nose." As mucous membranes in the nose accelerate their secretion of antibody-containing mucus, you get a runny nose. Congestion in the middle ear or sinuses can cause dizziness or a headache.

The "swollen glands" sometimes felt during a cold are actually swollen lymph nodes. The nodes swell as immune cells, including macrophages, T cells, and B cells, work overtime to fight pathogens. A sore throat can result from "postnasal drip" as the sinuses drain mucus into the throat. Throat tissue can also become dry and irritated from breathing through your mouth or coughing.

Like the nasal passages, the trachea and bronchial tubes become inflamed and produce extra secretions if invaded by the cold virus. A wheezing sound indicates airway congestion as mucus accumulates and restricts the flow of air. A "productive" cough assists the respiratory passages in getting rid of the mucus and the virus. A "nonproductive" or dry cough is usually caused by throat irritation. Your body may produce a fever to create an inhospitable climate for the virus. Most cold viruses prefer temperatures of 86 to 96°F (30 to 35.5°C). Since fever is a helpful part of your nonspecific resistance, medication should only be used if the fever exceeds 101.5°F (38.6°C) or is needed to treat accompanying aches and pains.

Cold Prevention

Despite its name, a cold is not caused by cold weather, wet feet, or getting cold, at least according to laboratory studies. Colds do occur more frequently in the winter than in the summer, but no one knows why. The incidence of colds usually rises sharply in the early fall and spring. Some believe that when children go back to school and are exposed to each other's viruses, they bring them home to their families.

Research indicates that most of the time cold viruses are transmitted from the hands of an infected person to the hands of a susceptible person. The virus can survive on the skin for only a few hours and must reach the nose in order to invade the body. On the face near the nose is no good, since the skin provides an effective barrier. The mucous membranes of the mouth are also an inhospitable environment; kissing seldom spreads colds.

If all goes well for the virus, eventually the hand delivers the virus to its new home, the person's respiratory system, by touching the mucous membranes of the nose or the eyes (the virus can travel down the tear duct to the upper nose). One study found that 40 to 90 percent of people with colds had rhinoviruses on their hands. The viruses were also found on about 15 percent of nearby objects, such as door knobs, telephones, and coffee cups.

It is not known what makes some people more susceptible to colds. Small children are the most susceptible, because their immune systems are still immature and haven't learned to recognize as many pathogens. People who are around children a lot also get colds more frequently. Smokers are more likely to catch colds than nonsmokers, partly because smoking inhibits the airway cilia that help move mucus. Some studies have shown that stress can decrease the effectiveness of the immune system, and many people believe that stress and fatigue increase their susceptibility to colds.

Given what we know about the transmission of colds, the single best way to prevent colds is frequent hand-washing, especially when you're around people who have colds. Avoid sharing telephones, glasses, towels, and other objects with a person who has a cold. And try not to touch your nose or eyes.

Getting enough rest, eating well, exercising moderately, and managing stress certainly won't hurt and may help keep your resistance up. If you're a smoker, cold prevention is yet another good reason to quit.

What about vitamin C? Studies have failed to show that vitamin C prevents colds, although some research has found that it may lessen the severity of cold symptoms. Vitamin C also increases the intactness of cell membranes, so it may make them harder for viruses to penetrate.

Cold Self-Care

Since a cure for colds continues to elude medical researchers, the best we can do is to treat the symptoms. It's been said that with aggressive medical treatment a cold will disappear in seven days, while if left alone a cold will last a week. Nevertheless, symptom treatment can at least make us feel better until the cold has run its course.

The first step in cold self-care is to decide whether your symptoms are those of a cold or something more serious requiring medical attention. People who have heart disease, emphysema, diabetes, or another health condition should get professional advice before initiating self-care, especially before taking over-the-counter medication. Pregnant and lactating women should also check with their doctors before taking any medication.

Symptoms that indicate your infection may be more than a cold include:

1. Oral temperature over 103°F (39.5°C).
2. Sore throat with temperature above 101°F (38.5°C) for over 24 hours.
3. Temperature over 100°F (37.5°C) for three days.
4. Severe pain in ears, head, chest, or stomach.
5. Symptoms that persist more than a week.
6. Enlarged lymph nodes.
7. In a child, difficulty breathing or greater than normal irritability or lethargy.

Once you decide you have a cold, there are several things you can do to help yourself feel better. They include the following:

1. Chicken soup, broth, or other hot drinks. Your mother was right: hot fluids help relieve congestion by increasing the flow of nasal secretions. They also soothe irritated throats.
2. Gargle with salt water (¼ teaspoon salt in 8 oz water) to soothe a sore throat.
3. Use a vaporizer or humidifier to increase humidity, especially if the air is very dry. Humid air is gentler on nose and throat.
4. Breathing steam gives your nose a temporary fever, creating an inhospitable environment for the virus. It also helps to thin the mucous causing a stuffy nose, and thus temporarily relieve conjestion. The steam may also feel soothing to irritated throats and nasal passages.
5. While rest may not hasten your recovery, it may help you feel better. It's good to stay out of circulation for the first few days of a cold to keep others from getting it.
6. Many over-the-counter cold medications are available. If you decide you need something, avoid combination drugs that contain several active ingredients to treat several symptoms. If, instead, you buy single drugs for the symptoms you wish to treat, you will avoid taking unnecessary drugs and decrease unpleasant side effects.

Should You Exercise When You Have a Cold?

Many people wonder whether they should continue their exercise programs when they have a cold. Some hope that the exertion will create a fever and "burn out the cold," while others believe that the added stress of exercise will only exhaust an already overwhelmed immune system. At this point, there is nothing but anecdotal evidence for these two beliefs. As long as symptoms are mild, exercising doesn't appear to either prolong the cold or hasten recovery.

It is important to recognize, however, that colds can sometimes lead to more serious complications, such as secondary bacterial infection in the middle ear, sinuses, or respiratory system. Medical authorities generally advise rest during the initial days of infection, just to be sure that what you are catching (or that what is catching you) is really just a cold. People who insist on exercising should start slowly. If they begin to feel better after five or ten minutes, then the exercise is probably not harmful.

FIGURE 17-11 **Role of B cells in humoral immunity.**

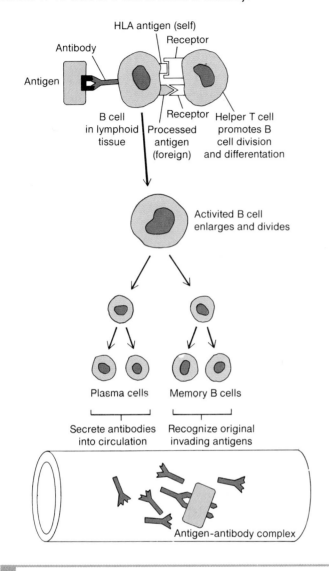

Distinguish between cellular and humoral immunity.

The activated B cells that do not differentiate into plasma cells remain as **memory B cells**, ready to respond more rapidly and forcefully should the same antigen appear at a future time.

Each specific antigen activates only those B cells already predetermined to secrete antibody specific to that antigen. The specific antigen selects a specific B cell because that particular B cell has surface antibodies that serve as receptor sites with which the antigen can combine. Antibodies produced by B cells then travel in lymph and blood to the invasion site and form antigen–antibody complexes with the foreign antigens. These antibodies activate complement enzymes for attack and fix the complement to the surface of the antigens.

The immune response is much more intense after a second or subsequent exposure to an antigen than after the initial exposure. After an initial contact with an antigen, there is a period of several days during which no antibody is present. Then there is a slow rise in the antibody level, followed by a gradual decline. This process is called the **primary response**. Every time the antigen is contacted again, there is an immediate proliferation of lymphocytes, and the plasma level of antibodies rises to higher levels than were initially produced. This accelerated, more intense response, called the **secondary response**, is produced by the lymphocytes formed during the primary response that remain as memory cells. These memory cells not only add to the pool of cells that can respond to the antigen but their response is more intense.

Actions of Antibodies

Once produced by plasma cells, antibodies may attack antigens directly or indirectly. Direct attack may cause lysis of a microbial cell, agglutination of cells, or precipitation of toxins so that they cannot circulate in the blood and are more easily removed by phagocytosis. Antibodies may adhere to antigenic cells (opsonization) or toxins (neutralization) so that they are weakened and more readily phagocytized.

Antigens may be attacked indirectly also. The combination of the antibody with the antigen may activate the complement system, which then works to remove the pathogens as previously described. IgM and IgG are antibodies that tend to promote complement fixation.

The secondary response provides the basis for immunization against certain diseases. When you receive the initial immunization, your body is sensitized. Should you encounter the pathogen again as an infecting microbe or "booster dose," your body responds by boosting the antibody to a higher level. However, booster doses of an immunizing agent must be given periodically, to maintain high antibody levels adequate for protection against the pathogenic organism.

A summary of the various cells involved in immune responses is presented in Exhibit 17-2.

THE SKIN AND IMMUNITY

The skin not only provides nonspecific resistance to disease, but it is also an active component of the immune system. Recall from Chapter 5 that the epidermis contains four kinds of cells: melanocytes, keratinocytes, and nonpigmented granular dendrocytes (Langerhans' cells and Granstein cells). Keratinocytes and nonpigmented granular dendrocytes assume a role in immunity according to the following mechanism.

If an antigen penetrates the epidermis, it binds to either Langerhans' or Granstein cells. Langerhans' cells present the antigen to helper T cells, which appear in the epidermis in response to injury to the skin, thus activating the T cells. At the same time, keratinocytes secrete interleukin-1, which stimulates T cells to produce interleukin-2, which, in turn, causes T cells to proliferate. The greatly increased number of T cells then enter the lymphatic system and spread throughout the body.

Just as Langerhans' cells interact with helper T cells, Granstein cells interact with suppressor T cells. Destruction of Langerhans' cells by ultraviolet radiation could result in a predominance of suppressor T cells, which could lessen the person's resistance to infection.

IMMUNOLOGY AND CANCER

When a normal cell becomes transformed into a cancer cell, the tumor cell develops *tumor-specific antigens* on its surface. It is believed that the immune system usually recognizes tumor-specific antigens as ''nonself'' and cytotoxic (killer) T cells, and possibly macrophages, destroy the cancer cells carrying them. This is called *immunologic surveillance*. Most investigators believe that cellular immunity is the basic mechanism involved in tumor destruction.

Despite immunologic surveillance, some cancer cells escape destruction, a phenomenon called *immunologic escape*. One possible explanation is that tumor cells shed their tumor-specific antigens, thus evading recognition. Another theory is that antibodies produced by plasma cells bind to tumor-specific antigens, preventing recognition by cytotoxic (killer) T cells. It is reasonble to assume that immunologic techniques using monoclonal antibodies (laboratory-produced pure antibodies formed by joining B cells and cancer cells) and toxins or radioactive compounds may be employed in the future to treat cancer. It is hoped that the monoclonal antibody would selectively locate the cancer cell, and the attached toxic agent or radioactive substance would destroy the cell but cause little or no damage to healthy tissue, an approach analogous to the use of antibiotics in treating infectious diseases.

EXHIBIT 17-2
Summary of Cells That Are Important in Immune Responses

Cell	Function	Cell	Function
Macrophage	Phagocytosis; processing and presentation of foreign antigens to T cells; secretion of interleukin-1 that stimulates secretion of interleukin-2 by helper T cells (stimulates proliferation of cytotoxic T cells) and induces proliferation of B cells; secretion of interferons that stimulate T cell growth.	Suppressor T Cell	Inhibits secretions by cytotoxic T cells and inhibits antibody production by plasma cells.
Cytotoxic (Killer) T Cell	Lysis of foreign cells by lymphotoxin and release of various other lymphokines that recruit and intensify cytotoxic T cell actions, attract macrophages, increase phagocytic activity of macrophages, prevent macrophage migration from site of action, and proliferation of uncommitted or nonsensitized lymphocytes. Cytotoxic T cells also secrete interferons.	Delayed Hypersensitivity T Cell	Secretes key substances related to hypersensitivity (allergy).
		Amplifier T Cell	Stimulates helper T cells, suppressor T cells, and B cells to higher levels of activity.
		Memory T Cell	Remains in lymphoid tissue and recognizes original invading antigens, even years after infection.
		Natural Killer (NK) Cell	Lymphocyte that destroys foreign cells by lysis and produces interferon.
		B Cell	Differentiates into antibody-producing plasma cell.
		Plasma Cell	Descendant of B cell that produces antibodies.
Helper T Cell	Cooperates with B cells to amplify antibody production by plasma cells and secretes interleukin-2, which stimulates proliferation of cytotoxic T cells.	Memory B Cell	Ready to respond more rapidly and forcefully than initially should the same antigen challenge the body in the future.

■ COMMON DISORDERS ■

Acquired Immune Deficiency Syndrome (AIDS)

Never before has science been confronted with an epidemic in which the primary disease only lowers the victim's immunity, and then a second unrelated disease produces the fatal symptoms. The primary disease is called *acquired immune deficiency syndrome (AIDS).* It is an illness characterized by a positive antibody test that indicates the presence of the AIDS virus and by the presence of *indicator diseases.* Indicator diseases are diseases that are found in AIDS sufferers much more frequently than in others and to which the AIDS sufferer is more than usually susceptible. One example is a cancer called Kaposi's sarcoma (KS), a deadly form of skin cancer prevalent in equatorial Africa but previously almost unknown in the United States. It produces painless purple or brownish lesions that resemble bruises on the skin or on the side of the mouth, nose, or rectum. There is some evidence that the AIDS virus may, by itself, be carcinogenic.

AIDS surfaced in June 1981 as a result of reports from the Los Angeles area to the Centers for Disease Control (CDC) of several cases of a very rare type of pneumonia called *Pneumocystis carinii* pneumonia (PCP), occurring among homosexual males. PCP causes shortness of breath, persistent dry cough, sharp chest pains, and difficulty in breathing. At about the same time, the CDC also received reports from New York and Los Angeles of an increase in the incidence of Kaposi's sarcoma among homosexual males. In 1983, a group of scientists in France and the United States isolated the virus causing AIDS, called *human immunodeficiency virus* (*HIV*).

The primary victims are homosexual males, intravenous drug users, and hemophilia patients. Other high-risk groups are the heterosexual partners of AIDS sufferers and transfusion patients. Most AIDS victims are between the ages of 20 and 49 years, and most are males. The average incubation period (time interval from infection with HIV to date of diagnosis) is nearly eight years. Forty-eight percent of known AIDS cases have died within two years of diagnosis.

All carriers of the virus, whether they have the disease or not, are assumed to be infected for life and are capable of transmitting the virus to others. The risk of developing AIDS increases yearly after infection with the virus; the longer a person is infected, the greater the chances of developing AIDS.

HIV: Structure and Infection

Viruses consist of a core of DNA or RNA covered by a protein coat. Some viruses also have an envelope (outer layer) composed of a double layer of lipid penetrated by proteins (Figure 17-12). Outside a living host cell, a virus has no life functions and is unable to replicate. However, once a virus makes contact with a host cell, the viral nucleic acid enters the host cell. Once inside, the viral nucleic acid uses the host cell's enzymes, ribosomes, nutrients, and other resources to make copies of itself and new protein coats and envelopes. As these components accumulate, they assemble into viruses that then leave the infected cell to infect other cells.

As viruses replicate, they damage or kill their host cells in various ways. Moreover, the body's own defenses, attacking the infected cells, can kill the cell as well as the viruses it harbors. Not only can the AIDS virus damage and kill host cells, it can also lie dormant for years, attacking more different types of host cells than imagined. Moreover, it has more complex means of destroying the immune system than suspected.

Infection with HIV begins when the virus binds to a protein receptor on the surface of helper T cells and other susceptible cells. Through very complicated mechanisms, helper T cells are killed, and as the process progresses, there is a decline in immune functioning. Recall that helper T cells cooperate with B cells to amplify antibody production. The reduction in the number of T cells inhibits antibody production by descendants of B cells (plasma cells) against HIV. Helper T cells also stimulate the destruction of infected cells by cytotoxic (killer) T cells and natural killer (NK) cells. Helper T cells also influence the activity of monocytes and macrophages, which engulf infected cells and foreign particles. Since helper T cells orchestrate a large portion of our immune defense, their destruction leads to collapse of the immune system and susceptibility to opportunistic infections such as Kaposi's sarcoma (KS) and *Pneumocystis carinii* pneumonia (PCP). Opportunistic infections are ones that take advantage of a person's compromised immunity. For about 60 percent of patients, PCP is the opportunistic infection that marks the progression from HIV infection to AIDS.

AIDS victims are also subject to tuberculosis, thrush, oral hairy leukoplakia, cytomegalovirus, a form of herpes that attacks the central nervous system, cryptococcal meningitis, central nervous system toxoplasmosis, oral and esophageal candidiasis, cryptosporidiosis, a severe form of arthritis called Reiter's syndrome, salmonellosis, and psoriasis. The AIDS virus also attacks macrophages; brain cells (where it causes AIDS dementia complex); endothelial cells that line various organs, body cavities, and blood vessels; bone marrow cells; colon cells; and skin cells. Heart and liver cells are also target cells. The virus might kill these cells or multiply in them, thus spreading the virus.

Presently, macrophages are receiving a great deal of attention as pivotal cells in the development of AIDS. Research suggests that macrophages may actually be the first and possibly the only cells invaded by the AIDS virus. It seems that once the virus invades macrophages, it then spreads to helper T cells. As a result, helper T cells are killed and macrophages do not function properly in the immune response. Macrophages serve as reservoirs for the virus, and even though the viruses multiply within the macrophages, they are not themselves killed. A very important consequence of this is that commonly used screening tests that detect antibodies to the AIDS virus in the blood may be negative even though the person has the AIDS virus hidden in macrophages. While stored in macrophages, the virus does not trigger the production of antibodies that would appear in blood. A new test, called the Cetus test, is used to detect the presence of AIDS viruses in macrophages, rather than AIDS antibodies in blood.

Soon after infection with the AIDS virus, the host develops antibodies against several proteins in the virus. The presence

FIGURE 17-12 Human immunodeficiency virus (HIV), the causative agent of AIDS.

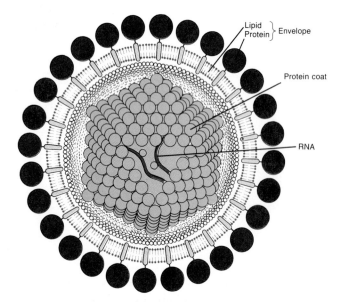

Lipid
Protein } Envelope

Protein coat

RNA

Describe the stages through which AIDS progresses.

of these antibodies in blood is used as a basis for diagnosing AIDS. Normally, antibodies so developed are protective, but in the case of the AIDS virus, this is not necessarily so.

Symptoms

The stages through which an individual passes correlate with decreasing helper T cell counts. Following exposure to HIV, there are usually no symptoms at first, and it may take from six weeks to one year before HIV is detected by standard antibody tests. Although most people have no symptoms when AIDS is first diagnosed, some develop mononucleosis-like symptoms such as fatigue, fever, swollen glands, headache, and even encephalitis. Within a few weeks, these symptoms disappear. In the next stage of the disease, which lasts from three to five years, chronically swollen lymph nodes develop in the neck, armpits, and groin. Following this, helper T cell counts decline even more, and the patients fail to respond to most skin tests that measure delayed hypersensitivity, a measure of the individual's ability to produce a cellular immune response against specific proteins injected under the skin. The next stage is characterized by a total absence of delayed hypersensitivity. It is during this stage that opportunistic infections develop (thrush, hairy leukoplakia, Kaposi's sarcoma, *Pneumocystis carinii* pneumonia, toxoplasmosis, candidiasis, and so on). In the terminal stages of HIV infection, many patients suffer from AIDS dementia complex, which is characterized by a progressive loss of motor function, cognition, and behavior.

HIV Outside the Body

Outside the body, the AIDS virus is fragile and can be eliminated in a number of ways. Dishwashing and clotheswashing by exposing the virus to 135°F (56°C) for 10 minutes will kill HIV. Chemicals such as hydrogen peroxide (H_2O_2), rubbing alcohol, Lysol, household bleach, and germicidal skin cleaner (such as Betadine and Hibiclens) are very effective against the virus. Standard chlorination is also sufficient to kill HIV in swimming pools and hot tubs.

Transmission

Although HIV has been isolated from a number of body fluids, such as blood, semen, vaginal fluid, cerebrospinal fluid, tears, and saliva, it appears that the fluids that provide sufficient virus for transmission are limited to blood, semen, and vaginal secretions. The virus is found in macrophages in these fluids. After exposure, it appears the infection establishes itself in the cardiovascular system, open wounds of the skin, the penis, vagina, and rectum.

HIV is effectively transmitted by sexual contact between males, from males to females, and from females to males through vaginal or anal intercourse with infected persons. HIV is also effectively transmitted through exchanges of blood, such as by contaminated hypodermic needles and needle stick, open wound, or mucous membrane exposure in healthcare workers. The virus is also transmitted from infected mothers to their infants before or during birth or through breast-feeding. It does not appear that individuals become infected as a result of routine, nonintimate contacts. Thousands of family members, co-workers, and friends of AIDS patients do not have AIDS.

No evidence exists that AIDS can be spread through *normal* kissing (some experts feel that prolonged, vigorous, wet deep kissing that results in breaks or tears in the lining of the mouth or if preexisting sores are present could theoretically transmit the virus). Also, although mosquitoes can carry the AIDS virus for several days, there is no evidence that the virus can multiply inside mosquitoes or that they are capable of transmitting the disease. It also appears that health-care personnel who take proper routine barrier precautions (gloves, masks, safety glasses) are not at risk.

Drugs and Vaccines Against HIV

Medical scientists are mounting what is probably the greatest concentrated effort ever to find a cure for a single virus disease. One of the problems is that HIV can lie dormant in body cells. In addition, HIV can infect a variety of cells, including those in the central nervous system that are protected by the blood–brain barrier. Added to this is the problem of opportunistic infections, which may be very difficult to treat. Antiviral therapy is aimed at disrupting various points in the viral life cycle. Some research seeks to prevent the attachment of the virus to the host cell plasma membrane. Other research is designed to counterattack the virus once it gets inside a host cell and takes over its machinery. Still other research aims to inhibit assembly of viruses in the host cell and their subsequent release. Most drugs now used against the AIDS virus are directed against reverse transcriptase.

The drug currently most widely tried against the AIDS virus and given approval by the Food and Drug Administration (FDA) is zidovudine (Retrovir), formerly known as azidothymidine (AZT). This drug prevents the virus from making DNA. Clinical improvements among patients taking zidovudine are weight gain, increased energy, and neurological improvements (reversal of loss of mental function and dementia). Side effects include severe bone marrow damage, anemia, and suppression of the immune system. Moreover, patients must take the medication every four hours and receive weekly or bimonthly transfusions. Researchers are now conducting human trials of a new drug that selectively kills HIV-infected macrophages. The drug is known as GLQ223, which is a protein derived from the root of a Chinese cucumber plant. Dideoxycytidine (ddC), a drug similar in action to zidovudine but with fewer toxic effects, is also in early human trials, as is phosphonoformate. Other drugs that also prevent DNA synthesis are dideoxyadenosine (ddA), dideoxyinosine (ddI), and rifabutin. Preliminary studies using ddI showed increases in immune system cells, decreases in HIV proteins in blood, and weight gain, all without toxic side effects. A drug called ribavirin (Virazole) appears to prevent the synthesis of viral proteins. It is not yet approved for AIDS patients in the United States. Alpha interferon is believed to exert its effect at the final stage of virus production. It is being tried in trials both alone and in combination with other drugs.

In addition to drug development and testing, considerable emphasis is also being given to developing a vaccine against AIDS. A vaccine would stimulate the production of antibodies that kill the virus directly and bolster immune defenses after the virus has invaded the body. The principal experimental vaccines under study make use of various subunits of the AIDS virus. In a process that is largely trial and error, researchers have selected different elements of the virus (proteins from the envelope or coat) that are believed to be most likely to produce the broadest range of antibodies. Some scientists propose using the entire killed AIDS virus in a vaccine to stimulate antibody production, but there is concern that some of the viruses might remain alive and thus cause disease.

The development of an AIDS vaccine has been impeded because it is not known which part of HIV elicits the most powerful immune response and because of the ability of HIV to mutate so quickly (it can even show dramatic variation within a single individual). In addition, since the AIDS virus can remain protected in certain cells, antibodies cannot attack it. Moreover, clinical trials have raised safety concerns, and there may be a shortage of volunteers in the United States.

At present, most studies are concentrated on proteins in the envelope of the HIV virus as the HIV component likely to elicit the most powerful immune response. Others focus on internal components of HIV rather than envelope antigens. Dr. Jonas Salk is using the whole virus minus its outer envelope in an attempt to trigger a protective immune response. This method is similar to the one used in the 1950s to develop the polio vaccine. In one experiment using whole viruses, researchers inoculated nine rhesus monkeys with a vaccine for the monkey version of AIDS and then exposed them to the active virus. Nine months later, none of the monkeys developed the disease and only one showed any signs of infection.

Prevention of Transmission

At the present time, there are no drugs or vaccines to prevent AIDS. The only means of prevention is to stop transmission of the virus. Sexual transmission of the AIDS virus can be prevented if infected persons do not have vaginal, oral, or anal intercourse with susceptible persons, or, if during intercourse, effective barrier techniques (condoms and spermicides such as nonoxynol 9) are used. Infection from donated blood and blood products can be prevented by testing blood for evidence of the AIDS virus. AIDS transmitted by needles and syringes could be avoided if the use of intravenous drugs is stopped or if only sterile paraphernalia is used. Mother-to-infant infections could be avoided if infected females would not become pregnant. All these measures must be part of an overall program involving education, counseling, screening high-risk individuals, tracing contacts, and modifying behavior.

Until there is effective drug therapy or an effective vaccine, preventing the spread of AIDS must rely on education and safer sexual practices.

Autoimmune Diseases

Under normal conditions, the body's immune mechanism is able to recognize its own tissues and chemicals and it does not produce T cells or B cells against its own substances. Such recognition of self is called *immunologic tolerance*. Although the mechanism of tolerance is not completely understood, it is believed that suppressor T cells may inhibit the differentiation of B cells into antibody-producing plasma cells or inhibit helper T cells that cooperate with B cells to amplify antibody production.

At times, however, immunologic tolerance breaks down, and this leads to an *autoimmune disease* (*autoimmunity*). For reasons still not understood, certain tissues undergo changes that cause the immune system to recognize them as foreign antigens and attack them. Among human autoimmune diseases are rheumatoid arthritis (RA), systemic lupus erythematosus (SLE), thyroiditis, rheumatic fever, glomerulonephritis, encephalomyelitis, hemolytic and pernicious anemias, Addison's disease, Graves' disease, type I diabetes mellitus, myasthenia gravis, and multiple sclerosis (MS).

Severe Combined Immunodeficiency (SCID)

Severe combined immunodeficiency (*SCID*) is a rare disease in which both B cells and T cells are missing or inactive. Perhaps the most famous patient with SCID was David, the ''bubble boy,'' who lived in a sterile plastic chamber for all but 15 days of his life; he died on February 22, 1984, at age 12. He was the oldest untreated survivor of SCID.

Hypersensitivity (Allergy)

A person who is overly reactive to an antigen is said to be *hypersensitive (allergic)*. Whenever an allergic reaction occurs, there is tissue injury. The antigens that induce an allergic reaction are called *allergens*. Almost any substance can be an allergen for some individual. Common allergens include certain foods (milk, peanuts, shellfish, eggs), antibiotics (penicillin, tetracycline), vitamins (thiamine, folic acid), protein drugs (insulin, ACTH, estradiol), vaccines (pertussis, typhoid), venoms (honeybee, wasp, snake), cosmetics, chemicals in plants such as poison ivy, pollens, dust, molds, iodine-containing dyes used in certain x-ray procedures, and even microbes.

Some allergic reactions, such as hives, eczema, swelling of the lips or tongue, abdominal cramps, and diarrhea, are referred to as *localized* (affecting one part or a limited area). Others are considered *systemic* (affecting several parts or the entire body). An example is acute anaphylaxis (anaphylactic shock), in which the person develops respiratory symptoms (wheezing and shortness of breath) as bronchioles constrict, usually accompanied by cardiovascular failure and collapse due to vasodilation and fluid loss from blood.

Tissue Rejection

Transplantation is the replacement of an injured or diseased tissue or organ. Usually, the body recognizes the proteins in the transplanted tissue or organ as foreign and produces antibodies against them. This is known as *tissue rejection*. Rejection can be somewhat reduced by matching donor and recipient HLA antigens and by administering drugs that inhibit the body's ability to form antibodies. Until recently, *immunosuppressive drugs* suppressed not only the recipient's immune rejection of the donor organ but also the immune response to all antigens as well. This makes patients susceptible to infectious diseases. A drug called *cyclosporine*, derived from a fungus, has largely overcome this problem for kidney, heart, and liver transplants.

Hodgkin's Disease (HD)

Hodgkins' disease (*HD*) is a form of cancer, usually arising in lymph nodes. It may arise from a combination of genetic predisposition, disturbance of the immune system, and an infectious agent (Epstein–Barr virus).

MEDICAL TERMINOLOGY AND CONDITIONS

Adenitis (ad′-e-NĪ-tis; *adeno* = gland; *itis* = inflammation of) Enlarged, tender, and inflamed lymph nodes resulting from an infection.

Elephantiasis (el′-e-fan-TĪ-a-sis) Long-standing edema of one or both lower extremities, and sometimes of the arms or other body parts, due to lymphatic obstruction. The involved part is tremendously swollen and hardened, and the skin surface folds and produces fissures, causing it to resemble the leg of an elephant. Elephantiasis may be due to a parasitic worm, congestive heart failure, or obstruction of the lymphatics.

Hypersplenism (hī′-per-SPLĒN-izm; *hyper* = over) Abnormal splenic activity due to splenic enlargement and associated with an increased rate of destruction of normal blood cells.

Lymphadenectomy (lim-fad′-e-NEK-tō-mē; *ectomy* = removal) Removal of a lymph node.

Lymphadenopathy (lim-fad′-e-NOP-a-thē; *patho* = disease) Enlarged, sometimes tender lymph glands.

Lymphangioma (lim-fan′-jē-Ō-ma; *angio* = vessel; *oma* = tumor). A benign tumor of the lymphatic vessels.

Lymphangitis (lim′-fan-JĪ-tis) Inflammation of lymphatic vessels.

Lymphedema (lim′-fe-DĒ-ma; *edema* = swelling) Accumulation of lymph fluid producing subcutaneous tissue swelling.

Lymphoma (lim′-FŌ-ma) Any tumor composed of lymphatic tissue.

Splenomegaly (splē′-nō-MEG-a-lē; *mega* = large) Enlarged spleen.

STUDY OUTLINE

Functions (p. 353)

1. The lymphatic system consists of lymph, lymphatic vessels, and structures and organs that contain lymphatic tissue (specialized reticular tissue containing large numbers of lymphocytes).
2. The lymphatic-tissue-containing components of the lymphatic system are diffuse lymphatic tissue, lymphatic nodules, and lymphatic organs (lymph nodes, spleen, and thymus gland).
3. The lymphatic system drains tissue spaces of excess fluid and returns proteins that have escaped from blood to the cardiovascular system, transports fats from the gastrointestinal tract to the blood, and provides immunity.

Lymph and Interstitial Fluid (p. 353)

1. Interstitial fluid and lymph are basically the same. When the fluid bathes body cells, it is called interstitial fluid; when it is found in lymph vessels, it is called lymph.
2. These fluids differ chemically from plasma in that both contain less protein.

Lymphatic Vessels (p. 354)

1. Lymphatic vessels begin as lymph capillaries in tissue spaces between cells.
2. Lymph capillaries merge to form larger vessels, called lymphatic vessels, which ultimately converge into the thoracic duct or right lymphatic duct.
3. Lymphatic vessels have thinner walls and more valves than veins.

Lymphatic Tissue (p. 356)

1. Lymph nodes are oval structures located along lymphatic vessels.
2. Lymph enters nodes through afferent lymphatic vessels and exits through efferent lymphatic vessels.
3. Lymph passing through the nodes is filtered. Lymph nodes also produce lymphocytes.
4. A tonsil is a group of large lymphatic nodules embedded in mucous membranes. The five tonsils are the pharyngeal, palatine, and lingual.
5. The spleen is the largest mass of lymphatic tissue in the body. It produces lymphocytes and antibodies, phagocytizes bacteria and worn-out red blood cells, and stores blood.
6. The thymus gland functions in immunity by producing T cells.

Lymph Circulation (p. 357)

1. The passage of lymph is from interstitial fluid, to lymph capillaries, to lymphatic vessels, to lymph trunks, to the thoracic duct or right lymphatic duct, to the subclavian veins.
2. Lymph flows as a result of skeletal muscle contractions, respiratory movements, and valves in the lymphatics.

Nonspecific Resistance to Disease (p. 358)

1. The ability to ward off disease using a number of defenses is called resistance. Lack of resistance is called susceptibility.
2. Nonspecific resistance refers to a wide variety of body responses against a wide range of pathogens.

3. Nonspecific resistance includes mechanical factors (skin, mucous membranes, lacrimal apparatus, saliva, mucus, cilia, epiglottis, and flow of urine), chemical factors (gastric juice, acid pH of skin, sebum, and perspiration), antimicrobial substances (interferon, complement, and properdin), phagocytosis, inflammation, and fever.

Immunity (Specific Resistance to Disease) (p. 362)

1. Specific resistance to disease involves the production of a specific lymphocyte or antibody against a specific antigen and is called immunity.
2. Acquired immunity is gained during life as a result of contact with an antigen. It may be obtained actively or passively or naturally or artificially.
3. Antigens (Ags) are chemical substances that, when introduced into the body, stimulate the production of antibodies that react with the antigen.
4. Examples of antigens are microbes (such as bacteria or viruses), pollen, incompatible blood cells, and transplants.
5. Antibodies (Abs) are proteins produced in response to antigens.
6. Based on chemistry and structure, antibodies are distinguished into five principal classes, each with specific biological roles (IgG, IgA, IgM, IgD, and IgE).
7. Cellular immunity refers to destruction of antigens by T cells.
8. Humoral immunity refers to destruction of antigens by antibodies, which are produced by descendants of B cells called plasma cells.
9. T cells are processed in the thymus gland; B cells may be processed in bone marrow, the fetal liver and spleen, or gut-associated lymphoid tissue.
10. Macrophages process and present antigens to T cells and B cells and secrete interleukin-1 that induces production of T cells and B cells.
11. There are six kinds of T cells: cytotoxic (killer) T cells secrete lymphotoxin that destroys antigens directly and other lymphokines that destroy antigens indirectly; helper T cells cooperate with B cells to amplify antibody production and secrete interleukin-2; suppressor T cells inhibit secretions by cytotoxic T cells and antibody production by plasma cells; delayed hypersensitivity T cells produce lymphokines and are important in allergic responses; amplifier T cells stimulate helper and suppressor T cells and plasma cells to higher levels of activity; and memory T cells recognize antigens at a later date.
12. Natural killer (NK) cells are lymphocytes that resemble cytotoxic T cells but can kill certain cells without interacting with lymphocytes or antigens.
13. B cells develop into antibody-producing plasma cells under the influence of thymic hormones and interleukin-1; memory B cells recognize the original, invading antigen.
14. The secondary response provides the basis for immunization against certain diseases.
15. The immunologic properties of the skin are due to keratinocytes and nonpigmented granular dendrocytes (Langerhans' cells and Granstein cells).
16. Cancer cells contain tumor-specific antigens and are usually destroyed by the body's immune system (immunologic surveillance); some cancer cells escape detection and destruction, a phenomenon called immunologic escape.

REVIEW QUESTIONS

1. Identify the components and functions of the lymphatic system. (p. 353)
2. Compare interstitial fluid and lymph with regard to location, chemical composition, and function. (p. 353)
3. How do lymphatic vessels originate? Compare veins and lymphatic vessels with regard to structure. (p. 354)
4. Describe the structure of a lymph node. What functions do lymph nodes serve? (p. 356)
5. Define edema. What are some of its causes? (p. 358)
6. Describe and locate the tonsils. (p. 357)
7. Describe the structure and functions of the spleen. (p. 357)
8. Describe the route of lymph circulation and explain the factors involved in maintaining lymph circulation. (p. 357)
9. Describe the role of the thymus gland in immunity. (p. 357)
10. Describe the mechanical and chemical factors involved in nonspecific resistance. (p. 358)
11. How do interferon (IFN), complement, and properdin function in nonspecific resistance? (p. 359)
12. What is phagocytosis? Describe the steps involved in adherence and ingestion. (p. 359)
13. Define immunity. Compare the four types of acquired immunity. (p. 362)
14. Distinguish between an antigen (Ag) and an antibody (Ab). (p. 363)
15. Where are T cells and B cells processed? How do they differ in function? (p. 364)
16. What is the function of macrophages in immunity? (p. 364)
17. Describe the role of T cells in cellular immunity. What are the specific functions of each type of T cell? (p. 364)
18. What are natural killer (NK) cells? How do they differ from cytotoxic (killer) T cells? (p. 364)
19. Describe the role of B cells in humoral immunity. (p. 365)
20. Discuss the importance of the secondary response of the body to an antigen. (p. 368)
21. Describe the role of the skin in immunity. (p. 368)
22. Explain how immunology is related to cancer. (p. 369)
23. Define the following: acquired immune deficiency syndrome (AIDS), autoimmune diseases, severe combined immunodeficiency (SCID), hypersensitivity (allergy), and Hodgkin's disease. (p. 369)
24. Refer to the glossary of medical terminology and conditions associated with the lymphatic system. Be sure that you can define each term. (p. 373)

The Digestive System

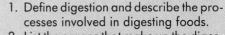

STUDENT OBJECTIVES

1. Define digestion and describe the processes involved in digesting foods.
2. List the organs that make up the digestive system.
3. Explain the role of the mouth in digestion.
4. Describe the phases involved in swallowing food.
5. Describe the functions of the stomach in digestion.
6. Explain how the pancreas, liver, and gallbladder participate in digestion.
7. Explain the role of the small intestine in digestion and absorption of food.
8. Describe the functions of the large intestine in digestion, feces formation, and defecation.
9. Define common disorders and medical terminology and conditions associated with the digestive system.

A LOOK AHEAD

DIGESTIVE PROCESSES
ORGANIZATION
   General Histology
     Mucosa
     Submucosa
     Muscularis
     Serosa
MOUTH (ORAL CAVITY)
   Tongue
   Salivary Glands
     Composition of Saliva
     Secretion of Saliva
   Teeth
   Digestion in the Mouth
     Mechanical
     Chemical
PHARYNX
ESOPHAGUS
   Deglutition
STOMACH
   Anatomy
   Digestion in the Stomach
     Mechanical
     Chemical
   Regulation of Gastric Secretion
     Stimulation
     Inhibition
   Regulation of Gastric Emptying
   Absorption
PANCREAS
   Anatomy
   Pancreatic Juice
   Regulation of Pancreatic Secretions
LIVER
   Anatomy
   Blood Supply
   Bile
   Regulation of Bile Secretion
   Functions of the Liver
GALLBLADDER (GB)
   Functions
   Emptying of the Gallbladder
SMALL INTESTINE
   Anatomy
   Intestinal Juice
   Digestion in the Small Intestine
     Mechanical
     Chemical
   Regulation of Intestinal Secretion
   Absorption
     Carbohydrates
     Proteins
     Lipids
     Water
     Electrolytes
     Vitamins
LARGE INTESTINE
   Anatomy
   Digestion in the Large Intestine
     Mechanical
     Chemical
   Absorption and Feces Formation
   Defecation
COMMON DISORDERS
MEDICAL TERMINOLOGY AND
     CONDITIONS

Food is vital for life because it is the source of energy that drives the chemical reactions occurring in every cell. Energy is needed for muscle contraction, conduction of nerve impulses, and secretory and absorptive activities of many cells. Food as it is consumed, however, is not in a state suitable for use. It must be broken down into molecules that can be absorbed through the wall of the gastrointestinal (GI) tract. The breaking down of food molecules for use by cells is called *digestion*, and the organs that collectively perform this function comprise the *digestive system*.

The medical specialty that deals with the structure, function, diagnosis, and treatment of diseases of the stomach and intestines is called *gastroenterology* (gas′-trō-en′-ter-OL-ō-jē; *gastro* = stomach; *enteron* = intestines).

DIGESTIVE PROCESSES

1. **Ingestion.** Taking food into the body (eating).
2. **Movement of food.** Passage of food along the gastrointestinal tract.
3. **Digestion.** The breakdown of food by chemical and mechanical processes.
4. **Absorption.** The passage of digested food from the gastrointestinal tract into the cardiovascular and lymphatic systems for distribution to cells.
5. **Defecation.** The elimination of indigestible substances from the gastrointestinal tract.

Chemical digestion is a series of reactions that break down the large, complex carbohydrate, lipid, and protein molecules that we eat into simple molecules small enough to pass through the walls of the digestive organs, into the blood and lymph capillaries, and eventually into the body's cells. *Mechanical digestion* consists of various movements that aid chemical digestion. Food is prepared by the teeth before it can be swallowed. The smooth muscles of the stomach and small intestine churn the food so it is thoroughly mixed with enzymes that digest foods.

ORGANIZATION

The organs of digestion may be divided into two main groups. First is the *gastrointestinal (GI) tract* or *alimentary canal*, a continuous tube that begins at the mouth and ends at the anus (Figure 19-1). Organs composing the gastrointestinal tract include the mouth, pharynx, esophagus, stomach, small intestine, and large intestine. The GI tract contains the food from the time it is eaten until it is digested and prepared for elimination. Muscular contractions in the wall of the GI tract break down the food physically by churning it. Secretions produced by cells along the tract break down the food chemically.

The second group of organs composing the digestive system are the *accessory structures*: the teeth, tongue, salivary glands, liver, gallbladder, and pancreas. Teeth protrude into the GI tract and aid in the physical breakdown of food. The other accessory structures, except for the tongue, lie totally outside the tract and produce or store secretions necessary for chemical digestion. These secretions are released into the tract through ducts.

GENERAL HISTOLOGY

The wall of the GI tract consists of four layers of tissue. From the inside out, they are the mucosa, submucosa, muscularis, and serosa (Figure 19-2).

Mucosa

The *mucosa*, or inner lining of the tract, is a mucous membrane. It is composed of a layer of *epithelium* in direct contact with the contents of the GI tract, loose connective tissue, and smooth muscle.

FIGURE 19-1 Organs of the digestive system and related structures.

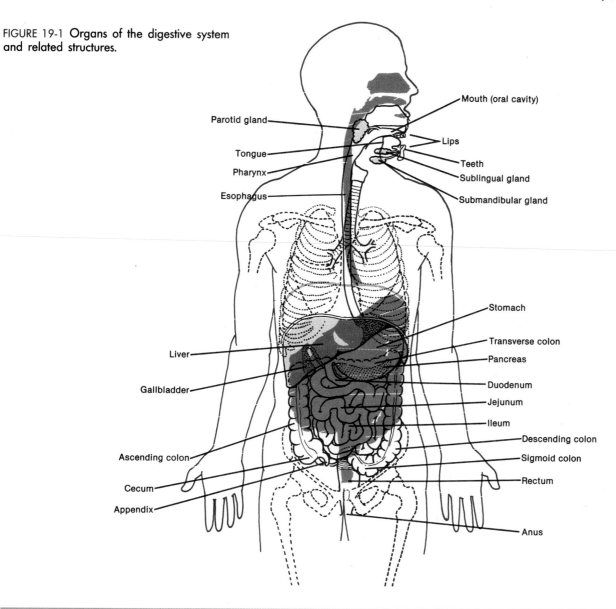

Put a check next to the organs that are part of the gastrointestinal tract and circle the organs that are classified as accessory structures.

Submucosa

The *submucosa* consists of loose connective tissue that binds the mucosa to the muscle layer. It contains numerous blood and lymphatic vessels for the absorption of the breakdown products of digestion and nerves that control GI tract secretions.

Muscularis

The *muscularis* of the mouth, pharynx, and upper esophagus consists in part of skeletal muscle that produces voluntary swallowing. In the rest of the tract, the muscularis consists of smooth muscle usually arranged as an inner sheet of circular fibers and an outer sheet of longitudinal fibers. Involuntary contractions of these smooth muscles help break down food physically, mix it with digestive secretions, and propel it through the tract. The muscularis also contains nerves that control GI tract movements.

Serosa

The *serosa* is the outermost layer of the GI tract. It secretes serous fluid that allows the tract to glide easily against other organs. This layer is also called the *visceral peritoneum*.

MOUTH (ORAL CAVITY)

The *mouth,* also referred to as the *oral* or *buccal* (BUK-al) *cavity,* is formed by the cheeks, hard and soft palates, and tongue (Figure 19-3).

The *lips* are fleshy folds around the opening of the mouth. They are covered on the outside by skin and on the inside by a mucous membrane. During chewing, the lips and cheeks help keep food between the upper and lower teeth. They also assist in speech.

Hanging from the soft palate is a projection called the *uvula*

FIGURE 19-2 Composite of various sections of the gastrointestinal tract seen in a three-dimensional drawing depicting the various layers and related structures.

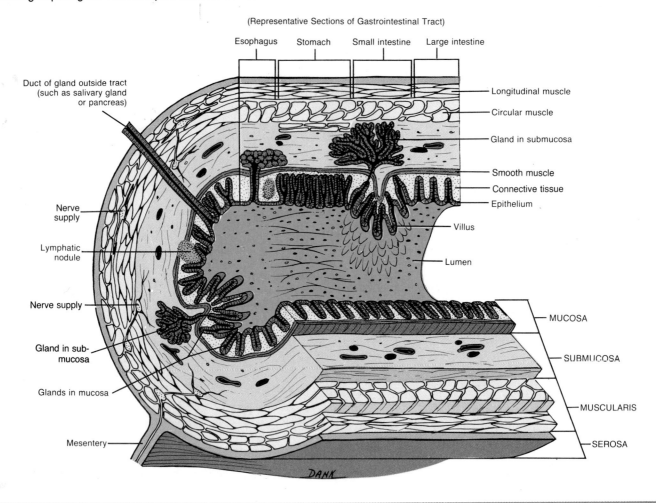

(Representative Sections of Gastrointestinal Tract)

Esophagus   Stomach   Small intestine   Large intestine

Duct of gland outside tract (such as salivary gland or pancreas)

Nerve supply

Lymphatic nodule

Nerve supply

Gland in submucosa

Glands in mucosa

Mesentery

Longitudinal muscle

Circular muscle

Gland in submucosa

Smooth muscle

Connective tissue

Epithelium

Villus

Lumen

MUCOSA

SUBMUCOSA

MUSCULARIS

SEROSA

DANK

What is the importance of the nerve supply in the submucosa? Muscularis?

(YOU-vyoo-la). At the back of the soft palate, the mouth opens into the oropharynx through an opening called the *fauces* (FAW-sēs). Just behind this is the palatine tonsils.

TONGUE

The *tongue* forms the floor of the oral cavity. It is an accessory structure of the digestive system composed of skeletal muscle covered with mucous membrane (see Figure 12-3). The tongue consists of extrinsic and intrinsic muscles.

The *extrinsic muscles* of the tongue originate outside the tongue and insert into it. They move the tongue from side to side and in and out. These movements maneuver food for chewing, shape the food into a rounded mass, called a *bolus*, and force the food to the back of the mouth for swallowing. The *intrinsic muscles* are within the tongue and alter the shape and size of the tongue for speech and swallowing. The *lingual frenulum*, a fold of mucous membrane in the midline of the undersurface of the tongue, limits

the movement of the tongue posteriorly (Figure 19-3). The lingual tonsil lies at the base of the tongue.

The upper surface and sides of the tongue are covered with projections called *papillae* (pa-PIL-ē). *Filiform papillae* are conical projections distributed over the front two-thirds of the tongue. They are whitish and contain no taste buds. *Fungiform papillae* are mushroomlike elevations distributed among the filiform papillae and are more numerous near the tip of the tongue. They appear as red dots on the surface of the tongue, and most of them contain taste buds. *Circumvallate papillae*, 10 to 12 in number, are arranged in the form of an inverted V on the back surface of the tongue, and all of them contain taste buds. Note the taste zones of the tongue in Figure 12-3b.

SALIVARY GLANDS

The fluid called saliva is secreted by three pairs of *salivary glands*, accessory structures that lie outside the mouth and pour their con-

FIGURE 19-3 **Structures of the mouth (oral cavity).**

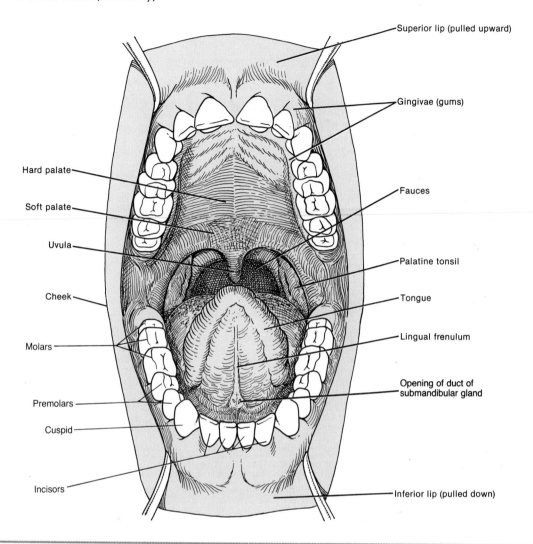

Which structure contains extrinsic and intrinsic muscles? Describe their functions.

tents into ducts emptying into the oral cavity. Their names are parotid, submandibular, and sublingual (Figure 19-4).

The *parotid glands* are located under and in front of the ears between the skin and the masseter muscle. They secrete into the mouth via ducts opposite the upper second molar tooth. The *submandibular glands* lie beneath the base of the tongue in the floor of the mouth. Their ducts enter the mouth just behind the central incisors. The *sublingual glands* lie in front of the submandibular glands, and their ducts open into the floor of the mouth.

Mumps is an inflammation and enlargement of the parotid glands accompanied by fever, malaise, a sore throat, and swelling on one or both sides of the face. In about 20 to 35 percent of males past puberty, the testes may also become inflamed, and although it rarely occurs, it may result in sterility.

Composition of Saliva

Chemically, *saliva* is 99.5 percent water and 0.5 percent solutes, the most important of which is the digestive enzyme salivary amy-

lase, which is described shortly. The water in saliva provides a medium for dissolving foods so they can be tasted and for initiating digestive reactions. Mucus in saliva lubricates food so it can be easily swallowed. The enzyme lysozyme destroys bacteria to protect the mucous membrane from infection and the teeth from decay.

Secretion of Saliva

Salivation is entirely under nervous control. Parasympathetic stimulation normally promotes continuous secretion of moderate amounts of saliva to keep the tongue and lips moist to facilitate speech. Sympathetic stimulation dominates during stress, resulting in dryness of the mouth. During dehydration, the salivary glands stop secreting to conserve water. The resulting dryness in the mouth contributes to the sensation of thirst. Drinking will then not only moisten the mouth but also restore the homeostasis of body water.

Food stimulates the glands to secrete heavily. Whenever food is tasted, smelled, touched by the tongue, or even thought about, parasympathetic stimulation activates increased secretion of saliva.

FIGURE 19-4 Location of the salivary glands.

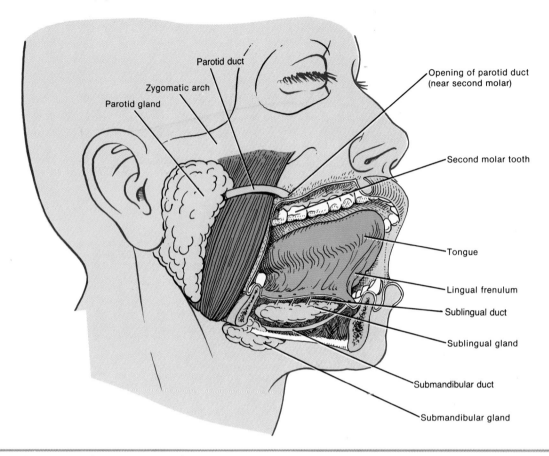

Describe how salivation is under nervous control.

Saliva continues to be secreted heavily some time after food is swallowed. This flow washes out the mouth and dilutes and buffers chemical remnants of any irritating substances.

TEETH

The *teeth (dentes)* are accessory structures of the digestive system located in bony sockets of the mandible and maxillae. The sockets are covered by the *gingivae* (jin-JI-vē) or gums and lined with the *periodontal ligament*, dense fibrous connective tissue that anchors the teeth to bone, keeps them in position, and acts as a shock absorber during chewing (Figure 19-5).

A typical tooth has three principal parts. The *crown* is the exposed portion above the level of the gums. The *root* consists of one to three projections embedded in the socket. The *neck* is the junction line of the crown and root, near the gum line.

Teeth are composed primarily of *dentin*, a bonelike substance that gives the tooth its basic shape and rigidity. The dentin encloses a cavity. The *pulp cavity* lies in the crown and is filled with *pulp*, a connective tissue containing blood vessels, nerves, and lymphatics. Narrow extensions of the pulp cavity run through the root of the tooth and are called *root canals*. Each root canal has an opening at its base through which enter blood vessels bearing

nourishment, lymphatic vessels affording protection, and nerves providing sensation. The dentin of the crown is covered by *enamel*, a bonelike substance which consists primarily of densely packed calcium phosphate and calcium carbonate. Enamel is the hardest substance in the body. It protects the tooth from the wear of chewing and is a barrier against acids that easily dissolve the dentin. The dentin of the root is covered by *cementum*, another bonelike substance that attaches the root to the periodontal ligament.

The branch of dentistry that is concerned with the prevention, diagnosis, and treatment of diseases that affect the pulp, root, periodontal ligament, and alveolar bone is known as *endodontics* (en′-dō-DON-tiks; *endo* = within; *odous* = tooth).

Humans have two sets of teeth. The *deciduous teeth* begin to erupt at about 6 months of age, with the incisors, and one pair appears about each month thereafter until all 20 are present. They are generally lost in the same sequence between 6 and 12 years of age. The *permanent teeth* appear between age 6 and adulthood. There are 32 teeth in a complete permanent set.

Humans also have different teeth for different functions (see Figure 19-3). Incisors are closest to the midline, are chisel-shaped, and adapted for cutting into food; cuspids (canines) are next to the incisors, have one pointed surface (cusp) to tear and shred food; premolars have two cusps to crush and grind food; and molars have more than two blunt cusps, to crush and grind food.

FIGURE 19-5 **Parts of a typical tooth as seen in a section of a mandibular (lower) molar.**

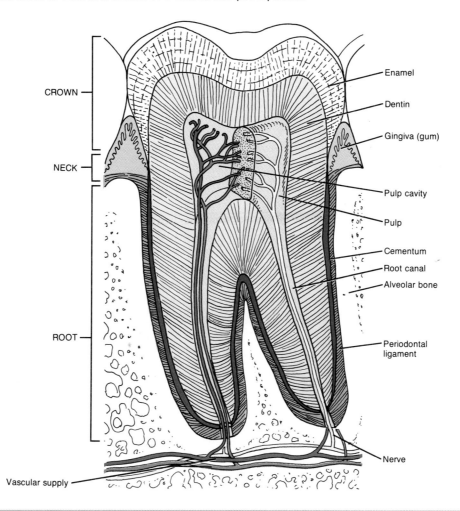

CROWN

NECK

ROOT

Vascular supply

Enamel

Dentin

Gingiva (gum)

Pulp cavity

Pulp

Cementum

Root canal

Alveolar bone

Periodontal ligament

Nerve

What bonelike substance gives shape to a tooth? How are teeth anchored in position? What connective tissue contains blood vessels, nerves, and lymphatic vessels? What are the first teeth to erupt?

DIGESTION IN THE MOUTH

Mechanical

In chewing, or *mastication*, the tongue manipulates food, the teeth grind it, and the food is mixed with saliva. As a result, the food is reduced to a soft bolus that is easily swallowed.

Chemical

The enzyme *salivary amylase* (AM-i-lās) initiates the breakdown of starch. This is the only chemical digestion that occurs in the mouth. Carbohydrates are either monosaccharide and disaccharide sugars or polysaccharide starches (see Chapter 2). Most of the carbohydrates we eat are polysaccharides. Since only monosaccharides can be absorbed into the bloodstream, ingested disaccharides and polysaccharides must be broken down. Salivary amylase breaks the chemical bonds between some of the monosaccharides in the starches to reduce the long-chain polysaccharides to the disaccharide maltose. Food usually is swallowed too quickly for all the starches to be reduced to disaccharides in the mouth, but salivary amylase in the swallowed food continues to act on starches in the stomach until the stomach acids eventually inactivate it.

PHARYNX

The *pharynx* is a funnel-shaped tube that extends from the internal nares to the esophagus posteriorly and the larynx anteriorly (see Figure 18-2). The pharynx is composed of skeletal muscle and lined by mucous membrane. Whereas the nasopharynx functions only in respiration, both the oropharynx and laryngopharynx have digestive as well as respiratory functions. Food that is swallowed passes from the mouth into the oropharynx and laryngopharynx before passing into the esophagus. Muscular contractions of the oropharynx and laryngopharynx help propel food into the esophagus.

ESOPHAGUS

The *esophagus* (e-SOF-a-gus) is a muscular, collapsible tube that lies behind the trachea (see Figure 18-2). It begins at the end of the laryngopharynx, passes through the mediastinum, pierces the diaphragm, and terminates at the top of the stomach. It transports food to the stomach and secretes mucus, which aids transport.

DEGLUTITION

Swallowing, or *deglutition* (dē-gloo-TISH-un), is a mechanism that moves food from the mouth to the stomach. It is helped by saliva and mucus and involves the mouth, pharynx, and esophagus. Swallowing is divided into three stages: (1) the voluntary stage, in which the bolus is moved into the oropharynx, (2) the pharyngeal stage, the involuntary passage of the bolus through the pharynx into the esophagus, and (3) the esophageal stage, the involuntary passage of the bolus through the esophagus into the stomach.

Swallowing starts when the bolus is forced to the back of the mouth cavity and into the oropharynx by the movement of the tongue upward and backward against the palate. This is the *voluntary stage*.

With the passage of the bolus into the oropharynx, the involuntary *pharyngeal stage* of swallowing begins (Figure 19-6b). The respira-tory passageways close and breathing is temporarily interrupted. The soft palate and uvula move upward to close off the nasopharynx and pull the larynx forward and upward under the tongue. As the larynx rises, it meets the epiglottis, which seals off the glottis, and it pulls the vocal cords together, further sealing off the respiratory tract and widening the opening between the laryngopharynx and esophagus. The respiratory passageways then reopen and breathing resumes.

The passage of food from the laryngopharynx into the esophagus is regulated by a sphincter (thick circle of muscle around an opening) at the entrance to the esophagus called the *upper esophageal* (e-sof'-a-JĒ-al) *sphincter*. The elevation of the larynx during the pharyngeal stage of swallowing causes the sphincter to relax and the bolus enters the esophagus.

In the *esophageal stage*, food is pushed through the esophagus by involuntary muscular movements called *peristalsis* (per'-i-STAL-sis). It occurs as follows (Figure 19-6c): In the section of esophagus just above the bolus, the circular muscle fibers contract. This constricts the esophageal wall and squeezes the bolus downward. Meanwhile, longitudinal fibers lying around the bottom of the bolus contract. This shortens the lower section, pushing its walls outward so it can receive the bolus. The contractions are repeated in a wave that moves down the esophagus, pushing the food toward the stomach.

Just above the diaphragm, the esophagus narrows slightly. This

FIGURE 19-6 Deglutition. (a) Position of structures prior to deglutition. (b) During the pharyngeal stage of deglutition, the tongue rises against the palate, the nasopharynx is closed off, the larynx rises, the epiglottis seals off the larynx, and the bolus is passed into the esophagus.

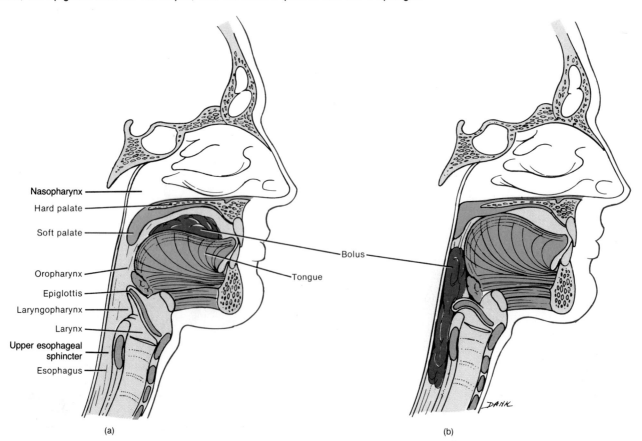

Nasopharynx
Hard palate
Soft palate
Oropharynx
Epiglottis
Laryngopharynx
Larynx
Upper esophageal sphincter
Esophagus

Bolus
Tongue

(a)

(b)

FIGURE 19-6 (*Continued*) (c) Esophageal stage of deglutition.

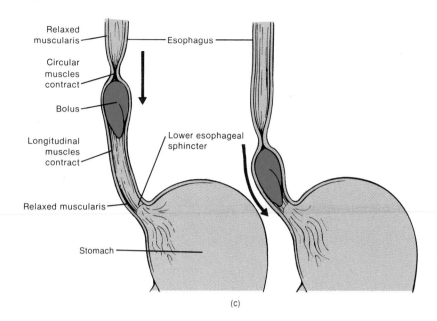

(c)

Describe the three stages of deglutition.

is the *lower esophageal sphincter*, which relaxes during swallowing and thus aids the passage of the bolus from the esophagus into the stomach.

If the lower esophageal sphincter fails to close adequately after food has entered the stomach, the stomach contents can enter the lower esophagus. Hydrochloric acid (HCl) from the contents can irritate the esophageal wall, resulting in a burning sensation known as *heartburn*, because it is experienced in the region near the heart, although it is not related to any cardiac problem. Heartburn can be treated by taking antacids that neutralize the hydrochloric acid and lessen the discomfort.

STOMACH

The *stomach* is a J-shaped region of the GI tract directly under the diaphragm. The superior portion of the stomach is a continuation of the esophagus. The inferior portion empties into the duodenum, the first part of the small intestine.

ANATOMY

The stomach is divided into four areas: cardia, fundus, body, and pylorus (Figure 19-7a). The *cardia* surrounds the lower esophageal sphincter. The rounded portion above and to the left of the cardia is the *fundus*. Below the fundus is the large central portion of the stomach, called the *body*. The narrow, inferior region is the *pylorus*. Between the pylorus and duodenum of the small intestine is the *pyloric sphincter*.

The stomach wall is composed of the same four basic layers as the rest of the GI tract, with certain modifications. When the stomach is empty, the mucosa lies in large folds, called *rugae*

(ROO-jē), that can be seen with the naked eye. Also, there is a layer of simple columnar epithelium with many narrow openings into the connective tissue of the mucosa. These pits—*gastric glands*—are lined with several kinds of secreting cells: peptic, parietal, mucous, and enteroendocrine (Figure 19-7b). The *peptic cells* secrete the principal gastric enzyme precursor, pepsinogen. The *parietal cells* secrete hydrochloric acid, involved in the conversion of pepsinogen to the active enzyme pepsin and intrinsic factor, involved in the absorption of vitamin B_{12} for red blood cell production. The *mucous cells* secrete mucus. Secretions from these three types of cells are collectively called *gastric juice*. The *enteroendocrine cells* secrete *stomach gastrin*, a hormone that stimulates secretion of hydrochloric acid and pepsinogen, contracts the lower esophageal sphincter, mildly increases motility of the GI tract, and relaxes the pyloric sphincter.

The muscularis, unlike that in other areas of the gastrointestinal tract, has three layers of smooth muscle: an outer longitudinal layer, a middle circular layer, and an inner oblique layer. This arrangement of fibers allows the stomach to contract in a variety of ways to churn food, break it into small particles, mix it with gastric juice, and pass it to the duodenum.

DIGESTION IN THE STOMACH

Mechanical

Several minutes after food enters the stomach, gentle, rippling, peristaltic movements called *mixing waves* pass over the stomach every 15 to 25 seconds. These waves macerate food, mix it with the secretions of the gastric glands, and reduce it to a thin liquid called *chyme* (kīm). As food reaches the pylorus, each mixing

FIGURE 19-7 Stomach. (a) External and internal anatomy of the stomach. (b) Gastric glands from the fundic wall of the stomach.

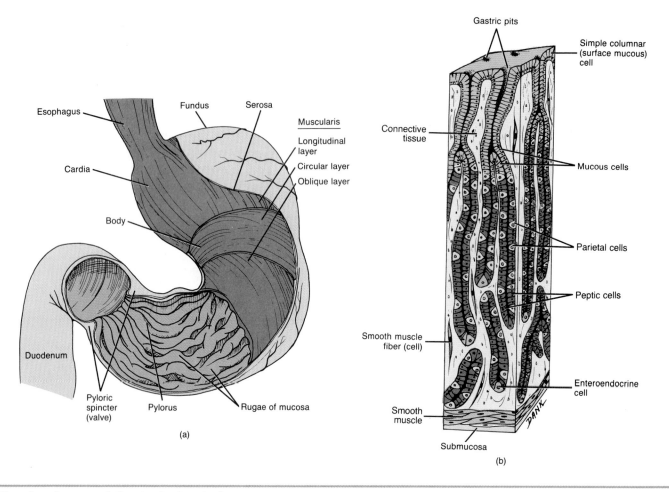

How does the stomach function in digestion?

wave forces a small amount of the gastric contents into the duodenum through the pyloric sphincter. Most of the food is forced back into the body of the stomach where it is subjected to further mixing. The next wave pushes it forward again and forces a little more into the duodenum. The forward and backward movements of the gastric contents are responsible for almost all the mixing in the stomach.

Chemical

The principal chemical activity of the stomach is to begin the digestion of proteins, which is achieved by *pepsin*. Pepsin breaks certain peptide bonds between the amino acids making up proteins. Thus, a protein chain of many amino acids is broken down into smaller fragments called *peptides*. Pepsin is most effective in the very acidic environment of the stomach (pH 2).

What keeps pepsin from digesting the protein in stomach cells along with the food? First, pepsin is secreted in an inactive form called *pepsinogen*, so it cannot digest the proteins in the peptic cells that produce it. It is not converted into active pepsin until it comes in contact with the hydrochloric acid secreted by the parietal cells. Second, the stomach cells are protected by mucus, which coats the mucosa to form a barrier between it and the gastric juices.

Gastric lipase, another stomach enzyme, aids in the breakdown of large lipids, such as butterfat molecules found in milk. This enzyme operates best at a pH of 5 to 6 and has a limited role in the adult stomach (pH 2). Fats are mainly digested in the more alkaline environment of the small intestine, to be described shortly.

The infant stomach also secretes *rennin*, which is important in the digestion of milk. (*Rennin* should not be confused with *renin*, the enzyme involved in the renin–angiotensin pathway in Chapter 13.) Rennin acts on the casein of milk to produce a curd that prevents too rapid a passage of milk from the stomach. Rennin is absent in gastric secretions of adults.

REGULATION OF GASTRIC SECRETION

Stimulation

The secretion of gastric juice is regulated by both nervous and hormonal mechanisms. Parasympathetic impulses from the medulla are transmitted via the vagus (X) nerves and stimulate the gastric glands to secrete pepsinogen, hydrochloric acid, mucus, and stomach gastrin.

The sight, smell, taste, or thought of food initiates a reflex that stimulates gastric secretion before food even enters the stomach. The gastric glands are stimulated by parasympathetic impulses that originate in the cerebral cortex or feeding center in the hypothalamus via the medulla.

Once food reaches the stomach, both nervous and hormonal mechanisms ensure that gastric secretion continues. Food of any kind causes stretching, which stimulates receptors in the wall of the stomach to send impulses to the medulla and back to the gastric glands. Partially digested proteins and caffeine also stimulate the secretion of gastric juice by triggering the release of stomach gastrin.

When partially digested proteins enter the duodenum, they stimulate the duodenal mucosa to release *enteric gastrin*, a hormone that stimulates the gastric glands to continue secretion. However, this mechanism produces relatively small amounts of gastric juice and chyme inhibits gastric secretion (described next), so the net effect of food in the small intestine is the *inhibition* of gastric secretion.

Inhibition

The presence of food in the small intestine ultimately initiates parasympathetic stimulation and stimulates sympathetic activity, thereby inhibiting gastric secretion. Negative emotions, such as anger, may slow down digestion by stimulating the sympathetic nervous system.

Several intestinal hormones also inhibit gastric secretion, the most important of which are *secretin* (se-KRĒ-tin) and *cholecystokinin* (kō-lē-sis′-tō-KĪN-in) or *CCK* (Exhibit 19-1). Secretin and cholecystokinin also help control secretion of bile and pancreatic and intestinal enzymes.

REGULATION OF GASTRIC EMPTYING

The stomach empties all its contents into the duodenum within 2 to 6 hours after ingestion. Foods rich in carbohydrate spend the least time in the stomach. Protein foods are somewhat slower, and emptying is slowest after a meal containing large amounts of fat.

Stomach emptying is regulated by both nervous and hormonal controls. The presence of chyme in the duodenum triggers a reflex that slows gastric emptying. It is also inhibited by secretin and cholecystokinin. These controls keep the rate of stomach emptying at the same rate the intestine can handle its final digestion of the chyme.

Vomiting is the forcible expulsion of the contents of the upper GI tract (stomach and sometimes duodenum) through the mouth. The strongest stimuli for vomiting are irritation and stretching of

---

EXHIBIT 19-1
Hormonal Control of Gastric Secretion, Pancreatic Secretion, and Secretion and Release of Bile

Hormone	Where Produced	Stimulant	Action
Stomach Gastrin	Pyloric mucosa.	Partially digested proteins and caffeine in stomach.	Stimulates secretion of gastric juice, constricts lower esophageal sphincter, increases movement of GI tract, and relaxes pyloric sphincter and ileocecal sphincter.
Enteric Gastrin	Intestinal mucosa.	Partially digested proteins in chyme in small intestine.	Same as above.
Secretin	Intestinal mucosa.	Acid, partially digested carbohydrates, proteins, and fats, or irritants in chyme that enter small intestine.	Inhibits secretion of gastric juice, decreases movement of the GI tract, stimulates secretion of pancreatic juice rich in sodium bicarbonate ions, and stimulates secretion of bile by hepatic cells of liver.
Cholecystokinin (CCK)	Intestinal mucosa.	Fats and proteins that enter small intestine.	Inhibits secretion of gastric juice, decreases movement of the GI tract, stimulates the secretion of pancreatic juice rich in digestive enzymes, causes ejection of bile from the gallbladder and opening of the sphincter at the common duct, and induces a feeling of satiety (feeling full to satisfaction).

the stomach. Other stimuli include unpleasant sights, dizziness, and certain drugs such as morphine. Impulses are transmitted to the vomiting center in the medulla and returning impulses to the upper GI tract organs, diaphragm, and abdominal muscles bring about the vomiting act. Basically, vomiting involves squeezing the stomach between the diaphragm and abdominal muscles and expelling the contents through the open esophageal sphincters. Prolonged vomiting, especially in infants and elderly people, can be serious because the loss of gastric juice and fluids leads to disturbances in fluid and acid–base balance.

ABSORPTION

The stomach wall is impermeable to the passage of most materials into the blood, so most substances are not absorbed until they reach the small intestine. However, the stomach does absorb some water, electrolytes, certain drugs (especially aspirin), and alcohol. The absorption of alcohol by the stomach of females is faster than that in males. The difference is attributed to smaller amounts of the enzyme alcohol dehydrogenase in the stomachs of females. The enzyme breaks down alcohol in the stomach, reducing the amount of alcohol that enters the blood.

PANCREAS

After the stomach, the next organ of the GI tract involved in the breakdown of food is the small intestine. Chemical digestion in the small intestine depends not only on its own secretions but also on activities of three accessory structures: the pancreas, liver, and gallbladder.

ANATOMY

The *pancreas* is a soft, oblong gland that lies behind the stomach (Figure 19-8). Secretions pass from the pancreas to the duodenum by the *pancreatic duct*, which unites with the common bile duct from the liver and gallbladder to enter the duodenum as a single duct.

The pancreas is made up of two types of cells. Small clusters of glandular epithelial cells, the *pancreatic islets (islets of Langerhans)*, constitute the endocrine portion of the pancreas, which produces the hormones glucagon and insulin (see Chapter 13). The *acini* (AS-i-nē) are exocrine glands that secrete a mixture of digestive enzymes called *pancreatic juice* (see Figure 13-17b).

PANCREATIC JUICE

Pancreatic juice is a clear, colorless liquid that consists mostly of water, some salts, sodium bicarbonate, and enzymes. The sodium bicarbonate gives pancreatic juice a slightly alkaline pH (7.1 to 8.2) that stops the action of pepsin from the stomach and creates the proper environment for the enzymes in the small intestine. The enzymes in pancreatic juice include a carbohydrate-digesting enzyme called *pancreatic amylase*; several protein-digesting enzymes called *trypsin* (TRIP-sin), *chymotrypsin* (kī′-mō-TRIP-sin),

and *carboxypeptidase* (kar-bok′-sē-PEP-ti-dās); the principal fat-digesting enzyme in adults called *pancreatic lipase*; and nucleic-acid-digesting enzymes called *ribonuclease* and *deoxyribonuclease*.

The protein-digesting enzymes are produced in inactive form to prevent them from digesting the pancreas itself. Generally, the inactive form is activated in the small intestine by an activating enzyme. For example, the activating enzyme called *enterokinase* (en′-ter-ō-KĪ-nās) activates *trypsinogen* (trip-SIN-ō-jen) in the small intestine into active trypsin, which in turn, activates the other two enzymes.

REGULATION OF PANCREATIC SECRETIONS

Pancreatic secretion is also regulated by both nervous and hormonal mechanisms. The secretion of pancreatic enzymes is also stimulated by parasympathetic fibers from the vagus (X) nerve and by the hormones secretin and cholecystokinin (see Exhibit 19-1).

LIVER

The *liver* weighs about 1.4 kg (about 3 lb) in the average adult and is located under the diaphragm, to the right. It is covered by a connective tissue capsule overlaid by the serous membrane that covers all the viscera, the peritoneum.

ANATOMY

The liver is divided by a ligament into two principal lobes: the *right lobe* and the *left lobe* (Figure 19-8).

The lobes are made up of functional units called *lobules* (Figure 19-9). A lobule consists of rows of *hepatic (liver) cells* arranged radially around a *central vein*. Hepatic cells produce bile, the function of which is described shortly. Between the rows of hepatic cells are spaces called *sinusoids*, through which blood passes. The sinusoids are partly lined with resident phagocytic *stellate reticuloendothelial (Kupffer's) cells* that destroy worn-out white and red blood cells, bacteria, and toxic substances. The liver contains sinusoids instead of typical capillaries.

Bile produced by the liver cells enters the *right* and *left hepatic ducts*, which unite to leave the liver as the *common hepatic duct* (Figure 19-8). The common hepatic duct then joins the *cystic duct* from the gallbladder and the two tubes become the *common bile duct*. The common bile duct and pancreatic duct then enter the duodenum in a common duct.

BLOOD SUPPLY

The liver receives a double supply of blood. From the hepatic artery it obtains oxygenated blood, and from the hepatic portal vein it receives deoxygenated blood containing newly absorbed nutrients. Branches of both the hepatic artery and the hepatic portal vein carry blood to the sinusoids of the lobules, where oxygen, most of the nutrients, and certain poisons are extracted by the hepatic cells. Reticuloendothelial (Kupffer's) cells remove microbes and foreign or dead matter from the blood. Nutrients are stored

FIGURE 19-8 Pancreas. Relation of pancreas to liver, gallbladder, and duodenum. The insert shows details of the common bile duct and pancreatic duct forming a single duct that empties into the duodenum.

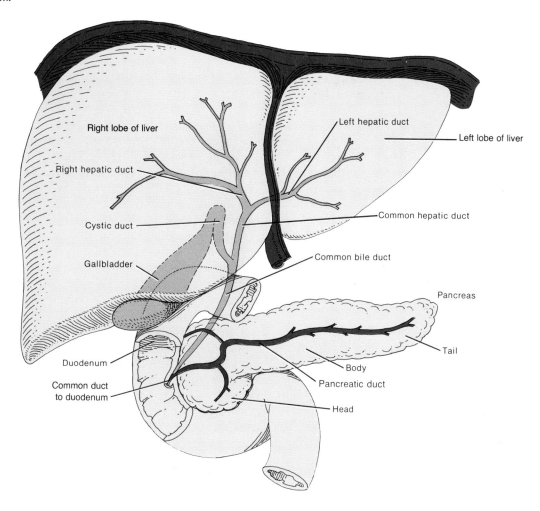

or used to make new materials. The poisons are stored or detoxified. Products manufactured by the hepatic cells and nutrients needed by other cells are secreted back into the blood. The blood then drains into the central vein and eventually leaves for the heart via the hepatic vein.

BILE

Hepatic cells secrete *bile*, a yellow, brownish, or olive-green liquid. It has a pH of 7.6 to 8.6. Bile consists mostly of water and bile salts, cholesterol, a phospholipid called lecithin, bile pigments, and several ions. It is produced in the liver but stored in the gallbladder.

Bile is partially an excretory product and partially a digestive secretion. Bile salts aid in *emulsification*, the breakdown of fat globules into a suspension of fat droplets, and absorption of fats following their digestion. The tiny fat droplets present a very large surface area for the action of pancreatic and intestinal lipase to

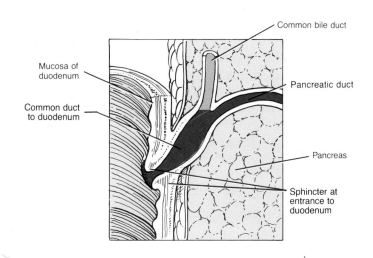

Using a pencil or pen, trace the flow of bile from the liver into the duodenum.

Alcohol: Use or Abuse?

What scenes come to mind when you hear the words "drug abuse"? Junkies shooting up heroin? Kids smoking crack? What about good old uncle Jim who drinks too much? After all, alcohol is the number one drug abuse and mental health problem in the United States. Far more people abuse alcohol (or should we say, abuse themselves with alcohol) than any other drug. The personal and social costs of alcohol abuse are high, since alcohol is our third leading cause of death. It is involved in a majority of traffic deaths, other fatal accidents, suicides, and homicides.

Although alcohol is a powerful, potentially fatal drug, the consumption of alcoholic beverages is socially sanctioned in many countries around the world when it is consumed in moderation and in culturally appropriate situations. We regard it as a "social lubricant," and about 70 percent of American adults drink alcohol at least occasionally. Some research has even suggested that moderate drinking may be healthful. Unfortunately, over 20 million Americans consume excessive amounts of alcohol (14 or more drinks a week). Fifty percent of all alcohol is consumed by 10 percent of all drinkers.

Physiological Effects

The digestive system is affected in several ways by alcohol overconsumption. Alcohol causes an increase in stomach secretions, which can lead to gastritis, or inflammation of the stomach lining. Alcohol also interferes with the ability of the small intestine to absorb and transport nutrients, especially the vitamins thiamin and folic acid and minerals. Impaired nutrient absorption and transport enhance the tendency of alcoholics to be malnourished; people who consume large amounts of alcohol tend to have a poor diet. This malnutrition occurs even though the alcoholic is consuming plenty of calories. Alcohol is calorically dense: 7 kcal/g, compared to 4 kcal/g for protein and carbohydrate and 9 kcal/g for fats. The calories in alcohol are empty, however. That is, they do not supply the protein, vitamins, fiber, or minerals of the foods they are replacing. Heavy drinking can also cause diarrhea, which decreases nutrient absorption from the small intestine.

It is malnutrition that is responsible for the liver damage that occurs in alcoholics. The liver is deprived of the carbohydrates (glycogen) and protein normally stored there and used by the liver for its own fuel. (Remember glycogenolysis and gluconeogenesis?) Fats replace carbohydrate stores, and the liver swells, a condition known as fatty liver. This is one of the early signs of liver damage.

If severe malnutrition continues, a more serious degeneration of liver tissue called *cirrhosis* results. The fatty liver shrinks and hardens as fats are used. Alcohol overconsumption is not the only cause of this disease, but cirrhosis is eight times more common in alcoholics than in nonalcoholics.

The digestive system is not the only site of alcohol damage. Ethanol, the alcohol found in alcoholic beverages, is a relatively small molecule, which diffuses easily from the gastrointestinal tract into the blood and then circulates to every organ in the body. It concentrates in body organs in proportion to the amount of water the organ contains. The brain has a relatively high water content since it has a high concentration of blood, which is 90 percent water, so quite a bit of alcohol ends up there, causing the psychological effects associated with alcohol consumption. Alcohol is physically harmful to body tissues, particularly because it is metabolized by the liver and body tissues into acetaldehyde, whose effects can be very damaging.

Alcohol consumption has been associated with hypertension. It is believed that about 5 to 10 percent of hypertension in men can be attributed to drinking. The effect of alcohol on blood pressure increases with the amount of alcohol consumed. Once the daily consumption of alcohol is reduced, blood pressure decreases as well. Since hypertension increases an individual's risk for stroke and heart disease, alcohol consumption is linked to those diseases as well. In addition to raising blood pressure, alcohol abuse also leads to the development of fat deposits in the heart muscle, which weaken the heart's contractile force. Alcohol consumption can also lead to heart arrhythmias, disturbances in the electrical conductivity of the heart beat.

Alcohol causes atrophy of the cytoplasm of cells of the nervous system. One of the results of nerve damage is decreased muscle tone, and associated muscle weakness and functional loss. Brain cells are also destroyed, and the alcoholic may experience memory loss, confusion, and even hallucinations and psychotic behavior.

The immune system may become suppressed if too much alcohol is consumed. Alcohol inhibits the bone marrow's ability to produce the white blood cells that destroy harmful bacteria. People whose immune systems are already weakened by other illness should be especially careful about alcohol use.

Alcohol use has been associated with several types of cancers, especially cancers of the stomach, liver, lung, pancreas, colon, and tongue. Alcohol and smoking together increase the risk for cancers of the mouth, larynx, and esophagus. Some evidence suggests a relationship between alcohol and breast cancer risk.

A woman who consumes alcohol while pregnant exposes the developing fetus to alcohol's damaging effects. Drinking while pregnant is one of the leading causes of birth defects in North America. Alcohol diffuses freely across the placenta, so the alcohol level in the fetus's bloodstream is the same as that in the mother's. Alcohol affects the fetus's rapidly growing tissues, especially the brain. Fetal alcohol syndrome is characterized by stunted growth, mental retardation, malformed facial features, and heart defects. The damage is irreversible. Fetal alcohol syndrome is found in babies of light drinkers as well as those of heavier abusers. Women consuming only one or two drinks a day during the first six weeks of pregnancy, a time during which many women may not yet realize they are pregnant, may have children with the some of the symptoms of fetal alcohol syndrome, especially impaired mental function.

How Much Is Too Much?

This stuff only happens to the wino on skid row, right? In fact, only about 5 percent of the alcoholics in the United States fit this description. Most hold jobs, have families, and participate in their communities.

It is difficult to say how much alcohol is too much. Individual tolerance varies greatly, so that two people consuming similar amounts may experience very different health effects. On the average, men who drink more than three drinks a day and women who drink more than one and a half drinks a day experience a higher incidence of liver disease. Rates of liver disease rise in proportion to the amount of alcohol consumed.

Women appear to be more susceptible to the effects of alcohol. This is due in part to differences in size; women are generally smaller than men, and a given amount of alcohol is more concentrated in a smaller body. But there's more to it than size. When men and women of the same size ingest the same amount of alcohol, the women end up with higher blood alcohol concentrations. Studies suggest that stomach differences between the sexes are the reason for this difference in alcohol tolerance. The digestion of alcohol begins in the stomach, with the action of an enzyme called alcohol dehydrogenase. Recent research has found that women do not digest alcohol as effectively as men do. Stomach enzymes of the women in this study broke down less than one-fourth as much of the alcohol as those of the men. Alcoholic subjects had an even lower stomach alcohol dehydrogenase enzyme activity. This research helps to explain why women develop liver disease at lower alcohol intakes.

Even relatively small amounts of alcohol (three or four drinks per week) have been associated with some cases of fetal alcohol syndrome, so pregnant women are generally advised to abstain during pregnancy.

Addiction

People prone to a drinking problem may find that no level of alcohol consumption is safe, if they are unable to stop after the first drink. There is no hard and fast standard of behavior that defines alcoholism. Alcoholism, or addiction to alcohol, is not so much a matter of how much people drink, when they drink, or how long they have been drinking. Addiction implies that alcohol is interfering with their sense of reality, relationships with family and friends, job performance, and ability to function in a self-reliant manner.

The National Council on Alcoholism suggests that if you think you or someone close to you may have a problem with alcohol, the following questions may help you decide. The more "yes" responses, the greater the chance there is a drinking problem.

1. When you have trouble or feel under pressure, do you drink more heavily than usual?
2. Have you noticed that you are able to handle more alcohol than you did when you were first drinking?
3. Do you sometimes gulp drinks?
4. When drinking with other people, do you try to have a few extra drinks when others will not know it?
5. Are there certain occasions when you feel uncomfortable if alcohol is not available?
6. Do you sometimes feel guilty about your drinking?
7. Have you tried switching brands or following different plans for controlling your drinking?
8. Do you eat very little or irregularly when you are drinking?

People who have lived with a friend or family member who has a drinking problem know that the health effects of alcoholism go far beyond its physiological consequences. Alcohol abuse can leave deep emotional scars on both the abuser and his/her loved ones.

FIGURE 19-9 Portion of a liver lobule.

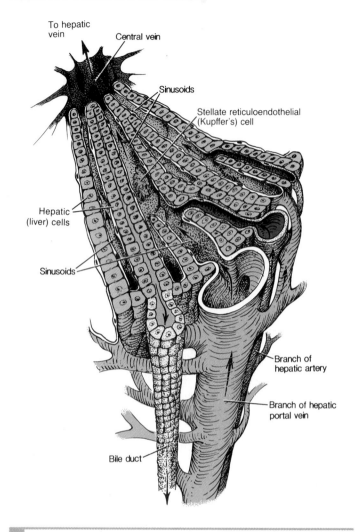

List nine functions of the liver.

rapidly digest fat. Cholesterol is made soluble in bile by bile salts and lecithin. The principal bile pigment is **bilirubin**. When worn-out red blood cells are broken down, iron, globin, and bilirubin (derived from heme) are released. The iron and globin are recycled, but some of the bilirubin is excreted in bile. Bilirubin eventually is broken down in the intestine, and one of its breakdown products (urobilinogen) gives feces their color (see Figure 14-4).

If the liver is unable to remove bilirubin from the blood because of increased rate of destruction of red blood cells or obstruction of bile ducts, the excess bilirubin collects in other tissues, giving the skin and eyes a yellow color. This condition is called **jaundice**.

REGULATION OF BILE SECRETION

The rate at which bile is secreted is determined by several factors. **Vagal** stimulation can increase the production of bile to more than twice the normal rate. Secretin also stimulates the secretion of bile (see Exhibit 19-1). Within limits, as blood flow through the liver increases, so does the secretion of bile.

FUNCTIONS OF THE LIVER

Many of the liver's vital functions are related to metabolism and are discussed in Chapter 20. Briefly, however, the main functions include the following:

1. **Carbohydrate metabolism.** The liver helps maintain a normal blood glucose level. The liver converts glucose to glycogen (glycogenesis) when blood sugar level is high and glycogen to glucose (glycogenolysis) when blood sugar level is low. The liver can also convert certain amino acids to glucose (gluconeogenesis) when blood sugar level is low, convert other sugars, such as fructose and galactose into glucose, and convert glucose to fats.
2. **Fat metabolism.** The liver breaks down fatty acids into acetyl coenzyme A, a process called beta oxidation, and converts excess acetyl coenzyme A into ketone bodies (ketogenesis).
3. **Protein metabolism.** Without the role of the liver in protein metabolism, death would occur in a few days. The liver deaminates (removes the amino group, NH_2) amino acids so that they can be used for energy or converted to carbohydrates or fats; converts ammonia (NH_3), a toxic substance, into the much less toxic urea for excretion in urine; synthesizes most plasma proteins, including alpha and beta globulins, albumin, prothrombin, and fibrinogen; and produces the anticoagulant heparin (together with mast cells).
4. **Removal of drugs and hormones.** The liver detoxifies or excretes into bile drugs such as penicillin and sulfonamides. It can also chemically alter or excrete steroid hormones, such as estrogens and aldosterone, and thyroxine.
5. **Excretion of bile.** Bilirubin, derived from the heme of worn-out red blood cells, is absorbed by the liver from the blood and secreted into bile.
6. **Synthesis of bile salts.** Bile salts are used in the small intestine for the emulsification and absorption of fats, cholesterol, phospholipids, and lipoproteins.
7. **Storage.** The liver stores glycogen, fat-soluble vitamins (A, D, E, and K), and minerals (iron and copper).
8. **Phagocytosis.** The stellate reticuloendothelial (Kupffer's) cells phagocytize worn-out red and white blood cells and some bacteria.
9. **Activation of vitamin D.** The liver and kidneys help activate vitamin D.

GALLBLADDER (GB)

The **gallbladder (GB)** is a pear-shaped sac about 7 to 10 cm (3 to 4 inches) long, located in a depression under the liver (see Figure 19-8). The muscularis, the middle coat of the wall, consists of smooth muscle fibers that contract following hormonal stimulation to eject the contents of the gallbladder into the **cystic duct**.

FUNCTIONS

The gallbladder concentrates and stores bile until it is needed in the small intestine. Bile enters the small intestine through the common bile duct. When the small intestine is empty, a valve around the common duct closes and bile backs up into the cystic duct to the gallbladder for storage.

EMPTYING OF THE GALLBLADDER

When fats enter the small intestine, cholecystokinin (CCK) is released to stimulate contraction of the gallbladder's muscular layer. Bile is then emptied into the common bile duct. CCK also relaxes the common duct's sphincter, which allows bile to flow into the intestine (see Exhibit 19-1).

SMALL INTESTINE

Most digestion and absorption occur in the *small intestine*. It is a tube that averages about 2.5 cm (1 inch) in diameter and 6.35 m (21 feet) in length.

ANATOMY

The small intestine is divided into three segments (see Figure 19-1). The *duodenum* (doo′-ō-DĒ-num), the shortest part (about 25 cm or 10 inches), originates at the pyloric sphincter of the stomach and merges with the jejunum. The *jejunum* (jē-JOO-num) is about 2.5 m (8 ft) long and extends to the ileum, the final portion. The *ileum* (IL-ē-um) is about 3.6 m (12 ft) long and joins the large intestine at the *ileocecal* (il′-ē-ō-SĒ-kal) *sphincter*.

Since almost all the absorption of nutrients occurs in the small intestine, its structure is specially adapted for this function. Its length, over 21 feet, provides a large surface area for absorption and that area is further increased by modifications in the structure of its wall.

The wall of the small intestine is composed of the same four layers that make up most of the GI tract. However, the mucosa and the submucosa are modified to allow the small intestine to complete the processes of digestion and absorption (Figure 19-10).

FIGURE 19-10 Small intestine. Shown are various structures that adapt the small intestine for digestion and absorption. (a) Villi in relation to the layers. (b) Enlarged aspect of several villi.

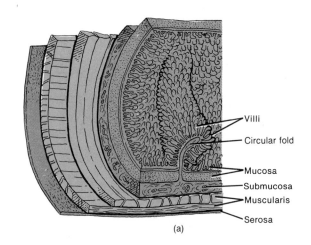

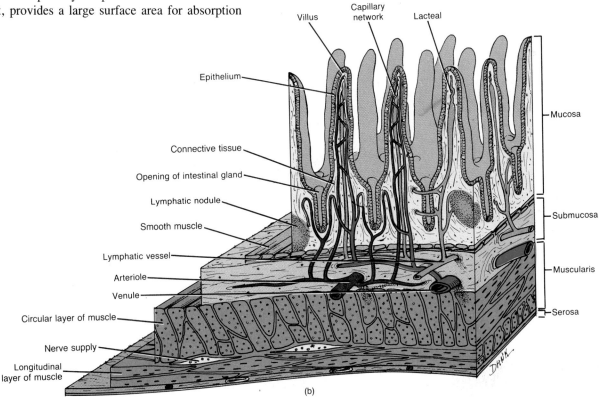

List five structures that adapt the small intestine for digestion and/or absorption.

The mucosa contains many *intestinal glands* that secrete intestinal juice. The submucosa of the duodenum contains *duodenal glands* that secrete an alkaline mucus to help neutralize acid in the chyme. The mucosa consists of simple columnar epithelium and contains goblet cells that secrete additional mucus and absorptive cells with microscopic *microvilli*, fingerlike projections of the plasma membrane that increase the surface area and allow larger amounts of digested nutrients to be digested and absorbed.

The mucosa also contains *villi*, projections found only in the small intestine, which give the intestinal mucosa its velvety appearance. Villi also increase the surface area of the epithelium available for digestion and absorption. Each villus consists of connective tissue, an arteriole, a venule, a capillary network, and *lacteal* (LAK-tē-al) or lymphatic vessel. Nutrients that diffuse through the epithelial cells covering the villus pass through the walls of the capillary or the lacteal to enter the cardiovascular or lymphatic systems.

In addition to the microvilli and villi, a third set of projections called *circular folds* further increases the surface area for absorption and digestion. The folds are permanent ridges in the mucosa. They enhance absorption by causing the chyme to spiral, rather than to move in a straight line, as it passes through the small intestine. The folds and villi decrease in size in the ileum, so most absorption occurs in the duodenum and jejunum.

There are numerous *solitary lymphatic nodules* and groups of lymphatic nodules referred to as *aggregated lymphatic follicles (Peyer's patches)* in the ileum. Their purpose is to prevent bacteria from entering the bloodstream.

INTESTINAL JUICE

Each day, the intestinal glands secrete about 2 to 3 liters (about 2 to 3 qt) of a clear yellow fluid which has a pH of 7.6 (slightly alkaline) and contains water and mucus. The juice is rapidly reabsorbed by the villi and provides a vehicle for the absorption of substances from chyme as they come in contact with the villi.

Intestinal enzymes are synthesized in the epithelial cells that line the villi and most digestion by enzymes of the small intestine occurs in or on the epithelial cells that line the villi, rather than in the lumen, as in other parts of the gastrointestinal tract.

DIGESTION IN THE SMALL INTESTINE

Mechanical

The movements of the small intestine are of two types: segmentation and peristalsis. *Segmentation* is the major movement. It is strictly a localized contraction in areas containing food that sloshes chyme back and forth, mixing it with digestive juices and bringing food particles into contact with the mucosa for absorption. It does not push the intestinal contents along the tract. Segmentation depends mainly on intestinal stretching, which initiates impulses to the central nervous system. Returning parasympathetic impulses increase motility. Sympathetic impulses decrease intestinal motility as stretching decreases.

Peristalsis propels the chyme through the intestinal tract. Peristaltic contractions in the small intestine are normally very weak compared to those in the esophagus or stomach. Peristalsis, like segmentation, is initiated by stretching and controlled by the autonomic nervous system.

Chemical

In the mouth, salivary amylase converts starch (polysaccharide) to maltose (disaccharide). In the stomach, pepsin converts proteins to peptides (small proteins). Thus, chyme entering the small intestine contains partially digested carbohydrates, partially digested proteins, and essentially undigested lipids. The completion of the digestion of carbohydrates, proteins, and lipids is a collective effort of pancreatic juice, bile, and intestinal juice in the small intestine.

■ **Carbohydrates** Starches not reduced to maltose by the time chyme leaves the stomach are broken down by *pancreatic amylase*, an enzyme in pancreatic juice that acts in the small intestine. Although amylase acts on both glycogen and starches, it does not act on the polysaccharide cellulose, an indigestible plant fiber.

Three enzymes in intestinal juice digest the disaccharides into monosaccharides: *maltase* splits maltose into two molecules of glucose; *sucrase* breaks sucrose into a molecule of glucose and a molecule of fructose; and *lactase* digests lactose into a molecule of glucose and a molecule of galactose. This completes the digestion of carbohydrates since monosaccharides are small enough to be absorbed.

■ **Proteins** Protein digestion starts in the stomach where *pepsin* fragments proteins into peptides. Enzymes found in pancreatic juice (trypsin, chymotrypsin, and carboxypeptidase) continue the digestion, though their actions differ somewhat since each splits peptide bonds between different amino acids. Protein digestion is completed by *peptidase*, enzymes produced by small intestinal cells. The final products of protein digestion are amino acids.

■ **Lipids** In an adult, almost all lipid digestion occurs in the small intestine. In the first step, bile salts emulsify globules of fat into small fat droplets. These droplets are neutral fats, called triglycerides because they consist of a molecule of glycerol and three molecules of fatty acid (see Figure 2-8). Now the fat-splitting enzyme can get at the lipid molecules. In the second step, *pancreatic lipase*, found in pancreatic juice, breaks down each fat molecule by removing two of the three fatty acids from glycerol; the third remains attached to the glycerol, thus forming monoglycerides, the end product of fat digestion.

■ **Nucleic Acids** Both intestinal juice and pancreatic juice contain *nucleases* that digest nucleotides into their constituent pentoses and nitrogenous bases. *Ribonuclease* acts on ribonucleic acid nucleotides, and *deoxyribonuclease* acts on deoxyribonucleic acid nucleotides.

A summary of digestive enzymes is presented in Exhibit 19-2.

REGULATION OF INTESTINAL SECRETION

Most small intestinal secretion is regulated by local reflexes in response to the presence of chyme.

EXHIBIT 19-2
Summary of Digestive Enzymes

Enzyme	Source	Substrate	Product
CARBOHYDRATE-DIGESTING			
Salivary Amylase	Salivary glands.	Starches (polysaccharides).	Maltose (disaccharide).
Pancreatic Amylase	Pancreas.	Starches (polysaccharides).	Maltose (disaccharide).
Maltase	Small intestine.	Maltose.	Glucose.
Sucrase	Small intestine.	Sucrose.	Glucose and fructose.
Lactase	Small intestine.	Lactose.	Glucose and galactose.
PROTEIN-DIGESTING			
Pepsin (activated from pepsinogen by hydrochloric acid)	Stomach (peptic cells).	Proteins.	Peptides.
Trypsin (activated from trypsinogen by enterokinase)	Pancreas.	Proteins.	Peptides.
Chymotrypsin (activated from chymotrypsinogen by trypsin)	Pancreas.	Proteins.	Peptides.
Carboxypeptidase (activated from procarboxypeptidase by trypsin)	Pancreas.	Terminal amino acid at carboxyl (acid) end of peptides.	Peptides and amino acids.
Peptidases	Small intestine.	Terminal amino acids at amino end of peptides and dipeptides.	Amino acids.
LIPID-DIGESTING			
Gastric Lipase	Stomach.	Butterfat in milk.	Fatty acids and monoglycerides.
Pancreatic Lipase	Pancreas.	Neutral fats (triglycerides) that have been emulsified by bile salts.	Fatty acids and monoglycerides.
NUCLEIC ACID–DIGESTING			
Ribonuclease	Pancreas and small intestine.	Ribonucleic acid nucleotides.	Pentoses and nitrogenous bases.
Deoxyribonuclease	Pancreas and small intestine.	Deoxyribonucleic acid nucleotides.	Pentoses and nitrogenous bases.

ABSORPTION

All the chemical and mechanical phases of digestion from the mouth down through the small intestine are directed toward changing food into forms that can pass through the mucosa into the underlying blood and lymphatic vessels. These forms are monosaccharides, amino acids, fatty acids, glycerol, and glycerides. Passage of these digested nutrients from the gastrointestinal tract into the blood or lymph is called *absorption*.

About 90 percent of all absorption takes place in the small intestine. The other 10 percent occurs in the stomach and large intestine. Any undigested or unabsorbed material left in the small intestine is passed on to the large intestine. Absorption in the small intestine occurs by diffusion, facilitated diffusion, osmosis, and active transport.

Carbohydrates

Essentially all carbohydrates are absorbed as monosaccharides. Glucose and galactose are transported into epithelial cells of the villi by active transport. Fructose is transported by facilitated diffusion. Transported monosaccharides then move out of the epithelial cells by diffusion and enter the capillaries of the villi. From here they are transported to the liver via the hepatic portal system, then through the heart and to general circulation (Figure 19-11).

Proteins

Most proteins are absorbed as amino acids and the process occurs mostly in the duodenum and jejunum. Amino acid transport into epithelial cells of the villi occurs by active transport and from there by diffusion into the bloodstream. They follow the same route as monosaccharides (Figure 19-11).

Lipids

As a result of emulsification and fat digestion, neutral fats (triglycerides) are broken down into monoglycerides and fatty acids. Short-chain fatty acids (fewer than 10 to 12 carbon atoms) pass into the epithelial cells by diffusion and follow the same route taken by monosaccharides and amino acids (Figure 19-11). Most fatty acids are long-chain fatty acids. They and the monoglycerides are transported with the help of bile salts. The bile salts form spherical aggregates called *micelles* (mī-SELZ). During fat digestion, fatty acids and monoglycerides dissolve in the center of the micelles, and it is in this form that they reach the epithelial cells of the villi.

Within the epithelial cells, many monoglycerides are further digested by lipase to glycerol and fatty acids, which are then recombined into triglycerides in the smooth endoplasmic reticulum of the epithelial cell. Here, the triglycerides are aggregated into globules called **chylomicrons** and coated with a protein that keeps them from sticking to each other. They leave the epithelial cell to enter the lacteal of a villus. From here, they move into the lymphatic system and, eventually, empty into the bloodstream by way of the subclavian veins (Figure 19-11).

The plasma lipids—fatty acids, triglycerides, cholesterol—are insoluble in water and body fluids. In order to be transported in blood and utilized by body cells, the lipids must be combined with protein transporters to make them soluble. The combination of lipid and protein is referred to as a **lipoprotein**. Chylomicrons are one example. Other lipoproteins are **high-density lipoproteins (HDLs)**, **low-density lipoproteins (LDLs)**, and **very low-density lipoproteins (VLDLs)**.

The function of VLDLs is to transport triglycerides that are synthesized in the liver to adipose tissue and muscle tissue where the VLDLs are catabolized and the triglycerides are deposited. The products of VLDL catabolism are LDLs, substances that are rich in cholesterol and also contain some phospholipids. LDLs contain approximately 60 to 70 percent of the total serum cholesterol. The function of LDLs is to transport cholesterol to peripheral tissues. In these tissues LDL is used for activities such as steroid hormone synthesis and the manufacture of cell membranes. High LDL levels are associated with the development of atherosclerosis since some of the cholesterol may be deposited in arteries ("bad" cholesterol). HDLs are rich in phospholipids and cholesterol and transport approximately 20 to 30 percent of the total serum cholesterol. Their function is to transport cholesterol from peripheral tissues to the liver. In the liver some of the HDL is catabolized to become a component of bile (bile salts). High HDL levels are associated with a decreased risk of cardiovascular disease since they may remove cholesterol from arterial walls ("good" cholesterol).

FIGURE 19-11 Absorption. (a) Movement of digested nutrients through epithelial cells of the villi. For simplicity, all digested foods are shown in the lumen of the small intestine, even though some nutrients are digested within or on the surface of small intestinal epithelial cells.

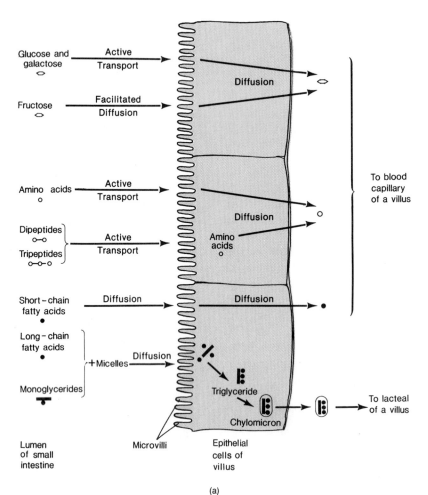

(a)

Water

The total volume of fluid that enters the small intestine each day is about 9 liters (about 9 qt). Nearly all this fluid is absorbed; the remainder, about 0.5 to 1.0 liter, passes into the large intestine where most of it is also absorbed. Water absorption occurs by osmosis.

Electrolytes

Electrolytes absorbed by the small intestine are mostly constituents of gastrointestinal secretions. Some are also components of ingested foods and liquids. Sodium ions move in and out of epithelial cells by diffusion. Chloride, iodide, and nitrate ions can passively follow sodium ions or be actively transported. Calcium, iron, potassium, magnesium, and phosphate ions also move by active transport.

Vitamins

Fat-soluble vitamins (A, D, E, and K) are absorbed along with ingested dietary fats in micelles. In fact, they cannot be absorbed unless they are ingested with some fat. Most water-soluble vitamins, such as the B vitamins and C, are absorbed by diffusion. Vitamin B_{12} must be combined with intrinsic factor produced by the stomach for its absorption.

LARGE INTESTINE

The overall functions of the large intestine are the completion of absorption, the manufacture of certain vitamins, the formation of feces, and the expulsion of feces from the body.

ANATOMY

The *large intestine* averages about 6.5 cm (2.5 inches) in diameter and about 1.5 m (5 feet) in length. It extends from the ileum to the anus and is attached to the posterior abdominal wall. The large intestine is divided into four principal regions: cecum, colon, rectum, and anal canal (Figure 19-12).

At the opening of the ileum into the large intestine is a fold

FIGURE 19-11 (*Continued*) (b) Movement of digested nutrients into the cardiovascular and lymphatic systems.

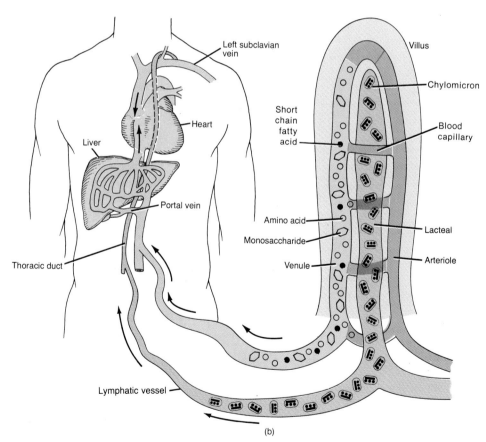

Name the enzyme(s) that digest starches, proteins, maltose, ribonucleic acid, and neutral fats.

FIGURE 19-12 Large intestine.

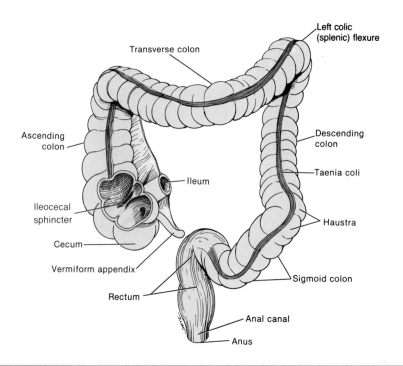

What movements characterize the large intestine?

of mucous membrane called the *ileocecal sphincter*. It allows materials from the small intestine to pass into the large intestine. Hanging below the ileocecal sphincter is a pouch called the *cecum*. Attached to the cecum is the *vermiform appendix* (*vermis* = worm), or just simply, the *appendix*.

The open end of the cecum merges with a long tube called the *colon*. The colon is divided into ascending, transverse, descending, and sigmoid portions. The *ascending colon* ascends on the right side of the abdomen, reaches the undersurface of the liver, and turns to the left. The colon continues across the abdomen to the left side as the *tranverse colon*. It curves beneath the lower end of the spleen on the left side and passes downward as the *descending colon*. The *sigmoid colon* begins near the left iliac crest and terminates as the *rectum*.

The terminal 2 to 3 cm (1 inch) of the rectum is called the *anal canal*. The mucous membrane of the anal canal is arranged in longitudinal folds containing arteries and veins. The opening of the anal canal to the exterior is called the *anus*. It has an internal sphincter of smooth (involuntary) muscle and an external sphincter of skeletal (voluntary) muscle. Normally, the anus is closed except during the elimination of the wastes (feces) of digestion.

The medical specialty that deals with the diagnosis and treatment of disorders of the rectum and anus is called *proctology* (prok-TOL-ō-jē; *proct* = rectum; *logos* = study of).

Varicosities in any veins involve inflammation and enlargement. Varicosities of the rectal veins are known as *hemorrhoids (piles)*. Initially contained within the anus (first degree), they gradually enlarge until they prolapse or extend outward on defecation (second degree) and finally remain prolapsed through the anal orifice (third degree). Hemorrhoids may be caused by constipation and pregnancy.

The wall of the large intestine differs from that of the small intestine in several respects. No villi or permanent circular folds are found in the mucosa, which does, however, contain simple columnar epithelium with numerous goblet cells that secrete mucus to lubricate the colonic contents. Solitary lymphatic nodules also are found in the mucosa. The muscularis consists of an external layer of longitudinal muscles and an internal layer of circular muscles. Unlike other parts of the gastrointestinal tract, portions of the longitudinal bands, the *taeniae coli* (TĒ-ni-ē KŌ-lī), run the length of most of the large intestine. Tonic contractions of the bands gather the colon into a series of pouches called *haustra* (HAWS-tra), which give the colon its puckered appearance.

DIGESTION IN THE LARGE INTESTINE

Mechanical

The passage of chyme from the ileum into the cecum is regulated by the ileocecal sphincter. The sphincter normally remains mildly contracted so the passage of chyme is usually a slow process, but immediately following a meal, there is a reflex action that intensifies peristalsis and any chyme in the ileum is forced into the cecum.

One movement characteristic of the large intestine is *haustral churning*. In this process, the haustra remain relaxed and distended while they fill up. When the distention reaches a certain point, the walls contract and squeeze the contents into the next haustrum.

Peristalsis also occurs, although at a slower rate than in other portions of the tract. The third type of movement is *mass peristalsis*, a strong peristaltic wave that begins in about the middle of the transverse colon and drives the colonic contents into the rectum. Food in the stomach initiates this reflex action. Thus, mass peristalsis usually takes place three or four times a day, during a meal or immediately after.

Chemical

The last stage of digestion occurs through bacterial, not enzymatic, action since no enzymes are secreted. Up to 40 percent of fecal mass is bacteria. Bacteria ferment any remaining carbohydrates and release hydrogen, carbon dioxide, and methane gas, which contribute to flatus (gas) in the colon. They also convert remaining proteins to amino acids and break down the amino acids into simpler substances: indole, skatole, hydrogen sulfide, and fatty acids. Some of the indole and skatole is carried off in the feces and contributes to their odor. The rest is absorbed and transported to the liver, converted to less toxic compounds, and excreted in the urine. Bacteria also decompose bilirubin to simpler pigments (urobilinogen), which give feces their brown color. Several vitamins needed for normal metabolism, including some B vitamins and vitamin K, are synthesized by bacterial action and absorbed.

When bacteria of the large intestine leave the intestine because of a perforation or ruptured appendix, they can cause *peritonitis*, an acute inflammation of the peritoneum. Complications of peritonitis include shock, renal failure, respiratory insufficiency, and liver failure.

ABSORPTION AND FECES FORMATION

By the time the chyme has remained in the large intestine 3 to 10 hours, it has become solid or semisolid as a result of absorption and is now known as *feces*. Chemically, feces consist of inorganic salts, sloughed off epithelial cells from the mucosa of the GI tract, bacteria, products of bacterial decomposition, undigested parts of food, and water.

Of the 0.5 to 1.0 liter (about ½ to 1 qt) of water that enters the large intestine each day, only about 100 ml (about 1/10 of a quart) remains in the feces; the rest is absorbed.

DEFECATION

Mass peristalsis from the sigmoid colon initiates the defecation reflex. Filling of the rectum stretches its wall and stimulates pressoreceptors, which send sensory impulses to the spinal cord. Reflex motor impulses return along parasympathetic fibers to the muscles of the lower colon, rectum, and anus. *Defecation*, or emptying of feces via the anus, results when the internal anal sphincter opens due to pressure caused by involuntary contraction of longitudinal rectal muscles and voluntary contractions of the diaphragm and abdominal muscles. The external sphincter is voluntarily controlled. If it is voluntarily relaxed, defecation occurs; if it is voluntarily constricted, defecation can be postponed after about age 3. If defecation does not occur, the feces move back into the sigmoid colon until the next wave of mass peristalsis again stimulates the pressure-sensitive receptors, creating the desire to defecate.

Diarrhea refers to frequent defecation of liquid feces caused by increased movement of the intestines. Since chyme passes too quickly through the small intestine and feces pass too quickly through the large intestine, there is not enough time for absorption. Like vomiting, diarrhea can result in dehydration and electrolyte imbalances. Diarrhea may be caused by stress and microbes that irritate the gastrointestinal mucosa.

Constipation refers to infrequent or difficult defecation. It is caused by decreased motility of the intestines, in which feces remain in the colon for prolonged periods of time. Water absorption continues and feces become dry and hard. Constipation may be caused by improper bowel habits, spasms of the colon, insufficient bulk in the diet, lack of exercise, and stress.

■ COMMON DISORDERS ■

Dental Caries

Dental caries, or tooth decay, involve a gradual demineralization (softening) of the enamel and dentin. If untreated, various microorganisms may invade the pulp, causing inflammation and infection with subsequent death of the pulp and abscess of the bone surrounding the root. Such teeth are treated by root canal therapy.

Dental caries begin when bacteria, acting on sugars, produce acids that demineralize the enamel. *Dextran*, a sticky polysaccharide produced from sucrose, causes the bacteria to stick to the teeth. Masses of bacterial cells, dextran, and other debris adhering to teeth constitute *dental plaque*. Saliva cannot reach the tooth surface to buffer the acid because the plaque covers the teeth. Brushing the teeth immediately after eating removes the plaque from flat surfaces before the bacteria have a chance to go to work. Dentists also suggest that the plaque between the teeth be removed every 24 hours with dental floss.

Periodontal Disease

Periodontal disease refers to a variety of conditions characterized by inflammation and degeneration of the gingivae, bone, periodontal ligament, and cementum.

Peptic Ulcers

An *ulcer* is a craterlike lesion in a membrane. Ulcers that develop in areas of the gastrointestinal tract exposed to acidic gastric juice (stomach and duodenum) are called *peptic ulcers*.

Appendicitis

Appendicitis is an inflammation of the vermiform appendix. Appendectomy (surgical removal of the appendix) is recommended in all suspected cases because it is safer to operate than to risk gangrene, rupture, and peritonitis.

Tumors

Both benign and malignant *tumors* can occur in all parts of the gastrointestinal tract. *Colorectal cancer* is one of the most common malignant diseases, ranking second to cancer of the lungs in males and breasts in females. Dietary fiber, retinoids, calcium, and selenium may be protective, whereas intake of animal fat and protein may cause an increase in the disease.

Diverticulitis

Diverticula are saclike outpouchings of the wall of the colon in places where the muscularis has become weak. The development of diverticula is called *diverticulosis*, and an inflammation in diverticula is known as *diverticulitis*.

Hepatitis

Hepatitis is an inflammation of the liver caused by viruses, drugs, or chemicals, including alcohol.

Hepatitis A (infectious hepatitis) is caused by the hepatitis A virus and is spread by fecal contamination of food, clothing, toys, eating utensils, and so forth (fecal–oral route). It does not cause lasting liver damage.

Hepatitis B (serum hepatitis) is caused by the hepatatitis B virus and is spread primarily by contaminated syringes and transfusion equipment. It can also be spread by any secretion of fluid by the body (tears, saliva, semen). Hepatitis B can produce chronic liver inflammation.

Non-A, non-B (NANB) hepatitis is a form of hepatitis that cannot be traced to either the hepatitis A or hepatitis B viruses. It is clinically similar to hepatitis B and is often spread by blood transfusions. The NANB hepatitis virus can cause cirrhosis and possibly liver cancer.

Cirrhosis

Cirrhosis refers to a distorted or scarred liver as a result of chronic inflammation. Cirrhosis may be caused by hepatitis, certain chemicals, parasites, and alcoholism.

Gallstones

If there are insufficient bile salts or lecithin in bile, or if there is excessive cholesterol, the cholesterol may crystallize to form *gallstones*. Once formed, gallstones grow in size and number and may cause minimal, intermittent, or complete obstruction to the flow of bile from the gallbladder into the duodenum.

Anorexia Nervosa

Anorexia nervosa is a chronic disorder characterized by self-induced weight loss, body-image and other perceptual disturbances, and physiologic changes that result from nutritional depletion. Patients with anorexia nervosa fixate on their weight and, often, insist on having a bowel movement every day despite the lack of adequate food intake. They abuse cathartics, which worsens the fluid/electrolyte/nutrient deficiencies. The disorder is found predominantly in young, single females and may be inherited.

Bulimia

A disorder that typically affects single, middle-class, young, white females is known as *bulimia* (*bous* = ox; *limos* = hunger), or *binge–purge syndrome*. It is characterized by overeating at least twice a week followed by purging by self-induced vomiting, strict dieting or fasting, vigorous exercise, or use of laxatives or diuretics. This binge–purge cycle occurs in response to fears of being overweight, stress, depression, and physiological disorders such as hypothalamic tumors.

MEDICAL TERMINOLOGY AND CONDITIONS

Borborygmus (bor′-bō-RIG-mus) A rumbling noise caused by the propulsion of gas through the intestines.

Canker (KANG-ker) *sore* Painful ulcer on the mucous membrane of the mouth that affects females more frequently than males and usually occurs between ages 10 to 40; may be an autoimmune reaction.

Cholecystitis (kō′-lē-sis-TĪ-tis; *chole* = bile; *kystis* = bladder; *itis* = inflammation of) Inflammation of the gallbladder that often leads to infection. Some cases are caused by obstruction of the cystic duct with bile stones.

Colitis (ko-LĪ-tis) Inflammation of the mucosa of the colon and rectum in which absorption of water and salts is reduced, producing watery, bloody feces and, in severe cases, dehydration and salt depletion. Spasms of the irritated muscularis produce cramps.

Colostomy (ko-LOS-tō-mē; *stomoun* = provide an opening) The diversion of the fecal stream through an opening in the colon, creating a surgical ''stoma'' (artificial opening) that is affixed to the exterior of the abdominal wall. This opening serves as a substitute anus through which feces are eliminated. A temporary colostomy may be done to allow a badly inflamed colon to rest and heal. If the rectum is removed for malignancy, the colostomy provides a permanent outlet for feces.

Dysphagia (dis-FĀ-jē-a; *dys* = abnormal; *phagein* = to eat) Difficulty in swallowing that may be caused by inflammation, paralysis, obstruction, or trauma.

Enteritis (en'-ter-Ī-tis; *enteron* = intestine) An inflammation of the intestine, particularly the small intestine.

Gastrectomy (gas-TREK-tō-mē; *gastro* = stomach; *tome* = excision) Removal of a portion of or the entire stomach.

Hernia (HER-nē-a) Protrusion of an organ or part of an organ through a membrane or cavity wall, usually the abdominal cavity.

Inflammatory bowel (in-FLAM-a-tō'-rē BOW-el) *disease* Disorder that exists in two forms: (1) *Crohn's disease* (inflammation of the gastrointestinal tract in which the inflammation may extend from the mucosa through the serosa) and (2) *ulcerative colitis* (inflammation of the mucosa of the gastrointestinal tract, usually limited to the large intestine and usually accompanied by rectal bleeding).

Irritable bowel (IR-i-ta-bul BOW-el) *syndrome* (*IBS*) Disease of the entire gastrointestinal tract in which persons with this condition may react to stress by developing symptoms such as cramping and abdominal pain associated with alternating patterns of diarrhea and constipation. Excessive amounts of mucus may appear in the stools, and other symptoms include flatulence, nausea, and loss of appetite. The condition is also known as *irritable colon* or *spastic colitis*.

Malocclusion (mal'-ō-KLOO-zhun; *mal* = disease; *occlusio* = to fit together) Condition in which the upper and lower teeth do not close together.

Nausea (NAW-sē-a; *nausia* = seasickness) Discomfort characterized by a loss of appetite and the sensation of impending vomiting. Its causes include local irritation of the gastrointestinal tract, a systemic disease, brain disease or injury, overexertion, or the effects of medication or drug overdosage.

Pancreatitis (pan'-krē-a-TĪ-tis) Inflammation of the pancreas, as may occur in association with mumps.

Traveler's diarrhea Infectious disease of the gastrointestinal tract that results in loose, urgent bowel movements, cramping, abdominal pain, malaise, nausea, and occasionally fever and dehydration. It is acquired through ingestion of food or water that has become contaminated with fecal material containing mostly bacteria.

STUDY OUTLINE

Digestive Processes (p. 400)

1. Food is prepared for use by cells by five basic activities: ingestion, movement, mechanical and chemical digestion, absorption, and defecation.
2. Chemical digestion is a series of reactions that break down the large carbohydrate, lipid, and protein molecules of food into molecules that are usable by body cells.
3. Mechanical digestion consists of movements that aid chemical digestion.
4. Absorption is the passage of nutrients from digested food in the digestive tract into blood or lymph for distribution to cells.
5. Defecation is emptying of the rectum.

Organization (p. 400)

1. The organs of digestion are usually divided into two main groups: those composing the gastrointestinal (GI) tract and accessory structures.
2. The GI tract is a continuous tube running from the mouth to the anus.
3. The accessory structures are the teeth, tongue, salivary glands, liver, gallbladder, and pancreas.
4. The basic arrangement of tissues in the alimentary canal from the inside outward is the mucosa, submucosa, muscularis, and serosa (peritoneum).

Mouth (Oral Cavity) (p. 401)

1. The mouth is formed by the cheeks, palates, lips, and tongue, which aid mechanical digestion.
2. The opening from the mouth to the throat is the fauces.

Tongue (p. 402)

1. The tongue forms the floor of the oral cavity. It is composed of skeletal muscle covered with mucous membrane.
2. The upper surface and sides of the tongue are covered with papillae. Some papillae contain taste buds.

Salivary Glands (p. 402)

1. The major portion of saliva is secreted by the salivary glands, which lie outside the mouth and pour their contents into ducts that empty into the oral cavity.
2. There are three pairs of salivary glands: the parotid, submandibular, and sublingual.
3. Saliva lubricates food and starts the chemical digestion of carbohydrates.
4. Salivation is entirely under nervous control.

Teeth (p. 404)

1. The teeth, or dentes, project into the mouth and are adapted for mechanical digestion.
2. A typical tooth consists of three principal portions: crown, root, and neck.
3. Teeth are composed primarily of dentin and are covered by enamel, the hardest substance in the body.
4. There are two dentitions: deciduous and permanent.

Digestion in the Mouth (p. 405)

1. Through mastication food is mixed with saliva and shaped into a bolus.
2. Salivary amylase converts polysaccharides (starches) to disaccharides (maltose).

Pharynx (p. 405)

1. Food that is swallowed passes from the mouth into the oropharynx.
2. From the oropharynx, food passes into the laryngopharynx.

Esophagus (p. 406)

1. The esophagus is a collapsible, muscular tube that connects the pharynx to the stomach.
2. It passes a bolus into the stomach by peristalsis.
3. It contains an upper and lower esophageal sphincter.

Deglutition (p. 406)

1. Deglutition or swallowing moves a bolus from the mouth to the stomach.
2. It consists of a voluntary stage, pharyngeal stage (involuntary), and esophageal stage (involuntary).

Stomach (p. 407)

Anatomy (p. 407)

1. The stomach begins at the bottom of the esophagus and ends at the pyloric sphincter.
2. The anatomic subdivisions of the stomach are the cardia, fundus, body, and pylorus.
3. Adaptations of the stomach for digestion include rugae; glands that produce mucus, hydrochloric acid, a protein-digesting enzyme (pepsin), intrinsic factor, and stomach gastrin; and a three-layered muscularis for efficient mechanical movement.

Digestion in the Stomach (p. 407)

1. Mechanical digestion consists of mixing waves.
2. Chemical digestion consists of the conversion of proteins into peptides by pepsin.
3. Mixing waves and gastric secretions reduce food to chyme.

Regulation of Gastric Secretion (p. 408)

1. Gastric secretion is regulated by nervous and hormonal mechanisms.
2. Stimulation occurs in three phases.

Regulation of Gastric Emptying (p. 409)

1. Gastric emptying is stimulated in response to stretching and stomach gastrin released in response to the presence of certain types of food.
2. Gastric emptying is inhibited by reflex action and hormones (secretin and CCK).

Absorption (p. 410)

1. The stomach wall is impermeable to most substances.
2. Among the substances absorbed are some water, certain electrolytes and drugs, and alcohol.

Pancreas (p. 410)

1. The pancreas is connected to the duodenum by the pancreatic duct.
2. Pancreatic islets (islets of Langerhans) secrete hormones and constitute the endocrine portion of the pancreas.
3. Acini cells secrete pancreatic juice; they constitute the exocrine portion of the pancreas.
4. Pancreatic juice contains enzymes that digest starch (pancreatic amylase), proteins (trypsin, chymotrypsin, and carboxypeptidase), fats (pancreatic lipase), and nucleic acids (nucleases).
5. Pancreatic secretion is regulated by nervous and hormonal mechanisms.

Liver (p. 410)

1. The liver has left and right lobes.
2. The lobes are made up of lobules that contain hepatocytes (liver cells), sinusoids, stellate reticuloendothelial (Kupffer's) cells, and a central vein.
3. Hepatocytes of the liver produce bile that is transported by a duct system to the gallbladder for storage.
4. Bile emulsifies neutral fats.
5. The liver also functions in carbohydrate, fat, and protein metabolism; excretion of bile; synthesis of bile salts; removal of drugs and hormones; storage of vitamins and minerals; phagocytosis; and activation of vitamin D.
6. Bile secretion is regulated by nervous and hormonal mechanisms.

Gallbladder (GB) (p. 414)

1. The gallbladder (GB) is a sac located in a depression under the liver.
2. The gallbladder stores and concentrates bile.
3. Bile is ejected into the common bile duct under the influence of cholecystokinin (CCK).

Small Intestine (p. 415)

Anatomy (p. 415)

1. The small intestine extends from the pyloric sphincter to the ileocecal sphincter.
2. It is divided into duodenum, jejunum, and ileum.
3. It is highly adapted for digestion and absorption. Its glands produce enzymes and mucus, and the microvilli, villi, and circular folds of its wall provide a large surface area for digestion and absorption.
4. Intestinal enzymes break down foods in or on epithelial cells of the mucosa.

Intestinal Juice; Digestion in the Small Intestine (p. 416)

1. Intestinal enzymes break down disaccharides to monosaccharides; protein digestion is completed by peptidase enzymes; lipids are broken down into monoglycerides by pancreatic lipase; and nucleases break down nucleic acids to pentoses and nitrogen bases.
2. Mechanical digestion in the small intestine involves segmentation and peristalsis.

Regulation of Intestinal Secretion (p. 416)

1. The most important mechanism is local reflexes.
2. They are initiated in the presence of chyme.

Absorption (p. 417)

1. Absorption is the passage of nutrients from digested food in the gastrointestinal tract into the blood or lymph.
2. Monosaccharides, amino acids, and short-chain fatty acids pass into the blood capillaries.
3. Long-chain fatty acids and monoglycerides are absorbed as part of micelles, resynthesized to triglycerides, and transported as chylomicrons.
4. Chylomicrons are taken up by the lacteal of a villus.
5. The small intestine also absorbs water, electrolytes, and vitamins.

Large Intestine (p. 419)

Anatomy (p. 419)

1. The large intestine extends from the ileocecal valve to the anus.
2. Its subdivisions include the cecum, colon, rectum, and anal canal.

3. The mucosa contains numerous goblet cells and the muscularis consists of taeniae coli.

Digestion in the Large Intestine (p. 420)

1. Mechanical movements of the large intestine include haustral churning, peristalsis, and mass peristalsis.
2. The last stages of chemical digestion occur in the large intestine through bacterial, rather than enzymatic, action. Substances are further broken down and some vitamins are synthesized.

Absorption and Feces Formation (p. 421)

1. The large intestine absorbs water, electrolytes, and vitamins.
2. Feces consist of water, inorganic salts, epithelial cells, bacteria, and undigested foods.

Defecation (p. 421)

1. The elimination of feces from the rectum is called defecation.
2. Defecation is a reflex action aided by voluntary contractions of the diaphragm and abdominal muscles.

REVIEW QUESTIONS

1. Define digestion. Distinguish between chemical and mechanical digestion. (p. 400)
2. Identify the organs of the gastrointestinal (GI) tract in sequence. How does the gastrointestinal tract differ from the accessory structures of digestion? (p. 400)
3. Describe the structure of each of the four coats of the gastrointestinal tract. (p. 400)
4. Make a simple diagram of the tongue. Indicate the types and locations of papillae. (p. 402)
5. Describe the location of the salivary glands and their ducts. (p. 403)
6. Describe the composition of saliva and the role of its components in digestion. (p. 403)
7. What are the principal portions of a typical tooth? What are the functions of each part? (p. 404)
8. Describe the location of the esophagus. What is its role in digestion? (p. 406)
9. Define deglutition. List the sequence of events involved in passing a bolus from the mouth to the stomach and discuss the voluntary, pharyngeal, and esophageal stages of swallowing. (p. 406)
10. Describe the location of the stomach. List and briefly explain the anatomical features of the stomach. (p. 407)
11. What is the importance of rugae, peptic cells, parietal cells, mucous cells, and enteroendocrine cells in the stomach? (p. 407)
12. What is the role of pepsin? Why is it secreted in an inactive form? (p. 408)
13. How is gastric emptying stimulated and inhibited? (p. 408)
14. What are pancreatic acini? Contrast their functions with those of the pancreatic islets (islets of Langerhans). (p. 410)
15. Describe the composition of pancreatic juice and the digestive functions of each component. (p. 410)
16. Where is the liver located? What are its principal functions? (p. 410)
17. What is the composition and function of bile? Where is it stored? (p. 411)
18. Where is the gallbladder located? How is it connected to the duodenum? Describe the function of the gallbladder. (p. 414)
19. What are the subdivisions of the small intestine? How are the mucosa and submucosa of the small intestine adapted for digestion and absorption? (p. 415)
20. Explain the function of each enzyme in intestinal juice. (p. 416)
21. Define absorption. How are the end products of carbohydrate and protein digestion absorbed? How are the end products of fat digestion absorbed? (p. 417)
22. What are the principal subdivisions of the large intestine? How does the muscularis of the large intestine differ from that of the rest of the gastrointestinal tract? What are haustra? (p. 419)
23. Explain the activities of the large intestine that change its contents into feces. (p. 421)
24. Define defecation. How does it occur? (p. 421)
25. Define the following: dental caries, periodontal disease, peptic ulcer, appendicitis, colorectal cancer, diverticulitis, hepatitis, cirrhosis, gallstones, anorexia nervosa, bulimia. (p. 421)
26. Refer to the glossary of medical terminology and conditions associated with the digestive system. Be sure that you can define each term. (p. 422)

Metabolism

STUDENT OBJECTIVES

STUDENT OBJECTIVES

1. Describe how food intake is regulated.
2. Define a nutrient and list the major groups.
3. Define metabolism and describe its importance to homeostasis.
4. Explain how carbohydrates, lipids, and proteins are used by the body.
5. Compare the sources, functions, and importance of minerals in metabolism.
6. Compare the sources, functions, and importance of the principal vitamins.
7. Explain how metabolism and body heat are related.
8. Describe how body heat is produced.
9. Define basal metabolic rate (BMR) and explain several factors that affect it.
10. Describe the various ways that body heat is lost.
11. Define common disorders associated with metabolism.

A LOOK AHEAD

REGULATION OF FOOD INTAKE
NUTRIENTS
METABOLISM
    Anabolism
    Catabolism
    Metabolism and Enzymes
    Oxidation–Reduction Reactions
CARBOHYDRATE METABOLISM
    Fate of Carbohydrates
    Glucose Catabolism
        Glycolysis
        Krebs Cycle
        Electron Transport Chain
    Glucose Anabolism
        Glucose Storage: Glycogenesis
        Formation of Glucose from
            Proteins and Fats:
            Gluconeogenesis
LIPID METABOLISM
    Fate of Lipids
    Fat Storage
    Lipid Catabolism
        Glycerol
        Fatty Acids
    Lipid Anabolism: Lipogenesis
PROTEIN METABOLISM
    Fate of Proteins
    Protein Catabolism
    Protein Anabolism
REGULATION OF METABOLISM
MINERALS
VITAMINS
METABOLISM AND BODY HEAT
    Measuring Heat
    Production of Body Heat
    Basal Metabolic Rate (BMR)
    Loss of Body Heat
        Radiation
        Conduction
        Convection
        Evaporation
    Body Temperature Regulation
        Hypothalamic Thermostat
        Mechanisms of Heat Production
        Mechanisms of Heat Loss
        Fever
COMMON DISORDERS

I n this chapter we will discuss how food intake is regulated; how each group of nutrients is used for energy, growth, and repair of the body; and how various factors affect the body's metabolic rate.

REGULATION OF FOOD INTAKE

In the hypothalamus of the brain are two groups of neurons related to food intake. When the **hunger center** is stimulated in animals, they begin to eat heartily, even if they are already full. When the **satiety center** is stimulated in animals, they stop eating, even if they have been starved for days. Apparently the hunger center is constantly active, but it is inhibited by the satiety center. Other parts of the brain that function in hunger and satiety are the brain stem and limbic system.

Food intake is also affected by blood glucose levels. When blood glucose levels are low, feeding increases. It is believed that low levels of blood glucose slow the rate of glucose utilization by neurons of the satiety center. As a result, neuronal activity decreases to the point where the satiety center can no longer inhibit the hunger center, and the individual eats. On the other hand, when blood glucose levels are high, the activity of the neurons of the satiety center is sufficiently high to inhibit the hunger center, and feeding is depressed. To a lesser extent, low levels of amino acids in the blood also enhance feeding, whereas high levels depress eating.

It has been observed that as the amount of adipose (fat) tissue increases in the body, the rate of feeding decreases. It is theorized that substances, perhaps fatty acids, are released from stored fats in proportion to the total fat content of the body and these substances then activate the satiety center.

Food intake is also affected by body temperature. Whereas a cold environment enhances eating, a warm environment depresses it. Food intake is also regulated by stretching of the gastrointestinal tract, particularly the stomach and duodenum. When these organs are stretched, a reflex is initiated that activates the satiety center and depresses the hunger center. The hormone cholecystokinin (CCK) also inhibits eating.

Psychological factors may override the usual intake mechanisms, for example, in obesity, anorexia nervosa, and bulimia.

NUTRIENTS

Nutrients are chemical substances in food that provide energy, form new body components, or assist in the functioning of various body processes. The six principal classes of nutrients are carbohydrates, lipids, proteins, minerals, vitamins, and water. Carbohydrates, proteins, and lipids are digested by enzymes in the gastrointestinal tract. Some of the end products of digestion are used to synthesize new structural molecules in cells or new regulatory molecules, such as hormones and enzymes. Most are used to produce energy to sustain life processes.

Some minerals and many vitamins are part of enzyme systems that catalyze metabolic reactions involving carbohydrates, proteins, and lipids.

Water has five major functions. It is an excellent solvent and suspending medium; it participates in hydrolysis reactions; it acts as a coolant; it lubricates; and it helps to maintain a constant body temperature.

METABOLISM

Metabolism (me-TAB-ō-lizm; *metabole* = change) refers to all the chemical reactions of the body. The body's metabolism may be thought of as an energy-balancing act between anabolic (synthesis) and catabolic (decomposition) reactions.

ANABOLISM

Chemical reactions that combine simple substances into more complex molecules are collectively known as **anabolism** (a-NAB-ō-lizm). Overall, anabolic reactions require energy, which is supplied by catabolic reactions (Figure 20-1). One example of an anabolic process is the formation of peptide bonds between amino acids, thereby building the amino acids into proteins. Fats are built into phospholipids and simple sugars into polysaccharides through anabolism.

CATABOLISM

The chemical reactions that break down complex organic compounds into simple ones are collectively known as **catabolism** (ka-TAB-ō-lizm). Catabolic reactions release the chemical energy in organic molecules. This energy is then stored in ATP until it is needed for anabolic reactions. The role of ATP in the relationship between catabolic and anabolic processes is shown in Figure 20-1.

Chemical digestion is a catabolic process in which the breaking of bonds of food molecules releases energy. Another example is oxidation (cellular respiration), to be described shortly.

METABOLISM AND ENZYMES

Chemical reactions occur when chemical bonds between substances are made or broken. Enzymes are proteins that serve as catalysts

FIGURE 20-1 Catabolism, anabolism, and ATP. When simple compounds are combined to form complex compounds (anabolism), ATP provides the energy for synthesis. When large compounds are split apart (catabolism), much of the energy is given off as heat. Some is transferred to and trapped in ATP and then utilized to drive anabolic reactions.

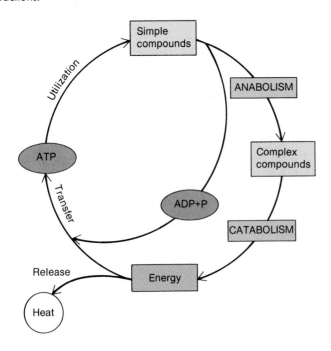

Define metabolism.

to speed up chemical reactions. A **catalyst** is a chemical substance that alters the rate of a chemical reaction without changing itself. That is, it does not become part of the products of the reaction and it is not used up.

Enzymes are very specific in the reactions they catalyze and the molecules with which they react are called **substrates**. The names of enzymes usually end in the suffix -ase, for example, salivary amylase.

Most enzymes work together with ions such as calcium, iron, and zinc that hold the enzyme to the substrate or with substances called **coenzymes** that accept or donate atoms to a substrate. Many coenzymes are derived from vitamins.

Enzymes are believed to work as follows (Figure 20-2):

1. A specific region on the surface of the enzyme molecule known as the **active site** makes contact with the surface of the substrate.
2. A temporary compound called an **enzyme–substrate complex** forms.
3. The substrate molecule becomes transformed. Existing atoms rearrange, or the substrate molecule breaks down completely, or several substrate molecules may recombine.
4. The transformed substrate molecules, now called the **products** of the reaction, move away from the enzyme molecule.
5. The unchanged enzyme is free to attach to another substrate molecule.

OXIDATION–REDUCTION REACTIONS

Oxidation refers to the removal of electrons (e^-) and hydrogen ions (H^+) from a molecule. This is equivalent to the removal of a hydrogen atom ($e^- + H^+ \rightarrow H$). An example of an oxidation reaction is the conversion of lactic acid into pyruvic acid:

$$
\begin{array}{ccc}
\text{COOH} & & \text{COOH} \\
| & \xrightarrow{\ -2H\ (oxidation)\ } & | \\
\text{CHOH} & & \text{C}=\text{O} \\
| & & | \\
\text{CH}_3 & & \text{CH}_3 \\
\text{Lactic acid} & & \text{Pyruvic acid}
\end{array}
$$

The two freed hydrogen atoms are transferred to another compound by coenzymes.

Reduction refers to the addition of electrons and hydrogen ions (hydrogen atoms) to a molecule. Reduction is the opposite of oxidation. An example of a reduction reaction is the conversion of pyruvic acid into lactic acid:

$$
\begin{array}{ccc}
\text{COOH} & & \text{COOH} \\
| & \xrightarrow{\ +2H\ (reduction)\ } & | \\
\text{C}=\text{O} & & \text{CHOH} \\
| & & | \\
\text{CH}_3 & & \text{CH}_3 \\
\text{Pyruvic acid} & & \text{Lactic acid}
\end{array}
$$

Oxidation and reduction reactions are always coupled; that is, whenever a substance is oxidized, another is reduced. This coupling of reactions is referred to as **oxidation–reduction**.

Oxidation is usually an energy-releasing reaction. Cells take nutrients (potential energy sources) and break them down from

FIGURE 20-2 Enzyme action in a decomposition reaction. The enzyme and substrate molecules combine to form an enzyme–substrate complex. During combination, the substrate is changed into products. Once the products are formed, the enzyme is recovered and may be used again to catalyze a similar reaction.

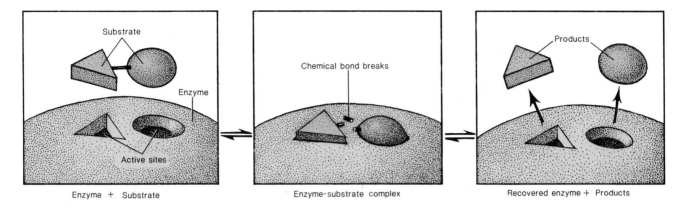

Enzyme + Substrate   Enzyme-substrate complex   Recovered enzyme + Products

What are five characteristics of enzymes?

compounds with many hydrogen atoms to compounds with many oxygen atoms or multiple bonds, thereby releasing energy. For example, when a cell oxidizes a molecule of glucose ($C_6H_{12}O_6$), the energy in the chemical bonds of the glucose molecule is removed and trapped by adenosine triphosphate (ATP). ATP is thus an energy source for energy-requiring reactions. Compounds such as glucose that have many hydrogen atoms are termed highly reduced compounds and contain more potential energy than oxidized compounds.

Oxidation–reduction reactions play important roles in carbohydrate, lipid, and protein metabolism. Also, most body heat is a result of oxidation of the food we eat.

CARBOHYDRATE METABOLISM

During digestion, polysaccharides and disaccharides are digested to monosaccharides—glucose, fructose, and galactose—which are absorbed in the small intestine and carried to the liver, where fructose and galactose are then converted to glucose (see Figure 19-11a). The liver is the only organ that has the necessary enzymes to make this conversion. Thus, the story of carbohydrate metabolism is really the story of glucose metabolism.

FATE OF CARBOHYDRATES

Since glucose is the body's preferred source of energy, the fate of absorbed glucose depends on the body cells' energy needs. If the cells require immediate energy, the glucose is oxidized by the cells. Each gram of carbohydrate produces about 4.0 kilocalories (kcal). (The kilocalorie content of a food is a measure of the heat it releases upon oxidation. Caloric value is described later in the chapter.) Glucose not needed for immediate use may be converted to glycogen by the liver (glycogenesis) and stored there or in skeletal muscle fibers (cells). If these glycogen storage areas

are filled up, the liver cells can transform the glucose to fat for storage in adipose tissue. When the cells need more energy, the glycogen and fat can be converted back to glucose.

Before glucose can be used by body cells, it must pass through the plasma membrane and enter the cytoplasm. This occurs through facilitated diffusion (see Chapter 3). The rate of glucose transport is greatly increased by insulin (see Chapter 13).

GLUCOSE CATABOLISM

The *oxidation* of glucose is also known as *cellular respiration*. It occurs in virtually every cell in the body and provides the cell's chief source of energy. The complete oxidation of one molecule of glucose yields 38 molecules of ATP. It occurs in three stages: glycolysis, the Krebs cycle, and the electron transport chain.

Glycolysis

Glycolysis (glī-KOL-i-sis; *glyco* = sugar; *lysis* = breakdown) refers to the chemical reactions that convert a molecule of glucose into two molecules of pyruvic acid (Figure 20-3). It takes place in the cytoplasm of a cell. Glycolysis is also called *anaerobic respiration* because it occurs without oxygen. In the process, several compounds are formed on the way to producing pyruvic acid. For each molecule of glucose that undergoes glycolysis two molecules of ATP are produced.

The fate of pyruvic acid depends on the availability of oxygen. If conditions remain anaerobic, as during strenuous exercise, pyruvic acid is converted into lactic acid. The lactic acid may be transported to the liver where it is converted back to pyruvic acid. Or, it may remain in the cells until aerobic (with oxygen) conditions are restored (recall the mechanisms for the repayment of the oxygen debt described in Chapter 8) and it is then converted to pyruvic acid in the cells.

FIGURE 20-3 Simplified version of glycolysis.

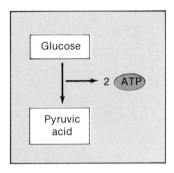

How does the fate of pyruvic acid differ under anaerobic and aerobic conditions?

Under anaerobic conditions, glucose is only partially oxidized. Complete oxidation of glucose requires the presence of oxygen and is referred to as *aerobic respiration*. During aerobic respiration, pyruvic acid enters mitochondria where it is oxidized to produce a great deal more energy (ATP) than anaerobic respiration. Aerobic respiration consists of a number of chemical reactions known as the Krebs cycle and electron transport chain.

Krebs Cycle

The *Krebs cycle* is a series of oxidation–reduction reactions that occur in the mitochondria of cells (Figure 20-4). In order for pyruvic acid to enter the Krebs cycle, it must first be changed into an acetyl group and combined with coenzyme A (CoA) to form a substance called *acetyl coenzyme A*. During the Krebs cycle, a series of oxidation–reduction reactions breaks down acetyl coenzyme A, releases carbon dioxide (CO_2), produces a substance called guanosine triphosphate (GTP), and reduces various coenzymes. The CO_2 is transported by the blood to the lungs where it is exhaled. The GTP is a high-energy compound that is used to change ADP into ATP (GTP is the equivalent of ATP). The reduced coenzymes now contain the stored (potential) energy that was in glucose, pyruvic acid, and acetyl coenzyme A. But, the reduced coenzymes must first go through the electron transport chain before the energy will be available for cellular activities.

Electron Transport Chain

The *electron transport chain* is also a series of oxidation–reduction reactions that take place in mitochondria and that transfer the energy that was stored in the reduced coenzymes to 36 molecules of ATP. This represents only 43 percent of the energy originally stored in glucose since the rest of the energy is released as heat. In the electron transport chain, oxygen is required as the final electron acceptor and water is produced.

The complete catabolism of a molecule of glucose can be summarized as follows:

Glucose + Oxygen → 38 ATP + Carbon dioxide + Water

A summary of the complete oxidation of glucose is shown in Figure 20-5.

Glycolysis, the Krebs cycle, and the electron transport chain provide all the ATP for cellular activities. And because the Krebs cycle and electron transport chain are aerobic processes, the cells cannot carry on their activities for long without sufficient oxygen.

GLUCOSE ANABOLISM

Most of the glucose in the body is catabolized to supply energy. However, some glucose participates in anabolic reactions. One is the manufacture of glucose from the breakdown products of proteins and lipids. Another is the synthesis of glycogen from glucose.

Glucose Storage: Glycogenesis

If glucose is not needed immediately for energy, it combines with many other molecules of glucose to form a long-chain molecule called glycogen (Figure 20-6). This process is called *glycogenesis* (glī-kō-JEN-e-sis; *glyco* = sugar; *genesis* = origin). The body can store about 500 g (about 1.1 lb) of glycogen, roughly 80 percent in the skeletal muscles and the rest in the liver. Liver and skeletal muscle fibers (cells) are very active in glycogenesis. Glycogenesis is stimulated by insulin from the pancreas.

Insulin stimulates glycogenesis as another way of decreasing blood sugar level. When blood sugar level drops too much, glucagon is released and it stimulates the liver to convert glycogen back into glucose, a process called *glycogenolysis* (glī-kō-je-NOL-i-sis; *lysis* = breakdown) (see Figure 20-6). Glycogenolysis usually occurs between meals.

Formation of Glucose from Proteins and Fats: Gluconeogenesis

When your liver runs low on glycogen, it is time to eat. If you don't, your body starts catabolizing fats and proteins. Actually, the body normally catabolizes some of its fats and a few of its proteins, but large-scale fat and protein catabolism does not happen unless you are starving, eating very few carbohydrates, or suffering from an endocrine disorder.

Fat molecules and protein molecules are converted in the liver to glucose (Figure 20-6). The process by which glucose is formed from noncarbohydrate sources is called *gluconeogenesis* (gloo'-kō-nē'-o-JEN-e-sis). Gluconeogenesis occurs when the liver is stimulated by cortisol, thyroxine, epinephrine, glucagon, and/or human growth hormone (hGH).

LIPID METABOLISM

Lipids are second to carbohydrates as a source of energy. Neutral fats (triglycerides) are ultimately digested into short-chain and long-chain fatty acids and monoglycerides.

FATE OF LIPIDS

Like carbohydrates, lipids such as fatty acids and glycerol may be oxidized to produce ATP. Each gram of fat produces about 9.0 kilocalories. If the body has no immediate need to utilize fats this way, they are stored in adipose tissue (fat depots) and in the liver.

FIGURE 20-4 Simplified version of the Krebs cycle. NADH + H⁺ and FADH₂ are reduced coenzymes that contain the stored energy that was in glucose, pyruvic acid, and acetyl coenzyme A. When they pass through the electron transport chain, the energy liberated will produce ATP.

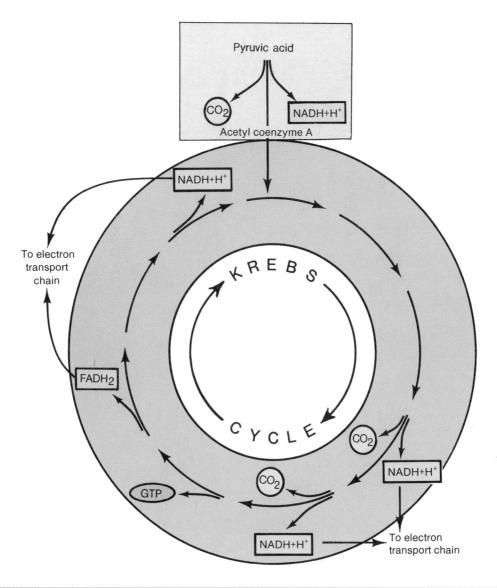

Why is the production of reduced coenzymes in the Krebs cycle important?

Other lipids are used as structural molecules or to synthesize other essential substances. For example, phospholipids are constituents of plasma membranes; lipoproteins transport cholesterol throughout the body; thromboplastin is needed for blood clotting; myelin sheaths speed up nerve impulse conduction; and cholesterol is used to synthesize bile salts and steroid hormones.

FAT STORAGE

The major function of adipose tissue is to store fats until they are needed for energy in other parts of the body. It also insulates and protects. About 50 percent of stored fat is deposited in subcutaneous tissue.

LIPID CATABOLISM

Fats stored in adipose tissue constitute the largest reserve of energy. The body can store much more fat than it can store glycogen. Moreover, the energy yield of fats is more than twice that of carbohydrates. Fats are nevertheless the body's second-favorite source of energy because they are more difficult to catabolize than carbohydrates.

FIGURE 20-5 Summary of the complete oxidation of glucose.

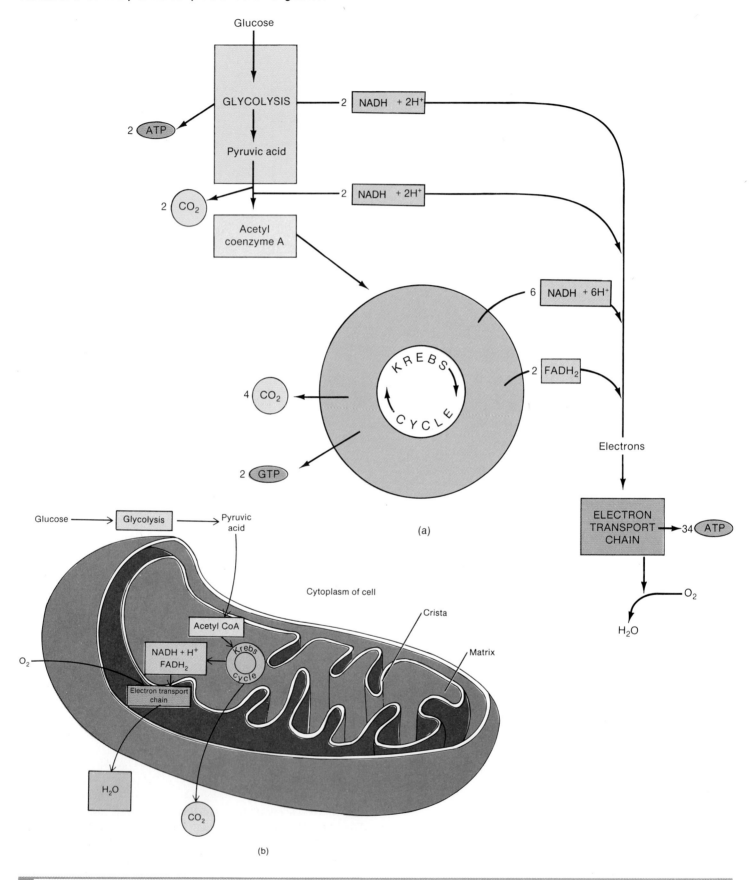

(a)

(b)

Write a summary equation for the complete oxidation of glucose.

FIGURE 20-6 Gluconeogenesis. This process involves the conversion of noncarbohydrate sources (amino acids and glycerol) into glucose. Note also where glucogenesis and glycogenolysis occur. Glyceraldehyde-3-phosphate is a substance formed during glycolysis.

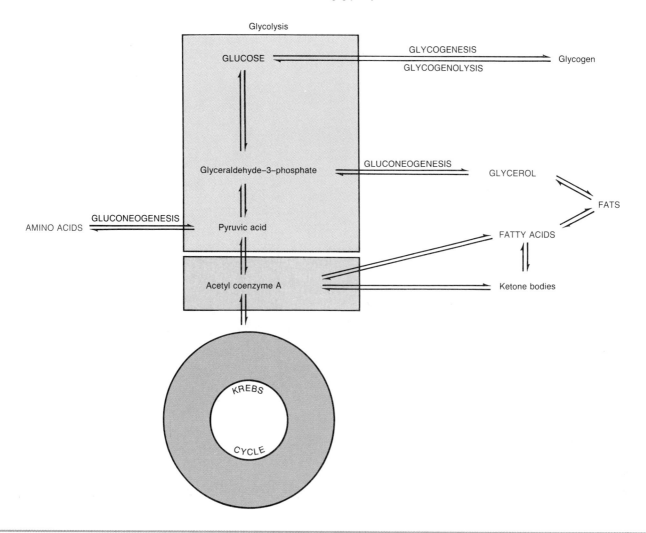

Name the process concerned with the formation of glucose for noncarbohydrate sources, conversion of glucose into glycogen, and conversion of glycogen into glucose.

Glycerol

Before fat molecules can be metabolized for energy, they must be split into glycerol and fatty acids. Then, the glycerol and fatty acids are catabolized separately (Figure 20-7).

Glycerol is converted to glyceraldehyde-3-phosphate, a compound also formed during the catabolism of glucose. It then continues the catabolic sequence to pyruvic acid or goes on to become glucose. This is one example of gluconeogenesis.

Fatty Acids

Fatty acids are too complex to be catabolized directly by body cells. They must first undergo a series of reactions in the liver called **beta oxidation**, which result in a number of molecules of **acetyl coenzyme A (CoA)**. The acetyl CoA may then enter the Krebs cycle in cells (Figure 20-7). In terms of energy yield, an 18-carbon fatty acid, such as palmitic acid, can yield 131 ATPs by the time it is completely oxidized via the Krebs cycle and electron transport chain.

As part of normal fatty-acid catabolism, the liver converts some acetyl CoA molecules into substances known as **ketone bodies**, a process known as **ketogenesis** (kē'-tō-JEN-e-sis) (Figure 20-7). They then leave the liver to enter body cells where they are broken down into acetyl CoA, which enters the Krebs cycle for oxidation.

Since the body prefers glucose as a source of energy, ketone bodies are generally produced in very small quantities. When the number of ketone bodies in the blood rises above normal—a condition called **ketosis**—the ketone bodies, most of which are acids,

FIGURE 20-7 Metabolism of lipids. Glycerol may be converted to glyceraldehyde-3-phosphate, which can then be converted to glucose or enter the Krebs cycle for oxidation. Fatty acids undergo beta oxidation and enter the Krebs cycle via acetyl coenzyme A. Fatty acids also can be converted into ketone bodies (ketogenesis). Lipogenesis is the synthesis of lipids from glucose or amino acids.

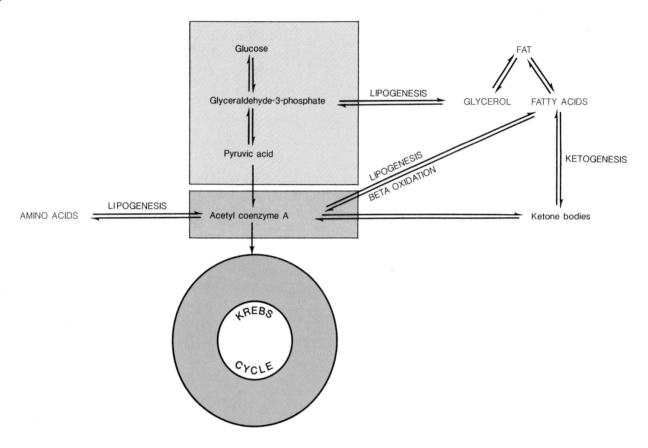

Why are fats a secondary source of energy even though they yield the greatest amount of energy?

must be buffered by the body. If too many accumulate, they use up the body's buffers and the blood pH falls, which leads to *acidosis*, or abnormally low blood pH, and possibly death.

LIPID ANABOLISM: LIPOGENESIS

Liver cells and fat cells can synthesize lipids from glucose or amino acids through a process called *lipogenesis* (lip′-ō-JEN-e-sis). Lipogenesis occurs when a greater quantity of carbohydrate enters the body than can be used for energy or stored as glycogen. The excess carbohydrates are synthesized into fats. The conversion of glucose to lipids involves the formation of glycerol and acetyl CoA, which can be converted to fatty acids (see Figure 20-6). The process is enhanced by insulin. Glycerol and fatty acids (1) can undergo anabolic reactions to become fat that can be stored or (2) can produce other lipids such as lipoproteins, phospholipids, and cholesterol.

Many amino acids can be converted into acetyl CoA, which can then be converted into fats (Figure 20-7). When there is more

protein in the diet than can be utilized as such, the excess is converted to and stored as fats.

PROTEIN METABOLISM

During digestion, proteins are broken down into their constituent amino acids. Although proteins may be catabolized for energy, amino acids are mainly used to synthesize new proteins for body growth and repair.

FATE OF PROTEINS

The amino acids that enter body cells are almost immediately synthesized into proteins. Proteins are used for structural components of the body (collagen, elastin, and keratin) or for functional purposes (actin and myosin for contraction, fibrinogen for clotting, hemoglobin for carrying oxygen, antibodies for fighting infection, and enzymes and hormones to alter the rate of metabolic processes).

Excess amino acids may be converted to fat or glycogen or glucose for energy. Each gram of protein produces about 4.0 kilocalories.

PROTEIN CATABOLISM

A certain amount of protein catabolism occurs in the body each day, though much of this is only partial catabolism. Proteins are extracted from worn-out cells, such as red blood cells, and broken down into free amino acids. Some amino acids are converted into other amino acids, peptide bonds are re-formed, and new proteins are made as part of the constant turnover in all cells.

If other energy sources are used up or inadequate and protein intake is high, the liver can convert protein to fat or glucose or oxidize it to carbon dioxide and water. Like fatty acids, the liver must process amino acids before they can be used by the cells for energy. The liver removes the nitrogen from amino acids in a process called **deamination** (dē-am'-i-NĀ-shun). The resulting keto acids can then be converted to acetyl CoA and used for energy in the cells' Krebs cycle. Amino acids can be altered in various ways to enter the Krebs cycle at various locations. The gluconeogenesis of amino acids into glucose may be reviewed in Figure 20-6. The conversion of amino acids into fatty acids or ketone bodies is also shown in Figure 20-6.

PROTEIN ANABOLISM

Protein anabolism involves the formation of peptide bonds between amino acids to produce new proteins. Protein anabolism, or synthesis, is carried out on the ribosomes of almost every cell in the body, directed by the cells' DNA and RNA (see Figure 3-16). The synthesized proteins function as enzymes, hormones, structural components of cells, and so forth. To balance protein catabolism and protein anabolism, dietary guidelines recommend 60 to 75 g (0.8 g/kg of body weight) of protein per day. However, this should be increased in times of growth or tissue repair such as during pregnancy or following surgery.

Of the naturally occurring amino acids, 10 are referred to as **essential amino acids** because they cannot be synthesized by the body. They are synthesized by plants or bacteria and so foods containing these amino acids are "essential" for human growth and must be part of the diet. **Nonessential amino acids** can be synthesized by body cells and the appropriate essential and nonessential amino acids must be present in cells before protein synthesis can occur.

A summary of carbohydrate, lipid, and protein metabolism is presented in Exhibit 20-1.

REGULATION OF METABOLISM

Absorbed nutrients may be oxidized for energy, stored, or converted into other molecules based on the body's needs. The pathway taken by a particular nutrient is controlled by enzymes and regulated by hormones. Hormones and enzymes are ineffective, however, without the proper minerals and vitamins. Some minerals and many vitamins are components of the enzymes that catalyze the metabolic reactions.

EXHIBIT 20-1
Summary of Metabolism

Process	Comment
CARBOHYDRATE METABOLISM	
Glucose Catabolism	Complete oxidation of glucose, also called aerobic respiration, is the chief source of energy in cells. In requires glycolysis, Krebs cycle, and electron transport chain. The complete oxidation of 1 molecule of glucose yields 38 molecules of ATP.
Glycolysis	Conversion of glucose into pyruvic acid produces some ATP. Reactions do not require oxygen and are thus termed anaerobic.
Krebs Cycle	Series of oxidation–reduction reactions in which coenzymes pick up hydrogen atoms from oxidized organic acids; some ATP is produced. CO_2, H_2O, and heat are by-products. Reactions are aerobic.
Electron Transport Chain	Third set of reactions in glucose catabolism is a series of oxidation–reduction reactions, in which most of the ATP is produced. Reactions are aerobic.
Glucose Anabolism	Some glucose is converted into glycogen (glycogenesis) for storage if not needed immediately for energy. Glycogen can be converted back to glucose (glycogenolysis) for energy. The conversion of fats and proteins into glucose is called gluconeogenesis.
LIPID METABOLISM	
Catabolism	Glycerol may be converted into glucose (gluconeogenesis) or catabolized via glycolysis. Fatty acids are catabolized via beta oxidation into acetyl CoA that is catabolized in the Krebs cycle. Acetyl CoA can also be converted into ketone bodies (ketogenesis).
Anabolism	Synthesis of lipids from glucose and amino acids is lipogenesis. Fats are stored in adipose tissue.
PROTEIN METABOLISM	
Catabolism	Amino acids are deaminated to enter the Krebs cycle and then are oxidized. Amino acids may be converted into glucose (gluconeogenesis) and fatty acids or ketone bodies.
Anabolism	Protein synthesis is directed by DNA and RNA.

MINERALS

Minerals are inorganic substances. They constitute about 4 percent of the total body weight and are concentrated most heavily in the skeleton. Minerals known to perform functions essential to life are calcium, phosphorus, sodium, chlorine, potassium, magnesium, iron, sulfur, iodine, manganese, cobalt, copper, zinc, selenium, and chromium. Other minerals—aluminum, silicon, arsenic, and nickel—are present in the body, but their functions have not yet been determined.

Some minerals, such as calcium, phosphorus, sodium, potassium, chlorine, magnesium, and sulfur are referred to as *macrominerals* because they are needed in the diet at levels of 100 mg per day or more. Other minerals, such as iron, iodine, copper, zinc, and fluorine, are called *microminerals* or *trace elements* because levels of under 100 mg per day are sufficient.

Exhibit 20-2 describes some vital minerals.

EXHIBIT 20-2
Minerals Vital to the Body

Mineral	Comments	Importance
MACROMINERALS		
Calcium	Most abundant cation in body. Appears in combination with phosphorus in ratio of 2:1.5. About 99 percent is stored in bone and teeth. Remainder stored in muscle, other soft tissues, and blood plasma. Blood calcium level controlled by calcitonin (CT) and parathyroid hormone (PTH). Absorption occurs only in the presence of vitamin D. Most is excreted in feces and small amount in urine. Sources are milk, egg yolk, shellfish, green leafy vegetables.	Formation of bones and teeth, blood clotting, muscle contraction and nerve activity, endocytosis and exocytosis, cellular motility, chromosome movement during cell division, glycogen metabolism, synthesis and release of neurotransmitters, and functions as a second messenger.
Phosphorus	About 80 percent found in bones and teeth. Remainder distributed in muscle, brain cells, blood. More functions than most other minerals. Blood phosphorus level controlled by calcitonin (CT) and parathyroid hormone (PTH). Most excreted in urine; small amount eliminated in feces. Sources are dairy products, meat, fish, poultry, nuts.	Formation of bones and teeth. Constitutes a major buffer system of blood. Plays important role in muscle contraction and nerve activity. Component of many enzymes. Involved in transfer and storage of energy (ATP). Component of DNA and RNA.
Sulfur	Constituent of many proteins (such as collagen and insulin) and some vitamins (thiamine and biotin). Excreted in urine. Sources include beef, liver, lamb, fish, poultry, eggs, cheese, beans.	As component of hormones and vitamins, regulates various body activities.
Sodium	Most found in extracellular fluids, some in bones. Excreted in urine and perspiration. Normal intake of NaCl (table salt) supplies required amounts.	As most abundant cation in extracellular fluid, strongly affects distribution of water through osmosis. Part of bicarbonate buffer system. Functions in nerve impulse conduction and muscle contraction.
Potassium	Principal cation in intracellular fluid. Most is excreted in urine. Normal food intake supplies required amounts.	Functions in transmission of nerve impulses and muscle contraction.
Chlorine	Found in extracellular and intracellular fluids. Principal anion of extracellular fluid. Most excreted in urine. Normal intake of NaCl supplies required amounts.	Assumes role in acid–base balance of blood, water balance, and formation of HCl in stomach.
Magnesium	Component of soft tissues and bone. Excreted in urine and feces. Widespread in various foods, such as green leafy vegetables, seafood, and whole-grain cereals.	Required for normal functioning of muscle and nervous tissue. Participates in bone formation. Constituent of many coenzymes. Deficiency linked to diabetes, hypertension, high blood levels of cholesterol, pregnancy problems, and spasms in blood vessels.
MICROMINERALS (TRACE ELEMENTS)		
Iron	About 66 percent found in hemoglobin of blood. Remainder distributed in skeletal muscles, liver, spleen, enzymes. Normal losses of iron occur by shedding of hair, epithelial cells, and mucosal cells, and in sweat, urine, feces, and bile. Sources are meat, liver, shellfish, egg yolk, beans, legumes, dried fruits, nuts, cereals.	As component of hemoglobin, carries O_2 to body cells. Involved in formation of ATP from catabolism. Large amounts of stored iron are associated with an increased risk of cancer. This may be related to the ability of iron to catalyze the production of oxygen radicals and serve as a nutrient for cancer cells.
Iodine	Essential component of thyroid hormones. Excreted in urine. Sources are seafood, cod-liver oil, iodized salt, and vegetables grown in iodine-rich soil.	Required by thyroid gland to synthesize thyroid hormones, hormones that regulate metabolic rate.

EXHIBIT 20-2 (Continued) Minerals Vital to the Body		
Mineral	**Comments**	**Importance**
Copper	Some stored in liver and spleen. Most excreted in feces. Sources include eggs, whole-wheat flour, beans, beets, liver, fish, spinach, asparagus.	Required with iron for synthesis of hemoglobin. Component of enzyme necessary for melanin pigment formation.
Zinc	Important component of certain enzymes. Widespread in many foods, especially meats.	Important in carbon dioxide metabolism. Necessary for normal growth and wound healing, proper functioning of prostate gland, normal taste sensations and appetite, and normal sperm counts in males. As a component of peptidases, it is involved in protein digestion. Deficiency may be involved in impaired immunity and slow learning. Excess may raise cholesterol level.
Fluorine	Component of bones, teeth, other tissues.	Appears to improve tooth structure and inhibit tooth decay. Inhibits bone resorption.
Manganese	Some stored in liver and spleen. Most excreted in feces.	Activates several enzymes. Needed for hemoglobin synthesis, urea formation, growth, reproduction, lactation, bone formation, and possibly production and release of insulin, and inhibiting cell damage.
Cobalt	Constituent of vitamin B_{12}.	As part of B_{12}, required for erythropoiesis.
Chromium	Found in high concentrations in brewer's yeast. Also found in wine and some brands of beer.	Necessary for the proper utilization of dietary sugars and other carbohydrates by optimizing the production and effects of insulin. Helps increase blood levels of HDL, while decreasing levels of LDL.
Selenium	Found in seafood, meat, chicken, grain cereals, egg yolk, milk, mushrooms, and garlic.	An antioxidant. Prevents chromosome breakage and may assume a role in preventing certain birth defects and certain types of cancer (esophageal).

VITAMINS

Vitamins are organic nutrients required in minute amounts to maintain growth and normal metabolism. Unlike carbohydrates, fats, or proteins, vitamins do not provide energy or serve as building materials. They regulate physiological processes, primarily by serving as coenzymes.

Most vitamins cannot be synthesized by the body; they must be ingested in foods or pills. (Vitamin D is an example of a vitamin made by the skin.) Other vitamins, such as vitamin K, are produced by bacteria in the gastrointestinal tract. The body can assemble some vitamins if the raw materials called *provitamins* are provided. Vitamin A is produced by the body from the provitamin carotene, a chemical present in spinach, carrots, liver, and milk. No single food contains all the required vitamins—one of the best reasons for eating a balanced diet.

On the basis of solubility, vitamins are divided into two principal groups: fat-soluble and water-soluble. *Fat-soluble* vitamins are absorbed along with ingested dietary fats by the small intestine. In fact, they cannot be absorbed unless they are ingested with some fat. Fat-soluble vitamins are generally stored in cells, particularly liver cells, so reserves can be built up. The fat-soluble vitamins are A, D, E, and K. *Water-soluble* vitamins, such as vitamins C and B, are absorbed with water in the gastrointestinal tract and dissolve in the body fluids. Excess quantities are excreted in the urine, not stored.

Exhibit 20-3 lists the principal vitamins, their sources, functions, and related disorders.

METABOLISM AND BODY HEAT

We will now consider the relationship of foods to body heat, heat gain and loss, and the regulation of body temperature.

MEASURING HEAT

Heat is a form of energy that can be measured as *temperature* and expressed in units called calories. A *kilocalorie (kcal)* is equal to 1000 cal and is defined as the amount of heat required to raise the temperature of 1000 g of water 1°C (1 kcal = 4.2 kilojoules or kJ, the metric measurement for the energy value of foods). The kilocalorie is the unit we use to express the heating value of foods and to measure the body's metabolic rate. Knowing the caloric value of foods is important; if we know the amount of energy the body uses for various activities, we can adjust our food intake. In this way we can control body weight by taking in only enough kilocalories to sustain our activities.

PRODUCTION OF BODY HEAT

Most of the heat produced by the body comes from oxidation of the food we eat. The rate at which this heat is produced—the *metabolic rate*—is measured in kilocalories. The following factors affect metabolic rate:

1. **Exercise.** During strenuous exercise, the metabolic rate

EXHIBIT 20-3
The Principal Vitamins

Vitamin	Comment and Source	Function	Deficiency Symptoms and Disorders
FAT-SOLUBLE			
A	Formed from provitamin carotene (and other provitamins) in GI tract. Requires bile salts and fat for absorption. Stored in liver. Sources of carotene and other provitamins include yellow and green vegetables; sources of vitamin A include fish-liver oils, milk, butter.	Maintains general health and vigor of epithelial cells. Potential role in cancer prevention is currently under investigation. Essential for formation of rhodopsin, light-sensitive chemical in rods of retina. Aids in growth of bones and teeth by apparently helping to regulate activity of osteoblasts and osteoclasts.	Deficiency results in atrophy and keratinization of epithelium, leading to dry skin and hair, increased incidence of infections, inability to gain weight, drying of cornea and ulceration, nervous disorders, and skin sores. *Night blindness* or decreased ability for dark adaptation. Slow and faulty development of bones and teeth.
D	In presence of sunlight, 7-dehydrocholesterol in the skin is converted to the active form of vitamin D. Dietary vitamin D requires moderate amounts of bile salts and fat for absorption. Stored in tissues to slight extent. Most excreted via bile. Sources include fish-liver oils, egg yolk, fortified milk.	Essential for absorption and utilization of calcium and phosphorus from gastrointestinal tract. May work with parathyroid hormone (PTH) that controls calcium metabolism. Necessary for leucocyte differentiation.	Defective utilization of calcium by bones leads to *rickets* in children and *osteomalacia* in adults. Possible loss of muscle tone.
E (Tocopherols)	Stored in liver, adipose tissue, and muscles. Requires bile salts and fat for absorption. Sources include fresh nuts and wheat germ, seed oils, green leafy vegetables.	Probably functions as an antioxidant. Believed to inhibit catabolism of certain fatty acids that help form cell membranes. Involved in formation of DNA, RNA, and red blood cells. May promote wound healing, contribute to the normal structure and functioning of the nervous system, prevent scarring, and reduce the severity of visual loss associated with retrolental fibroplasia (an eye disease in premature infants caused by too much oxygen in incubators) by functioning as an antioxidant. Believed to help protect liver from toxic chemicals like carbon tetrachloride.	May cause the oxidation of monounsaturated fats, resulting in abnormal structure and function of mitochondria, lysosomes, and plasma membranes, a possible consequence being hemolytic anemia. Deficiency also causes muscular dystrophy in monkeys and sterility in rats.
K	Produced in considerable quantities by intestinal bacteria. Requires bile salts and fat for absorption. Stored in liver and spleen. Other sources include spinach, cauliflower, cabbage, liver.	Coenzyme believed essential for synthesis of prothrombin by liver and several clotting factors. Also known as antihemorrhagic vitamin.	Delayed clotting time results in excessive bleeding.
WATER-SOLUBLE			
B_1 (Thiamine)	Rapidly destroyed by heat. Not stored in body. Excessive intake eliminated in urine. Sources include whole-grain products, eggs, pork, nuts, liver, yeast.	Acts as coenzyme for many different enzymes that break carbon-to-carbon bonds and are involved in carbohydrate metabolism of pyruvic acid to CO_2 and H_2O. Essential for synthesis of acetylcholine.	Deficiency leads to two syndromes: (1) *beriberi*—partial paralysis of smooth muscle of GI tract causing digestive disturbances, skeletal muscle paralysis, atrophy of limbs; (2) *polyneuritis*—due to degeneration of myelin sheaths; reflexes are impaired, impairment of sense of touch, decreased intestinal motility, stunted growth in children, and poor appetite.

EXHIBIT 20-3 (Continued)
The Principal Vitamins

Vitamin	Comment and Source	Function	Deficiency Symptoms and Disorders
B$_2$ (Riboflavin)	Not stored in large amounts in tissues. Most is excreted in urine. Small amounts supplied by bacteria of GI tract. Other sources include yeast, liver, beef, veal, lamb, eggs, whole-grain products, asparagus, peas, beets, peanuts.	Component of certain coenzymes, such as FAD and FMN, concerned with carbohydrate and protein metabolism, especially in cells of eye, integument, mucosa of intestine, blood.	Deficiency may lead to improper utilization of oxygen resulting in blurred vision, cataracts, and corneal ulcerations. Also dermatitis and cracking of skin, lesions of intestinal mucosa, and development of one type of anemia.
Niacin (Nicotinamide)	Derived from amino acid tryptophan. Sources include yeast, meats, liver, fish, whole-grain products, peas, beans, nuts.	Essential component of NAD and NADP coenzymes concerned with oxidation–reduction reactions. In lipid metabolism, inhibits production of cholesterol and assists in fat breakdown.	Principal deficiency is *pellagra*, characterized by dermatitis, diarrhea, and psychological disturbances.
B$_6$ (Pyridoxine)	Synthesized by bacteria of GI tract. Stored in liver, muscle, brain. Other sources include salmon, yeast, tomatoes, yellow corn, spinach, whole-grain products, liver, yogurt.	May function as coenzyme in fat metabolism. Essential coenzyme for normal amino acid metabolism. Assists production of circulating antibodies.	Most common deficiency symptom is dermatitis of eyes, nose, and mouth. Other symptoms are retarded growth and nausea.
B$_{12}$ (Cyanocobalamin)	Only B vitamin not found in vegetables; only vitamin containing cobalt. Absorption from GI tract depends on HCl and intrinsic factor secreted by gastric mucosa. Sources include liver, kidney, milk, eggs, cheese, meat.	Coenzyme necessary for red blood cell formation, formation of amino acid methionine, entrance of some amino acids into Krebs cycle, and manufacture of choline (chemical similar in function to acetylcholine).	Pernicious anemia, neuropsychiatric abnormalities (ataxia, memory loss, weakness, personality and mood changes, and abnormal sensations), and impaired osteoblast activity.
Pantothenic Acid	Stored primarily in liver and kidneys. Some synthesized by bacteria of GI tract. Other sources include kidney, liver, yeast, green vegetables, cereal.	Constituent of coenzyme A essential for transfer of pyruvic acid into Krebs cycle, conversion of lipids and amino acids into glucose, and synthesis of cholesterol and steroid hormones.	Experimental deficiency tests indicate fatigue, muscle spasms, neuromuscular degeneration, insufficient production of adrenal steroid hormones.
Folic Acid	Synthesized by bacteria of GI tract. Other sources include green leafy vegetables and liver.	Component of enzyme systems synthesizing purines and pyrimidines built into DNA and RNA. Essential for normal production of red and white blood cells.	Production of abnormally large red blood cells (macrocytic anemia).
Biotin	Synthesized by bacteria of GI tract. Other sources include yeast, liver, egg yolk, kidneys.	Essential coenzyme for conversion of pyruvic acid to oxaloacetic acid and synthesis of fatty acids and purines.	Mental depression, muscular pain, dermatitis, fatigue, nausea.
C (Ascorbic Acid)	Rapidly destroyed by heat. Some stored in glandular tissue and plasma. Sources include citrus fruits, tomatoes, green vegetables.	Exact role not understood. Promotes many metabolic reactions, particularly protein metabolism, including laying down of collagen in formation of connective tissue. As coenzyme may combine with poisons, rendering them harmless until excreted. Works with antibodies. Promotes wound healing. Role in cancer prevention is under investigation.	Scurvy; anemia; many symptoms related to poor connective tissue growth and repair including tender swollen gums, loosening of teeth (alveolar processes also deteriorate), poor wound healing, bleeding (vessel walls fragile because of connective tissue degeneration), and retardation of growth.

The Metabolic Realities of Weight Control

Weight control is a question of metabolic balance. If calories consumed equal calories expended, there is no net gain or loss in energy stores. Lipid and glycogen stores may be catabolized for energy production, but they are then replenished after the next meal, or in the case of heavy exercise, during the next 24 hours.

People interested in losing weight must manipulate this energy balance equation in some fashion so that the number of calories expended exceeds the number of calories consumed. An energy deficit means that energy stores (one hopes, fat cells) will be used and not replenished.

There are a number of important variables in this energy balance equation. Dieters, ever hopeful of defying thermodynamic laws, are easy prey for devices, drugs, and diets that promise to manipulate these variables in ways conducive to rapid weight loss. An understanding of these variables can help the dieter understand what behaviors truly promote healthful weight loss and why.

Magic Foods and Negative Calories

Some diet schemes suggest that certain foods, such as grapefruit, contain fewer calories than the number required to digest that food. In other words, the more you eat, the more weight you will lose. As you know, this is a metabolic impossibility. This is like the shopping episode from "I Love Lucy" in which the saleswoman keeps reminding Lucy of how much money she is saving as she buys items on sale. Lucy asks the saleswoman to please tell her when she has saved enough to pay for all her selections.

Research does suggest, however, that some calories are more fattening than others. A certain amount of energy is required to convert energy nutrients into storage fuels. For example, it is metabolically very efficient to convert triglyceride molecules from the diet into triglycer-ide molecules in fat cells; the conversion is relatively easy and requires little ATP. On the other hand, the conversion of dietary carbohydrate into adipose triglyceride is metabolically "costly"; it takes several steps and more ATP, so some energy is actually expended during this conversion process. Fats are more fattening, so a person trying to lose weight will benefit by selecting foods low in fat.

Some interesting weight-loss studies have compared the rate of weight loss and long-term success of various caloric-restriction programs. They have generally found that the dietary method does not make too much difference in the rate of weight loss when volunteers are followed for several months. In one study, people who consumed 1400 kcal per day of low-fat foods lost weight almost as quickly as people consuming only 800 kcal per day of a standard liquid weight-loss diet. Which would you rather eat? The people consuming 1400 kcal not only had a more interesting and palatable diet, but they were also learning which foods are nutritionally dense and low in fat. Although dietary method did not predict which subjects would maintain their weight loss, participation in regular exercise did. Those subjects who reported regular aerobic exercise were much more likely to stay at their target weight.

In one study, volunteers were told to eat as much as they wanted of low-fat, high-fiber foods, such as vegetables, fruits, grains, and legumes, while limiting calorically dense and processed foods. Even though no limit was placed on consumption, these volunteers naturally limited themselves to approximately 1400 kcal per day, and lost weight.

Take the Crash Out of Dieting

Diets very low in calories, commonly and appropriately known as crash diets, are not only ineffective weight-loss methods but are harmful to one's health.

These diets typically allow the dieter to consume a very limited and often bizarre selection of foods or food substitutes (usually purchased at a substantial price). Diets like these do nothing to educate the dieter about prudent food choices, so once the diet is over, the dieter usually gains back all the weight lost on the diet, and then some. In fact, while a large part of the weight loss is water and muscle tissue, a greater proportion of the weight regained is often fat, so the person may weigh the same before and after the crash diet but actually be relatively *fatter* after the diet.

Common side effects of very-low-calorie diets are a feeling of deprivation and depression and extremely powerful food cravings. These cravings can lead to food binges, where the dieter wolfs down mountains of forbidden foods. The binge is usually followed by even stricter dieting, greater cravings, and more misery. It is not uncommon for this harmful pattern to evolve into full-fledged eating disorders.

Much of the recent interest in caloric restriction and metabolic rate is a result of the desire to understand why some people have a great deal of difficulty losing weight, even though they may be eating very little food. According to standard metabolic calculations, they should be losing steadily. But they aren't.

It used to be that if you were one of these people who dieted but didn't lose weight, you were suspected of cheating or lying. You were accused of underestimating your food intake, fudging your calorie counts, and failing to follow orders.

But many of the people who do not lose weight on very-low-calorie diets may be experiencing a metabolic adaptation to what their bodies perceive as starvation. In the face of starvation, BMR may decline as the body attempts to conserve energy. This response appears to become stronger the more frequently a person undergoes severe caloric restriction.

Frequent dieting interspersed with periods of normal or greater-than-normal caloric intake are common in people attempting to lose weight. The same 10 or 20 pounds may be lost and regained many times, in what has become known as the rhythm method of girth control, or weight cycling. These people seem to be especially vulnerable to metabolic adaptation to a very-low-calorie diet.

You've heard it before: successful weight-control programs are based on behavior change, on developing nutritious, low-fat, lifelong eating habits you can live with. While some caloric restriction is necessary to create a negative energy balance, a person should not consume fewer than 10 kcal per pound of ideal weight, and never fewer than 1200 kcal. Moderately low-calorie diets do not seem to cause a drop in BMR for most people.

Are the Calories Burned During Exercise Worth All That Hard Work?

The health benefits of exercise arise from the fact that exercise provides a metabolic challenge to the body. The elevation in metabolic rate that occurs during aerobic exercise causes the training response that strengthens the heart, improves serum lipid profile, and burns calories. People unaccustomed to physical activity are sometimes lured into salons claiming that "passive exercise" can achieve the same effects. Machines that move, jiggle, or massage body parts do not elevate metabolic rate, cause the body to expend energy, or use fat.

Dieters are fond of reminding us that to burn off the calories in one piece of pecan pie you'd have to run four miles. All that work for 400 measly calories? Forget it. What dieters don't mention is the fact that exercise confers more benefits than simple caloric expenditure. Exercise helps a person build or maintain muscle tissue. Since muscle tissue is metabolically active, the more muscle you have, the higher your BMR. Crash diets that lead to loss of lean body mass result in a decline in BMR. A moderately low-calorie diet combined with a somewhat vigorous exercise program will result in a negative energy balance and no drop in resting energy expenditure.

Exercise also improves mood and self-concept and decreases one's risk of heart disease, hypertension, and other disorders associated with a sedentary life-style. Exercise is a better way to decrease the tension associated with stress, anxiety, and anger, which many overweight people try to decrease through inappropriate overindulgence in food. And as mentioned above, people who exercise are less likely to regain lost weight, and thus less likely to become yo-yo dieters.

Aerobic exercise is the best type of exercise for burning calories. Resistance exercise, such as weight lifting, does not burn as many calories as aerobic exercise, but since it increases muscle mass, it can help maintain metabolic rate. People who are very overweight should get their physician's exercise recommendations before embarking on a program of vigorous exercise.

increases. It may increase to as much as 15 times the normal rate in well-trained athletes.

2. **Nervous system.** In a stress situation, the sympathetic nervous system causes the release of norepinephrine (NE), which increases metabolic rate.

3. **Hormones.** In addition to norepinephrine (NE), several other hormones affect metabolic rate. Epinephrine is also secreted under stress. Increased secretions of thyroid hormones increase metabolic rate. Testosterone, glucocorticoids, and human growth hormone (hGH) also increase metabolic rate.

4. **Body temperature.** The higher the body temperature, the higher the metabolic rate. Metabolic rate may be substantially increased during fever and strenuous exercise.

5. **Ingestion of food.** The ingestion of food, especially proteins, can raise metabolic rate by as much as 10 to 20 percent.

6. **Age.** The metabolic rate of a child, in relation to its size, is about double that of an elderly person because of the high rates of reactions related to growth. Metabolic rate decreases with age.

7. **Others.** Other factors that affect metabolic rate are sex (lower in females, except during pregnancy and lactation), climate (higher in tropical regions), sleep (lower), and malnutrition (lower).

BASAL METABOLIC RATE (BMR)

Since many factors affect metabolic rate, it is measured under standard conditions designed to reduce or eliminate those factors as much as possible. These conditions of the body are called the **basal state** and the measurement obtained is the **basal metabolic rate (BMR).** Basal metabolic rate is a measure of the rate at which the body breaks down foods (and therefore releases heat). BMR is also a measure of how much thyroxine the thyroid gland is producing, since thyroxine regulates the rate of food breakdown and is not a controllable factor under basal conditions.

LOSS OF BODY HEAT

Body heat is produced by the oxidation of foods we eat. This heat must be removed continuously or body temperature would rise steadily. The principal routes of heat loss are radiation, conduction, convection, and evaporation.

Radiation

Radiation is the transfer of heat from one object to another without physical contact. Your body loses heat by the radiation of heat waves to cooler objects nearby such as ceilings, floors, and walls. If these objects are at a higher temperature, you absorb heat by radiation. About 60 percent of body heat is lost by radiation in a room at 21°C (70°F).

Conduction

In *conduction*, body heat is transferred to a substance or object in contact with the body, such as chairs, clothing, or jewelry. About 3 percent of body heat is lost via conduction.

Convection

Convection is the transfer of heat by the movement of a liquid or gas between areas of different temperature. When cool air makes contact with the body, it becomes warmed and is carried away by convection currents. Then, more cool air makes contact and is carried away. The faster the air moves, the faster the rate of convection. About 15 percent of body heat is lost to the air by convection.

Evaporation

Evaporation is the conversion of a liquid to a vapor. Every gram of water evaporating from the skin takes heat with it. About 22 percent of heat loss occurs through evaporation. Under extreme conditions, about 4 liters (1 gal) of perspiration is produced each hour and this volume can remove 2000 kilocalories of heat from the body. The higher the relative humidity, however, the lower the rate of evaporation.

BODY TEMPERATURE REGULATION

If the amount of heat production equals the amount of heat loss, you maintain a constant body temperature near 37°C (98.6°F). If your heat-producing mechanisms generate more heat than is lost by your heat-losing mechanisms, your body temperature rises. If your heat-losing mechanisms give off more heat than is generated by heat-producing mechanisms, your temperature falls. A too high temperature destroys body proteins, while a too low temperature causes cardiac arrhythmias; both can result in death.

Hypothalamic Thermostat

Body temperature is regulated by mechanisms that attempt to keep heat production and heat loss in balance. The control center for these mechanisms is a group of neurons in the hypothalamus called the preoptic area. If blood temperature rises, these neurons fire impulses more rapidly. If blood temperature goes down, the neurons fire impulses more slowly.

The impulses are sent to two other portions of the hypothalamus: the *heat-losing center*, which initiates a series of responses that lower body temperature, and the *heat-promoting center,* which initiates a series of responses that raise body temperature. The heat-losing center is mainly parasympathetic in function; the heat-promoting center is primarily sympathetic.

Mechanisms of Heat Production

The heat-promoting center discharges impulses that automatically set into operation responses designed to bring body temperature back to normal (Figure 20-8a). A negative feedback cycle operates to maintain homeostasis.

▪ **Vasoconstriction** Sympathetic nerves stimulate blood vessels of the skin to constrict. This decreases the flow of warm blood from the internal organs to the skin, thus decreasing the transfer of heat from the internal organs to the skin and raising internal body temperature.

FIGURE 20-8 Regulation of body temperature by the hypothalamus. (a) Responses to low body temperature. (b) Responses to high body temperature. The preoptic area is the hypothalamic thermostat.

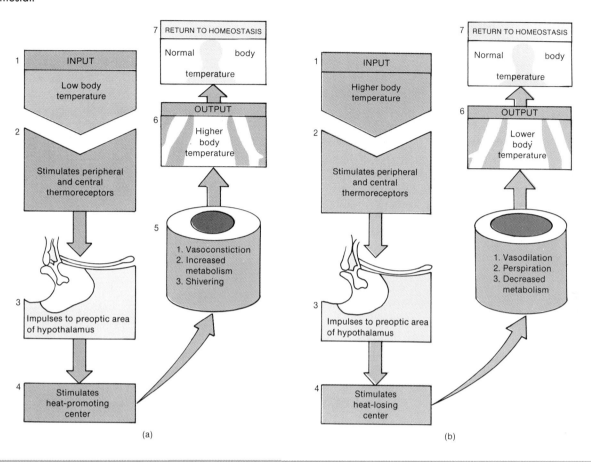

What is the source of most body heat?

■ **Sympathetic Stimulation of Metabolism** Sympathetic nerves stimulate the adrenal medulla to secrete epinephrine and norepinephrine (NE) into the blood. The hormones increase cellular metabolism, which, in turn, increases heat production.

■ **Thyroxine** As a result of a series of steps initiated by the hypothalamic control center, the thyroid gland releases thyroxine into the blood. Since increased levels of thyroxine increase the metabolic rate, body temperature is increased.

■ **Skeletal Muscles** Stimulation of the heat-promoting center causes stimulation of parts of the brain that increase muscle tone. As the muscle tone increases, the stretching of the agonist muscle initiates the stretch reflex and the muscle contracts. This contraction causes the antagonist muscle to stretch, and it too develops a stretch reflex. The repetitive cycle—called ***shivering***—increases the rate of heat production. During maximal shivering, body heat production can rise to about four times the normal rate.

Mechanisms of Heat Loss

When the heat-losing center is stimulated, it inhibits the heat-promoting center. Now, blood vessels dilate. Excess heat is lost to the environment as an increased volume of blood flows into the skin. At the same time, sweat glands produce perspiration, which evaporates and cools the skin, and the metabolic rate and shivering decrease. All these responses reverse the heat-promoting effects and bring body temperature down to normal. This negative feedback cycle attempts to lower body temperature to normal (Figure 20-8b).

Fever

A ***fever*** is an abnormally high body temperature. The most frequent cause of fever is infection from bacteria and viruses. Other causes are heart attacks, tumors, tissue destruction by x-rays, surgery, or trauma, and reactions to vaccines. The mechanism of fever production is believed to occur as follows. When phagocytes, namely, monocytes and macrophages, ingest certain bacteria, a portion of the cell wall of the bacteria is released, causing the phagocytes to secrete interleukin-1. Interleukin-1 circulates to the hypothalamus and induces neurons of the preoptic area to secrete prostaglandins, particularly of the E series. Prostaglandins reset the hypothalamic thermostat at a higher temperature, and temperature-regulating reflex mechanisms will then act to bring the core body temperature up to this new setting. Aspirin, acetaminophen

(Tylenol), and ibuprofen (Medipren) reduce fever by inhibiting synthesis of prostaglandins. (Fever may also be reduced by peripheral cooling in which a cooling blanket is used.)

Suppose that as a result of fever the body thermostat is set at 39.4°C (103°F). Now the heat-promoting mechanisms (vasoconstriction, increased metabolism, shivering) are operating at full force. Thus, even though body temperature is climbing higher than normal—say, 38.3°C (101°F)—the skin remains cold and shivering occurs. This condition, called a *chill*, is a definite sign that body temperature is rising. After several hours, body temperature reaches the setting of the thermostat and the chills disappear. But the body will continue to regulate temperature at 39.4°C (103°F) until the stress is removed. When the stress is removed, the thermostat is reset at normal—37°C (98.6°F). Since body temperature remains high in the beginning, the heat-losing mechanisms (vasodilation and sweating) go into operation to decrease body temperature.

The skin becomes warm and the person begins to sweat. This phase of the fever is called the *crisis* and indicates that body temperature is falling.

Up to a certain point, fever has a beneficial effect on the body. High body temperature is believed to inhibit the growth of some bacteria and viruses. Fever also increases heart rate so that white blood cells are delivered to sites of infection more rapidly and their secretions are increased. In addition, antibody production and T cell proliferation are increased. Also, heat speeds up the rate of chemical reactions that may help body cells repair themselves more quickly during a disease. Among the complications of fever are dehydration, acidosis, and permanent brain damage. As a rule, death results if body temperature rises to 44.4 to 45.5°C (112 to 114°F). At the other end of the scale, death usually results when body temperature falls to 21.2 to 23.9°C (70 to 75°F).

▪ COMMON DISORDERS ▪

Obesity

Obesity is defined as a body weight 10 to 20 percent above the desirable standard due to accumulation of fat. Even moderate obesity is hazardous to health. It is a risk factor in cardiovascular disease, hypertension, pulmonary disease, diabetes mellitus (type II), arthritis, certain cancers (uterine and colon), varicose veins, and gallbladder disease.

Starvation

Starvation is characterized by the loss of energy stores in the form of glycogen, fats, and proteins. It may arise from inadequate intake of nutrients or the inability to digest, absorb, or metabolize ingested nutrients. Starvation may be voluntary, as in fasting or anorexia nervosa, or involuntary, as in deprivation or disease, (such as diabetes mellitus and cancer).

Phenylketonuria (PKU)

Phenylketonuria (fen′-il-kē′-tō-NOO-rē-a) or *PKU* is a genetic error of metabolism characterized by elevated levels of the amino acid phenylalanine in the blood. It is frequently associated with mental retardation. The artificial sweetener aspartame contains phenylalanine; its consumption should be restricted in children with PKU since they cannot metabolize it and it may produce neuronal damage.

Cystic Fibrosis (CF)

Cystic fibrosis (*CF*) is an inherited disease of the exocrine glands that affects the pancreas, respiratory passageways, and salivary and sweat glands. It is the most common lethal genetic disease of Caucasians: 5 percent of the population are thought to be genetic carriers.

S TUDY OUTLINE

Regulation of Food Intake (p. 427)

1. Two centers in the hypothalamus relate to regulation of food intake, the hunger center and satiety center; the hunger center is constantly active but may be inhibited by the satiety center.
2. Among the stimuli that affect the hunger and satiety centers are glucose, amino acids, lipids, body temperature, distention, and cholecystokinin (CCK).

Nutrients (p. 427)

1. Nutrients are chemical substances in food that provide energy,

act as building blocks in forming new body components, or assist in the functioning of various body processes.
2. There are six major classes of nutrients: carbohydrates, lipids, proteins, minerals, vitamins, and water.

Metabolism (p. 427)

1. Metabolism refers to all chemical reactions of the body and has two phases: catabolism and anabolism.
2. Anabolism consists of a series of synthesis reactions whereby small molecules are built up into larger ones that form the

body's structural and functional components. Anabolic reactions use energy.

3. Catabolism refers to decomposition reactions that provide energy.
4. Anabolic reactions require energy, which is supplied by catabolic reactions.
5. Metabolic reactions are catalyzed by enzymes, proteins that speed up chemical reactions without themselves being changed.
6. Oxidation refers to the removal of electrons and hydrogen ions from a molecule; oxidation releases energy.
7. Reduction is the opposite of oxidation; oxidation and reduction reactions are always coupled.

Carbohydrate Metabolism (p. 429)

1. During digestion, polysaccharides and disaccharides are converted to monosaccharides, which are transported to the liver.
2. Carbohydrate metabolism is primarily concerned with glucose metabolism.

Fate of Carbohydrates (p. 429)
1. Some glucose is oxidized by cells to provide energy; it moves into cells by facilitated diffusion; insulin stimulates glucose movement into cells.
2. Excess glucose can be stored by the liver and skeletal muscles as glycogen or converted to fat.

Glucose Catabolism (p. 429)
1. Glucose oxidation is also called cellular respiration.
2. The complete oxidation of glucose to CO_2 and H_2O involves glycolysis, the Krebs cycle, and the electron transport chain.

Glycolysis (p. 429)
1. Glycolysis is also called anaerobic respiration because it occurs without oxygen.
2. Glycolysis refers to the breakdown of glucose into two molecules of pyruvic acid.
3. When oxygen is in short supply, pyruvic acid is converted to lactic acid; under aerobic conditions, pyruvic acid enters the Krebs cycle.
4. Glycolysis yields two molecules of ATP.

Krebs Cycle (p. 430)
1. The Krebs cycle begins when pyruvic acid is converted to acetyl coenzyme A.
2. Then a series of oxidations and reductions of various organic acids take place; coenzymes are reduced.
3. The energy originally in glucose and then pyruvic acid is now in the reduced coenzymes.

Electron Transport Chain (p. 430)
1. The electron transport chain is a series of oxidation–reduction reactions in which the energy in the coenzymes is liberated and transferred to ATP for storage.
2. The complete oxidation of glucose can be represented as follows:

$$\text{Glucose} + O_2 \rightarrow 38 \text{ ATP} + CO_2 + H_2O$$

Glucose Anabolism (p. 430)
1. The conversion of glucose to glycogen for storage in the liver and skeletal muscle is called glycogenesis. It occurs extensively in liver and skeletal muscle fibers (cells) and is stimulated by insulin.
2. The body can store about 500 g of glycogen.
3. The conversion of glycogen back to glucose is called glycogenolysis; it occurs between meals.
4. Gluconeogenesis is the conversion of fat and protein molecules into glucose.

Lipid Metabolism (p. 430)

1. Lipids are second to carbohydrates as a source of energy.
2. During digestion, fats are ultimately broken down into fatty acids and glycerol.

Fate of Lipids (p. 430)
1. Some fats may be oxidized to produce ATP.
2. Some fats are stored in adipose tissue.
3. Other lipids are used as structural molecules or to synthesize essential molecules. Examples include phospholipids of plasma membranes, lipoproteins that transport cholesterol, thromboplastin for blood clotting, and cholesterol used to synthesize bile salts and steroid hormones.

Fat Storage (p. 431)
1. Fats are stored in adipose tissue.
2. Most storage occurs in the subcutaneous layer.

Lipid Catabolism (p. 431)
1. Fat must be split into fatty acids and glycerol before it can be catabolized.
2. Glycerol can be converted into glucose by conversion into glyceraldehyde-3-phosphate.
3. Fatty acids are catabolized through beta oxidation, yielding acetyl coenzyme A, which can enter the Krebs cycle.
4. The formation of ketone bodies by the liver is a normal phase of fatty acid catabolism, but an excess of ketone bodies, called ketosis, may cause acidosis.

Lipid Anabolism: Lipogenesis (p. 434)
1. The conversion of glucose or amino acids into lipids is called lipogenesis.
2. The process is stimulated by insulin.

Protein Metabolism (p. 434)

1. During digestion, proteins are broken down into amino acids.
2. Protein anabolism and catabolism must be balanced through daily dietary intake to prevent protein depletion.

Fate of Proteins (p. 434)
1. Amino acids that enter cells are almost immediately synthesized into proteins.
2. Proteins function as enzymes, hormones, structural elements, and so forth; are stored as fat or glycogen; or are used for energy.

Protein Catabolism (p. 435)

1. Before amino acids can be catabolized, they must be converted to substances that can enter the Krebs cycle.
2. Amino acids may also be converted into glucose, fatty acids, and ketone bodies.

Protein Anabolism (p. 435)

1. Protein synthesis is directed by DNA and RNA and carried out on the ribosomes of cells.
2. Before protein synthesis can occur, all the essential and nonessential amino acids must be present.

Regulation of Metabolism (p. 435)

1. Absorbed nutrients may be oxidized, stored, or converted, based on the needs of the body.
2. The pathway taken by a particular nutrient is controlled by enzymes and regulated by hormones.

Minerals (p. 436)

1. Minerals are inorganic substances that help regulate body processes. They are classified as macrominerals and microminerals.
2. Minerals known to perform essential functions are calcium, phosphorus, sodium, chlorine, potassium, magnesium, iron, sulfur, iodine, manganese, cobalt, copper, zinc, selenium, and chromium. Their functions are summarized in Exhibit 20-2.

Vitamins (p. 437)

1. Vitamins are organic nutrients that maintain growth and normal metabolism. Many function in enzyme systems.
2. Fat-soluble vitamins (A, D, E, K) are absorbed with fats.
3. Water-soluble vitamins (B, C) are absorbed with water.
4. The functions and deficiency disorders of the principal vitamins are summarized in Exhibit 20-3.

Metabolism and Body Heat (p. 437)

1. A kilocalorie (kcal) is the amount of energy required to raise the temperature of 1000 g of water 1°C from 14 to 15°C.

2. The kilocalorie is the unit of heat used to express the caloric value of foods and to measure the body's metabolic rate.

Production of Body Heat (p. 437)

1. Most body heat is a result of oxidation of the food we eat. The rate at which this heat is produced is known as the metabolic rate.
2. Metabolic rate is affected by exercise, the nervous system, hormones, body temperature, ingestion of food, age, sex, climate, sleep, and malnutrition.
3. Measurement of the metabolic rate under conditions designed to minimize influential factors is called the basal metabolic rate (BMR).

Loss of Body Heat (p. 442)

1. Radiation is the transfer of heat from one object to another without physical contact.
2. Conduction is the transfer of body heat to a substance or object in contact with the body.
3. Convection is the transfer of body heat by the movement of air that has been warmed by the body.
4. Evaporation is the conversion of a liquid to a vapor, as in perspiration.

Body Temperature Regulation (p. 442)

1. A normal body temperature is maintained by a delicate balance between heat-producing and heat-losing mechanisms.
2. The hypothalamic thermostat exists as a group of neurons (preoptic area) that stimulate the heat-losing and heat-promoting centers, also in the hypothalamus.
3. Mechanisms that produce heat are vasoconstriction, sympathetic stimulation, skeletal muscle contraction, and thyroxine production.
4. Mechanisms of heat loss include vasodilation, decreased metabolic rate, decreased skeletal muscle contraction, and perspiration.
5. Fever is an abnormally high body temperature.
6. Fever is caused by bacteria, viruses, heart attacks, tumors, tissue destruction by x-rays and trauma, and reactions to vaccines.

REVIEW QUESTIONS

1. How is food intake regulated? (p. 427)
2. Define a nutrient. List the six classes of nutrients and indicate the function of each. (p. 427)
3. What is metabolism? Distinguish between anabolism and catabolism and give examples of each. (p. 427)
4. How does ATP couple anabolism and catabolism? (p. 428)
5. List and explain the characteristics of an enzyme. (p. 428)
6. Define oxidation and reduction and explain their importance. (p. 428)
7. How are carbohydrates absorbed and what are their fates in the body? How does glucose move into body cells? (p. 429)
8. Define glycolysis. Describe its principal events and outcome. (p. 429)
9. Outline the principal events and outcomes of the Krebs cycle. (p. 430)
10. Explain what happens in the electron transport chain. (p. 430)
11. Define glycogenesis and glycogenolysis. Under what circumstances does each occur? (p. 430)

A LOOK AHEAD

FLUID COMPARTMENTS AND FLUID
    BALANCE
WATER
    Fluid Intake and Output
    Regulation of Intake
    Regulation of Output
ELECTROLYTES
    Concentration
    Distribution
    Functions and Regulation
        Sodium
        Potassium
        Calcium
        Magnesium
        Chloride
        Phosphate
MOVEMENT OF BODY FLUIDS
    Between Plasma and Interstitial Com-
        partments
    Between Interstitial and Intracellular
        Compartments
ACID–BASE BALANCE
    Buffer Systems
        Carbonic Acid–Bicarbonate
        Phosphate
        Hemoglobin
        Protein
    Respiration
    Kidney Excretion
ACID–BASE IMBALANCES

The term **body fluid** refers to body water and its dissolved substances. Fluid composes 45 to 75 percent of body weight.

FLUID COMPARTMENTS AND FLUID BALANCE

About two-thirds of body fluid is within cells; it is called **intracellular fluid (ICF)**. The other third, called **extracellular fluid (ECF)**, includes all other body fluids (see Figure 3-3). ECF includes interstitial fluid; plasma; lymph; cerebrospinal fluid; gastrointestinal tract fluids; synovial fluid; the fluids of the eyes and ears; pleural, pericardial, and peritoneal fluids; and glomerular filtrate.

Body fluids are separated into distinct compartments by the selectively permeable membranes of cells, but they are in constant motion from one compartment to another. Also, homeostatic mechanisms keep the fluid volume in each compartment relatively stable.

The term *fluid balance* means that the proportion of water in each fluid compartment is kept in homeostasis. Fluid balance is essential for health because water is the main component of all body fluids.

Water constitutes the bulk of all the body fluids. When we say that the body is in fluid balance, we mean that the required amount of water has been distributed to the various compartments according to their needs.

Osmosis is the primary way in which water moves in and out of the body compartments. The concentration of solutes in the fluids is therefore a major determinant of fluid balance. Most solutes in body fluids are electrolytes—substances that dissociate into ions. Fluid balance, then, means water balance, but it also implies electrolyte balance. The two are inseparable.

WATER

Water makes up from 45 to 75 percent of total body weight. The percentage depends on age and the amount of body fat. Since fat is basically water-free, lean people have a greater proportion of water than fat people. Water proportion also decreases with age. An infant has the highest amount of water per body weight. In normal adult males, water averages about 65 percent of body weight. Because females have more subcutaneous fat, their water averages about 55 percent. The functions of water may be reviewed in Chapters 2 and 20.

FLUID INTAKE AND OUTPUT

The primary source of body fluid is derived from ingested liquids (1600 ml) and foods (700 ml) that have been absorbed from the gastrointestinal tract. This is called **preformed water** and amounts to about 2300 ml/day. Another source is **metabolic water**, the water produced through catabolism which amounts to about 200 ml/day. Thus, total fluid input averages about 2500 ml/day (Figure 22-1).

Under normal conditions, fluid intake equals fluid output, so the body maintains a constant volume. The approximate amounts of water lost from each avenue of fluid output are 1500 ml/day by the kidneys as urine, 500 ml/day by the skin most through evaporation (400 ml/day) and some through perspiration (100 ml/day), 300 ml/day by the lungs as water vapor, and 200 ml/day by the gastrointestinal tract as feces. Therefore, total fluid output is about 2500 ml/day (Figure 22-1).

REGULATION OF INTAKE

Fluid intake is regulated by thirst. When water loss is greater than water intake, the resulting *dehydration* stimulates thirst and the desire to drink through both local and general responses. Locally, it leads to a decrease in saliva that produces

FIGURE 22-1 Summary of fluid intake and output per day under normal conditions.

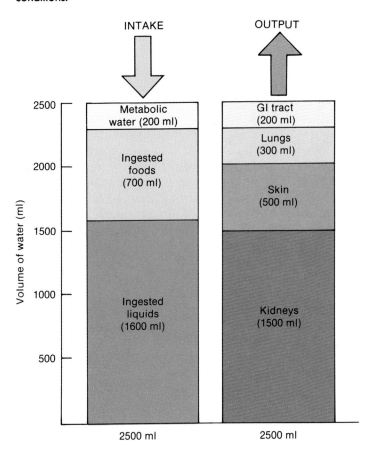

How would each of the following affect fluid balance: Hyperventilation? Vomiting? Fever? Diuretics?

dryness of the mouth (Figure 22-2). Dryness is interpreted by the brain as thirst. Dehydration also raises blood osmotic pressure, which stimulates receptors (osmoreceptors) in the thirst center of the hypothalamus to initiate impulses that are also interpreted as thirst. The subsequent desire to drink can result in balancing the fluid loss.

When one drinks, the sensation of thirst is then inhibited by (1) wetting the oral mucosa and (2) inhibiting the thirst center in the hypothalamus. The latter is caused by both decreased blood osmotic pressure and impulses from stretch receptors in the stomach and intestinal walls.

REGULATION OF OUTPUT

Under normal circumstances, fluid output is adjusted by antidiuretic hormone (ADH) and aldosterone, both of which regulate urine production (see Chapter 21). Under abnormal conditions, other factors may influence output. If the body is dehydrated, blood pressure falls, the glomerular filtration rate decreases accordingly, and water is conserved. Conversely, excessive blood volume results in an increase of blood pressure, glomerular filtration rate, and

fluid output. Hypertension (high blood pressure) produces the same effect. Fluid loss may be accelerated by hyperventilation (rapid breathing), vomiting, diarrhea, excess perspiration, hemorrhage, or burns that permit excess fluid loss. Lack of ADH results in diabetes insipidus, in which there is excessive loss of urine.

ELECTROLYTES

Body fluids contain a variety of dissolved chemicals. Some are compounds with covalent bonds; that is, the atoms that compose the molecule share electrons and do not form ions. They are called **nonelectrolytes** and include most organic compounds, such as glucose, urea, and creatine.

Other compounds, called **electrolytes**, have at least one ionic bond. When they dissolve in a fluid, they dissociate into positive ions (**cations**) and negative ions (**anions**). Acids, bases, and salts are electrolytes. Most electrolytes are inorganic compounds, but a few are organic. For example, when some proteins are put in solution, the ion detaches and the rest of the protein molecule carries the opposite charge.

Electrolytes serve three general functions in the body. First, many are essential minerals. Second, they control the osmosis of water between body compartments. Third, they help maintain the acid–base balance required for normal cellular activities.

CONCENTRATION

In osmosis, water moves from an area with fewer particles in solution to an area with more particles in solution. A particle may be a whole molecule or an ion. An electrolyte exerts a far greater effect on osmosis than a nonelectrolyte because an electrolyte breaks apart into at least two particles, both of them charged. Suppose the nonelectrolyte glucose and two different electrolytes are placed in solution:

FIGURE 22-2 Regulation of fluid volume by the adjustment of intake to output.

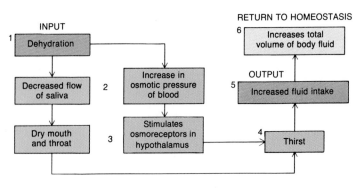

What body mechanisms prevent overhydration by inhibiting thirst and the desire to drink?

$$C_6H_{12}O_6 \xrightarrow{H_2O} C_6H_{12}O_6 = \text{one particle}$$

$$\underset{\substack{\text{Sodium} \\ \text{chloride}}}{NaCl} \xrightarrow{H_2O} Na^+ + Cl^- = \text{two particles}$$

$$\underset{\substack{\text{Calcium} \\ \text{chloride}}}{CaCl_2} \xrightarrow{H_2O} Ca^{2+} + Cl^- + Cl^- = \text{three particles}$$

Because glucose does not break apart when dissolved in water, a molecule of glucose contributes only one particle to the solution. Sodium chloride, on the other hand, contributes two particles, or ions, and calcium chloride contributes three. Thus, calcium chloride has three times as great an effect on solute concentration as glucose.

Just as important, once the electrolyte breaks apart, its ions can attract other ions of the opposite charge. If equal amounts of

Ca^{2+} and Na^+ are placed in solution, the calcium ions will attract about twice as many chloride ions to their area as the sodium ions.

To determine how much effect an electrolyte has on concentration, it is necessary to know its concentration in a solution. The concentration of an ion is commonly expressed in *milliequivalents per liter (mEq/liter)*— the number of electrical charges in each liter of solution. The mEq/liter equals the number of ions in solution times the number of charges the ions carry.

DISTRIBUTION

Figure 22-3 compares the principal chemical constituents of plasma, interstitial fluid, and intracellular fluid. The chief difference between plasma and interstitial fluid is that plasma contains many protein anions, whereas interstitial fluid has hardly any. Since capillary

FIGURE 22-3 Comparison of electrolyte concentrations in plasma, interstitial fluid, and intracellular fluid. The height of each column represents the total electrolyte concentration.

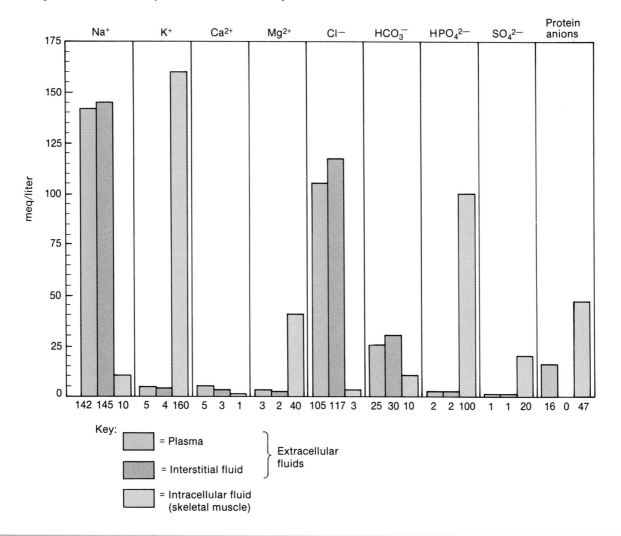

What are three major differences between intracellular and extracellular fluid?

membranes are practically impermeable to protein, the protein stays in the plasma and does not move out of the blood into the interstitial fluid. Plasma also contains more Na^+ ions and fewer Cl^- ions than interstitial fluid. In other respects the two fluids are similar.

Intracellular fluid differs considerably from extracellular fluid, however. In extracellular fluid, the most abundant cation is sodium (Na^+) and the most abundant anion is chloride (Cl^-). In intracellular fluid, the most abundant cation is potassium (K^+) and the most abundant anion is phosphate (HPO_4^{2-}). Also, there are many more protein anions in intracellular fluid than in extracellular fluid.

FUNCTIONS AND REGULATION

Sodium

Sodium (Na^+), the most abundant extracellular ion, represents about 90 percent of extracellular cations. Normal blood (serum) sodium is about 142 mEq/liter. Sodium is necessary for the transmission of action potentials in nervous and muscle tissue. It is a component of buffer systems and it plays a significant role in fluid and electrolyte balance by creating most of the osmotic pressure of extracellular fluid (ECF).

The kidneys normally excrete excess sodium and conserve it during periods of sodium restriction. The sodium level in the blood is controlled primarily by the hormone aldosterone from the adrenal cortex (see Figure 13-16). Aldosterone acts on the distal convoluted tubules of the nephron of the kidneys, causing them to increase their reabsorption of sodium. The sodium thus moves from the filtrate back into the blood and establishes an osmotic gradient that causes water to follow from the filtrate back into the blood as well. Aldosterone is secreted in response to reduced blood volume or cardiac output, decreased extracellular sodium, increased extracellular potassium, and physical stress. A decrease in sodium concentration inhibits the release of antidiuretic hormone (ADH) by the pituitary gland, which, in turn, permits more water to be excreted in urine and the sodium level to be restored to normal. The atria of the heart produce the hormone atrial natriuretic factor (ANF) that also increases sodium and water excretion by the kidneys.

Sodium may be lost from the body through excessive perspiration, vomiting, diarrhea, and burns. A lower than normal blood sodium level can result in *hyponatremia* (hī′-pō-na-TRĒ-mē-a; *natrium* = sodium). It is characterized by muscular weakness, dizziness, headache, hypotension, tachycardia, and shock. Severe sodium loss can result in mental confusion, stupor, and coma. A higher-than-normal blood sodium level, called *hypernatremia*, may occur because of excess water loss, water deprivation, or sodium intake. The average daily intake of sodium far exceeds the body's normal daily requirements. Since sodium is the major determinant of the osmotic pressure of ECF, too high sodium levels cause water to move out of body cells into ECF, resulting in cellular dehydration. Symptoms include intense thirst, fatigue, restlessness, agitation, and coma.

Potassium

Potassium (K^+) is the most abundant cation in intracellular fluid. Normal blood (serum) potassium is about 5.0 mEq/liter. Potassium

assumes a key role in the functioning of nervous and muscle tissue. Abnormal serum potassium levels adversely affect neuromuscular and cardiac function. Potassium helps maintain fluid volume in cells and regulate extracellular fluid pH. When potassium ions move out of the cell, they are replaced by sodium and hydrogen ions; it is the shift of hydrogen ions that affects pH.

The blood level of potassium is under the control of mineralocorticoids, mainly aldosterone. When potassium concentration is high, more aldosterone is secreted and more potassium is excreted. This process occurs in the distal convoluted tubules of the kidneys by way of tubular secretion.

A higher-than-normal blood potassium level, called *hyperkalemia*, is characterized by irritability, anxiety, abdominal cramping, diarrhea, weakness (especially of the lower extremities), and paresthesia (abnormal sensation, such as burning or prickling). Hyperkalemia can also cause death from fibrillation of the heart.

A lower-than-normal level of potassium, called *hypokalemia* (hī′-pō-ka-LĒ-mē-a; *kalium* = potassium), may result from vomiting, diarrhea, high sodium intake, kidney disease, or taking certain diuretics. Symptoms include cramps, fatigue, flaccid paralysis, nausea, vomiting, mental confusion, increased urine output, shallow respirations, and changes in the electrocardiogram.

Calcium

Calcium (Ca^{2+}) is the most abundant ion in the body; it is principally an extracellular electrolyte. Most calcium in the body is combined with phosphate (HPO_4^{2-}) as a mineral salt. Normal blood calcium is about 5.0 mEq/liter. About 98 percent of the calcium in the adult is present in the skeleton (and teeth) as calcium and phosphorus salts. The remaining calcium is found in extracellular fluid and cells of all tissues, especially skeletal muscle. Calcium is not only a structural component of bones and teeth and a second messenger, it is also involved in blood clotting, neurotransmitter release, neuromuscular conduction, maintenance of muscle tone, and excitability of nervous and muscle tissue.

Calcium blood levels are regulated principally by parathyroid hormone (PTH) and calcitonin (CT) (see Figure 13-13). PTH is released when calcium blood level is low. PTH stimulates osteoclasts to release calcium (and phosphate) into the blood. PTH also helps to increase absorption of calcium from the gastrointestinal tract and increases the rate at which calcium is reabsorbed in the kidneys and returned to the blood. CT, from the thyroid gland, is released when blood calcium levels are high. It decreases blood level of calcium by stimulating osteoblasts and inhibiting osteoclasts. In the presence of CT, osteoblasts remove calcium (and phosphate) from blood and deposit them in bone.

An abnormally low level of calcium is called *hypocalcemia* (hī′-pō-kal-SĒ-mē-a). It may be due to increased calcium loss, reduced calcium intake, elevated levels of phosphate (as one goes up, the other goes down), or altered regulation as might occur in hypoparathyroidism. It is characterized by numbness and tingling of the fingers, hyperactive reflexes, muscle cramps, tetany, and convulsions; it may also lead to bone fractures. Hypocalcemia may also cause spasms of laryngeal muscles that can cause death by asphyxiation.

Hypercalcemia, an abnormally high level of calcium, is characterized by lethargy, weakness, anorexia, nausea, vomiting, polyuria, itching, bone pain, depression, confusion, paresthesia, stupor, and coma.

Magnesium

Magnesium (Mg^{2+}) is primarily an intracellular electrolyte. Normal blood (serum) magnesium is about 3.0 mEq/liter. In an adult, about 50 percent of the body's magnesium is in bone; about 45 percent is in intracellular fluid; and about 5 percent is in extracellular fluid. Magnesium activates enzymes for the metabolism of carbohydrates and proteins, triggers the sodium–potassium pump, and preserves the structure of DNA, RNA, and ribosomes. It is also important in neuromuscular activity, neural transmission in the central nervous system (CNS), and myocardial functioning.

Magnesium level is regulated by aldosterone. When magnesium concentration is low, increased aldosterone secretion acts on the kidneys so that more magnesium is reabsorbed.

Magnesium deficiency, called *hypomagnesemia* (hī'- pō-mag'-ne-SĒ-mē-a), may be caused by malabsorption, diarrhea, alcoholism, malnutrition, excessive lactation, diabetes mellitus, or diuretics. Symptoms include weakness, irritability, tetany, delirium, convulsions, confusion, anorexia, nausea, vomiting, paresthesia, and cardiac arrhythmias.

Hypermagnesemia, or magnesium excess, occurs almost exclusively in persons with renal failure who take medications containing magnesium. Other causes include Addison's disease, acute diabetic acidosis, severe dehydration, and hypothermia. Symptoms include flaccidity, hypotension, muscle weakness or paralysis, nausea, vomiting, and altered mental functioning.

Chloride

Chloride (Cl^-) is the major extracellular anion. However, it easily diffuses between extracellular and intracellular compartments. This makes chloride important in regulating osmotic pressure differences between the compartments. Also, in the gastric mucosal glands, chloride combines with hydrogen to form hydrochloric acid (HCl). Normal blood (serum) chloride is about 105 mEq/liter.

Part of chloride regulation is indirectly under the control of aldosterone. Aldosterone regulates sodium reabsorption, and chloride follows sodium passively since the negatively charged Cl^- ions follow the positively charged Na^+ ions.

An abnormally low level of chloride in the blood, called *hypochloremia* (hī'-pō-klō-RĒ-mē-a), may be caused by excessive vomiting, dehydration, and certain diuretics. Symptoms include muscle spasms, alkalosis, depressed respirations, and even coma.

Phosphate

Phosphate (HPO_4^{2-}) is principally an intracellular electrolyte. Normal blood (serum) phosphate is about 2.0 mEq/liter. About 85 percent of the phosphate in an adult is in bones and teeth as calcium phosphate salts. The remainder is mostly combined with lipids (phospholipids), proteins, carbohydrates, and other organic molecules to form cellular membranes and to synthesize nucleic acids (DNA and RNA) and high-energy compounds such as adenosine triphosphate (ATP). In addition, it plays an important role in buffering reactions (the phosphate buffer system is discussed later in the chapter).

Phosphate blood levels are regulated by PTH and CT. PTH stimulates osteoclasts to release phosphate from mineral salts of bone matrix and causes renal tubular cells to excrete phosphate. CT lowers high phosphate levels by stimulating osteoblasts and inhibiting osteoclasts. In the presence of CT, osteoblasts remove phosphate from blood, combine it with calcium, and deposit it in bone.

An abnormally low level of phosphate, called *hypophosphatemia* (hī'-pō-fos'-fa-TĒ-mē-a), may occur from polyuria, decreased intestinal absorption, increased utilization of phosphate, or alcoholism. Symptoms include confusion, seizures, coma, chest and muscle pain, numbness and tingling of the fingers, lack of coordination, memory loss, and lethargy.

Hyperphosphatemia occurs most often in response to renal insufficiency because the kidneys increase the intake of phosphates or fail to excrete excess phosphate. The primary complication is the accumulation of calcium phosphate in soft tissues, joints, and arteries. Symptoms include anorexia, nausea, vomiting, muscular weakness, hyperreflexes, tetany, and tachycardia.

MOVEMENT OF BODY FLUIDS

BETWEEN PLASMA AND INTERSTITIAL COMPARTMENTS

The movement of fluid between plasma and interstitial compartments occurs across capillary membranes. This movement was discussed in general in Chapter 16. At this point, we will discuss it in more detail.

The movement of water and dissolved substances, except proteins, through capillary walls is mostly by diffusion and is due to several opposing forces, or pressures. Some forces push fluid out of capillaries into the surrounding interstitial (tissue) spaces. To prevent fluid from moving in one direction only and accumulating in interstitial spaces, opposing forces push fluid from interstitial spaces into blood capillaries. The pressures involved are called hydrostatic and osmotic pressures (Figure 22-4).

Hydrostatic pressures are due to the pressure of water in the fluids. Blood pressure in capillaries, called *blood hydrostatic pressure (BHP)*, tends to move fluids out of capillaries into interstitial fluid. Opposing BHP is the pressure of the interstitial fluid, called *interstitial fluid hydrostatic pressure (IFHP),* which tends to move fluid out of interstitial spaces into capillaries.

Osmotic pressures are due to the presence of proteins that are too large to diffuse in blood and interstitial fluid. *Blood osmotic pressure (BOP)* tends to move fluid from interstitial spaces into capillaries. Opposing BOP is *interstitial fluid osmotic pressure (IFOP),* which tends to move fluid out of capillaries into interstitial fluid.

Whether fluids enter or leave capillaries depends on how these four pressures relate to each other. The difference between the two forces that move fluid out of plasma and the two forces that push it into plasma is the *net filtration pressure (NFP)*:

$$NFP = (BHP + IFOP) - (IFHP + BOP)$$

The NFP at the arterial end of a capillary is 8 mm Hg; at the

Endurance Exercise: A Challenge to Fluid and Electrolyte Balance

We've come a long way since the days when the coach handed out salt tablets to the players and warned them not to drink any water during practice. Heavy and/or prolonged physical activity can lead to dehydration and disrupt fluid and electrolyte homeostasis, so coaches are right to be concerned. But our methods of preventing and treating exercise-induced dehydration have improved considerably, thanks to the research of exercise physiologists. Their laboratory techniques have allowed us to closely monitor changes in plasma volume and electrolyte concentrations under a variety of exercise conditions.

During exercise, a great deal of heat is generated because of muscle contraction. In fact, heat production by active muscles can be up to 100 times greater than that of muscles at rest. Recall from Chapter 20 that the body can get rid of this extra heat by increasing blood flow to the skin, where heat is given off by radiation and convection, and by activating the sweat glands to increase heat loss by evaporation. In very hot weather, radiation and convection do not work, so the body must rely primarily on evaporation to cool itself. Strenuous exercise in hot weather may cause the loss of over 2 liters of water per hour from the skin and lungs. Such losses can lead to dehydration and hyperthermia if fluids are not replaced.

Dehydration

Dehydration refers to a loss of body fluid that amounts to 1 percent or more of total body weight. It is most common during exercise in the heat but can also occur during very low levels of physical activity in a hot environment or during strenuous exercise in a thermally neutral environment. Fluid deficits of 5 percent are common in such athletic events as football, soccer, tennis, and distance running. At this level, symptoms include irritability, fatigue, and loss of appetite. Dehydration levels greater than 7 percent may cause prostration.

As dehydration occurs, water is lost from all body compartments. The decrease in blood volume has deleterious effects on athletic performance, since it decreases the amount of blood the heart can pump per beat. Muscles need oxygen to work, so performance decreases as cardiac output is reduced. The body tries to maintain blood volume to the muscles by constricting vessels in the skin, so less heat is lost and body temperature rises. Intracellular electrolyte changes may also occur with dehydration and interfere with optimal performance.

Thirst is the body's signal that its water level is getting too low, and it motivates a person to drink. Unfortunately, thirst is not a reliable indicator of fluid needs. People tend to drink just enough to relieve their parched throats. This is especially true in hot weather, when more fluid than usual is lost through sweating. The thirst mechanism is especially unreliable in children and older adults. Aging also decreases the kidney's ability to retain water when the body needs fluids, which increases the susceptibility of older people to dehydration.

Rehydration

Nutritionists advise that most people can meet their daily fluid needs with 6 to 8 cups of various combinations of water, juices, and milk. Alcoholic and caffeinated beverages act as diuretics and thus *increase* fluid needs, so these should not be counted as part of the recommended intake. One of the ways to tell whether or not you are drinking enough fluid is to check the color of your urine. Dilute urine is very pale, and indicates sufficient fluid intake.

People experiencing dehydration may find that plain water is not the optimal solution. Studies have shown that when a dehydrated person consumes water, the water dilutes the blood as plasma volume is replenished.

This removes the feeling of thirst and protects against low plasma electrolyte levels. In other words, as electrolyte levels drop, the sensation of thirst goes away, so the blood will not become any more dilute. Water moves from the plasma into compartments outside the blood, since they contain less water and more electrolytes. The kidney senses the increase in fluids in renal tubules and begins to excrete water. In laboratory studies, subjects consuming plain water after dehydration needed to urinate long before hydration was complete. In other words, although plasma volume increased to some extent when water was consumed, it did not return to its desirable level, and subjects did not rehydrate all body cells and extracellular compartments.

Enter Gatorade. When sodium is taken along with water, dehydrated subjects rehydrate to a greater level than subjects taking only water. Sodium helps to restore plasma volume and to retain water in the blood without inhibiting thirst. This observation has prompted extensive research into the ideal sports drink, designed to help the athlete recover fluid and electrolyte homeostasis.

A typical 8-ounce sports drink contains about 50 to 100 mg of sodium. While some sodium is lost in sweat, the amount is quite small, especially if the person is acclimated to the heat. The National Research Council recommends a daily sodium intake of 1100 to 3300 mg for most adults. Some nutritionists have expressed concern that since most Americans consume 10 to 60 times this daily sodium requirement, sports drinks add insult to injury. Sports drinks certainly aren't necessary for the recreational athlete who plays a leisurely game of tennis doubles or a person who walks briskly for half an hour.

You are a candidate for such a drink if you exercise to the point of dehydration. Your body weight is a good measure of hydration. Weigh yourself before and after exercise. For each pound of body weight lost, drink 2 cups of fluids. Use sports drinks only if you lose more than 1 percent of your body weight, and you are not on a restricted salt diet.

Sports drinks also contain some amount of carbohydrate. You'll recall from Chapter 20 that glycogen is a preferred source of energy during physical activity. Glycogen stores need to be replenished following prolonged exercise, so the sports drinks people add carbohydrate to some drinks to supply energy. Many of these drinks contain a glucose polymer, an easily digestible form of complex carbohydrate. People performing heavy physical activity for longer than 60 to 90 minutes experience depletion of muscle glycogen. They are able to work better and longer when they consume sport drinks containing carbohydrate. Needless to say, an athlete should never try something new on the day of a race or contest. Some people have reported gastric distress after consuming some drinks. Therefore, athletes should use these drinks regularly if dehydration is a problem, and not just on race day.

Carbohydrate supplies calories, and extra calories are made into fat, so exercisers who are watching their calorie intake should not forget to count the calories in these drinks.

Carbohydrate also increases the palatability of the drink. One study showed that when subjects were given unlimited access to either a carbohydrate drink or plain water during recovery from exercise, subjects tended to drink more of the former. If it tastes good, we're likely to drink more of it.

Recommendations

Dehydration is easily prevented. Athletes and people who work outdoors in the heat should avoid the hottest part of the day and dress appropriately. People adjusted to the heat are less likely to suffer from dehydration. If weight drops more than 2 to 3 percent after physical activity, the person should not continue working or working out until rehydrated.

People adjusted to the heat are less likely to suffer from dehydration. Physical conditioning helps to prevent dehydration, since training increases blood volume. A higher blood volume means that a worker or athlete can lose more water before dehydration affects cardiac output. A greater blood volume also permits a better heat transfer from the working muscles to the environment during physical activity, since more blood is available for circulation to the skin. Physical conditioning also causes changes in the function of sweat glands. The sweat produced is more dilute, and less sodium is lost. Sodium conservation improves the body's ability to retain water, and maintain fluid and electrolyte homeostasis.

Athletes and others performing prolonged, heavy physical activity should drink about 2 cups of fluids 2 hours before physical activity, and another cup or so 15 minutes beforehand. They should continue to drink every 15 to 20 minutes during practice or competition. If dehydration occurs, sports drinks can help the athlete recover fluid and electrolyte homeostasis.

FIGURE 22-4 Fluid movement between plasma and interstitial compartments. Shown are hydrostatic and osmotic pressure forces involved in moving fluid out of capillaries (filtration) and into capillaries (reabsorption).

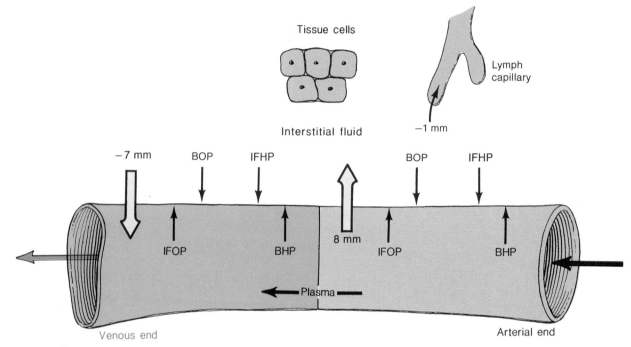

Key:

BHP = Blood hydrostatic pressure
IFHP = Interstitial fluid hydrostatic pressure
BOP = Blood osmotic pressure
IFOP = Interstitial fluid osmotic pressure

venous end it is −7 mm Hg. Thus, at the arterial end of a capillary, fluid moves out (filtered) from plasma into the interstitial compartment at a pressure of 8 mm Hg; at the venous end of a capillary, fluid moves in (reabsorbed) from the interstitial to the plasma compartment at a pressure of −7 mm Hg. Thus, not all the fluid filtered at one end of the capillary is reabsorbed at the other. The fluid not reabsorbed and any proteins that escape from capillaries pass into lymph capillaries. From here, the fluid (lymph) moves through lymphatic vessels to the thoracic duct or right lymphatic duct to enter the cardiovascular system (see Figure 17-3b). Under normal conditions, there is a state of balance at the arterial and venous ends of a capillary in which filtered fluid and absorbed fluid, as well as fluid picked up by the lymphatic system, are nearly equal.

BETWEEN INTERSTITIAL AND INTRACELLULAR COMPARTMENTS

Intracellular fluid has a higher osmotic pressure than interstitial fluid. In addition, the principal cation inside the cell is K^+, whereas the principal cation outside is Na^+ (see Figure 22-3). Normally, the higher intracellular osmotic pressure is balanced by forces that move water out of the cell, so the amount of water inside the cell does not change. When a fluid imbalance between these two compartments occurs, it is usually because the Na^+ or K^+ concentration has changed.

As previously discussed, ADH and aldosterone normally regulate extracellular sodium but lower than normal blood sodium level may result from a variety of causes. A decrease in sodium concentration in the interstitial fluid lowers the interstitial fluid osmotic pressure and establishes a filtration pressure gradient between the interstitial fluid and the intracellular fluid (Figure 22-5). Water moves from the interstitial fluid into the cells, with two potentially serious results.

First, an increase in intracellular water concentration, called **overhydration**, disrupts nerve cell function. In fact, severe overhydration, or **water intoxication**, produces symptoms ranging from disoriented behavior to convulsions, coma, and even death. The second result of the fluid shift is a loss of interstitial fluid volume that leads to a decrease in the interstitial fluid hydrostatic pressure. Consequently, water moves out of the plasma, blood volume goes down, and in a severe case, circulatory shock may result.

ACID–BASE BALANCE

Before reading this section, you might want to review the discussion of acids, bases, and pH in Chapter 2. In addition to controlling

FIGURE 22-5 Interrelations between fluid imbalance and electrolyte imbalance.

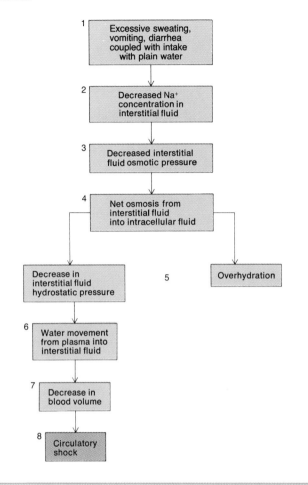

1. Excessive sweating, vomiting, diarrhea coupled with intake with plain water
2. Decreased Na$^+$ concentration in interstitial fluid
3. Decreased interstitial fluid osmotic pressure
4. Net osmosis from interstitial fluid into intracellular fluid
5. Overhydration / Decrease in interstitial fluid hydrostatic pressure
6. Water movement from plasma into interstitial fluid
7. Decrease in blood volume
8. Circulatory shock

What are the effects of overhydration?

water movement, electrolytes help regulate the body's acid—base balance. The overall acid—base balance depends on the hydrogen ion (H$^+$) concentration of body fluids, particularly extracellular fluid. Recall that acids ionize to produce hydrogen ions. The majority of hydrogen ions are produced as a result of the metabolic activities of cells. For example, during the aerobic respiration of glucose, carbon dioxide (CO_2) is produced. It combines with water in extracellular fluid to form carbonic acid (H_2CO_3), which breaks down into hydrogen ions and bicarbonate ions (HCO_3^-):

$$CO_2 + H_2O \rightarrow H_2CO_3 \rightarrow H^+ + HCO_3^-$$

Also, during the anaerobic respiration of glucose (glycolysis), the lactic acid produced is a source of hydrogen ions. During fatty acid catabolism, ketone bodies, most of which are acids, are produced and they contribute to the pool of hydrogen ions. When sulfur-containing amino acids are catabolized, they produce sulfuric acid (H_2SO_4) and when nucleoproteins are catabolized, they produce phosphoric acid (H_3PO_4). Both sulfuric and phosphoric acid produce hydrogen ions when they ionize. Recall that as part of digestion

the stomach produces hydrochloric acid (HCl), which could also seriously decrease blood pH if not buffered.

In a healthy person, the pH of the extracellular fluid is between 7.35 and 7.45. Homeostasis of this narrow range is essential to survival and depends on three major mechanisms: buffer systems, respiration, and kidney excretion.

BUFFER SYSTEMS

Most human **buffer systems** consist of a weak acid and a weak base that function to prevent drastic changes in the pH of a body fluid by rapidly changing strong acids and bases into weak acids and bases. A strong acid dissociates into H$^+$ ions more easily than a weak acid. Strong acids, therefore, lower pH more than weak ones because strong acids contribute more H$^+$ ions. Similarly, strong bases raise pH more than weak ones because strong bases dissociate more easily into OH$^-$ ions. The principal buffer systems are the carbonic acid—bicarbonate system, the phosphate system, the hemoglobin—oxyhemoglobin system, and the protein system.

Carbonic Acid—Bicarbonate

As a result of metabolic processes, the body normally produces more acids than bases. A strong acid like HCl can be buffered by a weak base, sodium bicarbonate ($NaHCO_3$), to produce a weaker acid, carbonic acid (H_2CO_3), and a salt (NaCl):

$$HCl + NaHCO_3 \rightleftharpoons H_2CO_3 + NaCl$$

For example, this reaction occurs when alkaline secretions from the pancreas mix with acidic chyme from the stomach.

Although the body needs more bicarbonate to neutralize acids, strong bases must also be buffered to prevent the blood pH from rising too high. In this case weak acids such as carbonic acid can neutralize strong bases such as sodium hydroxide (NaOH) to weaker bases like sodium bicarbonate:

$$NaOH + H_2CO_3 \rightleftharpoons NaHCO_3 + H_2O$$

Therefore, the **carbonic acid—bicarbonate buffer system** is an important regulator of blood pH.

Phosphate

The **phosphate buffer system** helps regulate pH in red blood cells but is especially important in kidney tubular fluids. With the use of this system, the kidneys help maintain normal blood pH by the acidification of urine. The two phosphate buffers act in the same manner as the bicarbonate buffer system. Strong acids are converted to weaker ones by sodium monohydrogen phosphate (Na_2HPO_4) and strong bases are buffered to weaker ones by sodium dihydrogen phosphate (NaH_2PO_4).

Hemoglobin

The **hemoglobin buffer system** buffers carbonic acid in the blood. When blood moves from the arterial end of a capillary to the venous end, the carbon dioxide given up by body cells enters the erythrocytes and combines with water to form carbonic acid (Figure 22-6). Simultaneously, oxyhemoglobin gives up its oxygen to the body cells, and some becomes reduced hemoglobin, which carries a negative charge. The hemoglobin anion attracts the hydrogen ion (H^+) from the carbonic acid and becomes an acid that is even weaker than carbonic acid.

Protein

The *protein buffer system* is the most abundant buffer in body cells and plasma. The amino acids in proteins contain at least one carboxyl group (COOH) and one amine group (NH_2). In solution, a carboxyl group may become ionized (COOH $\rightarrow$ COO^- + H^+) and buffer bases by combining its H^+ ions with excess hydroxide ions (OH^-) to form water (H^+ + OH^- $\rightarrow$ H_2O). In the presence of excess H^+ ions, free COO^- will recombine with H^+ to form COOH, thereby raising the pH of the fluids again. Thus, the carboxyl group can buffer both acids and bases.

The amine group (NH_2) can also act as an acid or a base. By accepting free H^+ ions (NH_3^+) it removes them from a body fluid. By releasing H^+ from NH_3^+ in the presence of excess OH^- ions to form water, it lowers the pH again.

RESPIRATION

Respiration also plays a role in maintaining blood pH. An increase in the carbon dioxide concentration in body fluids as a result of

cellular respiration lowers pH (makes the fluid more acid). This is illustrated by the following equation:

$$CO_2 + H_2O \rightleftharpoons H_2CO_3 \rightleftharpoons H^+ + HCO_3^-$$

Carbon dioxide   Water   Carbonic acid   Hydrogen ion   Bicarbonate ion

Conversely, a decrease in the carbon dioxide concentration of body fluids raises the pH.

The pH of body fluids may be adjusted by a change in the rate and depth of breathing, an adjustment that usually takes from 1 to 3 minutes. If the rate and depth of breathing are increased, more carbon dioxide is exhaled and blood pH rises. A slowed respiratory rate means less carbon dioxide is exhaled, causing blood pH to fall. Breathing rate can be altered up to eight times the normal rate; thus, respiration can greatly influence the pH of body fluids.

The pH of body fluids, in turn, affects the rate of breathing (Figure 22-7). If, for example, the blood becomes more acidic, the increase in hydrogen ions stimulates the respiratory center in the medulla, and respirations increase in rate and depth. The same effect occurs if the blood concentration of carbon dioxide increases. Increased respiration removes more carbon dioxide from blood to reduce the hydrogen ion concentrations. On the other hand, if the pH of the blood increases, or carbon dioxide concentration decreases, the respiratory center is inhibited and respirations decrease. The decreased respirations allow carbon dioxide to accumulate in blood and the hydrogen ion concentration to increase.

The respiratory mechanism normally can eliminate more acid or base than all the buffers combined.

KIDNEY EXCRETION

The role of the kidneys in maintaining pH was discussed in Chapter 21. This is done principally by the secretion of hydrogen (H^+) ions into filtrate.

ACID–BASE IMBALANCES

In *acidosis*, blood pH ranges from 7.35 to 6.80 or lower. In *alkalosis*, pH ranges from 7.45 to 8.00 or higher. A change in blood pH that leads to acidosis or alkalosis can be compensated to return pH to normal. *Compensation* refers to the physiological response to an acid–base imbalance. If a person has an altered pH due to metabolic causes, respiratory mechanisms (hyperventilation or hypoventilation) can help compensate for the alteration. Respiratory compensation occurs within minutes and is maximized within hours. Conversely, if a person has an altered pH due to respiratory causes, metabolic mechanisms (kidney excretion) can compensate for the alteration. Metabolic compensation may begin in minutes but takes days to be maximized.

Acidosis and alkalosis affect the body in different ways. The principal physiological effect of acidosis is depression of the central nervous system through depression of synaptic transmission. Blood pH below 7 causes disorientation and coma. In fact, patients with severe acidosis usually die in a state of coma. On the other hand, the major physiological effect of alkalosis is overexcitability of

FIGURE 22-6 Hemoglobin buffer system. When CO_2 enters the blood, it combines with water to form carbonic acid (H_2CO_3), which dissociates into bicarbonate (HCO_3^-) ions and hydrogen (H^+) ions. Oxyhemoglobin gives up its oxygen and combines with excess H^+ ions to form HbH (a weak acid). The bicarbonate may move out of the cell and combine with sodium (Na^+) in the plasma. In this way, much of the carbon dioxide is carried back to the lungs in the form of sodium bicarbonate ($NaHCO_3$).

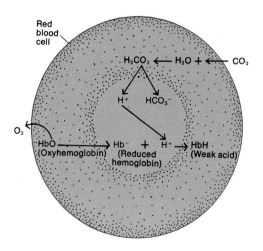

What do all buffers have in common?

FIGURE 22-7 Relationship between pH and respirations.

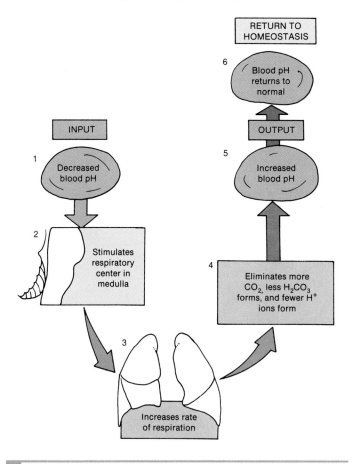

Does an increase in carbon dioxide concentration raise or lower pH? As the pH of the blood becomes more acid, do respirations increase or decrease in rate and depth?

the nervous system through facilitation of synaptic transmission. The overexcitability occurs both in the central nervous system and in peripheral nerves. Nerves conduct impulses repetitively even when not stimulated, causing nervousness, muscle spasms, and even convulsions.

Respiratory acidosis is caused by hypoventilation, a rate of breathing below normal. It occurs as a result of any condition that decreases the movement of carbon dioxide from the blood to the alveoli of the lungs and therefore causes a buildup of carbon dioxide, carbonic acid, and hydrogen ions. Such conditions include emphysema, pulmonary edema, injury to the respiratory center of the medulla, or disorders of the muscles involved in breathing. Treatment is aimed at increasing the exhalation of carbon dioxide.

Respiratory alkalosis is caused by hyperventilation. It occurs as a result of any condition that stimulates the respiratory center, such as oxygen deficiency at high altitude, severe anxiety, and aspirin overdose. Therefore, treatment is aimed at increasing the level of carbon dioxide in the blood. One corrective measure is to have the patient breathe repeatedly into a paper bag and then rebreathe the exhaled mixture of CO_2 and oxygen from the bag.

Metabolic acidosis results from an abnormal increase in acidic metabolic products other than carbon dioxide and from the loss of bicarbonate ions from the body. Ketosis is a good example. Loss of bicarbonate may occur with diarrhea and renal tubular dysfunction. Treatment consists of intravenous solutions of sodium bicarbonate and correcting the primary cause.

Metabolic alkalosis is caused by a nonrespiratory loss of acid by the body or excessive intake of alkaline drugs. Excessive vomiting of gastric contents results in a substantial loss of hydrochloric acid and is probably the most frequent cause of metabolic alkalosis. Treatment consists of giving a medication containing chloride ions and correcting the primary cause.

STUDY OUTLINE

Fluid Compartments and Fluid Balance (p. 471)

1. Body fluid refers to water and its dissolved substances.
2. About two-thirds of the body's fluid is located inside cells and is called intracellular fluid (ICF).
3. The other third is called extracellular fluid (ECF). It includes interstitial fluid, plasma, lymph, cerebrospinal fluid, gastrointestinal tract fluids, synovial fluid, fluids of the eyes and ears, glomerular filtrate, and pleural, pericardial, and peritoneal fluids.
4. Fluid balance means that the proportion of water in each fluid compartment is kept in homeostasis.
5. Fluid balance and electrolyte balance are inseparable.

Water (p. 471)

1. Water is the largest single constituent in the body, varying

from 45 to 75 percent of body weight depending on the amount of fat present and age.
2. Primary sources of fluid intake are ingested liquids and foods and water produced by catabolism.
3. Avenues of fluid output are the kidneys, skin, lungs, and gastrointestinal tract.
4. The stimulus for fluid intake is dehydration, which causes thirst sensations. Under normal conditions, fluid output is adjusted by aldosterone and antidiuretic hormone (ADH).

Electrolytes (p. 472)

1. Electrolytes are chemicals that dissolve in body fluids and dissociate into either cations (positive ions) or anions (negative ions) because of their ionic bonds.

2. Electrolyte concentration is expressed in milliequivalents per liter (mEq/liter).

3. Electrolytes have a greater effect on osmosis than nonelectrolytes.

4. Plasma, interstitial fluid, and intracellular fluid contain varying kinds and amounts of electrolytes.

5. Electrolytes serve three functions in the body: they are needed for normal metabolism, proper fluid movement between compartments, and regulation of pH.

6. Sodium (Na^+) is the most abundant extracellular ion. It is involved in nerve impulse transmission, muscle contraction, and creates most of the osmotic pressure of ECF. Its level is controlled by aldosterone.

7. Potassium (K^+) is the most abundant cation in intracellular fluid. It is involved in maintaining fluid volume, nerve impulse conduction, muscle contraction, and regulating pH. Its level is controlled by aldosterone.

8. Calcium (Ca^{2+}), the most abundant ion in the body, is principally an extracellular ion that is a structural component of bones and teeth. It also functions in blood clotting, neurotransmitter release, muscle contraction, secretion, and heartbeat. Its level is controlled by parathyroid hormone (PTH) and calcitonin (CT).

9. Magnesium (Mg^{2+}) is primarily an intracellular electrolyte that activates several enzyme systems. Its level is controlled by aldosterone.

10. Chloride (Cl^-) is the major extracellular anion. It assumes a role in regulating osmotic pressure and forming HCl. Its level is controlled indirectly by aldosterone.

11. Phosphate (HPO_4^{2-}) is principally an intracellular ion that is a structural component of bones and teeth. It is also required for the synthesis of nucleic acids and ATP and for buffer reactions. Its level is controlled by PTH and CT.

Movement of Body Fluids (p. 475)

1. At the arterial end of a capillary, fluid moves from plasma into interstitial fluid (filtration). At the venous end, fluid moves in the opposite direction (reabsorption).

2. There is a state of near equilibrium at the arterial and venous ends of a capillary between filtered fluid and absorbed fluid plus that picked up by the lymphatic system.

3. Fluid movement between interstitial and intracellular compartments depends on the movement of sodium and potassium and the secretion of aldosterone and ADH.

4. Fluid imbalance may lead to overhydration (water intoxication).

Acid—Base Balance (p. 478)

1. The overall acid—base balance of the body is maintained by controlling the H^+ concentration of body fluids, especially extracellular fluid.

2. The normal pH of extracellular fluid is 7.35 to 7.45.

3. Homeostasis of pH is maintained by buffers, respirations, and kidney excretion.

4. The important buffer systems include carbonic acid–bicarbonate, phosphate, hemoglobin, and protein.

5. An increase in rate of respiration increases pH; a decrease in rate decreases pH.

Acid—Base Imbalances (p. 480)

1. Acidosis is a blood pH between 7.35 and 6.80 and lower. Its principal effect is depression of the central nervous system (CNS).

2. Alkalosis is a blood pH between 7.45 and 8.00 and higher. Its principal effect is overexcitability of the CNS.

3. Respiratory acidosis is characterized by high CO_2 levels and is caused by hypoventilation; metabolic acidosis is characterized by a decreased bicarbonate level and results from an abnormal increase in acid metabolic products (other than CO_2) and loss of bicarbonate.

4. Respiratory alkalosis is characterized by low CO_2 levels and is caused by hyperventilation; metabolic alkalosis is characterized by increased bicarbonate and results from nonrespiratory loss of acid or excess intake of alkaline drugs.

5. Metabolic acidosis or alkalosis is compensated by respiratory mechanisms; respiratory acidosis or alkalosis is compensated by renal mechanisms.

REVIEW QUESTIONS

1. Define body fluid and give examples. Name the three major fluid compartments and describe how they are separated. (p. 471)

2. What is meant by fluid balance? How are fluid balance and electrolyte balance related? (p. 471)

3. Describe the avenues of fluid intake and fluid output. Indicate volumes in each case. (p. 471)

4. Discuss the role of thirst in regulating fluid intake. (p. 471)

5. Explain how aldosterone and antidiuretic hormone (ADH) adjust normal fluid output. What are some abnormal routes of fluid output? (p. 472)

6. Define a nonelectrolyte and an electrolyte. Give examples of each. (p. 472)

7. Describe the functions of electrolytes in the body. (p. 472)

8. Describe the major differences in electrolytic concentration of the three major fluid compartments. (p. 473)

9. Name three important extracellular electrolytes and three important intracellular electrolytes. (p. 473)

10. Indicate the function and regulation of each of the following electrolytes: sodium, potassium, calcium, magnesium, chloride, and phosphate. (p. 474)

11. Describe the physiological effects of too low and too high concentrations of sodium, potassium, calcium, magnesium, chloride, and phosphate. (p. 474)

12. What factors are involved in interstitial and intracellular fluid movement? (p. 475)

13. Explain how the following buffer systems help maintain the pH of body fluids: carbonic acid–bicarbonate, phosphate, hemoglobin, and protein. (p. 479)

14. Describe how respiration helps maintain pH. (p. 480)

15. Briefly discuss the role of the kidneys in maintaining pH. (p. 480)

16. Define acidosis and alkalosis. Distinguish between respiratory and metabolic acidosis and alkalosis. (p. 480)

17. Describe the principal physiological effects of acidosis and alkalosis and treatment for each. (p. 480)

23

The Reproductive Systems

STUDENT OBJECTIVES

1. Define reproduction and explain its significance.
2. Classify the male organs of reproduction by function.
3. Describe how sperm cells are produced and transported to the exterior.
4. Explain the functions of the male reproductive hormones.
5. Classify the female organs of reproduction by function.
6. Describe how ova are produced.
7. Explain the roles of the uterine (Fallopian) tubes, uterus, and vagina in reproduction.
8. Describe the structure and functions of the mammary glands.
9. Define the menstrual and ovarian cycles and explain how they are related.
10. Explain the functions of the female reproductive hormones.
11. Define common disorders and medical terminology and conditions associated with the reproductive systems.

A LOOK AHEAD

MALE REPRODUCTIVE SYSTEM
  Scrotum
  Testes
    Spermatogenesis
    Spermatozoa
    Testosterone and Inhibin
  Male Puberty
  Ducts
    Ducts of the Testis
    Epididymis
    Ductus (Vas) Deferens
    Ejaculatory Duct
    Urethra
  Accessory Sex Glands
  Semen (Seminal Fluid)
  Penis
FEMALE REPRODUCTIVE SYSTEM
  Ovaries
    Oogenesis
  Uterine (Fallopian) Tubes
  Uterus
  Vagina
  Vulva
  Perineum
  Mammary Glands
  Female Puberty
FEMALE REPRODUCTIVE CYCLE (FRC)
  Hormonal Regulation
  Menstrual Phase (Menstruation)
  Preovulatory Phase
  Ovulation
  Postovulatory Phase
  Menopause
COMMON DISORDERS
MEDICAL TERMINOLOGY AND
    CONDITIONS

Reproduction is the mechanism by which life is sustained. In one sense, reproduction is the process by which a single cell duplicates its genetic material, allowing an organism to grow and repair itself; thus, reproduction maintains the life of the individual. But reproduction is also the process by which genetic material is passed from generation to generation. In this regard, reproduction maintains the continuation of the species.

The organs of the male and female reproductive systems may be grouped by function. The testes and ovaries, also called *gonads* (*gonos* = seed), produce gametes—sperm cells and ova, respectively, and secrete hormones. The *ducts* transport, receive, and store gametes. Other reproductive organs, called *accessory sex glands,* produce materials that support gametes.

MALE REPRODUCTIVE SYSTEM

The organs of the male reproductive system are the testes (male gonads), which produce sperm and hormones; a number of ducts that either store or transport sperm to the exterior; accessory sex glands that secrete semen; and several supporting structures, including the penis (Figure 23-1).

SCROTUM

The *scrotum* is a pouch that supports the testes; it consists of loose skin, superficial fascia, and smooth muscle fibers (Figure 23-1). Internally, it is divided by a septum into two sacs, each containing a single testis.

The location of the scrotum and contraction of its muscle fibers regulate the temperature of the testes. The production and survival of sperm require a lower than normal blood temperature. Because the scrotum is outside the body cavities, it supplies an environment about 3°C below body temperature. On exposure to cold and during sexual arousal, skeletal muscles contract to elevate the testes, moving them closer to the pelvic cavity where they can absorb body heat. Exposure to warmth reverses the process.

TESTES

The *testes*, or *testicles*, are paired oval glands that develop from an embryonic tissue called *mesoderm,* high on the embryo's posterior abdominal wall and usually begin their descent into the scrotum in the seventh month of fetal development. When the testes do not descend, the condition is referred to as *cryptorchidism* (krip-TOR-ki-dizm).

The testes are covered by a dense white fibrous capsule that extends inward and divides each testis into internal compartments called *lobules* (Figure 23-2a). Each of the 200 to 300 lobules contains one to three tightly coiled *seminiferous tubules* that produce sperm by a process called spermatogenesis. This process is considered shortly.

Seminiferous tubules are lined with spermatogenic cells in various stages of development (Figure 23-2b). The most immature spermatogenic cells, the *spermatogonia* (sper′-ma-tō-GŌ-nē-a), lie against the basement membrane toward the outside of the tubules. Toward the lumen of the tube are layers of progressively more mature cells in order of advancing maturity: primary spermatocytes, secondary spermatocytes, and spermatids. By the time a *sperm cell*, or *spermatozoon* (sper′-ma-tō-ZŌ-on), has nearly reached maturity, it is in the lumen of the tubule and begins to be moved through a series of ducts.

Between the developing sperm cells in the tubules are *sustentacular* (sus-ten-TAK-yoo-lar) or *Sertoli cells*. These cells support, protect, and nourish developing spermatogenic cells; phagocytize degenerating spermatogenic cells; and secrete the hormone inhibin that helps regulate sperm production. Between the seminiferous

485

FIGURE 23-1 **Male organs of reproduction and surrounding structures seen in sagittal section.**

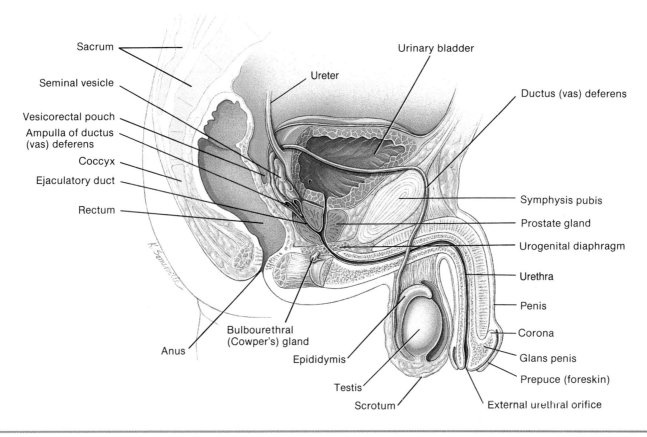

Which of the structures shown are the gonads? Ducts? Accessory sex glands?

tubules are clusters of ***interstitial endocrinocytes (interstitial cells of Leydig)***. These cells secrete the male hormone testosterone, the most important androgen (AN-drō-jen), a substance producing male characteristics.

Spermatogenesis

The process by which the seminiferous tubules of the testes produce spermatozoa is called ***spermatogenesis*** (sper′-ma-tō-JEN-e-sis). It consists of three stages: meiosis I, meiosis II, and spermiogenesis.

■ **Meiosis: Overview** In sexual reproduction, a new organism is produced by the union and fusion of two different cells, one produced by each parent. The sex cells, called ***gametes***, are the ovum produced in the female gonads (ovaries) and the sperm produced in the male gonads (testes). The union and fusion of gametes is called ***fertilization*** and the cell thus produced is known as a ***zygote***. The zygote contains a mixture of chromosomes (DNA) from the two parents and, through its repeated mitotic division, develops into a new organism.

Gametes differ from all other body cells (somatic cells) in that they contain the ***haploid*** (one-half) ***chromosome number***, 23, a single set of chromosomes. It is symbolized by ***n***. Somatic cells, such as brain, stomach, kidney, and so on, contain 46 chromosomes in their nuclei. Of the 46 chromosomes, 23 are a complete set from one parent that contains one copy of all the genes necessary for carrying out the activities of the cell. The other 23 chromosomes are another set from the other parent, which codes for the same traits. Since somatic cells contain two sets of chromosomes, they are referred to as ***diploid*** (DIP-loyd; *di* = two) ***cells***, symbolized as ***2n***. In a diploid cell, two chromosomes that belong to a pair are called ***homologous*** (hō-MOL-ō-gus) ***chromosomes***, or ***homologues***.

If gametes had the same number of chromosomes as somatic cells, the zygote formed from their fusion would have double the diploid number, or 92, and with every succeeding generation, the number of chromosomes would continue to double and normal development would not occur. The chromosome number does not double with each generation because of a special nuclear division called ***meiosis*** that occurs only in the production of gametes. In meiosis, a developing spermatozoon or ovum relinquishes its duplicate set of chromosomes so that the mature gamete has only 23.

Review the process of mitosis in Chapter 3. A basic difference between mitosis and meiosis is that in mitosis, a parent cell divides into two daughter cells, each of which receives the same chromosome number as the parent cell. In meiosis, the daughter cells have only half as many chromosomes as the parent cell. Let us see how this happens.

■ **Stages of Spermatogenesis** Spermatogenesis begins during puberty and continues through life. It takes about 74 days. The sperma-

FIGURE 23-2 Testes. (a) Internal structure. Sagittal section illustrating internal anatomy.

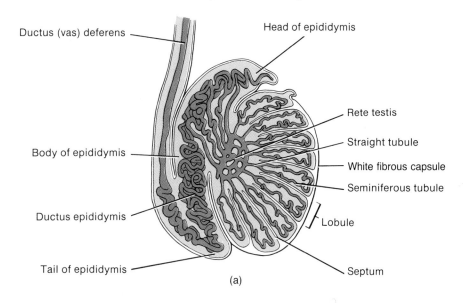

Ductus (vas) deferens

Head of epididymis

Body of epididymis

Rete testis

Straight tubule

White fibrous capsule

Seminiferous tubule

Ductus epididymis

Lobule

Tail of epididymis

Septum

(a)

Trace a sperm cell from its origin in a seminiferous tubule into the ductus (vas) deferens.

FIGURE 23-2 (*Continued*) (b) Microscopic cross section of a portion of a seminiferous tubule showing the stages of spermatogenesis.

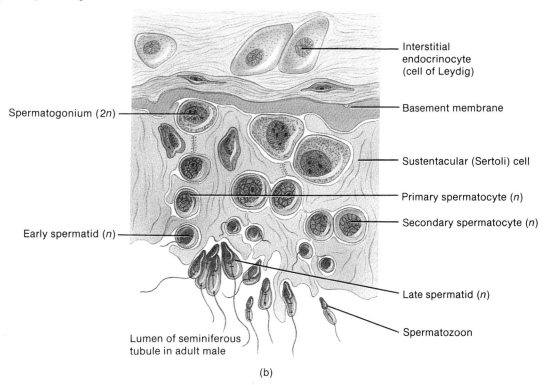

Interstitial endocrinocyte (cell of Leydig)

Basement membrane

Spermatogonium (2n)

Sustentacular (Sertoli) cell

Primary spermatocyte (n)

Secondary spermatocyte (n)

Early spermatid (n)

Late spermatid (n)

Spermatozoon

Lumen of seminiferous tubule in adult male

(b)

Which cell in a seminiferous tubule is least mature? Most mature?

togonia or sperm stem cells (Figure 23-3) that line the seminiferous tubules contain the diploid chromosome number (46). Following division, some spermatogonia remain near the basement membrane to prevent depletion of the cell population. Other spermatogonia undergo certain developmental changes and become ***primary spermatocytes*** (SPER-ma-tō-sitz′). Like spermatogonia, they are diploid (2*n*).

1. **Meiosis I (reduction division).** Each primary spermatocyte enlarges before dividing. Then, two nuclear divisions take place as part of meiosis. In the first, DNA is replicated and 46 chromosomes (each made up of two identical chromatids) form and

move toward the center of the nucleus where they line up as 23 homologous pairs of chromosomes. The four chromatids of each homologous pair then twist around each other to form a ***tetrad***. In a tetrad, portions of one chromatid may be exchanged with portions of another (***crossing-over***), which permits recombination of genes. Thus, the spermatozoa eventually produced may be genetically unlike each other and unlike the cell that produced them.

Next the pairs separate and one member of each pair migrates to opposite poles of the dividing nucleus. The cells formed by this first nuclear division are called ***secondary spermatocytes***. Each cell has 23 chromosomes—the haploid number. Each chro-

FIGURE 23-3 **Spermatogenesis. The designation 2*n* means diploid; *n* means haploid.**

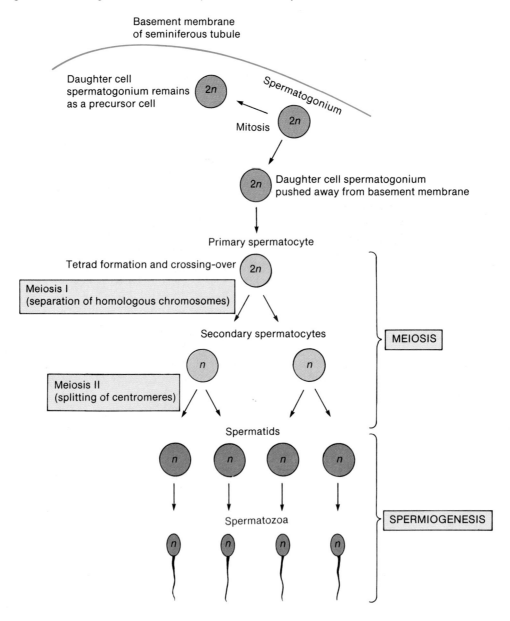

Why is meiosis I also called reduction division and meiosis II also called equatorial division?

mosome of the secondary spermatocytes, however, is made up of two identical chromatids, but the genes may be rearranged as a result of crossing-over.

2. **Meiosis II (equatorial division).** The second nuclear division of meiosis is equatorial division. There is no replication of DNA. The chromosomes (each composed of two identical chromatids) line up in single file at the equator and the chromatids of each chromosome separate from each other. The cells formed from the equatorial division are called *spermatids*. Each contains half the original chromosome number, 23, and is haploid. Each primary spermatocyte therefore produces four spermatids by meiosis.

3. **Spermiogenesis.** In the final stage of spermatogenesis, called *spermiogenesis* (sper′-mē-ō-JEN-e-sis), spermatids mature into spermatozoa. Each spermatid develops a head and a flagellum (tail). The developing spermatids are then nourished by sustentacular (Sertoli) cells (see Figure 23-2b). Since there is no cell division in spermiogenesis, each spermatid develops into a single *spermatozoon* (*sperm cell*).

Spermatozoa (plural of spermatozoon) enter the lumen of the seminiferous tubule and migrate to the ductus epididymis, where in 10 to 14 days they complete their maturation and become capable of fertilizing ova. Spermatozoa are also stored in the ductus (vas) deferens. Here, they can retain their fertility for up to several months.

Spermatozoa

Spermatozoa are produced or matured at the rate of about 300 million per day and, once ejaculated, have a life expectancy of about 48 hours in the female reproductive tract. A spermatozoon is composed of a head, a midpiece, and a tail (Figure 23-4). In

FIGURE 23-4 **Parts of a spermatozoon.**

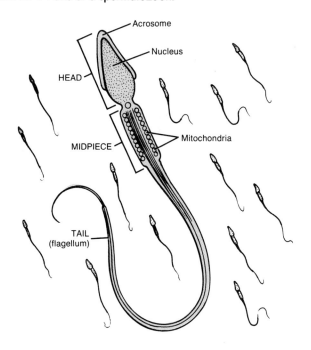

What are the functions of each part of a spermatozoon?

the *head* are the nuclear material and a dense granule called the *acrosome*, which contains enzymes that help the sperm cell penetrate a secondary oocyte (potential ovum). Mitochondria in the *midpiece* carry on the metabolism that provides energy for locomotion. The *tail*, a typical flagellum, propels the spermatozoon.

Testosterone and Inhibin

At the onset of puberty, the anterior pituitary starts to secrete two hormones that have profound effects on male reproductive organs: follicle-stimulating hormone (FSH) and luteinizing hormone (LH). Their release is controlled from the hypothalamus by gonadotropin releasing hormone (GnRH). FSH acts on the seminiferous tubules to initiate spermatogenesis. LH stimulates the interstitial endocrinocytes (interstitial cells of Leydig) to secrete the hormone testosterone (tes-TOS-te-rōn).

Testosterone is synthesized from cholesterol or acetyl coenzyme A in the testes. It has a number of effects on the male body:

1. During prenatal development, it facilitates the development of male internal genitals.
2. It stimulates the descent of the testes just prior to birth.
3. At puberty, it brings about development and enlargement of the male sex organs and the development of male secondary sex characteristics, including pubic, axillary, facial, and chest hair (within hereditary limits), temporal hairline recession, thickening of the skin, increased sebaceous (oil) gland secretion, growth of skeletal muscles and bones, and enlargement of the larynx and deepening of the voice.
4. It stimulates metabolic rate and is the basis for sex drive.
5. It is thought to stimulate formation of spermatogonia and meiosis II.

Testosterone production is controlled by a negative feedback system with both the hypothalamus and the anterior pituitary (see Chapter 13).

Inhibin, a hormone secreted by sustentacular (Sertoli) cells, inhibits the secretion of FSH. Once the degree of spermatogenesis required for male reproductive functions has been achieved, sustentacular cells secrete inhibin. Inhibin feeds back negatively to the anterior pituitary to inhibit FSH and thus decrease spermatogenesis. If spermatogenesis is proceeding too slowly, lack of inhibin production permits FSH secretion and an increased rate of spermatogenesis.

MALE PUBERTY

Puberty (PŪ-ber-tē; *puber* = marriageable age) refers to the period of time when secondary sex characteristics begin to develop and the potential for sexual reproduction is reached. Male puberty begins at an average age of 10 to 11 and ends at an average age of 15 to 17. The factors that determine the onset of puberty are poorly understood, but the sequence of events is well established. During prepubertal years, levels of LH, FSH, and testosterone are low. At around age six or seven, a prepubertal growth spurt occurs that is probably related to secretion of adrenal androgens and human growth hormone (hGH).

The onset of puberty is signaled by sleep-associated surges in LH and, to a lesser extent, FSH secretion. As puberty advances, elevated LH and FSH levels are present throughout the day and are accompanied by increased levels of testosterone. The rise in LH and FSH are believed to result from increased GnRH secretion and enhanced responsiveness of the anterior pituitary to GnRH. With sexual maturity, the hypothalamic–pituitary system becomes less sensitive to the feedback inhibition of testosterone on LH and FSH secretion.

The changes in the testes that occur during puberty include maturation of sustentacular cells and initiation of spermatogenesis. The anatomical and functional changes associated with puberty are the result of increased testosterone secretion. Usually, the first sign is enlargement of the testes. About a year later, the penis increases in size. The prostate gland, seminal vesicles, bulbourethral (Cowper's) gland, and epididymis increase in size over a period of several years. Development of the secondary sex characteristics occurs and a growth spurt takes place as elevated testosterone levels increase both bone and muscle growth.

DUCTS

Ducts of the Testis

Following their production, spermatozoa are moved through the seminiferous tubules to the *straight tubules* (see Figure 23-2a). The straight tubules lead to a network of ducts in the testis called the *rete* (RĒ-tē) *testis*. Some of the cells lining the rete testis possess cilia that probably help move the sperm along. The sperm are next transported out of the testis into an adjacent organ, the epididymis (see Figure 23-2a).

Epididymis

The *epididymis* (ep'-i-DID-i-mis; *epi* = above; *didymos* = testis) is a comma-shaped organ that lies along the posterior border of the testis (see Figures 23-1 and 23-2a) and consists mostly of a tightly coiled tube, the *ductus epididymis*. The ductus epididymis is the site of sperm maturation. The ductus epididymis also stores spermatozoa for up to four weeks, after which they are expelled or reabsorbed.

Ductus (Vas) Deferens

Within the epididymis, the ductus epididymis becomes less convoluted, its diameter increases, and at this point it is referred to as the *ductus (vas) deferens* (see Figure 23-2a). The ductus (vas) deferens ascends along the posterior border of the testis, penetrates the inguinal canal (a passageway in the anterior abdominal wall), and enters the pelvic cavity, where it loops over the side and down the posterior surface of the urinary bladder (see Figure 23-1). The ductus (vas) deferens has a heavy coat of three layers of muscle. It stores sperm for up to several months and propels them toward the urethra during ejaculation by peristaltic contractions of its muscular coat.

One method of sterilization of males is called *vasectomy*. It is a relatively uncomplicated procedure typically performed under local anesthesia in which a portion of each ductus (vas) deferens is removed. In the procedure, an incision is made in the scrotum, the ducts are tied in two places, and the portion between the ties is removed. Although sperm production continues in the testes, the sperm cannot reach the exterior because the ducts are cut; the sperm degenerate and are destroyed by phagocytosis. Vasectomy has no effect on sexual desire and performance, and if performed correctly, is virtually 100 percent effective. The procedure is reversible.

Traveling with the ductus (vas) deferens as it ascends in the scrotum are blood vessels, autonomic nerves, lymphatic vessels, and the cremaster muscle. These structures together make up the *spermatic cord*.

Ejaculatory Duct

Behind the urinary bladder are the *ejaculatory* (e- JAK-yoo-la-tō'-rē) *ducts* (Figure 23-5), formed by the union of the duct from the seminal vesicle (to be described shortly) and ductus (vas) deferens. The ejaculatory ducts eject spermatozoa into the urethra.

Urethra

The *urethra* is the terminal duct of the system, serving as a passageway for spermatozoa or urine. In the male, the urethra passes through the prostate gland, urogenital diaphragm, and penis. The opening of the urethra to the exterior is called the *external urethral orifice*.

ACCESSORY SEX GLANDS

Whereas the ducts of the male reproductive system store and transport sperm cells, the *accessory sex glands* secrete the liquid portion of semen. The paired *seminal vesicles* (VES-i-kuls) are pouchlike structures, lying at the base of the urinary bladder in front of the rectum (Figure 23-5). They secrete an alkaline, viscous fluid that is rich in fructose and pass it into the ejaculatory duct. The fructose provides energy for the sperm. It constitutes about 60 percent of the volume of semen. The alkaline nature of the fluid helps neutralize acid in the female tract that would otherwise inactivate and kill sperm.

The *prostate* (PROS-tāt) *gland* is a single, doughnut-shaped gland about the size of a chestnut (Figure 23-5). It is below the urinary bladder and surrounds the superior portion of the urethra. The prostate secretes a slightly alkaline fluid rich in citric acid and prostaglandins into the prostatic urethra. The prostatic secretion constitutes 13 to 33 percent of the volume of semen and contributes to sperm motility and viability. The prostate gland slowly increases in size from birth to puberty, and then a rapid growth spurt occurs. The size attained by the third decade remains stable until about age 45, when enlargement may occur.

The paired *bulbourethral* (bul'-bō-yoo-RĒ-thral) or *Cowper's glands* are about the size of peas. They are located beneath the prostate on either side of the urethra and within the urogenital diaphragm (Figure 23-5). The bulbourethral glands secrete an alkaline substance that protects sperm by neutralizing the acid environment of the urethra and mucus that lubricates the end of the penis during sexual intercourse.

FIGURE 23-5 Male reproductive organs in relation to surrounding structures seen in posterior view.

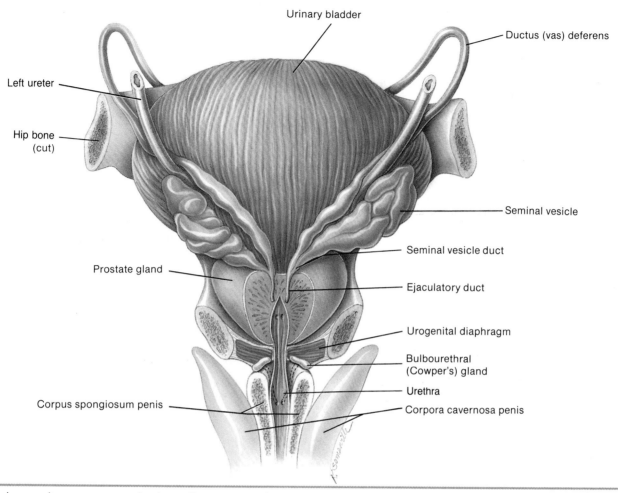

Urinary bladder

Ductus (vas) deferens

Left ureter

Hip bone (cut)

Seminal vesicle

Seminal vesicle duct

Prostate gland

Ejaculatory duct

Urogenital diaphragm

Bulbourethral (Cowper's) gland

Urethra

Corpus spongiosum penis

Corpora cavernosa penis

What does each accessory sex gland contribute to semen?

SEMEN (SEMINAL FLUID)

Semen (*seminal fluid*) is a mixture of sperm and the secretions of the seminal vesicles, prostate gland, and bulbourethral glands. The average volume of semen for each ejaculation is 2.5 to 5 ml, and the average range of spermatozoa ejaculated is 50 to 150 million/ml. When the number of spermatozoa falls below 20 million/ml, the male is likely to be infertile. The very large number is required because only a small percentage eventually reach the ovum. Also, though only a single spermatozoon fertilizes an ovum, fertilization requires the combined action of a larger number of spermatozoa to digest the intercellular material covering the ovum. The acrosome of a spermatozoon produces enzymes that dissolve the barrier but a passageway through which one sperm cell may enter can be created only by the enzymatic actions of many sperm cells.

Semen has a slightly alkaline pH of 7.20 to 7.60. The prostatic secretion gives semen a milky appearance, and fluids from the seminal vesicles and bulbourethral glands give it a mucoid consistency. Semen provides spermatozoa with a transportation medium and nutrients. It neutralizes the acid environment of the male urethra

and the female vagina. It also contains enzymes that activate sperm after ejaculation, and an antibiotic that kills bacteria in semen and the female reproductive tract.

Once ejaculated into the vagina, liquid semen coagulates rapidly because of a clotting enzyme produced by the prostate gland that acts on a substance produced by the seminal vesicle. This clot liquefies in about 5 to 20 minutes because of another enzyme produced by the prostate gland. Abnormal or delayed liquefaction of coagulated semen may cause complete or partial immobilization of spermatozoa, thus inhibiting their movement through the cervix of the uterus.

PENIS

The *penis* is used to introduce spermatozoa into the vagina (Figure 23-6). The penis is a cylinder-shaped organ that consists of a body, root, and glans penis. It is composed of three masses of erectile tissue. The two dorsal masses are called the *corpora cavernosa penis*. The ventral mass, the *corpus spongiosum penis*, contains

FIGURE 23-6 Internal structure of the penis. (a) Coronal section. (b) Cross section. The insert shows details of the skin and fascia.

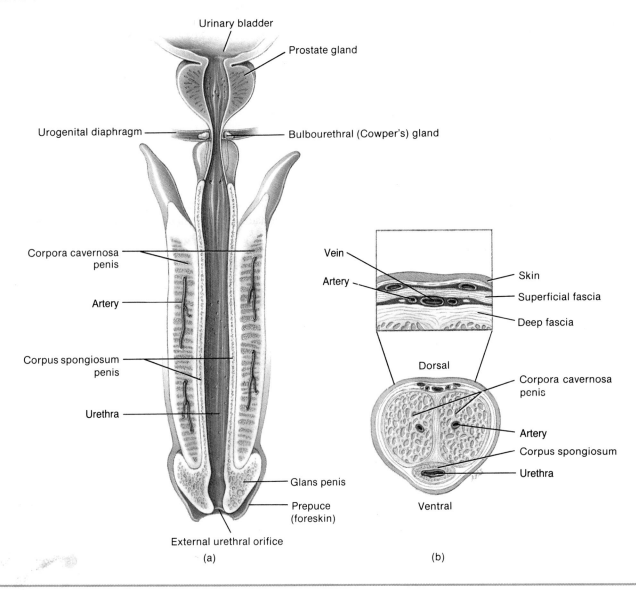

What happens in the penis when an erection occurs?

the urethra. All three masses are enclosed by fascia and skin and consist of erectile tissue containing blood sinuses. Under the influence of sexual stimulation, the arteries supplying the penis dilate, and large quantities of blood enter the sinuses, resulting in an *erection*, a parasympathetic reflex. Details of male sexual responses are presented in the next chapter. A smooth-muscled sphincter at the base of the urinary bladder closes during ejaculation to prevent the mixing of acidic urine with semen in the urethra which could immobilize sperm. The sphincter also keeps semen from entering the urinary bladder.

The distal end of the corpus spongiosum penis is a slightly enlarged region called the *glans penis*, which means shaped like an acorn. In the glans penis is a slitlike opening of the urethra (the external urethral orifice) to the exterior. Covering the glans

penis is the loosely fitting *prepuce* (PRĒ-pyoos), or *foreskin*.

Circumcision (*circumcido* = to cut around) is a surgical procedure in which part or all of the prepuce is removed. It is usually performed on the third or fourth day after birth or on the eighth day as part of a Jewish religious rite.

FEMALE REPRODUCTIVE SYSTEM

The female organs of reproduction include the ovaries (female gonads), which produce secondary oocytes (cells that develop into mature ova, or eggs, following fertilization) and the female sex hormones progesterone, estrogens, and relaxin; the uterine (Fallopian) tubes that transport ova to the uterus (womb); the vagina;

and external organs—the vulva, or pudendum (Figure 23-7). The mammary glands also are considered part of the female reproductive system.

The specialized branch of medicine that deals with the diagnosis and treatment of diseases of the female reproductive system is **gynecology** (gī'-ne-KOL-ō-jē; *gyneco* = woman).

OVARIES

The **ovaries** (*ovarium* = egg receptacle) are paired organs that in fetal life arise from the same embryonic tissue (mesoderm) as the testes. They are the size and shape of almonds. One lies on each side of the pelvic cavity, held in place by broad, ovarian, and suspensory ligaments (Figure 23-8). Each contains a **hilus** where nerves, blood, and lymphatic vessels enter. They consist of the following parts (Figure 23-9):

1. **Germinal epithelium**. A surface layer of simple cuboidal epithelium.
2. **Tunica albuginea**. A capsule of connective tissue immediately beneath the germinal epithelium.
3. **Stroma**. A central region of connective tissue composed of an outer dense layer (the *cortex*) that contains ovarian follicles and an inner loose layer (the *medulla*) that contains the nerve, blood, and lymph supply.

4. **Ovarian follicles**. Immature ova (oocytes) lie within surrounding protective glandular tissues called follicles. Both are at various stages of development. A relatively large, fluid-filled follicle is called the vesicular ovarian (Graafian) follicle. Follicles secrete estrogens.
5. **Corpus luteum**. A mature vesicular ovarian follicle that has ruptured to expel a secondary oocyte (potential mature ovum), a process called ovulation. It continues to produce the hormones progesterone, estrogens, relaxin, and inhibin until it degenerates and turns to fibrous tissue (corpus albicans).

Oogenesis

The formation of a haploid ovum in the ovary is called **oogenesis** (ō'-ō-JEN-e-sis). Oogenesis occurs in essentially the same manner as spermatogenesis. It involves meiosis and maturation.

■ **Meiosis I (Reduction Division)** During early fetal development, germ cells in the ovaries differentiate into **oogonia** (ō'-o-GŌ-n-a; *oo* = egg), cells that can give rise to cells that develop into ova (Figure 23-10). Oogonia are diploid cells that divide mitotically. At about the third month of prenatal development, oogonia develop into larger diploid cells called **primary oocytes** (Ō-o-sītz). They are enclosed in a follicle and migrate into the cortex where they remain at this stage until puberty, when they are stimulated by

FIGURE 23-7 **Female organs of reproduction and surrounding structures seen in sagittal section.**

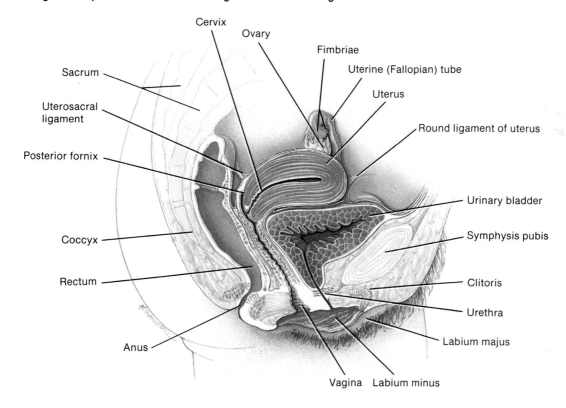

Which male structures are equivalent to the ovaries? Clitoris? Paraurethral (Skene's) glands? Greater vestibular (Bartholin's) glands?

FIGURE 23-8 Uterus and associated structures seen in posterior view. The left side of the figure has been sectioned to show internal structures.

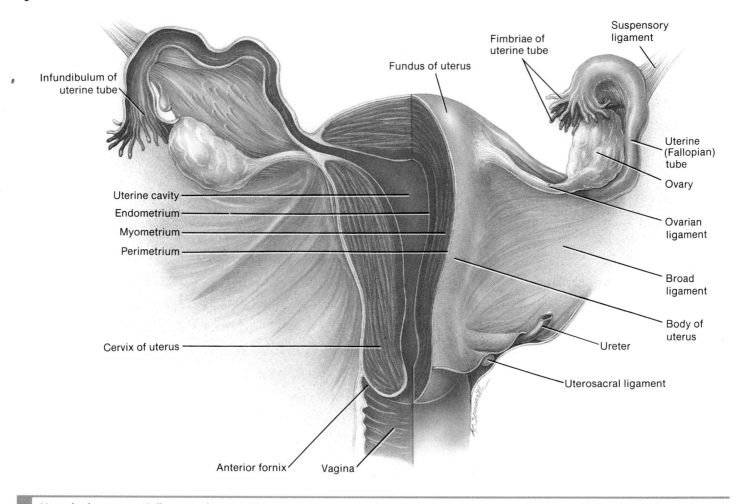

Infundibulum of uterine tube

Fimbriae of uterine tube

Suspensory ligament

Fundus of uterus

Uterine cavity

Endometrium

Myometrium

Perimetrium

Uterine (Fallopian) tube

Ovary

Ovarian ligament

Broad ligament

Body of uterus

Cervix of uterus

Ureter

Uterosacral ligament

Anterior fornix   Vagina

How do the uterine (Fallopian) tubes contribute to transportation of an ovum?

follicle-stimulating hormone (FSH) from the anterior pituitary gland, which, in turn, has responded to gonadotropin releasing hormone (GnRH) from the hypothalamus. All the ova a woman will ever produce are present at birth.

With puberty, several primary follicles respond each month to the rising level of FSH. When luteinizing hormone (LH) is secreted from the anterior pituitary, one of the primary follicles is selected to continue to develop. Two cells of unequal size, both with 23 chromosomes of two chromatids each, are produced. The smaller cell, called the *first polar body*, is essentially a packet of discarded nuclear material. The larger cell, the *secondary oocyte*, receives most of the cytoplasm and is surrounded by layers of epithelial cells.

■ **Meiosis II (Equatorial Division)** At ovulation, the secondary oocyte is discharged. It enters the uterine (Fallopian) tube, and if spermatozoa are present and fertilization occurs, the second division, meiosis II, is completed.

■ **Maturation** The secondary oocyte produces two cells of unequal

size, both of them haploid. The larger cell eventually develops into an *ovum*, or mature egg; the smaller is the *second polar body*.

All polar bodies disintegrate. Thus, each oogonium produces a single secondary oocyte, whereas each spermatocyte produces four spermatozoa. Spermatogenesis and oogenesis differ in other ways as well. Spermatogenesis is a continuous process that begins in puberty and continues throughout life; oogenesis begins at the first menstruation and ends at menopause. Also, spermatozoa are quite small, have flagella for locomotion, and contain few nutrients; a secondary oocyte is larger, lacks flagella, and contains more nutrients for nourishment until implantation occurs in the uterus.

UTERINE (FALLOPIAN) TUBES

The female body contains two *uterine (Fallopian) tubes* that extend laterally from the uterus and transport the ova from the ovaries to the uterus (see Figure 23-8). The open funnel-shaped end of each tube, the *infundibulum*, lies close to the ovary and is surrounded by fingerlike projections called *fimbriae* (FIM-brē-ē),

FIGURE 23-9 Parts of an ovary seen in sectional view. The arrows indicate the sequence of developmental stages that occur as part of the ovarian cycle.

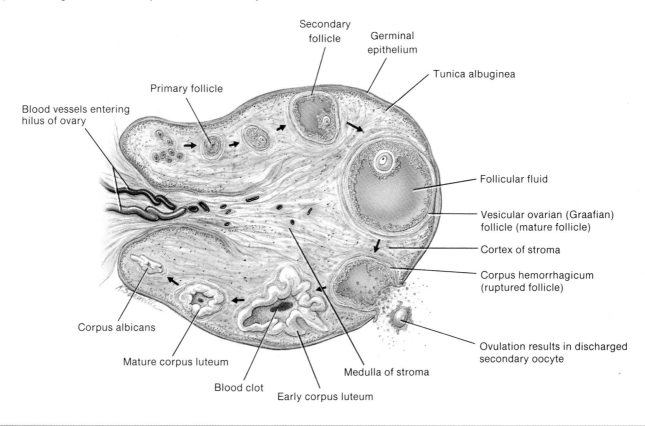

Secondary follicle

Germinal epithelium

Primary follicle

Tunica albuginea

Blood vessels entering hilus of ovary

Follicular fluid

Vesicular ovarian (Graafian) follicle (mature follicle)

Cortex of stroma

Corpus hemorrhagicum (ruptured follicle)

Ovulation results in discharged secondary oocyte

Corpus albicans

Mature corpus luteum

Medulla of stroma

Blood clot

Early corpus luteum

What happens at each stage in the production and discharge a secondary oocyte?

which help gather ova into the tube following ovulation. From the infundibulum the uterine tube extends across and down to the upper outer corners of the uterus.

About once a month a vesicular ovarian (Graafian) follicle (developed from a secondary follicle) ruptures, releasing a secondary oocyte, a process called **ovulation**. The oocyte is swept into the uterine tube by the ciliary action of the epithelium of the infundibulum. The oocyte is then moved along the tube by the cilia of the tube's mucous lining and the peristaltic contractions of the muscle layer.

If the oocyte is fertilized by a spermatozoon, it usually occurs in the uterine tube. Fertilization may occur any time up to about 24 hours following ovulation. The fertilized ovum (zygote) descends into the uterus within 7 days. An unfertilized secondary oocyte disintegrates.

Ectopic (*ektopos* = displaced) **pregnancy (EP)** refers to the development of an embryo or fetus outside the uterine cavity. Most occur in the uterine tube. Some occur in the ovaries, abdomen, uterine cervix, and broad ligaments. The basic cause of a tubal pregnancy is impaired passage of the fertilized ovum through the uterine tube, which might be due to pelvic inflammatory disease (PID), previous uterine tube surgery, previous ectopic pregnancy, repeated elective abortions, pelvic tumors, or developmental abnormalities. Ectopic pregnancy is characterized by one or two missed periods, followed by vaginal bleeding and acute pelvic pain.

UTERUS

The **uterus** is the site of menstruation, implantation of a fertilized ovum, development of the fetus during pregnancy, and labor. It is situated between the urinary bladder and the rectum and is shaped like an inverted pear (see Figures 23-7 and 23-8).

Parts of the uterus include the dome-shaped portion above the uterine tubes called the **fundus**, the tapering central portion called the **body**, and the narrow portion opening into the vagina called the **cervix**. The interior of the body is called the **uterine cavity**.

The uterus is supported and held in position by the broad, uterosacral, cardinal, and round ligaments (see Figures 23-7 and 23-8). The latter maintain the uterus in an anteflexed position over the bladder. The broad ligament also forms part of the outer layer of the uterus, the **perimetrium.**

The middle muscular layer of the uterus, the **myometrium**, forms the bulk of the uterine wall. It consists of smooth muscle and is thickest in the fundus and thinnest in the cervix. During childbirth, coordinated contractions of the muscles help expel the fetus.

The innermost part of the uterine wall, the **endometrium**, is a mucous membrane composed of two layers. The **stratum basalis**, a permanent layer lying next to the myometrium, is supplied with blood by **straight arterioles**. The **stratum functionalis** surrounds the uterine cavity. It nourishes a growing fetus or is shed each month during menstruation if fertilization does not occur. It is

FIGURE 23-10 Oogenesis. The designation 2n means diploid; n means haploid.

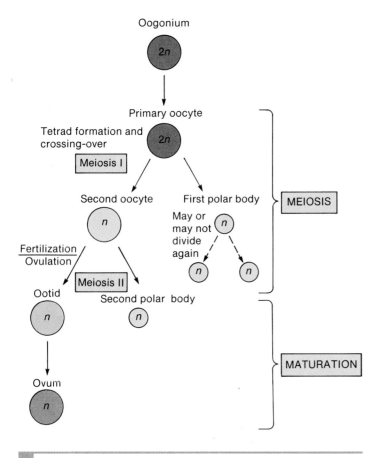

How do spermatogenesis and oogenesis differ?

nourished by *spiral arterioles* that undergo marked changes during the menstrual cycle (explained shortly). Following menstruation, it is replaced by the stratum basalis. The endometrium contains many glands whose secretions nourish sperm and the zygote. The mucosa of the cervix secretes mucus that changes in character radically during various stages of the menstrual cycle.

Cancer of the uterus can be diagnosed through the *Papanicolaou* (pap'-a-NIK-ō-la-oo) *test*, or *Pap smear*. In this generally painless procedure, a few cells from the part of the vagina surrounding the cervix and from the cervix are removed with a swab and examined microscopically. Malignant cells have a characteristic appearance that allows diagnosis even before symptoms occur. Estimates indicate that the Pap smear is more than 90 percent reliable in detecting cancer of the cervix.

VAGINA

The *vagina* serves as a passageway for menstrual flow and childbirth. It is also the receptacle for the penis during sexual intercourse (see Figures 23-7 and 23-8). It is situated between the urinary bladder and the rectum. A recess, called the *fornix*, surrounds the vaginal attachment to the cervix. The fornix makes it possible for a woman to use contraceptive diaphragms.

The mucosa of the vagina is continuous with that of the uterus and cervix and lies in a series of transverse folds, the *rugae*. The muscular layer is composed of smooth muscle that can stretch considerably to receive the penis during intercourse and allow for birth of the fetus. At the vaginal opening, the *vaginal orifice*, there may be a thin fold of vascularized mucous membrane called the *hymen*, which partially covers it (see Figure 23-1).

During the childbearing years, the mucosa of the vagina has an acid environment that retards microbial growth. However, the acidity is also injurious to sperm cells. Semen neutralizes the acidity of the vagina to ensure survival of the sperm.

VULVA

The term *vulva* (VUL-va), or *pudendum* (pyoo-DEN-dum), refers to the external genitalia of the female (Figure 23-11). Its components are as follows.

The *mons pubis* is an elevation of adipose tissue covered by coarse pubic hair which cushions the symphysis pubis during sexual intercourse. From the mons pubis, two longitudinal folds of skin, the *labia majora* (LĀ-bē-a ma-JŌ-ra), extend down and back. The labia majora and scrotum are equivalent structures. The labia majora contain adipose tissue and sebaceous (oil) and sudoriferous (sweat) glands; they are covered by pubic hair. Inside the labia majora are two folds of skin called the *labia minora* (MĪ-nō-ra). The labia minora do not contain pubic hair or fat and have few sudoriferous (sweat) glands; they do, however, contain numerous sebaceous (oil) glands.

The *clitoris* (KLI-to-ris) is a small, cylindrical mass of erectile tissue and nerves. It is located at the front junction of the labia minora. A layer of skin called the *prepuce* (foreskin) is formed at a point where the labia minora unite and cover the body of the clitoris. The exposed portion of the clitoris is the *glans*. The clitoris and glans penis of the male are equivalent structures. Like the penis, the clitoris is capable of enlargement upon tactile stimulation and assumes a role in sexual excitement.

The cleft between the labia minora is called the *vestibule*. In the vestibule are the hymen (if present), *vaginal orifice*, the opening of the vagina to the exterior; *external urethral orifice*, the opening of the urethra to the exterior; and on either side of the vaginal orifice, the openings of the ducts of the *paraurethral (Skene's) glands*. These glands are in the wall of the urethra and secrete mucus. The paraurethral glands and the male prostate are equivalent structures. On either side of the vaginal orifice itself are the *greater vestibular (Bartholin's) glands*, which produce a mucoid secretion that supplements lubrication during sexual intercourse. The greater vestibular glands and the male bulbourethral glands are equivalent structures.

PERINEUM

The *perineum* (per'-i-NĒ-um) is the diamond-shaped area between the thighs and buttocks of both males and females (Figure 23-11). It contains the external genitals and anus.

FIGURE 23-11 **Components of the vulva.**

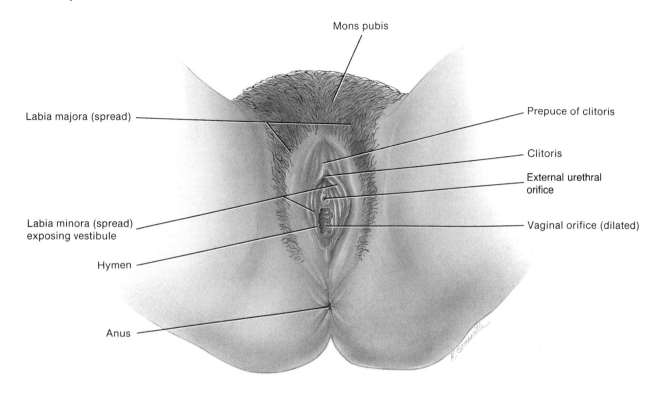

In addition to the vulva, what are the other female reproductive organs?

In the female, the region between the vagina and anus is known as the ***clinical perineum***. If the vagina is too small to accommodate the head of an emerging fetus, the skin, vaginal epithelium, subcutaneous fat, and muscle of the clinical perineum may tear. Moreover, the tissues of the rectum may be damaged. To avoid this, a small incision called an ***episiotomy*** (e-piz'-ē-OT-ōmē) is made in the perineal skin and underlying tissues just prior to delivery. After delivery the episiotomy is sutured in layers.

MAMMARY GLANDS

The ***mammary glands*** are modified sudoriferous (sweat) glands that lie over the pectoralis major and serratus anterior muscles and are attached to them by a layer of connective tissue (Figure 23-12). Internally, each mammary gland consists of 15 to 20 ***lobes*** arranged radially and separated by adipose tissue and strands of connective tissue, called ***suspensory ligaments of the breast (Cooper's ligaments)***, which support the breast. In each lobe are smaller ***lobules***, in which milk-secreting cells referred to as ***alveoli*** are found. Breast size has nothing to do with the amount of milk produced. Milk secretion by alveoli is transported through ducts that terminate in the ***nipple***. The circular pigmented area of skin surrounding the nipple is called the ***areola*** (a-RĒ-ō-la). It appears rough because it contains modified sebaceous (oil) glands.

At birth, both male and female mammary glands are undeveloped and appear as slight elevations on the chest. With the onset of puberty under the influence of estrogens and progesterone, the female breasts begin to develop; the duct system matures, and fat is deposited, which increases breast size to varying degrees. The areola and nipple also grow and become pigmented.

The essential function of the mammary glands is milk secretion and ejection, together called ***lactation.*** The secretion of milk following delivery is due largely to the hormone prolactin (PRL). The ejection of milk occurs in the presence of oxytocin (OT). Lactation is considered in detail in the next chapter.

Breast self-examination (BSE) and mammography are the most effective method of detecting ***breast cancer*** early. It is estimated that 95 percent of breast cancer is first detected by women themselves. Each month after the menstrual period the breasts should be thoroughly examined for lumps, puckering of the skin, or discharge.

Mammography is the most effective screening technique for routinely detecting tumors less than 1.27 cm (0.5 inch) in diameter. Mammography can also detect small calcium deposits, called microcalcifications, in breast tissue which frequently indicate the presence of a tumor.

One of the most recent breast-cancer-detecting procedures is ***ultrasound***, which can be used to determine whether a lump is a benign cyst or a malignant tumor.

Another technique combines computed tomography with mammography (CT/M) and is based on the fact that breast carcinoma

FIGURE 23-12 Mammary glands. (a) Sagittal section. (b) Anterior view of right mammary gland, partially sectioned.

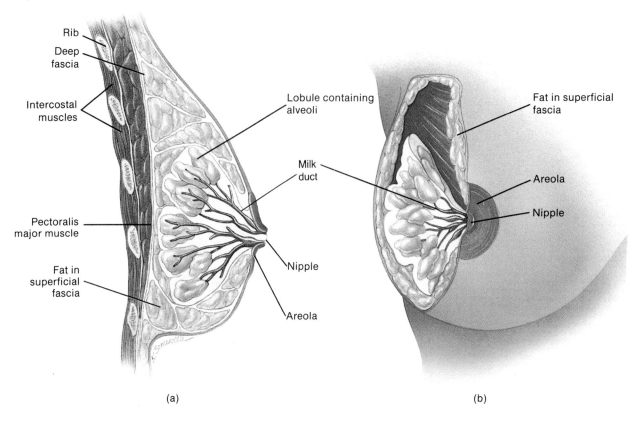

Rib
Deep fascia
Intercostal muscles
Pectoralis major muscle
Fat in superficial fascia

Lobule containing alveoli
Milk duct
Nipple
Areola

Fat in superficial fascia
Areola
Nipple

(a)   (b)

What hormones regulate lactation?

has an abnormal affinity for iodide. CT/M affords definitive diagnostic help when mammographic and physical examinations are inconclusive and appears to be a significantly improved method of breast cancer diagnosis.

FEMALE PUBERTY

The factors that determine the onset of female puberty are no better understood than those that determine male puberty. Prepubertal levels of LH, FSH, and estrogens are low. At around age seven or eight, the secretion of adrenal androgens begins to increase. The onset of puberty is signaled by sleep-associated surges in LH and FSH. As puberty progresses, LH and FSH levels increase throughout the day. The rising levels stimulate the ovaries to secrete estrogens, which bring about development of the secondary sex characteristics. These include fat distribution to the breasts, abdomen, mons pubis, and hips; increased vascularization of the skin; voice pitch; broad pelvis; and hair pattern. (Budding of the breasts is the first observable sign of puberty.) Estrogens also stimulate the growth of the uterine (Fallopian) tubes, uterus, and vagina

and bring on *menarche* (me-NAR-kē), the onset of menstruation, at an average age of 12. But, the first ovulation does not take place until 6 to 9 months after menarche because the positive feedback of estrogens on LH and FSH is the last step in the maturation of the hypothalamic–pituitary– ovarian loop, which is discussed next.

FEMALE REPRODUCTIVE CYCLE (FRC)

The general term *female reproductive cycle* (**FRC**) refers to the menstrual and ovarian cycles and the hormonal cycles that regulate them. It also includes other cyclic changes in female reproductive organs, such as those that relate to the breasts and cervical secretions. At this point, we will consider the menstrual and ovarian cycles and their hormonal regulation.

The *menstrual* (*mens* = monthly) *cycle* is a series of changes in the endometrium of a nonpregnant female. Each month the endometrium is prepared to receive a fertilized ovum. If no fertilization occurs, the stratum functionalis of the endometrium is shed. The *ovarian cycle* is a monthly series of events associated with the maturation of an ovum.

HORMONAL REGULATION

The menstrual cycle, ovarian cycle, and other changes associated with puberty in the female are regulated by gonadotropin releasing hormone (GnRH) from the hypothalamus (Figure 23-13). GnRH stimulates the release of follicle-stimulating hormone (FSH) from the anterior pituitary. FSH stimulates the initial development of the ovarian follicles and their secretion of estrogens. GnRH also stimulates the release of luteinizing hormone (LH) by the anterior pituitary, which stimulates the further development of ovarian follicles, brings about ovulation, and stimulates the production of estrogens, progesterone, inhibin, and relaxin by the corpus luteum.

Estrogens develop and maintain female reproductive structures, especially the endometrial lining of the uterus, and secondary sex characteristics, including the breasts.

Estrogens also control fluid and electrolyte balance and, with human growth hormone (hGH), increase protein anabolism. High levels of estrogens in the blood inhibit the release of GnRH by the hypothalamus, which, in turn, inhibits the secretion of FSH by the anterior pituitary gland. This inhibition provides the basis for action of one kind of contraceptive pill.

Progesterone (PROG) works with estrogens to prepare the endometrium for implantation of a fertilized ovum and the mammary glands for milk secretion. High levels of progesterone also inhibit GnRH and prolactin (PRL).

Inhibin is secreted by the corpus luteum and sustentacular (Sertoli) cells. In the female, it inhibits secretion of FSH and, to a lesser extent, LH, and may decrease secretion of FSH and LH toward the end of the menstrual cycle.

Relaxin is produced by the corpus luteum and placenta during pregnancy. It facilitates delivery by relaxing the symphysis pubis and helping to dilate the cervix. Relaxin also helps increase sperm motility.

MENSTRUAL PHASE (MENSTRUATION)

The menstrual cycle ranges from 24 to 35 days, with an average of 28 days. The menstrual cycle may be divided into three phases: the menstrual phase, the preovulatory phase, and the postovulatory phase (Figure 23-14).

The *menstrual phase*, or *menstruation,* is the periodic discharge

FIGURE 23-13 Secretion and physiological effects of estrogens, progesterone, and relaxin.

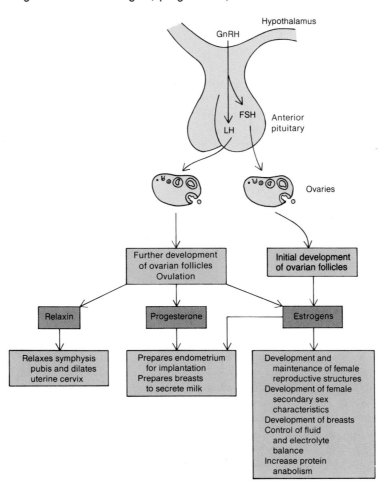

What are the effects of FSH and LH?

FIGURE 23-14 Correlation of menstrual and ovarian cycles with the hypothalamic and anterior pituitary gland hormones. In the cycle shown, fertilization and implantation have not occurred.

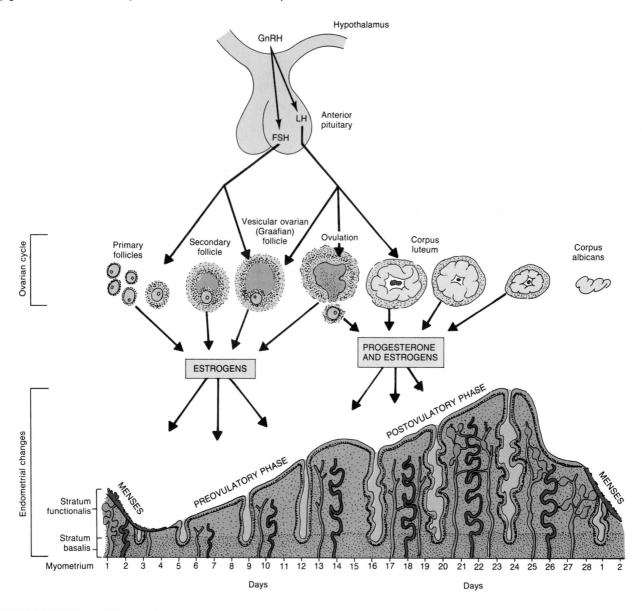

What brings about ovulation? A new menstrual cycle?

of 25 to 65 ml of blood, tissue fluid, mucus, and epithelial cells, collectively called **menses**. It is caused by a sudden reduction in estrogens and progesterone and lasts for approximately the first 5 days of the cycle. The first day of menstruation is designated as the first day of the cycle. Menstruation begins when the stratum functionalis layer degenerates, patchy areas of bleeding develop, and small areas of the stratum functionalis detach one at a time (total detachment would result in hemorrhage). When the entire stratum functionalis has been shed, the endometrium is very thin because only the stratum basalis remains.

During the menstrual phase, the ovarian cycle is also in operation. Ovarian follicles, called **primary follicles**, begin their development. At birth, each ovary contains about 200,000 follicles, each containing a primary oocyte. During the early part of each menstrual phase, 20 to 25 primary follicles start to produce very low levels of estrogens. Toward the end of the menstrual phase (days 4 and 5), about 20 primary follicles develop into **secondary follicles**. As a secondary follicle grows, a fluid is secreted that forces the secondary oocyte to the edge of the follicle and fills the follicular cavity. The secondary follicles also produce estrogens. Ovarian follicle development is the result of GnRH secretion by the hypothalamus, which, in turn, stimulates FSH production by the anterior pituitary. During this part of the cycle, FSH secretion is relatively high. Although about 20 follicles begin development each cycle, usually only 1 attains maturity and the rest degenerate.

PREOVULATORY PHASE

The second or *preovulatory phase* of the menstrual cycle is the time between menstruation and ovulation. This phase is more variable than the other phases, lasting from days 6 to 13 in a 28-day cycle.

FSH and LH stimulate the ovarian follicles to produce more estrogens, which, in turn, stimulate cells of the stratum basalis to undergo mitosis and produce a new stratum functionalis. Estrogens also cause spiral arterioles to coil and lengthen to nourish the stratum functionalis. Endometrial glands develop and the thickness of the endometrium approximately doubles. Estrogens are the dominant hormones during this phase of the menstrual cycle (Figure 23-15).

It is in the preovulatory phase that one of the secondary follicles matures into a *vesicular ovarian (Graafian) follicle,* a follicle ready for ovulation. As the follicle matures, it increases its estrogen production. Early in the preovulatory phase, FSH is the dominant hormone of the anterior pituitary, but close to the time of ovulation, LH is secreted in increasing quantities (Figure 23-15). Also, small amounts of progesterone may be produced by the vesicular ovarian (Graafian) follicle a day or two before ovulation.

OVULATION

Ovulation, the rupture of the vesicular ovarian (Graafian) follicle with release of the secondary oocyte into the pelvic cavity, usually

occurs on day 14 in a 28-day cycle. During ovulation, the secondary oocyte remains surrounded by a thick transparent membrane, the *zona pellucida,* and a covering of follicle cells directly around it. These follicle cells are referred to as the *corona radiata*. Just prior to ovulation, the high level of estrogens that developed during the preovulatory phase acts directly on the hypothalamus to secrete GnRH and indirectly on the anterior pituitary to release a surge of LH. Without this surge of LH, ovulation will not occur. (An over-the-counter home test that detects the LH surge a day in advance of ovulation is now available.) FSH also increases at this time, but not as dramatically as LH because FSH is stimulated only by the increase in GnRH. Following ovulation, the vesicular ovarian (Graafian) follicle collapses, and blood from the ruptured wall forms a clot *(corpus hemorrhagicum)* that is eventually absorbed by the remaining follicular cells. Under the influence of LH, the follicular cells enlarge, change character, and form the *corpus luteum*, or yellow body, which also secretes estrogens and progesterone (see Figure 23-9).

POSTOVULATORY PHASE

The *postovulatory phase* of the menstrual cycle is the most constant in duration and lasts from days 15 to 28 in a 28-day cycle. It is the time between ovulation and the onset of the next menses. The corpus luteum now secretes increasing quantities of estrogens and progesterone to prepare the endometrium to receive a fertilized ovum. Preparatory activities include stimulating the endometrial glands to secrete; increasing growth of arterioles into the new stratum functionalis; thickening of the endometrium; glycogen storage; and increasing the amount of tissue fluid. These preparatory changes are maximal about 1 week after ovulation when the arrival of the fertilized ovum is anticipated. During the postovulatory phase, FSH secretion again gradually increases and LH secretion decreases. The functionally dominant ovarian hormone during this phase is progesterone (Figure 23-15).

If fertilization occurs, the developing placenta secretes a hormone (human chorionic gonadotropin) that maintains the production of estrogens and progesterone by the corpus luteum and prevents menstruation. If fertilization does not occur, the corpus luteum degenerates into the *corpus albicans* (white body) and the production of estrogens and progesterone drops. As these hormone levels decrease, the spiral arterioles shrink, causing ischemia and necrosis (death) of the stratum functionalis of the endometrium. Eventually, part of the degenerated portion sloughs off and a new cycle commences with menstruation.

Meanwhile, the decreased levels of progesterone and estrogens bring about a new output of the anterior pituitary hormones—especially FSH in response to an increased output of GnRH by the hypothalamus. Thus, a new ovarian cycle is initiated. A summary of these hormonal interactions is presented in Figure 23-16.

MENOPAUSE

The menstrual cycle normally occurs once each month from menarche to *menopause* (*mens* = monthly; *pause* = to stop), the last menses. The advent of menopause is signaled by the *climacteric* (klī-MAK-ter-ik): menstrual cycles become less frequent. The cli-

FIGURE 23-15 Relative concentrations of anterior pituitary hormones (FSH and LH) and ovarian hormones (estrogens and progesterone) during a normal menstrual cycle.

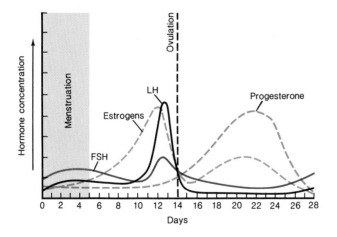

Why is LH secreted in increasing quantities close to the time of ovulation?

macteric typically begins between ages 40 and 50. Some women experience hot flashes, copious sweating, headache, insomnia, and emotional instability. In postmenopausal women there is some atrophy of the ovaries, uterine (Fallopian) tubes, uterus, vagina, external genitalia, and breasts.

Menopause occurs because of degeneration of the aging ovaries. Over the years of menstrual cycling, the number of primary follicles diminishes, causing the production of ova, estrogens, and progesterone by the ovary to decrease.

FIGURE 23-16 Summary of hormonal interactions of the menstrual and ovarian cycles.

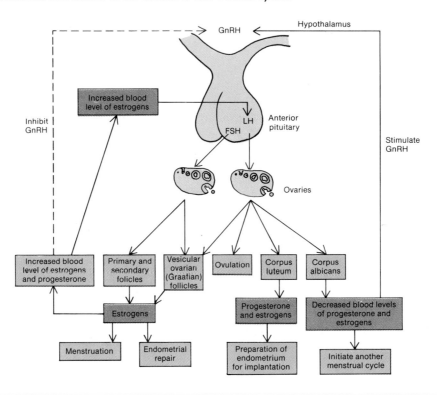

What is the dominant ovarian hormone in each phase of the menstrual cycle?

▪ COMMON DISORDERS ▪

Sexually Transmitted Diseases (STDs)

The general term *sexually transmitted disease* (*STD*) applies to any of the large groups of diseases that are contracted sexually or may be contracted otherwise but are then transmitted to a sexual partner. The group includes conditions traditionally specified as *venereal disease* (*VD*) (from Venus, goddess of love), such as gonorrhea, syphilis, and genital herpes, as well as other conditions.

Gonorrhea

Gonorrhea (or "*clap*") is an infectious sexually transmitted disease that primarily affects the mucous membrane of the urogenital tract, the rectum, and occasionally the eyes. It is caused by the bacterium *Neisseria gonorrhoeae*. Discharges from the infected mucous membranes transmit the bacteria by direct contact, usually sexual, or during passage of a newborn through the birth canal. Males usually suffer inflammation of the urethra with pus and painful urination. In females, infection may occur in the urethra, vagina, and cervix, often with a discharge of pus. However, frequently there are no noticeable symptoms. Treatment is with antibiotics, mainly penicillin or tetracyclines. Untreated gonorrhea in females may lead to sterility as a result of damage to the uterine (Fallopian) tubes.

Syphilis

Syphilis is a sexually transmitted disease caused by the bacterium *Treponema pallidum*. It is transmitted through sexual contact or through the placenta to a fetus. The disease progresses through several stages: primary, secondary, latent, and sometimes tertiary. During the *primary stage,* the chief symptom is an open sore, called a *chancre* (SHANKG-ker), at the point of contact. The chancre heals in 1 to 5 weeks. From 6 to 24 weeks later, symptoms such as a skin rash, fever, and aches in the joints and muscles usher in the *secondary stage.* These symptoms also eventually disappear in about 4 to 12 weeks and the disease ceases to be infectious, but a blood test for the presence of the bacteria generally remains positive. During this "symptomless" period, called the *latent stage*, the bacteria may invade body structures. When other structures become infected (usually the brain, heart valves, and aorta), the disease is said to be in the *tertiary stage*. Treatment is with penicillin.

Genital Herpes

The sexually transmitted disease **genital herpes** is common in North America and, unlike syphilis and gonorrhea, genital herpes is incurable. Type I herpes simplex virus causes most infections above the waist, such as cold sores. Type II herpes simplex virus causes most infections below the waist, such as painful genital blisters on the prepuce, glans penis, and penile shaft in males and on the vulva or sometimes high up in the vagina in females. The blisters disappear and reappear in most patients, but the virus itself remains in the body and causes recurrence of symptoms such as fever, enlarged and sometimes tender lymph nodes, and numerous clusters of genital blisters.

Trichomoniasis

The microorganism *Trichomonas vaginalis,* a flagellated protozoan, causes **trichomoniasis,** an inflammation of the mucous membrane of the vagina in females and the urethra in males. Symptoms include a yellow vaginal discharge and severe vaginal itch in women. Flagyl is used for treatment.

Chlamydia

Chlamydia (kla-MID-ē-a), a sexually transmitted disease caused by the bacterium *Chlamydia trachomatis,* currently is the most prevalent and damaging of the sexually transmitted diseases. In males, it is characterized by frequent, painful or burning urination and low back pain. In females, urethritis may spread through the reproductive tract and develop into inflammation of the uterine (Fallopian) tubes, which increases the risk of ectopic pregnancy and sterility. As in gonorrhea, infant's eyes may become infected during birth.

Testicular Cancer

Testicular cancer is associated with males with a history of late-descended or undescended testes. Most testicular cancers arise from the sperm-producing cells. Early signs of testicular cancer are a mass in the testis, often with pain or discomfort. Treatment involves removal of the diseased testis.

Prostate Disorders

The prostate gland is susceptible to infection, enlargement, and benign and malignant tumors. Because the prostate surrounds the urethra, any of these disorders can obstruct the flow of urine and result in serious changes in the urinary bladder, ureters, and kidneys and may perpetuate urinary tract infections. Therefore, if the obstruction cannot be relieved by other means, surgical removal (prostatectomy) of part of or the entire gland is necessary.

Prostate cancer is the second leading cause of death from cancer in men in the United States. Its incidence is related to age, race, occupation, geography, and ethnic origin. Both benign and malignant growths are common in elderly men. Both types of tumors put pressure on the urethra, making urination painful and difficult. At times, the excessive back pressure destroys kidney tissue and gives rise to an increased susceptibility to infection.

Sexual Functional Abnormalities

Impotence (*impotenia* = lack of strength) is the inability of an adult male to attain or hold an erection long enough for sexual intercourse. Impotence may be the result of diabetes mellitus, physical abnormalities of the penis, systemic disorders such as syphilis, vascular disturbances (arterial or venous obstructions), neurological disorders, testosterone deficiency, drugs (alcohol, antidepressants, antihistamines, antihypertensives, narcotics, nicotine, and tranquilizers), or psychic factors.

Male infertility (*sterility*) is an inability to fertilize the ovum. It does not imply impotence. Male fertility requires production of adequate amounts of viable, normal spermatozoa by the testes, unobstructed transportation of sperm through the seminal tract, and satisfactory deposition in the vagina.

Female Disorders

Menstrual Abnormalities

Amenorrhea (ā-men′-ō-RĒ-a; *a* = without; *men* = month; *rhein* = to flow) is the absence of menstruation. It can be caused by endocrine disorders, congenitally abnormal development of the ovaries or uterus, changes in body weight, or continuous, rigorous athletic training.

Dysmenorrhea (dis′-men-ō-RĒ-a; *dys* = difficult) is painful menstruation caused by forceful contraction of the uterus. It is often accompanied by nausea, vomiting, diarrhea, headache, fatigue, and nervousness. Some cases are caused by pathological conditions such as uterine tumors, ovarian cysts, endometriosis, and pelvic inflammatory disease (PID). However, other cases of dysmenorrhea are not related to any pathologies.

Abnormal uterine bleeding includes menstruation of excessive duration or excessive amount, too frequent menstruation, intermenstrual bleeding, and postmenopausal bleeding. These abnormalities may be caused by improper hormonal regulation, emotional factors, fibroid tumors of the uterus, or systemic diseases.

Premenstrual syndrome (*PMS*) refers to severe physical and emotional distress occurring late in the postovulatory phase of the menstrual cycle and sometimes overlapping with menstruation. Symptoms usually increase in severity until the onset of menstruation and then dramatically disappear. Among the symptoms are edema, weight gain, breast swelling and tenderness, abdominal distention, backache, joint pain, constipation, skin eruptions, fatigue and lethargy, greater need for sleep, depression or anxiety, irritability, mood swings, headache, poor coordina-

Life-Style and the Menstrual Cycle

Gone are the days when girls and women were forced into confinement during their menstrual periods. Who can afford to take that much time off? Besides, we now acknowledge that menstruation is a normal, healthy, biological function that is part of the menstrual cycle and need not interfere with a woman's daily life.

But while a woman's life goes on whether or not she is having her period, few women in their reproductive years would say that their lives are unaffected by the reproductive cycle. Dysmenorrhea and premenstrual syndrome are two of the most common complaints. And while to many women the absence of the menstrual cycle might sound like a blessing, amenorrhea and oligomenorrhea (irregular and infrequent menstrual periods) are a cause for concern.

Dysmenorrhea

Many women experience some dysmenorrhea, especially during the first one to three days of menstruation. The dysmenorrhea may range from occasional, mild cramps to severe pain. Dysmenorrhea is caused by uterine spasms, which deprive the uterine muscle of oxygen, rather like angina pectoris, which results from lack of oxygen in the heart muscle. Uterine spasms are caused by prostaglandins, whose levels increase with the rise in progesterone following ovulation. Women who experience dysmenorrhea should check with their physicians to rule out underlying problems such as pelvic inflammatory disease and endometriosis.

Some women notice an association between various life-style factors, such as exercise and diet, and the degree of dysmenorrhea they experience. These responses vary considerably from person to person, however, so it is difficult to generalize. For example, some women report that they experience less dysmenorrhea after beginning a program of regular exercise, while others find that exercise has no effect. Some find that mild exercise, such as walking, during menstruation relieves some of the cramping, while others say exercise depletes their already low energy levels.

The dietary factor most frequently linked to dysmenorrhea is salt. Water retention is a common complaint during the day or two preceding menstruation, as well as during the first day or two of the period. Many women find that a high salt intake, especially at these times, increases fluid retention and dysmenorrhea, but others find that decreasing their salt intake brings no relief.

Home remedies for the treatment of dysmenorrhea abound. These include stretching exercises, especially for the lower back area, hot drinks, massage, and the use of hot water bottles and heating pads applied to the lower back and abdominal area.

The most effective treatment for dysmenorrhea is the use of inexpensive, over-the-counter drugs that suppress prostaglandins. These include aspirin and ibuprofen. Research has shown that these medications relieve dysmenorrhea in about 80 percent of the women tested. Women who experience cramps with each menstrual period are advised to take the medication the day before the period is due and for the first one or two days of the period.

Oral contraceptives often diminish dysmenorrhea, since they prevent ovulation and thus also prevent high levels of progesterone and prostaglandins. They are most effective when taken regularly as prescribed, not just when dysmenorrhea occurs.

Premenstrual Syndrome (PMS)

Like dysmenorrhea, PMS varies in its severity and its relationship to life-style factors and requires consultation with a gynecologist to rule out any underlying pathology.

Women who find PMS to be a significant burden sometimes find keeping a diary of life-style factors and PMS symptom severity helpful. For example, if salt intake is kept low for several months and PMS symptoms improve, then salt intake is allowed to rise and PMS symptoms observed. If they become worse again, the woman can be fairly sure that the low-salt diet was the cause of the previous months' symptom relief. Other dietary culprits that may increase PMS symptoms include foods high in sugar or fat, alcohol, and caffeine.

The prostaglandin-suppressing drugs that help relieve dysmenorrhea often help relieve symptoms of PMS as well. Diuretics and vitamin B_6 have both been found helpful by some women as well.

One study found that some women experienced fewer PMS symptoms after beginning a jogging program. Subjects in this study averaged only one and a half miles per day but reported improvement in several symptoms such as breast tenderness and irritability.

Menstrual Irregularity

Amenorrhea means an absence of menstrual periods, and oligomenorrhea refers to irregular periods. A woman with oligomenorrhea may have a few regular periods, then go for over 40 days with no period, start menstruating again, and then have no period for two months. It's the irregularity of the cycle that defines a woman as oligomenorrheic.

The exact definition of these terms is not so important to the woman who is trying to figure out whether or not to notify her doctor. Physicians usually recommend that if a girl or woman stops having menstrual periods, or starts menstruating irregularly, she should consult her doctor. Such changes in the menstrual cycle sometimes signal serious underlying physiological disorders, which should be diagnosed and treated immediately.

In a majority of cases, menstrual irregularity is not caused by any discernible physiological disorder. Some women experience a temporary change in menstrual schedule when they travel or undergo other major changes in routine and life-style. More serious and sustained menstrual irregularity can be caused by a number of life-style factors (besides pregnancy). Vigorous exercise, low caloric intake, eating disorders, low levels of body fat, and psychological stress can all cause amenorrhea. Amenorrhea seems to occur more frequently in younger women than in women who have had a regular menstrual cycle for a number of years.

When menstrual irregularity is apparently caused by one or more of these life-style factors and is not associated with another physical disorder, it may still signal a hormonal imbalance. While short periods of menstrual irregularity may cause no lasting harm, long-term hormonal imbalance may be accompanied by a decrease in bone density, which can lead to irreversible bone damage. If menstrual irregularity is accompanied by a lack of ovulation, then a significant source of estrogen production is absent. This is a concern not only for women wishing to conceive, because estrogen has several important health effects in addition to those related to reproductive function. For one thing, estrogen helps bones retain calcium. That's why many women develop osteoporosis after menopause, when estrogen levels drop.

Amenorrheic athletes have provided interesting examples of what can happen to bones when estrogen levels remain low for an extended period of time. Exercise has been shown to increase bone density, especially if the exercise involves bone stress, as running does. Thus, researchers were surprised to discover that many amenorrheic runners suffered from a significant degree of osteoporosis. These athletes experienced frequent injuries and sustained many stress fractures. Bone density measurements revealed that many of these young women had the bone density of women in their fifties. Particularly disturbing were changes in the density of the vertebrae. Some subjects were already showing evidence of crush fractures, which are not reversible.

Estrogen has also been associated with low-risk blood cholesterol profiles. Premenopausal women generally have higher levels of HDL cholesterol than do men. One study of amenorrheic runners found that these women had much lower levels of this protective lipoprotein than runners with normal menstrual periods.

Menstrual irregularity related to life-style factors is usually reversible once indicated changes are made. For example, treatment of eating disorders eventually results in the resumption of a normal cycle. Decreasing activity level usually brings back a regular cycle in female athletes. If bone density is a concern, the physician may recommend more aggressive treatment, such as hormone replacement therapy.

tion and clumsiness, and cravings for sweet or salty foods. The basic cause of PMS is unknown.

Toxic Shock Syndrome (TSS)

Toxic shock syndrome (TSS) is primarily a disease of previously healthy young, menstruating females who use tampons, but it is also recognized in males, children, and nonmenstruating females. Clinically, TSS is characterized by high fever up to 40.6°C (105°F), sore throat or very tender mouth, headache, fatigue, irritability, muscle soreness and tenderness, conjunctivitis, diarrhea and vomiting, abdominal pain, vaginal irritation, and rash. Other symptoms include lethargy, unresponsiveness, memory loss, hypotension, peripheral vasoconstriction, respiratory distress syndrome, intravascular coagulation, decreased platelet count, renal failure, circulatory shock, and liver involvement.

Toxin-producing strains of the bacterium *Staphylococcus aureus* are necessary for development of the disease. Actually, it appears that a virus has become incorporated into *S. aureus*, causing the bacterium to produce the toxins. Although all tampon users are at some risk for developing TSS, the risk is increased considerably by females who use highly absorbent tampons. Apparently, high-absorbency tampons provide a substrate on which the bacteria grow and produce toxins. There is some evidence that the absorption of magnesium by the tampon fibers is also a factor. As the metal is absorbed, its absence in the reproductive tract stimulates the staphylococci to produce toxin. TSS can also occur as a complication of influenza and influenza-like illness and use of contraceptive sponges.

Ovarian Cysts

Ovarian cysts are fluid-containing tumors of the ovary. Follicular cysts may occur in the ovaries of elderly women, in ovaries that have inflammatory diseases, and in menstruating females. They have thin walls and contain a serous albuminous material. Cysts may also arise from the corpus luteum or the endometrium.

Endometriosis

Endometriosis (en'-dō-mē-trē-Ō-sis; *endo* = within; *metri* = uterus; *osis* = condition) is a benign condition characterized by the growth of endometrial tissue outside the uterus. The tissue enters the pelvic cavity via the open uterine (Fallopian) tubes and may be found on the ovaries, surface of the uterus, sigmoid colon, pelvic and abdominal lymph nodes, cervix, abdominal wall, kidneys, and urinary bladder. Symptoms include premenstrual or unusual menstrual pain. The unusual pain is caused by the displaced tissue sloughing off at the same time the normal uterine endometrium is being shed during menstruation. Infertility can be a consequence. Endometriosis disappears at menopause or when the ovaries are removed.

Female Infertility

Female infertility, or the inability to conceive, occurs in about 10 percent of married females in the United States. Once it is established that ovulation occurs regularly, the reproductive tract is examined for functional and anatomical disorders to determine the possibility of union of the spermatozoon and the ovum in the uterine tube. Female infertility may be caused by obstruction in the uterine (Fallopian) tubes, ovarian disease, and certain conditions of the uterus. An upset in hormone balance, so that the endometrium is not adequately prepared to receive the fertilized ovum, may also be the problem.

Disorders Involving the Breasts

In the female, benign **fibroadenoma** is a common tumor of the breast. It occurs most frequently in young women. Fibroadenomas have a firm rubbery consistency and are easily moved about within the mammary tissue. The usual treatment is excision of the growth.

Breast cancer has one of the highest fatality rates of all cancers affecting women, but it is rare in men. In the female, breast cancer is rarely seen before age 30, and its occurrence rises rapidly after menopause. Breast cancer is generally not painful until it becomes quite advanced, so often it is not discovered early or, if noted, is ignored. Any lump, no matter how small, should be reported to a doctor at once. New evidence links some breast cancers to loss of protective antioncogenes.

Among the factors that clearly increase the risk of breast cancer development are (1) a family history of breast cancer, especially in a mother or sister; (2) never having a child or having a first child after age 34; (3) previous cancer in one breast; (4) exposure to ionizing radiation; and (5) excessive fat and alcohol intake. Females who take birth control pills do not have a higher risk of developing breast cancer than females who do not. Cigarette smoking may increase the incidence of breast cancer, especially in postmenopausal females.

The American Cancer Society recommends the following steps to help diagnose breast cancer as early as possible:

1. A mammogram should be taken between the ages of 35 and 39, to be used later for comparison (baseline mammogram).
2. A physician should examine the breasts every three years when a female is between the ages of 20 and 40, and every year after 40.
3. Females with no symptoms should have a mammogram every year or two between ages 40 and 49, and every year after 50.
4. Females of any age with a history of breast cancer, a strong family history of the disease, or other risk factors should consult a physician to determine a schedule for mammography.
5. All females over 20 should develop the habit of monthly breast self-examination (BSE).

Cervical Cancer

Another common disorder of the female reproductive tract is **cervical cancer**, carcinoma of the cervix of the uterus. The condition starts with **cervical dysplasia** (dis-PLĀ-sē-a), a change in the shape, growth, and number of the cervical cells. If the condition is minimal, the cells may regress to normal. If it is severe, it may progress to cancer. Cervical cancer may be detected in most cases in its earliest stages by a Pap smear. There is some evidence linking cervical cancer to penile virus (papillomavirus) infections of male sexual partners. Depending on the progress of the disease, treatment may consist of excision of lesions, radiotherapy, chemotherapy, and hysterectomy.

Pelvic Inflammatory Disease (PID)

Pelvic inflammatory disease (PID) is a collective term for any extensive bacterial infection of the pelvic organs, especially the uterus, uterine (Fallopian) tubes, or ovaries. PID is most commonly caused by the bacterium that causes gonorrhea, but any bacterium can trigger infection. Often the early symptoms of PID, which include increased vaginal discharge and pelvic pain, occur just after menstruation. As infection spreads, fever may develop in advanced cases along with painful abscesses of the reproductive organs. Early treatment with antibiotics (tetracycline or penicillin) can stop the spread of PID.

MEDICAL TERMINOLOGY AND CONDITIONS

Culdoscopy (kul-DOS-kō-pē; *skopein* = to examine) A procedure in which a culdoscope (endoscope) is used to view the pelvic cavity. The approach is through the vagina.

Hermaphroditism (her-MAF-rō-di-tizm′) Presence of both male and female sex organs in one individual.

Hypospadias (hī′-pō-SPĀ-dē-as; *hypo* = below; *span* = to draw) A displaced urethral opening. In the male, the opening may be on the underside of the penis, at the penoscrotal junction, between the scrotal folds, or in the perineum. In the female, the urethra opens into the vagina.

Leukorrhea (loo′-kō-RĒ-a; *leuco* = white; *rrhea* = discharge) A nonbloody vaginal discharge that may occur at any age and affects most women at some time.

Salpingectomy (sal′-pin-JEK-tō-mē; *salpingo* = tube) Excision of a uterine (Fallopian) tube.

Smegma (SMEG-ma; *smegma* = soap) A secretion, consisting principally of desquamated epithelial cells, found chiefly about the external genitalia and especially under the foreskin of the male.

Vaginitis (vaj′-i′-NĪ-tis) Inflammation of the vagina.

S TUDY OUTLINE

Male Reproductive System (p. 485)

1. Reproduction is the process by which genetic material is passed on from one generation to the next.
2. The organs of reproduction are grouped as gonads (produce gametes), ducts (transport and store gametes), and accessory sex glands (produce materials that support gametes).
3. The male structures of reproduction include the testes, ductus epididymis, ductus (vas) deferens, ejaculatory duct, urethra, seminal vesicles, prostate gland, bulbourethral (Cowper's) glands, and penis.

Scrotum (p. 485)

1. The scrotum supports the testes.
2. It regulates the temperature of the testes by elevating them closer to the pelvic cavity.

Testes (p. 485)

1. The testes are oval-shaped glands (gonads) in the scrotum containing seminiferous tubules, in which sperm cells are made; sustentacular (Sertoli) cells, which nourish sperm cells and produce inhibin; and interstitial endocrinocytes (cells of Leydig), which produce the male sex hormone testosterone.
2. Ova and sperm are collectively called gametes, or sex cells, and are produced in gonads.
3. Somatic cells divide by mitosis, the process in which each daughter cell receives the full complement of 23 chromosome pairs (46 chromosomes). Somatic cells are therefore diploid.
4. Immature gametes divide by meiosis in which the pairs of chromosomes are split so that the mature gamete has only 23 chromosomes. Sperm are therefore haploid.
5. Spermatogenesis occurs in the testes. It results in the formation of four haploid spermatozoa.
6. Spermatogenesis consists of meiosis I, meiosis II, and spermiogenesis.
7. Mature spermatozoa consist of a head, midpiece, and tail. Their function is to fertilize an ovum.
8. At puberty, gonadotropin releasing hormone (GnRH) stimulates anterior pituitary secretion of FSH and LH. FSH initiates spermatogenesis, and LH assists spermatogenesis and stimulates production of testosterone.

9. Testosterone controls the growth, development, and maintenance of sex organs; stimulates bone growth, protein anabolism, and sperm maturation; and stimulates development of male secondary sex characteristics.
10. Inhibin is produced by sustentacular (Sertoli) cells. Its inhibition of FSH helps regulate the rate of spermatogenesis.
11. Puberty refers to the period of time when secondary sex characteristics begin to develop and the potential for sexual reproduction is reached.
12. The onset of male puberty is signaled by increased levels of LH, FSH, and testosterone.

Ducts (p. 490)

1. The duct system of the testes includes the seminiferous tubules, straight tubules, and rete testis.
2. Sperm are transported out of the testes into an adjacent organ, the epididymis.
3. The ductus epididymis is the site of sperm maturation and storage.
4. The ductus (vas) deferens stores sperm and propels them toward the urethra during ejaculation.
5. Alteration of the ductus (vas) deferens to prevent fertilization is called vasectomy.
6. The ejaculatory ducts are formed by the union of the ducts from the seminal vesicles and ductus (vas) deferens and eject spermatozoa into the urethra.
7. The male urethra passes through the prostate gland, urogenital diaphragm, and penis.

Accessory Sex Glands (p. 490)

1. The seminal vesicles secrete an alkaline, viscous fluid that constitutes most of the volume of semen and contributes to sperm viability.
2. The prostate gland secretes an alkaline fluid that contributes to sperm motility.
3. The bulbourethral (Cowper's) glands secrete mucus for lubrication and a substance that neutralizes acid.
4. Semen (seminal fluid) is a mixture of spermatozoa and accessory gland secretions that provides the fluid in which spermatozoa are transported, provides nutrients, and neutralizes the acidity of the male urethra and female vagina.

Penis (p. 491)

1. The penis is the male organ of copulation.
2. Expansion of its blood sinuses under the influence of sexual excitation is called erection.

Female Reproductive System (p.492)

1. The female organs of reproduction include the ovaries (gonads), uterine (Fallopian) tubes, uterus, vagina, and vulva.
2. The mammary glands are considered part of the reproductive system.

Ovaries (p. 493)

1. The ovaries are female gonads located in the upper pelvic cavity on either side of the uterus.
2. They produce secondary oocytes, discharge secondary oocytes (ovulation), and secrete estrogens, progesterone, inhibin, and relaxin.

3. Oogenesis occurs in the ovaries. It results in the formation of a single haploid ovum.
4. Oogenesis consists of meiosis I, meiosis II, and maturation.

Uterine (Fallopian) Tubes (p. 494)

1. The uterine (Fallopian) tubes transport ova from the ovaries to the uterus and are the normal sites of fertilization.
2. Implantation outside the uterus is called an ectopic pregnancy.

Uterus (p. 495)

1. The uterus is an inverted, pear-shaped organ that functions in menstruation, implantation of a fertilized ovum, development of a fetus during pregnancy, and labor.
2. The innermost layer of the uterine wall is the endometrium, which undergoes marked changes during the menstrual cycle.

Vagina (p. 496)

1. The vagina is a passageway for the menstrual flow, the receptacle for the penis during sexual intercourse, and the lower portion of the birth canal.
2. It is capable of considerable stretching to accomplish its functions.

Vulva (p. 496)

1. The vulva is a collective term for the external genitals of the female.
2. It consists of the mons pubis, labia majora, labia minora, clitoris, vestibule, vaginal and urethral orifices, and the paraurethral (Skene's) and greater vestibular (Bartholin's) glands.

Perineum (p. 496)

1. The perineum is a diamond-shaped area at the inferior end of the trunk between the thighs and buttocks.
2. An incision in the perineal skin prior to delivery is called an episiotomy.

Mammary Gland (p. 497)

1. The mammary glands are modified sweat glands located over the pectoralis major muscles. Their function is to secrete and eject milk (lactation).
2. Mammary gland development depends on estrogens and progesterone.
3. Milk secretion is due mainly to the hormone prolactin (PRL).

Female Reproductive Cycle (FRC) (p. 498)

1. The purpose of the menstrual cycle is to prepare the endometrium each month for the reception of a fertilized egg. The ovarian cycle is associated with the maturation of an ovum each month.
2. The menstrual and ovarian cycles are regulated by GnRH, which stimulates the release of FSH and LH.
3. FSH stimulates the initial development of ovarian follicles and secretion of estrogens by the ovaries. LH stimulates further development of ovarian follicles, ovulation, and the secretion of estrogens and progesterone by the ovaries.
4. Estrogens stimulate the growth, development, and maintenance

of female reproductive structures; stimulate the development of secondary sex characteristics; regulate fluid and electrolyte balance; and stimulate protein anabolism.

5. Progesterone (PROG) works with estrogens to prepare the endometrium for implantation and the mammary glands for milk secretion.

6. Relaxin relaxes the symphysis pubis, helps dilate the uterine cervix to facilitate delivery, and increases sperm motility.

7. During the menstrual phase, the stratum functionalis layer of the endometrium is shed with a discharge of blood, tissue fluid, mucus, and epithelial cells. Primary follicles develop into secondary follicles.

8. During the preovulatory phase, endometrial repair occurs. A secondary follicle develops into a vesicular (Graafian) follicle. Estrogens are the dominant ovarian hormones.

9. Ovulation is the rupture of a vesicular (Graafian) follicle and the release of a secondary oocyte (immature ovum) into the pelvic cavity brought about by inhibition of FSH and a surge of LH.

10. During the postovulatory phase, the endometrium thickens in anticipation of implantation. Progesterone is the dominant ovarian hormone.

11. If fertilization and implantation do not occur, the corpus luteum degenerates and low levels of estrogens and progesterone initiate another menstrual and ovarian cycle.

12. If fertilization and implantation do occur, the corpus luteum is maintained by a placental hormone until the placenta itself takes over secretion of estrogens and progesterone.

13. Climacteric is a period of infrequent menstrual cycles; menopause is the cessation of the sexual cycles.

14. The onset of female puberty is signaled by increased levels of LH, FSH, and estrogens.

REVIEW QUESTIONS

1. Define reproduction. Describe how the reproductive organs are classified and list the male and female organs of reproduction. (p. 485)

2. Describe the internal structure of a testis. Where are sperm cells made? (p. 485)

3. Describe the principal events of spermatogenesis. Why is meiosis important? Distinguish between haploid and diploid cells. (p. 486)

4. Explain the effects of FSH and LH on the male reproductive system. How are these hormones controlled by GnRH? (p. 489)

5. Describe the physiological effects of testosterone and inhibin on the male reproductive system. How is testosterone level controlled? (p. 489)

6. How is male puberty controlled? What changes are associated with male puberty? (p. 489)

7. Describe the functions of the ductus epididymis, ductus (vas) deferens, and ejaculatory duct. (p. 490)

8. Trace the course of spermatozoa through the male system of ducts from the seminiferous tubules through the urethra. (p. 490)

9. Explain the functions of the seminal vesicles, prostate gland, and bulbourethral (Cowper's) glands. (p. 490)

10. How is the penis structurally adapted as an organ of copulation. How does an erection occur? (p. 491)

11. Describe the parts of an ovary. What are the functions of the ovaries? (p. 493)

12. Describe the principal events that take place during oogenesis. (p. 493)

13. Where are the uterine (Fallopian) tubes located? What is their function? What is an ectopic pregnancy? (p. 494)

14. Describe the functions of the uterus. (p. 495)

15. What is the function of the vagina? (p. 496)

16. List the parts of the vulva and the functions of each part. (p. 496)

17. What is the perineum? Define episiotomy. (p. 496)

18. Describe the structure of the mammary glands. Describe the passage of milk from the alveolar cells of the mammary gland to the nipple. (p. 497)

19. How is female puberty controlled? What changes are associated with female puberty? (p. 498)

20. Define the menstrual cycle and ovarian cycle. What is the function of each of the following in the menstrual and ovarian cycles: GnRH, FSH, LH, estrogens, progesterone (PROG), and relaxin? (p. 498)

21. Define the following: gonorrhea, syphilis, genital herpes, trichomoniasis, chlamydia, testicular cancer, impotence, infertility, amenorrhea, dysmenorrhea, premenstrual syndrome (PMS), toxic shock syndrome (TSS), ovarian cysts, endometriosis, breast cancer, and pelvic inflammatory disease (PID). (p. 502)

22. Refer to the glossary of medical terminology and conditions associated with reproduction. Be sure that you can define each term. (p. 507)

24
Development and Inheritance

STUDENT OBJECTIVES

1. Describe the roles of the male and female in sexual intercourse.
2. Explain how an ovum is fertilized and how it implants in the uterus.
3. Discuss alternative procedures to natural fertilization and implantation.
4. Describe the principal events associated with embryonic and fetal development.
5. Explain the functions of the hormones secreted during pregnancy.
6. Describe some of the changes in the mother's body associated with pregnancy.
7. Describe the stages of labor.
8. Discuss how lactation occurs and how it is controlled.
9. Compare the various kinds of birth control (BC) methods and their effectiveness.
10. Describe the basic principles of inheritance.
11. Define medical terminology and conditions associated with development and inheritance.

A LOOK AHEAD

SEXUAL INTERCOURSE
  Male Sexual Act
    Erection
    Lubrication
    Orgasm
  Femal Sexual Act
    Erection
    Lubrication
    Orgasm (Climax)
DEVELOPMENT DURING PREGNANCY
  Fertilization
  Formation of the Morula
  Development of the Blastocyst
  Implantation
EMBRYONIC DEVELOPMENT
  Beginnings of Organ Systems
    Embryonic Membranes
    Placenta and Umbilical Cord
    Fetal Circulation
FETAL GROWTH
HORMONES OF PREGNANCY
GESTATION
PRENATAL DIAGNOSTIC TECHNIQUES
  Amniocentesis
  Chorionic Villus Sampling (CVS)
PARTURITION AND LABOR
LACTATION
BIRTH CONTROL (BC)
  Sterilization
  Hormonal
  Intrauterine Devices (IUDs)
  Barrier
  Chemical
  Physiologic (Natural)
  Coitus Interruptus (Withdrawal)
  Induced Abortion
INHERITANCE
  Genotype and Phenotype
  Genes and the Environment
  Inheritance of Sex
  Color Blindness and X-Linked
    Inheritance
MEDICAL TERMINOLOGY AND
    CONDITIONS

Now that you have learned about the structure and function of the female and male reproductive systems, we will examine the physiology of sexual intercourse, the principal events of pregnancy and lactation, methods of birth control, and the basic principles of inheritance.

SEXUAL INTERCOURSE

Sexual intercourse, or *copulation* (in humans, called *coitus*; KŌ-i-tus), is the process by which spermatozoa are deposited in the vagina.

MALE SEXUAL ACT

Erection

The male role in the sexual act starts with *erection*, the enlargement and stiffening of the penis. An erection may be initiated in the cerebrum by stimuli such as anticipation, memory, and visual sensation, or it may be a reflex brought on by stimulation of the touch receptors in the penis, especially in the glans. In either case, parasympathetic impulses from the sacral portion of the spinal cord to the penis cause dilation of the arteries to the penis, filling spaces in the corpora cavernosa and corpus spongiosum with blood, resulting in erection.

Lubrication

Parasympathetic impulses from the sacral cord also cause the bulbourethral (Cowper's) glands to secrete mucus through the urethra, which affords some *lubrication* for intercourse. However, the female's cervical mucus provides most of the lubrication. Without satisfactory lubrication, the male sexual act is difficult; unlubricated intercourse causes pain and inhibits coitus.

Orgasm

Tactile stimulation of the penis brings about emission and ejaculation. When sexual stimulation becomes intense, rhythmic sympathetic impulses leave the upper lumbar spinal cord and stimulate peristaltic contractions of the ducts of the testes, epididymides, and ductus (vas) deferens that propel spermatozoa into the urethra—a process called *emission*. Simultaneously, peristaltic contractions of the seminal vesicles and prostate expel seminal and prostatic fluid along with the spermatozoa. All these mix with the mucus of the bulbourethral glands to produce semen. Other rhythmic impulses sent from the sacral spinal cord stimulate skeletal muscles at the base of the penis which then expels the semen. The propulsion of semen from the urethra to the exterior constitutes an *ejaculation*.

A number of sensory and motor responses accompany ejaculation, including rapid heart rate, increased blood pressure, increased respiration, and pleasurable sensations. These responses, together with the muscular events involved in ejaculation, constitute an *orgasm*.

FEMALE SEXUAL ACT

Erection

The female role in the sex act also involves erection, lubrication, and orgasm. Stimulation of the female also depends on both psychic and tactile responses. Under appropriate conditions, stimulation of the female genitalia results in clitoral erection and widespread sexual arousal. This response is also controlled by parasympathetic impulses from the sacral spinal cord.

511

Lubrication

Parasympathetic impulses from the sacral spinal cord also stimulate *lubrication* of the vagina. The cervical mucosa produces a mucoid fluid and some mucus is also produced by the greater vestibular (Bartholin's) glands.

Orgasm (Climax)

When genital stimulation reaches maximum intensity, it initiates reflexes that produce an *orgasm (climax)*. Female orgasm also causes increased respirations, pulse, and blood pressure and muscular contractions of the reproductive organs which may aid fertilization of an ovum (to be described shortly). However, unlike males who experience a single climax with ejaculation, females may have multiple orgasms of varying intensity with accompanying pleasurable sensations, but no ejaculation.

About 15 minutes after semen is deposited in the vagina, it liquefies to allow sperm to swim into the cervix aided by perineal muscle contractions. Prostaglandins in semen and oxytocin (OT) released from the posterior pituitary during female orgasm may stimulate uterine contractions to further aid the movement of sperm toward the uterine (Fallopian) tubes.

DEVELOPMENT DURING PREGNANCY

Once spermatozoa and ova are developed through meiosis and maturation and the spermatozoa are deposited in the vagina, pregnancy can occur. *Pregnancy* is the sequence of events that includes fertilization, implantation, and embryonic and fetal growth that terminates in birth. We will now examine these events in detail.

FERTILIZATION

The term *fertilization* refers to the penetration of a secondary oocyte by a spermatozoon and the union of the chromosomes from the spermatozoon and oocyte to form a zygote. Of the hundreds of millions of sperm cells introduced into the vagina, less than 1 percent contact the oocyte. Fertilization normally occurs in the first third of the uterine (Fallopian) tube within 24 hours after ovulation.

In order for sperm to fertilize a secondary oocyte, they must shed the protective covering of the acrosome during their ascent through the female reproductive tract, in a process called *capacitation* (ka'-pas'-i-TĀ- shun). This allows enzymes from the acrosome to penetrate the *corona radiata*, several layers of follicular cells around the oocyte, and *zona pellucida* (pe-LOO-si-da), a jellylike layer under the corona radiata (Figure 24-1). The enzymes from many sperm are needed before one spermatozoon's head can finally break through. Once this occurs, the zona pellucida undergoes cellular changes that block the entry of other sperm.

When a spermatozoon penetrates a secondary oocyte, it triggers the completion of meiosis II and oogenesis. This produces a second tiny polar body (which also disintegrates) and a larger ovum with 23 chromosomes. The nuclear membranes of the ovum and spermatozoon disintegrate: the haploid chromosomes of each cell then combine to form a diploid cell called the fertilized ovum or *zygote*.

FIGURE 24-1 Fertilization. Diagram of a spermatozoon moving through the corona radiata and zona pellucida on its way to reach the nucleus of a secondary oocyte.

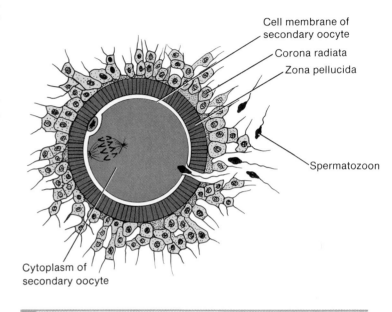

Cell membrane of secondary oocyte
Corona radiata
Zona pellucida
Spermatozoon
Cytoplasm of secondary oocyte

What is capacitation?

FORMATION OF THE MORULA

Immediately after fertilization, rapid cell division of the zygote takes place. This early division of the zygote is called *cleavage*. Although cleavage increases the number of cells, it does not result in an increase in the size of the zygote, which is still contained inside the zona pellucida. Successive cleavages produce a solid mass of tiny cells, the *morula* (MOR-yoo-la) or mulberry, within a few days after fertilization.

DEVELOPMENT OF THE BLASTOCYST

As the number of cells in the morula increases, it continues to move down through the uterine (Fallopian) tube and enters the uterine cavity. By this time, the morula forms a hollow ball of cells, now referred to as a *blastocyst* (Figure 24-2).

The blastocyst consists of an outer layer, the *trophoblast* (TRŌF-ō-blast; *troph* = nourish), which surrounds a fluid-filled cavity (*blastocoele*), and an *inner cell mass*, which develops into the *embryo* by the end of the second week. The trophoblast will form the chorion and placenta, which protect and nourish the growing organism.

IMPLANTATION

During the next 2 to 4 days, the zona pellucida disintegrates and the blastocyst is nourished by "uterine milk," secretions from endometrial glands. The attachment of the blastocyst to the endometrium is called *implantation* (Figure 24-3). In the process, the trophoblast secretes enzymes that digest the uterine lining and allow the blastocyst to bury into the endometrium during the second

FIGURE 24-2 **Blastocyst. (a) External view. (b) Internal view.**

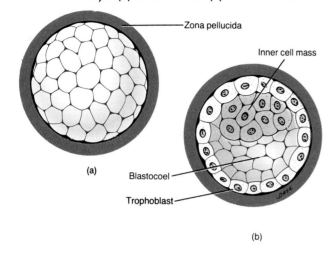

(a)

(b)

What is the next developmental stage of the trophoblast? Of the inner cell mass?

week after fertilization. The placenta will develop between the inner cell mass and the endometrial wall to provide nutrients for the growth of the embryo (more detail on this shortly). By now, the trophoblast has begun to secrete **human chorionic gonadotropin (hCG),** a hormone that maintains the corpus luteum and is the basis of positive pregnancy tests. The hormone may cause the woman to feel nauseated (often called **morning sickness**).

EMBRYONIC DEVELOPMENT

The first two months of development are considered the **embryonic period**. During this time, the developing human is called an **embryo** (*bryein* = grow). The study of development from the fertilized egg through the eighth week in utero is termed **embryology** (em-brē-OL-ō-jē). The months of development after the second month are considered the **fetal period**, and during this time the developing human is called a **fetus** (*feo* = to bring forth). By the end of the embryonic period, the rudiments of all principal adult organs are present, the embryonic membranes are developed, and by the end of the third month, the placenta is functioning.

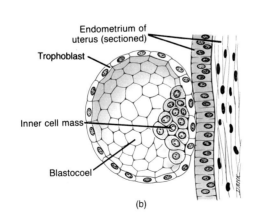

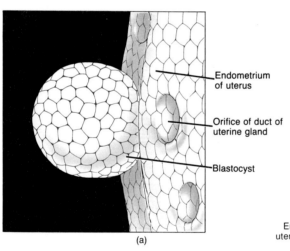

(a)

(b)

FIGURE 24-3 **Implantation. (a) External view of the blastocyst in relation to the endometrium of the uterus about 5 days after fertilization. (b) Internal view of the blastocyst in relation to the endometrium about 6 days after fertilization. (c) Internal view of the blastocyst at implantation about 7 days after fertilization.**

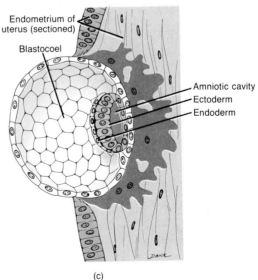

(c)

What allows the blastocyst to merge with the endometrium?

BEGINNINGS OF ORGAN SYSTEMS

Following implantation, the inner cell mass of the blastocyst begins to differentiate into the three *primary germ layers:* ectoderm, endoderm, and mesoderm. They are the embryonic tissues from which all tissues and organs of the body develop.

The germ layers form so quickly it is difficult to determine the exact sequence of events. Within 8 days of fertilization, the top layer of the inner cell mass proliferates to form the *ectoderm* and separates to form a fluid-filled space, the *amniotic cavity* (Figure 24-4). Another layer of the inner cell mass differentiates into the *endoderm*, which also separates to form another fluid-filled space, the *yolk sac* (Figure 24-4). Finally, a third layer forms between these two, the *mesoderm*. The ectoderm, mesoderm, and endoderm now are attached to the trophoblast (now called the *chorion*) by a structure formed by the mesoderm called the *body stalk* (future umbilical cord). Early in embryonic life the yolk sac merges with the body stalk and becomes nonfunctional.

The endoderm becomes the epithelial lining of the gastrointestinal tract, respiratory tract, and a number of other organs. The mesoderm forms the peritoneum, muscle, bone, and other connective tissue. The ectoderm develops into the skin and nervous system. See Exhibit 24-1 for more detail.

Embryonic Membranes

During the embryonic period, the *embryonic membranes* form (Figure 24-5). These membranes lie outside the embryo and protect and nourish the embryo and, later, the fetus. The principal membranes are the amnion and chorion.

The *amnion* is a thin, protective membrane that surrounds the embryo and becomes filled with *amniotic fluid* (*AF*), which serves as a shock absorber for the fetus. At birth, the amnion ruptures and amniotic fluid flows out of the vagina (sometimes referred to as breaking the "bag of waters").

The *chorion* (KŌ-rē-on) surrounds the embryo and amnion and forms part of the placenta.

Placenta and Umbilical Cord

The *placenta* (pla-SEN-ta) begins development with the differentiation of the trophoblast and is fully developed by the third month of pregnancy. It is shaped like a flat cake and is formed by the inner portion of the chorion where it contacts a portion of the mother's endometrium (Figure 24-6). It allows the fetus and mother to exchange nutrients and wastes and secretes hormones necessary to maintain pregnancy. In fact, the placenta directs many of the functions normally associated with the pituitary gland.

During embryonic life, fingerlike projections of the chorion, called *chorionic villi* (kō'-rē-ON-ik VIL-ē), grow into the endometrium of the uterus (Figure 24-6). They contain fetal blood vessels of the body stalk. They are bathed in maternal blood sinuses called *intervillous* (in-ter-VIL-us) *spaces*. Thus, maternal and fetal blood vessels are brought close together, but maternal and fetal blood do not normally mix. Oxygen and nutrients from the mother's blood diffuse into capillaries of the villi. From the capillaries the nutrients circulate into the *umbilical vein*. Wastes leave the fetus through the *umbilical arteries*, pass into the capillaries of the villi, and diffuse into the maternal blood (see Figure 24-7). The *umbilical* (um-BIL-i-kul) *cord* consists of the umbilical arteries, umbilical vein, and supporting connective tissue called Wharton's jelly (Figure 24-6). By the tenth day after birth, the umbilical cord shrivels and drops off, leaving a scar called the *umbilicus* (*navel*).

Fetal Circulation

The cardiovascular system of a fetus, called *fetal circulation*, differs from an adult's because the lungs, kidneys, and gastrointestinal tract of a fetus are not functioning. The fetus derives its oxygen

---

EXHIBIT 24-1
Structures Produced by the Three Primary Germ Layers

Endoderm	Mesoderm	Ectoderm
Epithelium of gastrointestinal tract (except the oral cavity and anal canal) and the epithelium of its glands.	All skeletal, most smooth, and all cardiac muscle.	All nervous tissue.
Epithelium of urinary bladder, gallbladder, and liver.	Cartilage, bone, and other connective tissues.	Epidermis of skin.
Epithelium of pharynx, auditory (Eustachian) tube, tonsils, larynx, trachea, bronchi, and lungs.	Blood, bone marrow, and lymphoid tissue. Endothelium of blood vessels and lymphatic vessels.	Hair follicles, arrector pili muscles, nails, and epithelium of skin glands (sebaceous and sudoriferous).
Epithelium of thyroid, parathyroid, pancreas, and thymus glands.	Dermis of skin. Fibrous tunic and vascular tunic of eye.	Lens, cornea, and internal eye muscles. Internal and external ear.
Epithelium of prostate and bulbourethral (Cowper's) glands, vagina, vestibule, urethra, and associated glands, such as the greater (Bartholin's) vestibular and lesser vestibular glands.	Middle ear. Mesothelium of ventral body and joint cavities. Epithelium of kidneys and ureters. Epithelium of adrenal cortex. Epithelium of gonads and genital ducts.	Neuroepithelium of sense organs. Epithelium of oral cavity, nasal cavity, paranasal sinuses, salivary glands, and anal canal. Epithelium of pineal gland, pituitary gland, and adrenal medulla.

FIGURE 24-4 Formation of the primary germ layers and associated structures. (a) Internal view of the developing embryo about 12 days after fertilization. (b) Internal view of the developing embryo about 14 days after fertilization. (c) External view of the developing embryo about 25 days after fertilization.

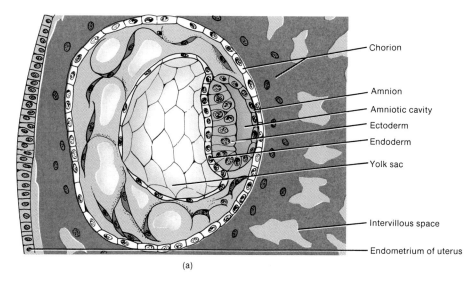

(a)

Chorion

Amnion

Amniotic cavity

Ectoderm

Endoderm

Yolk sac

Intervillous space

Endometrium of uterus

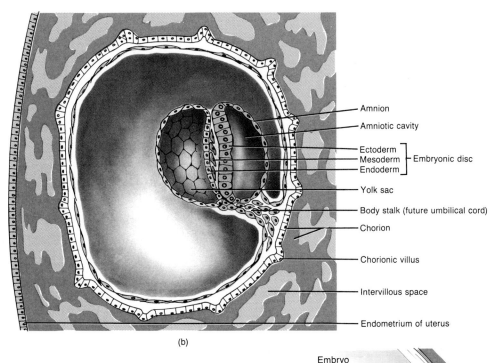

(b)

Amnion

Amniotic cavity

Ectoderm ⎤
Mesoderm ⎬ Embryonic disc
Endoderm ⎦

Yolk sac

Body stalk (future umbilical cord)

Chorion

Chorionic villus

Intervillous space

Endometrium of uterus

What are some of the structures formed by the ectoderm, mesoderm, and endoderm?

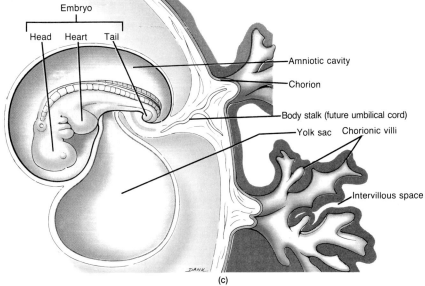

(c)

Embryo

Head   Heart   Tail

Amniotic cavity

Chorion

Body stalk (future umbilical cord)

Yolk sac   Chorionic villi

Intervillous space

FIGURE 24-5 Embryonic membranes.

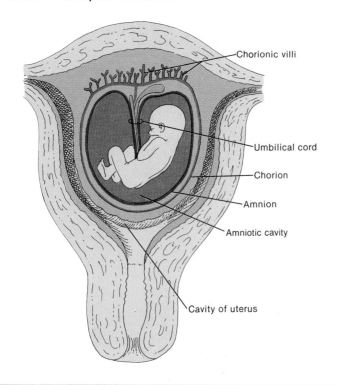

How do the amnion and chorion differ in function?

and nutrients from the maternal blood and eliminates its carbon dioxide and wastes into maternal blood.

The exchange of materials between fetal and maternal circulation occurs through the placenta. It is attached to the umbilicus of the fetus by the umbilical cord and communicates with the mother through countless small blood vessels that emerge from the uterine wall. The umbilical cord contains blood vessels that branch into capillaries in the placenta.

Blood passes from the fetus to the placenta via two **umbilical arteries** in the umbilical cord (Figure 24-7). At the placenta, the blood picks up oxygen and nutrients and eliminates carbon dioxide and wastes. The oxygenated blood returns from the placenta via a single **umbilical vein,** which ascends to the liver of the fetus, where it divides into two branches. Some blood flows through the branch that joins the hepatic portal vein and enters the liver. The fetal liver manufactures red blood cells but does not function in digestion. Therefore, most of the blood flows into the second branch, the **ductus venosus** (DUK-tus ve-NŌ-sus). The ductus venosus eventually passes its blood to the inferior vena cava, bypassing the liver.

In general, circulation through other portions of the fetus is not unlike postnatal circulation. Deoxygenated blood returning from the lower regions is mingled with oxygenated blood from the ductus venosus in the inferior vena cava. This mixed blood then enters the right atrium. The circulation of blood through the upper portion of the fetus is also similar to postnatal flow. Deoxygenated blood returning from the upper regions of the fetus is collected by the superior vena cava, and it also passes into the right atrium.

Most of the blood does not pass through the right ventricle to the lungs, as it does in postnatal circulation, since the fetal lungs do not operate. In the fetus, an opening called the **foramen ovale** (fō-RĀ-men ō-VAL-ē) exists in the septum between the right and left atria. A valve in the inferior vena cava directs about a third of the blood through the foramen ovale so that it may be sent directly into systemic circulation. The blood that does descend into the right ventricle is pumped into the pulmonary trunk, but little of this blood actually reaches the lungs. Most is sent through the **ductus arteriosus** (ar-tē-rē-Ō-sus), a small vessel connecting the pulmonary trunk with the aorta that allows most blood to bypass the fetal lungs. The blood in the aorta is carried to all parts of the fetus through its systemic branches. When the common iliac arteries branch into the external and internal iliacs, part of the blood flows into the internal iliacs. It then goes to the umbilical arteries and back to the placenta for another exchange of materials. The only fetal vessel that carries fully oxygenated blood is the umbilical vein.

At birth, when lung, renal, digestive, and liver functions are established, the special structures of fetal circulation are no longer needed. For example, the placenta is delivered by the mother as the "**afterbirth**"; the foramen ovale normally closes shortly after birth to become the **fossa ovalis**, a depression in the interatrial septum; the ductus venosus, ductus arteriosus, and umbilical vessels constrict and eventually turn into ligaments.

FETAL GROWTH

During the **fetal period**, organs established by the primary germ layers grow rapidly and the organism takes on a human appearance. A summary of changes associated with the embryonic and fetal period is presented in Exhibit 24-2.

HORMONES OF PREGNANCY

The corpus luteum is maintained for at least the first three or four months of pregnancy, during which time it continues to secrete **estrogens** and **progesterone (PROG).** The amounts secreted, however, are only slightly more than those produced after ovulation in a normal menstrual cycle. The high levels of estrogens and progesterone needed to maintain pregnancy and develop the mammary glands for lactation are provided by the placenta from the third month through the rest of the gestation period.

The chorion of the placenta secretes **human chorionic gonadotropin (hCG),** which stimulates continued production of progesterone by the corpus luteum. Progesterone is necessary for the implantation of the blastocyst to the lining of the uterus (Figure 24-8). hCG is excreted in the urine of pregnant women from about the eighth day of pregnancy, peaking about the ninth week. Recall that excretion of hCG in the urine serves as the basis for most home pregnancy tests. hCG can be detected in blood even before menstruation is missed.

The placenta begins to secrete estrogens after the third or fourth week and progesterone by the sixth week of pregnancy. The secretion of hCG is then greatly reduced because the secretions of the corpus luteum are no longer essential. Placental hormones are secreted in increasing quantities until delivery.

FIGURE 24-6 **Structure of the placenta and umbilical cord.**

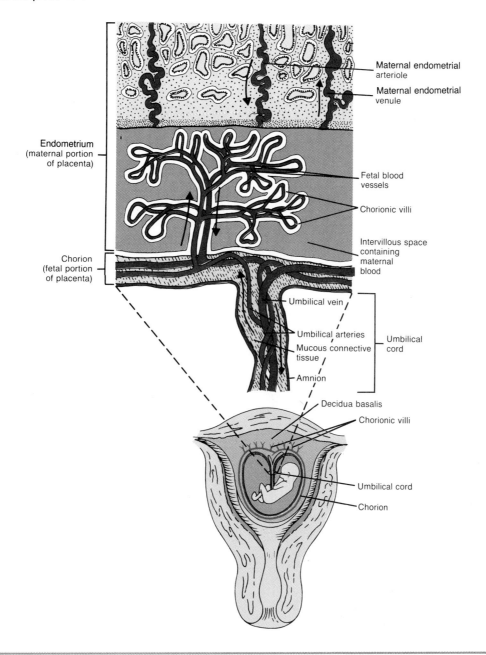

Maternal endometrial
arteriole

Maternal endometrial
venule

Endometrium
(maternal portion
of placenta)

Fetal blood
vessels

Chorionic villi

Intervillous space
containing
maternal
blood

Chorion
(fetal portion
of placenta)

Umbilical vein

Umbilical arteries

Umbilical
cord

Mucous connective
tissue

Amnion

Decidua basalis

Chorionic villi

Umbilical cord

Chorion

What is the function of the placenta?

The chorion also produces **human chorionic somatomammotropin (hCS).** hCS is believed to stimulate development of breast tissue for lactation. It also decreases maternal metabolism of glucose, thus making more available for fetal metabolism while promoting the release of fatty acids from fat depots as an alternative energy source for the mother.

Relaxin, produced by the placenta and ovaries, relaxes the symphysis pubis and helps dilate the uterine cervix toward the end of pregnancy to ease delivery.

Inhibin, produced by the ovaries and testes, has also been found in the human placenta at term. It inhibits secretion of follicle-stimulating hormone (FSH) and might regulate secretion of hCG.

GESTATION

The period of time a zygote, embryo, and fetus is carried in the female reproductive tract is called **gestation** (jes-TĀ-shun). The human gestation period is about 280 days, from the beginning of the last menstrual period. The specialized branch of medicine that deals with pregnancy, labor, and the period of time immediately following delivery is called **obstetrics** (ob-STET-riks; *obstetrix* = midwife).

By about the end of the third month of gestation, the uterus occupies most of the pelvic cavity, and as the fetus continues to grow, the uterus extends higher and higher into the abdominal

FIGURE 24-7 **Fetal circulation.** The various colors represent different degrees of oxygenation of blood ranging from greatest (red) to intermediate (two shades of purple) to least (blue).

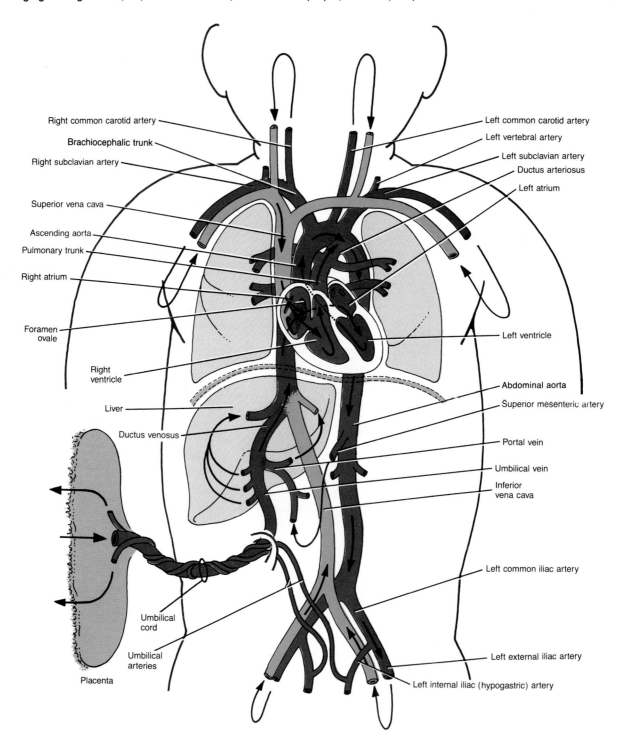

Right common carotid artery

Brachiocephalic trunk

Right subclavian artery

Superior vena cava

Ascending aorta

Pulmonary trunk

Right atrium

Foramen ovale

Right ventricle

Liver

Ductus venosus

Umbilical cord

Umbilical arteries

Placenta

Left common carotid artery

Left vertebral artery

Left subclavian artery

Ductus arteriosus

Left atrium

Left ventricle

Abdominal aorta

Superior mesenteric artery

Portal vein

Umbilical vein

Inferior vena cava

Left common iliac artery

Left external iliac artery

Left internal iliac (hypogastric) artery

Which four structures are found in fetal circulation, but not in postnatal (after birth) circulation?

EXHIBIT 24-2
Changes Associated with Embryonic
and Fetal Growth

End of Month	Approximate Size and Weight	Representative Changes	End of Month	Approximate Size and Weight	Representative Changes
1	0.6 cm (³⁄₁₆ inch)	Eyes, nose, and ears not yet visible. Backbone and vertebral canal form. Small buds that will develop into arms and legs form. Heart forms and starts beating. Body systems begin to form. The central nervous system appears at the start of the third week.			and joints begin to form. Continued development of body systems.
			5	25–30 cm (10–12 inches) 227–454 g (½–1 lb)	Head less disproportionate to rest of body. Fine hair (lanugo) covers body. Skin still bright pink. Rapid development of body systems.
2	3 cm (1¼ inches) 1 g (¹⁄₃₀ oz)	Eyes far apart, eyelids fused, nose flat. Ossification begins. Limbs become distinct as arms and legs. Digits are well formed. Major blood vessels form. Many internal organs continue to develop.	6	27–35 cm (11–14 inches) 567–681 g (1¼–1½ lb)	Head becomes even less disproportionate to rest of body. Eyelids separate and eyelashes form. Skin wrinkled and pink.
			7	32–42 cm (13–17 inches) 1135–1362 g (2½–3 lb)	Head and body become more proportionate. Skin wrinkled and pink. Seven-month fetus (premature baby) is capable of survival.
3	7.5 cm (3 inches) 28 g (1 oz)	Eyes almost fully developed but eyelids still fused, nose develops bridge, and external ears are present. Ossification continues. Appendages are fully formed and nails develop. Heartbeat can be detected. Body systems continue to develop.	8	41–45 cm (16½–18 inches) 2043–2270 g (4½–5 lb)	Subcutaneous fat deposited. Skin less wrinkled. Testes descend into scrotum. Bones of head are soft. Chances of survival much greater at end of eighth month.
4	18 cm (6½–7 inches) 113 g (4 oz)	Head large in proportion to rest of body. Face takes on human features and hair appears on head. Skin bright pink. Many bones ossified,	9	50 cm (20 inches) 3178–3405 g (7–7½ lb)	Additional subcutaneous fat accumulates. Lanugo shed. Nails extend to tips of fingers and maybe beyond.

cavity. By the end of a full-term pregnancy, the uterus practically fills all the abdominal cavity, causing displacement of the maternal intestines, liver, and stomach upward, elevation of the diaphragm, and widening of the thoracic cavity. In the pelvic cavity, there is compression of the ureters and urinary bladder.

In addition to the anatomical changes associated with pregnancy, there are also certain pregnancy-induced physiological changes. For example, cardiac output (CO) rises by 20 to 30 percent by the twenty-seventh week due to increased maternal blood flow to the placenta and increased metabolism.

Pulmonary function is altered in several ways. Tidal volume increases, expiratory reserve decreases, and shortness of breath occurs during the last few months.

A general decrease in gastrointestinal tract motility can result in constipation and a delay in gastric emptying time. Nausea, vomiting, and heartburn also occur.

Pressure on the urinary bladder by the enlarging uterus can produce urinary symptoms, such as frequency, urgency, and stress incontinence.

Changes in the skin during pregnancy include increased pigmentation around the eyes and cheekbones in a masklike pattern, in the areolae of the breasts, and in the linea alba of the lower abdomen. Striae (stretch marks) over the abdomen occur as the uterus enlarges, and hair loss also increases.

PRENATAL DIAGNOSTIC TECHNIQUES

AMNIOCENTESIS

In **amniocentesis**, a sample of amniotic fluid is used to diagnose genetic disorders or determine fetal maturity or well-being. By using ultrasound and palpation, the positions of the fetus and pla-

FIGURE 24-8 Hormones of pregnancy.

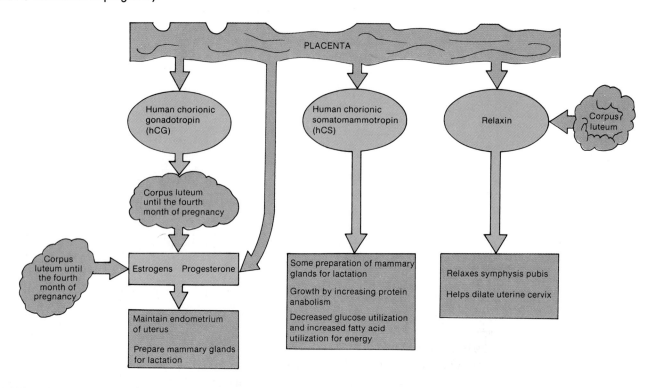

Which hormone serves as the basis for home pregnancy tests?

centa are first determined. About 10 to 20 ml of fluid containing sloughed-off fetal cells are removed by hypodermic needle puncture of the uterus, usually 14 to 16 weeks after conception. The fluid and cells are then examined. Close to 300 chromosomal disorders and over 50 inheritable biochemical defects can be detected through amniocentesis, including Down's syndrome, hemophilia, certain muscular dystrophies, Tay–Sachs disease, sickle-cell anemia, and cystic fibrosis. When both parents are known or suspected to be genetic carriers of any one of these disorders, or when maternal age approaches 35, amniocentesis is advisable.

CHORIONIC VILLUS SAMPLING (CVS)

A new test called *chorionic villus sampling* (*CVS*) to detect prenatal genetic defects is now available. It picks up the same defects as amniocentesis but can be performed during the first trimester of pregnancy, at 8 to 10 weeks gestation, and results are available within 24 hours. Moreover, the procedure does not require penetration of the abdominal wall, uterine wall, or amniotic cavity.

In CVS, a catheter (tube) is placed through the vagina into the uterus and then to the chorionic villi under ultrasound guidance. About 30 mg of tissue are suctioned out and subjected to chromosomal analysis. Chorion cells and fetal cells contain identical genetic information. The procedure is believed to be as safe as amniocentesis.

PARTURITION AND LABOR

The term *parturition* (par′-too-RISH-un) refers to birth. Parturition is accompanied by a sequence of events called *labor*.

The onset of labor appears to be hormonally directed. Just prior to birth, the muscles of the uterus contract rhythmically and forcefully. Both placental and ovarian hormones seem to play a role in these contractions. At the end of gestation, the level of estrogens in the mother's blood is sufficient to overcome the effects of progesterone, which initiates uterine contractions, and labor commences. Oxytocin (OT) also stimulates uterine contractions (see Figure 13-9). Relaxin assists by relaxing the symphysis pubis and helping to dilate the uterine cervix. Prostaglandins may also play a role in labor.

Uterine contractions occur in waves that start at the top of the uterus and move downward. These waves expel the fetus.

Labor can be divided into three stages (Figure 24-9).

1. The *stage of dilation* is the time from the onset of labor to the complete dilation of the cervix. During this stage there are regular contractions of the uterus and complete dilation (10 cm) of the cervix. If the amniotic sac does not rupture spontaneously, it is done artificially.
2. The *stage of expulsion* is the time from complete cervical dilation to delivery.

FIGURE 24-9 Parturition. (a) Fetal position prior to birth. (b) Dilation. Protrusion of amnionic sac through partly dilated cervix (left). Amnionic sac ruptured and complete dilation of cervix (right). (c) Stage of expulsion. (d) Placental stage.

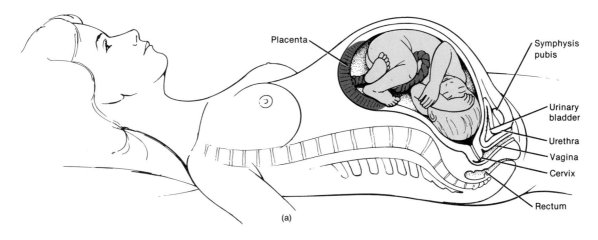

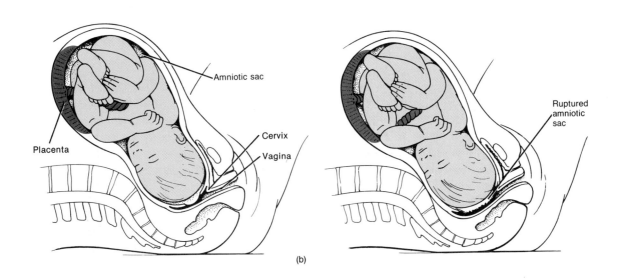

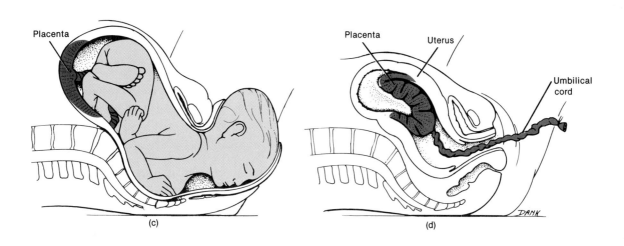

What is the purpose of doing amniocentesis and chorionic villus sampling (CVS)?

Eating for Two

During pregnancy, many of the habits a woman has taken for granted have to be reevaluated as she realizes that she has become responsible for shaping a new life. Some of the most important habits that deserve special consideration are those affecting the nutritional status of the mother-to-be and her developing fetus. Numerous research studies have shown that a woman's diet and exercise habits during pregnancy have important effects, both on the mental and physical development of the fetus and on her own health as well.

Eating for two means more than taking double portions of dessert. At issue during pregnancy is quality of the diet, rather than quantity. A pregnant woman is responsible for providing the developing fetus with nutrients, the raw materials necessary for the creation of a new life. Needs for many nutrients increase substantially during pregnancy, while caloric needs increase by only 300 to 500 kcal per day. In other words, there's still not much room for empty calories. Those calories need to be "spent" on food choices that provide the nutrients needed for fetal growth and to support the physical changes that accompany pregnancy.

People used to believe that the developing fetus was like a parasite that could take whatever nutrients it needed from its host, the mother. That's not a very nice analogy, really, and not true, either. While the placenta is capable of manufacturing some nutrients such as glycogen and cholesterol, most nutrients are brought to the fetus by the mother's blood. If a nutrient is missing from her bloodstream, fetal development may be compromised. For example, if the amino acids needed for brain development are absent, brain growth is not normal.

Maternal malnourishment is a major cause of low birth weight. Babies who weigh less than 5½ pounds at birth are 10 times more likely to be mentally impaired than bigger babies and have a much higher risk of illness and death. It's up to the mother to provide the best possible environment for her developing baby.

Nutrient Needs

The following are some of the nutrients considered especially vital to the growth of the fetus and the physiological changes occurring in the pregnant woman.

Protein

Amino acids are the stuff of which living structures are made. Protein requirements increase to 75 to 100 g during pregnancy. Pregnant women are advised to consume at least three servings of protein foods daily, in addition to their four servings of milk.

Iron

Both the fetus and the mother need iron to make red blood cells. It is almost impossible to obtain the 30 to 60 mg of iron required during pregnancy from food alone, so iron is a recommended prenatal supplement.

Folic Acid

Folic acid is a vitamin required for cell division and development of the fetal nervous system. Research has shown that women who receive extra folic acid a month before conception and for the first few months of fetal development have a much lower risk of giving birth to a baby with neural-tube defects, such as spinal bifida. Folic acid requirements during pregnancy are 400 to 800 micrograms, and because of its importance, folic acid is another recommended prenatal supplement.

Calcium

Calcium is needed for the growth of bones and teeth. Calcium level is regulated by a number of hormones (see Chapter 6) so the level of calcium in the blood remains fairly constant. If her dietary calcium is insufficient, calcium will be released from the mother's bones. That's why a woman is advised to consume four servings of milk and dairy products daily during pregnancy. Women who don't like milk often learn to disguise it in custards and milk-based soups. Teenage mothers, especially those who are still growing themselves, need six milk servings per day.

Fiber

Although not considered a nutrient, an adequate intake of fiber is important during pregnancy to prevent constipation, which typically increases because of the action of progesterone, which relaxes smooth muscle, and compression of the colon by the growing uterus. Iron supplements, often prescribed during pregnancy, frequently contribute to constipation.

Fluids

As metabolic rate and blood volume increase, so does a woman's need for fluids. Six to eight glasses of fluid daily are recommended.

Other Considerations

A woman's usual eating practices are often disrupted by pregnancy. Hormonal changes may cause nausea and vomiting (usually called "morning sickness") during the first three months. Since little weight gain occurs during the first trimester, many women don't realize how much the embryo is growing, and assume good nutrition is not yet a concern. Even though the embryo is still very tiny, good nutrition is as important as ever. Many women find that many small snacks work better than big meals, and manage to get some good nutrition that way.

Toward the end of the pregnancy, the stomach becomes compressed as the uterus pushes up against the diaphragm, and big meals become impossible. Five or six small but nutritious meals usually work best during this time.

Gestational diabetes refers to poor blood sugar control that develops during pregnancy. Women who develop high blood sugar are told to avoid foods with simple sugars such as sodas and fruit juice.

Alcohol and drugs may cause fetal abnormalities and should be avoided during pregnancy. Smoking during pregnancy has also been found to be a cause of low birth weight in newborns.

What About Exercise

Physical activity during pregnancy may indirectly affect the nutritional status of the fetus by affecting the source of the fetus's food supply: blood flow to the uterus, and the intrauterine environment in general. Researchers have been concerned that exercise raises core temperature, and may thus have an adverse feverlike effect on the fetus. In addition, they have wondered whether exercise might "use up" fuel at the expense of fetal metabolism. This is why not so long ago, pregnant women were urged to quit working, rest, and take it easy.

Most of the research on the physiological effects of exercise during pregnancy has been done on laboratory animals, especially sheep. In those studies, blood flow to the uterus did decrease during vigorous exercise, and fetal temperature did not return to normal for one hour following exercise. But, in general, results have been mixed, and are difficult to apply to humans. Sheep are not known for their desire to perform endurance-type exercise; they eliminate heat through panting, not sweating; and they always exercise in wool coats. Nevertheless, most physicians agree that women should avoid very vigorous exercises and large increases in body temperature (more than 2°F, or 1°C).

A moderate amount of physical activity does not appear to harm the developing fetus or result in low-birth-weight babies. Besides, most women have no choice but to remain at least somewhat active during pregnancy, as the responsibilities of life don't stop for nine months. Housekeeping, shopping, and childcare all take quite a bit of energy. Job demands continue for women who are employed. And most women want to stay involved in the things that are important to them. Many studies have shown that some exercise can improve the health of a pregnant woman, prevent excessive weight gain, relieve some of the physical discomforts associated with pregnancy, such as backache, and also help her bounce back more quickly after the birth of her baby.

A pregnant woman should always check with her doctor about continuing or beginning to exercise, since exercise is contraindicated for many problems that can occur during pregnancy. Once she has the go-ahead, the following recommendations should be observed.

1. Heart rate should remain below 140 beats per minute, and aerobic exercise should not last more than about 15 minutes. This guideline is designed to prevent increases in body temperature. Workouts in hot, humid weather should be avoided.
2. Avoid supine (back-lying) exercise after the fourth month, because in this position the large veins returning blood to the heart are compressed.
3. Maintain an adequate caloric intake to achieve satisfactory weight gain. Consume plenty of fluids before and after exercise.
4. Avoid movements that require jumping, bouncing, and fast changes in direction, because pregnancy decreases joint stability by making the ligaments relax. Eliminate sports that might involve abdominal injury.
5. Supervised weight training may be continued during pregnancy as long as there is no straining or breath-holding and good back care is observed.

Warning signs include feelings of fatigue that continue more than an hour after exercise. A woman who experiences pain or bleeding should stop exercising and consult her doctor immediately.

3. The *placental stage* is the time after delivery until the placenta or ''afterbirth'' is expelled by powerful uterine contractions. These contractions also constrict blood vessels that were torn during delivery to prevent hemorrhage.

During the six week period of time following delivery of the baby and placenta, called the *puerperium* (pyoo′-er-PŌ-rē-um), the reproductive organs and maternal physiology return to the pre-pregnancy state.

LACTATION

The term *lactation* refers to the secretion and ejection of milk by the mammary glands. The major hormone in promoting lactation is *prolactin* (PRL) from the anterior pituitary gland. It is released in response to the hypothalamic prolactin releasing factor (PRF). Even though PRL levels increase as the pregnancy progresses, there is no milk secretion because estrogens and progesterone inhibit the PRL. Following delivery, the levels of estrogens and progesterone in the mother's blood decrease following delivery of the placenta and the inhibition is removed.

The principal stimulus in maintaining prolactin secretion during lactation is the sucking action (suckling) of the infant. Sucking initiates impulses from the receptors in the nipples to the hypothalamus and PRL is released. The sucking action also initiates impulses to the hypothalamus which stimulate the release of *oxytocin* (*OT*) by the posterior pituitary gland. Oxytocin promotes *milk letdown*, the process by which milk is moved from the alveoli of the mammary gland into the ducts where it can be sucked.

During late pregnancy and the first few days after birth, the mammary glands secrete a cloudy fluid called *colostrum* (first milk). It is not as nutritious as true milk because it contains less lactose and virtually no fat, but it serves until the appearance of true milk on about the fourth day. Colostrum and maternal milk contain antibodies that protect the infant during the first few months of life.

Following birth, PRL level starts to diminish but each time the mother nurses the infant, nerve impulses from the nipples to the hypothalamus cause the release of PRF and a 10-fold increase in PRL that lasts about an hour, which provides milk for the next nursing period. If this surge of PRL is blocked by injury or disease or if nursing is discontinued, the mammary glands lose their ability to secrete milk in a few days.

BIRTH CONTROL (BC)

Several types of contraceptive methods, or *birth control* (*BC*), are available, each with its own advantages and disadvantages. The methods discussed here are sterilization, hormonal, intra-uterine, barrier, chemical, physiologic (natural), coitus interruptus (withdrawal), and induced abortion.

STERILIZATION

One means of *sterilization* of males is *vasectomy* (discussed in Chapter 23). Sterilization in females is usually by *tubal ligation*

(lī-GĀ-shun). In one procedure, an incision is made into the abdominal cavity, the uterine (Fallopian) tubes are squeezed, and a small loop called a knuckle is made. A suture is tied tightly at the base of the knuckle, and the knuckle is then cut. After four or five days, the suture is digested by body fluids, and the two severed ends of the tube separate. The ovum is thus prevented from passing to the uterus, and the sperm cannot reach the ovum. Tubal ligation may also be performed by *laparoscopy*. Following local or general anesthesia, a small incision is made in the abdominal wall. A needle is inserted through the incision, and gas is injected to inflate the abdomen and create a space for better visualization and easier manipulation of instruments. A *laparoscope* (lighted tube) is then inserted so the physician can view the abdominal and pelvic viscera. Tubal ligation, like vasectomy, is reversible.

HORMONAL

The *hormonal method* is also called *oral contraception* (*OC*) or *the pill*. Although several pills are available, the one most commonly used contains a high concentration of progesterone and a low concentration of estrogens (combination pill). These two hormones decrease the secretion of FSH and LH by inhibiting GnRH by the hypothalamus. Low levels of FSH and LH usually prevent ovulation, and thus pregnancy cannot occur. Oral contraceptives also alter cervical mucus so that it makes the endometrium less receptive to implantation if ovulation occurs, which it does in some cases.

Oral contraceptives should not be used by women with a predisposition to blood clotting, cerebral blood vessel damage, hypertension, liver malfunction, heart disease, or cancer of the breast or reproductive system.

Recent reports link pill users with an increased risk of infertility. Also, about 40 percent of all pill users experience side effects—generally minor problems such as nausea, weight gain, headache, irregular menses, spotting between periods, and amenorrhea. The statistics on the life-threatening conditions associated with the pill such as blood clots, heart attacks, liver tumors, and gallbladder disease indicate, for all the problems combined, fewer than 3 deaths per 100,000 in users under age 30, 4 among those 30 to 35, 10 among those 35 to 39, and 18 among women over 40. Women who take the pill and smoke face far higher odds of developing heart attack and stroke than do nonsmoking pill users.

The quest for an effective male oral contraceptive has been disappointing. Trials are now under way to test inhibin, which inhibits FSH secretion. Modified versions of GnRH have been shown to inhibit testosterone production, with a resultant decrease in sperm count and motility, but impotence is a side effect, making this approach unpopular.

INTRAUTERINE DEVICES (IUDs)

An *intrauterine device* (*IUD*) is a small plastic, copper, or stainless steel object that is inserted into the cavity of the uterus. It is not clear how IUDs operate, but some investigators believe they cause changes in the uterine lining, which, in turn, produce a substance that destroys either the sperm or fertilized ovum. IUDs can cause pelvic inflammatory disease (PID), infertility, and excess menstrual bleeding and pain. Consequently, IUDs are rapidly declining in popularity. Lawsuits resulting from damage claims have caused most manufacturers of copper IUDs in the United States to stop production and sales. To use an IUD, a woman must sign a consent form to indicate she is aware of the risks involved.

BARRIER

Barrier methods are designed to prevent spermatozoa from gaining access to the uterine cavity and uterine (Fallopian) tubes. Condoms, diaphragms, and cervical caps are barrier methods of contraception.

The *condom* is a nonporous, elastic (latex or similar material) covering placed over the penis that prevents sperm from being deposited in the female reproductive tract. Condoms, especially when used with sperm-killing chemicals, can also reduce, but not eliminate, the risk of sexually transmitted diseases (STDs). Individuals should be aware, too, that a condom cannot completely eliminate the risk of transmission of the AIDS virus to themselves or to others since condoms have a failure rate of up to 10 percent.

The *diaphragm* is a dome-shaped structure that fits over the cervix and is generally used in conjunction with a sperm-killing chemical. The diaphragm stops the sperm from passing into the cervix and the chemical kills the sperm cells. Toxic shock syndrome (TSS) and urinary tract infections are associated with diaphragm use in some females.

The *cervical cap* is a thimble-shaped contraceptive device made of latex or plastic that fits over the cervix of the uterus and is held in position by suction. Like the diaphragm, the cervical cap is used with a spermicide. Also, like the diaphragm, it must be fitted initially by a physician. The cervical cap has some advantages over the diaphragm. It can be worn up to 48 hours versus 24 hours for the diaphragm, and since the cap fits tightly and rarely leaks, it is not necessary to reintroduce spermicide before intercourse. The cervical cap is prescribed only to females with normal Pap smears, and users should undergo a follow-up Pap smear after three months. The cervical cap is not recommended for females with toxic shock syndrome (TSS), known or suspected cervical or uterine malignancies, and current vaginal or cervical infections.

CHEMICAL

Chemical methods of contraception refer to various foams, creams, jellies, suppositories, and douches that contain spermicidal agents, thus making the vagina and cervix unfavorable for sperm survival. Most spermicides kill spermatozoa by rupturing their plasma membranes. Spermicides are most effective when used in conjunction with a diaphragm. Studies have found no increase in the overall frequency of birth defects in association with the use of spermicides. A recent spermicidal development is the *contraceptive sponge*. It is a nonprescription polyurethane sponge that contains the spermicide nonoxynol-9. This spermicide decreases the incidence of chlamydia and gonorrhea but slightly increases the risk of vaginal infections caused by *Candida*. The sponge is placed in the vagina where it releases spermicide for up to 24 hours and also acts as a physical barrier to sperm. The sponge is equal to the diaphragm in effectiveness. Some cases of toxic shock syndrome (TSS) have been reported among users.

PHYSIOLOGIC (NATURAL)

Physiologic (natural) methods are based on knowledge of the menstrual cycle. In females with normal menstrual cycles, it is possible to predict during which days of the cycle ovulation is likely to occur.

The *rhythm method* takes advantage of the fact that a secondary oocyte is fertilizable for only 24 hours and is available only during a period of three to five days in each menstrual cycle. During this time, the couple refrains from intercourse (three days before ovulation, the day of ovulation, and three days after ovulation). This method is limited by the fact that few women have absolutely regular cycles.

Another natural family planning system is the *symptothermal method*. Couples are instructed to refrain from sexual intercourse when certain signs of ovulation are present, such as increased basal body temperature, clear, stretchy and abundant cervical mucus, elevated and softened cervix, and pain associated with ovulation (mittelschmerz).

COITUS INTERRUPTUS (WITHDRAWAL)

Coitus (KŌ-i-tus; *coitio* = coming together) *interruptus* refers to withdrawal of the penis from the vagina just prior to ejaculation. Failures with this method are related to either failure to withdraw before ejaculation or pre-ejaculatory escape of sperm-containing fluid from the urethra.

INDUCED ABORTION

Abortion refers to the premature expulsion from the uterus of the products of conception. An abortion may be spontaneous (naturally occurring) or induced (intentionally performed). When birth control methods are not practiced or are not successful, *induced abortion* may be performed. Procedures include vacuum aspiration (suction), a saline solution, surgical evacuation (scraping), or use of drugs. (See the Medical Terminology list also.)

A summary of methods of birth control is presented in Exhibit 24-3.

INHERITANCE

Inheritance is the passage of hereditary traits from one generation to another. The branch of biology that studies inheritance is *genetics* (je-NET-iks).

GENOTYPE AND PHENOTYPE

The nuclei of all human cells except gametes contain 23 pairs of chromosomes. One chromosome from each pair comes from the mother, and the other comes from the father. *Homologous chromosomes*, the two chromosomes of a pair, contain genes that control the same traits. If a chromosome contains a gene for height, its homologous chromosome will contain a gene for height.

The relationship of genes to heredity is illustrated by the disorder phenylketonuria or PKU, the inability to manufacture the enzyme

EXHIBIT 24-3
Summary of Birth Control (BC) Methods

Method	Comments	Method	Comments
Sterilization	Procedure involving severing ductus (vas) deferens in males (vasectomy) and uterine (Fallopian) tubes in females (tubal ligation). Failure rate is less than 1 percent.[a]		be left in place as long as 24 hours. Must be fitted by physician and refitted every two years and after each pregnancy. Offers high level of protection if used with spermicide. Occasional failures caused by improper insertion or displacement during sexual intercourse. Failure rate is about 20 percent. A cervical cap is a thimble-shaped latex device that fits snugly over the cervix. Used with a spermicide. Must be fitted by a physician. May be left in place for up to 48 hours, and it is not necessary to reintroduce spermicide before sexual intercourse. Prescribed for females with normal Pap smears, and users should undergo a follow-up after three months. Failure rate is about 10 to 20 percent.
Hormonal	Except for total abstinence or surgical sterilization, oral contraception (the pill) is the most effective method. Side effects include nausea, occasional bleeding between periods, breast tenderness or enlargement, fluid retention, and weight gain. Should not be used by women who have cardiovascular conditions (blood clot disorders, cerebrovascular disease, heart disease, hypertension), liver malfunction, cancer or neoplasia of breast or reproductive organs, or by women who smoke. Pill users may have an increased risk of infertility. Failure rate is about 2 percent.		
Intrauterine Device (IUD)	Small object (loop, coil, T, or 7) made of plastic, copper, or stainless steel that is inserted into uterus by physician. May be left in place for long periods of time (some must be changed every 2–3 years). Some women cannot use them because of expulsion, bleeding, or discomfort. Not recommended for women who have not had children because uterus is too small and cervical canal too narrow. Use of IUDs has diminished because of problems experienced by some females such as pelvic inflammatory disease (PID) and infertility. Para Gard model T380A is now available in the United States, but a consent form indicating the potential risks must be signed. Failure rate is about 5 percent.	Chemical	Sperm-killing chemicals inserted into vagina to coat vaginal surfaces and cervical opening. Provide protection for about one hour. Effective when used alone but significantly more effective when used with diaphragm or condom. The contraceptive sponge is made of polyurethane and releases a spermicide for up to 24 hours. A few cases of toxic shock syndrome (TSS) have been reported among users of the sponge. Failure rate is about 20 percent.
Barrier	A condom is a thin, strong sheath of rubber or similar material worn by the male to prevent sperm from entering vagina. Failure is caused by sheath tearing or slipping off after climax or by not putting the sheath on soon enough. If used correctly and consistently, especially with a spermicide, effectiveness is similar to that of diaphragm. Failure rate is about 10 percent. A diaphragm is a flexible rubber dome inserted into vagina to cover cervix, creating barrier to sperm. Usually used with spermicidal cream or jelly. Must be left in place at least 6 hours after intercourse and may	Physiologic (Natural)	In the rhythm method, sexual intercourse is avoided just before and just after ovulation (about seven days). Failure rate is about 20 percent even in females with regular menses. In the symptothermal method, signs of ovulation are noted (increased basal body temperature, stretchy and abundant cervical mucus, elevation and softening of the cervix, and pain associated with ovulation), and sexual intercourse is avoided. Failure rate is about 20 percent.
		Coitus Interruptus	Withdrawal of penis from vagina before ejaculation occurs. Failure rate is 20 to 25 percent.
		Induced Abortion	Involves vacuum aspiration (suction), saline solution, surgical evacuation (scraping) to remove prematurely products of conception, or drugs such as mifepristone (RU486) and epostane.

[a] Failure rates are based on estimated pregnancy rates (percentage) in the first year of use, assuming variations in consistency of use.

phenylalanine hydroxylase. It is believed PKU is caused by an abnormal gene, which we symbolize as p (see Figure 24-10). The normal gene is symbolized as P. The chromosome concerned with directions for phenylalanine hydroxylase production has either p or P on it. Its homologous chromosome also has p or P. Thus, every individual will have one of the following genetic makeups, or *genotypes* (JĒ-nō-tīps): PP, Pp, or pp. Although people with genotypes of Pp have the abnormal gene, only those with genotype pp suffer from the disorder because the normal gene, when present, dominates over the abnormal one. A gene that dominates is called the *dominant gene*, and the trait expressed is said to be a dominant trait. The gene that is inhibited is called the *recessive gene*, and the trait it controls is called the recessive trait. Genes that control the same inherited trait, for example, height, eye color, or hair color, and that occupy the same position on homologous (paired) chromosomes are known as *alleles*.

By tradition, we symbolize the dominant gene with a capital letter and the recessive one with a lowercase letter. An individual with the same genes on homologous chromosomes (for example, PP or pp) is said to be *homozygous* for the trait. An individual with different genes on homologous chromosomes (for example, Pp), is said to be *heterozygous* for the trait. *Phenotype* (FĒ-nō-tīp) refers to how the genetic makeup is expressed in the body. A person with Pp has a different genotype from one with PP, but both have the same phenotype—which in this case is normal production of phenylalanine hydroxylase. It should be noted that individuals who carry a recessive gene but do not express it (Pp) can

pass the gene on to their offspring. Such individuals are called *carriers*.

To determine how gametes containing haploid chromosomes unite to form diploid fertilized eggs, special charts called *Punnett squares* are used. Usually, the male gametes (sperm cells) are placed at the side of the chart and the female gametes (ova) at the top (Figure 24-10). The four spaces on the chart represent the possible combinations (genotypes) of male and female gametes that could form fertilized eggs.

Exhibit 24-4 lists some of the inherited structural and functional traits in humans.

Genes for severe disorders are more frequently recessive than dominant, but this is not always true. For example, the neurological disease **Huntington's chorea** is caused by a dominant gene. Huntington's chorea is characterized by degeneration of nervous tissue, usually leading to mental disturbances and death. The first signs of Huntington's chorea do not occur until adulthood, very often after the person has already produced offspring. A genetic test is

FIGURE 24-10 Inheritance of phenylketonuria (PKU).

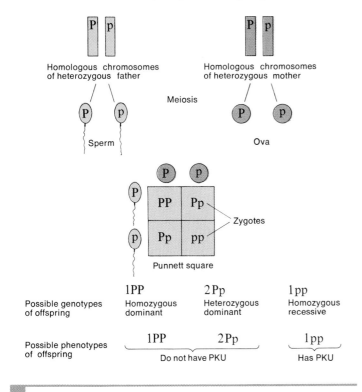

What is a carrier?

EXHIBIT 24-4 *Hereditary Traits in Humans*	
Dominant	**Recessive**
Dark brown hair	All other colors
Coarse body hair	Fine body hair
Pattern baldness (dominant in males)	Baldness (recessive in females)
Widow's peak	Straight hairline
Normal skin pigmentation	Albinism
Freckles	Absence of freckles
Brown eyes	Blue or gray eyes
Astigmatism	Normal vision
Nearsightedness or farsightedness	Normal vision
Free earlobes	Attached earlobes
Normal hearing	Deafness
Normal color vision	Color blindness
Broad lips	Thin lips
Tongue roller	Inability to roll tongue (into a U shape)
PTC taster	PTC nontaster
Large eyes	Small eyes
Polydactylism (extra digits)	Normal digits
Brachydactylism (short digits)	Normal digits
Syndactylism (webbed digits)	Normal digits
Feet with normal arches	Flat feet
Hypertension	Normal blood pressure
Diabetes insipidus	Normal excretion
Huntington's chorea	Normal nervous system
Normal mentality	Schizophrenia
Migraine headaches	Normal
Normal resistance to disease	Susceptibility to disease
Enlarged spleen	Normal spleen
Enlarged colon	Normal colon
A or B blood factor	O blood factor
Rh blood factor	No Rh blood factor

now available to determine whether or not an individual carries the gene for the disorder.

Until recently, it was believed that a gene contributed by one parent functioned the same as the identical gene contributed by the other parent. It has been shown, however, that this is not the case. A gene inherited from one parent may be expressed differently from the same gene inherited from the other parent. Apparently, during spermatogenesis and oogenesis some genes combine with chemicals called methyl groups and are inactivated. This inactivation process is called *genomic imprinting* and may explain why, for example, patients who inherit the gene for Huntington's chorea from their father sometimes acquire a more severe form of the disease that begins earlier in life than those who inherit it from their mother.

GENES AND THE ENVIRONMENT

Most patterns of inheritance are not simple *dominant–recessive inheritance* in which only dominant and recessive genes interact. In fact, the phenotypic expression of a particular gene is influenced not only by the alleles of the gene present but also by other genes and the environment. Put another way, a given phenotype is the result of both genotype and environment. For example, height, along with body build, hair color, and skin color, among other traits, is controlled by the combined effects of many genes. At the same time, even though a person inherits genes for tallness, full potential may not be reached due to environmental factors, such as disease or malnutrition during the growth years.

Exposure of a developing embryo or fetus to certain environmental factors can damage the developing organism or even cause death. A *teratogen* (*terato* = monster) is any agent or influence that causes physical defects in the developing embryo. Examples are pesticides, defoliants, industrial chemicals, some hormones, antibiotics, oral anticoagulants, anticonvulsants, antitumor agents, thyroid drugs, thalidomide, diethylstilbestrol (DES), LSD, marijuana, cocaine, alcohol, nicotine, and ionizing radiation. Thalidomide, a sedative in use in the 1960s, affected limb development in fetuses of pregnant women. A pregnant female who uses cocaine subjects the fetus to risks of underweight, retarded growth, attention and orientation problems, alteration in visual processing, hyperirritability, a tendency to stop breathing, crib death, malformed or missing organs, strokes, and seizures. The risks of spontaneous abortion (miscarriage), premature birth, and stillbirth also increase from fetal exposure to cocaine.

Alcohol is another teratogen. Intrauterine exposure to alcohol results in *fetal alcohol syndrome* (*FAS*). Studies indicate that in the general population FAS may be by far the number one fetal teratogen, affecting 1 per 1000 live births. FAS is one of the major causes of mental retardation in the United States. Other symptoms include slow growth before and after birth, poor eyesight, small head, facial irregularities such as narrow eye slits and sunken nasal bridge, defective heart and other organs, malformed arms and legs, genital abnormalities, mental retardation, and learning disabilities, plus behavioral problems, such as hyperactivity, extreme nervousness, and poor attention span.

The latest evidence not only indicates a causal relationship between cigarette smoking during pregnancy and low infant birth weight but also points to a strong probable association between smoking and cardiac abnormalities, anencephaly (an absence of neural tissue in the cranium), and higher fetal and infant mortality rate. Maternal smoking also appears to be a significant factor in the development of cleft lip and palate and has been tentatively linked with sudden infant death syndrome (SIDS). Infants nursing from smoking mothers have been found to have an increased incidence of gastrointestinal disturbances. Finally, infants of smoking mothers have an increased incidence of respiratory problems during the first year of life.

Ionizing radiations are potent teratogens. Treatment of pregnant mothers with large doses of x-rays and radium during the embryo's susceptible period of development may cause microcephaly (small size of head in relation to the rest of the body), mental retardation, and skeletal malformations. Caution is advised for diagnostic x-rays during the first trimester of pregnancy.

INHERITANCE OF SEX

One pair of chromosomes in cells differs in males and females (Figure 24-11a). In females, the pair consists of two rod-shaped chromosomes designated as X chromosomes. One X chromosome is presnt in males, but its mate is a chromosome called a Y chromosome. The XX pair in the female and the XY pair in the male are called the *sex chromosomes*. All other chromosomes are called *autosomes*.

The sex chromosomes are responsible for the sex of the individual (Figure 24-11b). When a spermatocyte undergoes meiosis to reduce its chromosome number, one daughter cell will contain the X chromosome and the other contain the Y chromosome. Oocytes produce only X-containing ova. If the secondary oocyte is subsequently fertilized by an X-bearing spermatozoon, the offspring will be female (XX). Fertilization by a Y spermatozoon produces a male (XY). Thus, sex is determined at fertilization.

Both female and male embryos develop identically until about seven weeks after fertilization. At that point, one particular gene sets into motion a series of events that leads to the development of a male. In the absence of the gene, a female develops. The gene is called *testis-determining factor* (*TDF*) and it is contained in a fragment of the Y chromosome. The TDF gene apparently produces a protein that binds to DNA to control the activity of other genes. If the TDF gene is present in a fertilized ovum, the fetus develops testes and differentiates into a male; in the absence of TDF, the fetus develops ovaries and differentiates into a female.

COLOR BLINDNESS AND X-LINKED INHERITANCE

The sex chromosomes also are responsible for the transmission of a number of nonsexual traits. Genes for these traits appear on X chromosomes, but many of these genes are absent from Y chromosomes. This feature produces a pattern of heredity that is different from the pattern described earlier. Let us consider color blindness. The gene for *color blindness* is a recessive one designated c. Normal color vision, designated C, dominates. The C/c genes are located on the×chromosome. The Y chromosome does not contain the segment of DNA that programs this aspect of vision. Thus, the ability to see colors depends entirely on the X chromosomes. The genetic possibilities are

$X^C X^C$ Normal female
$X^C X^c$ Normal female carrying the recessive gene
$X^c X^c$ Color-blind female
$X^C Y$ Normal male
$X^c Y$ Color-blind male

FIGURE 24-11 Inheritance of sex. (a) Normal human male chromosomes. Sex chromosomes are the twenty-third pair, indicated in the blue colored box. (b) Sex determination.

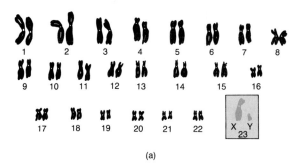

Only females who have two X^c chromosomes are color-blind. In $X^C X^c$ females, the trait is inhibited by the normal, dominant gene. Males, on the other hand, do not have a second X chromosome that would inhibit the trait. Therefore, all males with an X^c chromosome will be color-blind. The inheritance of color blindness is illustrated in Figure 24-12.

Traits inherited in the manner just described are called *X-linked traits*. Another X-linked trait is **hemophilia**—a condition in which the blood fails to clot or clots very slowly after an injury (see Chapter 14). Like the trait for color blindness, hemophilia is caused by a recessive gene. If H represents normal clotting and h represents abnormal clotting, then $X^h X^h$ females will have the disorder. Males with $X^H Y$ will be normal; males with $X^h Y$ will be hemophiliac. Actually, clotting time varies somewhat among hemophiliacs, so the condition may be affected by other genes as well.

A few other X-linked traits in humans are nonfunctional sweat glands, certain forms of diabetes, some types of deafness, uncontrollable rolling of the eyeballs, absence of central incisors, night blindness, one form of cataract, white forelocks, juvenile glaucoma, and juvenile muscular dystrophy.

FIGURE 24-12 Inheritance of color blindness.

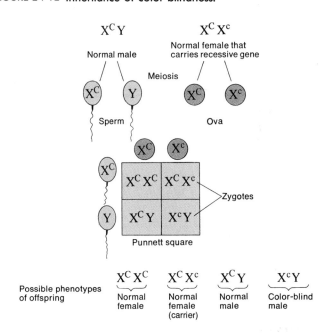

How does testis-determining factor (TDF) affect sex determination?

MEDICAL TERMINOLOGY AND CONDITIONS

Abortion (a-BOR-shun) Premature expulsion from the uterus of the products of conception—embryo or nonviable fetus. May be caused by abnormal development of embryo, placental abnormalities, endocrine disturbances, certain diseases, trauma, and stress. ***Induced (nonspontaneous) abortions*** are brought on intentionally by methods such as vacuum aspiration (suction curettage), up to the twelfth week of pregnancy; dilation and evacuation (D&E), commonly used between the thirteenth and fifteenth weeks of pregnancy and sometimes up to twenty weeks; and use of saline or prostaglandin preparations to induce labor and delivery, usually after the fifteenth week of pregnancy. **Spontaneous abortions (miscarriages)** occur without apparent cause. Recent evidence suggests that as many as one-third of all successful fertilizations end in spontaneous abortions, most of which occur even before a female or her physician is aware that she is pregnant.

Breech presentation A malpresentation in which the fetal buttocks or lower extremities, instead of the head, enter the birth canal first; most common cause is prematurity.

Mutation (myoo-TĀ-shun; *mutare* = change) A permanent heritable change in a gene that causes it to have a different

function than it had previously.

Preeclampsia (prē′-e-KLAMP-sē-a) A syndrome characterized by sudden hypertension, large amounts of protein in urine, and generalized edema; possibly related to an autoimmune or allergic reaction to the presence of a fetus; when the condition is also associated with convulsions and coma, it is referred to as **eclampsia.**

Puerperal (pyoo-ER-per-al; *puer* = child; *parere* = to bring forth) **fever** Infectious disease of childbirth that originates in the birth canal and affects the endometrium. It may spread to other pelvic structures and lead to septicemia.

STUDY OUTLINE

Sexual Intercourse (p. 511)

1. The role of the male in the sex act involves erection, lubrication, and orgasm.
2. The female role also involves erection, lubrication, and orgasm (climax).

Development During Pregnancy (p. 512)

1. Pregnancy is a sequence of events that includes fertilization, implantation, embryonic growth, fetal growth, and birth.
2. Its various events are hormonally controlled.

Fertilization; Implantation (p. 512)

1. Fertilization refers to the penetration of a secondary oocyte by a sperm cell and the subsequent union of the spermatozoon and oocyte chromosomes to form a zygote.
2. Penetration is facilitated by enzymes produced by sperm.
3. Normally only one spermatozoon fertilizes a secondary oocyte.
4. Early rapid cell division of a zygote is called cleavage.
5. The solid mass of tiny cells produced by cleavage is a morula.
6. The morula develops into a blastocyst, a hollow ball of cells differentiated into a trophoblast (future embryonic membranes) and inner cell mass (future embryo).
7. The attachment of a blastocyst to the endometrium is called implantation; it occurs by enzymatic degradation of the endometrium.

Embryonic Development (p. 513)

1. During embryonic growth, the primary germ layers and embryonic membranes are formed and the placenta is functioning.
2. The primary germ layers—ectoderm, mesoderm, and endoderm—form all tissues of the developing organism.
3. Embryonic membranes include the amnion and chorion.
4. Fetal and maternal materials are exchanged through the placenta.
5. Fetal circulation involves the exchange of materials between fetus and mother.

6. The fetus derives its oxygen and nutrients and eliminates its carbon dioxide and wastes through the maternal blood supply by means of the placenta.
7. At birth, when lung, digestive, and liver functions are established, the special structures of fetal circulation are no longer needed.

Fetal Growth (p. 516)

1. During the fetal period, organs established by the primary germ layers grow rapidly.
2. The principal changes associated with fetal growth are summarized in Exhibit 24-2.

Hormones of Pregnancy (p. 516)

1. Pregnancy is maintained by human chorionic gonadotropin (hCG), estrogens, and progesterone (PROG).
2. Human chorionic somatomammotropin (hCS) facilitates breast development, decreases glucose catabolism, and promotes release of fatty acids as an alternate energy source.
3. Relaxin relaxes the symphysis pubis and helps dilate the uterine cervix toward the end of pregnancy.
4. Inhibin inhibits secretion of FSH and might regulate secretion of hGH.

Gestation (p. 517)

1. The time an embryo or fetus is carried in the uterus is called gestation.
2. Human gestation lasts about 280 days from the beginning of the last menstrual period.
3. During gestation, the mother undergoes anatomical and physiological changes in preparation for birth.

Prenatal Diagnostic Techniques (p. 519)

1. Amniocentesis is the withdrawal of amniotic fluid to diagnose biochemical defects and chromosomal disorders, such as hemo-

philia, Tay–Sachs disease, sickle-cell anemia, and Down's syndrome.

2. Chorionic villus sampling (CVS) involves withdrawal of chorionic villi for chromosomal analysis.

3. CVS can be done earlier than amniocentesis and the results are available in a shorter time.

Parturition and Labor (p. 520)

1. Parturition refers to birth and is accompanied by a sequence of events called labor.

2. The birth of a baby involves dilation of the cervix, expulsion of the fetus, and delivery of the placenta.

Lactation (p. 524)

1. Lactation refers to the secretion and ejection of milk by the mammary glands.

2. Secretion is influenced by prolactin (PRL), estrogens, and progesterone.

3. Ejection is influenced by oxytocin (OT).

Birth Control (BC) (p. 524)

1. Methods include sterilization (vasectomy, tubal ligation), hormonal, intrauterine devices, barriers (condom, diaphragm, cervical cap), chemical (spermicides), physiologic (rhythm, symptothermal method), coitus interruptus, and induced abortion.

2. See Exhibit 24-3 for a summary of birth control methods.

Inheritance (p. 525)

1. Inheritance is the passage of hereditary traits from one generation to another.

2. The genetic makeup of an organism is its genotype. The traits expressed are its phenotype.

3. Dominant genes control a particular trait; expression of recessive genes is inhibited by dominant genes.

4. Most patterns of inheritance do not conform to the simple dominant–recessive patterns.

5. A given phenotype is the result of the interactions of genotype and the environment.

6. Teratogens are agents that cause physical defects in developing embryos. They include thalidomide, drugs, alcohol, nicotine, and ionizing radiation.

7. Sex is determined by the Y chromosome of the male at fertilization.

8. Color blindness and hemophilia primarily affect males because there are no counterbalancing dominant genes on the Y chromosomes.

REVIEW QUESTIONS

1. Explain the roles of the male's erection, lubrication, and orgasm and the female's erection, lubrication, and orgasm (climax) in the sex act. (p. 511)

2. Define fertilization. Where does it normally occur? How is a morula formed? What are the parts of a blastocyst? (p. 512)

3. What is implantation? How does the fertilized ovum implant itself? (p. 512)

4. Define the embryonic period and the fetal period. (p. 513)

5. List several body structures formed by the endoderm, mesoderm, and ectoderm. (p. 514)

6. What is an embryonic membrane? Describe the functions of the embryonic membranes. (p. 514)

7. Explain the importance of the placenta and umbilical cord to fetal growth. (p. 514)

8. Discuss the physiology of fetal circulation. Indicate the function of the umbilical arteries, umbilical vein, ductus venosus, foramen ovale, and ductus arteriosus. (p. 514)

9. Outline some of the major developmental changes during fetal growth. (p. 516)

10. List the hormones involved in pregnancy and describe the functions of each. (p. 516)

11. Define gestation and parturition. (p. 517)

12. Describe several anatomical and physiological changes that occur during gestation. (p. 517)

13. What is amniocentesis? What is its value? Describe chorionic villus sampling (CVS). What are its advantages over amniocentesis? (p. 519)

14. Describe the three stages of labor. (p. 520)

15. Define lactation. How is it controlled? (p. 524)

16. Briefly describe the various methods of birth control (BC) and the effectiveness of each. (p. 524)

17. Define inheritance. What is genetics? (p. 525)

18. Define the following terms: homologous chromosomes, genotype, phenotype, allele, dominant, recessive. (p. 527)

19. List several dominant and recessive traits inherited in humans. (p. 527)

20. How do genes and the environment determine phenotype? (p. 527)

21. Set up Punnett squares to show the inheritance of the following traits: sex, color blindness, hemophilia. (p. 527)

22. Refer to the glossary of medical terminology associated and conditions with development and inheritance. Be sure you can define each term. (p. 529)

Bibliography

CHAPTER 1

Clemente, C. D. *Anatomy: A Regional Atlas of the Human Body*, 3rd ed. Baltimore: Urban & Schwarzenberg, 1987.

Geller, S. A. "Autopsy," *Scientific American*, March 1983.

Gosling, J. A., P. F. Harris, J. R. Humpherson, I. Whitmore, and P. L. T. Willan. *Atlas of Human Anatomy*. London: Gower Medical, 1985.

Keiffer, S. A., and E. R. Heitzman. *An Atlas of Cross-Sectional Anatomy*. New York: Harper & Row, 1979.

Netter, F. H. *Atlas of Human Anatomy*. Summit, NJ: CIBA, 1989.

Rohen, J. W., and C. Yokochi, *Color Atlas of Anatomy*. Tokyo, New York: Igaku-Shoin, Ltd., 1983.

Sochurek, H. "Medicine's New Vision," *National Geographic*, January 1987.

Tortora, G. J. *Principles of Human Anatomy*, 5th ed. New York: Harper & Row, 1989.

Yokochi, C., and J. W. Rohen. *Photographic Anatomy of the Human Body*, 2nd ed. Tokyo, New York: Igaku-Shoin, Ltd., 1978.

CHAPTER 2

Darnell, J. E. "RNA," *Scientific American*, October 1985.

Dickerson, R. E., and I. Geis. *Chemistry, Matter, and the Universe*. Menlo Park, CA: Benjamin-Cummings, 1976.

Doolittle, R. F. "Proteins," *Scientific American,* October 1985.

Felsenfeld, G. "DNA," *Scientific American,* October 1985.

Frieden, E. "The Chemical Elements of Life," *Scientific American*, July 1972.

Marx, J. L. "Limits to DNA Fingerprinting," *Science*, 24 March 1989.

Sharon, N. "Carbohydrates," *Scientific American,* November 1980.

Weinberg, R. A. "The Molecular Basis of Life," *Scientific American*, October 1985.

CHAPTER 3

Begley, S., and M. Hager. "Brave New Gene Therapy," *Newsweek*, 13 February 1989.

Berridge, M. J. "The Molecular Basis of Communication Within the Cell," *Scientific American,* October 1985.

Bretscher, M. S. "The Molecules of the Cell Membrane," *Scientific American*, October 1985.

Cairns, J. "The Treatment of Diseases and the War Against Cancer," *Scientific American*, November 1985.

Dautry-Varsat, A., and H. F. Lodish. "How Receptors Bring Proteins and Particles into Cells," *Scientific American*, May 1984.

deDuve, C. *A Guided Tour of the Living Cell*. New York: Scientific American Books, 1984.

Feldman, M., and L. Eisenbach. "What Makes a Tumor Cell Metastatic?" *Scientific American*, November 1988.

Kartner, N., and V. Ling. "Multidrug Resistance in Cancer," *Scientific American*, March 1989.

Marx, J. L. "How DNA Viruses May Cause Cancer," *Science*, 24 February 1989.

McIntosh, R., and K. L. McDonald. "The Mitotic Spindle," *Scientific American*, October 1989.

Porter, K. R., and J. B. Tucker. "The Ground Substance of the Living Cell," *Scientific American*, March 1981.

Ptashne, M. "How Gene Activators Work," *Scientific American*, January 1989.

Radman, M., and R. Wagner. "The High Fidelity of DNA Duplication," *Scientific American*, August 1988.

Rothman, J. E. "The Compartmental Organization of the Golgi Apparatus," *Scientific American*, September 1985.

Unwin, N., and R. Henderson, "The Structure of Proteins in Biological Membranes," *Scientific American*, February 1984.

Weinberg, R. A. "Finding the Anti-Oncogene," *Scientific American*, September 1988.

CHAPTER 4

Cormack, D. H. *Ham's Histology*, 9th ed. Philadelphia: Lippincott, 1987.

Fawcett, D. W. *Bloom and Fawcett: A Textbook of Histology*, 11th ed. Philadelphia: Saunders, 1986.

Kelly, D. E., R. L. Wood, and A. C. Enders. *Bailey's Textbook of Microscopic Anatomy*, 18th ed. Baltimore: Williams & Wilkins, 1984.

Miller, J. "The Connective Tissue Perspective," *Science News*, 20 February 1982.

Synder, S. H. "The Molecular Basis of Communication Between Cells," *Scientific American*, October 1985.

CHAPTER 5

Dahl, M. V. "Acne: How It Happens and How It's Treated," *Modern Medicine*, September 1982.

Edelson, L. E., and J. M. Fink. "The Immunologic Function of Skin," *Scientific American*, June 1985.

Flanagan, B. P. "Skin Cancer: An Illustrated Guide to Early Diagnosis," *Modern Medicine*, October 1985.

Goldberg, L. H., and H. A. Rubin. "Management of Basal Cell Carcinoma," *Postgraduate Medicine*, January 1989.

Martin, L. M. "Nursing Implications of Today's Burn Care Techniques," *RN*, May 1989.

Montagna, W. "The Skin," *Scientific American*, June 1989.

Silverberg, N., and L. Silverberg. "Aging and the Skin," *Postgraduate Medicine*, July 1989.

Sober, A. "Malignant Melanoma: A Guide to Early, Accurate Diagnosis," *Modern Medicine*, July 1988.

Wachtel, T. L. "Major Burns," *Postgraduate Medicine*, January 1989.

CHAPTER 6

Baran, D. "Diagnosis and Management of Osteoporosis," *Modern Medicine*, March 1989.

Bock, H. (ed.). "Tricky Bone Healing Technique From the Soviet Union," *Medical Update*, March 1989.

Gamble, J. G. *The Musculoskeletal System: Physiological Basis.* New York: Raven Press, 1988.

Gosling, J. A., P. F. Harris, J. R. Humpherson, I. Whitmore, and P. L. T. Willon. *Atlas of Human Anatomy*, Philadelphia: Lippincott, 1985.

Hogan, M. J. (ed.). "TMJ Disorder," *Mayo Clinic Health Letter*, October 1988.

Langone, J. "Back Surgery Without Stitches," *Time*, 5 September 1988.

Netter, F. H. *Musculoskeletal System, Part I, Anatomy, Physiology, and Metabolic Disorders.* Summit, NJ: CIBA, 1987.

Ravnikav, V. "Clinical Considerations in the Diagnosis of Osteoporosis," *Modern Medicine*, May 1988.

Rohen, J. W., and C. Yokochi. *Color Atlas of Anatomy.* Tokyo, New York: Igaku-Shoin, Ltd., 1983.

Rudy, D. R. "Osteoporosis," *Postgraduate Medicine*, August 1989.

Shipman, P., A. Walker, and D. Bichell. *The Human Skeleton.* Cambridge: Harvard University Press, 1985.

Steele, D. G., and C. A. Bramblett. *The Anatomy and Biology of the Human Skeleton.* College Station: Texas A&M University Press, 1988.

CHAPTER 7

Allman, W. F. "The Knee," *Science 83*, November 1983.

Arehart-Trechel, J. "The Joint Destroyers," *Science News*, September 1982.

Beck, M., M. Hager, and V. E. Smith. "Living with Arthritis," *Newsweek*, 20 March 1989.

Bertram, Z., and M. Adams. "Knee Injuries in Sports," *New England Journal of Medicine*, 14 April 1988.

Bienenstock, H. "Diagnosis: Arthritis," *Hospital Medicine*, October 1984.

Edwards, D. D. "Severe Arthritis Under Attack," *Science News*, 19 October 1985.

Epstein, F. H. "The Biology of Osteoarthritis," *New England Journal of Medicine*, 18 May 1989.

Hogan, M. J. (ed.). "Arthroscopy," *Mayo Clinic Health Letter*, March 1989.

———. "Artificial Joints," *Mayo Clinic Health Letter*, Three parts. November and December 1988 and January 1989.

Kiley, J. M. (ed.). "Rheumatoid Arthritis," *Mayo Clinic Health Letter*, December 1986.

Wilson, J. D. "Tiny Tick, Big Worry," *Newsweek*, 22 May 1989.

CHAPTER 8

Allman, W. F. "Weight-Lifting: Inside the Pumphouse," *Science 84*, September 1984.

Bohigian, G. M. "Drug Abuse in Athletes," *Journal of the American Medical Association*, 18 March 1988.

Brody, D. M. "Running Injuries: Prevention and Management," *CIBA-GEIGY Clinical Symposia*, 39(3), 1987.

Gamble, J. G. *The Musculoskeletal System: Physiological Basics.* New York: Raven Press, 1988.

Gosling, J. A., P. F. Harris, J. R. Humpherson, I. Whitmore, and P. L. T. Willan. *Atlas of Human Anatomy.* Philadelphia: Lippincott, 1985.

Hinson, M. M. *Kinesiology.* Dubuque, IA: Wm. C. Brown, 1981.

Huxley, H. E. "The Contraction of Muscle," *Scientific American*, November 1968.

Johnson, G. T. (ed.). "Steroids: A Primer," *Harvard Medical School Health Letter*, October 1983.

Netter, F. H. *Musculoskeletal System: Anatomy, Physiology, and Metabolic Disorders.* Summit, NJ: CIBA-GEIGY Corporation, 1987.

Ravits, J. "Myasthenia Gravis," *Postgraduate Medicine*, January 1988.

Wilson, F. C. *The Musculoskeletal System: Basic Processes and Disorders*, 2nd ed. Philadelphia: Lippincott, 1983.

Windsor, R. E., and D. Dumitru. "Anabolic Steroid Use by Athletes," *Postgraduate Medicine*, September 1988.

CHAPTER 9

Barr, M. L., and J. A. Kiernan. *The Human Nervous System*, 5th ed. Philadelphia: Lippincott, 1988.

Bloom, F. E. "Neuropeptides," *Scientific American*, October 1981.

Byrne, J. H., and S. G. Schultz. *An Introduction to Membrane Transport and Bioelectricity.* New York: Raven Press, 1988.

Dunant, Y., and M. Israël. "The Release of Acetylcholine," *Scientific American*, March 1985.

Kalil, R. E. "Synapse Formation in the Developing Brain," *Scientific American*, December 1989.

Llinas, R. R. "Calcium in Synaptic Transmission," *Scientific American*, October 1982.

Miller, J. A. "Cell Communication Equipment: Do-It-Yourself," *Science News*, 14 April 1984.

Morell, P., and W. T. Norton, "Myelin," *Scientific American*, May 1980.

Netter, F. H. *Nervous System: Anatomy and Physiology*, Summit, NJ: CIBA-GEIGY Corporation, 1983.

Patton, H. D., A. F. Fuchs, B. Hille, A. M. Scher, and R. Steiner. *Textbook of Physiology: Excitable Cells and Neurophysiology*, 21st ed. Philadelphia: Saunders, 1989.

Stevens, C. F. "The Neuron," *Scientific American*, September 1979.

CHAPTER 10

Barr, M. L., and J. A. Kiernan. *The Human Nervous System*, 5th ed. Philadelphia: Lippincott, 1988.

Begley, S., J. Carey, and R. Sawhill. "How the Brain Works," *Newsweek*, 7 February 1983.

"The Brain," *Scientific American*, September 1979. (Entire issue devoted to the brain and the human nervous system.)

Brown, M. R., and L. A. Fisher. "Brain Peptides as Intercellular Messengers," *Journal of the American Medical Association*, 9 March 1984.

Bruno, M., and S. Katz. "New Hope for Hurt Nerves," *Newsweek*, 7 October 1985.

Burden, N. "Regional Anesthesia: What Patients and Nurses Need to Know," *RN*, May 1988.

Cowley, G. "Medical Mystery Tour," *Newsweek*, 18 December 1989.

———. "Parkinson's Breakthrough," *Newsweek*, 27 November 1989.

Edwards, D. D. "A Common Medical Denominator," *Science News*, 25 January 1986.

Goldfinger, S. E. (ed). "Alzheimer's Disease," *Harvard Medical School Health Letter*, April 1988.

_____. "Shingles," *Harvard Medical School Health Letter*, June 1984.

Gorelick, P. B. "Clues to the Mystery of Multiple Sclerosis," *Postgraduate Medicine*, March 1989.

Greenspan, J. "Carpal Tunnel Syndrome," *Postgraduate Medicine*, November 1988.

Kiley, M. J. (ed.). "Parkinson's Disease," *Mayo Clinic Health Letter*, May 1989.

_____. "A Long Goodbye," *Mayo Clinic Health Letter*, January 1988.

Kimelberg, H. K., and M. D. Norenberg. "Astrocytes," *Scientific American*, April 1989.

Koller, W. C. "Diagnosis and Treatment of Parkinson's Disease," *Modern Medicine*, May 1989.

Mattewson, J. "Alzheimer's Disease: Source Searching," *Science News*, 13 July 1985.

Miller, J. A. "Grow Nerves Grow," *Science News*, 29 March 1986.

Nauta, W. J. H., and M. Feirtag. *Fundamental Neuroanatomy*. New York: W. H. Freeman, 1985.

Romeo, J. H. "Spinal Cord Injury," *RN*, May 1988.

Thompson, R. F. *The Brain: An Introduction to Neuroscience*. New York: W. H. Freeman, 1985.

Weiss, R. "Muscular Dystrophy Protein Identified," *Science News*, 2 January 1988.

Wurtman, R. J. "Alzheimer's Disease," *Scientific American*, January 1985.

CHAPTER 11

Agras, W. S. "Relaxation Therapy in Hypertension," *Hospital Practice*, May 1983.

Carney, R. M. "Clinical Applications of Relaxation Training," *Hospital Practice*, July 1983.

Carpenter, M. B. *Core Text of Neuroanatomy*, 3rd ed. Baltimore: Williams & Wilkins, 1985.

Nobach, C. R., and R. J. Demarest. *The Nervous System: Introduction and Review*, 3rd ed. New York: McGraw-Hill, 1986.

CHAPTER 12

Franklin, D. "Crafting Sound from Silence," *Science News*, 20 October 1984.

Goldfinger, S. E. (ed.). "Pain Control," *Harvard Medical School Health Letter*, June 1989.

_____. "Hearing Loss and Hearing Aids," *Harvard Medical School Health Letter*, April 1989.

Hogan, M. J. (ed.). "Color Blindness," *Mayo Clinic Health Letter*, March 1989.

Hudspeth, A. J. "The Hair Cells of the Inner Ear," *Scientific American*, January 1983.

Kiely, J. M. (ed.). "Cataracts," *Mayo Clinic Health Letter*, February 1989.

_____. "Retinal Detachment," *Mayo Clinic Health Letter*, January 1988.

_____. "Your Eyes Can Be Windows to Your Health," *Mayo Clinic Health Letter*, March 1985.

Koretz, J. F., and G. H. Handelman. "How the Human Eye Focuses," *Scientific American*, July 1988.

Loeb, G. E. "The Functional Replacement of the Ear," *Scientific American*, February 1985.

Morrison, A. R. "A Window on the Sleeping Brain," *Scientific American*, April 1983.

Von Bekesy, G. "The Ear," *Scientific American*, August 1975.

Wiet, R. J. "Help for the Hearing-Impaired," *Postgraduate Medicine*, November 1988.

CHAPTER 13

Clark, M. "The Power of Hormones," *Newsweek on Health*, Spring 1987.

Crapo, L. *Hormones: Messengers of Life*. New York: W. H. Freeman, 1985.

Fellman, B. "A Clockwork Gland," *Science 85*, May 1985.

Fry, W. F. "Benefits of Stress," *Healthline*, January 1984.

Goldfinger, S. E. (ed.). "Diabetes: The Long Run," *Harvard Medical School Health Letter*, April 1985.

_____. "Insulin and the Diabetic," *Harvard Medical School Health Letter*, March 1985.

_____. "The Diseases Called Sugar Diabetes" (Part 1), *Harvard Medical School Health Letter*, February 1985.

Goodman, H. M. *Basic Medical Endocrinology*. New York: Raven Press, 1988.

Griffin, J. E., and S. R. Ojeda. *Textbook of Endocrine Physiology*. New York: Oxford University Press, 1988.

Huzar, J. G., and P. L. Cerrato. "The Role of Diet and Drugs in Diabetes," *RN*, April 1989.

Miller, J. "Eye to (Third) Eye," *Science News*, 9 November 1985.

Molitch, M. E. "Diabetes Mellitus," *Postgraduate Medicine*, March 1989.

Muldoon, T. G., and A. C. Evans. "Hormones and Their Receptors," *Archives of Internal Medicine*, April 1988.

Orci, L., J. D. Vassalli, and A. Perrelet. "The Insulin Factory," *Scientific American*, September 1988.

Seyle, H. *Stress in Health and Disease*. London: Butterworths, 1976.

Williams, R. *Textbook of Endocrinology*, 7th ed. Philadelphia: Saunders, 1985.

CHAPTER 14

Aledort, L. M. "Current Concepts in Diagnosis and Management of Hemophilia," *Hospital Practice*, October 1982.

Bennett, W. I. "Blood and Safety," *Harvard Medical School Health Letter*, November 1989.

Bloom, M. (ed.). "New Jobs for Man-Made Blood," *Physician's Weekly*, 17 February 1986.

Bock, H. (ed.). "New Method for Saving Your Own Blood," *Medical Update*, October 1988.

Dixon, B. "Of Different Bloods," *Science 84*, November 1984.

Doolittle, R. F. "Fibrinogen and Fibrin," *Scientific American*, December 1981.

England, J. A. "The Many Faces of Epstein–Barr Virus," *Postgraduate Medicine*, February 1988.

Froberg, J. H. "The Anemias: Causes and Course of Action," *RN*, January, March, and May 1989.

Golde, D. W., and J. C. Ganon. "Hormones that Stimulate the Growth of Blood Cells," *Scientific American*, July 1988.

Goldfinger, S. E. (ed.). "Chronic Fatigue Syndrome," *Harvard Medical School Health Letter*, July 1988.

Hogan, M. J. (ed.). "Blood Transfusions," *Mayo Clinic Health Letter*, July 1989.

Kiley, J. M. (ed.). "Chronic Mononucleosis," *Mayo Clinic Health Letter*, June 1988.

Langone, J. "New Methods for Saving Blood," *Time*, 5 December 1988.

Lefant, C. "Dissolving Blood Clots," *Medical Update*, March 1988.

Sabetta, J. R. "Diagnosis: Infectious Mononucleosis," *Hospital Medicine*, March 1984.

Silberner, J. "Clot Dissolver May Save Heart Tissue," *Science News*, 26 November 1983.

Spivak, J. *Fundamentals of Clinical Hematology*, 2nd ed. New York: Harper & Row, 1984.

Thorup, O. A., et al. *Leavell and Thorup's Fundamentals of Clinical Hematology*, 5th ed. Philadelphia: Saunders, 1987.

Vichinsky, E. P., D. Hurst, and B. Lubin. "Sickle Cell Disease: Basic Concepts," *Hospital Medicine*, September 1983.

Weiss, R. "Postponing Red-Cell Retirement," *Science News*, 23 and 30 December 1989.

Zucker, M. D. "The Functioning of Blood Platelets," *Scientific American*, June 1980.

CHAPTER 15

Alpert, J. S. "Management of Acute and Chronic Myocardial Ischemia," *Modern Medicine*, April 1989.

Becker, B. L. "Cholesterol," *Heartbeat*, June 1988.

Brown, M. S., and J. L. Goldstein. "How LDL Receptors Influence Cholesterol and Atherosclerosis," *Scientific American*, November 1984.

Comroe, J. H. "Doctor, You Have Six Minutes," *Science 84*, January/February 1984.

DeVries, W. C. "The Permanent Artificial Heart," *Journal of the American Medical Association*, 12 February 1988.

Dolan, B., and J. M. Nash. "Searching for Life's Elixir," *Newsweek*, 12 December 1988.

Eisenber, M. S., L. Bergner, A. P. Hallstrom, and R. O. Cummins. "Sudden Cardiac Death," *Scientific American*, May 1986.

Escher, D. J. W. "Use of Cardiac Pacemakers," *Hospital Practice*, September 1981.

Falk, J. S., B. Kaufman, and M. H. Weil. "Cardiopulmonary Resuscitation: An Update," *Hospital Medicine*, January 1984.

Fowler, R. E. "Acute Myocardial Infarction," *Postgraduate Medicine*, November 1988.

Goldfinger, S. E. (ed.). "Balloons: Expanding," *Harvard Medical School Health Letter*, May 1987.

_____. "Keeping Up with Cholesterol," *Harvard Medical School Health Letter*, June 1985.

Gwynne, J. T., and M. K. Lawrence, "Current Concepts in the Evaluation and Treatment of Hypercholesterolemia," *Modern Medicine*, March 1989.

Hogan, M. J. (ed.). "Coronary Atherectomy," *Mayo Clinic Health Letter*, April 1989.

_____. "Cholesterol," *Mayo Clinic Health Letter*, March 1988.

Johnson, R. "A Consumer's Guide to Coronary Angioplasty," *Cardiac Alert*, February 1989.

Leaf, D. A. "Omega-3 Fatty Acids and Coronary Artery Disease," *Postgraduate Medicine*, June 1989.

McIntyre, K. "Cardiac Arrest," *Hospital Medicine*, November 1984.

O'Keefe, J. H., C. J. Lavie, and J. O. O'Keefe. "Dietary Prevention of Coronary Artery Disease," *Postgraduate Medicine*, May 1989.

O'Toole, M. T., and A. R. Waldman. "Chest Pain," *RN*, April 1989.

Rios, J. C. (ed.). "Breakthrough Discoveries on How to Avoid a Fatal Heart Attack," *Cardiac Alert*, 1988.

Rogers, W. J. "Diagnosis: Coronary Artery Disease," *Hospital Medicine*, August 1984.

Silberner, J. "Anatomy of Atherosclerosis," *Science News*, 16 March 1985.

Toufexis, A. "The Latest Word on What to Eat," *Time* 13 March 1989.

Wehrmacher, W. H. "Acute Myocardial Infarction," *Postgraduate Medicine*, February 1989.

CHAPTER 16

Blake, P. "Precision Moves that Counter Cardiogenic Shock," *RN*, May 1989.

Bock, H. (ed.). "New Ways to Control Blood Pressure," *Medical Update*, July 1988.

Briening, E. P. "Septic Shock," *RN*, September 1988.

Cressman, M. D., and R. W. Gifford. "Pharmacologic Management of Hypertension," *Postgraduate Medicine*, June 1989.

Garvas, I., M. Bursztyn, and H. Garvas. "Changing Trends in Hypertension Therapy," *Modern Medicine*, October 1988.

Goldfinger, S. E (ed.). "High Blood Pressure: Newer Treatments," *Harvard Medical School Health Letter*, January 1989.

_____. "High Blood Pressure: A New Look," *Harvard Medical School Health Letter*, December 1988.

_____. "Exercise and Well-Being," *Harvard Medical School Health Letter*, February 1985.

Johnson, G. T. (ed.). "The Ups and Downs of Blood Pressure Numbers," *Harvard Medical School Health Letter*, November 1987.

Rios, J. C. (ed.). "Exercise and the Heart," *Cardiac Alert*, February 1984.

Rippe, J. M., A. Ward, J. P. Pocari, and P. S. Freedson. "Walking for Health and Fitness," *Journal of the American Medical Association*, 13 May 1988.

CHAPTER 17

Barnes, D. M. "Obstacles to an AIDS Vaccine," *Science*, 6 May 1988.

Bolognesi, D. P. "Progress in Vaccines Against AIDS," *Science*, 8 December 1989.

_____. "Prospects for Prevention of and Early Intervention Against HIV," *Journal of the American Medical Association*, 26 May 1989.

Cohen, I. R. "The Self, the World and Immunity," *Scientific American*, April 1988.

Ding, E. Y., and Z. A. Cohn. "How Killer Cells Kill," *Scientific American*, January 1988.

Frolich, E. D. "Research into AIDS: Status, Prospects, By-products," *Hospital Medicine*, August 1988.

Gallo, R. C., and L. Montagnier. "AIDS in 1988," *Scientific American*, October 1988.

Haseltine, W. A., and F. Wong-Staal. "The Molecular Biology of the AIDS Virus," *Scientific American*, October 1988.

Henry, K., J. Thurn, and D. Anderson. "Testing for Human Immunodeficiency Virus," *Postgraduate Medicine*, January 1989.

Heyward, W. L., and J. W. Curran. "The Epidemiology of AIDS in the U.S.," *Scientific American*, October 1988.

Jaroff, L. "Stop that Germ," *Time*, 23 May 1988.

Johnson, R. B. "The Therapeutic Potential of Macrophage Activation," *Modern Medicine*, July 1989.

_____. "Current Concepts: Immunology," *New England Journal of Medicine*, 24 March 1988.

Koop, C. E. "Surgeon General's Report on Acquired Immune Deficiency Syndrome," *U.S. Public Health Service Office*, 22 October 1986.

Marx, J. L. "AIDS Drugs—Coming but Not Here," *Science*, 21 April 1989.

Matthews, T. J., and D. P. Bolognese. "AIDS Vaccines," *Scientific American*, October 1988.

Palca, J. "New AIDS Drugs Take Careful Aim," *Science*, 22 December 1989.

Redfield, R. R., and D. S. Burke. "HIV Infections: The Clinical Picture," *Scientific American*, October 1988.

Robertson, S. "Drugs That Keep AIDS Patients Alive," *RN*, February 1989.

Scutchfield, F. D., and A. S. Benenson. "AIDS Update," *Postgraduate Medicine*, March 1989.

Silberner, J. "AIDS: Disease, Research Efforts Advance," *Science News*, 27 April 1985.

Soto-Aguilar, M. C., R. D. de Shazo, and N. P. Waring. "Anaphylaxis," *Postgraduate Medicine*, October 1987.

Springate, C. F., M. A. Flaum, and E. S. Cooper. "The Therapeutic Potential of Interleukins," *Modern Medicine*, October 1987.

Tonegawa, S. "The Molecules of the Immune System," *Scientific American*, October 1985.

Weber, J. N., and R. A. Weiss. "HIV Infection: The Cellular Picture," *Scientific American*, October 1988.

Weiss, R. "AIDS Vaccine: Preliminary but Promising," *Science News*, 17 June 1989.

_____. "HIV Can Linger Years with No Antibodies," *Science News*, 3 June 1989.

Yarchoan, R., and S. Broder, "AIDS Therapies," *Scientific American*, October 1988.

CHAPTER 18

Amin, N. M. "Evaluation and Treatment of Pneumonia in Adult Patients," *Modern Medicine*, January 1989.

Carr, D. T. "Malignant Lung Disease," *Hospital Practice*, January 1981.

Eberhart, J. "Common Cold Preventive," *Science News*, 11 January 1986.

Edwards, D. D. "Nicotine: A Drug of Choice?" *Science News*, 18 January 1986.

Heimlich, H. J. "A Life-Saving Maneuver to Prevent Food Choking," *Journal of the American Medical Association*, 27 October 1975.

Kiley, J. (ed.). "Sudden Infant Death Syndrome," *Mayo Clinic Health Letter*, April 1989.

Mines, A. H. *Respiratory Physiology*, 2nd ed. New York: Raven Press, 1986.

Niewoehner, D. E. "Diagnosis, Management, and Monitoring of Chronic Pulmonary Obstructive Disease," *Modern Medicine*, February 1989.

Petty, T. L. "Prevention of Emphysema," *Postgraduate Medicine*, November 1989.

Weiss, R. "TB Troubles," *Science News*, 6 February 1988.

Wenger, N. K. "Pulmonary Embolism," *Postgraduate Medicine*, August 1988.

CHAPTER 19

Bruckstein, A. H. "Chronic Hepatitis," *Postgraduate Medicine*, May 1989.

Council on Scientific Affairs. "Dietary Fiber and Health," *Journal of the American Medical Association*, 28 July 1989.

Cunha, B. A. "The Yellow Patient," *Postgraduate Medicine*, October 1988.

Dolan, B. "A Mother's Gift of Life," *Time*, 11 December 1989.

Farley, D. "Today's Dentistry: A Mouthful of Marvels," *Healthline*, October 1985.

Farley, P. C., and K. H. McFaden, "Colorectal Cancer," *Postgraduate Medicine*, November 1989.

Friedman, G. "Peptic Ulcer Disease," *Clinical Symposia*, 40, (5), 1988.

Goldfinger, S. E. (ed.). "Shock Waves for Gallstones," *Harvard Medical School Health Letter*, July 1988.

_____. "Screening for Bowel Cancer," *Harvard Medical School Health Letter*, April 1986.

Hogan, M. J. (ed.). "Gallstones," *Mayo Clinic Health Letter*, May 1989.

_____. "Plaque and Tartar," *Mayo Clinic Health Letter*, January 1989.

Jenkins, E. "Large Bowel Exam: When to Use What Test," *Modern Medicine*, March 1984.

Johnson, D. A. "Fecal Occult Blood Testing," *Postgraduate Medicine*, April 1989.

Johnson, G. T. (ed.). "The Tragedy of Anorexia Nervosa," *Harvard Medical School Health Letter*, December 1981.

Leff, E. "Hemorrhoids," *Postgraduate Medicine*, November 1987.

Levine, M. P. "Eating Disorders," *Postgraduate Medicine*, November 1988.

Rios, J. E. (ed.). "Which Kind of Fiber Will Help You?" *Cardiac Alert*, December 1989.

Steinberg, S. "Cancer and Cuisine," *Science News*, 1 October 1983.

Uvnäs-Moberg, K. "The Gastrointestinal Tract in Growth and Reproduction," *Scientific American*, July 1989.

CHAPTER 20

Ackerman, S. "The Management of Obesity," *Hospital Practice*, March 1983.

Cunha, B. A. "The Patient with Fever," *Postgraduate Medicine*, April 1989.

Friedman, R. B. "Very Low-Calorie Diets: How Successful?" *Postgraduate Medicine*, May 1988.

Goldfinger, S. E. "The Overweight Problem," *Harvard Medical School Health Letter*, September 1986.

Hogan, M. J. (ed.). "Liquid Diet Supplements," *Mayo Clinic Health Letter*, January 1989.

Johnson, G. T. (ed.). "Weight Control," *Harvard Medical School Health Letter*, December 1980.

Kiester, E. "A Little Fever Is Good for You," *Science 84*, November 1984.

Miller, J. A. "Obesity: If the Gene Fits . . . ," *Science News*, 25 January 1986.

Rios, J. C. (ed.). "Reviewing the Most Popular Diets," *Cardiac Alert*, February 1984.

Rosenblatt, E. "Weight-Loss Programs," *Postgraduate Medicine*, May 1988.

Rosenthal, T. C., and D. A. Silverstein. "Fever," *Postgraduate Medicine*, June 1988.

Sachs, A. "Drinking Yourself Silly," *Time*, 19 December 1988.

Schimke, R. N. "Metabolic Diseases," *Hospital Practice*, January 1983.

Scott, J. "Heat-Related Illnesses," *Postgraduate Medicine*, June 1989.

CHAPTER 21

Cochran, J. S., S. N. Robinson, V. S. Crane, and D. G. Jones. "Extracorporeal Shock Wave Lithotripsy," *Postgraduate Medicine*, May 1988.

Dretler, S. P. "Stone Crushers," *Harvard Medical School Health Letter*, December 1985.

Nolph, K. D., A. S. Lindblad, and J. W. Novak. "Continuous Ambulatory Peritoneal Dialysis," *New England Journal of Medicine*, 16 June 1988.

Reilly, N. J., and L. C. Torosian. "The New Wave in Lithotripsy," *RN*, March 1988.

Smith, D. A. "Incontinence," *RN*, March 1989.

Spiegel, D. M., M. Burnier, and R. W. Schrier. "Acute Renal Failure," *Postgraduate Medicine*, September 1987.

Twardowski, Z. J. "Peritoneal Dialysis," *Postgraduate Medicine*, April 1989.

Vander, A. J. *Renal Physiology*, 2nd ed. New York: McGraw-Hill, 1980.

Townsend, C. M. "Management of Breast Cancer," *CIBA-GEIGY Clinical Symposia*, 39(4), 1987.

CHAPTER 22

Arieff, A., and R. A. DeFonzo. *Fluid Electrolyte, and Acid–Base Disorders*. New York: Churchill Livingstone, 1985.

Barta, M. A. "Correcting Electrolyte Imbalances," *RN*, February 1987.

Halperin, M. L., and M. C. Goldstein. *Fluid, Electrolyte, and Acid–Base Emergencies*. Philadelphia: Saunders, 1988.

Hays, R. M. "Principles of Ion and Water Transport," *Hospital Practice*, March 1978.

Hobbs, J. "Disturbances in Acid–Base Metabolism," *Postgraduate Medicine*, February 1988.

Horne, M. M., and P. L. Swearinger. *Pocket Guide to Fluids and Electrolytes*, St. Louis: C. V. Mosby, 1989.

Kassirer, J. P., and N. E. Madias. "Respiratory Acid–Base Disorders," *Hospital Practice*, February 1980.

Kopp, U. C., and G. F. DiBona. "Diagnosis: Edema," *Hospital Medicine*, February 1982.

Mueller, K. D. "Keeping Your Patient's Water Level Up," *RN*, July 1989.

Tepper, D., and R. S. Aronson. "Effects of Potassium and Calcium Abnormalities," *Hospital Medicine*, February 1984.

CHAPTER 23

Bennett, I. W. "Estrogen Replacement and Breast Cancer," *Harvard Medical School Health Letter*, October 1989.

Copeland, L. J. "Screening for Cervical Cancer: The Role of the Pap Test," *Modern Medicine*, January 1987.

Danforth, D. N., and J. R. Scott (eds.). *Obstetrics and Gynecology*, 5th ed. Philadelphia: Lippincott, 1986.

Devalon, M. L., and J. W Bachman. "Premenstrual Syndrome," *Postgraduate Medicine*, November 1989.

Goldfinger, S. E. (ed.). "Circumcision and Urinary Tract Infections," *Harvard Medical School Health Letter*, April 1989.

_____. "Prostate Cancer," *Harvard Medical School Health Letter*, September 1988.

_____. "Breast Cancer: Early Decisions," *Harvard Medical School Health Letter*, March 1988.

Hogan, M. J. (ed.). "Impotence," *Mayo Clinic Health Letter*, December 1989.

Johnson, G. T. (ed.). "Cesarean Section," *Harvard Medical School Health Letter*, June 1981.

_____. "Sexually Transmitted Diseases," *Harvard Medical School Health Letter*, April 1981.

Kiely, J. M. (ed.). "Endometriosis," *Mayo Clinic Health Letter*, March 1987.

Langone, J. "Hard Looks at Hormones," *Time*, 14 August 1989.

McElhose, P. "The Other STDs: As Dangerous as Ever," *RN*, June 1988.

McKeon, V. A. "Cruel Myths and Clinical Facts About Menopause," *RN*, June 1989.

Nero, F. A. "When Couples Ask About Infertility," *RN*, November 1988.

Robinson, D. S., N. Love, and J. G. Schwade. "Breast Cancer Screening and Early Diagnosis," *Postgraduate Medicine*, October 1989.

Sweet, R. L. "Pelvic Inflammatory Disease," *Modern Medicine*, April 1989.

CHAPTER 24

Begley, S., and J. Carey. "How Human Life Begins," *Newsweek*, 11 January 1982.

Carlson, B. M. *Patten's Foundations of Human Embryology*, 4th ed. New York: McGraw-Hill, 1981.

Cole, H. M. "Intrauterine Devices," *Journal of the American Medical Association*, 14 April 1989.

Dorfman, A. "Alcohol's Youngest Victims," *Time*, 28 August 1989.

England, M. A. *Color Atlas of Life Before Birth*. Chicago: YearBook Medical Publishers, 1983.

Fuchs, F. "Genetic Amniocentesis," *Scientific American*, June 1980.

Gehring, W. J. "The Molecular Basis of Development," *Scientific American*, October 1985.

Gold, M. "The Baby Makers," *Science 85*, April 1985.

Goldfinger, S. E. (ed.). "Pregnancy: Age and Outcome," *Harvard Medical School Health Letter*, October 1985.

Grimes, D. A. "Reversible Contraception for the 1980's," *Journal of the American Medical Association*, 3 January 1986.

Grobstein, C. "External Human Fertilization," *Scientific American*, June 1979.

Hacker, N. F., and J. G. Moore. *Essentials of Obstetrics and Gynecology*. Philadelphia: Saunders, 1986.

Hogan, M. J. (ed.). "Cesarean Birth," *Mayo Clinic Health Letter*, February 1988.

Holliday, R. "A Different Kind of Inheritance," *Scientific American*, June 1989.

Kantrowitz, B., P. Wingert, and M. Hager. "Preemies," *Newsweek*, 16 May 1988.

Lawn, R. M., and G. A. Vehar, "The Molecular Genetics of Hemophilia," *Scientific American*, March 1986.

Leaf, D. A. "Exercise During Pregnancy," *Postgraduate Medicine*, January 1989.

Lemonick, M. D. "Trying to Fool the Infertile," *Time*, 13 March 1989.

McKusick, V. "Mapping and Sequencing the Human Genome," *New England Journal of Medicine*, 6 April 1989.

Miller, J. A. "Window on the Womb," *Science News*, 2 February 1985.

Mishell, D. R. "Contraception," *New England Journal of Medicine*, 23 March 1989.

Moore, K. L. *Essentials of Human Embryology*. Toronto: B. C. Decker, 1988.

Orshan, S. A. "The Pill, the Patient, and You," *RN*, July 1988.

Scott, W. C. "Medical Applications of Fetal Tissue Transplantation," *Journal of the American Medical Association*, 26 January 1990.

Seibel, M. M. "A New Era in Reproductive Technology," *New England Journal of Medicine*, 31 March 1988.

Silberner, J. "Babymaking: Expanding Horizons," *Science News*, 14 December 1985.

Singer, S. *Human Genetics*. New York: W. H. Freeman, 1985.

Wassarman, P. M. "Fertilization in Mammals," *Scientific American*, December 1988.

Weiss, R. "A Genetic Gender Gap," *Science News*, 20 May 1989.

Zuckerman, B. "Effects of Maternal Marijuana and Cocaine Use on Fetal Growth," *New England Journal of Medicine*, 23 March 1989.

Glossary of Combining Forms, Word Roots, Prefixes, and Suffixes

PRONUNCIATION KEY

1. The strongest accented syllable appears in capital letters, for example, bilateral (bī-LAT-er-al) and diagnosis (dī-ag-NŌ-sis).

2. If there is a secondary accent, it is noted by a single quote mark ('), for example, constitution (kon'-sti-TOO-shun) and physiology (fiz'-ē-OL-ō-jē). Any additional secondary accents are also noted by a single quote mark, for example, decarboxylation (dē'-kar-bok'-si-LĀ-shun).

3. Vowels marked with a line above the letter are pronounced with the long sound as in the following common words:

ā as in *māke*
ē as in *bē*
ī as in *īvy*
ō as in *pōle*

4. Vowels not so marked are pronounced with the short sound, as in the following words:

e as in *bet*
i as in *sip*
o as in *not*
u as in *bud*

5. Other phonetic symbols are used to indicate the following sounds:

a as in *above*

oo as in *sue*
yoo as in *cute*
oy as in *oil*

Many medical terms are ''compound'' words; that is, they are made up of one or more word roots or combining forms of word roots with prefixes or suffixes. For example, *leucocyte* (white blood cell) is a combination of *leuco,* the combining form for the word root meaning ''white,'' and *cyt,* the word root meaning ''cell.'' Learning the medical meanings of the fundamental word parts will enable you to analyze many long, complicated terms.

The following list includes some of the most commonly used combining forms, word roots, prefixes, and suffixes used in making medical terms and an example for each.

COMBINING FORMS AND WORD ROOTS

Acou-, Acu- hearing Acoustics (a-KOO-stiks), the science of sounds or hearing.
Acr-, Acro- extremity Acromegaly (ak'-rō-MEG-a-lē), hyperplasia of the nose, jaws, fingers, and toes.
Aden-, Adeno- gland Adenoma (ad-en'-Ō-ma), a tumor with a glandlike structure.

G-1

Alg-, Algia- pain Neuralgia (nyoo-RAL-ja), pain along the course of a nerve.

Angi- vessel Angiocardiography (an'-jē-ō-kard-ē-OG-ra-fē), roentgenography of the great blood vessels and heart after intravenous injection of radiopaque fluid.

Arthr-, Arthro- joint Arthropathy (ar-THORP-a-thē), disease of a joint.

Aut-, Auto- self Autolysis (aw-TOL-i-sis), destruction of cells of the body by their own enxymes, even after death.

Bio- life, living Biopsy (BĪ-op-sē), examination of tissue removed from a living body.

Blast- germ, bud Blastocyte (BLAS-tō-sīt), an embryonic or undifferentiated cell.

Blephar- eyelid Blepharitis (blef-a-RĪT-is), inflammation of the eyelids.

Brachi- arm Brachialis (brā-kē-AL-is), muscle that flexes the forearm.

Bronch- trachea, windpipe Bronchoscopy (bron-KOS-kō-pē), direct visual examination of the bronchi.

Bucc-, cheek Buccocervical (bū-kō-SER-vi-kal), pertaining to the cheek and neck.

Capit- head Decapitate (dē-KAP-i-tāt), to remove the head.

Carcin- cancer Carcinogenic (kar-sin-ō-JEN-ik), causing cancer.

Cardi-, Cardia-, Cardio- heart Cardiogram (KARD-ē-o-gram), a recording of the force and form of the heart's movements.

Cephal- head Hydrocephalus (hī-drō-SEF-a-lus), enlargement of the head due to an abnormal accumulation of fluid.

Cerebro- brain Cerebrospinal (se-rē'-brō-SPĪN-al) fluid, fluid contained within the cranium and spinal canal.

Cheil- lip Cheilosis (kī-LŌ-sis), dry scaling of the lips.

Chole- bile, gall Cholecystogram (kō-lē-SIS-tō-gram), roentgenogram of the gallbladder.

Chondr-, Chondri-, Chondrio- cartilage Chondrocyte (KON-drō-sīt), a cartilage cell.

Chrom-, Chromat-, Chromato- color Hyperchromic (hī-per-KRŌ-mik), highly colored.

Cili- eyelash Supercilia (soo'-per-SIL-ē-a), eyebrow (hairs above eyelash).

Colpo- vagina Colpotomy (kol-POT-ō-mē), incision into the wall of the vagina.

Cor-, coron- heart Coronary (KOR-ō-na-rē), arteries supplying blood to the heart muscle.

Cost- rib Costal (KOS-tal), pertaining to a rib.

Crani- skull Craniotomy (krā-nē-OT-ō-mē), surgical opening of the skull.

Cry-, Cryo- cold Cryosurgery (krī-ō-SERJ-e-rē), surgical procedure using a very cold liquid nitrogen probe.

Cut- skin Subcutaneous (sub-kyoo-TĀ-nē-us), under the skin.

Cysti-, Cysto- sac, bladder Cystoscope (SIS-tō-skōp), instrument for interior examination of the urinary bladder.

Cyt-, Cyto-, Cyte- cell Cytology (sī-TOL-ō-jē), the study of cells.

Dactyl-, Dactylo- digits (usually fingers, but sometimes toes) Polydactylism (pol-ē-DAK-til-ism), above normal number of fingers or toes.

Derma-, Dermato- skin Dermatosis (der-ma-TŌ-sis), any skin disease.

Dura- hard Dura mater (DYOO-ra MĀ-ter), outer membrane covering brain and spinal cord.

Entero- intestine Enteritis (ent-e-RĪT-is), inflammation of the intestine.

Erythro- red Erythrocyte (e-RITH-rō-sīt), red blood cell.

Galacto- milk Galactose (ga-LAK-tōse), a milk sugar.

Gastr- stomach Gastrointestinal (gas'-trō-in-TES-tin-al), pertaining to the stomach and intestine.

Gloss-, Glosso- tongue Hypoglossal (hī'-pō-GLOS-al), located under the tongue.

Glyco- sugar Glycosuria (glī'-kō-SUR-ē-a), sugar in the urine.

Gravid- pregnant Gravidity (gra-VID-i-tē), condition of being pregnant.

Gyn-, Gyne-, Gynec- female, women Gynecology (gīn'-e-KOL-ō-jē), the medical specialty dealing with disorders of the female reproductive system.

Hem-, Hemat- blood Hematoma (hē'-ma-TŌ-ma), a tumor or swelling filled with blood.

Hepar-, Hepato- liver Hepatitis (hep-a-TĪT-is), inflammation of the liver.

Hist-, Histio- tissue Histology (his-TOL-ō-jē), the study of tissues.

Hydr- water Hydrocele (HĪ-drō-sēl), accumulation of fluid in a saclike cavity.

Hyster- uterus Hysterectomy (his'-te-REK-tō-mē), surgical removal of the uterus.

Ileo- ileum Ileocecal (il'-ē-ō-SĒ-kal) valve, folds at the opening between ileum and cecum.

Ilio- ilium Iliosacral (il'-ē-ō-SĀ-kral), pertaining to ilium and sacrum.

Kines- motion Kinesiology (ki-nē-sē-OL-ō-jē), study of movement of body parts.

Labi- lip Labial (LĀ-bē-al), pertaining to a lip.

Lachry-, Lacri- tears Nasolacrimal (nā-zō-LAK-rim-al), pertaining to the nose and lacrimal apparatus.

Laparo- loin, flank, abdomen Laparoscopy (lap'-a-ROS-kō-pē), examination of the interior of the abdomen by means of a laparoscope.

Leuco-, Leuko- white Leucocyte (LYOO-kō-sīt), white blood cell.

Lingua- tongue Lingual (LIN-gwal), pertaining to the tongue.

Lip-, Lipo fat Lipoma (lī-PŌ-ma), a fatty tumor.

Lith- stone Lithiasis (li-THĒ-a-sis), the formation of stones.

Lumbo- lower back, loin Lumbar (LUM-bar), pertaining to the loin.

Macul- spot, blotch Macula (MAK-yoo-la), spot or blotch.

Malign- bad, harmful Malignant (ma-LIG-nant), condition that gets worse and results in death.

Mamm- breast Mammography (ma-MOG-ra-fē), roentgenography of the mammary gland.

Mast- breast Mastitis (ma-STĪT-is), inflammation of the mammary gland.

Meningo- membrane Meningitis (men-in-JĪT-is), inflammation of the membranes of spinal cord and brain.

Metro- uterus Endometrium (en'-dō-MĒ-trē-um), lining of the uterus.

Morpho- form, shape Morphology (mor-FOL-o-jē), the study of the form and structure of things.

Myelo- marrow, spinal cord Poliomyelitis (pō'-lē-ō-mī'-a-LĪT-is), inflammation of the gray matter of the spinal cord.

Myo- muscle Myocardium (mī-ō-KARD-ē-um), heart muscle.

Necro- corpse, dead Necrosis (ne-KRŌ-sis), death of areas of tissue surrounded by healthy tissue.

Nephro- kidney Nephrosis (ne-FRŌ-sis), degeneration of kidney tissue.

Neuro- nerve Neuroblastoma (nyoor'-ō-blas-TŌ-ma), malignant tumor of the nervous system composed of embryonic nerve cells.

Oculo- eye Binocular (bī-NOK-yoo-lar), pertaining to the two eyes.

Odont- tooth Orthodontic (or-thō-DONT-ik), pertaining to the proper positioning and relationship of the teeth.

Onco- mass, tumor Oncology (ong-KOL-ō-jē), study of tumors.

Oo- egg Oocyte (Ō-ō-sīt), original egg cell.

Oophor- ovary, egg carrier Oophorectomy (ō'-ōf-o-REK-tō-mē), surgical removal of ovaries.

Ophthalm- eye Ophthalmology (of'-thal-MOL-ō-jē), the study of the eye and its diseases.

Or- mouth Oral (Ō-ral), pertaining to the mouth.

Orchido- testicle Orchidectomy (or'-ki-DEK-tō-me), surgical removal of a testicle.

Osmo- odor, sense of smell Anosmia (an-OZ-mē-a), absence of sense of smell.

Oss-, Osseo-, Osteo- bone Osteoma (os-tē-Ō-ma), bone tumor.

Oto- ear Otosclerosis (ō'-tō-skle-RŌ-sis), formation of bone in the labyringh of the ear.

Palpebr- eyelid Palpebra (PAL-pe-bra), eyelid.

Part- birth, delivery, labor Parturition (par'-too-RISH-un), act of giving birth.

Patho- disease Pathogenic (path'-ō-JEN-ik), causing disease.

Ped- children Pediatrician (pēd-ē-a-TRISH-an), medical specialist in the treatment of children.

Peps- digest Peptic (PEP-tik), pertaining to digestion.

Phag-, Phago- to eat Phagocytosis (fag'-ō-sī-TŌ-sis), the process by which cells ingest particulate matter.

Philic-, Philo- to like, have an affinity for Hydrophilic (hī-drō-FIL-ik), having an affinity for water.

Phleb- vein Phlebitis (fle-BĪT-is), inflammation of the veins.

Phon- voice, sound Phonogram (FŌ-nō-gram), record made of sound.

Phren- diaphragm Phrenic (FREN-ik), pertaining to the diaphragm.

Pilo- hair Depilatory (de-PIL-a-tō-re), hair remover.

Pneumo- lung, air Pneumothorax (nyoo-mō-THŌR-aks), air in the thoracic cavity.

Pod- foot Podiatry (po-DĪ-a-trē), the diagnosis and treatment of foot disorders.

Procto- anus, rectum Proctoscopy (prok-TOS-kō-pē), instrumental examination of the rectum.

Psycho- soul, mind Psychiatry (sī-KĪ-a-trē), treatment of mental disorders.

Pulmon- lung Pulmonary (PUL-mō-ner'-ē), pertaining to the lungs.

Pyle-, Pyloro opening, passage Pyloric (pī-LOR-ik), pertaining to the pylorus of the stomach.

Pyo- pus Pyuria (pī-YOOR-ē-a), pus in the urine.

Ren- kidneys Renal (RĒ-nal), pertaining to the kidney.

Rhin- nose Rhinitis (ri-NĪT-is), inflammation of nasal mucosa.

Salpingo- uterine (Fallopian) tube Salpingitis (sal'-pin-JĪ-tis), inflammation of the uterine (Fallopian) tubes.

Scler-, Sclero- hard Atherosclerosis (ath'-er-ō-skle-RŌ-sis), hardening of the arteries.

Sep-, Septic- toxic condition due to microorganisms Septicemia (sep'-ti-SĒ-mē-a), presence of bacterial toxins in the blood (poisoning).

Soma-, Somato- body Somatotropic (sō-mat-ō-TRŌ-pik), having a stimulating effect on body growth.

Somni- sleep Insomnia (in-SOM-nē-a), inability to sleep.

Stasis-, Stat- stand still Homeostasis (hō'-mē-ō-STĀ-sis), achievement of a steady state.

Sten- narrow Stenosis (ste-NŌ-sis), narrowing of a duct or canal.

Tegument- skin, covering Integumentary (in-teg-yoo-MEN-ta-rē), pertaining to the skin.

Therm- heat Thermometer (ther-MOM-et-er), instrument used to measure and record heat.

Thromb- clot, lump Thrombus (THROM-bus), clot in a blood vessel or heart.

Tox-, Toxic- poison Toxemia (tok-SĒ-mē-a), poisonous substances in the blood.

Trich- hair Trichosis (trik-Ō-sis), disease of the hair.

Tympan- eardrum Tympanic (tim-PAN-ik) membrane, eardrum.

Vas- vessel, duct Cerebrovascular (se-rē-brō-VAS-kyoo-lar), pertaining to the blood vessels of the cerebrum of the brain.

Viscer- organ Visceral (VIS-e-ral), pertaining to the abdominal organs.

Zoo- animal Zoology (zō-OL-o-jē), the study of animals.

PREFIXES

A-, An- without, lack of, deficient Anesthesia (an'-es-THĒ-zha), without sensation.

Ab- away from, from Abnormal (ab-NOR-mal), away from normal.

Ad- to, near, toward Adduction (a-DUK-shun), movement of an extremity toward the axis of the body.

Alb- white Albino (Al-BĪ-no), person whose skin, hair, and eyes lack the pigment melanin.

Alveol- cavity, socket Alveolus (al-VĒ-ō-lus), air sac in the lung.

Ambi- both sides Ambidextrous (am'-bi-DEK-strus), able to use either hand.

Ambly- dull Amblyaphia (am-blē-A-fē-a), dull sense of touch.

Andro- male, masculine Androgen (AN-drō-jen), male sex hormone.

Ankyl(o)- bent, fusion Ankylosed (ANG-ki-lōsd), fused joint.

Ante- before Antepartum (ant-ē-PAR-tum), before delivery of a baby.

Anti- against Anticoagulant (an-tī-ko-AG-yoo-lant), a substance that prevents coagulation of blood.

Basi- base, foundation Basal (BĀ-sal), located near the base.

Bi- two, double, both Biceps (BĪ-seps), a muscle with two heads of origin.

Bili- bile, gall Biliary (BIL-ē-er-ē), pertaining to bile, bile ducts, or gallbladder.

Brachy- short Brachyesophagus (brā-kē-e-SOF-a-gus), short esophagus.

Brady- slow Bradycardia (brād'-ē-KARD-ē-a), abnormal slowness of the heartbeat.

Cata- down, lower, under, against Catabolism (ka-TAB-a-lizm), metabolic breakdown into simpler substances.

Circum- around Circumrenal (ser-kum-RĒ-al), around the kidney.

Cirrh- yellow Cirrhosis (si-RŌ-sis), liver disorder that causes yellowing of skin.

Co-, Con-, Com- with, together Congenital (kon-JEN-i-tal), existing at birth.

Contra- against, opposite Contraception (kon-tra-SEP-shun), the prevention of conception.

Crypt- hidden, concealed Cryptorchidism (krip-TOR-ka-dizm'), undescended or hidden testes.

Cyano- blue Cyanosis (sī-a-NŌ-sis), bluish discoloration due to inadequate oxygen.

De- down, from Decay (de-KĀ), waste away from normal.

Demi-, hemi- half Hemiplegia (hem'-ē-PLĒ-jē-a), paralysis on one side of the body.

Di-, Diplo- two Diploid (DIP-loyd), having double the haploid number of chromosomes.

Dis- separation, apart, away from Disarticulate (dis'-ar-TIK-yoo-lāt'), to separate at a joint.

Dys- painful, difficult Dyspnea (disp-NĒ-a), difficult breathing.

E-, Ec-, Ex- out from, out of Eccentric (ek-SEN-trik), not located at the center.

Ecto-, Exo- outside Ectopic (ek-TOP-ik) pregnancy, gestation outside the uterine cavity.

Em-, En- in, on Empyema (em'-pī-Ē-ma), pus in a body cavity.

En-, Endo- inside Endocardium (en'dō-KARD-ē-um), membrane lining the inner surface of the heart.

Epi- upon, on, above Epidermis (ep'i-DER-mis), outermost layer of skin.

Eu- well Eupnea (YOOP-nē-a), normal breathing.

Ex-, Exo- out, away from Exocrine (EK-sō-krin), excreting outwardly or away from.

Extra- outside, beyond, in addition to Extracellular (ek'-stra-SEL-yoo-lar), outside of the cell.

Fore- before, in front of Forehead (FOR-hed), anterior part of head.

Gen- originate, produce, form Pathogen (PATH-ō-jen), disease producer.

Gingiv- gum Gingivitis (jin'-je-VĪ-tus), inflammation of the gums.

Hemi- half Hemiplegia (hem-ē-PLĒ-jē-a), paralysis of only half of the body.

Heter, Hetero- other, different Heterogeneous (het'-e-rō-JEN-ē-us), composed of different substances.

Homeo, Homo- unchanging, the same, steady Homeostasis (hō'-mē-ō-STĀ-sis), achievement of a steady state.

Hyper- beyond, excessive Hyperglycemia (hī-per-glī-SĒ-mē-a), excessive amount of sugar in the blood.

Hypo- under, below, deficient Hypodermic (hī-pō-DER-mik), below the skin or dermis.

Idio- self, one's own, separate Idiopathic (id'-ē-ō-PATH-ik), a disease without recognizable cause.

In-, Im- in, inside, not Incontinent (in-KON-ti-nent), not able to retain urine or feces.

Infra- beneath Infraorbital (in'-fra-OR-bi-tal), beneath the orbit.

Inter- among, between Intercoastal (int'-er-KOS-tal), between the ribs.

Intra within, inside Intracellular (in'-tra-SEL-yoo-lar), inside the cell.

Iso- equal, like Isogenic (ī-sō-JEN-ik), alike in morphological development.

Later- side Lateral (LAT-er-al), pertaining to a side or farther from the midline.

Lepto- small, slender, thin Leptodermic (lep'-tō-DER-mik), having thin skin.

Macro- large, great Macrophage (MAK-rō-fāj), large phagocytic cell.

Mal- bad, abnormal Malnutrition (mal'-noo-TRISH-un), lack of necessary food substances.

Medi-, Meso- middle Medial (MĒD-ē-al), nearer to midline.

Mega, Megalo- great, large Megakaryocyte (meg'-a-KAR-ē-ō-sīt), giant cell of bone marrow.

Melan- black Melanin (MEL-a-nin), black or dark brown pigment found in skin and hair.

Meta- after, beyond Metacarpus (met'-a-KAR-pus), the part of the hand between the wrist and fingers.

Micro- small Microtome (MĪ-krō-tōm), instrument for preparing very thin slices of tissue for microscopic examination.

Mono- one Monorchid (mon-OR-kid), having one testicle.

Neo- new Neonatal (nē-ō-NĀT-al), pertaining to the first 4 weeks after birth.

Noct(i)- night Nocturia (nok-TOO-rē-a), urination occurring at night.

Null(i)- none Nullipara (nu-LIP-a-ra), woman with no children.

Nyct- night Nyctalopia (nik'-ta-LŌ-pē-a), night blindness.

Oligo- small, deficient Oliguria (ol-ig-YOO-rē-a), abnormally small amount of urine.

Ortho- straight, normal Orthopnea (or-thop-NĒ-a), inability to breathe in any position except when straight or erect.

Pan- all Pancarditis (pan-kar-DĪ-tis), inflammation of the entire heart.

Para- near, beyond, apart from, beside Paranasal (par-a-NĀ-zal), near the nose.

Per- through Percutaneous (per'-kyoo-TĀ-nē-us), through the skin.

Peri- around Pericardium (per'-i-KARD-ē-um), membrane or sac around the heart.

Poly- much, many Polycythemia (pol'-i-sī-THĒ-mē-a), an excess of red blood cells.

Post- after, beyond Postnatal (pōst-NAT-al), after birth.

Pre-, Pro- before, in front of Prenatal (prē-NAT-al), before birth.

Prim- first Primary (PRĪ-me-rē), first in time or order.

Proto- first Protocol (PRŌ-tō-kol), clinical report made from first notes taken.

Pseud-, Pseudo- false Pseudoangina (soo'-dō-an-JĪ-na), false angina.

Retro- backward, located behind Retroperitoneal (re'-trō-per'-it-on-Ē-al), located behind the peritoneum.

Schizo- split, divide Schizophrenia (skiz'-ō-FRE-nē-a), split personality mental disorder.

Semi- half Semicircular (semi'i-SER-kyoo-lar) canals, canals in the shape of a half circle.

Sub- under, beneath, below Submucosa (sub'-myoo-KŌ-sa), tissue layer under a mucous membrane.

Super- above, beyond Superficial (soo-per-FISH-al), confined to the surface.

Supra- above, over Suprarenal (soo-pra-RĒN-al), adrenal gland above the kidney.

Sym- Syn- with, together, joined Syndrome (SIN-drōm), all the symptoms of a disease considered as a whole.

Tachy- rapid Tachycardia (tak'-i-KARD-ē-a), rapid heart action.

Terat(o)- malformed fetus Teratogen (TER-a-tō-jen), an agent that caused development of a malformed fetus.

Tetra-, quadra- four Tetrad (TET-rad), group of four with something in common.

Trans- across, through, beyond Transudation (trans-yoo-DĀ-shun), oozing of a fluid through pores.

Tri- three Trigone (TRĪ-gon), a triangular space, as at the base of the bladder.

SUFFIXES

-able capable of, having ability to Viable (VĪ-a-bal), capable of living.

-ac, -al pertaining to Cardiac (KARD-ē-ak), pertaining to the heart.

-agra severe pain Myagra (mī-AG-ra), severe muscle pain.

-an, -ian pertaining to Circadian (ser-KĀ-dē-an), pertaining to a cycle of active and inactive periods.

-ant having the characteristic of Malignant (ma-LIG-nant), having the characteristic of badness.

-ary connected with Ciliary (SIL-ē-ar-ē), resembling any hairlike structure.

-asis, -asia, -esis, -osis condition or state of Hemostasis (hē-mō-STĀ-sis), stopping of bleeding or circulation.

-asthenia weakness Myasthenia (mi-as-THĒ-nē-a), weakness of skeletal muscles.

-ation process, action, condition Inspiration (in-spu-RĀ-shun), process of drawing air into lungs.

-cel, -cele swelling, an enlarged space or cavity Meningocele (men-IN-gō-sēl), enlargement of the meninges.

-centesis puncture, usually for drainage Amniocentesis (am'-nē-ō-sen-TĒ-sis), withdrawal of amniotic fluid.

-cid, -cide, -cis cut, kill, destroy Germicide (jer-mi-SĪD), a substance that kills germs.

-ectasia, -ectasis stretching, dilation Bronchiectasis (bron-kē-EK-ta-sis), dilation of a bronchus or bronchi.

-ectomize, ectomy excision of, removal of Thyroidectomy (thī-royd-EK-tō-mē), surgical removal of a thyroid gland.

-ema swelling, distension Emphysema (em'-fi-SĒ-ma), swelling of air sacs in lungs.

-emia condition of blood Lipemia (lip-Ē-mē-a), abnormally high concentration of fat in the blood.

-esis condition, process Enuresis (en'-yoo-RĒ-sis), condition of involuntary urination.

-esthesia sensation, feeling Anesthesia (an'-es-THĒ-zē-a), total or partial loss of feeling.

-ferent carry Efferent (EF-e-rent), carrying away from a center.

-form shape Fusiform (FYOO-zi-form), spindle-shaped.

-gen agent that produces or orginates Pathogen (PATH-ō-jen), microorganism or substance capable of producing a disease.

-genic produced from, producing Pyogenic (pī-ō-JEN-ik), producing pus.

-gram record, that which is recorded Electrocardiogram (e-lek'-trō-KARD-ē-ō-gram), record of heart action.

-graph instrument for recording Electroencephalograph (e-lek'-grō-en-SEF-a-lō-graf), instrument for recording electrical activity of the brain.

-ia state, condition Hypermetropia (hī'-per-me-TRŌ-pē-a), condition of farsightedness.

-iatrics, iatry medical practice specialities Pediatrics (pēd-ē-A-triks), medical science relating to care of children and treatment of their diseases.

-ician person associated with Technician (tek-NISH-an), person skilled in a technical field.

-ics art or science of Optics (OP-tiks), science of light and vision.

-ion action, condition resulting from action Incision (in-SIZH-un), act or result of cutting into flesh.

-ism, condition, state Rheumatism (ROO-ma-tizm), inflammation, especially of muscles and joints.

-ist one who practices Internist (in-TER-nist), one who practices internal medicine.

-itis inflammation Neuritis (nyoo-RĪT-is), inflammation of a nerve or nerves.

-ive relating to Sedative (SED-a-tive), relating to a pain or tension reliever.

-logy, -ology the study or science of Physiology (fez-ē-OL-ō-jē), the study of function of body parts.

-lyso, -lysis solution, dissolve, loosening Hemolysis (hē-MOL-i-sis), dissolution of red blood cells.

-malacia softening Osteomalacia (os'-tē-ō-ma-LĀ-shē-a), softening of bone.

-megaly enlarged Cardiomegaly (kar'-dē-ō-MEG-a-lē), enlarged heart.

-oid resembling Lipoid (li-POYD), resembling fat.

-ologist specialist Dermatologist (der-ma-TOL-ō-gist), specialist in the study of the skin.

-oma tumor Fibroma (fi-BRŌ-ma), tumor composed mostly of fibrous tissue.

-ory pertaining to Sensory (SENS-o-rē), pertaining to sensation.

-ose full of Adipose (AD-i-pōz), characterized by presence of fat.

-osis condition, disease Necrosis (ne-KRŌ-sis), condition of death of cells.

-ostomy create an opening Colostomy (kō-LOS-tō-me), surgical creation of an opening between the colon and body surface.

-otomy surgical incision Tracheotomy (trā-kē-OT-ō-me), surgical incision of the trachea.

-pathy disease Neuropathy (nyoo-ROP-a-thē), disease of the peripheral nervous system.

-penia deficiency Thrombocytopenia (throm'-bō-sīt'-o-PĒ-nē-a), deficiency of thrombocytes in the blood.

-phobe, -phobia fear of, aversion to Hydrophobia (hī-drō-FŌ-bē-a), fear of water.

-plasia, -plasty development, formation Rhinoplasty (RĪ-nō-plas-tē), surgical reconstruction of the nose.

-plegia, -plexy stroke, paralysis Apoplexy (AP-ō-plek-sē), sudden loss of consciousness and paralysis.

-pnea to breathe Apnea (AP-nē-a), temporary absence of respiration, following a period of overbreathing.

-poiesis production Hematopoiesis (he-mat'-a-poy-Ē-sis), formation and development of red blood cells.

-ptosis falling, sagging Blepharoptosis (blef'-a-rō-TŌ-sis), dropping of upper eyelid.

-rrhage bursting forth, abnormal discharge Hemorrhage (HEM-or-rij), bursting forth of blood.

-rrhea flow, discharge Diarrhea (dī-a-RĒ-a), abnormal frequency of bowel evacuation, the stools with a more or less fluid consistency.

-scope instrument for viewing Bronchoscope (BRON-kō-skōp), instrument used to examine the interior of a bronchus.

-stomy creation of a mouth or artificial opening Tracheostomy (trā-kē-OST-ō-mē), creation of an opening in the trachea.

-tic, -ulnar pertaining to Diagnostic (dī-ag-NOS-tik), pertaining to diagnosis.

-tomy cutting into, incision into Laparatomy (lap-a-ROT-ō-mē), an abdominal incision to gain access to the peritoneal cavity.

-tripsy crushing Lithotripsy (LITH-ō-trip'-sē), crushing of a calculus (stone).

-trophy state relating to nutrition or growth Hypertrophy (hī-PER-trō-fē), excessive growth of an organ or part.

-tropic turning toward, influencing, changing Gonadotropic (gō-nad-a-TRŌ-pic), influencing the gonads.

-uria urine Polyuria (pol-ē-YOOR-ē-a), excessive secretion of urine.

Glossary of Terms

PRONUNCIATION KEY

1. The strongest accented syllable appears in capital letters, for example, bilateral (bī-LAT-er-al) and diagnosis (dī-ag-NŌ-sis).

2. If there is a secondary accent, it is noted by a single quote mark ('), for example, constitution (kon'-sti-TOO-shun) and physiology (fiz'-ē-OL-ō-mē). Any additional secondary accents are also noted by a single quote mark, for example, decarboxylation (dē'-kar-bok'-si-LĀ-shun).

3. Vowels marked with a line above the letter are pronounced with the long sound, as in the following common words:

ā as in *māke*
ē as in *bē*
ī as in *īvy*
ō as in *pōle*

4. Vowels not so marked are pronounced with the short sound, as in the following words:

e as in *bet*
i as in *sip*
o as in *not*
u as in *bud*

5. Other phonetic symbols are used to indicate the following sounds:

a as in *above*
oo as in *sue*
yoo as in *cute*
oy as in *oil*

Abatement (a-BĀT-ment)   A decrease in the seriousness of a disorder or in the severity of pain or other symptoms.

Abdomen (ab-DŌ-men or AB-dō-men)   The area between the diaphragm and pelvis.

Abdominal (ab-DŌM-i-nal) *cavity*   Superior portion of the abdominopelvic cavity that contains the stomach, spleen, liver, gallbladder, pancreas, small intestine, and most of the large intestine.

Abdominal thrust maneuver   A first-aid procedure for choking. Employs a quick, upward thrust against the diaphragm that forces air out of the lungs with sufficient force to eject any lodged material. Also called the *Heimlich* (HĪM-lik) *maneuver.*

Abdominopelvic (ab-dom'-i-nō-PEL-vic) *cavity*   Inferior component of the ventral body cavity that is subdivided into an upper abdominal cavity and a lower pelvic cavity.

Abduction (ab-DUK-shun)   Movement away from the axis or midline of the body or one of its parts.

Abortion (a-BOR-shun)   The premature loss (spontaneous) or removal (induced) of the embryo or nonviable fetus; any failure in the normal process of developing or maturing.

Abrasion (a-BRĀ-shun)   A portion of skin that has been scraped away.

Abscess (AB-ses)   A localized collection of pus and liquefied tissue in a cavity.

Absorption (ab-SORP-shun)   The taking up of liquids by solids or of gases by solids or liquids; intake of fluids or other substances by cells of the skin or mucous membranes; the passage of digested foods from the gastrointestinal tract into blood or lymph.

Absorptive state   Metabolic state during which ingested nutrients are being absorbed by the blood or lymph from the gastrointestinal tract.

Accessory duct   A duct of the pancreas that empties into the duodenum about 2.5 cm (1 in.) superior to the ampulla of Vater (hepatopancreatic ampulla). Also called the *duct of Santorini* (san'-tō-RE-ne).

Accommodation (a-kom-ō-DĀ-shun)   A change in the curvature of the eye lens to adjust for vision at various distances; focusing.

Acetabulum (as'-e-TAB-yoo-lum)   The rounded cavity on the external surface of the coxal (hip) bone that receives the head of the femur.

Acetylcholine (as'-ē-til-KŌ-lēn) *(ACh)*   A neurotransmitter, liberated at synapses in the central nervous system, that stimulates skeletal muscle contraction.

Achille's tendon   *See Calcaneal tendon.*

Acid (AS-id)   A proton donor, or substance that dissociates into hydrogen ions (H^+) and anions, characterized by an excess of hydrogen ions and a pH less than 7.

Acidosis (as-i-DŌ-sis)   A condition in which blood pH ranges from 7.35 to 6.80 or lower.

Acini (AS-i-nē)   Masses of cells in the pancreas that secrete digestive enzymes.

Acne (AK-nē)   Inflammation of sebaceous (oil) glands that usually begins at puberty; the basic acne lesions in order of increasing severity are comedones, papules, pustules, and cysts.

Acoustic (a-KOOS-tik)   Pertaining to sound or the sense of hearing.

Acquired immune deficiency syndrome (AIDS)   A disorder characterized by a positive HIV-antibody test and certain indicator diseases (Kaposi's sarcoma, *Pneumocytis carinii* pneumonia, tuberculosis, fungus diseases, etc.). A deficiency of helper T cells and a reversed ratio of helper T cells to suppressor T cells that results in fever or night sweats, coughing, sore throat, fatigue, body aches, weight loss, and enlarged lymph nodes. Caused by a virus called human immunodeficiency virus (HIV).

Acromegaly (ak'-rō-MEG-a-lē)   Condition caused by hypersecretion of human growth hormone (hGH) during adulthood characterized by thickened bones and enlargement of other tissues.

Acrosome (AK-rō-sōm)   A dense granule in the head of a spermatozoon that contains enzymes that facilitate the penetration of a spermatozoon into a secondary oocyte.

Actin (AK-tin)   The contractile protein that makes up thin myofilaments in muscle fiber (cell).

Action potential   A wave of negativity that self-propagates along the outside surface of the membrane of a neuron or muscle fiber (cell); a rapid change in membrane potential that involves a depolarization following a repolarization. Also called a *nerve action potential nerve impulse* as it relates to a neuron and a *muscle action potential* as it relates to a muscle fiber (cell).

Active transport   The movement of substances, usually ions, across cell membranes, against a concentration gradient, requiring the expenditure of energy (ATP).

Acuity (a-KYOO-i-tē)   Clearness or sharpness, usually of vision.

Acupuncture (AK-ū-punk'-chur)   The insertion of a needle into a tissue for the purpose of drawing fluid or relieving pain. It is also an ancient Chinese practice employed to cure illnesses by inserting needles into specific locations of the skin.

Acute (a-KYOOT)   Having rapid onset, severe symptoms, and a short course; not chronic.

Adam's apple   *See Thyroid cartilage.*

Adaptation (ad'-ap-TĀ-shun)   The adjustment of the pupil of the eye to light variations. The property by which a neuron relays a decreased frequency of action potentials from a receptor even though the strength of the stimulus remains constant. The decrease in perception of a sensation over time while the stimulus is still present.

Addison's (AD-i-sonz) *disease*   Disorder caused by hyposecretion of glucocorticoids (and aldosterone) characterized by muscular weakness, hypoglycemia, mental lethargy, anorexia, nausea and vomiting, weight loss, low blood pressure, dehydration, and excessive skin and mucous membrane pigmentation.

Adduction (ad-DUK-shun)   Movement toward the axis or midline of the body or one of its parts.

Adenohypophysis (ad'-e-nō-hī-POF-i-sis)   The anterior portion of the pituitary gland.

Adenoids (AD-e-noyds)   The pharyngeal tonsils.

Adenosine triphosphate (a-DEN-ō-sēn trī-FOS-fāt) (*ATP*)   The universal energy-carrying molecule manufactured in all living cells as a means of capturing and storing energy. It consists of the purine base *adenine* and the five-carbon sugar *ribose,* to which are added, in linear array, three *phosphate* molecules.

Adenylate cyclase (a-DEN-i-lāt SĪ-klās)   An enzyme in the postsynaptic membrane that is activated when certain neurotransmitters bind to their receptors; the enzyme converts ATP into cyclic AMP.

Adhesion (ad-HĒ-zhun)   Abnormal joining of parts to each other.

Adipocyte (AD-ipō-sīt)   Fat cell, derived from a fibroblast.

Adrenal cortex (a-DRĒ-nal KOR-teks)   The outer portion of an adrenal gland, divided into three zones, each of which has a different cellular arrangement and secretes different hormones.

Adrenal (a-DRĒ-nal) *glands*   Two glands located superior to each kidney. Also called the *suprarenal* (soo'-pra-RĒ-nal) *glands.*

Adrenal medulla (me-DUL-a)   The inner portion of an adrenal gland, consisting of cells that secrete epinephrine and norepinephrine (NE) in response to the stimulation of preganglionic sympathetic neurons.

Adrenergic (ad'-ren-ER-jik) *fiber*   A nerve fiber that when stimulated releases norepinephrine (noradrenaline) at a synapse.

Adrenocorticotropic   (ad-rē'-nō-kor-ti-kō-TRŌ-ik)   *hormone (ACTH)*   A hormone produced by the adenohypophysis (anterior lobe) of the pituitary gland that influences the production and secretion of certain hormones of the adrenal cortex.

Adrenoglomerulotropin (a-drē'-nō-glō-mer'-yoo-lō-TRŌ-pin)   A hormone secreted by the pineal gland that may stimulate aldosterone secretion.

Adventitia (ad-ven-TISH-ya)   The outermost covering of a structure or organ.

Aerobic (air-Ō-bik)   Requiring molecular oxygen.

Afferent arteriole (AF-er-ent ar-TĒ-rē-ōl)   A blood vessel of a kidney that breaks up into the capillary network called a glomerulus; there is one afferent arteriole for each glomerulus.

Afferent neuron (NOO-ron)   A neuron that carries a nerve impulse toward the central nervous system. Also called a *sensory neuron.*

Afterimage   Persistence of a sensation even though the stimulus has been removed.

Agglutination (a-gloo'-ti-NĀ-shun)   Clumping of microorganisms or blood corpuscles; typically an antigen–antibody reaction.

Agglutinin (a-GLOO-ti-nin)   A specific principle or antibody in blood serum capable of causing the clumping of bacteria, blood corpuscles, or particles. Also called an *isoantibody.*

Agglutinogen (ag'-loo-TIN-ō-gen)   A genetically determined antigen located on the surface of erythrocytes; basis for the ABO grouping and Rh system of blood classification. Also called an *isoantigen.*

Aggregated lymphatic follicles   Aggregated lymph nodules that are most numerous in the ileum. Also called *Peyer's* (PĪ-erz) *patches.*

Aging   Normal process accompanied by a progressive alteration of the body's homeostatic adaptive responses.

Agnosia (ag-NŌ-zē-a)   A loss of the ability to recognize the meaning of stimuli from the various senses (visual, auditory, touch).

Agraphia (a-GRAF-ē-a)   An inability to write.

Albinism (AL-bin-izm)   Abnormal, nonpathological, partial or total absence of pigment in skin, hair, and eyes.

Albumin (al-BYOO-min)   The most abundant (60 percent) and smallest of the plasma proteins, which functions primarily to regulate osmotic pressure of plasma.

Albuminuria (al-byoo'-min-UR-ēa)   Presence of albumin in the urine.

Aldosterone (al-do-STER-ōn)   A mineralocorticoid produced by the adrenal cortex that brings about sodium and water reabsorption and potassium excretion.

Aldosteronism (al'do-STER-ōn-izm')   Condition caused by hypersecretion of aldosterone that results in increased sodium concentration and decreased potassium concentration in blood and characterized by muscular paralysis, high blood pressure, and edema.

Alimentary (al-i-MEN-ta-rē)   Pertaining to nutrition.

Alkaline (AL-ka-līn)   Containing more hydroxyl ions (OH⁻) than hydrogen ions (H⁺) to produce a pH of more than 7.

Alkalosis (al-ka-LŌ-sis)   A condition in which blood pH ranges from 7.45 to 8.00 or higher.

Allele (a-LĒL)   Genes that control the same inherited trait (such as height or eye color) that are located on the same position (locus) on homologous chromosomes.

Allergen (AL-er-jen)   An antigen that evokes a hypersensitivity reaction.

Allergic (a-LER-jik)   Pertaining to or sensitive to an allergen.

All-or-none principle   In muscle physiology, muscle fibers (cells) of a motor unit contract to their fullest extent or not at all. In neuron physiology, if a stimulus is strong enough to initiate an action potential, a nerve impulse is transmitted along the entire neuron at a constant and minimum strength.

Alpha (AL-fa) *cell*   A cell in the pancreatic islets (islets of Langerhans) in the pancreas that secretes glucagon.

Alpha receptor   Receptor found on visceral effectors innervated by most sympathetic postganglionic axons; in general, stimulation of alpha receptors leads to excitation.

Alveolar-capillary (al-VĒ-ō-lar) *membrane*   Structure in the lungs consisting of the alveolar wall and basement membrane and a capillary endothelium and basement membrane through which the diffusion of respiratory gases occurs. Also called the *respiratory membrane*.

Alveolar duct   Branch of a respiratory bronchiole around which alveoli and alveolar sacs are arranged.

Alveolar macrophage (MAK-rō-fāj)   Cell found in the alveolar walls of the lungs that is highly phagocytic. Also called a *dust cell*.

Alveolar sac   A collection or cluster of alveoli that share a common opening.

Alveolus (al-VĒ-ō-lus)   A small hollow or cavity; an air sac in the lungs; milk-secreting portion of a mammary gland.

Alzheimer's (ALTZ-hī-merz) *disease* (*AD*)   Disabling neurological disorder characterized by dysfunction and death of specific cerebral neurons resulting in widespread intellectual impairment, personality changes, and fluctuations in alertness.

Amenorrhea (ā-men-ō-RĒ-a)   Absence of menstruation.

Amino acid   An organic acid, containing a carboxyl group (COOH) and an amino group (NH₂), that is the building unit from which proteins are formed.

Amnesia (am-NĒ-zē-a)   A lack or loss of memory.

Amniocentesis (am'-nē-ō-sen-TĒ-sis)   Removal of amniotic fluid by inserting a needle transabdominally into the amniotic cavity.

Amnion (AM-nē-on)   The innermost fetal membrane; a thin transparent sac that holds the fetus suspended in amniotic fluid. Also called the *"bag of waters."*

Amniotic (am'-nē-OT-ik) *fluid*   Fluid in the amniotic cavity, the space between the developing embryo (or fetus) and amnion; the fluid is initially produced as a filtrate from maternal blood and later from fetal urine.

Amphiarthrosis (am'-fē-ar-THRŌ-sis)   Articulation midway between diarthrosis and synarthrosis, in which the articulating bony surfaces are separated by an elastic susbstnce to which both are attached, so that the mobility is slight.

Ampulla of Vater   See **Hepatopancreatic ampulla.**

Amyotrophic (a-mē-ō-TROF-ik) *lateral sclerosis* (*ALS*)   Progressive neuromuscular disease characterized by degeneration of motor neurons in the spinal cord that leads to muscular weakness. Also called *Lou Gehrig's disease.*

Anabolism (a-NAB-ō-lizm)   Synthetic energy-requiring reactions whereby small molecules are built up into larger ones.

Anaerobic (an-AIR-ō-bik)   Not requiring molecular oxygen.

Anal (Ā-nal) *canal*   The terminal 2 or 3 cm (1 in.) of the rectum; opens to the exterior of the anus.

Anal column   A longitudinal fold in the mucous membrane of the anal canal that contains a network of arteries and veins.

Analgesia (an-al-JĒ-zē-a)   Pain relief.

Anal triangle   The subdivision of the female or male perineum that contains the anus.

Anamnestic (an'-am-NES-tik) *response*   Accelerated, more intense production of antibodies upon a subsequent exposure to an antigen after the initial exposure.

Anaphase (AN-a-fāz)   The third stage of mitosis in which the chromatids that have separated at the centromeres move to opposite poles of the cell.

Anaphylaxis (an'-a-fi-LAK-sis)   Against protection; a hyper-sensitivity (allergic) reaction in which IgE antibodies attach to mast cells and basophils, causing them to produce mediators of anaphylaxis (histamine, leukotrines, kinins, and prostaglandins) that bring about increased blood permeability, increased smooth muscle contraction, and increased mucus production. Examples are hayfever, hives, and anaphylactic shock.

Anastomosis (a-nas-tō-MŌ-sis)   An end-to-end union or joining together of blood vessels, lymphatics, or nerves.

Anatomic dead space   The volume of air that is inhaled but remains in spaces in the upper respiratory system and does not reach the alveoli to participate in gas exchange; about 150 ml.

Anatomical (an'-a-TOM-i-kal) *position*   A position of the body universally used in anatomical descriptions in which the body is erect, facing the observer, the upper extremities are at the sides, the palms of the hands are facing forward, and the feet are on the floor.

Anatomy (a-NAT-ō-mē)   The structure or study of structure of the body and the relation of its parts to each other.

Androgen (AN-drō-jen)   Substance producing or stimulating male characteristics, such as the male hormone testosterone.

Anemia (a-NĒ-mē-a)   Condition of the blood in which the number of functional red blood cells or their hemoglobin content is below normal.

Anesthesia (an'-es-THĒ-zha)   A total or partial loss of feeling or sensation, usually defined with respect to loss of pain sensation; may be general or local.

Aneurysm (AN-yoo-rizm)   A saclike enlargement of a blood vessel caused by a weakening of its wall.

Angina pectoris (an-JĪ-na *or* AN-ji-na PEK-tō-ris)   A pain in the chest related to reduced coronary circulation that may or may not involve heart or artery disease.

Angiotensin (an-jē-ō-TEN-sin) Either of two forms of a protein associated with regulation of blood pressure. Angiotensin I is produced by the action of renin on angiotensinogen and is converted by the action of a plasma enzyme into angiotensin II, which stimulates aldosterone secretion by the adrenal cortex.

Anion (AN-ī-on) A negatively charged ion. An example is the chloride ion (Cl^-).

Ankyloglossia (ang'-ki-lō-GLOSS-ē-a) "Tongue-tied"; restriction of tongue movements by a short lingual frenulum.

Anomaly (a-NOM-a-lē) An abnormality that may be a developmental (congenital) defect; a variant from the usual standard.

Anorexia nervosa (an-ō-REK-sē-a ner-VŌ-sa) A chronic disorder characterized by self-induced weight loss, body-image and other perceptual disturbances, and physiologic changes that result from nutritional depletion.

Anoxia (an-OK-sē-a) Deficiency of oxygen.

Antagonist (an-TAG-ō-nist) A muscle that has an action opposite that of the prime mover (agonist) and yields to the movement of the prime mover.

Antepartum (an-tē-PAR-tum) Before delivery of the child; occurring (to the mother) before childbirth.

Anterior (an-TĒR-ē-or) Nearer to or at the front of the body. Also called *ventral.*

Anterior root The structure composed of axons of motor or efferent fibers that emerges from the anterior aspect of the spinal cord and extends laterally to join a posterior root, forming a spinal nerve. Also called a *ventral root.*

Antibiotic (an'-ti-bī-OT-ik) Literally, "antilife"; a chemical produced by a microorganism that is able to inhibit the growth of or kill other microorganisms.

Antibody (AN-ti-bod'-ē) A protein produced by certain cells in the body in the presence of a specific antigen; the antibody combines with that antigen to neutralize, inhibit, or destroy it.

Anticoagulant (an-tī-cō-AG-yoo-lant) A substance that is able to delay, suppress, or prevent the clotting of blood.

Antidiuretic (an'-ti-dī-yoo-RET-ik) Substance that inhibits urine formation.

Antidiuretic hormone (ADH) Hormone produced by neurosecretory cells in the paraventricular and supraoptic nuclei of the hypothalamus that stimulates water reabsorption from kidney cells into the blood and vasoconstriction of arterioles.

Antigen (AN-ti-jen) Any substance that when introduced into the tissues or blood induces the formation of antibodies and reacts only with its specific antibodies.

Anti-oncogene (ONG-kō-jēn) A gene that can cause cancer when inappropriately inactivated.

Anulus fibrosus (AN-yoo-lus fī-BRŌ-sus) A ring of fibrous tissue and fibrocartilage that encircles the pulpy substance (nucleus pulposus) of an intervertebral disc.

Anuria (a-NOO-rē-a) A daily urine output of less than 50 ml.

Anus (Ā-nus) The distal end and outlet of the rectum.

Aorta (ā-OR-ta) The main systemic trunk of the arterial system of the body; emerges from the left ventricle.

Aortic (ā-OR-tik) *body* Receptor on or near the arch of the aorta that responds to alterations in blood levels of oxygen, carbon dioxide, and hydrogen ions.

Aortic reflex A reflex concerned with maintaining normal general systemic blood pressure.

Aphasia (a-FĀ-zē-a) Loss of ability to express oneself properly through speech or loss of verbal comprehension.

Apnea (ap-NĒ-a) Temporary cessation of breathing.

Apneustic (ap-NOO-stik) *area* Portion of the respiratory center in the pons that sends stimulatory nerve impulses to the inspiratory area that activate and prolong inspiration and inhibit expiration.

Apocrine (AP-ō-krin) *gland* A type of gland in which the secretory products gather at the free end of the secreting cell and are pinched off, along with some of the cytoplasm, to become the secretion, as in mammary glands.

Aponeurosis (ap'-ō-noo-RŌ-sis) A sheetlike tendon joining one muscle with another or with bone.

Appendage (a-PEN-dij) A structure attached to the body.

Appendicitis (a-pen-di-SĪ-tis) Inflammation of the vermiform appendix.

Aqueous humor (AK-wē-us HYOO-mor) The watery fluid, similar in composition to cerebrospinal fluid, that fills the anterior cavity of the eye.

Arachnoid (a-RAK-noyd) The middle of the three coverings (meninges) of the brain or spinal cord.

Arachnoid villus (VIL-us) Berrylike tuft of arachnoid that protrudes into the superior sagittal sinus and through which cerebrospinal fluid is reabsorbed into the bloodstream.

Arch of the aorta (ā-OR-ta) The most superior portion of the aorta, lying between the ascending and descending segments of the aorta.

Areflexia a'-rē-FLEK-sē-a) Absence of reflexes.

Areola (a-RĒ-ō-la) Any tiny space in a tissue. The pigmented ring around the nipple of the breast.

Arm The portion of the upper extremity from the shoulder to the elbow.

Arrector pili (a-REK-tor PI-lē) Smooth muscles attached to hairs; contraction pulls the hairs into a more vertical position, resulting in "goose bumps."

Arrhythmia (a-RITH-mē-a) Irregular heart rhythm. Also called a *dysrhythmia.*

Arteriole (ar-TĒ-rē-ōl) A small, almost microscopic, artery that delivers blood to a capillary.

Artery (AR-ter-ē) A blood vessel that carries blood away from the heart.

Arthritis (ar-THRĪ-tis) Inflammation of a joint.

Arthrology (ar-THROL-ō-jē) The study or description of joints.

Arthroscopy (ar-THROS-co-pē) A procedure for examining the interior of a joint, usually the knee, by inserting an arthroscope into a small incision; used to determine extent of damage, remove torn cartilage, repair cruciate ligaments, and obtain samples for analysis.

Articular (ar-TIK-yoo-lar) *capsule* Sleevelike structure around a synovial joint composed of a fibrous capsule and a synovial membrane.

Articular cartilage (KAR-ti-lij) Hyaline cartilage attached to articular bone surfaces.

Articular disc Fibrocartilage pad between articular surfaces of bones of some synovial joints. Also called a *meniscus* (men-IS-cus).

Articulation (ar-tik'-yoo-LĀ-shun) A joint; a point of contact between bones, cartilage and bones, or teeth and bones.

Artificial pacemaker A device that generates and delivers electrical signals to the heart to maintain a regular heart rhythm.

Ascending colon (KŌ-lon) The portion of the large intestine that passes upward from the cecum to the lower edge of the liver where it bends at the right colic (hepatic) flexure to become the transverse colon.

Ascites (as-SĪ-tēz) Serous fluid in the peritoneal cavity.

Aseptic (ā-SEP-tik) Free from any infectious or septic material.

Asphyxia (as-FIX-ē-a) Unconsciousness due to interference with the oxygen supply of the blood.

Aspiration (as'-pi-RĀ-shun) Inhalation of a foreign substance (water, food, or foreign body) into the bronchial tree; drainage of a substance in or out by suction.

Association area A portion of the cerebral cortex connected by many motor and sensory fibers to other parts of the cortex. The association areas are concerned with motor patterns, memory, concepts of word-hearing and word-seeing, reasoning, will, judgment, and personality traits.

Association neuron (NOO-ron) A nerve cell lying completely within the central nervous system that carries nerve impulses from sensory neurons to motor neurons. Also called a *connecting neuron.*

Astereognosis (as-ter'-ē-ōg-NŌ-sis) Inability to recognize objects or forms by touch.

Asthenia (as-THĒ-nē-a) Lack or loss of strength; debility.

Astigmatism (a-STIG-ma-tizm) An irregularity of the lens or cornea of the eye causing the image to be out of focus and producing faulty vision.

Astrocyte (AS-trō-sīt) A neuroglial cell having a star shape that supports neurons in the brain and spinal cord and attaches the neurons to blood vessels.

Ataxia (a-TAK-sē-a) A lack of muscular coordination, lack of precision.

Atelectasis (at'-ē-LEK-ta-sis) A collapsed or airless state of all or part of the lung, which may be acute or chronic.

Atherosclerosis (ath'-er-ō-skle-RŌ-sis) A process in which fatty substances (cholesterol and triglycerides) are deposited in the walls of medium and large arteries in response to certain stimuli (hypertension, carbon monoxide, dietary cholesterol). Following endothelial damage, monocytes stick to the tunica interna, develop into macrophages, and take up cholesterol and low-density lipoproteins. Smooth muscle fibers (cells) in the tunica media ingest cholesterol. This results in the formation of an atherosclerotic plaque that decreases the size of the arterial lumen.

Atom Unit of matter that comprises a chemical element; consists of a nucleus and electrons.

Atomic number Number of protons in an atom.

Atomic weight Total number of protons and neutrons in an atom.

Atresia (a-TRĒ-zē-a) Abnormal closure of a passage, or absence of a normal body opening.

Atrial fibrillation (Ā-trē-al fib-ri-LĀ-shun) Asynchronous contraction of the atria that results in the cessation of atrial pumping.

Atrial natriuretic (na'-trē-yoo-RET-ik) *factor (ANF)* Peptide hormone produced by the atria of the heart in response to stretching that inhibits aldosterone production and thus lowers blood pressure.

Atrioventricular (AV) (ā'-trē-ō-ven-TRIK-yoo-lar) *bundle* The portion of the conduction system of the heart that begins at the atrioventricular (AV) node, passes through the cardiac skeleton separating the atria and the ventricles, then runs a short distance down the interventricular septum before splitting into right and left bundle branches. Also called the *bundle of His (HISS).*

Atrioventricular (AV) node The portion of the conduction system of the heart made up of a compact mass of conducting cells located near the orifice of the coronary sinus in the right atrial wall.

Atrioventricular (AV) valve A structure made up of membranous flaps or cusps that allows blood to flow in one direction only, from an atrium into a ventricle.

Atrium (Ā-trē-um) A superior chamber of the heart.

Atrophy (AT-rō-fē) Wasting away or decrease in size of a part, due to a failure, abnormality of nutrition, or lack of use.

Auditory ossicle (AW-di-tō-rē OS-si-kul) One of the three small bones of the middle ear called the malleus, incus, and stapes.

Auditory tube The tube that connects the middle ear with the nose and nasopharynx region of the throat. Also called the *Eustachian* (yoo-STĀ-kē-an) *tube.*

Auscultation (aws-kul-TĀ-shun) Examination by listening to sounds in the body.

Autoimmunity An immunologic response against a person's own tissue antigens.

Autologous transfusion (aw-TOL-ō-gus trans-FYOO-zhun) Donating one's own blood up to six weeks before elective surgery to ensure an abundant supply and reduce transfusion complications such as those that may be associated with diseases such as AIDS and hepatitis. Also called *predonation.*

Autolysis (aw-TOL-i-sis) Spontaneous self-destruction of cells by their own digestive enzymes upon death or a pathological process or during normal embryological development.

Autonomic ganglion (aw'-tō-NOM-ik GANG-lē-on) A cluster of sympathetic or parasympathetic cell bodies located outside the central nervous system.

Autonomic nervous system (ANS) Visceral efferent neurons, both sympathetic and parasymphetic, that transmit nerve impulses from the central nervous system to smooth muscle, cardiac muscle, and glands; so named because this portion of the nervous system was thought to be self-governing or spontaneous.

Autonomic plexus (PLEK-sus) An extensive network of sympathetic and parasympathetic fibers; the cardiac, celiac, and pelvic plexuses are located in the thorax, abdomen, and pelvis, respectively.

Autophagy (aw-TOF-a-jē) Process by which worn-out organelles are digested within lysosomes.

Autopsy (AW-top-sē) The examination of the body after death.

Autoregulation (aw-tō-reg-yoo-LĀ-shun) A local, automatic adjustment of blood flow in a given region of the body in response to tissue needs.

Autosome (AW-tō-sōm) Any chromosome other than the pair of sex chromosomes.

Avitaminosis (ā-vī-ta-min-Ō-sis) A deficiency of a vitamin in the diet.

Axilla (ak-SIL-a) The small hollow beneath the arm where it joins the body at the shoulders. Also called the *armpit.*

Axon (AK-son)   The usually single, long process of a nerve cell that carries a nerve impulse away from the cell body.

Axon terminal   Terminal branch of an axon and its collateral. Also called a **telodendrium** (tel-ō-DEN-drē-um).

Babinski (ba-BIN-skē) **sign**   Extension of the great toe, with or without fanning of the other toes, in response to stimulation of the outer margin of the sole of the foot; normal up to 1½ years of age.

Back   The posterior part of the body; the dorsum.

Ball-and-socket joint   A synovial joint in which the rounded surface of one bone moves within a cup-shaped depression or fossa of another bone, as in the shoulder or hip joint. Also called a **spheroid** (SFĒ-roid) **joint.**

Baroreceptor (bar'-ō-re-SEP-tor)   Nerve cell capable of responding to changes in blood pressure. Also called a **pressoreceptor.**

Bartholin's glands   *See Greater vestibular glands.*

Basal ganglia (GANG-glē-a)   Paired clusters of cell bodies that make up the central gray matter in each cerebral hemisphere, including the caudate nucleus, lentiform nucleus, claustrum, and amygdaloid body. Also called **cerebral nuclei** (SER-e-bral NOO-klē-ī).

Basal metabolic (BĀ-sal met'-a-BOL-ik) **rate** (**BMR**)   The rate of metabolism measured under standard or basal conditions.

Base   The broadest part of a pyramidal structure. A nonacid or a proton acceptor, characterized by excess of hydroxide ions (OH⁻) and a pH greater than 7. A ring-shaped, nitrogen-containing organic molecule that is one of the components of a nucleotide, for example, adenine, guanine, cytosine, thymine, and uracil.

Basement membrane   Thin, extracellular layer consisting of basal lamina secreted by epithelial cells and reticular lamina secreted by connective tissue cells.

Basilar (BAS-i-lar) **membrane**   A membrane in the cochlea of the inner ear that separates the cochlear duct from the scala tympani and on which the spiral organ (organ of Corti) rests.

Basophil (BĀ-sō-fil)   A type of white blood cell characterized by a pale nucleus and large granules that stain readily with basic dyes.

B cell   A lymphocyte that develops into a plasma cell that produces antibodies or a memory cell.

Belly   The abdomen. The gaster or prominent, fleshy part of a skeletal muscle.

Benign (be-NĪN)   Not malignant; favorable for recovery; a mild disease.

Beta (BĀ-ta) **cell**   A cell in the pancreatic islets (islets of Langerhans) in the pancreas that secretes insulin.

Beta receptor   Receptor found on visceral effectors innervated by most sympathetic postganglionic axons; in general, stimulation of beta receptors leads to inhibition.

Bicuspid (bī-KUS-pid) **valve**   Atrioventricular (AV) valve on the left side of the heart. Also called the **mitral valve.**

Bilateral (bī-LAT-er-al)   Pertaining to two sides of the body.

Bile (bīl)   A secretion of the liver consisting of water, bile salts, bile pigments, cholesterol, lecithin, and several ions; it assumes a role in emulsification of fats prior to their digestion.

Biliary calculi (BIL-ē-er-ē CAL-kyoo-lē)   Gallstones formed by the crystallization of cholesterol in bile.

Bilirubin (bil-ē-ROO-bin)   A red pigment that is one of the end products of hemoglobin breakdown in the liver cells and is excreted as a waste material in the bile.

Bilirubinuria (bil-ē-roo-bi-NOO-rē-a)   The presence of above-normal levels of bilirubin in urine.

Biliverdin (bil-ē-VER-din)   A green pigment that is one of the first products of hemoglobin breakdown in the liver cells and is converted to bilirubin or excreted as a waste material in bile.

Biofeedback   Process by which an individual gets constant signals (feedback) about various visceral biological functions.

Biopsy (BĪ-op-sē)   Removal of tissue or other material from the living body for examination, usually microscopic.

Blastocoel (BLAS-tō-sēl)   The fluid-filled cavity within the blastocyst.

Blastocyst (BLAS-tō-sist)   In the development of an embryo, a hollow ball of cells that consists of a blastocoel (the internal cavity), trophoblast (outer cells), and inner cell mass.

Blastomere (BLAS-tō-mer)   One of the cells resulting from the cleavage of a fertilized ovum.

Blastula (BLAS-tyoo-la)   An early stage in the development of a zygote.

Bleeding time   The time required for the cessation of bleeding from a small skin puncture; identifies platelet function defects and integrity of small blood vessels; ranges from 4 to 8 minutes.

Blepharism (BLEF-a-rizm)   Spasm of the eyelids; continuous blinking.

Blind spot   Area in the retina at the end of the optic (II) nerve in which there are no light receptor cells.

Blood   The fluid that circulates through the heart, arteries, capillaries, and veins and that constitutes the chief means of transport within the body.

Blood–brain barrier (**BBB**)   A special mechanism that prevents the passage of materials from the blood to the cerebrospinal fluid and brain.

Blood pressure (**BP**)   Pressure exerted by blood as it presses against and attempts to stretch blood vessels, especially arteries; the force is generated by the rate and force of heartbeat; clinically, a measure of the pressure in arteries during ventricular systole and ventricular diastole.

Blood reservoir (REZ-er-vwar)   Systemic veins that contain large amounts of blood that can be moved quickly to parts of the body requiring the blood.

Blood–testis barrier   A barrier formed by sustentacular (Sertoli) cells that prevents an immune response against antigens produced by spermatozoa and developing cells by isolating the cells from the blood.

Body cavity   A space within the body that contains various internal organs.

Bolus (BŌ-lus)   A soft, rounded mass, usually food, that is swallowed.

Bony labyrinth (LAB-i-rinth)   A series of cavities within the petrous portion of the temporal bone forming the vestibule, cochlea, and semicircular canals of the inner ear.

Bowman's capsule   *See Glomerular capsule.*

Brachial plexus (BRĀ-kē-al PLEK-sus)   A network of nerve fibers of the anterior rami of spinal nerves C5, C6, C7, C8, and T1. The nerves that emerge from the brachial plexus supply the upper extremity.

Bradycardia (brād'-i-KAR-dē-a) A slow heartbeat or pulse rate.

Brain A mass of nervous tissue located in the cranial cavity.

Brain electrical activity mapping (*BEAM*) Noninvasive procedure that measures and displays the electrical activity of the brain; used primarily to diagnose epilepsy.

Brain stem The portion of the brain immediately superior to the spinal cord, made up of the medulla oblongata, pons, and midbrain.

Brain wave Electrical activity produced as a result of action potentials of brain cells.

Bright's disease *See Glomerulonephritis.*

Broca's (BRŌ-kaz) *area* Motor area of the brain in the frontal lobe that translates thoughts into speech. Also called the *motor speech area.*

Bronchi (BRONG-kē) Branches of the respiratory passageway including primary bronchi (the two divisions of the trachea), secondary or lobar bronchi (divisions of the primary that are distributed to the lobes of the lung), and tertiary or segmental bronchi (divisions of the secondary that are distributed to bronchopulmonary segments of the lung).

Bronchial asthma (BRONG-kē-al AZ-ma) Usually allergic reaction characterized by smooth muscle spasms in bronchi resulting in wheezing and difficult breathing.

Bronchial tree The trachea, bronchi, and their branching structures.

Bronchiectasis (brong'-kē-EK-ta-sis) A chronic disorder in which there is a loss of the normal tissue and expansion of lung air passages; characterized by difficult breathing, coughing, expectoration of pus, and foul breath.

Bronchiole (BRONG-kē-ol) Branch of a tertiary bronchus further dividing into terminal bronchioles (distributed to lobules of the lung), which divide into respiratory bronchioles (distributed to alveolar sacs).

Bronchitis (brong-KĪ-tis) Inflammation of the bronchi characterized by hypertrophy and hyperplasia of seromucous glands and goblet cells that line the bronchi and resulting in a productive cough.

Bronchogenic carcinoma (brong'-kō-JEN-ik kar'-si-NŌ-ma) Cancer originating in the bronchi.

Bronchogram (BRONG-kō-gram) A roentgenogram of the bronchial tree.

Bronchography (bron-KOG-ra-fē) Technique for examining the bronchial tree in which an opaque contrast medium is introduced into the trachea for distribution to the bronchial branches. The roentgenogram produced is called a bronchogram.

Bronchopulmonary (brong'-kō-PUL-mō-ner-ē) *segment* One of the smaller divisions of a lobe of a lung supplied by its own branches of a bronchus.

Bronchoscope (BRONG-kō-skōp) An instrument used to examine the interior of the bronchi of the lungs.

Bronchoscopy (brong-KOS-kō-pē) Visual examination of the interior of the trachea and bronchi with a bronchoscope to biopsy a tumor, clear an obstruction, take cultures, stop bleeding, or deliver drugs.

Bronchus (BRONG-kus) One of the two large branches of the trachea. *Plural, bronchi* (BRONG-kē).

Brunner's gland *See Duodenal gland.*

Buccal (BUK-al) Pertaining to the cheek or mouth.

Buffer (BUF-er) *system* A pair of chemicals, one a weak acid and one the salt of the weak acid, which functions as a weak base, that resists changes in pH.

Bulbourethral (bul'-bō-yoo-RĒ-thral) *gland* One of a pair of glands located inferior to the prostate gland on either side of the urethra that secretes an alkaline fluid into the cavernous urethra. Also called a *Cowper's* (KOW-perz) *gland.*

Bulimia (boo-LIM-ē-a) A disorder characterized by overeating, at least twice a week, followed by purging by self-induced vomiting, strict dieting or fasting, vigorous exercise, or use of laxatives.

Bullae (BYOOL-ē) Blisters beneath or within the epidermis.

Bundle branch One of the two branches of the atrioventricular (AV) bundle made up of specialized muscle fibers (cells) that transmit electrical impulses to the ventricles.

Bundle of His *See Atrioventricular (AV) bundle.*

Bunion (BUN-yun) Lateral deviation of the great toe that produces inflammation and thickening of the bursa, bone spurs, and calluses.

Burn An injury in which proteins are destroyed (denatured) as a result of heat (fire, steam), chemicals, electricity, or the ultraviolet rays of the sun.

Bursa (BUR-sa) A sac or pouch of synovial fluid located at friction points, especially about joints.

Bursitis (bur-SĪ-tis) Inflammation of a bursa.

Buttocks (BUT-oks) The two fleshy masses on the posterior aspect of the lower trunk, formed by the gluteal muscles.

Cachexia (kah-KEK-sē-ah) A state of ill health, malnutrition, and wasting.

Calcaneal tendon The tendon of the soleus, gastrocnemius, and plantaris muscles at the back of the heel. Also called the *Achilles* (a-KIL-ēz) *tendon.*

Calcify (KAL-si-fī) To harden by deposits of calcium salts.

Calcitonin (kal-si-TŌ-nin) (*CT*) A hormone produced by the thyroid gland that lowers the calcium and phosphate levels of the blood by inhibiting bone breakdown and accelerating calcium absorption by bones.

Calculus (KAL-kyoo-lus) A stone, or insoluble mass of crystallized salts or other material, formed within the body, as in the gallbladder, kidney, or urinary bladder.

Callus (KAL-lus) A growth of new bone tissue in and around a fractured area, ultimately replaced by mature bone. An acquired, localized thickening.

Calorie (KAL-ō-rē) A unit of heat. A calorie (cal) is the standard unit and is the amount of heat necessary to raise 1 g of water 1°C from 14° to 15°C. The kilocalorie (kcal), used in metabolic and nutrition studies, is the amount of heat necessary to raise 1,000 g of water 1°C and is equal to 1,000 cal.

Calyx (KĀl-iks) Any cuplike division of the kidney pelvis. *Plural, calyces* (KĀ-li-sēz).

Canaliculus (kan'-a-LIK-yoo-lus) A small channel or canal, as in bones, where they connect lacunae. *Plural, canaliculi* (kan'-a-LIK-yoo-lī).

Canal of Schlemm *See Scleral venous sinus.*

Cancer (KAN-ser) A malignant tumor of epithelial origin tending to infiltrate and give rise to new growths or metastases. Also called *carcinoma* (kar'-si-NŌ-ma).

Canker (KANG-ker) **sore**   Painful ulcer on the mucous membrane of the mouth that may result from an autoimmune response.

Capacitation (ka'-pas-i-TĀ-shun)   The functional changes that sperm undergo in the female reproductive tract that allow them to fertilize a secondary oocyte.

Capillary (KAP-i-lar'-ē)   A microscopic blood vessel located between an arteriole and venule through which materials are exchanged between blood and body cells.

Carbohydrate (kar'-bō-HĪ-drāt)   An organic compound containing carbon, hydrogen, and oxygen in a particular amount and arrangement and comprised of sugar subunits; usually has the formula $(CH_2O)_n$.

Carbon monoxide (CO) poisoning   Hypoxia due to increased levels of carbon monoxide as a result of its preferential and tenacious combination with hemoglobin rather than with oxygen.

Carcinogen (kar-SIN-ō-jen)   Any substance that causes cancer.

Carcinoma (kar'-si-NŌ-ma)   A malignant tumor consisting of epithelial cells.

Cardiac (KAR-dē-ak) **arrest**   Cessation of an effective heartbeat in which the heart is completely stopped or in ventricular fibrillation.

Cardiac catheterization (KAR-dē-ak kath'-e-ter-i-ZĀ-shun)   Introduction of a catheter into the heart and/or its blood vessels to measure pressure; assess left ventricular function and cardiac output; measure blood flow, oxygen content of blood, and the status of valves and conduction system; and identify valvular and septal defects.

Cardiac (KAR-dē-ak) **cycle**   A complete heartbeat consisting of systole (contraction) and diastole (relaxation) of both atria plus systole and diastole of both ventricles.

Cardiac muscle   An organ specialized for contraction, composed of striated muscle fibers (cells), forming the wall of the heart, and stimulated by an intrinsic conduction system and visceral efferent neurons.

Cardiac notch   An angular notch in the anterior border of the left lung.

Cardiac output (CO)   The volume of blood pumped from one ventricle of the heart (usually measured from the left ventricle) in 1 min; about 5.2 liters/min under normal resting conditions

Cardiac reserve   The maximum percentage that cardiac output can increase above normal.

Cardiac tamponade (tam'-pon-ĀD)   Compression of the heart due to excessive fluid or blood in the pericardial sac that could result in cardiac failure.

Cardinal ligament   A ligament of the uterus, extending laterally from the cervix and vagina as a continuation of the broad ligament.

Cardioacceleratory (kar-dē-ō-ak-SEL-er-a-tō-rē) **center (CAC)**   A group of neurons in the medulla from which cardiac nerves (sympathetic) arise; nerve impulses along the nerves release epinephrine that increases the rate and force of heartbeat.

Cardioinhibitory (kar-dē-ō-in-HIB-i-tō-rē) **center (CIC)**   A group of neurons in the medulla from which parasympathetic fibers that reach the heart via the vagus (X) nerve arise; nerve impulses along the nerves release acetylcholine that decreases the rate and force of heartbeat.

Cardiology (kar-dē-OL-ō-jē)   The study of the heart and diseases associated with it.

Cardiopulmonary resuscitation (rē-sus-i-TĀ-shun) **(CPR)**   A technique employed to restore life or consciousness to a person apparently dead or dying; includes external respiration (exhaled air respiration) and external cardiac massage.

Carina (ka-RĪ-na)   A ridge on the inside of the division of the right and left primary bronchi.

Carotid (ka-ROT-id) **body**   Receptor on or near the carotid sinus that responds to alterations in blood levels of oxygen, carbon dioxide, and hydrogen ions.

Carotid sinus   A dilated region of the internal carotid artery immediately above the bifurcation of the common carotid artery that contains receptors that monitor blood pressure.

Carotid sinus reflex   A reflex concerned with maintaining normal blood pressure in the brain.

Carpus (KAR-pus)   A collective term for the eight bones of the wrist.

Cartilage (KAR-ti-lij)   A type of connective tissue consisting of chondrocytes in lacunae embedded in a dense network of collagenous and elastic fibers and a matrix of chondroitin sulfate.

Cartilaginous (kar'-ti-LAJ-i-nus) **joint**   A joint without a synovial (joint) cavity where the articulating bones are held tightly together by cartilage, allowing little or no movement.

Cast   A small mass of hardened material formed within a cavity in the body and then discharged from the body; can originate in different areas and be composed of various materials.

Castration (kas-TRĀ-shun)   The removal of the testes.

Catabolism (ka-TAB-ō-lizm)   Chemical reactions that break down complex organic compounds into simple ones with the release of energy.

Cataract (KAT-a-rakt)   Loss of transparency of the lens of the eye or its capsule or both.

Catheter (KATH-i-ter)   A tube that can be inserted into a body cavity through a canal or into a blood vessel; used to remove fluids, such as urine and blood, and to introduce diagnostic materials or medication.

Cation   A positively charged ion. An example is a sodium ion (na^+).

Cauda equina (KAW-da ē-KWĪ-na)   A taillike collection of roots of spinal nerves at the inferior end of the spinal canal.

Cecum (SĒ-kum)   A blind pouch at the proximal end of the large intestine to which the ileum is attached.

Cell   The basic structural and functional unit of all organisms; the smallest structure capable of performing all the activities vital to life.

Cell division   Process by which a cell reproduces itself that consists of a nuclear division (mitosis) and a cytoplasmic division (cytokinesis); types include somatic and reproductive cell division.

Cell inclusion   A lifeless, often temporary, constituent in the cytoplasm of a cell as opposed to an organelle.

Cellular immunity   That component of immunity in which specially sensitized lymphocytes (T cells) attach to antigens to destroy them. Also called **cell-mediated immunity**.

Cementum (se-MEN-tum)   Calcified tissue covering the root of a tooth.

Center   An area in the brain where a particular function is localized.

Center of ossification (os'-i-fi-KĀ-shun)   An area in the cartilage model of a future bone where the cartilage cells hypertrophy and then secrete enzymes that result in the calcification of their matrix, resulting in the death of the cartilage cells, followed by the invasion of the area by osteoblasts that then lay down bone.

Central canal A circular channel running longitudinally in the center of an osteon (Haversian system) of mature compact bone, containing blood and lymphatic vessels and nerves. Also called a **Haversian** (ha-VĒR-shun) **canal.** A microscopic tube running the length of the spinal cord in the gray commissure.

Central fovea (FŌ-vē-a) A cuplike depression in the center of the macula lutea of the retina, containing cones only; the area of clearest vision.

Central nervous system (CNS) That portion of the nervous system that consists of the brain and spinal cord.

Centrioles (SEN-trē-ōlz) Paired, cylindrical structures within a centrosome, each consisting of a ring of microtubules and arranged at right angles to each other; function in cell division.

Centromere (SEN-trō-mēr) The clear, constricted portion of a chromosome where the two chromatids are joined; serves as the point of attachment for the chromosomal microtubules.

Centrosome (SEN-trō-sōm) A rather dense area of cytoplasm, near the nucleus of a cell, containing a pair of centrioles.

Cephalic (se-FAL-ik) Pertaining to the head; superior in position.

Cerebellar peduncle (ser-e-BEL-ar pe-DUNG-kul) A bundle of nerve fibers connecting the cerebellum with the brain stem.

Cerebellum (ser-e-BEL-um) The portion of the brain lying posterior to the medulla and pons, concerned with coordination of movements.

Cerebral aqueduct (SER-ē-bral AK-we-dukt) A channel through the midbrain connecting the third and fourth ventricles and containing cerebrospinal fluid.

Cerebral arterial circle A ring of arteries forming an anastomosis at the base of the brain between the internal carotid and basilar arteries and arteries supplying the brain. Also called the **circle of Willis.**

Cerebral cortex The surface of the cerebral hemispheres, 2–4 mm thick, consisting of six layers of nerve cell bodies (gray matter) in most areas.

Cerebral palsy (PAL-zē) A group of motor disorders resulting in muscular uncoordination and loss of muscle control and caused by damage to motor areas of the brain (cerebral cortex, basal ganglia, and cerebellum) during fetal life, birth, or infancy.

Cerebral peduncle (pe-DUNG-kul) One of a pair of nerve fiber bundles located on the ventral surface of the midbrain, conducting nerve impulses between the pons and the cerebral hemispheres.

Cerebrospinal (se-rē'-brō-SPĪ-nal) **fluid (CSF)** A fluid produced in the choroid plexuses and ependymal cells of the ventricles of the brain that circulates in the ventricles and the subarachnoid space around the brain and spinal cord.

Cerebrovascular (se-rē'-brō-VAS-kyoo-lar) **accident (CVA)** Destruction of brain tissue (infarction) resulting from disorders of blood vessels that supply the brain. Also called a **stroke.**

Cerebrum (SER-ē-brum) The two hemispheres of the forebrain, making up the largest part of the brain.

Ceruminous (se-ROO-mi-nus) **gland** A modified sudoriferous (sweat) gland in the external auditory meatus that secretes cerumen (ear wax).

Cervical dysplasia (dis-PLĀ-sē-a) A change in the shape, growth, and number of cervical cells of the uterus that, if severe, may progress to cancer.

Cervical mucus A mixture of water, glycoprotein, serum-type proteins, lipids, enzymes, and inorganic salts produced by secreting cells of the mucosa of the cervix.

Cervical plexus (PLEK-sus) A network of neuron fibers formed by the anterior rami of the first four cervical nerves.

Cervix (SER-viks) Neck; any constricted portion of an organ, such as the lower cylindrical part of the uterus.

Cesarean (se-SA-rē-an) **section** Procedure in which a low, horizontal incision is made through the abdominal wall and uterus for removal of the baby and placenta. Also called a **C-section.**

Chemical bond Force of attraction in a molecule that holds atoms together. Examples include ionic, covalent, and hydrogen bonds.

Chemical element Unit of matter that cannot be decomposed into a simpler substance by ordinary chemical reactions. Examples include hydrogen (H), carbon (C), and oxygen (O).

Chemical reaction The combination or breaking apart of atoms in which chemical bonds are formed or broken and new products with different properties are produced.

Chemoreceptor (kē'-mō-rē-SEP-tor) Receptor outside the central nervous system on or near the carotid and aortic bodies that detects the presence of chemicals.

Chemotaxis (kē-mō-TAK-sis) Attraction of phagocytes to microbes by a chemical stimulus.

Chemotherapy (kē-mō-THER-a-pē) The treatment of illness or disease by chemicals.

Chiropractic (kī'-rō-PRAK-tik) A system of treating disease by using one's hands to manipulate body parts, mostly the vertebral column.

Cholecystectomy (kō'-lē-sis-TEK-tō-mē) Surgical removal of the gallbladder.

Cholesterol (kō-LES-te-rol) Classified as a lipid, the most abundant steroid in animal tissues; located in cell membranes and used for the synthesis of steroid hormones and bile salts.

Cholinergic (kō'-lin-ER-jik) **fiber** A nerve ending that liberates acetylcholine at a synapse.

Chondrocyte (KON-drō-sīt) Cell of mature cartilage.

Chondroitin (kon-DROY-tin) **sulfate** An amorphous matrix material found outside the cell.

Chordae tendineae (KOR-dē TEN-di-nē-ē) Tendonlike, fibrous cords that connect the heart valves with the papillary muscles.

Chorion (KŌ-rē-on) The outermost fetal membrane that becomes the principal embryonic portion of the placenta; serves a protective and nutritive function.

Chorionic villus (kō'-rē-ON-ik VIL-lus) Fingerlike projection of the chorion that grows into the decidua basalis of the endometrium and contains fetal blood vessels.

Chorionic villus sampling (CVS) The removal of a sample of chorionic villus tissue by means of a catheter to analyze the tissue for prenatal genetic defects.

Choroid (KŌ-royd) One of the vascular coats of the eyeball.

Choroid plexus (PLEK-sus) A vascular structure located in the roof of each of the four ventricles of the brain; produces cerebrospinal fluid.

Chromaffin (krō-MAF-in) **cell** Cell that has an affinity for chrome salts, due in part to the presence of the precursors of the neurotransmitter epinephrine; found, among other places, in the adrenal medulla.

Chromatid (KRŌ-ma-tid) One of a pair of identical connected nucleoprotein strands that are joined at the centromere and separate during cell division, each becoming a chromosome of one of the two daughter cells.

Chromatin (KRŌ-ma-tin) The threadlike mass of the genetic material consisting principally of DNA, which is present in the nucleus of a nondividing or interphase cell.

Chromatolysis (krō'-ma-TOL-i-sis) The breakdown of chromatophilic substance (Nissl bodies) into finely granular masses in the cell body of a central or peripheral neuron whose process (axon or dendrite) has been damaged.

Chromatophilic substance Rough endoplasmic reticulum in the cell bodies of neurons that functions in protein synthesis. Also called *Nissl bodies.*

Chromosomal microtubule (mī-krō-TOOB-yool) Microtubule formed during prophase of mitosis that originates from centromeres, extends from a centromere to a pole of the cell, and assists in chromosomal movement; constitutes a part of the mitotic spindle.

Chromosome (KRŌ-mō-sōm) One of the 46 small, dark-staining bodies that appear in the nucleus of a human diploid (*2n*) cell during cell division.

Chronic (KRON-ik) Long term or frequently recurring; applied to a disease that is not acute.

Chyme (kīm) The semifluid mixture of partly digested food and digestive secretions found in the stomach and small intestine during digestion of a meal.

Cicatrix (SIK-a-triks) A scar left by a healed wound.

Ciliary (SIL-ē-ar'-ē) *body* One of the three portions of the vascular tunic of the eyeball, the others being the choroid and the iris; includes the ciliary muscle and the ciliary processes.

Cilium (SIL-ē-um) A hair of hairlike process projecting from a cell that may be used to move the entire cell or to move substances along the surface of the cell.

Circadian (ser-KĀ-dē-an) *rhythm* A cycle of active and nonactive periods in organisms determined by internal mechanisms and repeating about every 24 hours.

Circle of Willis *See Cerebral arterial circle.*

Circular folds Permanent, deep, transverse folds in the mucosa and submucosa of the small intestine that increase the surface area for absorption. Also called *plicae circulares* (PLĪ-kē SER-kyoo-lar-ēs).

Circulation time Time required for blood to pass from the right atrium, through pulmonary circulation, back to the left ventricle, through systemic circulation to the foot, and back again to the right atrium; normally about 1 min.

Circumcision (ser'-kum-SIZH-un) Surgical removal of the foreskin (prepuce), the fold of skin over the glans penis.

Circumduction (ser'-kum-DUK-shun) A movement at a synovial joint in which the distal end of a bone moves in a circle while the proximal end remains relatively stable.

Circumvallate papilla (ser'-kum-VAL-āt pa-PIL-a) One of the circular projections that is arranged in an inverted V-shaped row at the posterior portion of the tongue; the largest of the elevations on the upper surface of the tongue containing taste buds.

Cirrhosis (si-RŌ-sis) A liver disorder in which the parenchymal cells are destroyed and replaced by connective tissue.

Cisterna chyli (sis-TER-na KĪ-lē) The origin of the thoracic duct.

Cleavage The rapid mitotic divisions following the fertilization of a secondary oocyte, resulting in an increased number of progressively smaller cells, called blastomeres, so that the overall size of the zygote remains the same.

Cleft palate Condition in which the palatine processes of the maxillae do not unite before birth; cleft lip, a split in the upper lip, is often associated with cleft palate.

Climacteric (klī-mak-TER-ik) Cessation of the reproductive function in the female or diminution of testicular activity in the male.

Climax The peak period of moments of greatest intensity during sexual excitement.

Clitoris (KLI-to-ris) An erectile organ of the female located at the anterior junction of the labia minora that is homologous to the male penis.

Clot The end result of a series of biochemical reactions that changes liquid plasma into a gelatinous mass; specially, the conversion of fibrinogen into a tangle of polymerized fibrin molecules.

Clot retraction (rē-TRAK-shun) The consolidation of a fibrin clot to pull damaged tissue together.

Coagulation (cō-ag-yoo-LĀ-shun) Process by which a blood clot is formed.

Coarctation (kō'-ark-TĀ-shun) *of the aorta* Congenital condition in which the aorta is too narrow and results in reduced blood supply, increased ventricular pumping, and high blood pressure.

Coccyx (KOK-six) The fused bones at the end of the vertebral column.

Cochlea (KŌK-lē-a) A winding, cone-shaped tube forming a portion of the inner ear and containing the spiral organ (organ of Corti).

Cochlear duct The membranous cochlea consisting of a spirally arranged tube enclosed in the bony cochlea and lying along its outer wall. Also called the *scala media* (SCA-la MĒ-dē-a).

Coenzyme A type of cofactor; a nonprotein organic molecule that is associated with and activates an enzyme; many are derived from vitamins. An example is nicotinamide adenine dinucleotide (NAD), derived from the B vitamin niacin.

Coitus (KŌ-i-tus) Sexual intercourse. Also called *copulation* (cop-yoo-LĀ-shun).

Colitis (ko-LĪ-tis) Inflammation of the mucosa of the colon and rectum in which absorption of water and salts is reduced, producing watery, bloody feces, and, in severe cases, dehydration and salt depletion. Spasms of the irritated muscularis produce cramps.

Collagen (KOL-a-jen) A protein that is the main organic constituent of connective tissue.

Collateral circulation The alternate route taken by blood through an anastomosis.

Collision theory Theory that explains how chemical reactions occur. It states that all atoms, ions, and molecules are constantly moving and colliding, and the energy transferred during collision might disrupt their electron structures so that chemical bonds are broken or formed.

Colon The division of the large intestine consisting of ascending, transverse, descending, and sigmoid portions.

Colony-stimulating factor (CSF) One of a group of hematopoietins that stimulates development of white blood cells. Examples are macrophage CSF and granulocyte CSF.

Color blindness Any deviation in the normal perception of colors, resulting from the lack of usually one of the photopigments of the cones.

Colostomy (kō-LOS-tō-mē) The diversion of feces through an opening in the colon, creating a surgical opening at the exterior of the abdominal wall.

Colostrum (kō-LOS-trum) A thin, cloudy fluid secreted by the mammary glands a few days prior to or after delivery before true milk is secreted.

Colposcopy (kol-POS-kō-pē) Direct examination of the vaginal and cervical mucosa using a magnifying device; frequently the first procedure performed following an abnormal Pap smear.

Coma (KŌ-ma) Final stage of brain failure that is characterized by total unresponsivenss to all external stimuli.

Common bile duct A tube formed by the union of the common hepatic duct and the cystic duct that empties bile into the duodenum at the hepatopancreatic ampulla (ampulla of Vater).

Compact (dense) bone Bone tissue with no apparent spaces in which the layers of lamellae are fitted tightly together. Compact bone is found immediately deep to the periosteum and external to spongy bone.

Complement (KOM-ple-ment) A group of at least 20 proteins found in serum that forms a component of nonspecific resistance and immunity by bringing about cytolysis, inflammation, and opsonization.

Complete blood count (CBC) Hematology test that usually includes hemoglobin determination, hematocrit, red and white blood cell count, differential white blood cell count, and platelet count.

Compound A substance that can be broken down into two or more other substances by chemical means.

Computed tomography (tō-MOG-ra-fē) **(CT)** X-ray technique that provides a cross-sectional image of any area of the body. Also called **computed axial tomography (CAT).**

Concha (KONG-ka) A scroll-like bone found in the skull. *Plural*, **conchae** (KONG-kē).

Concussion (kon-KUSH-un) Traumatic injury to the brain that produces no visible bruising but may result in abrupt, temporary loss of consciousness.

Conduction myofiber Muscle fiber (cell) in the subendocardial tissue of the heart specialized for conducting an action potential to the myocardium; part of the conduction system of the heart. Also called a **Purkinje** (pur-KIN-jē) **fiber.**

Conduction system An intrinsic regulating system composed of specialized muscle tissue that generates and distributes electrical impulses that stimulate cardiac muscle fibers (cells) to contract.

Conductivity (kon'-duk-TIV-i-tē) The ability to carry the effect of a stimulus from one part of a cell to another; highly developed in nerve and muscle fibers (cells).

Cone The light-sensitive receptor in the retina concerned with color vision.

Congenital (kon-JEN-i-tal) Present at the time of birth.

Congestive heart failure (CHF) Chronic or acute state that results when the heart is not capable of supplying the oxygen demands of the body.

Conjunctiva (kon'-junk-TĪ-va) The delicate membrane covering the eyeball and lining the eyes.

Conjunctivits (kon-junk'-ti-VĪ-tis) Inflammation of the conjunctiva, the delicate membrane covering the eyeball and lining the eyelids.

Connective tissue The most abundant of the four basis tissue types in the body, performing the functions of binding and supporting; consists of relatively few cells in a great deal of intercellular substance.

Constipation (con-sti-PĀ-shun) Infrequent or difficult defecation caused by decreased motility of the intestines.

Contact inhibition Phenomenon by which migration of a growing cell is stopped when it makes contact with another cell of its own kind.

Continuous microtubules (mī-krō-TOOB-yoolz) Microtubules formed during prophase of mitosis that originate from the vicinity of the centrioles, grow toward each other, extend from one pole of the cell to another, and assist in chromosomal movement; constitute a part of the mitotic spindle.

Contraception (kon'-tra-SEP-shun) The prevention of conception or impregnation without destroying fertility.

Contractility (kon'-trak-TIL-i-tē) The ability of cells or parts of cells actively to generate force to undergo shortening and change form for purposeful movements. Muscle fibers (cells) exhibit a high degree of contractility.

Contralateral (kon'-tra-LAT-er-al) On the opposite side; affecting the opposite side of the body.

Contusion (kon-TOO-shun) Condition in which tissue below the skin is damaged, but the skin is not broken.

Conus medullaris (KŌ-nus med-yoo-LAR-is) The tapered portion of the spinal cord below the lumbar enlargement.

Convergence (con-VER-jens) An anatomical arrangement in which the synaptic end bulbs of several presynaptic neurons terminate on one postsynaptic neuron. The medial movement of the two eyeballs so that both are directed toward a near object being viewed in order to produce a single image.

Convulsion (con-VUL-shun) Violent, involuntary, tetanic contractions of an entire group of muscles.

Cornea (KOR-nē-a) The nonvasclar, transparent fibrous coat through which the iris can be seen.

Coronal (kō-RŌ-nal) **plane** A plane that runs vertical to the ground and divides the body into anterior and posterior portions. Also called **frontal plane.**

Corona radiata Several layers of follicle cells surrounding a secondary oocyte.

Coronary (KOR-ō-na-rē) **artery bypass grafting (CABG)** Surgical procedure in which a portion of a blood vessel is removed from another part of the artery and grafted onto a coronary artery so as to bypass an obstruction in the coronary artery.

Coronary (KOR-ō-na-rē) **artery disease (CAD)** A condition in which the heart muscle receives inadequate blood due to an interruption of its blood supply.

Coronary artery spasm A condition in which the smooth muscle of a coronary artery undergoes a sudden contraction, resulting in vasoconstriction.

Coronary circulation The pathway followed by the blood from the ascending aorta through the blood vessels supplying the heart and returning to the right atrium. Also called *cardiac circulation.*

Coronary sinus (SĪ-nus)   A wide venous channel on the posterior surface of the heart that collects the blood from the coronary circulation and returns it to the right atrium.

Corpora quadrigemina (KOR-por-a kwad-ri-JEM-in-a) Four small elevations (superior and inferior colliculi) on the dorsal region of the midbrain concerned with visual and auditory functions.

Cor pulmonale (kor pul-mōn-ALE) **(CP)**   Right ventricular hypertrophy from disorders that bring about hypertension in pulmonary circulation.

Corpus (KOR-pus)   The principal part of any organ; any mass or body.

Corpus albicans (KOR-pus AL-bi-kanz)   A white fibrous patch in the ovary that forms after the corpus luteum regresses.

Corpus callosum (kal-LŌ-sum)   The great commissure of the brain between the cerebral hemispheres.

Corpuscle of touch   The sensory receptor for the sensation of touch; found in the dermal papillae, especially in palms and soles. Also called a *Meissner's* (MĪS-nerz) *corpuscle.*

Corpus luteum (LOO-tē-um)   A yellow endocrine gland in the ovary formed when a follicle has discharged its secondary oocyte; secretes estrogens, progesterone, and relaxin.

Cortex (KOR-teks)   An outer layer of an organ. The convoluted layer of gray matter covering each cerebral hemisphere.

Costal cartilage (KOS-tal KAR-ti lij)   Hyaline cartilage that attaches a rib to the sternum.

Cowper's gland   *See Bulbourethral gland.*

Cramps   A spasmodic, especially a tonic, contraction of one of many muscles, usually painful.

Cranial (KRĀ-nē-al) *cavity*   A subdivision of the dorsal body cavity formed by the cranial bones and containing the brain.

Cranial nerve   One of 12 pairs of nerves that leave the brain, pass through foramina in the skull, and supply the head, neck, and part of the trunk; each is designated by a Roman numeral and a name.

Cranium (KRĀ-nē-um)   The skeleton of the skull that protects the brain and the organs of sight, hearing, and balance; includes the frontal, parietal, temporal, occipital, sphenoid, and ethmoid bones.

Crenation (krē-NĀ-shun)   The shrinkage of red blood cells into knobbed, starry forms when placed in a hypertonic solution.

Cretinism (KRĒ-tin-izm)   Severe congenital thyroid deficiency during childhood leading to physical and mental retardation.

Crista (KRIS-ta)   A crest or ridged structure. A small elevation in the ampulla of each semicircular duct that serves as a receptor for dynamic equilibrium.

Crossing-over   The exchange of a portion of one chromatid with another in a tetrad during meiosis. It permits an exchange of genes among chromatids and is one factor that results in genetic variation.

Cryosurgery (KRĪ-ō-ser-jer-ē)   The destruction of tissue by application of extreme cold.

Crypt of Lieberkühn   *See Intestinal gland.*

Cryptorchidism (krip-TOR-ki-dizm)   The condition of undescended testes.

Cupula (KUP-yoo-la)   A mass of gelatinous material covering the hair cells of a crista, a receptor in the ampulla of a semicircular canal stimulated when the head moves.

Curvature (KUR-va-tūr)   A nonangular deviation of a straight line, as in the greater and lesser curvatures of the stomach. Abnormal curvatures of the vertebral column include kyphosis, lordosis, and scoliosis.

Cushing's syndrome   Condition caused by a hypersecretion of glucocorticoids characterized by spindly legs, ''moon face,'' ''buffalo hump,'' pendulous abdomen, flushed facial skin, poor wound healing, hyperglycemia, osteoporosis, weakness, hypertension, and increased susceptibility to disease.

Cutaneous (kyoo-TĀ-nē-us)   Pertaining to the skin.

Cyanosis (sī-a-NŌ-sis)   Reduced hemoglobin (unoxygenated) concentration of blood of more than 5 g/dl that results in a blue or dark purple discoloration that is most easily seen in nail beds and mucous membranes.

Cyclic AMP (cyclic adenosisn-3',5'-monophosphate)   Molecule formed from ATP by the action of the enzyme adenylate cyclase; serves as an intracellular messenger (second messenger) for some hormones.

Cyst (SIST)   A sac with a distinct connective tissue wall, containing a fluid or other material.

Cystic (SIS-tik) *duct*   The duct that transports bile from the gallbladder to the common bile duct.

Cystitis (sis-TĪ-tis)   Inflammation of the urinary bladder.

Cystoscope (SIS-to-skōp)   An instrument used to examine the inside of the urinary bladder.

Cytochrome (SĪ-tō-krōm)   A protein with an iron-containing group (heme) capable of alternating between a reduced form (Fe^{2+}) and an oxidized form (Fe^{3+}).

Cytokinesis (sī-tō-ki-NĒ-sis)   Division of the cytoplasm.

Cytology (sī-TOL-ō-jē)   The study of cells.

Cytoplasm (SĪ-tō-plazm)   Substance that surrounds organelles and located within a cell's plasma membrane and external to its nucleus. Also called *protoplasm.*

Cytoskeleton   Complex internal structure of cytoplasm consisting of microfilaments, microtubules, and intermediate filaments.

Dartos (DAR-tōs)   The contractile tissue under the skin of the scrotum.

Deafness   Lack of the sense of hearing or a significant hearing loss.

Debility (dē-BIL-i-tē)   Weakness of tonicity in functions or organs of the body.

Decibel (DES-i-bel) **(db)**   A unit that measures relative sound intensity (loudness).

Decidua (dē-SID-yoo-a)   That portion of the endometrium of the uterus (all but the deepest layer) that is modified for pregnancy and shed after childbirth.

Deciduous (dē-SID-yoo-us)   Falling off or being shed seasonally or at a particular stage of development. In the body, referring to the first set of teeth.

Decompression sickness   A condition characterized by joint pains and neurologic symptoms; follows from a too-rapid reduction of environmental pressure or decompression, so that nitrogen that dissolved in body fluid under pressure comes out of solution as bubbles that form air emboli and occlude blood vessels. Also called *caisson* (KĀ-son) *disease* or *bends.*

Decubitus (dē-KYOO-bi-tus) *ulcer* Tissue destruction due to a constant deficiency of blood to tissues overlying a bony projection that has been subjected to prolonged pressure against an object such as a bed, cast, or splint. Also called *bedsore, pressure sore,* or *trophic ulcer.*

Decussation (dē'-ku-SA-shun) A crossing-over, usually refers to the crossing of most of the fibers in the large motor tracts to opposite sides in the medullary pyramids.

Deep Away from the surface of the body.

Deep fascia (FASH-ē-a) A sheet of connective tissue wrapped around a muscle to hold it in place.

Deep inguinal (IN-gwi-nal) *ring* A slitlike opening in the aponeurosis of the transversus abdominis muscle that represents the origin of the inguinal canal.

Deep-venous thrombosis (DVT) The presence of a thrombus in a vein, usually a deep vein of the lower extremities.

Defecation (def-e-KA-shun) The discharge of feces from the rectum.

Defibrillation (dē-fib-ri-LA-shun) Delivery of a very strong electrical current to the heart in an attempt to stop ventricular fibrillation.

Degeneration (dē-jen-er-A-shun) A change from a higher to a lower state; a breakdown in structure.

Deglutition (dē-gloo-TISH-un) The act of swallowing.

Dehydration (dē-hī-DRA-shun) Excessive loss of water from the body or its parts.

Delirium (de-LIR-ē-um) A transient disorder of abnormal cognition (perception, thinking, and memory) and disordered attention that is accompanied by disturbances of the sleep-wake cycle and psychomotor behavior (hyperactivity or hypoactivity of movements and speech). Also called *acute confusional state (ACS).*

Delta cell A cell in the pancreatic islets (islets of Langerhans) in the pancreas that secretes somatostatin.

Dementia (de-MEN-shē-a) An organic mental disorder that results in permanent or progressive general loss of intellectual abilities such as impairment of memory, judgment, and abstract thinking and changes in personality; most common cause is Alzheimer's disease.

Demineralization (de-min'-er-al-i-ZA-shun) Loss of calcium and phosphorus from bones.

Dendrite (DEN-drīt) A nerve cell process that carries a nerve impulse toward the cell body.

Dental caries (KA-rēz) Gradual demineralization of the enamel and dentin of a tooth that may invade the pulp and alveolar bone. Also called *tooth decay.*

Dentin (DEN-tin) The osseous tissues of a tooth enclosing the pulp cavity.

Dentition (den-TI-shun) The eruption of teeth. The number, shape, and arrangement of teeth.

Deoxyribonucleic (dē-ok'-sē-ri'-bō-nyoo-KLE-ik) *acid (DNA)* A nucleic acid in the shape of a double helix constructed of nucleotides consisting of one of four nitrogenous bases (adenine, cytosine, guanine, or thymine), deoxyribose, and a phosphate group; encoded in the nucleotides is genetic information.

Depolarization (dē-pē-lar-i-ZA-shun) Used in neurophysiology to describe the reduction of voltage across a cell membrane; expressed as a movement toward less negative (more positive) voltages on the interior side of the cell membrane.

Depression (dē-PRESS-shun) Movement in which a part of the body moves downward.

Dermal papilla (pa-PILL-a) Fingerlike projection of the papillary region of the dermis that may contain blood capillaries or corpuscles of touch (Meissner's corpuscles).

Dermatology (der-ma-TOL-ō-jē) The medical specialty dealing with diseases of the skin.

Dermatome (DER-ma-tōm) An instrument for incising the skin or cutting thin transplants of skin. The cutaneous area developed from one embryonic spinal cord segment and receiving most of its innervation from one spinal nerve.

Dermis (DER-mis) A layer of dense connective tissue lying deep to the epidermis; the true skin or corium.

Descending colon (KŌ-lon) The part of the large intestine descending from the left colic (splenic) flexure to the level of the left iliac crest.

Developmental anatomy The study of development from the fertilized egg to the adult form. The branch of anatomy called embryology is generally restricted to the study of development from the fertilized egg through the eighth week in utero.

Diabetes insipidus (dī-a-BE-tēz in-SIP-i-dus) Condition caused by hyposecretion of antidiuretic hormone (ADH) and characterized by thirst and excretion of large amounts of urine.

Diabetes mellitus (MEL-i-tus) Hereditary condition caused by hyposecretion of insulin and characterized by hyperglycemia, increased urine production, excessive thirst, and excessive eating.

Diagnosis (dī-ag-NO-sis) Distinguishing one disease from another or determining the nature of a disease from signs and symptoms by inspection, palpation, laboratory tests, and other means.

Dialysis (dī-AL-i-sis) The process of separating small molecules from large by the difference in their rates of diffusion through a selectively permeable membrane.

Diapedesis (dī-a-pe-DE-sis) The passage of white blood cells through intact blood vessel walls.

Diaphragm (DĪ-a-fram) Any partition that separates one area from another, especially the dome-shaped skeletal muscle between the thoracic and abdominal cavities. Also a dome-shaped structure that fits over the cervix, usually with a spermicide, to prevent conception.

Diaphysis (dī-AF-i-sis) The shaft of a long bone.

Diarrhea (dī-a-RE-a) Frequent defecation of liquid feces caused by increased motility of the intestines.

Diarthrosis (dī-'ar-THRO-sis) Articulation in which opposing bones move freely, as in a hinge joint.

Diastole (dī-AS-tō-lē) In the cardiac cycle, the phase of relaxation or dilation of the heart muscle, especially of the ventricles.

Diastolic (di-as-TOL-ik) *blood pressure* The force exerted by blood on arterial walls during ventricular relaxation; the lowest blood pressure measured in the large arteries, about 80 mm Hg under normal conditions for a young, adult male.

Diencephalon (dī'-en-SEF-a-lon) A part of the brain consisting primarily of the thalamus and the hypothalamus.

Differential (dif-fer-EN-shal) *white blood cell count* Determination of the number of each kind of white blood cell in a sample of 100 cells for diagnostic purposes.

Differentiation (dif'-e-ren'-shē-Ā-shun) Acquisition of specific functions different from those of the original general type.

Diffusion (dif-YOO-zhun) A passive process in which there is a net or greater movement of molecules or ions from a region of high concentration to a region of low concentration until equilibrium is reached.

Digestion (dī-JES-chun) The mechanical and chemical breakdown of food to simple molecules that can be absorbed and used by body cells.

Dilate (DĪ-lāte) To expand or swell.

Dilation (dī-LĀ-shun) ***and curettage*** (ku-re-TAZH) Following dilation of the urterine cervix, the uterine endometrium is scraped with a curette (spoon-shaped instrument). Also called a ***D and C.***

Diploid (DIP-loyd) Having the number of chromosomes characteristically found in the somatic cells of an organism. Symbolized $2n$.

Diplopia (di-PLŌ-pē-a) Double vision.

Disease Any change from a state of health.

Dislocation (dis-lō-KĀ-shun) Displacement of a bone from a joint with tearing of ligaments, tendons, and articular capsules. Also called ***luxation*** (luks-Ā-shun).

Dissect (DIS-sekt) To separate tissues and parts of a cadaver (corpse) or an organ for anatomical study.

Dissociation (dis'-sō-sē-Ā-shun) Separation of inorganic acids, bases, and salts into ions when dissolved in water. Also called ***ionization*** (ī-on-i-ZĀ-shun).

Distal (DIS-tal) Farther from the attachment of an extremity to the trunk or a structure; farther from the point of origin.

Diuretic (dī-yoo-RET-ik) A chemical that inhibits sodium reabsorption, reduces antidiuretic hormone (ADH) concentration, and increases urine volume by inhibiting facultative reabsorption of water.

Divergence (di-VER-jens) An anatomical arrangement in which the synaptic end bulbs of one presynaptic neuron terminate on several postsynaptic neurons.

Diverticulitis (dī-ver-tik-yoo-LĪ-tis) Inflammation of diverticula, saclike outpouchings of the colonic wall, when the muscularis becomes weak.

Diverticulum (dī-ver-TIK-yoo-lum) A sac or pouch in the wall of a canal or organ, especially in the colon.

Dominant gene A gene that is able to override the influence of the complementary gene on the homologous chromosome; the gene that is expressed.

Donor insemination (in-sem'-i-NĀ-shun) The deposition of seminal fluid within the vagina or cervix at a time during the menstrual cycle when pregnancy is most likley to occur. It may be homologous (using the husband's semen) or heterologous (using a donor's semen). Also called ***artificial insemination.***

Dorsal body cavity Cavity near the dorsal surface of the body that consists of a cranial cavity and vertebral canal.

Dorsiflexion (dor'-si-FLEK-shun) Bending the foot in the direction of the dorsum (upper surface).

Down's syndrome (DS) An inherited defect due to an extra copy of chromosome 21. Symptoms include mental retardation; a small skull, flattened from front to back; a short, flat nose; short fingers; and a widened space between the first two digits of the hand and foot. Also called ***trisomy 21.***

Duct of Santorini   *See* ***Accessory duct.***

Duct of Wirsung   *See* ***Pancreatic duct.***

Ductus arteriosus (DUK-tus ar-tē-rē-Ō-sus) A small vessel connecting the pulmonary trunk with the arota; found only in the fetus.

Ductus (vas) deferens (DEF-er-ens) The duct that conducts spermatozoa from the epididymis to the ejaculatory duct. Also called the ***seminal duct.***

Ductus epididymis (ep'-i-DID-i-mis) A tightly coiled tube inside the epididymis, distinguished into a head, body, and tail, in which spermatozoa undergo maturation.

Ductus venosus (ve-NŌ-sus) A small vessel in the fetus that helps the circulation bypass the liver.

Duodenal (doo-ō-DĒ-nal) ***gland*** Gland in the submucosa of the duodenum that secretes an alkaline mucus to protect the lining of the small intestine from the action of enzymes and to help neutralize the acid in chyme. Also called ***Brunner's*** (BRUN-erz) ***gland.***

Duodenal papilla (pa-PILL-a) An elevation on the duodenal mucosa that receives the hepatopancreatic ampulla (ampulla of Vater).

Duodenum (doo'-ō-DĒ-num) The first 25 cm (10 in.) of the small intestine.

Dura mater (DYOO-ra MĀ-ter) The outer membrane (meninx) covering the brain and spinal cord.

Dynamic equilibrium (ē-kwi-LIB-rē-um) The maintenance of body position, mainly the head, in response to sudden movements such as rotation.

Dysfunction (dis-FUNK-shun) Absence of complete normal function.

Dyslexia (dis-LEX-sē-a) Impairment of the brain's ability to translate images received from the eyes or ears into understandable language.

Dysmenorrhea (dis'-men-ō-RĒ-a) Painful menstruation.

Dysphagia (dis-FĀ-jē-a) Difficulty in swallowing.

Dysplasia (dis-PLĀ-zē-a) Change in the size, shape, and organization of cells due to chronic irritation or inflammation; may revert to normal if stress is removed or progresses to neoplasia.

Dyspnea (DISP-nē-a) Shortness of breath.

Dystocia (dis-TŌ-sē-a) Difficult labor due to factors such as pelvic deformities, malpositioned fetus, and premature rupture of fetal membranes.

Dystrophia (dis-TRŌ-fē-a) Progressive weakening of a muscle.

Dysuria (dis-SOO-rē-a) Painful urination.

Ectoderm The outermost of the three primary germ layers that gives rise to the nervous system and the epidermis of skin and its derivatives.

Ectopic (ek-TOP-ik) Out of the normal location, as in ectopic pregnancy.

Eczema (EK-ze-ma) A skin rash characterized by itching, swelling, blistering, oozing, and scaling of the skin.

Edema (e-DĒ-ma) An abnormal accumulation of interstitial fluid.

Effective filtration pressure (**Peff**) Net pressure that expresses the relationship between the force that promotes glomerular filtration and the forces that oppose it.

Effector (e-FEK-tor) The organ of the body, either a muscle or a gland, that responds to a motor neuron impulse.

Efferent arteriole (EF-er-ent ar-TĒ-rē-ōl) A vessel of the renal vascular system that transports blood from the glomerulus to the peritubular capillary.

Efferent (EF-er-ent) *ducts* A series of coiled tubes that transport spermatozoa from the rete testis to the epididymis.

Efferent neuron (NOO-ron) A neuron that conveys nerve impulses from the brain and spinal cord to effectors that may be either muscles or glands. Also called a *motor neuron.*

Effusion (e-FYOO-zhun) The escape of fluid from the lymphatics or blood vessels into a cavity or into tissues.

Ejaculation (e-jak-yoo-LĀ-shun) The reflex ejection or expulsion of semen from the penis.

Ejaculatory (e-JAK-yoo-la-tō′-rē) *duct* A tube that transports spermatozoa from the ductus (vas) deferens to the prostatic urethra.

Elasticity (e-las-TIS-i-tē) The ability of tissue to return to its original shape after contraction or extension.

Electrocardiogram (e-lek′-trō-KAR-dē-ō-gram) *(ECG* or *EKG)* A recording of the electrical changes that accompany the cardiac cycle and can be recorded on the surface of the body; may be resting, stress, or ambulatory.

Electroencephalogram (e-lek′-trō-en-SEF-a-lō-gram) *(EEG)* A recording of the electrical impulses of the brain to diagnose certain diseases (such as epilepsy), furnish information regarding sleep and wakefulness, and confirm brain death.

Electrolyte (ē-LEK-trō-līt) Any compound that separates into ions when dissolved in water and is able to conduct electricity.

Electron transport chain A series of oxidation-reduction reactions in the catabolism of glucose that occur on the inner mitochrondrial membrane and in which energy is released and transferred for storage to ATP.

Elevation (el-e-VĀ-shun) Movement in which a part of the body moves upward.

Ellipsoidal (e-lip-SOY-dal) *joint* A synovial joint structured so that an oval-shaped condyle of one bone fits into an elliptical cavity of another bone, permitting side-to-side and back-and-forth movements, as at the joint at the wrist between the radius and carpals. Also called a *condyloid* (KON-di-loid) *joint.*

Embolism (EM-bō-lizm) Obstruction of closure of a vessel by an embolus.

Embolus (EM-bō-lus) A blood clot, bubble of air, fat from broken bones, mass of bacteria, or other debris or foreign material transported by the blood.

Embryo (EM-brē-ō) The young of any organism in an early stage of development; in humans, the developing organism from fertilization to the end of the eighth week in utero.

Embryology (em′-brē-OL-ō-jē) The study of development from the fertilized egg to the end of the eighth week in utero.

Embryo transfer A type of *in vitro* fertilization in which semen is used to artificially inseminate a fertile secondary oocyte donor and the morula or blastocyst is then transferred from the donor to the infertile woman, who then carries it to term.

Emesis (EM-e-sis) Vomiting.

Emmetropia (em′-e-TRŌ-pē-a) The ideal optical condition of the eyes.

Emphysema (em′-fi-SĒ-ma) A swelling or inflation of air passages due to loss of elasticity in the alveoli.

Emulsification (ē-mul′-si-fi-KĀ-shun) The dispersion of large fat globules to smaller uniformly distributed particles in the presence of bile.

Enamel (e-NAM-el) The hard, white substance covering the crown of a tooth.

Endocardium (en-dō-KAR-dē-um) The layer of the heart wall, composed of endothelium and smooth muscle, that lines the inside of the heart and covers the valves and tendons that hold the valves open.

Endochondral ossification (en′-dō-KON-dral os′-i-fi-KĀ-shun) The replacement of cartilage by bone. Also called *intracartilaginous* (in′-tra-kar′-ti-LAJ-i-nus) *ossification.*

Endocrine (EN-dō-krin) *gland* A gland that secretes hormones into the blood; a ductless gland.

Endocrinology (en′-dō-kri-NOL-ō-jē) The science concerned with the structure and functions of endocrine glands and the diagnosis and treatment of disorders of the endocrine system.

Endocytosis (en′-dō-sī-TŌ-sis) The uptake into a cell of large molecules and particles in which a segment of plasma membrane surrounds the substance, encloses it, and brings it in; includes phagocytosis, pinocytosis, and receptor-mediated endocytosis.

Endoderm The innermost of the three primary germ layers of the developing embryo that gives rise to the gastrointestinal tract, urinary bladder and urethra, and respiratory tract.

Endodontics (en′-dō-DON-tiks) The branch of dentistry concerned with the prevention, diagnosis, and treatment of diseases that affect the pulp, root, periodontal ligament, and alveolar bone.

Endogenous (en-DOJ-e-nus) Growing from or beginning within the organism.

Endolymph (En-dō-lymf) The fluid within the membranous labyrinth of the inner ear.

Endometriosis (en′-dō-MĒ-trē-ō′-sis) The growth of endometrial tissue outside the uterus.

Endometrium (en′-dō-MĒ-trē-um) The mucous membrane lining the uterus.

Endomysium (em′-dō-MĪZ-ē-um) Invagination of the perimysium separating each individual muscle fiber (cell).

Endoneurium (en′-dō-NYOO-rē-um) Connective tissue wrapping around individual nerve fibers (cells).

Endoplasmic reticulum (en′-do-PLAZ-mik re-TIK-yoo-lum) *(ER)* A network of channels running through the cytoplasm of a cell that serves in intracellular transportation, support, storage, synthesis, and packaging of molecules. Portions of ER where ribosomes are attached to the outer surface are called *granular* or *rough reticulum;* portions that have no ribosomes are called *agranular* or *smooth reticulum.*

End organ of Ruffini See *Type II cutaneous mechanoreceptor.*

Endorphin (en-DOR-fin) A neuropeptide in the central nervous system that acts as a painkiller.

Endoscope (EN-dō-skōp′) An illuminated tube with lenses used to look inside hollow organs such as the stomach (gastroscope) or urinary bladder (cystoscope).

Endoscopy (en-DOS-kō-pē) The visual examination of any cavity of the body using an endoscope, an illuminated tube with lenses.

Endosteum (en-DOS-tē-um) The membrane that lines the medullary cavity of bones, consisting of osteoprogenitor cells and scattered osteoclasts.

Endothelial-capsular (en-dō-THĒ-lē-al) *membrane* A filtration membrane in a nephron of a kidney consisting of the endothelium and basement membrane of the glomerulus and the epithelium of the visceral layer of the glomerular (Bowman's) capsule.

Endothelium (en'-dō-THĒ-lē-um)   The layer of simple squamous epithelium that lines the cavities of the heart and blood and lymphatic vessels.

Energy   The capacity to do work.

Enkephalin (en-KEF-a-lin)   A peptide found in the central nervous system that acts as a painkiller.

Enteroendocrine (en-ter-ō-EN-dō-krin) *cell*   A stomach cell that secretes the hormone stomach gastrin.

Enterogastric (en-te-rō-GAS-trik) *reflex*   A reflex that inhibits gastric secretion; initiated by food in the small intestine.

Enuresis (en'-yoo-RĒ-sis)   Involuntary discharge of urine, complete or partial, after age 3.

Enzyme (EN-zīm)   A substance that affects the speed of chemical changes; an organic catalyst, usually a protein.

Eosinophil (ē'-ō-SIN-ō-fil)   A type of white blood cell characterized by granular cytoplasm readily stained by eosin.

Ependyma (e-PEN-de-ma)   Neuroglial cells that line ventricles of the brain and probably assist in the circulation of cerebrospinal fluid (CSF). Also called *ependymocytes* (e-PEN-di-mō-sītz).

Epicardium (ep'-i-KAR-dē-um)   The thin outer layer of the heart wall, composed of serous tissue and mesothelium. Also called the *visceral pericardium.*

Epidemic (ep'-i-DEM-ik)   A disease that occurs above the expected level among individuals in a population.

Epidemiology (ep'-i-dē-mē-OL-ō-jē)   Medical science concerned with the occurrence and distribution of disease in human populations.

Epidermis (ep-i-DERM-is)   The outermost, thinner layer of skin, composed of stratified squamous epithelium.

Epididymis (ep'-i-DID-i-mis)   A comma-shaped organ that lies along the posterior border of the testis and contains the ductus epididymis, in which sperm undergo maturation. *Plural, epididymides* (ep'-i-DID-i-mi-dēz).

Epidural (ep'-i-DOO-ral) *space*   A space between the spinal dura mater and the vertebral canal, containing loose connective tissue and a plexus of veins.

Epiglottis (ep'-i-GLOT-is)   A large, leaf-shaped piece of cartilage lying on top of the larynx, with its "stem" attached to the thyroid cartilage and its "leaf" portion unattached and free to move up and down to cover the glottis (vocal folds and rima glottidis).

Epilepsy (EP-i-lep'-sē)   Neurological disorder characterized by short, periodic attacks of motor, sensory, or psychological malfunction.

Epimysium (ep'-i-MĪZ-ē-um)   Fibrous connective tissue around muscles.

Epinephrine (ep-ē-NEF-rin)   Hormone secreted by the adrenal medulla that produces actions similar to those that result from sympathetic stimulation. Also called *adrenaline* (a-DREN-a-lin).

Epineurium (ep'-i-NYOO-rē-um)   The outermost covering around the entire nerve.

Epiphyseal (ep'-i-FIZ-ē-al) *line*   The remnant of the epiphyseal plate in a long bone.

Epiphyseal (ep'-i-FIZ-ē-al) *plate*   The cartilaginous plate between the epiphysis and diaphysis that is responsible for the lengthwise growth of long bones.

Epiphysis (ē-PIF-i-sis)   The end of a long bone, usually larger in diameter than the shaft (diaphysis).

Episiotomy (e-piz'-ē-OT-ō-mē)   A cut made with surgical scissors to avoid tearing of the perineum at the end of the second stage of labor.

Epistaxis (ep'-i-STAK-sis)   Loss of blood from the nose due to trauma, infection, allergy, neoplasm, and bleeding disorders. Also called *nosebleed.*

Epithelial (ep'-i-THĒ-lē-al) *tissue*   The tissue that forms glands or the outer part of the skin and lines blood vessels, hollow organs, and passages that lead externally from the body.

Eponychium (ep'-ō-NIK-ē-um)   Narrow band of stratum corneum at the proximal border of a nail that extends from the margin of the nail wall. Also called the *cuticle.*

Erection (ē-REK-shun)   The enlarged and stiff state of the penis (or clitoris) resulting from the engorgement of the spongy erectile tissue with blood.

Eructation (e-ruk'-TĀ-shun)   The forceful expulsion of gas from the stomach. Also called *belching.*

Erythema (er'-e-THĒ-ma)   Skin redness usually caused by engorgement of the capillaries in the lower layers of the skin.

Erythematosus (er-i'-them-a-TŌ-sus)   Pertaining to redness.

Erythrocyte (e-RITH-rō-sīt)   Red blood cell.

Erythropoiesis (e-rith'-rō-poy-Ē-sis)   The process by which erythrocytes (red blood cells) are formed.

Erythropoietin (e-rith'-rō-POY-ē-tin)   A hormone formed from a plasma protein that stimulates erythrocyte (red blood cell) production.

Esophagus (e-SOF-a-gus)   A hollow muscular tube connecting the pharynx and the stomach.

Essential amino acids   Those 10 amino acids that cannot be synthesized by the human body at an adequate rate to meet its needs and therefore must be obtained from the diet.

Estrogens (ES-tro-jens)   Female sex hormones produced by the ovaries concerned with the development and maintenance of female reproductive structures and secondary sex characteristics, fluid and electrolyte balance, and protein anabolism. Examples are β-estradiol, estrone, and estriol.

Etiology (ē'-tē-OL-ō-jē)   The study of the causes of disease, including theories of origin and the organisms, if any, involved.

Euphoria (yoo-FŌR-ē-a)   A subjectively pleasant feeling of well-being marked by confidence and assurance.

Eupnea (yoop-NĒ-a)   Normal quiet breathing.

Eustachian tube   *See Auditory tube.*

Euthanasia (yoo'-tha-NĀ-zē-a)   The practice of ending a life in case of incurable diseases.

Eversion (ē-VER-zhun)   The movement of the sole outward at the ankle joint.

Exacerbation (eg-zas'-er-BĀ-shun)   An increase in the severity of symptoms or of disease.

Excitability (ek-sīt'-a-BIL-i-tē)   The ability of muscle tissue to receive and respond to stimuli; the ability of nerve cells to respond to stimuli and convert them into nerve impulses.

Excrement (EKS-kre-ment)   Material cast out from the body as waste, especially fecal matter.

Excretion (eks-KRĒ-shun)   The process of eliminating waste products from a cell, tissue, or the entire body; or the products excreted.

Exocrine (EK-sō-krin) ***gland*** A gland that secretes substances into ducts that empty at covering or lining epithelium or directly onto a free surface.

Exocytosis (ex'-ō-sī-TŌ-sis) A process of discharging cellular products too big to go through the membrane. Particles for export are enclosed by Golgi membranes when they are synthesized. Vesicles pinch off from the Golgi complex and carry the enclosed particles to the interior surface of the cell membrane, where the vesicle membrane and plasma membrane fuse and the contents of the vesicle are discharged.

Exogenous (ex-SOJ-e-nus) Originating outside an organ or part.

Exophthalmic goiter (ek'-sof-THAL-mik GOY-ter) An autoimmune disease that may result in hypersecretion of thyroid hormones characterized by protrusion of the eyeballs (exophthalmos) and an enlarged thyroid (goiter). Also called ***Graves disease.***

Expiration (ek-spi-RĀ-shun) Breathing out; expelling air from the lungs into the atmosphere. Also called ***exhalation.***

Expiratory (eks-PĪ-ra-tō-rē) ***reserve volume*** The volume of air in excess of tidal volume that can be exhaled forcibly; about 1,200 ml.

Extensibility (ek-sten'-si-BIL-i-tē) The ability of muscle tissue to be stretched when pulled.

Extension (ek-STEN-shun) An increase in the angle between two bones; restoring a body part to its anatomical position after flexion.

External Located on or near the surface.

External auditory (AW-di-tōr-ē) ***canal*** or ***meatus*** (mē-Ā-tus) A curved tube in the temporal bone that leads to the middle ear.

External ear The outer ear, consisting of the pinna, external auditory canal, and tympanic membrane or eardrum.

External nares (NA-rēz) The external nostrils, or the openings into the nasal cavity on the exterior of the body.

External respiration The exchange of respiratory gases between the lungs and blood.

Exteroceptor (eks'-ter-ō-SEP-tor) A receptor adapted for the reception of stimuli from outside the body.

Extracellular fluid (ECF) Fluid outside body cells, such as interstitial fluid and plasma.

Extrinsic (ek-STRIN-sik) Of external origin.

Extrinsic clotting pathway Sequence of reactions leading to blood clotting that is initiated by the release of a substance (tissue factor or thromboplastin) *outside* blood itself, from damaged blood vessels or surrounding tissues.

Exudate (EKS-yoo-dāt) Escaping fluid or semifluid material that oozes from a space that may contain serum, pus, and cellular debris.

Eyebrow The hairy ridge above the eye.

Face The anterior aspect of the head.

Facilitated diffusion (fa-SIL-i-tā-ted dif-YOO-zhun) Diffusion in which a substance not soluble by itself in lipids is transported across a selectively permeable membrane by combining with a carrier substance.

Facilitation (fa-sil-i-TĀ-shun) The process in which a nerve cell membrane is partially depolarized by a subliminal stimulus so

that a subsequent subliminal stimulus can further depolarize the membrane to reach the threshold of nerve impulse initiation.

Facultative (FAK-ul-tā-tive) ***water reabsorption*** The absorption of water from distal convoluted tubules and collecting tubules of nephrons in response to antidiuretic hormone (ADH).

Falciform ligament (FAL-si-form LIG-a-ment) A sheet of parietal peritoneum between the two principal lobes of the liver. The ligamentum teres, or remnant of the umbilical vein, lies within its fold.

Fallopina tube *See Uterine tube.*

Falx cerebelli (FALKS ser'-e-BEL-lē) A small triangular process of the dura mater attached to the occipital bone in the posterior cranial fossa and projecting inward between the two cerebellar hemispheres.

Falx cerebri (SER-e-brē) A fold of the dura mater extending down into the longitudinal fissure between the two cerebral hemispheres.

Fascia (FASH-ē-a) A fibrous membrane covering, supporting, and separating muscles.

Fascicle (FAS-i-kul) A small bundle or cluster, especially of nerve or muscle fibers (cells). Also called a ***fasciculus*** (fa-SIK-yoo-lus). *Plural,* ***fasciculi*** (fa-SIK-yoo-lī).

Fat A lipid compound formed from one molecule of glycerol and three molecules of fatty acids; the body's most highly concentrated source of energy. Adipose tissue, composed of adipocytes specialized for fat storage and present in the form of soft pads between various organs for support, protection, and insulation.

Fauces (FAW-sēz) The opening from the mouth into the pharynx.

Febrile (FĒ-bril) Feverish; pertaining to a fever.

Feces (FĒ-sēz) Material discharged from the rectum and made up of bacteria, excretions, and food residue. Also called ***stool.***

Feedback system A circular sequence of events in which information about the status of a situation is continually reported (fed back) to a central control region.

Fertilization (fer'-ti-li-ZĀ-shun) Penetration of a secondary oocyte by a spermatozoon and subsequent union of the nuclei of the cells.

Fetal (FĒ-tal) ***alcohol syndrome (FAS)*** Term applied to the effects of intrauterine exposure to alcohol, such as slow growth, defective organs, and mental retardation.

Fetal circulation The cardiovascular system of the fetus, including the placenta and special blood vessels involved in the exchange of materials between fetus and mother.

Fetus (FĒ-tus) The latter stages of the developing young of an animal; in humans, the developing organism in utero from the beginning of the third month to birth.

Fever An elevation in body temperature above its normal temperature of 37°C (98.6°F).

Fibrillation (fi-bri-LĀ-shun) Involuntary brief twitch of a muscle that is not visible under the skin and is not associated with movement of the affected muscle.

Fibrin (FĪ-brin) An insoluble protein that is essential to blood clotting; formed from fibrinogen by the action of thrombin.

Fibrinogen (fī-BRIN-ō-jen) A high-molecular-weight protein in the blood plasma that by the action of thrombin is converted to fibrin.

Fibrinolysis (fī-brin-OL-i-sis) Dissolution of a blood clot by the action of a proteolytic enzyme that converts insoluble fibrin into a soluble substance.

Fibroblast (FĪ-brō-blast) A large, flat cell that forms collagenous and elastic fibers and intercellular substance of loose connective tissue.

Fibromyalgia (fī-bro-mī-AL-jē-a) Groups of common nonarticular rheumatic disorders characterized by pain, tenderness, and stiffness of muscles, tendons, and surrounding tissue. Examples of fibromyalgia are lumbago and charleyhorse.

Fibroplasia (fī-brō-PLĀ-zē-a) Period of scar tissue formation.

Fibrosis (fī-BRŌ-sis) Abnormal formation of fibrous tissue.

Fibrous (FĪ-burs) *joint* A joint that allows little or no movement, such as a suture and syndesmosis.

Fibrous tunic (TOO-nik) The outer coat of the eyeball, made up of the posterior sclera and the anterior cornea.

Fight-or-flight response The effect of the stimulation of the sympathetic division of the autonomic nervous system.

Filiform papilla (FIL-i-form pa-PIL-a) One of the conical projections that are distributed in parallel rows over the anterior two-thirds of the tongue and contain no taste buds.

Filtrate (fil-TRĀT) The fluid produced when blood is filtered by the endothelial-capsular membrane.

Filtration (fil-TRĀ-shun) The passage of a liquid through a filter or membrane that acts like a filter.

Filtration fraction The percentage of plasma entering the nephrons that actually becomes glomerular filtrate.

Filum terminale (FĪ-lum ter-mi-NAL-ē) Nonnervous fibrous tissue of the spinal cord that extends inferiorly from the conus medullaris to the coccyx.

Fimbriae (FIM-brē-ē) Fingerlike structures, expecially the lateral ends of the uterine (Fallopian) tubes.

Fissue (FISH-ur) A groove, fold, or slit that may be normal or abnormal.

Fistula (FIS-choo-la) An abnormal passage between two organs or between an organ cavity and the outside.

Fixator A muscle that stabilizes the origin of the prime mover so that the prime mover can act more efficiently.

Fixed macrophage (MAK-rō-fāj) Stationary phagocytic cell found in the liver, lungs, brain, spleen, lymph nodes, subcutaneous tissue, and the bone marrow. Also called a *histiocyte* (HIS-tē-ō-sīt).

Flaccid (FLAS-sid) Relaxed, flabby, or soft; lacking muscle tone.

Flagellum (fla-JEL-um) A hairlike, motile process on the extremity of a bacterium of protozoan. *Plural, flagella* (fla-JEL-a).

Flatfoot A condition in which the ligaments and tendons of the arches of the foot are weakened and the height of the longitudinal arch decreases.

Flatus (FLĀ-tus) Air (gas) in the stomach or intestines, commonly used to denote passage of gas rectally.

Flexion (FLEK-shun) A folding movement in which there is a decrease in the angle between two bones.

Flexor reflex A protective reflex in which flexor muscles are stimulated while extensor muscles are inhibited.

Fluoroscope (FLOOR-ō-skōp) An instrument for visual observation of the body by means of x-ray.

Follicle-stimulating (FOL-i-kul) *hormone* (*FSH*) Hormone secreted by the adenohypophysis (anterior lobe) of the pituitary gland that initiates development of ova and stimulates the ovaries to secrete estrogens in females and initiates sperm production in males.

Fontanel (fon'-ta-NEL) A membrane-covered spot where bone formation is not yet complete, especially between the cranial bones of an infant's skull.

Foot The terminal part of the lower extremity.

Foramen (fo-RĀ-men) A passage or opening; a communication between two cavities of an organ or a hole in a bone for passage of vessels or nerves.

Foramen ovale (ō-VAL-ē) An opening in the fetal heart in the septum between the right and left atria. A hole in the greater wing of the sphenoid bone that transmits the mandibular branch of the trigeminal (V) nerve.

Forearm (FOR-arm) The part of the upper extremity between the elbow and the wrist.

Fossa (FOS-a) A furrow or shallow depression.

Fourth ventricle (VEN-tri-kul) A cavity within the brain lying between the cerebellum and the medulla and pons.

Fracture (FRAK-chur) Any break in a bone.

Fragile X syndrome Inherited disorder characterized by learning difficulties, mental regardation, and physical abnormalities; due to a defective gene on the X chromosome.

Frenulum (FREN-yoo-lum) A small fold of mucous membrane that connects two parts and limits movement.

Frontal plane A plane at a right angle to a midsagittal plane that divides the body or organs into anterior and posterior portions. Also called a *coronal* (kō-RŌ-nal) *plane.*

Fulminate (FUL-mi-nāt') To occur suddenly with great intensity.

Functional residual (re-ZID-yoo-al) *volume* The sum of residual volume plus expiratory reserve volume; about 2,400 ml.

Fungiform papilla (FUN-ji-form pa-PIL-a) A mushroomlike elevation on the upper surface of the tongue appearing as a red dot; papillae contain taste buds.

Furuncle (FYOOR-ung-kul) A boil; painful nodule caused by bacterial infection and inflammation of a hair follicle or sebaceous (oil) gland.

Gallbladder A small pouch that stores bile, located under the liver, which is filled with bile and emptied via the cystic duct.

Gallstone A concretion, usually consisting of cholesterol, formed anywhere between bile canaliculi in the liver and the hepatopancreatic ampulla (ampulla of Vater), where bile enters the duodenum. Also called a *biliary calculus.*

Gamete (GAM-ēt) A male or female reproductive cell; the spermatozoon or ovum.

Gamete intrafallopian transfer (*GIFT*) A type of *in vitro* fertilization in which aspirated secondary oocytes are combined with a solution containing sperm outside the body and then the secondary oocytes are inserted into the uterine (Fallopian) tubes.

Ganglion (GANG-glē-on) A group of nerve cell bodies that lie outside the central nervous system. *Plural, ganglia* (GANG-glē-a).

Gangrene (GANG-rēn) Death and rotting of a considerable mass of tissue that usually is caused by interruption of blood supply followed by bacterial (*Clostridium*) invasion.

Gastroenterology (gas'-trō-en'ter-OL-ō-jē) The medical specialty that deals with the structure, function, diagnosis, and treatment of diseases of the stomach and intestines.

Gastrointestinal (gas-trō-in-TES-ti-nal) *(GI) tract*   A continuous tube running through the ventral body cavity extending from the mouth to the anus. Also called the *alimentary* (al'-i-MEN-tar-ē) *canal.*

Gastrulation (gas'-troo-LĀ-shun)   The various movements of groups of cells that lead to the establishment of the primary germ layers.

Gavage (ga-VAZH)   Feeding through a tube passed through the esophagus and into the stomach.

Gene (jēn)   Biological unit of heredity; an ultramicroscopic, self-reproducing DNA particle located in a definite position on a particular chromosome.

General adaptation syndrome (GAS)   Wide-ranging set of bodily changes triggered by a stressor that gears the body to met an emergency.

Generator potential   The graded depolarization that results in a change in the resting membrane potential in a receptor (specialized neuronal ending).

Genetic engineering   The manufacture and manipulation of genetic material.

Genetics   The study of heredity.

Genital herpes (JEN-i-tal HER-pēz)   A sexually transmitted disease caused by type II herpes simplex virus.

Genitalia (jen'-i-TAL-ya)   Reproductive organs.

Genome (JĒ-nōm)   The complete gene complement of an organism.

Genotype (JĒ-nō-tīp)   The total hereditary information carried by an individual; the genetic makeup of an organism.

Geriatrics (jer'-ē-AT-riks)   The branch of medicine devoted to the medical problems and care of elderly persons.

Germinal (JER-mi-nal) *epithelium*   A layer of epithelial cells that covers the ovaries and lines the seminiferous tubules of the testes.

Germinativum (jer'-mi-na-TĒ-vum)   Skin layers where new cells are germinated.

Gestation (jes-TĀ-shun)   The period of intrauterine fetal development.

Giantism (GĪ-an-tizm)   Condition caused by hypersecretion of human growth hormone (hGH) during childhood characterized by excessive bone growth and body size. Also called *gigantism.*

Gingivae (jin-JI-vē)   Gums. They cover the alveolar processes of the mandible and maxilla and extend slightly into each socket.

Gingivitis (jin'-je-VĪ-tis)   Inflammation of the gums.

Gland   Single or group of specialized epithelial cells that secrete substances.

Glans penis (glanz PĒ-nis)   The slightly enlarged region at the distal end of the penis.

Glaucoma (glaw-KŌ-ma)   An eye disorder in which there is increased intraocular pressure due to an excess of aqueous humor.

Gliding joint   A synovial joint having articulating surfaces that are usually flat, permitting only side-to-side and back-and-forth movements, as between carpal bones, tarsal bones, and the scapula and clavicle. Also called an *arthrodial* (ar-THRŌ-dē-al) *joint.*

Glomerular (glō-MER-yoo-lar) *capsule*   A double-walled globe at the proximal end of a nephron that encloses the glomerulus. Also called *Bowman's* (BŌ-manz) *capsule.*

Glomerular filtration   The first step in urine formation in which substances in blood are filtered at the endothelial-capsular membrane and the filtrate enters the proximal convoluted tubule of a nephron.

Glomerular filtration rate (GFR)   The total volume of fluid that enters all the glomerular (Bowman's) capsules of the kidneys in 1 min; about 125 ml/min.

Glomerulonephritis (glō-mer-yoo-lō-nef-RĪ-tis)   Inflammation of the glomeruli of the kidney that increases the permeability of the endothelial-capsular membrane and permits blood cells and proteins to enter the filtrate. Also called *Bright's disease.*

Glomerulus (glō-MER-yoo-lus)   A rounded mass of nerves or blood vessels, especially the microscopic tuft of capillaries that is surrounded by the glomerular (Bowman's) capsule of each kidney tubule.

Glottis (GLOT-is)   The vocal folds (true vocal cords) in the larynx and the space between them (rima glottidis).

Glucagon (GLOO-ka-gon)   A hormone produced by the alpha cells of the pancreas that increases the blood glucose level.

Glucocorticoids (gloo-kō-KOR-ti-koyds)   A group of hormones of the adrenal cortex.

Gluconeogenesis (gloo'-kō-nē'-ō-JEN-e-sis)   The conversion of a substance other than carbohydrate into glucose.

Glucose (GLOO-kōs)   A six-carbon sugar, $C_6H_{12}O_6$; the major energy source for every cell type in the body. Its metabolism is possible by every known living cell for the production of ATP.

Glycogen (GLĪ-kō-jen)   A highly branched polymer of glucose containing thousands of subunits; functions as a compact store of glucose molecules in liver and muscle fibers (cells).

Glycogenesis (glī'-kō-JEN-e-sis)   The process by which many molecules of glucose combine to form a molecule called glycogen.

Glycogenolysis (glī-kō-je-NOL-i-sis)   The process of converting glycogen to glucose.

Glycolysis (glī-KŌL-i-sis)   A series of chemical reactions that break down glucose into pyruvic acid, with a net gain of two molecules of ATP.

Glycosuria (glī-kō-SOO-rē-a)   The presence of glucose in the urine; may be temporary or pathological. Also called *glucosuria.*

Gnostic (NOS-tik)   Pertaining to the faculties of perceiving and recognizing.

Gnostic area   Sensory area of the cerebral cortex that receives and integrates sensory input from various parts of the brain so that a common thought can be formed.

Goblet cell   A goblet-shaped unicellular gland that secretes mucus. Also called a *mucus cell.*

Goiter (GOY-ter)   An enlargement of the thyroid gland.

Golgi (GOL-jē) *complex*   An organelle in the cytoplasm of cells consisting of four to eight flattened channels, stacked upon one another, with expanded areas at their ends; functions in packaging secreted proteins, lipid secretion, and carbohydrate synthesis.

Golgi tendon organ   See Tendon organ.

Gomphosis (gom-FŌ-sis)   A fibrous joint in which a cone-shaped peg fits into a socket.

Gonad (GŌ-nad)   A gland that produces gametes and hormones; the ovary in the female and the testis in the male.

Gonadocorticoids (gō-na-dō-KOR-ti-koydz)   Sex hormones secreted by the adrenal cortex.

Gonadotropic (gō'-nad-ō-TRŌ-pik) *hormone* A hormone that regulates the functions of the gonads.

Gonorrhea (gon'-ō-RĒ-a) Infectious, sexually transmitted disease caused by the bacterium *Neisseria gonorrhoeae* and characterized by inflammation of the urogenital mucosa, discharge of pus, and painful urination.

Gout (gowt) Hereditary condition associated with excessive uric acid in the blood; the acid crystallizes and deposits in joints, kidneys, and soft tissue.

Graafian follicle *See Vesicular ovarian follicle.*

Gray commissure (KOM-i-shur) A narrow strip of gray matter connecting the two lateral gray masses within the spinal cord.

Gray matter Area in the central nervous system and ganglia consisting of nonmyelinated nerve tissue.

Greater omentum (ō-MEN-tum) A large fold in the serosa of the stomach that hangs down like an apron over the front of the intestines

Greater vestibular (ves-TIB-yoo-lar) *glands* A pair of glands on either side of the vaginal orifice that open by a duct into the space between the hymen and the labia minora. Also called **Bartholin's** (BAR-to-linz) *glands.*

Groin (groyn) The depression between the thigh and the trunk, the inguinal region.

Gross anatomy The branch of anatomy that deals with structures that can be studied without using a microscope. Also called **macroscopic anatomy.**

Growth An increase in size due to an increase in the number of cells or an increase in the size of existing cells as internal components increase in size or an increase in the size of intercellular substances.

Gustatory (GUS-ta-tō'-rē) Pertaining to taste.

Gynecology (gī'-ne-KOL-ō-jē) The branch of medicine dealing with the study and treatment of disorders of the female reproductive system.

Gynecomastia (gīn'-e-kō-MAS-té-a) Excessive growth (benign) of the male mammary glands due to secretion of sufficient estrogens by an adrenal gland tumor (feminizing adenoma).

Gyrus (JĪ-rus) One of the folds of the cerebral cortex of the brain. *Plural,* **gyri** (JĪ-rī). Also called a **convolution.**

Hair A threadlike structure produced by hair follicles that develops in the dermis. Also called *pilus* (PI-lus).

Hair follicle (FOL-li-kul) Structure composed of epithelium surrounding the root of a hair from which hair develops.

Hair root plexus (PLEK-sus) A network of dendrites arranged around the root of a hair as free or naked nerve endings that are stimulated when a hair shaft is moved.

Hallucination (ha-loo'-si-NĀ-shun) A sensory perception of something that does not really exist in the world, that is, a sensory experience created from within the brain.

Hand The terminal portion of an upper extremity, including the carpus, metacarpus, and phalanges.

Haploid (HAP-loyd) Having half the number of chromosomes characteristically found in the somatic cells of an organism; characteristic of mature gametes. Symbolized *n*.

Hard palate (PAL-at) The anterior portion of the roof of the mouth, formed by the maxillae and palatine bones and lined by mucous membrane.

Haustra (HAWS-tra) The sacculated elevations of the colon.

Haversian canal *See Central canal.*

Haversian system *See Osteon.*

Head The superior part of a human, cephalic to the neck. The superior or proximal part of a structure.

Heart A hollow muscular organ lying slightly to the left of the midline of the chest that pumps the blood through the cardiovascular system.

Heart block An arrhythmia (dysrhythmia) of the heart in which the atria and ventricles contract independently because of a blocking of electrical impulses through the heart at a critical point in the conduction system.

Heartburn Burning sensation in the esophagus due to reflux of hydrochloric acid (HCl) from the stomach.

Heart-lung machine A device that pumps blood, functioning as a heart, and removes carbon dioxide from blood and oxygenates it, functioning as lungs; used during heart transplantation, open-heart surgery, and coronary artery bypass grafting.

Heart murmur (MER-mer) An abnormal sound that consists of a flow noise that is heard before the normal lubb-dupp or that may mask normal heart sounds.

Heat exhaustion Condition characterized by cool, clammy skin, profuse perspiration, and fluid and electrolyte (especially salt) loss that results in muscle cramps, dizziness, vomiting, and fainting. Also called **heat prostration.**

Heatstroke Condition produced when the body cannot easily lose heat and characterized by reduced perspiration and elevated body temperature. Also called **sunstroke.**

Heimlich maneuver *See Abdominal thrust maneuver.*

Hematocrit (hē-MAT-ō-krit) *(Hct)* The percentage of blood made up of red blood cells. Usually calculated by centrifuging a blood sample in a graduated tube and then reading off the volume of red blood cells and total blood.

Hematology (hē'-ma-TOL-ō-jē) The study of blood.

Hematoma (hē'-ma-TŌ-ma) A tumor or swelling filled with blood.

Hematopoiesis (hem'-a-tō-poy-Ē-sis) Blood cell production occurring in the red marrow of bones. Also called **hemopoiesis** (hē-mō-poy-Ē-sis).

Hematuria (hē'-ma-TOOR-ē-a) Blood in the urine.

Hemiballismus (hem'-i-ba-LIZ-mus) Violent muscular restlessness of half of the body, especially of the upper extremity.

Hemiplegia (hem-i-PLĒ-jē-a) Paralysis of the upper extremity, trunk, and lower extremity on one side of the body.

Hemocytoblast (hē'-mō-SĪ-tō-blast) Immature stem cell in bone marrow that develops along different lines into all the different mature blood cells.

Hemodialysis (hē'-mō-dī-AL-i-sis) Filtering of the blood by means of an artificial device so that certain substances are removed from the blood as a result of the difference in rates of their diffusion through a selectively permeable membrane while the blood is being circulated outside the body.

Hemodynamics (hē-mō-dī-NAM-iks) The study of factors and forces that govern the flow of blood through blood vessels.

Hemoglobin (hē'-mō-GLŌ-bin) *(Hb)* A substance in erythrocytes (red blood cells) consisting of the protein globin and the iron-containing red pigment heme and constituting about 33 percent of the cell volume; involved in the transport of oxygen and carbon dioxide.

Hemolysis (hē-MOL-i-sis) The escape of hemoglobin from the interior of the red blood cells into the surrounding medium; results from disruption of the integrity of the cell membrane by toxins or drugs, freezing or thawing, or hypotonic solutions.

Hemolytic disease of the newborn A hemolytic anemia of a newborn child that results from the destruction of the infant's red blood cells by antibodies produced by the mother; usually the antibodies are due to an RH blood type incompatibility. Also called **erythroblastosis fetalis** (e-rith'-rō-blas-TŌ-sis fe-TAL-is).

Hemophilia (hē'-mō-FĒL-ē-a) A hereditary blood disorder where there is a deficient production of certain factors involved in blood clotting, resulting in excessive bleeding into joints, deep tissues, and elsewhere.

Hemoptysis (hē-MOP-ti-sis) Spitting of blood from the respiratory tract.

Hemorrhage (HEM-or-rij) Bleeding; the escape of blood from blood vessels, especially when it is profuse.

Hemorrhoids (HEM-ō-royds) Dilated or varicosed blood vessels (usually veins) in the anal region. Also called **piles.**

Hemostasis (hē-MŌS-tā-sis) The stoppage of bleeding.

Hemostat (HĒ-mō-stat) An agent or instrument used to prevent the flow or escape of blood.

Hepatic (he-PAT-ik) Refers to the liver.

Hepatic duct A duct that receives bile from the bile capillaries. Small hepatic ducts merge to form the larger right and left hepatic ducts that unite to leave the liver as the common hepatic duct.

Hepatic portal circulation The flow of blood from the gastrointestinal organs to the liver before returning to the heart.

Hepatitis (hep-a-TĪ-tis) Inflammation of the liver due to a virus, drugs, and chemicals.

Hepatopancreatic (hep'-a-tō-pan'-krē-A-tik) **ampulla** A small, raised area in the duodenum where the combined common bile duct and main pancreatic duct empty into the duodenum. Also called the **ampulla of Vater** (VA-ter).

Hering-Breuer reflex *See **Inflation reflex.***

Hernia (HER-nē-a) The protrusion or projection of an organ or part of an organ through a membrane or cavity wall, usually the abdominal cavity.

Herniated (her'-nē-Ā-ted) **disc** A rupture of an intervertebral disc so that the nucleus pulposus protrudes into the vertebral cavity. Also called a **slipped disc.**

Heterozygous (he-ter-ō-ZĪ-gus) Possessing a pair of different genes on homologous chromosomes for a particular hereditary characteristic.

High altitude sickness Disorder caused by decreased levels of alveolar PO_2 as altitude increases and characterized by headache, fatigue, insomnia, shortness of breath, nausea, and dizziness. Also called **acute mountain sickness.**

Hilus (HĪ-lus) An area, depression, or pit where blood vessels and nerves enter or leave an organ. Also called a **hilum.**

Hinge joint A synovial joint in which a convex surface of one bone fits into a concave surface of another bone, such as the elbow, knee, ankle, and interphalangeal joints. Also called a **ginglymus** (JIN-gli-mus) **joint.**

Hirsutism (HER-soot-izm) An excessive growth of hair in females

and children, with a distribution similar to that in adult males, due to the conversion of vellus hairs into large terminal hairs in response to higher-than-normal levels of androgens.

Histamine (HISS-ta-mēn) Substance found in many cells, especially mast cells, basophils, and platelets, released when the cells are injured; results in vasodilation, increased permeability of blood vessels, and bronchiole constriction.

Histology (hiss-TOL-ō-jē) Microscopic study of the structure of tissues.

Hives (HĪVZ) Condition of the skin marked by reddened elevated patches that are often itchy; may be caused by infections, trauma, medications, emotional stress, food additives, and certain foods.

Hodgkin's disease (HD) A malignant disorder, usually arising in lymph nodes.

Holocrine (HŌL-ō-krin) **gland** A type of gland in which the entire secreting cell, along with its accumulated secretions, makes up the secretory product of the gland, as in the sebaceous (oil) glands.

Holter monitor Electrocardiograph worn by a person while going about everyday routines.

Homeostasis (hō'-mē-ō-STĀ-sis) The condition in which the body's internal environment remains relatively constant, within physiological limits.

Homologous (hō-MOL-ō-gus) Correspondence of two organs in structure, position, and origin.

Homologous chromosomes Two chromosomes that belong to a pair. Also called **homologues.**

Horizontal plane A plane that runs parallel to the ground and divides the body or organs into superior and inferior portions. Also called a **transverse plane.**

Hormone (HOR-mōn) A secretion of endocrine tissue that alters the physiological activity of target cells of the body.

Horn Principal area of gray matter in the spinal cord.

Human chorionic gonadotropin (hCG) (kō-rē-ON-ik gō-nad-ō-TRŌ-pin) A hormone produced by the developing placenta that maintains the corpus luteum.

Human chorionic somatomammotropin (sō-mat-ō-mam-ō-TRŌ-pin) **(hCS)** A hormone produced by the chorion of the placenta that may stimulate breast tissue for lactation, enhance body growth, and regulate metabolism.

Human growth hormone (hGH) Hormone secreted by the adenohypophysis (anterior lobe) of the pituitary that brings about growth of body tissues, especially skeletal and muscular. Also known as **somatotropin** and **somatotropic hormone (STH).**

Human leucocyte associated (HLA) antigens Surface proteins on white blood cells and other nucleated cells that are unique for each person (except for identical twins) and are used to type tissues and help prevent rejection.

Humoral (YOO-mor-al) **immunity** That component of immunity in which lymphocytes (B cells) develop into plasma cells that produce antibodies that destroy antigens. Also called **antibody-mediated immunity.**

Hunger center A cluster of neurons in the lateral nuclei of the hypothalamus that, when stimulated, brings about feeding.

Hyaluronic (hī'-a-loo-RON-ik) **acid** A viscous, amorphous extracellular material that binds cells together, lubricates joints, and maintains the shape of the eyeballs.

Hydrocephalus (hī-drō-SEF-a-lus) Abnormal accumulation of cerebrospinal fluid on the brain.

Hydrophobia (hī'-drō-FŌ-bē-a) Rabies; a condition characterized by severe muscle spasms when attempting to drink water. Also, an abnormal fear of water.

Hymen (HĪ-men) A thin fold of vascularized mucous membrane at the vaginal orifice.

Hyperbaric oxygenation (hī'-per-BA-rik ok'-sē-je-NĀ-shun) (*HBO*) Use of pressure supplied by a hyperbaric chamber to cause more oxygen to dissolve in blood to treat patients infected with anaerobic bacteria (tetanus and gangrene bacteria). Also used to treat carbon monoxide poisoning, asphyxia, smoke inhalation, and certain heart disorders.

Hypercalcemia (hī'-per-kal-SĒ-mē-a) An excess of calcium in the blood.

Hypercapnia (hī'-per-KAP-nē-a) An abnormal increase in the amount of carbon dioxide in the blood.

Hyperemia (hī'-per-Ē-mē-a) An excess of blood in an area or part of the body.

Hyperextension (hī'-per-ek-STEN-shun) Continuation of extension beyond the anatomical position, as in bending the head backward.

Hyperglycemia (hī'-per-glī-SĒ-mē-a) An elevated blood sugar level.

Hypermetropia (hī'-per-mē-TRŌ-pē-a) A condition in which visual images are focused behind the retina with resulting defective vision of near objects; farsightedness.

Hyperphosphatemia (hī-per-fos'-fa-TĒ-mē-a) An abnormally high level of phosphate in the blood.

Hyperplasia (hī'-per-PLĀ-zē-a) An abnormal increase in the number of normal cells in a tissue or organ, increasing its size.

Hyperpolarizatin (hī'-per-PŌL-a-ri-zā'-shun) Increase in the internal negativity across a cell membrane, thus increasing the voltage and moving it farther away from the threshold value.

Hypersecretion (hī'-per-se-KRĒ-shun) Overactivity of glands resulting in excessive secretion.

Hypersensitivity (hī'-per-sen-si-TI-vi-tē) Overreaction to an allergen that results in pathological changes in tissues. Also called *allergy.*

Hypertension (hī'-per-TEN-shun) High blood pressure.

Hyperthermia (hī'-per-THERM-ē-a) An elevated body temperature.

Hypertonia (hī-per-TŌ-nē-a) Increased muscle tone that is expressed as spasticity or rigidity.

Hypertonic (hī'-per-TON-ik) Having an osmotic pressure greater than that of a solution with which it is compared.

Hypertrophy (hī-PER-trō-fē) An excessive enlargement or overgrowth of tissue without cell division.

Hyperventilation (hī'-per-ven-ti-LĀ-shun)) A rate of respiration higher than that required to maintain a normal level of plasma PCO_2.

Hypervitaminosis (hī'-per-vī'-ta-min-Ō-sis) An excess of one or more vitamins.

Hypocalcemia (hī'-pō-kal-SĒ-mē-a) A below normal level of calcium in the blood.

Hypochloremia (hī'-pō-klō-RĒ-mē-a) Deficiency of chloride in the blood.

Hypoglycemia (hī'-pō-glī-SĒ-mē-a) An abnormally low concentration of glucose in the blood; can result from excess insulin (injected or secreted).

Hypokalemia (hī'-pō-kā-LĒ-mē-a) Deficiency of potassium in the blood.

Hypomagnesemia (hī'-pō-mag'-ne-SĒ-mē-a) Deficiency of magnesium in the blood.

Hyponatremia (hī'-pō-na-TRĒ-mē-a) Deficiency of sodium in the blood.

Hyponychium (hī-pō-NIK-ē-um) Free edge of the fingernail.

Hypophosphatemia (hī-pō-fos'-fa-TĒ-mē-a) An abnormally low level of phosphate in the blood.

Hypophysis (hī-POF-i-sis) Pituitary gland.

Hypoplasia (hī-pō-PLĀ-zē-a) Defective development of tissue.

Hyposecretion (hī'-pō-se-KRĒ-shun) Underactivity of glands resulting in diminished secretion.

Hypospadias (hī'-pō-SPĀ-dē-as) A displaced urethral opening. In the male, the opening may be on the underside of the penis, at the penoscrotal junction, between the scrotal folds, or in the perineum. In the female, the urethra opens into the vagina.

Hypothalamic-hypophyseal (hī'-pō-thal-AM-ik hī'-po-FIZ-ē-al) *tract* A bundle of nerve processes made up of fibers that have their cell bodies in the hypothalamus but release their neurosecretions in the posterior pituitary gland or neurohypophysis.

Hypothalamus (hī'-pō-THAL-a-mus) A portion of the diencephalon, lying beneath the thalamus and forming the floor and part of the wall of the third ventricle.

Hypothermia (hī-pō-THER-mē-a) Lowering of body temperature below 35°C (95°F); in surgical procedures, it refers to deliberate cooling of the body to slow down metabolism and reduce oxygen needs of tissues.

Hypotonia (hī'-pō-TŌ-nē-a) Decreased or lost muscle tone in which muscles appear flaccid.

Hypotonic (hī'-pō-TON-ik) Having an osmotic pressure lower than than of a solution with which it is compared.

Hypoventilation (hī-pō-ven-ti-LĀ-shun) A rate of respiration lower than that required to maintain a normal level of plasma PCO_2.

Hypoxia (hī-POKS-ē-a) Lack of adequate oxygen at the tissue level.

Hysterectomy (his-te-REK-tō-mē) The surgical removal of the uterus.

Ileocecal (il'-ē-ō-SĒ-kal) *sphincter* A fold of mucous membrane that guards the opening from the ileum into the large intestine. Also called the *ileocecal valve.*

Ileum (IL-ē-um) The terminal portion of the small intestine.

Immunity (i-MYOON-i-tē) The state of being resistant to injury, particularly by poisons, foreign proteins, and invading parasites, due to the presence of antibodies.

Immunogenicity (im-yoo-nō-jen-IS-it-ē) Ability of an antigen to stimulate antibody production.

Immunoglobulin (im-yoo-nō-GLOB-yoo-lin) (*IG*) An antibody synthesized by plasma cells derived from B lymphocytes in response to the introduction of antigen. Immunoglobulins are divided into five kinds (IgG, IgM, IgA, IgD, IgE) based primarily on the larger protein component present in the immunoglobulin.

Immunology (im'-yoo-NOL-ō-jē) The branch of science that

deals with the responses of the body when challenged by antigens.

Immunosuppression (im'-yoo-nō-su-PRESH-un) Inhibition of the immune response.

Immunotherapy (im-yoo-nō-THER-a-pē) Attempt to induce the immune system to mount an attack against cancer cells.

Impetigo (im'-pe-TĪ-go) A contagious skin disorder characterized by pustular eruptions.

Implantation (im-plan-TĀ-shun) The insertion of a tissue or a part into the body. The attachment of the blastocyst to the lining of the uterus 7–8 days after fertilization.

Impotence (IM-pō-tens) Weakness; inability to copulate; failure to maintain an erection long enough for sexual intercourse.

Incontinence (in-KON-ti-nens) Inability to retain urine, semen, or feces, through loss of sphincter control.

Infant respiratory distress syndrome (RDS) A disease of newborn infants, especially premature ones, in which insufficient amounts of surfactants are produced and breathing is labored. Also called ***hyaline*** (HĪ-a-lin) ***membrane disease*** (***HMD***).

Infarction (in-FARK-shun) The presence of a localized area of necrotic tissue, produced by inadequate oxygenation of the tissue.

Infection (in-FEK-shun) Invasion and multiplication of microorganisms in body tissues, which may be inapparent or characterized by cellular injury.

Infectious mononucleosis (mon-ō-nook'-lē-Ō-sis) (***IM***) Contagious disease caused by the Epstein-Barr virus (EBV) and characterized by an elevated mononucleocyte and lymphocyte count, fever, sore throat, stiff neck, cough, and malaise.

Inferior (in-FĒR-ē-or) Away from the head or toward the lower part of a structure. Also called ***caudad*** (KAW-dad).

Inferior vena cava (VĒ-na CĀ-va) (***IVC***) Large vein that collects blood from parts of the body inferior to the hear and returns it to the right atrium.

Infertility Inability to conceive or to cause conception. Also called ***sterility.***

Inflammation (in'-fla-MĀ-shun) Localized, protective response to tissue injury designed to destroy, dilute, or wall off the infecting agent or injured tissue; characterized by redness, pain, heat, swelling, and sometimes loss of function.

Inflammatory bowel (in-FLAM-a-tō'-rē BOW-el) ***disease*** Disorder that exists in two forms: (1) Crohn's disease (inflammation of the gastrointestinal tract, especially the distal ileum and proximal colon, in which the inflammation may extend from the mucosa through the serosa); and (2) ulcerative colitis (inflammation of the mucosa of the gastrointestinal tract, usually limited to the large intestine and usually accompanied by rectal bleeding).

Inflation reflex Reflex that prevents overinflation of the lungs. Also called ***Hering-Breuer reflex.***

Infundibulum (in'-fun-DIB-yoo-lum) The stalklike structure that attaches the pituitary gland (hypophysis) to the hypothalamus of the brain. The funnel-shaped, open, distal end of the uterine (Fallopian) tube.

Ingestion (in-JES-chun) The taking in of food, liquids, or drugs, by mouth.

Inheritance The acquisition of body characteristics and qualities by transmission of genetic information from parents to offspring.

Inhibin A male sex hormone secreted by sustentacular (Sertoli) cells that inhibits FSH release by the adenohypophysis (anterior pituitary) and thus spermatogenesis.

Inner cell mass A region of cells of a blastocyst that differentiates into the three primary germ layers—ectoderm, mesoderm, and endoderm—from which all tissues and organs develop; also called an ***embryoblast.***

Inorganic (in'-or-GAN-ik) ***compound*** Compound that usually lacks carbon, usually small, and contains ionic bonds. Examples include water and many acids, bases, and salts.

Insertion (in-SER-shun) The manner or place of attachment of a muscle to the bone that it moves.

Insomnia (in-SOM-nē-a) Difficulty in falling asleep and, usually, frequent awakening.

Inspiration (in-spi-RĀ-shun) The act of drawing air into the lungs.

Inspiratory (in-SPĪ-ra-tō-rē) ***capacity*** The total inspiratory ability of the lungs; the sun of tidal volume and inspiratory reserve; about 3,600 ml.

Inspiratory reserve volume The volume of air in excess of tidal volume that can be inhaled by forced inspiration; about 3,100 ml.

Insula (IN-su-la) A triangular area of cerebral cortex that lies deep within the lateral cerebral fissure, under the parietal, frontal, and temporal lobes, and cannot be seen in an external view of the brain. Also called the ***island*** or ***isle of Reil*** (RĪL).

Insulin (IN-su-lin) A hormone produced by the beta cells of the pancreas that decreases the blood glucose level.

Integumentary (in-teg'-yoo-MEN-tar-ē) Relating to the skin.

Intercalated (in-TER-ka-lāt-ed) ***disc*** An irregular transverse thickening of sarcolemma that contains desmosomes that hold cardiac muscle fibers (cells) together and gap junctions that aid in conduction of muscle action potentials.

Intercostal (in'-ter-KOS-tal) ***nerve*** A nerve supplying a muscle located between the ribs.

Interferon (in'-ter-FĒR-on) (***IFN***) Three principal types of protein (alpha, beta, gamma) naturally produced by virus-infected host cells that induce uninfected cells to synthesize antiviral proteins (AVPs) that inhibit intracellular viral replication in uninfected host cells; artificially synthesized through recombinant DNA techniques.

Intermediate Between two structures, one of which is medial and one of which is lateral.

Internal Away from the surface of the body.

Internal capsule A tract of projection fibers connecting various parts of the cerebral cortex and lying between the thalamus and the caudate and lentiform nuclei of the basal ganglia.

Internal ear The inner ear or labyrinth, lying inside the temporal bone, containing the organs of hearing and balance.

Internal nares (NA-rēz) The two openings posterior to the nasal cavities opening into the nasopharynx. Also called the ***choanae*** (kō-A-nē).

Internal respiration The exchange of respiratory gasses between blood and body cells.

Interphase (IN-ter-fāz) The period during its life cycle when a cell is carrying on every life process except division; the stage between two mitotic divisions. Also called ***metabolic phase.***

Interstitial cell of Leydig *See* ***Interstitial endocrinocyte.***

Interstitial (in'-ter-STISH-al) **endocrinocyte** A cell located in the connective tissue between semiinferous tubules in a mature testis that secretes testosterone. Also called an **interstitial cell of Leydig** (LĪ-dig).

Interstitial (in'-ter-STISH-al) **fluid** The portion of extracellular fluid that fills the microscopic spaces between the cells of tissues; the internal environment of the body. Also called **intercellular** or **tissue fluid.**

Intervertebral (in'-ter-VER-te-bral) **disc** A pad of fibrocartilage located between the bodies of two vertebrae.

Intestinal gland Simple tubular gland that opens onto the surface of the intestinal mucosa and secretes digestive enzymes. Also called a **crypt of Lieberkühn** (LĒ-ber-kyoon).

Intracellular (in'-tra-SEL-yoo-lar) **fluid** (**ICF**) Fluid located within cells.

Intramembranous ossification (in'-tra-MEM-bra-nus os'-i'-fi-KĀ-shun) The method of bone formation in which the bone is formed directly in membranous tissue.

Intraocular (in-tra-OC-yoo-lar) **pressure** (**IOP**) Pressure in the eyeball, produced mainly by aqueous humor.

Intrapleural pressure Air pressure between the two pleural layers of the lungs, usually subatmospheric. Also called **intrathoracic pressure.**

Intrapulmonic pressure Air pressure within the lungs. Also called **intraalveolar pressure.**

Intrauterine device (IUD) A small metal or plastic object inserted into the uterus for the purpose of preventing pregnancy.

Intrinsic clotting pathway Sequence of reactions leading to blood clotting that is initiated by the release of a substance (tissue factor or thromboplastin) contained *within* blood itself or cells in direct contact with blood.

Intrinsic factor (**IF**) A glycoprotein synthesized and secreted by the parietal cells of the gastric mucosa that facilitates vitamin B_{12} absorption.

Intubation (in'-too-BĀ-shun) Insertion of a tube through the nose or mouth into the larynx and trachea for entrance of air or to dilate a stricture.

In utero (YOO-ter-ō) Within the uterus.

Invagination (in-vaj'-i-NĀ-shun) The pushing of the wall of a cavity into the cavity itself.

Inversion (in-VER-zhun) The movement of the sole inward at the ankle joint.

In vitro (VĒ-trō) Literally, in glass; outside the living body and in artificial environment such as a laboratory test tube.

In vivo (VĒ-vō) In the living body.

Ion (Ī-on) Any charged particle or group of particles; usually formed when a substance, such as a salt, dissolves and dissociates.

Ipsilateral (ip'-si-LAT-er-al) On the same side, affecting the same side of the body.

Iris The colored portion of the eyeball seen through the cornea that consists of circular and radial smooth muscle; the black hole in the center of the iris is the pupil.

Irritable bowel (IR-i-ta-bul BOW-el) **syndrome** (**IBS**) Disease of the entire gastrointestinal tract in which persons with the condition may react to stress by developing symptoms such as cramping and abdominal pain associated with alternating patterns of diarrhea and constipation. Excessive amounts of mucus may appear in the stools, and other symptoms include flatulence, nausea, and loss of appetitie. The condition is also known as **irritable colon** or **spastic colitis.**

Ischemia (is-KĒ-mē-a) A lack of sufficient blood to a part due to obstruction of circulation.

Island of Reil *See* **Insula.**

Islet of Langerhans *See* **Pancreatic islet.**

Isometric contraction A muscle contraction in which tension on the muscle increases, but there is only minimal muscle shortening so that no movement is produced.

Isotonic (ī'-sō-TON-ik) Having equal tension or tone. Having equal osmotic pressure between two different solutions or between two elements in a solution.

Isotonic contraction A muscle contraction in which tension remains constant, but the muscle shortens and pulls on another structure to produce movement.

Isotope (Ī-sō-tōpe') A chemical element that has the same atomic number as another but a different atomic weight. Radioactive isotopes change into other elements with the emission of certain radiations.

Isthmus (IS-mus) A narrow strip of tissue or narrow passage connecting two larger parts.

Jaundice (JAWN-dis) A condition characterized by yellowness of skin, white of eyes, mucous membranes, and body fluids because of a buildup of bilirubin.

Jejunum (jē-JOO-num) The middle portion of the small intestine.

Joint kinesthetic (kin-es-THET-ik) **receptor** A proprioceptive receptor located in a joint, stimulated by joint movement.

Juxtaglomerular (juks-ta-glō-MER-yoo-lar) **apparatus** (**JGA**) Consists of the macula densa (cells of the distal convoluted tubule adjacent to the afferent and efferent arteriole) and juxtaglomerular cells (modified cells of the afferent and sometimes efferent arteriole); secretes renin when blood pressure starts to fall.

Karyotype (KAR-ē-ō-tīp) An arrangement of chromosomes based on shape, size, and position of centromeres.

Keratin (KER-a-tin) An insoluble protein found in the hair, nails, and other keratinized tissues of the epidermis.

Keratinocyte (ker-A-tin'-ō-sīt) The most numerous of the epidermal cells that function in the production of keratin.

Keratohyalin (ker'-a-tō-HĪ-a-lin) A compound involved in the formation of keratin.

Keratosis (ker'-a-TŌ-sis) Formation of a hardened growth of tissue.

Ketone (KĒ-ton) **bodies** Substances produced primarily during excessive fat metabolism, such as acetone, acetoacetic acid, and β-hydroxybutyric acid.

Ketosis (kē-TŌ-sis) Abnormal condition marked by excessive production of ketone bodies.

Kidney (KID-nē) One of the paired reddish organs located in the lumbar region that regulates the composition and volume of blood and produces urine.

Kidney stone A concentration, usually consisting of calcium oxalate, uric acid, and calcium phosphate crystals, that may form in any portion of the urinary tract. Also called a **renal calculus.**

Kilocalorie (KIL-ō-kal'-ō-rē) (**kcal**) The amount of heat required to raise the temperature of 1,000 g of water 1°C; the unit used to express the heating value of foods and to measure metabolic rate.

Kinesiology (ki-nē'-sē-OL-ō-jē) The study of the movement of body parts.

Kinesthesia (kin-is-THĒ-szē-a) Ability to perceive extent, direction, or weight of movement; muscle sense.

Korotkoff (kō-ROT-kof) *sounds* The various sounds that are heard while taking blood pressure.

Krebs cycle A series of energy-yielding chemical reactions that occur in the matrix of mitochondria in which energy is transferred to carrier molecules for subsequent liberation and carbon dioxide is formed. Also called the *citric acid cycle* and *tricarboxylic acid (TCA) cycle*.

Kupffer's cell *See Stellate reticuloendothelial cell.*

Kyphosis (kī-FŌ-sis) An exaggeration of the thoracic curve of the vetebral column, resulting in a ''round-shouldered'' or hunchback appearance.

Labial frenulum (LĀ-bē-al FREN-yoo-lum) A medial fold of mucous membrane between the inner surface of the lip and the gums.

Labia majora (LĀ-bē-a ma-JO-ra) Two longitudinal folds of skin extending downward and backward from the mons pubis of the female.

Labia minora (min-OR-a) Two small folds of mucous membrane lying medial to the labia majora of the female.

Labium (LĀ-bē-um) A lip. A liplike structure. *Plural, labia* (LĀ-bē-a).

Labor The process by which the product of conception is expelled from the uterus through the vagina.

Labyrinth (LAB-i-rinth) Intricate communicating passageway, especially in the internal ear.

Labyrinthine (lab-i-RIN-thēn) *disease* Malfunction of the internal ear characterized by deafness, tinnitus, vertigo, nausea, and vomiting.

Laceration (las'-er-Ā-shun) Wound or irregular area of the skin.

Lacrimal (LAK-ri-mal) Pertaining to tears.

Lacrimal (LAK-ri-mal) *canal* A duct, one on each eyelid, commencing at the punctum at the medial margin of an eyelid and conveying tears medially into the nasolacrimal sac.

Lacrimal gland Secretory cells located at the superior anterolateral portion of each orbit that secrete tears into excretory ducts that open onto the surface of the conjunctiva.

Lacrimal sac The superior expanded portion of the nasolacrimal duct that receives the tears from a lacrimal canal.

Lactation (lak-TĀ-shun) The secretion and ejection of milk by the mammary glands.

Lacteal (LAK-tē-al) One of many intestinal lymphatic vessels in villi that absorb fat from digested food.

Lacuna (la-KOO-na) A small, hollow space, such as that found in bones in which the osteoblasts lie. *Plural, lacunae* (la-KOO-nē).

Lambdoidal (lam-DOY-dal) *suture* The line of union in the skull between the parietal bones and the occipital bone; sometimes contains sutural (Wormian) bones.

Lamellae (la-MEL-ē) Concentric rings found in compact bone.

Lamellated corpuscle Oval pressure receptor located in subcutaneous tissue and consisting of concentric layers of connective tissue wrapped around an afferent nerve fiber. Also called a *Pacinian* (pa-SIN-ē-an) *corpuscle.*

Lamina propria (PRO-prē-a) The connective tissue layer of a mucous membrane.

Lanugo (lan-YOO-gō) Fine downy hairs that cover the fetus.

Large intestine The portion of the gastrointestinal tract extending from the ileum of the small intestine to the anus, divided structurally into the cecum, colon, rectum, and anal canal.

Laryngitis (la-rin-JĪ-tis) Inflammation of the mucous membrane lining the larynx.

Laryngopharynx (la-rin-gō-FAR-inks) The inferior portion of the pharynx, extending downward from the level of the hyoid bone to divide posteriorly into the esophagus and anteriorly into the larynx.

Laryngoscope (la-RIN-gō-skōp) An instrument for examining the larynx.

Larynx (LAR-inks) The voice box, a short passageway that connects the pharynx with the trachea.

Lateral (LAT-er-al) Farther from the midline of the body or a structure.

Lateral ventricle (VEN-tri-kul) A cavity within a cerebral hemisphere that communicates with the lateral ventricle in the other cerebral hemisphere and with the third ventricle by way of the interventricular foramen.

Learning The ability to acquire knowledge or a skill through instruction or experience.

Leg The part of the lower extremity between the knee and the ankle.

Lens A transparent organ constructed of proteins (crystallins) lying posterior to the pupil and iris of the eyeball and anterior to the vitreous body.

Lesion (LĒ-zhun) Any localized, abnormal change in tissue formation.

Lesser omentum (ō-MEN-tum) A fold of the peritoneum that extends from the liver to the lesser curvature of the stomach and the commencement of the duodenum.

Lesser vestibular (ves-TIB-yoo-lar) *gland* One of the paired mucus-secreting glands that have ducts that open on either side of the urethral orifice in the vestibule of the female.

Lethargy (LETH-ar-jē) A condition of drowsiness or indifference.

Leucocyte (LOO-kō-sīt) A white blood cell. Also spelled *leukocyte.*

Leucocytosis (loo'-kō-sī-TŌ-sis) An increase in the number of white blood cells, characteristic of many infections and other disorders.

Leucopenia (loo-kō-PĒ-nē-a) A decrease of the number of white blood cells below 5,000/mm^3.

Leukemia (loo-KĒ-mē-a) A malignant disease of the blood-forming tissues characterized by either uncontrolled production and accumulation of immature leucocytes in which many cells fail to reach maturity (acute) or an accumulation of mature leucocytes in the blood because they do not die at the end of their normal life span (chronic).

Leukoplakia (loo-kō-PLĀ-kē-a) A disorder in which there are white patches in the mucous membranes of the tongue, gums, and cheeks.

Libido (li-BĒ-dō)   The sexual drive, conscious or unconscious.

Ligament (LIG-a-ment)   Dense, regularly arranged connective tissue that attaches bone to bone.

Limbic system   A portion of the forebrain, sometimes termed the visceral brain, concerned with various aspects of emotion and behavior, that includes the limbic lobe, dentate gyrus, amygdaloid body, septal nuclei, mammillary bodies, anterior thalamic nucleus, olfactory bulbs, and bundles of myelinated axons.

Lingual frenulum (LIN-gwal FREN-yoo-lum)   A fold of mucous membrane that connects the tongue to the floor of the mouth.

Lipase (LĪ-pās)   A fat-splitting enzyme.

Lipid   an organic compound composed of carbon, hydrogen, and oxygen that is usually insoluble in water, but soluble in alcohol, ether, and chloroform; examples include fats, phospholipids, steroids, and prostaglandins.

Lipid profile   Blood test that measures total cholesterol, high-density lipoprotein, low-density lipoprotein, and triglycerides, to assess risk for cardiovascular disease.

Lipogenesis (li-pō-GEN-e-sis)   The synthesis of lipids from glucose or amino acids by liver cells.

Lipoma (li-PŌ-ma)   A fatty tissue tumor, usually benign.

Lipoprotein (lip'-ō-PRŌ-tēn)   Protein containing lipid that is produced by the liver and combines with cholesterol and triglycerides to make it water-soluble for transportation by the cardiovascular system; high levels of low-density lipoproteins (LDL) are associated with increased risk of atherosclerosis, while high levels of high-density lipoproteins (HDL) are associated with decreased risk of atherosclerosis.

Lithotripsy (LITH-ō-trip'-sē)   A noninvasive procedure in which shock waves generated by a lithotriptor are used to pulverize kidney stones or gallstones.

Liver   Large gland under the diaphragm that occupies most of the right hypochondriac region and part of the epigastric region; functionally, it produces bile salts, heparin, and plasma proteins; converts one nutrient into another; detoxifies substances; stores glycogen, minerals, and vitamins; carries on phagocytosis of blood cells and bacteria; and helps activate vitamin D.

Lordosis (lor-DŌ-sis)   An exaggeration of the lumbar curve of the vertebral column.

Lou Gehrig's disease   *See Amyotrophic lateral sclerosis*

Lower extremity   The appendage attached at the pelvic (hip) girdle, consisting of the thigh, knee, leg, ankle, foot, and toes.

Lumbar (LUM-bar)   Region of the back and side between the ribs and pelvis; loin.

Lumbar plexus (PLEK-sus)   A network formed by the anterior branches of spinal nerves L1 through L4.

Lumen (LOO-men)   The space within an artery, vein, intestine, or a tube.

Lung   One of the two main organs of respiration, lying on either side of the heart in the thoracic cavity.

Lunula (LOO-nyoo-la)   The moon-shaped white area at the base of a nail.

Luteinizing (LOO-tē-in'-īz-ing) *hormone* (*LH*)   A hormone secreted by the adenohypophysis (anterior lobe) of the pituitary gland that stimulates ovulation, progesterone secretion by the corpus luteum, and readies the mammary glands for milk secretion in females and stimulates testosterone secretion by the testes in males.

Lymph (limf)   Fluid confined in lymphatic vessels and flowing through the lymphatic system to be returned to the blood.

Lymphatic tissue   A specialized form of reticular tissue that contains large numbers of lymphocytes.

Lymphatic (lim-FAT-ik) *vessel*   A large vessel that collects lymph from lymph capillaries and converges with other lymphatic vessels to form the thoracic and right lymphatic ducts.

Lymph capillary   Blind-ended microscopic lymph vessel that begins in spaces between cells and converges with other lymph capillaries to form lymphatic vessels.

Lymph node   An oval or bean-shaped structure located along lymphatic vessels.

Lymphocyte (LIM-fō-sīt)   A type of white blood cell, found in lymph nodes, associated with the immune system.

Lymphokines (LIM-fō-kīns)   Powerful proteins secreted by T cells that endow T cells with their ability to assist in immunity.

Lysosome (LĪ-sō-sōm)   An organelle in the cytoplasm of a cell, enclosed by a single membrane and containing powerful digestive enzymes.

Lysozyme (LĪ-sō-zīm)   A bactericidal enzyme found in tears, saliva, and perspiration.

Macrophage (MAK-rō-fāj)   Phagocytic cell derived from a monocyte. May be fixed or wandering.

Macula (MAK-yoo-la)   A discolored spot or a colored area. A small, thickened region on the wall of the utricle and saccule that serves as a receptor for static equilibrium.

Macula lutea (LOO-tē-a)   The yellow spot in the center of the retina.

Magnetic resonance imaging (*MRI*)   A diagnostic procedure that focuses on the nuclei of atoms of a single element in a tissue, usually hydrogen, to determine if they behave normally in the presence of an external magnetic force; used to indicate the biochemical activity of a tissue. Formerly called *nuclear magnetic resonance* (*NMR*).

Malaise (ma-LĀYZ)   Discomfort, uneasiness, and indisposition, often indicative of infection.

Malignant (ma-LIG-nant)   Referring to diseases that tend to become worse and cause death; especially the invasion and spreading of cancer.

Malignant melanoma (mel'-a-NŌ-ma)   A usually dark, malignant tumor of the skin containing melanin.

Malnutrition (mal'-nu-TRISH-un)   State of bad or poor nutrition that may be due to inadequate food intake, imbalance of nutrients, malabsorption of nutrients, improper distribution of nutrients, increased nutrient requirements, increased nutrient losses, or overnutrition.

Mammary (MAM-ar-ē) *gland*   Modified sudoriferous (sweat) gland of the female that secretes milk for the nourishment of the young.

Mammillary (MAM-i-ler-ē) *bodies*   Two small rounded bodies posterior to the tuber cinereum that are involved in reflexes related to the sense of smell.

Mammography (mam-OG-ra-fē)   Procedure for imaging the breasts (xeromammography or film-screen mammography) to evaluate for breast disease or screen for breast cancer.

Marfan (MAR-fan) *syndrome*   Inherited disorder that results in abnormalities of connective tissue, especially in the skeleton, eyes, and cardiovascular system.

Margination (mar'-ji-NĀ-shun)   Accumulation and adhesion of neutrophils to the endothelium of blood vessels at the site of injury during the early stages of inflammation.

Marrow (MAR-ō)   Soft, spongelike material in the cavities of bone. Red marrow produces blood cells; yellow marrow, formed mainly of fatty tissue, has no blood-producing function.

Mast cell   A cell found in loose connective tissue along blood vessels that produces heparin, an anticoagulant. The name given to a basophil after it has left the bloodstream and entered the tissues.

Mastectomy (mas-TEK-tō-mē)   Surgical removal of breast tissue.

Mastication (mas'-ti-KĀ-shun)   Chewing.

Maximal oxygen uptake   Maximum rate of oxygen consumption during aerobic catabolism of pyruvic acid that is determined by age, sex, and body size.

Meatus (mē-Ā-tus)   A passage or opening, especially the external portion of a canal.

Mechanoreceptor (me-KAN-ō-rē'-sep-tor)   Receptor that detects mechanical deformation of the receptor itself or adjacent cells; stimuli so detected include those related to touch, pressure, vibration, proprioception, hearing, equilibrium, and blood pressure.

Medial (MĒ-dē-al)   Nearer the midline of the body or a structure.

Medial lemniscus (lem-NIS-kus)   A flat band of myelinated nerve fibers extending through the medulla, pons, and midbrain and terminating in the thalamus on the same side. Sensory neurons in this tract transmit impulses for proprioception, fine touch, pressure, and vibration sensations.

Median plane   A vertical plane dividing the body into right and left halves. Situated in the middle.

Mediastinum (mē'-dē-as-TĪ-num)   A broad, median partition, actually a mass of tissue found between the pleurae of the lungs that extends from the sternum to the vertebral column.

Medulla (me-DULL-la)   An inner layer of an organ, such as the medulla of the kidneys.

Medulla oblongata (ob'-long-GA-ta)   The most inferior part of the brain stem.

Medullary (MED-yoo-lar'-ē) *cavity*   The space within the diaphysis of a bone that contains yellow marrow. Also called the **marrow cavity.**

Medullary rhythmicity (rith-MIS-i-tē) *area*   Portion of the respiratory center in the medulla that controls the basic rhythm of respiration.

Meibomian gland   See Tarsal gland.

Meiosis (mē-Ō-sis)   A type of cell division restricted to sex-cell production involving two successive nuclear divisions that result in daughter cells with the haploid (*n*) number of chromosomes.

Meissner's corpuscle   See Corpuscle of touch.

Melanin (MEL-a-nin)   A dark black, brown, or yellow pigment found in some parts of the body such as the skin.

Melanocyte (MEL-a-nō-sīt')   A pigmented cell located between or beneath cells of the deepest layer of the epidermis that synthesizes melanin.

Melanocyte-stimulating hormone (MSH)   A hormone secreted by the adenohypophysis (anterior lobe) of the pituitary gland that stimulates the dispersion of melanin granules in melanocytes

in amphibians; continued administration produces darkening of skin in humans.

Melatonin (mel-a-TŌN-in)   A hormone secreted by the pineal gland that may inhibit reproductive activities.

Membrane   A thin, flexible sheet of tissue composed of an epithelial layer and an underlying connective tissue layer, as in an epithelial membrane, or of loose connective tissue only, as in a synovial membrane.

Membranous labyrinth (mem-BRA-nus LAB-i-rinth)   The portion of the labyrinth of the inner ear that is located inside the bony labyrinth and separated from it by the perilymph; made up of the membranous semicircular canals, the saccule and utricle, and the cochlear duct.

Memory   The ability to recall thoughts; commonly classified as short term (activated) and long term.

Menarche (me-NAR-kē)   Beginning of the menstrual function.

Ménière's (men-YAIRZ) *syndrome*   A type of labyrinthine disease characterized by fluctuating loss of hearing, vertigo, and tinnitus due to an increased amount of endolymph that enlarges the labyrinth.

Meninges (me-NIN-jēz)   Three membranes covering the brain and spinal cord, called the dura mater, arachnoid, and pia mater. *Singular,* **meninx** (MEN-inks).

Meningitis (men-in-JĪ-tis)   Inflammation of the meninges, most commonly the pia mater and arachnoid.

Menopause (MEN-ō-pawz)   The termination of the menstrual cycles.

Menstrual (MEN-stroo-al) *cycle*   A series of changes in the endometrium of a nonpregnant female that prepares the lining of the uterus to receive a fertilized ovum.

Menstruation (men'-stroo-Ā-shun)   Periodic discharge of blood, tissue fluid, mucus, and epithelial cells that usually lasts for 5 days; caused by a sudden reduction in estrogens and progesterone. Also called the **menstrual phase** or **menses.**

Merocrine (MER-ō-krin) *gland*   A secretory cell that remains intact throughout the process of formation and the discharge of the secretory product, as in the salivary and pancreatic glands.

Mesenchyme (MEZ-en-kīm)   An embryonic connective tissue from which all other connective tissues arise.

Mesentery (MEZ-en-ter'-ē)   A fold of peritoneum attaching the small intestine to the posterior abdominal wall.

Mesocolon (mez'-ō-KŌ-lon)   A fold of peritoneum attaching the colon to the posterior abdominal wall.

Mesoderm   The middle of the three primary germ layers that gives rise to connective tissues, blood and blood vessels, and muscles.

Mesothelium (mez'-ō-THĒ-lē-um)   The layer of simple squamous epithelium that lines serous cavities.

Mesovarium (mez'-ō-VAR-ē-um)   A short fold of peritoneum that attaches an ovary to the broad ligament of the uterus.

Metabolism (me-TAB-ō-lizm)   The sum of all the biochemical reactions that occur within an organism, including the synthetic (anabolic) reactions and decomposition (catabolic) reactions.

Metacarpus (met'-a-KAR-pus)   A collective term for the five bones that make up the palm of the hand.

Metaphase (MET-a-phāz)   The second stage of mitosis in which chromatid pairs line up on the equatorial plane of the cell.

Metaphysis (me-TAF-i-sis)   Growing portion of a bone.

Metaplasia (met'-a-PLĀ-zē-a) The transformation of one cell into another.

Metastasis (me-TAS-ta-sis) The spread of cancer to surrounding tissues (local) or to other body sites (distant).

Metatarsus (met'-a-TAR-sus) A collective term for the five bones located in the foot between the tarsals and the phalanges.

Micelle (mī-SEL) A spherical aggregate of bile salts that dissolves fatty acids and monoglycerides so that they can be transported into small intestinal epithelial cells.

Microcephalus (mi-krō-SEF-a-lus) An abnormally small head; premature closing of the anterior fontanel so that the brain has insufficient room for growth, resulting in mental retardation.

Microfilament (mī-krō-FIL-a-ment) Rodlike cytoplasmic structure about 6 nm in diameter; comprises contractile units in muscle fibers (cells) and provides support, shape, and movement in nonmuscle cells.

Microglia (mī-krō-GLĒ-a) Neuroglial cells that carry on phagocytosis. Also called *brain macrophages* (MAK-rō-fāj-ez).

Microphage (MĪK-rō-fāj) Granular leucocyte that carries on phagocytosis, especially neutrophils and eosinophils.

Microtrabecular (mī-krō-tra-BEK-yoo-lar) *lattice* (LAT-is) Collective term for microfilaments, microtubules, and intermediate filaments held together by microtrabeculae in cytoplasm.

Microtubule (mī-krō-TOOB-yool') Cylindrical cytoplasmic structure, ranging in diameter from 18 to 30 nm, consisting of the protein tubulin; provides support, structure, and transportation.

Microvilli (mī'-krō-VIL-ē) Microscopic, fingerlike projections of the cell membranes of small intestinal cells that increase surface area for absorption.

Micturition (mik'-too-RISH-un) The act of expelling urine from the urinary bladder. Also called *urination* (yoo-ri-NĀ-shun).

Midbrain The part of the brain between the pons and the diencephalon. Also called the *mesencephalon* (mes'-en-SEF-a-lon).

Middle ear A small, epithelial-lined cavity hollowed out of the temporal bone, separated from the external ear by the eardrum and from the internal ear by a thin bony partition containing the oval and round windows; extending across the middle ear are the three auditory ossicles. Also called the *tympanic* (tim-PAN-ik) *cavity.*

Midline An imaginary vertical line that divides the body into equal left and right sides.

Midsagittal plane A vertical plane through the midline of the body that divides the body or organs into *equal* right and left sides. Also called a *median plane.*

Milk let-down reflex Contraction of alveolar cells to force milk into ducts of mammary glands, stimulated by oxytocin (OT), which is released from the posterior pituitary in response to suckling action.

Mineral Inorganic, homogeneous solid substance that may perform a function vital to life; examples include calcium, sodium, potassium, iron, phosphorus, and chlorine.

Mineralocorticoids (min'-er-al-ō-KOR-ti-koyds) A group of hormones of the adrenal cortex.

Minimal volume The volume of air in the lungs even after the thoracic cavity has been opened forcing out some of the residual volume.

Minute volume of respiration (MVR) Total volume of air taken to the lungs per minute; about 6,000/ml.

Mitochondrion (mī'tō-KON-drē-on) A double-membraned organelle that plays a central role in the production of ATP; known as the ''powerhouse'' of the cell.

Mitosis (mī-TŌ-sis) The orderly division of the nucleus of a cell that ensures that each new daughter nucleus has the same number and kind of chromosomes as the original parent nucleus. The process includes the replication of chromosomes and the distribution of the two sets of chromosomes into two separate and equal nuclei.

Mitotic apparatus Collective term for continuous and chromosomal microtubules and centrioles; involved in cell division.

Mitotic spindle The combination of continuous and chromosomal microtubules, involved in chromosomal movement during mitosis.

Mitral (MĪ-tral) *stenosis* (ste-NŌ-sis) Narrowing of the mitral valve by scar formation or a congential defect.

Mitral (MĪ-tral) *valve prolapse* (PRŌ-laps) or *MVP* An inherited disorder in which a portion of a mitral valve is pushed back too far (prolapsed) during contraction due to expansion of the cusps and elongation of the chordae tendineae.

Mittelschmerz (MIT-el-shmerz) Abdominopelvic pain that supposedly indicates the release of a secondary oocyte from the ovary.

Modality (mō-DAL-i-tē) Any of the specific sensory entities, such as vision, smell, or taste.

Molecule (MOL-e-kyool) The chemical combination of two or more atoms.

Monoclonal antibody (MAb) Antibody produced by *in vitro* clones of B cells hybridized with cancerous cells.

Monocyte (MON-ō-sīt') A type of white blood cell characterized by agranular cytoplasm; the largest of the leucocytes.

Monosaturated fat A fat that contains one double covalent bond between its carbon atoms; it is not completely saturated with hydrogen atoms. Examples are olive and peanut oil.

Mons pubis (monz PYOO-bis) The rounded, fatty prominence over the symphysis pubis, covered by coarse pubic hair.

Morbid (MOR-bid) Diseased; pertaining to disease.

Morula (MOR-yoo-la) A solid mass of cells produced by successive cleavages of a fertilized ovum a few days after fertilization.

Motor area The region of the cerebral cortex that governs muscular movement, particularly the precentral gyrus of the frontal lobe.

Motor end plate Portion of the sarcolemma of a muscle fiber (cell) in close approximation with an axon terminal.

Motor unit A motor neuron together with the muscle fibers (cells) it stimulates.

Mucous (MYOO-kus) *cell* A unicellular gland that secretes mucus. Also called a *goblet cell.*

Mucous membrane A membrane that lines a body cavity that opens to the exterior. Also called the *mucosa* (myoo-KŌ-sa).

Mucus The thick fluid secretion of mucous glands and mucous membranes.

Multiple sclerosis (skler-Ō-sis) Progressive destruction of myelin sheaths of neurons in the central nervous system, short-circuiting conduction pathways.

Mumps Inflammation and enlargement of the parotid glands

accompanied by fever and extreme pain during swallowing.

Muscarinic (mus'-ka-RIN-ik) *receptor*   Receptor found on all effectors innervated by parasympathetic postganglionic axons and some effectors innervated by sympathetic postganglionic axons; so named because the actions of acetylcholine (ACh) on such receptors are similar to those produced by muscarine.

Muscle   An organ composed of one of three types of muscle tissue (skeletal, cardiac, or visceral), specialized for contraction to produce voluntary or involuntary movement of parts of the body.

Muscle action potential   A stimulating impulse that travels along a sarcolemma and then into transverse tubules; it is generated by acetylcholine from synaptic vesicles which alters permeability of the sarcolemma to sodium (Na^+) ions.

Muscle fatigue (fa-TĒG)   Inability of a muscle to maintain its strength of contraction or tension; may be related to insufficient oxygen, depletion of glycogen, and/or lactic acid buildup.

Muscle spindle   An encapsulated receptor in a skeletal muscle, consisting of specialized muscle fiber (cell) and nerve endings, stimulated by changes in length or tension of muscle fibers; a proprioceptor. Also called a *neuromuscular* (noo-rō-MUS-kyoo-lar) *spindle.*

Muscle tissue   A tissue specialized to produce motion in response to muscle action potentials by its qualities of contractility, extensibility, elasticity, and excitability.

Muscle tone   A sustained, partial contraction of portions of a skeletal muscle in response to activation of stretch receptors.

Muscular dystrophies (DIS-trō-fēz')   Inherited muscle-destroying diseases, characterized by degeneration of the individual muscle fibers (cells), which leads to progressive atrophy of the skeletal muscle.

Muscularis (MUS-kyoo-la'-ris)   A muscular layer (coat or tunic) of an organ.

Mutation (myoo-TĀ-shun)   Any change in the sequence of bases in the DNA molecule resulting in a permanent alteration in some inheritable characteristic.

Myasthenia (mī-as-THĒ-nē-a) *gravis*   Weakness of skeletal muscles caused by antibodies directed against acetylcholine receptors that inhibit muscle contraction.

Myelin (MĪ-e-lin) *sheath*   A white, phospholipid, segmented covering, formed by neurolemmocytes (Schwann cells), around the axons and dendrites of many peripheral neurons.

Myenteric plexus   A network of nerve fibers from both autonomic divisions located in the muscularis coat of the small intestine. Also called the *plexus of Auerbach* (OW-er-bak).

Myocardial infarction (mī'-ō-KAR-dē-al in-FARK-shun) *(MI)*   Gross necrosis of myocardial tissue due to interrupted blood supply. Also called a *heart attack.*

Myocardium (mī'-ō-KAR-dē-um)   The middle layer of the heart wall, made up of cardiac muscle, comprising the bulk of the heart, and lying between the epicardium and the endocardium.

Myofibril (mī'-ō-FĪ-bril)   A threadlike structure, running longitudinally through a muscle fiber (cell) consisting mainly of thick myofilaments (myosin) and thin myofilaments (actin).

Myoglobin (mī-ō-GLŌ-bin)   The oxygen-binding, iron-containing conjugated protein complex present in the sarcoplasm of muscle fibers (cells); contributes the red color to muscle.

Myogram (MĪ-ō-gram)   The record or tracing produced by the myograph, the apparatus that measures and records the effects of muscular contractions.

Myology (mī-OL-ō-jē)   The study of the muscles.

Myometrium (mī'-ō-MĒ-trē-um)   The smooth muscle layer of the uterus.

Myopia (mī-Ō-pē-a)   Defect in vision so that objects can be seen distinctly only when very close to the eyes; nearsightedness.

Myosin (MĪ-ō-sin)   The contractile protein that makes up the thick myofilaments of muscle fibers (cells)

Myotonia (mī-ō-TŌ-nē-a)   A continuous spasm of muscle; increased muscular irritability and tendency to contract, and less ability to relax

Myxedema (mix-e-DĒ-ma)   Condition caused by hypothyroidism during the adult years characterized by swelling of facial tissues.

Nail   A hard plate, composed largely of keratin, that develops from the epidermis of the skin to form a protective covering on the dorsal surface of the distal phalanges of the fingers and toes.

Nail matrix (MĀ-triks)   The part of the nail beneath the body and root from which the nail is produced.

Nasal (NĀ-zal) *cavity*   A mucosa-lined cavity on either side of the nasal septum that opens onto the face at an external naris and into the nasopharynx at an internal naris.

Nasal septum (SEP-tum)   A vertical partition composed of bone (perpendicular plate of ethmoid and vomer) and cartilage, covered with a mucous membrane, separating the nasal cavity into left and right sides.

Nasolacrimal (nā'-zō-LAK-ri-mal) *duct*   A canal that transports the lacrimal secretion (tears) from the nasolacrimal sac into the nose.

Nasopharynx (nā'-zō-FAR-inks)   The uppermost portion of the pharynx, lying posterior to the nose and extending down to the soft palate.

Nausea (NAW-sē-a)   Discomfort characterized by loss of appetite and sensation of impending vomiting.

Nebulization (neb'-yoo-li-ZĀ-shun)   Administration of medication to selected portions of the respiratory tract by droplets suspended in air.

Neck   The part of the body connecting the head and the trunk. A constricted portion of an organ such as the neck of the femur or uterus.

Necrosis (ne-KRŌ-sis)   Death of a cell or group of cells as a result of disease or injury.

Negative feedback   The principle governing most control systems; a mechanism of response in which a stimulus initiates actions that reverse or reduce the stimulus.

Neonatal (nē'-ō-NĀ-tal)   Pertaining to the first 4 weeks after birth.

Neoplasm (NĒ-ō-plazm)   A new growth that may be benign or malignant.

Nephritis (ne-FRĪT-is)   Inflammation of the kidney.

Nephron (NEF-ron)   The functional unit of the kidney.

Nephrotic (ne-FROT-ik) *syndrome*   A condition in which the endothelial-capsular membrane leaks, allowing large amounts of protein to escape into urine.

Nerve   A cordlike bundle of nerve fibers (axons and/or dendrites) and their associated connective tissue coursing together outside the central nervous system.

Nerve impulse A wave of negativity (depolarization) that self-propagates along the outside surface of the plasma membrane of a neuron; also called a **nerve action potential.**

Nervous tissue Tissue that initiates and transmits nerve impulses to coordinate homeostasis.

Neuralgia (noo-RAL-jē-a) Attacks of pain along the entire course or branch of a peripheral sensory nerve.

Neuritis (noo-RĪ-tis) Inflammation of a single nerve, two or more nerves in separate areas, or many nerves simultaneously.

Neuroeffector (noo-rō-e-FEK-tor) **junction** Collective term for neuromuscular and neuroglandular junctions.

Neurofibral (noo-rō-FĪ-bral) **node** A space, along a myelinated nerve fiber, between the individual neurolemmocytes (Schwann cells) that form the myelin sheath and the neurolemma. Also called **node of Ranvier** (ron-VĒ-ā).

Neurofibril (noo-rō-FĪ-bril) One of the delicate threads that forms a complicated network in the cytoplasm of the cell body and processes of a neuron.

Neuroglandular (noo-rō-GLAND-yoo-lar) **junction** Area of contact between a motor neuron and a gland.

Neuroglia (noo-RŌG-lē-a) Cells of the nervous system that are specialized to perform the functions of connective tissue. The neuroglia of the central nervous system are the astrocytes, oligodendrocytes, microglia, and ependyma; neuroglia of the peripheral nervous system include the neurolemmocytes (Schwann cells) and the ganglion satellite cells. Also called **glial** (GLĒ-al) **cells.**

Neurohypophysis (noo-rō-hī-POF-i-sis) The posterior lobe of the pituitary gland.

Neurolemma (noo-rō-LEM-ma) The peripheral, nucleated cytoplasmic layer of the neurolemmocyte (Schwann cell). Also called **sheath of Schwann** (SCHVON).

Neurolemmocyte A neuroglial cell of the peripheral nervous system that forms the myelin sheath and neurolemma of a nerve fiber by wrapping around a nerve fiber in a jelly-roll fashion. Also called a **Schwann** (SCHVON) **cell.**

Neurology (noo-ROL-ō-jē) The branch of science that deals with the normal functioning and disorders of the nervous system.

Neuromuscular (noo-rō-MUS-kyoo-lar) **junction** The area of contact between the axon terminal of a motor neuron and a portion of the sarcolemma of a muscle fiber (cell). Also called a **myoneural** (mi-o-NOO-ral) **junction.**

Neuron (NOO-ron) A nerve cell, consisting of a cell body, dendrites, and an axon.

Neuropeptide (noo-rō-PEP-tīd) Chain of 2 to about 40 amino acids that occurs naturally in the brain that acts primarily to modulate the response of or to a neurotransmitter. Examples are enkephalins and endorphins.

Neurosecretory (noo-rō-SĒC-re-tō-rē) **cell** A cell in a nucleus (paraventricular and supraoptic) in the hypothalamus that produces oxytocin (OT) or antidiuretic hormone (ADH), hormones stored in the neurohypophysis of the pituitary gland.

Neurosyphilis (noo-rō-SIF-i-lis) A form of the tertiary stage of syphilis in which various types of nervous tissue are attacked by bacteria and degenerate.

Neurotransmitter One of a variety of molecules synthesized within the nerve axon terminals, released into the synaptic cleft in response to a nerve impulse, and affecting the membrane potential of the postsynaptic neuron. Also called a **transmitter substance.**

Neutrophil (NOO-trō-fil) A type of white blood cell characterized by granular cytoplasm that stains as readily with acid or basic dyes.

Nicotinic (nik'-ō-TIN-ik) **receptor** Receptor found on both sympathetic and parasympathetic postganglionic neurons so named because the actions of acetylcholine (ACh) in such receptors are similar to those produced by nicotine.

Night blindness Poor or no vision in dim light or at night although good vision is present during bright illumination; frequently caused by a deficiency of vitamin A. Also referred to as **nyctalopia** (nik'-ta-LŌ-pē-a).

Nipple A pigmented, wrinkled projection on the surface of the mammary gland that is the location of the openings of the lactiferous ducts for milk release.

Nissl bodies See **Chromatophilic substance.**

Nociceptor (nō'-sē-SEP-tor) A free (naked) nerve ending that detects pain.

Node of Ranvier See **Neurofibral node.**

Nonessential amino acid An amino acid that can be synthesized by body cells through transamination, the transfer of an amino group from an amino acid to another substance.

Nonpigmented granular dendrocytes Two distinct cell types found in the epidermis, formerly known as **Langerhans cells** and **Granstein cells,** that differ in their sensitivity to damage by ultraviolent (UV) radiation and their functions in immunity.

Norepinephrine (nor'-ep-ē-NEF-rin) (**NE**) A hormone secreted by the adrenal medulla that produces actions similar to those that result from sympathetic stimulation. Also called **noradrenaline** (nor-a-DREN-a-lin).

Nuclear medicine The branch of medicine concerned with the use of radioisotopes in the diagnosis of disease and therapy.

Nuclease (NOO-klē-ās) An enzyme that breaks nucleotides into pentoses and nitrogenous bases; examples are ribonuclease and deoxyribonuclease.

Nucleic (noo-KLĒ-ic) **acid** An organic compound that is a long polymer of nucleotides, with each nucleotide containing a pentose sugar, a phosphate group, and one of four possible nitrogenous bases (adenine, cytosine, guanine, and thymine or uracil).

Nucleolus (noo-KLĒ-ō-lus) Nonmembranous spherical body within the nucleus composed of protein, DNA, and RNA that functions in the synthesis and storage of ribosomal RNA.

Nucleus (NOO-klē-us) A spherical or oval organelle of a cell that contains the hereditary factors of the cell, called genes. A cluster of unmyelinated nerve cell bodies in the central nervous system. The central portion of an atom made up of protons and neutrons.

Nucleus pulposus (pul-PŌ-sus) A soft, pulpy, highly elastic substance in the center of an intervertebral disc, a remnant of the notochord.

Nutrient A chemical substance in food that provides energy, forms new body components, or assists in the functioning of various body processes.

Nystagmus (nis-TAG-mus) Rapid, involuntary, rhythmic movement of the eyeballs; horizontal, rotary, or vertical.

Obesity (ō-BĒS-i-tē) Body weight 10–20 percent over a desirable standard as a result of excessive accumulation of fat. Types of obesity are hypertrophic (adult-onset) and hyperplastic (lifelong).

Obligatory water reabsorption The absorption of water from proximal convoluted tubules of nephrons as a function of osmosis.

Obstetrics (ob-STET-riks) The specialized branch of medicine that deals with pregnancy, labor, and the period of time immediately following delivery.

Occlusion (ō-KLOO-zhun) The act of closure or state of being closed.

Occult (o-KULT) Obscure or hidden from view, as for example, occult blood in stools or urine.

Olfactory (ōl-FAK-tō-rē) Pertaining to smell.

Olfactory bulb A mass of gray matter at the termination of an olfactory (I) nerve, lying beneath the frontal lobe of the cerebrum on either side of the crista galli of the ethmoid bone.

Olfactory cell A bipolar neuron with its cell body lying between supporting cells located in the mucous membrane lining the upper portion of each nasal cavity.

Olfactory tract A bundle of axons that extends from the olfactory bulb posteriorly to the olfactory portion of the cortex.

Oligodendrocyte (o-lig-ō-DEN-drō-sīt) A neuroglial cell that supports neurons and produces a phospholipid myelin sheath around axons of neurons of the central nervous system.

Oliguria (ol'-i-GYOO-rē-a) Daily urinary output usually less than 250 ml.

Oncogene (ONG-kō-jēn) Gene that has the ability to transform a normal cell into a cancerous cell.

Oncology (ong-KOL-ō-jē) The study of tumors.

Oogenesis (ō'-ō-JEN-e-sis) Formation and development of the ovum.

Oophorectomy (ō'-of-ō-REK-tō-mē) The surgical removal of the ovaries.

Ophthalmologist (of'-thal-MOL-ō-jist) A physician who specializes in the diagnosis and treatment of eye disorders with drugs, surgery, and corrective lenses.

Ophthalmology (of'-thal-MOL-ō-jē) The study of the structure, function, and diseases of the eye.

Ophthalmoscopy (of'-thal-MOS-kō-pē) Examination of the interior fundus of the eyeball to detect retinal changes associated with hypertension, diabetes mellitus, atherosclerosis, and increased intracranial pressure.

Opsonization (op-sō-ni-ZĀ-shun) The action of some antibodies that renders bacteria and other foreign cells more susceptible to phagocytosis. Also called *immune adherence.*

Optic (OP-tik) Refers to the eye, vision, or properties of light.

Optic chiasma (kī-AZ-ma) A crossing point of the optic (II) nerves, anterior to the pituitary gland.

Optic disc A small area of the retina containing openings through which the fibers of the ganglion neurons emerge as the optic (II) nerve. Also called the *blind spot.*

Optician (op-TISH-an) A technician who fits, adjusts, and dispenses corrective lenses on prescription of an ophthalmologist or optometrist.

Optic tract A bundle of axons that transmits nerve impulses from the retina of the eye between the optic chiasma and the thalamus.

Optometrist (op-TOM-e-trist) Specialist with a doctorate degree in optometry who is licensed to examine and test the eyes and treat visual defects by prescribing corrective lenses.

Oral contraceptive (*OC*) A hormone compound, usually a high concentration of progesterone and a low concentration of estrogens, that is swallowed and prevents ovulation, and thus pregnancy. Also called **"the pill."**

Ora serrata (Ō-ra ser-RĀ-ta) The irregular margin of the retina lying internal and slightly posterior to the junction of the choroid and ciliary body.

Orbit (OR-bit) The bony, pyramid-shaped cavity of the skull that holds the eyeball.

Organ A structure composed of two or more different kinds of tissues with a specific function and usually a recognizable shape.

Organelle (or-gan-EL) A permanent structure within a cell with characteristic morphology that is specialized to serve a specific function in cellular activities.

Organic (or-GAN-ik) *compound* Compound that always contains carbon and hydrogen and the atoms are held together by covalent bonds. Examples include carbohydrates, lipids, protein, and nucleic acids (DNA and RNA).

Organism (OR-ga-nizm) A total living form; one individual.

Orgasm (OR-gazm) Sensory and motor events involved in ejaculation for the male and involuntary contraction of the perineal muscles in the female at the climax of sexual intercourse.

Orifice (OR-i-fis) Any aperture or opening.

Origin (OR-i-jin) The place of attachment of a muscle to the more stationary bone, or the end opposite the insertion

Oropharynx (or'-ō-FAR-inks) The second portion of the pharynx, lying posterior to the mouth and extending from the soft palate down to the hyoid bone.

Orthopedics (or'-thō-PĒ-diks) The branch of medicine that deals with the perservation and restoration of the skeletal system, articulations, and associated structures.

Orthopnea (or'-thop-NĒ-a) Dyspnea that occurs in the horizontal position.

Osmoreceptor (oz'-mō-re-CEP-tor) Receptor in the hypothalamus that is sensitive to changes in blood osmotic pressure and, in response to high osmotic pressure (low water concentration), causes synthesis and release of antidiuretic hormone (ADH).

Osmosis (os-MŌ-sis) The net movement of water molecules through a selectively permeable membrane from an area of high water concentration to an area of lower water concentration until an equilibrium is reached.

Osmotic pressure The pressure required to prevent the movement of pure water into a solution containing solutes when the solutions are separated by a selectively permeable membrane.

Osseous (OS-ē-us) Bony.

Ossicle (OS-si-kul) Small bone, as in the middle ear (malleus, incus, stapes).

Ossification (os'-i-fi-KĀ-shun) Formation of bone. Also called *osteogenesis.*

Osteoblast (OS-tē-ō-blast') Cell formed from an osteoprogenitor cell that participates in bone formation by secreting some organic components and inorganic salts.

Osteoclast (OS-tē-ō-clast') A large multinuclear cell that develops from a monocyte and destroys or resorbs bone tissue.

Osteocyte (OS-tē-ō-sīt') A mature bone cell that maintains the daily activities of bone tissue.

Osteogenic (os'-tē-ō-JEN-ik) *layer* The inner layer of the periosteum that contains cells responsible for forming a new bone during growth and repair.

Osteology (os'-tē-OL-ō-jē) The study of bones.

Osteomalacia (os'-tē-ō-ma-LĀ-shē-a) A deficiency of vitamin D in adults causing demineralization and softening of bone.

Osteomyelitis (os'-tē-ō-mī-i-LĪ-tis) Inflammation of bone marrow or of the bone and marrow.

Osteon The basic unit of structure in adult compact bone, consisting of a central (Haversian) canal with its concentrically arranged lamellae, lacunae, osteocytes, and canaliculi. Also called a *Haversian* (ha-VĒR-shun) *system.*

Osteoporosis (os'-tē-ō-pō-RŌ-sis) Age-related disorder characterizied by decreased bone mass and increased susceptibility to fractures as a result of decreased levels of estrogens.

Osteoprogenitor (os'-tē-ō-prō-JEN-i-tor) *cell* Stem cell derived from mesenchyme that has mitotic potential and the ability to differentiate into an osteoblast.

Otalgia (ō-TAL-jē-a) Pain in the ear; earache.

Otic (Ō-tik) Pertaining to the ear.

Otitis media (ō-TĪ-tus MĒ-dē-a) Acute infection of the middle ear characterized by pain, malaise, fever, and an inflamed tympanic membrane, subject to rupture.

Otolith (Ō-tō-lith) A particle of calcium carbonate embedded in the otolithic membrane that functions in maintaining static equilibrium.

Otolithic (ō-tō-LITH-ik) *membrane* Thick, gelatinous, glycoprotein layer located directly over hair cells of the macula in the saccule and utricle of the inner ear.

Otorhinolaryngology (ō'-tō-rī-nō-lar'-in-GOL-ō-jē) The branch of medicine that deals with the diagnosis and treatment of diseases of the ears, nose, and throat.

Oval window A small opening between the middle ear and inner ear into which the footplate of the stapes fit. Also called the *fenestra vestibuli* (fe-NES-tra ves-TIB-yoo-lē).

Ovarian (ō-VAR-ē-an) *cycle* A monthly series of events in the ovary associated with the maturation of an ovum.

Ovarian follicle (FOL-i-kul) A general name for oocytes (immature ova) in any stage of development, along with their surrounding epithelial cells.

Ovarian ligament (LIG-a-ment) A rounded cord of connective tissue that attaches the ovary to the uterus.

Ovary (Ō-var-ē) Female gonad that produces ova and the hormones estrogens, progesterone, and relaxin.

Ovulation (ō-vyoo-LĀ-shun) The rupture of a vesicular ovarian (Graafian) follicle with discharge of a secondary oocyte into the pelvic cavity.

Ovum (Ō-vum) The female reproductive or germ cell; an egg cell.

Oxidation (ok-si-DĀ-shun) The removal of electrons and hydrogen ions (hydrogen atoms) from a molecule or, less commonly, the addition of oxygen to a molecule that results in a decrease in the energy content of the molecule. The oxidation of glucose in the body is also called *cellular respiration.*

Oxygen debt The volume of oxygen required to oxidize the lactic acid produced by muscular exercise.

Oxyhemoglobin (ok'-sē-HĒ-mō-glō-bin) (*HbO₂*) Hemoglobin combined with oxygen.

Oxyphil cell A cell found in the parathyroid gland that secretes parathyroid hormone (PTH).

Oxytocin (ok'-sē-TŌ-sin) (*OT*) A hormone secreted by neurosecretory cells in the paraventricular and supraoptic nuclei of the hypothalamus that stimulates contraction of the smooth muscle fibers (cells) in the pregnant uterus and contractile cells around the ducts of mammary glands.

Pacinian corpuscle See Lamellated corpuscle.

Paget's (PAJ-ets) *disease* A disorder characterized by a greatly accelerated remodeling process in which osteoclastic resorption is massive and new bone formation by osteoblasts is extensive. As a result, there is an irregular thickening and softening of the bones.

Palate (PAL-at) The horizontal structure separating the oral and the nasal cavities; the roof of the mouth.

Palliative (PAL-ē-a-tiv) Serving to relieve or alleviate without curing.

Palpate (PAL-pāt) To examine by touch; to feel.

Palpitation (pal'-pi-TĀ-shun) A fluttering of the heart or abnormal rate or rhythm of the heart.

Pancreas (PAN-krē-as) A soft, oblong organ lying along the greater curvature of the stomach and connected by a duct to the duodenum. It is both exocrine (secreting pancreatic juice) and endocrine (secreting insulin, glucagon, and somatostatin).

Pancreatic (pan'-krē-AT-ik) *duct* A single, large tube that unites with the common bile duct from the liver and gallbladder and drains pancreatic juice into the duodenum at the hepatopancreatic ampulla (ampulla of Vater). Also called the *duct of Wirsung.*

Pancreatic islet A cluster of endocrine gland cells in the pancreas that secretes insulin, glucagon, and somatostatin. Also called an *islet of Langerhans* (LANG-er-hanz).

Papanicolaou (pap'-a-NIK-ō-la-oo) *test* A cytological staining test for the detection and diagnosis of premalignant and malignant conditions of the female genital tract. Cells scraped from the genital epithelium are smeared, fixed, stained, and examined microscopically. Also called a *Pap smear.*

Papilla (pa-PIL-a) A small nipple-shaped projection or elevation.

Paralysis (pa-RAL-a-sis) Loss or impairment of motor function due to a lesion of nervous or muscular origin.

Paranasal sinus (par'-a-NĀ-zal SĪ-nus) A mucus-lined air cavity in a skull bone that communicates with the nasal cavity. Paranasal sinuses are located in the frontal, maxillary, ethmoid, and sphenoid bones.

Paraplegia (par-a-PLĒ-jē-a) Paralysis of both lower extremities.

Parasagittal plane A vertical plane that does not pass through the midline and that divides the body or organs into *unequal* left and right portions.

Parasympathetic (par'-a-sim-pa-THET-ik) *division* One of the two subdivisions of the autonomic nervous system, having cell bodies of preganglionic neurons in nuclei in the brain stem and in the lateral gray matter of the sacral portion of the spinal cord; primarily concerned with activities that conserve and restore body energy. Also called the *craniosacral* (krā-nē-ō-SĀ-kral) *division.*

Parathyroid (par'-a-THĪ-royd) *gland* One of four small endocrine glands embedded on the posterior surfaces of the lateral lobes of the thyroid gland.

Parathyroid hormone (PTH)   A hormone secreted by the parathyroid glands that decreases blood phosphate level and increases blood calcium level.

Paraurethral (par'-a-yoo-RĒ-thral) **gland**   Gland embedded in the wall of the urethra whose duct opens on either side of the urethral orifice and secretes mucus. Also called **Skene's** (SKĒNZ) **gland.**

Parenchyma (par-EN-ki-ma)   The functional parts of any organ, as opposed to tissue that forms its stroma or framework.

Parenteral (par-EN-ter-al)   Situated or occurring outside the intestines; referring to introduction of substances into the body other than by way of the intestines such as intradermal, subcutaneous, intramuscular, intravenous, or intraspinal.

Parietal (pa-RĪ-e-tal)   Pertaining to or forming the outer wall of a body cavity.

Parietal cell   The secreting cell of a gastric gland that produces hydrochloric acid and intrinsic factor. Also called an **oxyntic cell.**

Parietal pleura (PLOO-ra)   The ourter layer of the serous pleural membrane that encloses and protects the lungs; the layer that is attached to the wall of the pleural cavity.

Parkinson's disease   Progressive degeneration of the basal ganglia and substantia nigra of the cerebrum resulting in decreased production of dopamine (DA) that leads to tremor, slowing of voluntary movements, and muscle weakness. Also called **Parkinsonism.**

Parotid (pa-ROT-id) **gland**   One of the paired salivary glands located inferior and anterior to the ears connected to the oral cavity via a duct (Stensen's) that opens into the inside of the cheek opposite the upper second molar tooth.

Paroxysm (PAR-ok-sizm)   A sudden periodic attack or recurrence of symptoms of a disease.

Parturition (par'-too-RISH-un)   Act of giving birth to young; childbirth, delivery.

Patellar (pa-TELL-ar) **reflex**   Extension of the leg by contraction of the quadriceps femoris muscle in response to tapping the patellar ligament. Also called the **knee jerk.**

Patent ductus arteriosus   Congenital anatomical heart defect in which the fetal connection between the aorta and pulmonary trunk remains open instead of closing completely after birth.

Pathogen (PATH-ō-jen)   A disease-producing organism.

Pathogenesis (path'-ō-JEN-e-sis)   The development of disease or a morbid or pathological state.

Pathological (path'-ō-LOJ-i-kal)   Pertaining to or caused by disease.

Pathological (path'-ō-LOJ-i-kal) **anatomy**   The study of structural changes caused by disease.

Pectinate (PEK-ti-nāt) **muscles**   Projecting muscle bundles of the anterior atrial walls and the lining of the auricles.

Pectoral (PEK-tō-ral)   Pertaining to the chest or breast.

Pediatrician (pē'-dē-a-TRISH-un)   A physician who specializes in the care and treatment of children and their illnesses.

Pedicel (PED-i-sel)   Footlike structure, as on podocytes of a glomerulus.

Pedicle (PED-i-kul)   A short, thick process found on vertebrae.

Pelvic (PEL-vik) **cavity**   Inferior portion of the abdominopelvic cavity that contains the urinary bladder, sigmoid colon, rectum, and internal female and male reproductive structures.

Pelvic inflammatory disease (PID)   Collective term for any extensive bacterial infection of the pelvic organs, especially the uterus, uterine (Fallopian) tubes, and ovaries.

Pelvimetry (pel-VIM-e-trē)   Measurement of the size of the inlet and outlet of the birth canal.

Pelvis   The basinlike structure formed by the two pelvic (hip) bones, the sacrum, and the coccyx. The expanded, proximal portion of the ureter, lying within the kidney and into which the major calyces open.

Penis (PĒ-nis)   The male copulatory organ, used to introduce spermatozoa into the female vagina.

Pepsin   Protein-digesting enzyme secreted by zymogenic (chief) cells of the stomach as the inactive form pepsinogen, which is converted to active pepsin by hydrochloric acid.

Peptic ulcer   An ulcer that develops in areas of the gastrointestinal tract exposed to hydrochloric acid; classified as a gastric ulcer if in the lesser curvature of the stomach and as a duodenal ulcer if in the first part of the duodenum.

Percussion (per-KUSH-un)   The act of striking (percussing) an underlying part of the body with short, sharp blows as an aid in diagnosing the part by the quality of the sound produced.

Perforating canal   A minute passageway by means of which blood vessels and nerves from the periosteum penetrate into compact bone. Also called **Volkmann's** (FŌLK-manz) **canal.**

Pericardial (per'-i-KAR-dē-al) **cavity**   Small potential space between the visceral and parietal layers of the serous pericardium that contains pericardial fluid.

Pericardium (per'-i-KAR-dē-um)   A loose-fitting membrane that encloses the heart, consisting of an outer fibrous layer and an inner serous layer.

Perichondrium (per'-i-KON-drē-um)   The membrane that covers cartilage.

Perikaryon (per'-i-KAR-ē-on)   The nerve cell body that contains the nucleus and other organelles.

Perilymph (PER-i-lymf)   The fluid contained between the bony and membranous labyrinths of the inner ear.

Perimetrium (per-i-MĒ-trē-um)   The serosa of the uterus.

Perimysium (per'-i-MIZ-ē-um)   Invagination of the epimysium that divides muscles into bundles.

Perineum (per'-i-NĒ-um)   The pelvic floor; the space between the anus and the scrotum in the male and between the anus and the vulva in the female.

Perineurium (per'-i-NYOO-rē-um)   Connective tissue wrapping around fascicles in a nerve.

Periodontal (per-ē-ō-DON-tal) **disease**   A collective term for conditions characterized by degeneration of gingivae, alveolar bone, periodontal ligament, and cementum.

Periodontal membrane   The periosteum lining the alveoli (sockets) for the teeth in the alveolar processes of the mandible and maxillae.

Periosteum (per'-ē-OS-tē-um)   The membrane that covers bone and consists of connective tissue, osteoprogenitor cells, and osteoblasts and is essential for bone growth, repair, and nutrition.

Peripheral (pe-RIF-er-al)   Located on the outer part or a surface of the body.

Peripheral nervous system (PNS)   The part of the nervous system that lies outside the central nervous system—nerves and ganglia.

Peripheral resistance (pe-RIF-er-al re-ZIS-tans)   Resistance (impedance) to blood flow as a result of the force of friction between blood and the walls of blood vessels that is related to viscosity of blood and blood vessel length and diameter.

Periphery (pe-RIF-er-ē)   Outer part or a surface of the body; part away from the center.

Peristalsis (per'-i-STAL-sis)   Successive muscular contractions along the wall of a hollow muscular structure.

Peritoneum (per'-i-tō-NĒ-um)   The largest serous membrane of the body that lines the abdominal cavity and covers the viscera.

Peritonitis (per'-i-tō-NĪ-tis)   Inflammation of the peritoneum.

Peroxisome (pe-ROKS-i-sōm)   Organelle similar in structure to a lysosome that contains enzymes related to hydrogen peroxide metabolism; abundant in liver cells.

Perspiration   Substance produced by sudoriferous (sweat) glands containing water, salts, urea, uric acid, amino acids, ammonia, sugar, lactic acid, and ascorbic acid; helps maintain body temperature and eliminate wastes.

Peyer's patches   See Aggregated lymphatic follicles.

pH   A symbol of the measure of the concentration of hydrogen ions in a solution. The pH scale extends from 0 to 14, with a value of 7 expressing neutrality, values lower than 7 expressing increasing acidity, and values higher than 7 expressing increasing alkalinity.

Phagocytosis (fag'-ō-sī-TŌ-sis)   The process by which cells (phagocytes) ingest particulate matter; especially the ingestion and destruction of microbes, cell debris, and other foreign matter.

Phalanx (FĀ-lanks)   The bone of a finger or toe. *Plural, phalanges* (fa-LAN-jēz).

Phantom pain   A sensation of pain as originating in a limb that has been amputated.

Pharmacology (far'-ma-KOL-ō-jē)   The science that deals with the effects and uses of drugs in the treatment of disease.

Pharynx (FAR-inks)   The throat; a tube that starts at the internal nares and runs partway down the neck where it opens into the esophagus posteriorly and the larynx anteriorly.

Phenotype (FĒ-nō-tīp)   The observable expression of genotype; physical characteristics of an organism determined by genetic makeup and influenced by interaction between genes and internal and external environmental factors.

Phenylketonuria (fen'-il-kē'-tō-NOO-rē-a) (*PKU*)   A disorder characterized by an elevation of the amino acid phenylalanine in the blood.

Pheochromocytoma (fē-ō-krō'-mō-sī-TŌ-ma)   Tumor of the chromaffin cells of the adrenal medulla that results in hypersecretion of medullary hormones.

Phlebitis (fle-BĪ-tis)   Inflammation of a vein, usually in the lower extremities.

Phosphocreatine (fos'-fō-KRĒ-a-tin)   High-energy molecule in skeletal muscle fibers (cells) that is used to generate ATP rapidly; upon decomposition, phosphocreatine breaks down into creatine, phosphate, and energy—the energy is used to generate ATP from ADP.

Phospholipid (fos'-fō-LIP-id) *bilayer*   Arrangement of phospholipid molecules in two parallel rows in which the hydrophilic "heads" face outward and the hydrophobic "tails" face inward.

Photopigment   A substance that can absorb light and undergo structural changes that can lead to the development of a receptor potential. An example of rhodopsin.

Photoreceptor   Receptor that detects light on the retina of the eye.

Physiology (fiz'-ē-OL-ō-jē)   Science that deals with the functions of an organism or its parts.

Pia mater (PĪ-a MĀ-ter)   The inner membrane (meninx) covering the brain and spinal cord.

Piezoelectric (pē-e-zō-e-LEK-trik) *effect*   Response of bone, mainly collagen, to stress in which very minute currents of electricity are produced; believed to stimulate osteoblasts to make new bone cells.

Pineal (PĪN-ē-al) *gland*   The cone-shaped gland located in the roof of the third ventricle. Also called the *epiphysis cerebri* (ē-PIF-i-sis se-RĒ-brē).

Pinealocyte (pin-ē-AL-ō-sīt)   Secretory cell of the pineal gland that produces hormones.

Pinna (PIN-na)   The projecting part of the external ear composed of elastic cartilage and covered by skin and shaped like the flared end of a trumpet. Also called the *auricle* (OR-i-kul).

Pinocytosis (pi'-nō-sī-TŌ-sis)   The process by which cells ingest liquid.

Pituicyte (pi-TOO-i-sīt)   Supporting cell of the posterior lobe of the pituitary gland.

Pituitary (pi-TOO-i-tar'-ē) *dwarfism*   Condition caused by hyposecretion of human growth hormone (hGH) during the growth years and characterized by childlike physical traits in an adult.

Pituitary gland   A small endocrine gland lying in the sella turcica of the sphenoid bone and attached to the hypothalamus by the infundibulum; nicknamed the "master gland." Also called the *hypophysis* (hī-POF-i-sis).

Pivot joint   A synovial joint in which a rounded, pointed, or conical surface of one bone articulates with a ring formed partly by another bone and partly by a ligament, as in the joint between the atlas and axis and between the proximal ends of the radius and ulna. Also called a *trochoid* (TRŌ-koid) *joint.*

Placenta (pla-SEN-ta)   The special structure through which the exchange of materials between fetal and maternal circulations occurs. Also called the *afterbirth.*

Plantar flexion (PLAN-tar FLEK-shun)   Bending the foot in the direction of the plantar surface (sole).

Plaque (plak)   A cholesterol-containing mass in the tunica media of arteries. A mass of bacterial cells, dextran (polysaccharide), and other debris that adheres to teeth.

Plasma (PLAZ-ma)   The extracellular fluid found in blood vessels; blood minus the formed elements.

Plasma cell   Cell that produces antibodies and develops from a B cell (lymphocyte).

Plasma (cell) membrane   Outer, limiting membrane that separates the cell's internal parts from extracellular fluid and the external environment.

Plasmapheresis (plaz'-ma-fe-RĒ-sis)   A procedure in which blood is withdrawn from the body, its components are selectively separated, the undesirable component causing disease is removed, and the remainder is returned to the body. Among the substances removed are toxins, metabolic substances, and antibodies. Also called *therapeutic plasma exchange (TPE).*

Platelet plug   Aggregation of thrombocytes at a damaged blood vessel to prevent blood loss.

Pleura (PLOOR-a)   The serous membrane that covers the lungs and lines the walls of the chest and diaphragm.

Pleural cavity   Small potential space between the visceral and parietal pleurae.

Plexus (PLEK-sus)   A network of nerves, veins, or lymphatic vessels.

Plexus of Auerbach   *See Myenteric plexus.*

Plexus of Meissner   *See Submucosal plexus.*

Pneumonia (noo-MŌ-nē-a)   Acute infection or inflammation of the alveoli of the lungs.

Pneumotaxic (noo-mō-TAK-sik) *area*   Portion of the respiratory center in the pons that continually sends inhibitory nerve impulses to the inspiratory area that limit inspiration and facilitate expiration.

Podiatry (pō-DĪ-a-trē)   The diagnosis and treatment of foot disorders.

Polar body   The smaller cell resulting from the unequal division of cytoplasm during the meiotic divisions of an oocyte. The polar body has no function and is resorbed.

Polarized   A condition in which opposite effects or states exist at the same time. In electrical contexts, having one portion negative and another positive; for example, a polarized nerve cell membrane has the outer surface positively charged and the inner surface negatively charged.

Poliomyelitis (pō'-lē-ō-mī-e-LĪ-tis)   Viral infection marked by fever, headache, stiff neck and back, deep muscle pain and weakness, and loss of certain somatic reflexes; a serious form of the disease, *bulbar polio,* results in destruction of motor neurons in anterior horns of spinal nerves that leads to paralysis.

Polycythemia (pol'-ē-sī-THĒ-mē-a)   Disorder characterized by a hematocrit above the normal level of 55 in which hypertension, thrombosis, and hemorrhage occur.

Polyp (POL-ip)   A tumor on a stem found especially on a mucous membrane.

Polysaccharides (pol'-ē-SAK-a-rīds)   A carbohydrate in which three or more monosaccharides are joined chemically.

Polyunsaturated fat   A fat that contains more than one double covalent bond between its carbon atoms; examples are corn oil, safflower oil, and cottonseed oil.

Polyuria (pol'-ē-YOO-rē-a)   An excessive production of urine.

Pons (ponz)   The portion of the brain stem that forms a ''bridge'' between the medulla and the midbrain, anterior to the cerebellum.

Positron emission tomography (*PET*)   A type of radioactive scanning based on the release of gamma rays when positrons collide with negatively charged electrons in body tissues; it indicates where radioisotopes are used in the body.

Postabsorptive state   Metabolic state during which absorption is complete and energy needs of the body must be satisfied.

Postcentral gyrus   A gyrus immediately posterior to the central sulcus that contains the general sensory area of the cerebral cortex.

Posterior (pos-TĒR-ē-or)   Nearer to or at the back of the body. Also called *dorsal.*

Posterior root   The structure composed of afferent (sensory) fibers lying between a spinal nerve and the dorsolateral aspect of the spinal cord. Also called the *dorsal (sensory) root.*

Posterior root ganglion   A group of cell bodies of sensory (afferent) neurons and their supporting cells located along the posterior root of a spinal nerve. Also called a *dorsal (sensory) root ganglion* (GANG-glē-on).

Postganglionic neuron (pōst'-gang-lē-ON-ik NOO-ron)   The second visceral efferent neuron in an autonomic pathway, having its cell body and dendrites located in an autonomic ganglion and its unmyelinated axon ending at cardiac muscle, smooth muscle, or a gland.

Postpartum (pōst-PAR-tum)   After parturition; occurring after the delivery of a baby.

Postsynaptic (pōst-sin-AP-tik) *neuron*   The nerve cell that is activated by the release of a neurotransmitter substance from another neuron and carries nerve impulses away from the synapse.

Pouch of Douglas   *See Rectouterine pouch.*

Precapillary sphincter (SFINGK-ter)   A ring of smooth muscle fibers (cells) at the site of origin of true capillaries that regulate blood flow into true capillaries.

Precentral gyrus   A gyrus immediately anterior to the central sulcus that contains the primary motor area of the cerebral cortex.

Preeclampsia (pre'-e-KLAMP-sē-a)   A syndrome characterized by sudden hypertension, large amounts of protein in urine, and generalized edema; it might be related to an autoimmune or allergic reaction due to the presence of a fetus.

Preganglionic (prē'-gang-lē-ON-ik) *neuron*   The first visceral efferent neuron in an autonomic pathway, with its cell body and dendrites in the brain or spinal cord and its myelinated axon ending at an autonomic ganglion, where it synapses with a postganglionic neuron.

Pregnancy   Sequence of events that normally includes fertilization, implantation, embryonic growth, and fetal growth that terminates in birth.

Premenstrual syndrome (*PMS*)   Severe physical and emotional stress occurring late in the postovulatory phase of the menstrual cycle and sometimes overlapping with menstruation.

Premonitory (prē-MON-i-tō-rē)   Giving previous warning; as premonitory symptoms.

Prepuce (PRĒ-pyoos)   The loose-fitting skin covering the glans of the penis and clitoris. Also called the foreskin.

Presbyopia (prez-bē-Ō-pē-a)   A loss of elasticity of the lens of the eye due to advancing age with resulting inability to focus clearly on near objects.

Presynaptic (prē-sin-AP-tik) *neuron*   A nerve cell that carries nerve impulses toward a synapse.

Prevertebral ganglion (prē-VERT-e-bral GANG-lē-on)   A cluster of cell bodies of postganglionic sympathetic neurons anterior to the spinal column and close to large abdominal arteries. Also called a *collateral ganglion.*

Primary germ layer   One of three layers of embryonic tissue, called ectoderm, mesoderm, and endoderm, that give rise to all tissues and organs of the organism.

Primary motor area   A region of the cerebral cortex in the precentral gyrus of the frontal lobe of the cerebrum that controls specific muscles or groups of muscles.

Primary somesthetic (sō-mes-THET-ik) *area*   A region of the cerebral cortex posterior to the central sulcus in the postcentral gyrus of the parietal lobe of the cerebrum that localizes exactly the points of the body where sensations originate.

Prime mover   The muscle directly responsible for producing the desired motion. Also called an *agonist* (AG-ō-nist).

Primigravida (prī-mi-GRAV-i-da)   A woman pregnant for the first time.

Principal cell   Cell found in the parathyroid glands that secretes parathyroid hormone (PTH). Also called a *chief cell.*

Proctology (prok-TOL-ō-jē)   The branch of medicine that treats the rectum and its disorders.

Progeny (PROJ-e-nē)   Refers to offspring or descendants.

Progesterone (prō-JES-te-rōn) (*PROG*)   A female sex hormone produced by the ovaries that helps prepare the endometrium for implantation of a fertilized ovum and the mammary glands for milk secretion.

Prognosis (prog-NŌ-sis)   A forecast of the probable results of a disorder; the outlook for recovery.

Projection (prō-JEK-shun)   The process by which the brain refers sensations to their point of stimulation.

Prolactin (prō-LAK-tin) (*PRL*)   A hormone secreted by the adenohypophysis (anterior lobe) of the pituitary gland that initiates and maintains milk secretion by the mammary glands.

Prolapse (PRŌ-laps)   A dropping or falling down of an organ, especially the uterus or rectum.

Proliferation (pro-lif'-er-Ā-shun)   Rapid and repeated reproduction of new parts, especially cells.

Pronation (prō-NĀ-shun)   A movement of the forearm in which the palm of the hand is turned posteriorly or inferiorly.

Properdin (prō-PER-din)   A protein found in serum capable of destroying bacteria and viruses.

Prophase (PRŌ-fāz)   The first stage of mitosis during which chromatid pairs are formed and aggregate around the equatorial plane region of the cell.

Proprioception (prō-pre-ō-SEP-shun)   The receipt of information from muscles, tendons, and the labyrinth that enables the brain to determine movements and position of the body and its parts. Also called *kinesthesia* (kin'-es-THĒ-zē-a).

Proprioceptor (prō-'prē-ō-SEP-tor)   A receptor located in muscles, tendons, or joints that provides information about body position and movements.

Prostaglandin (pros'-ta-GLAN-din) (*PG*)   A membrane-associated lipid composed of 20-carbon fatty acids with 5 carbon atoms joined to form a cyclopentane ring; synthesized in small quantities and basically mimics the activities of hormones.

Prostatectomy (pros'-ta-TEK-tō-mē)   The surgical removal of part of or the entire prostate gland.

Prostate (PROS-tāt) *gland*   A doughnut-shaped gland inferior to the urinary bladder that surrounds the superior portion of the male urethra and secretes a slightly acid solution that contributes to sperm motility and viability.

Prosthesis (pros-THĒ-sis)   An artificial device to replace a missing body part.

Protein   An organic compound consisting of carbon, hydrogen, oxygen, nitrogen, and sometimes sulfur and phosphorus, and made up of amino acids linked by peptide bonds.

Prothrombin (prō-THROM-bin)   An inactive protein synthesized by the liver, released into the blood, and converted to active thrombin in the process of blood clotting.

Proto-oncogene (prō'-tō-ONG-kō-jēn)   Gene responsible for some aspect of normal growth and development; it may transform into an oncogene, a gene capable of causing cancer.

Protraction (prō-TRAK-shun)   The movement of the mandible or shoulder girdle forward on a plane parallel with the ground.

Proximal (PROK-si-mal)   Nearer the attachment of an extremity to the trunk or a structure; nearer to the point of origin.

Pruritus (proo'-RĪ-tus)   Itching.

Pseudopodia (soo'-dō-PŌ-dē-a)   Temporary, protruding projections of cytoplasm.

Psoriasis (sō-RĪ-a-sis)   Chronic skin disease characterized by reddish plaques or papules covered with scales.

Psychosomatic (sī'-kō-sō-MAT-ik)   Pertaining to the relation between mind and body. Commonly used to refer to those physiological disorders thought to be caused entirely or partly by emotional disturbances.

Ptosis (TŌ-sis)   Drooping, as of the eyelid or the kidney (nephrotosis).

Puberty (PYOO-ber-tē)   The time of life during which the secondary sex characteristics begin to appear and the capability for sexual reproduction is possible; usually between the ages of 10 and 17.

Puerperium (pyoo'-er-PER-ē-um)   The state immediately after childbirth, usually 4–6 weeks.

Pulmonary (PUL-mo-ner'-ē)   Concerning or affected by the lungs.

Pulmonary circulation   The flow of deoxygenated blood from the right ventricle to the lungs and the return of oxygenated blood from the lungs to the left atrium.

Pulmonary edema (e-DĒ-ma)   An abnormal accumulation of interstitial fluid in the tissue spaces and alveoli of the lungs due to increased pulmonary capillary permeability or increased pulmonary capillary pressure.

Pulmonary embolism (EM-bō-lizm) (*PE*)   The presence of a blood clot or other foreign substance in a pulmonary arterial blood vessel that obstructs circulation to lung tissue.

Pulmonary ventilation   The inflow (inspiration) and outflow (expiration) of air between the atmosphere and the lungs. Also called *breathing.*

Pulp cavity   A cavity within the crown and neck of a tooth, filled with pulp, a connective tissue containing blood vessels, nerves, and lymphatics.

Pulse pressure   The difference between the maximum (systolic) and minimum (diastolic) pressures; normally a value of about 40 mm Hg.

Pupil   The hole in the center of the iris, the area through which light enters the posterior cavity of the eyeball.

Purkinje fiber   See Conduction myofiber.

Pus   The liquid product of inflammation containing leucocytes or their remains and debris of dead cells.

P wave   The deflection wave of an electrocardiogram that records atrial depolarization (contraction).

Pyelitis (pī'-e-LĪ-tis)   Inflammation of the kidney pelvis and its calyces.

Pyemia (pī-Ē-mēa)   Infection of the blood, with multiple abscesses, caused by pus-forming microorganisms.

Pyloric (pī-LOR-ik) *sphincter*   A thickened ring of smooth muscle through which the pylorus of the stomach communicates with the duodenum. Also called the *pyloric valve.*

Pyogenesis (pi'-ō-JEN-e-sis)   Formation of pus.

Pyorrhea (pī-ō-RĒ-a)   A discharge or flow of pus, especially in the alveoli (sockets) and the tissues of the gums.

Pyramid (PIR-a-mid)   A pointed or cone-shaped structure; one of two roughly triangular structures on the ventral side of the medulla composed of the largest motor tracts that run from the cerebral cortex to the spinal cord; a triangular-shaped structure in the renal medulla composed of the straight segments of renal tubules.

Pyramidal (pi-RAM-i-dal) *pathways* Collections of motor nerve fibers arising in the brain and passing down through the spinal cord to motor cells in the anterior horns.

Pyuria (pī-YOO-rē-a) The presence of leucocytes and other components of pus in urine.

QRS wave The deflection wave of an electrocardiogram that records ventricular depolarization (contraction) and atrial relaxation.

Quadrant (KWOD-rant) One of four parts.

Quadriplegia (kwod'-ri-PLĒ-jē-a) Paralysis of the two upper and two lower extremities.

Radiographic (rā'-dē-ō-GRAF-ic) *anatomy* Diagnostic branch of anatomy that includes the use of x rays.

Rales (RALS) Sounds sometimes heard in the lungs that resemble bubbling or rattling due to the presence of an abnormal amount or type of fluid or mucus inside the bronchi or alveoli, or to bronchoconstriction so that air cannot enter or leave the lungs normally.

Rami communicantes (RĀ-mē ko-myoo-ni-KAN-tēz) Branches of a spinal nerve. *Singular, ramus communicans* (RĀ-mus ko-MYOO-ni-kans).

Rapid eye movement (REM) sleep A level of sleep characterized by symmetrical flutter of the eyes and eyelids and brain wave patterns similar to those of an awake person.

Raynaud's (rā-NOZ) *disease* A vascular disorder, primarily of females, characterized by bilateral attacks of ischemia, usually of the fingers and toes, in which the skin becomes pale and exhibits burning and pain; it is brought on by cold or emotional stimuli.

Reactivity (rē-ak-TI-vi-tē) Ability of an antigen to react specifically with the antibody whose formation it induced.

Receptor A specialized cell or a nerve cell terminal modified to respond to some specific sensory modality, such as touch, pressure, cold, light, or sound, and convert it to a nerve impulse by way of a generator or receptor potential. A specific molecule or arrangement of molecules organized to accept only molecules with a complementary shape.

Receptor-mediated endocytosis A highly selective process in which cells take up large molecules or particles (ligands). In the process, successive compartments called vesicles, endosomes, and CURLs form. Ligands are eventually broken down by enzymes in lysosomes.

Receptor potential Depolarization of the plasma membrane of a receptor cell which stimulates release of neurotransmitter from the cell; if the neuron connected to the receptor cell becomes depolarized to threshold, a nerve action potential (nerve impulse) is triggered.

Recessive gene A gene that is not expressed in the presence of a dominant gene on the homologous chromosome.

Reciprocal innervation (re-SIP-rō-kal in-ner-VĀ-shun) The phenomenon by which action potentials stimulate contraction of one muscle and simultaneously inhibit contraction of antagonistic muscles.

Recombinant DNA Synthetic DNA, formed by joining a fragment of DNA from one source to a portion of DNA from another.

Recruitment (rē-KROOT-ment) The process of increasing the number of active motor units. Also called *motor unit summation*.

Rectouterine pouch A pocket formed by the parietal peritoneum as it moves posteriorly from the surface of the uterus and is reflected onto the rectum; the lowest point in the pelvic cavity. Also called the *pouch* or *cul de sac of Douglas*.

Rectum (REK-tum) The last 20 cm (7 in.) of the gastrointestinal tract, from the sigmoid colon to the anus.

Recumbent (re-KUM-bent) Lying down.

Red nucleus A cluster of cell bodies in the midbrain, occupying a large portion of the tectum and sending fibers into the rubroreticular and rubrospinal tracts.

Reduction The addition of electrons and hydrogen ions (hydrogen atoms) to a molecule or, less commonly, the removal of oxygen from a molecule that results in an increase in the energy content of the molecule.

Referred pain Pain that is felt at a site remote from the place of origin.

Reflex Fast response to a change (stimulus) in the internal or external environment that attempts to restore homeostasis; passes over a reflex arc.

Reflex arc The most basic conduction pathway through the nervous system, connecting a receptor and an effector and consisting of a receptor, a sensory neuron, a center in the central nervous system for a synapse, a motor neuron, and an effector.

Refraction (rē-FRAK-shun) The bending of light as it passes from one medium to another.

Refractory (re-FRAK-to-rē) *period* A time during which an excitable cell cannot respond to a stimulus that is usually adequate to evoke an action potential.

Regeneration (rē-jen'-er-Ā-shun) The natural renewal of a structure.

Regimen (REJ-i-men) A strictly regulated scheme of diet, exercise, or activity designed to achieve certain ends.

Regional anatomy The division of anatomy dealing with a specific region of the body, such as the head, neck, chest, or abdomen.

Regulating factor Chemical secretion of the hypothalamus whose structure is unknown that can either stimulate or inhibit secretion of hormones of the adenohypophysis (anterior pituitary).

Regulating hormone Chemical secretion of the hypothalamus whose structure is known that can either stimulate or inhibit secretion of hormones of the adenohypophysis (anterior pituitary).

Regurgitation (rē-gur'-ji-TĀ-shun) Return of solids or fluids to the mouth from the stomach; flowing backward of blood through incompletely closed heart valves.

Relapse (RĒ-laps) The return of a disease weeks or months after its apparent cessation.

Relaxin (RLX) A female hormone produced by the ovaries that relaxes the symphysis pubis and helps dilate the uterine cervix to facilitate delivery.

Remodeling Replacement of old bone by new bone tissue.

Renal (RĒ-nal) Pertaining to the kidney.

Renal corpuscle (KOR-pus'-l)   A glomerular (Bowman's) capsule and its enclosed glomerulus.

Renal erythropoietic (ē-rith'-rō-poy-Ē-tik) **factor**   An enzyme released by the kidneys and liver in response to hypoxia that acts on a plasma protein to bring about the production of erythropoietin, a hormone that stimulates red blood cell production.

Renal failure   Inability of the kidneys to function properly, due to abrupt failure (acute) or progressive failure (chronic).

Renal pelvis   A cavity in the center of the kidney formed by the expanded, proximal portion of the ureter, lying within the kidney, and into which the major calyces open.

Renal pyramid   A triangular structure in the renal medulla composed of the straight segments of renal tubules.

Renin (RĒ-nin)   An enzyme released by the kidney into the plasma where it converts angiotensinogen into angiotensin I.

Renin-angiotensin (an'-jē-ō-TEN-sin) **pathway**   A mechanism for the control of aldosterone secretion by angiotensin II, initiated by the secretion of renin by the kidney in response to low blood pressure.

Reproduction (rē'-prō-DUK-shun)   Either the formation of new cells for growth, repair, or replacement, or the production of a new individual.

Reproductive cell division   Type of cell division in which sperm and egg cells are produced; consists of meiosis and cytokinesis.

Residual (re-ZID-yoo-al) **volume**   The volume of air still contained in the lungs after a maximal expiration; about 1,200 ml.

Resistance   Ability to ward off disease. The hindrance encountered by an electrical charge as it moves through a substance from one point to another. The hindrance encountered by blood as it flows through the vascular system or by air through respiratory passageways.

Respiration (res-pi-RĀ-shun)   Overall exchange of gases between the atmosphere, blood, and body cells consisting of pulmonary ventilation, external respiration, and internal respiration.

Respirator (RES-pi-rā'-tor)   An apparatus fitted to a mask over the nose and mouth, or hooked directly to an endotracheal or tracheotomy tube, that is used to assist or support ventilation or to provide nebulized medication to the air passages under positive pressure.

Respiratory center   Neurons in the reticular formation of the brain stem that regulate the rate of respiration.

Respiratory distress syndrome (RDS) of the newborn   A disease of newborn infants, especially premature ones, in which insufficient amounts of surfactant are produced and breathing is labored. Also called **hyaline** (HĪ-a-lin) **membrane disease (HMD).**

Respiratory failure   Condition in which the respiratory system cannot supply sufficient oxygen to maintain metabolism or eliminate enough carbon dioxide to prevent respiratory acidosis.

Resting membrane potential   The voltage that exists between the inside and outside of a cell membrane when the cell is not responding to a stimulus; about −70 to −90 mV, with the inside of the cell negative.

Resuscitation (rē-sus'-i-TĀ-shun)   Act of bringing a person back to full consciousness.

Retention (rē-TEN-shun)   A failure to void urine due to obstruction, nervous contraction of the urethra, or absence of sensation of desire to urinate.

Reticular (re-TIK-yoo-lar) **activating system (RAS)**   An extensive network of branched nerve cells running through the core of the brain stem. When these cells are activated, a generalized alert or arousal behavior results.

Reticular formation   A network of small groups of nerve cells scattered among bundles of fibers beginning in the medulla as a continuation of the spinal cord and extending upward through the central part of the brain stem.

Reticulocyte (re-TIK-yoo-lō-sīt)   An immature red blood cell.

Reticulocyte (re-TIK-yoo-lō-sīt) **count**   Examination of a stained sample of blood to determine the percentage of reticulocytes in the total number of red blood cells; used to evaluate rate of erythropoiesis and monitor treatment for anemia.

Retina (RET-i-na)   The inner coat of the eyeball, lying only in the posterior portion of the eye and consisting of nervous tissue and a pigmented layer comprised of epithelial cells lying in contact with the choroid. Also called the **nervous tunic** (TOO-nik).

Retinal (RE'-ti-nal)   The pigment portion of the photopigment rhodopsin. Also called **visual yellow.**

Retraction (rē-TRAK-shun)   The movement of a protracted part of the body backward on a plane parallel to the ground, as in pulling the lower jaw back in line with the upper jaw.

Retroflexion (re-trō-FLEK-shun)   A malposition of the uterus in which it is tilted posteriorly.

Retrograde degeneration (RE-trō-grād dē-jen-er-Ā-shun) Changes that occur in the proximal portion of a damaged axon only as far as the first neurofibral node (node of Ranvier); similar to changes that occur during Wallerian degeneration.

Retroperitoneal (re'-trō-per-i-tō-NĒ-al)   External to the peritoneal lining of the abdominal cavity.

Rheumatism (ROO-ma-tizm')   Any painful state of the supporting structures of the body—bones, ligaments, joints, tendons, or muscles.

RH factor   An inherited agglutinogen (antigen) on the surface of red blood cells.

Rhinology (rī-NOL-ō-jē)   The study of the nose and its disorders.

Rhinoplasty (RĪ-nō-plas'-tē)   Surgical procedure in which the structure of the external nose is altered.

Rhodopsin (rō-DOP-sin)   A photopigment in rods of the retina, consisting of a protein scotopsin plus retinal, that is sensitive to low levels of illumination. Also called **visual purple.**

Ribonucleic (rī'-bō-nyoo-KLĒ-ik) **acid (RNA)**   A single-stranded nucleic acid constructed of nucleotides consisting of one of four possible nitrogenous bases (adenine, cytosine, guanine, or uracil), ribose, and a phosphate group; three types are messenger RNA (mRNA), transfer RNA (tRNA), and ribosomal RNA (rRNA), each of which cooperates with DNA for protein synthesis.

Ribosome (RĪ-bō-sōm)   An organelle in the cytoplasm of cells, composed of ribosomal RNA and ribosomal proteins, that synthesizes proteins; nicknamed the "protein factory."

Rickets (RIK-ets)   Condition affecting children characterized by soft and deformed bones resulting from inadequate calcium metabolism due to a vitamin D deficiency.

Right heart (atrial) reflex   A reflex concerned with maintaining normal venous blood pressure.

Right lymphatic (lim-FAT-ik) **duct**   A vessel of the lymphatic system that drains lymph from the upper right side of the body and empties it into the right subclavian vein.

Rigidity (ri-JID-i-tē) Hypertonia characterized by increased muscle tone, but reflexes are not affected.

Rigor mortis State of partial contraction of muscles following death due to lack of ATP that causes cross bridges of thick myofilaments to remain attached to thin myofilaments, thus preventing relaxation.

Rod A visual receptor in the retina of the eye that is specialized for vision in dim light.

Roentgen (RENT-gen) The international unit of radiation; a standard quantity of x or gamma radiation.

Root canal A narrow extension of the pulp cavity lying within the root of a tooth.

Root of penis Attached portion of penis that consists of the bulb and crura.

Rotation (rō-TĀ-shun) Moving a bone around its own axis, with no other movement.

Round ligament (LIG-a-ment) A band of fibrous connective tissue enclosed between the folds of the broad ligament of the uterus, emerging from a point on the uterus just below the uterine (Fallopian) tube, extending laterally along the pelvic wall, and penetrating the abdominal wall through the deep inguinal ring to end in the labia majora.

Round window A small opening between the middle and inner ear, directly below the oval window, covered by the secondary tympanic membrane. Also called the *fenestra cochlea* (fe-NES-tra KŌK-lē-a).

Rugae (ROO-jē) Large folds in the mucosa of an empty hollow organ, such as the stomach and vagina.

Saccule (SAK-yool) The lower and smaller of the two chambers in the membranous labyrinth inside the vestibule of the inner ear containing a receptor organ for static equilibrium.

Sacral plexus (PLEK-sus) A network formed by the anterior branches of spinal nerves L4 through S3.

Sacral promontory (PROM-on-tor'-ē) The superior surface of the body of the first sacral vertebra that projects anteriorly into the pelvic cavity; a line from the sacral promontory to the superior border of the symphysis pubis divides the abdominal and pelvic cavities.

Saddle joint A synovial joint in which the articular surface of one bone is saddle shaped and the articular surface of the other bone is shaped like a rider sitting in the saddle, as in the joint between the trapezium and the metacarpal of the thumb. Also called a *sellaris* (sel-LA-ris) *joint.*

Sagittal (SAJ-i-tal) *plane* A vertical plane that divides the body or organs into left and right portions. Such a plane may be *midsagittal (median),* in which the divisions are equal, or *parasagittal,* in which the divisions are unequal.

Saliva (sa-LĪ-va) A clear, alkaline, somewhat viscous secretion produced by the three pairs of salivary glands; contains various salts, mucin, lysozyme, and salivary amylase.

Salivary amylase (SAL-i-ver-ē AM-i-lās) An enzyme in saliva that initiates the chemical breakdown of starch, mostly in the mouth.

Salivary gland One of three pairs of glands that lie outside the mouth and pour their secretory product (called saliva) into ducts that empty into the oral cavity; the parotid, submandibular, and sublingual glands.

Salpingitis (sal'-pin-JĪ-tis) Inflammation of the urterine (Fallopian) or auditory (Eustachian) tube.

Saltatory (sal-ta-TŌ-rē) *conduction* The propagation of an action potential (nerve impulse) along the exposed portions of a myelinated nerve fiber. The action potential appears at successive neurofibral nodes (nodes of Ranvier) and therefore seems to jump or leap from node to node.

Sarcolemma (sar'-kō-LEM-ma) The cell membrane of muscle fiber (cell), especially of a skeletal muscle fiber.

Sarcoma (sar-KŌ-ma) A connective tissue tumor, often highly malignant.

Sarcomere (SAR-kō-mēr) A contractile unit in a striated muscle fiber (cell) extending from one Z line to the next Z line.

Sarcoplasm (SAR-kō-plazm) The cytoplasm of a muscle fiber (cell).

Sarcoplasmic reticulum (sar'-kō-PLAZ-mik re-TIK-yoo-lum) A network of saccules and tubes surrounding myofibrils of a muscle fiber (cell), comparable to endoplasmic reticulum; functions to reabsorb calcium ions during relaxation and to release them to cause contraction.

Satiety (sa-TĪ-e-tē) Fullness or gratification, as of hunger or thirst.

Satiety center A collection of nerve cells located in the ventromedial nuclei of the hypothalamus that, when stimulated, brings about the cessation of eating.

Saturated fat A fat that contains no double bonds between any of its carbon atoms; all are single bonds and all carbon atoms are bonded to the maximum number of hydrogen atoms; found naturally in animal foods such as meat, milk, milk products, and egges.

Scala tympani (SKA-la TIM-pan-ē) The lower spiral-shaped channel of the bony chochlea, filled with perilymph.

Scala vertibuli (ves-TIB-yoo-lē) The upper spiral-shaped channel of the bony cochlea, filled with perilymph.

Schwann cell *See Neurolemmocyte.*

Sciatica (sī-AT-i-ka) Inflammation and pain along the sciatic nerve; felt at the back of the thigh running down the inside of the leg.

Sclera (SKLE-ra) The white coat of fibrous tissue that forms the outer protective covering over the eyeball except in the most anterior portion; the posterior portion of the fibrous tunic.

Scleral venous sinus A circular venous sinus located at the junction of the sclera and the cornea through which aqueous humor drains from the anterior chamber of the eyeball into the blood. Also called the *canal of Schlemm* (SHLEM).

Sclerosis (skle-RŌ-sis) A hardening with loss of elasticity of tissues.

Scoliosis (skō'-lē-Ō-sis) An abnormal lateral curvature from the normal vertical line of the backbone.

Scotoma (skō-TŌ-ma) An area of depressed or lost vision within the visual field.

Scotopsin (skō-TOP-sin) The protein portion of the visual pigment rhodopsin found in rods of the retina.

Scrotum (SKRŌ-tum) A skin-covered pouch that contains the testes and their accessory structures.

Sebaceous (se-BĀ-shus) Secreting oil.

Sebaceous (se-BĀ-shus) *gland* An exocrine gland in the dermis of the skin, almost always associated with a hair follicle, that secretes sebum. Also called an *oil gland.*

Sebum (SĒ-bum) Secretion of sebaceous (oil) glands.

Secondary sex characteristic A feature characteristic of the male or female body that develops at puberty under the stimulation of sex hormones but is not directly involved in sexual reproduction, such as distribution of body hair, voice pitch, body shape, and muscle development.

Secretion (se-KRĒ-shun) Production and release from a gland cell of a fluid, especially a functionally useful product as opposed to a waste product.

Selectively permeable membrane A membrane that permits the passage of certain substances, but restricts the passage of others. Also called a *semipermeable* (sem'-ē-PER-mē-a-bl) *membrane.*

Sella turcica (SEL-a TUR-si-ka) A depression on the superior surface of the sphenoid bone that houses the pituitary gland.

Semen (SĒ-men) A fluid discharged at ejaculation by a male that consists of a mixture of spermatozoa and the secretions of the seminal vesicles, prostate gland, and bulbourethral (Cowper's) glands. Also called *seminal* (SEM-i-nal) *fluid.*

Semicircular canals Three bony channels (anterior, posterior, lateral), filled with perilymph, in which lie the membranous semicircular canals filled with endolymph. They contain receptors for equilibrium.

Semicircular ducts The membranous semicircular canals filled with endolymph and floating in the perilymph of the bony semicircular canals. They contain cristae that are concerned with dynamic equilibrium.

Semilunar (sem'-ē-LOO-nar) *valve* A valve guarding the entrance into the aorta or the pulmonary trunk from a ventricle of the heart.

Seminal vesicle (SEM-i-nal VES-i-kul) One of a pair of convoluted, pouchlike structures, lying posterior and inferior to the urinary bladder and anterior to the rectum, that secrete a component of semen into the ejaculatory ducts.

Seminiferous tubule (sem'-i-NI-fer-us TOO-byool) A tightly coiled duct, located in a lobule of the testis, where spermatozoa are produced.

Senescence (se-NES-ens) The process of growing old; the period of old age.

Senile macular (MAK-yoo-lar) *degeneration (SMD)* A disease in which blood vessels grow over the macula lutea.

Senility (se-NIL-i-tē) A loss of mental or physical ability due to old age.

Sensation A state of awareness of external or internal conditions of the body.

Sensory area A region of the cerebral cortex concerned with the interpretation of sensory impulses.

Septal defect An opening in the septum (interatrial or interventricular) between the left and right sides of the heart.

Septicemia (sep'-ti-SĒ-mē-a) Toxins or disease-causing bacteria in blood. Also called *"blood poisoning."*

Septum (SEP-tum) A wall dividing two cavities.

Serosa (ser-Ō-sa) Any serous membrane. The outermost layer of an organ formed by a serous membrane. The membrane that lines the pleural, pericardial, and peritoneal cavities.

Serous (SIR-us) *membrane* A membrane that lines a body cavity that does not open to the exterior. Also called the *serosa* (se-RŌ-sa).

Serum Plasma minus its clotting proteins.

Sesamoid bones (SES-a-moyd) Small bones usually found in tendons.

Sex chromosomes The twenty-third pair of chromosomes, designated X and Y, which determine the genetic sex of an individual; in males, the pair is XY; in females, XX.

Sexual intercourse The insertion of the erect penis of a male into the vagina of a female. Also called *coitus* (KŌ-i-tus) or *copulation.*

Sexually transmitted disease (STD) General term for any of a large number of diseases spread by sexual contact. Also called a *venereal disease (VD).*

Sheath of Schwann *See Neurolemma.*

Shingles Acute infection of the peripheral nervous system caused by a virus.

Shinsplints Soreness or pain along the tibia probably caused by inflammation of the periosteum brought on by repeated tugging of the muscles and tendons attached to the periosteum. Also called *tibia stress syndrome.*

Shivering Involuntary contraction of a muscle that generates heat.

Shock Failure of the cardiovascular system to deliver adequate amounts of oxygen and nutrients to meet the metabolic needs of the body due to inadequate cardiac output. It is characterized by hypotension; clammy, cool, and pale skin; sweating; reduced urine formation; altered mental state; acidosis; tachycardia; weak, rapid pulse; and thirst. Types include hypovolemic, cardiogenic, obstructive, neurogenic, and septic.

Shoulder A synovial or diarthrotic joint where the humerus joins the scapula.

Sigmoid colon (SIG-moyd KŌ-lon) The S-shaped portion of the large intestine that begins at the level of the left iliac crest, projects inward to the midline, and terminates at the rectum at about the level of the third sacral vertebral.

Sign Any objective evidence of disease that can be observed or measured such as a lesion, swelling, or fever.

Sinoatrial (si-nō-Ā-trē-al) *(SA) node* A compact mass of cardiac muscle fibers (cells) specialized for conduction, located in the right atrium beneath the opening of the superior vena cava. Also called the *sinuatrial node* or *pacemaker.*

Sinus (SĪ-nus) A hollow in a bone (paranasal sinus) or other tissue; a channel for blood (vascular sinus); any cavity having a narrow opening.

Sinusitis (sīn-yoo-SĪT-is) Inflammation of the mucous membrane of a paranasal sinus.

Sinusoid (SĪN-yoo-soyd) A microscopic space or passage for blood in certain organs such as the liver or spleen.

Skeletal muscle An organ specialized for contraction, composed of striated muscle fibers (cells), supported by connective tissue, attached to a bone by a tendon or an aponeurosis, and stimulated by somatic efferent neurons.

Skene's gland *See Paraurethral gland.*

Skull The skeleton of the head consisting of the cranial and facial bones.

Sliding-filament theory The most commonly accepted explanation for muscle contraction in which actin and myosin myofilaments move into interdigitation with each other, decreasing the length of the sarcomeres.

Small intestine A long tube of the gastrointestinal tract that begins at the pyloric sphincter of the stomach, coils through the central and lower part of the abdominal cavity, and ends at the large intestine, divided into three segments: duodenum, jejunum, and ileum.

Smooth muscle An organ specialized for contraction, composed of smooth muscle fibers (cells), located in the walls of hollow internal structures, and innervated by a visceral efferent neuron.

Sodium-potassium pump An active transport system located in the cell membrane that transports sodium ions out of the cell and potassium ions into the cell at the expense of cellular ATP. It functions to keep the ionic concentrations of these elements at physiological levels.

Soft palate (PAL-at) The posterior portion of the roof of the mouth, extending posteriorly from the palatine bones and ending at the uvula. It is a muscular partition lined with mucous membrane.

Solution A homogeneous molecular or ionic dispersion of one or more substances (solutes) in a usually liquid-disssolving medium (solvent).

Somatic cell division Type of cell division in which a single starting cell (parent cell) duplicates itself to produce two identical cells (daughter cells); consists of mitosis and cytokinesis.

Somatic (sō-MAT-ik) **nervous system (SNS)** The portion of the peripheral nervous system made up of the somatic efferent fibers that run between the central nervous system and the skeletal muscles and skin.

Somatomedin (sō'-ma-tō-MĒ-din) Small protein produced by the liver in response to stimulation by human growth hormone (hGH) that mediates most of the effects of human growth hormone.

Somesthetic (sō-mes-THET-ik) Pertaining to sensations and sensory structures of the body.

Spasm (spazm) A sudden, involuntary contraction of large groups of muscles.

Spastic (SPAS-tik) An increase in muscle tone (stiffness) associated with an increase in tendon reflexes and abnormal reflexes (Babinski sign).

Spasticity (spas-TIS-t-tē) Hypertonia characterized by increased muscle tone, increased tendon reflexes, and pathological reflexes (Babinski sign).

Spermatic (sper-MAT-ik) **cord** A supporting structure of the male reproductive system, extending from a testis to the deep inguinal ring, that includes the ductus (vas) deferens, arteries, veins, lymphatics, nerves, cremaster muscle, and connective tissue.

Spermatogenesis (sper'-ma-tō-JEN-e-sis) The formation and development of spermatozoa in the seminiferous tubules of the testes.

Spermatozoon (sper'-ma-tō-ZŌ-on) A mature sperm cell.

Spermicide (SPER-mi-sīd') An agent that kills spermatozoa.

Spermiogenesis (sper'-mē-ō-JEN-e-sis) The maturation of spermatids into spermatozoa.

Sphincter (SFINGK-ter) A circular muscle constricting an orifice.

Sphincter of Oddi See **Sphincter of the hepatopancreatic ampulla.**

Sphincter of the hepatopancreatic ampulla A circular muscle at the opening of the common bile and main pancreatic ducts in the duodenum. Also called the **sphincter of Oddi** (OD-ē).

Sphygmomanometer (sfig'-mō-ma-NOM-e-ter) An instrument for measuring arterial blood pressure.

Spina bifida (SPĪ-na BIF-i-da) A congential defect of the vertebral column in which the halves of the neural arch of a vertebra fail to fuse in the midline.

Spinal (SPĪ-nal) **cord** A mass of nerve tissue located in the vertebral canal from which 31 pairs of spinal nerves originate.

Spinal nerve One of the 31 pairs of nerves that originate on the spinal cord from posterior and anterior roots.

Spinal shock A period of time, from several days to several weeks, following transection of the spinal cord and characterized by the abolition of all reflex activity.

Spinal (lumbar) tap (puncture) Withdrawal of some of the cerebrospinal fluid from the subarachnoid space in the lumbar region for diagnostic purposes, introduction of various substances, and evaluation of the effects of treatment.

Spinous (SPĪ-nus) **process** A sharp or thornlike process or projection. Also called **spine.** A sharp ridge running diagonally across the posterior surface of the scapula.

Spiral organ The organ of hearing, consisting of supporting cells and hair cells that rest on the basilar membrane and extend into the endolymph of the cochlear duct. Also called the **organ of Corti** (KOR-tē).

Spirometer (spī-ROM-e-ter) An apparatus used to measure air capacity of the lungs.

Spleen (SPLĒN) Large mass of lymphatic tissue between the fundus of the stomach and the diaphragm that functions in phagocytosis, production of lymphocytes, and blood storage.

Sprain Forcible wrenching or twisting of a joint with partial rupture or other injury to its attachments without dislocation.

Sputum (SPYOO-tum) Substance ejected from the mouth containing saliva and mucus.

Squamous (SKWĀ-mus) Scalelike.

Starling's law of the capillaries The movement of fluid between plasma and interstitial fluid is in a state of near equilibrium at the arterial and venous ends of a capillary; that is, filtered fluid and absorbed fluid plus that returned to the lymphatic system are nearly equal.

Starling's law of the heart The force of muscular contraction is determined by the length of the cardiac muscle fibers (cells); the greater the length of stretched fibers, the stronger the contraction.

Starvation (star-VĀ-shun) The loss of energy stores in the form of glycogen, fats, and proteins due to inadequate intake of nutrients or inability to digest, absorb, or metabolize ingested nutrients.

Stasis (STĀ-sis) Stagnation or halt of normal flow of fluids, as blood, urine, or of the intestinal mechanism.

Statis equilibrium (ē-kwi-LIB-rē-um) The maintenance of posture in response to changes in the orientation of the body, mainly the head, relative to the ground.

Stellate reticuloendothelial (STEL-āte re-tik'-yoo-lō-en'-dō-THĒ-lē-al) **cell** Phagocytic cell that lines a sinusoid of the liver. Also called a **Kupffer's** (KOOP-ferz) **cell.**

Stenosis (sten-Ō-sis) An abnormal narrowing or constriction of a duct or opening.

Sterocilia (ste'-rē-ō-SIL-ē-a) Groups of extremely long, slender, nonmotile microvilli projecting from epithelial cells lining the epididymis.

Stereognosis (ste'-rē-og-NŌ-sis) The ability to recognize the size, shape, and texture of an object by touch.

Sterile (STE-ril) Free from any living microorganisms. Unable to conceive or produce offspring.

Sterilization (ster'-i-li-ZĀ-shun) Elimination of all living microorganisms. The rendering of an individual incapable of reproduction (e.g., castration, vasectomy, hysterectomy).

Sternal puncture Introduction of a wide-bore needle into the marrow cavity of the sternum for aspiration of a sample of red bone marrow.

Stimulus Any change in the environment capable of altering the membrane potential.

Stomach The J-shaped enlargement of the gastrointestinal tract directly under the diaphragm in the epigastric, umbilical, and left hypochondriac regions of the abdomen, between the esophagus and small intestine.

Strabismus (stra-BIZ-mus) A condition in which the visual axes of the two eyes differ, so that they do not fix on the same object.

Straight tubule (TOO-byool) A duct in a testis leading from a convoluted seminiferous tubule to the rete testis.

Stratum basalis (STRĀ-tum ba-SAL-is) The outer layer of the endometrium, next to the myometrium, that is maintained during menstruation and gestation and produces a new functionalis following menstruation or parturition.

Stratum functionalis (funk'-shun-AL-is) The inner layer of the endometrium, the layer next to the uterine cavity, that is shed during menstruation and that forms the maternal portion of the placenta during gestation.

Stressor A stress that is extreme, unusual, or long-lasting and triggers the general adaptation syndrome.

Stretch receptor Receptor in the walls of bronchi, bronchioles, and lungs that sends impulses to the respiratory center that prevents overinflation of the lungs.

Stretch reflex A monosynaptic reflex triggered by a sudden stretch of a muscle and ending with a contraction of that same muscle. Also called a **tendon jerk.**

Stricture (STRIK-cher) A local construction of a tubular structure.

Stroke volume The volume of blood ejected by either ventricle in one systole; about 70 ml.

Stroma (STRŌ-ma) The tissue that forms the ground substance, foundation, or framework of an organ, as opposed to its functional parts.

Stupor (STOO-por) Unresponsiveness from which a patient can be aroused only briefly and by vigorous and repeated stimulation.

Subarachnoid (sub'-a-RAK-noyd) **space** A space between the arachnoid and the pia mater that surrounds the brain and spinal cord and through which cerebrospinal fluid circulates.

Subcutaneous (sub'-kyoo-TĀ-nē-us) Beneath the skin. Also called **hypodermic** (hī-pō-DER-mik).

Subcutaneous layer A continuous sheet of loose connective tissue and adipose tissue between the dermis of the skin and the deep fascia of the muscles. Also called the **superficial fascia** (FASH-ē-a).

Subdural (sub-DOO-ral) **space** A space between the dura mater and the arachnoid of the brain and spinal cord that contains a small amount of fluid.

Sublingual (sub-LING-gwal) **gland** One of a pair of salivary glands situated in the floor of the mouth under the mucous membrane and to the side of the lingual frenulum, with a duct (Rivinus's) that opens into the floor of the mouth.

Submandibular (sub'-man-DIB-yoo-lar) **gland** One of a pair of salivary glands found beneath the base of the tongue under the mucous membrane in the posterior part of the floor of the mouth, posterior to the sublingual glands, with a duct (Wharton's) situated to the side of the lingual frenulum. Also called the **submaxillay** (sub'-MAK-si-ler-ē) **gland.**

Submucosa (sub-myoo-KŌ-sa) A layer of connective tissue located beneath a mucous membrane, as in the gastrointestinal tract or the urinary bladder; the submucosa connects the mucosa to the muscularis tunic.

Submucosal plexus A network of autonomic nerve fibers located in the outer portion of the submucous layer of the small intestine. Also called the **plexus of Meissner** (MĪS-ner).

Substrate A substance with which an enzyme reacts.

Subthreshold stimulus A stimulus of such weak intensity that it cannot initiate an action potential (nerve impulse). Also called a **subliminal stimulus.**

Sudoriferous (soo'-dor-IF-er-us) **gland** An apocrine or eccrine exocrine gland in the dermis or subcutaneous layer that produces perspiration. Also called a **sweat gland.**

Sulcus (SUL-kus) A groove or depression between parts, especially between the convolutions of the brain. *Plural,* **sulci** (SUL-sē).

Summation (sum-MĀ-shun) The algebraic addition of the excitatory and inhibitory effects of many stimuli applied to a nerve cell body. The increased strength of muscle contraction that results when stimuli follow in rapid succession.

Superficial (soo'-per-FISH-al) Located on or near the surface of the body.

Superficial fascia (FASH-ē-a) A continuous sheet of fibrous connective tissue between the dermis of the skin and the deep fascia of the muscles. Also called **subcutaneous** (sub'-kyoo-TĀ-nē-us) **layer.**

Superior (soo-PER-eē-or) Toward the head or upper part of a structure. Also called **cephalad** (SEF-a-lad) or **craniad.**

Superior vena cava (VĒ-na CĀ-va) (**SVC**) Large vein that collects blood from parts of the body superior to the heart and returns it to the right atrium.

Supination (soo-pī-NĀ-shun) A movement of the forearm in which the palm of the hand is turned anteriorly or superiorly.

Suppuration (sup'-yoo-RĀ-shun) Pus formation and discharge.

Surface anatomy The study of the structures that can be identified from the outside of the body.

Surfactant (sur-FAK-tant) A phospholipid substance produced by the lungs that decreases surface tension.

Susceptibility (sus-sep'-ti-BIL-i-tē) Lack of resistance of a body to the deleterious or other effects of an agent such as pathogenic microorganisms.

Suspensory ligament (sus-PEN-so-rē LIG-a-ment) A fold of peritoneum extending laterally from the surface of the ovary to the pelvic wall.

Sustentacular (sus'-ten-TAK-yoo-lar) **cell** A supporting cell of seminiferous tubules that produces secretions for supplying nutrients to spermatozoa and the hormone inhibin. Also called a *Sertoli* (ser-TŌ-lē) *cell.*

Sutural (SOO-cher-al) *bone* A small bone located within a suture between certain cranial bones. Also called *Wormian* (WER-mē-an) *bone.*

Suture (SOO-cher) An immovable fibrous joint in the skull where bone surfaces are closely united.

Sympathetic (sim'-pa-THET-ik) *division* One of the two subdivisions of the autonomic nervous system, having cell bodies of preganglionic neurons in the lateral gray columns of the thoracic segment and first two or three lumbar segments of the spinal cord; primarily concerned with processes involving the expenditure of energy. Also called the *thoracolumbar* (thō'-ra-kō-LUM-bar) *division.*

Sympathetic trunk ganglion (GANG-glē-on) A cluster of cell bodies of postganglionic sympathetic neurons lateral to the vertebral column, close to the body of a vertebra. These ganglia extend downward through the neck, thorax, and abdomen to the coccyx on both sides of the vertebral column and are connected to one another to form a chain on each side of the vertebral column. Also called *lateral,* or *sympathetic, chain* or *vertebral chain ganglia.*

Sympathomimetic (sim'-pa-thō-mi-MET-ik) Producing effects that mimic those brought about by the sympathetic division of the autonomic nervous system.

Symphysis (SIM-fi-sis) A line of union. A slightly movable cartilaginous joint such as the symphysis pubis between the anterior surfaces of the coxal (hip) bones.

Symphysis pubis (PYOO-bis) A slightly movable cartilaginous joint between the anterior surfaces of the coxal (hip) bones.

Symptom (SIMP-tum) A subjective change in body function not apparent to an observer, such as fever or nausea, that indicates the presence of a disease or disorder of the body.

Synapse (SIN-aps) The junction between the process of two adjacent neurons, the place where the activity of one neuron affects the activity of another; may be electrical or chemical.

Synapsis (sin-AP-sis) The pairing of homologous chromosomes during prophase I of meiosis.

Synaptic (sin-AP-tik) *cleft* The narrow gap that separates the axon terminal of one nerve cell from another nerve cell or muscle fiber (cell) and across which a neurotransmitter diffuses to affect the postsynaptic cell.

Synaptic delay The length of time between the arrival of the action potential at the axon terminal and the membrane potential (IPSP or EPSP) change on the postsynaptic membrane; usually about 0.5 msec.

Synaptic end bulb Expanded distal end of an axon terminal that contains synaptic vesicles. Also called *synaptic knob* or *end foot.*

Synaptic gutter Invaginated portion of a sarcolemma under an axon terminal. Also called a *synaptic trough* (TROF).

Synaptic vesicle Membrane-enclosed sac in a synaptic end bulb that stores neurotransmitters.

Synarthrosis (sin'-ar-THRŌ-sis) An immovable joint.

Synchondrosis (sin'-kon-DRŌ-sis) A cartilaginous joint in which the connecting material is hyaline cartilage.

Syncope (SIN-kō-pē) Faint; a sudden temporary loss of consciousness associated with loss of postural tone and followed by spontaneous recovery; most commonly caused by cerebral ischemia.

Syndesmosis (sin'-dez-MŌ-sis) A fibrous joint in which articulating bones are united by dense fibrous tissue.

Syndrome (SIN-drōm) A group of signs and symptoms that occur together in a pattern that is characteristic of a particular disease or abnormal condition.

Syneresis (si-NER-e-sis) The process of clot retraction.

Synergist (SIN-er-jist) A muscle that assists the prime mover by reducing undesired action of unnecessary movement.

Synergistic (syn-er-GIS-tik) *effect* A hormonal interaction in which the effects of two or more hormones complement each other so that the target cell responds to the sum of the hormones involved. An example is the combined actions of estrogens, progesterone (PROG), prolactin (PRL), and oxytocin (OT) necessary for lactation.

Synostosis (sin'-os-TŌ-sis) A joint in which the dense fibrous connective tissue that unites bones at a suture has been replaced by bone, resulting in a complete fusion across the suture line.

Synovial (si-NŌ-vē-al) *cavity* The space between the articulating bones of a synovial (diarthrotic) joint, filled with synovial fluid. Also called a *joint cavity.*

Synovial fluid Secretion of synovial membranes that lubricates joints and nourishes articular cartilage.

Synovial joint A fully movable or diarthrotic joint in which a synovial (joint) cavity is present between the two articulating bones.

Synovial membrane The inner of the two layers of the articular capsule of a synovial joint, composed of loose connective tissue that secretes synovial fluid into the synovial (joint) cavity.

Syphilis (SIF-i-lis) A sexually transmitted disease caused by the bacterium *Treponema pallidum.*

System An association of organs that have a common function.

Systemic (sis-TEM-ik) Affecting the whole body; generalized.

Systemic anatomy The study of particular systems of the body, such as the skeletal, muscular, nervous, cardiovascular, or urinary systems.

Systemic circulation The routes through which oxygenated blood flows from the left ventricle through the aorta to all the organs of the body and deoxygenated blood returns to the right atrium.

Systemic lupus erythematosus (er-i-them-a-TŌ-sus) (*SLE*) An autoimmune, inflammatory disease that may affect every tissue of the body.

Systole (SIS-tō-lē) In the cardiac cycle, the phase of contraction of the heart muscle, especially of the ventricles.

Systolic (sis-TO-lik) *blood pressure* The force exerted by blood on arterial walls during ventricular contraction; the highest pressure measured in the large arteries, about 120 mm Hg under normal conditions for a young, adult male.

Tachycardia (tak'-i-KAR-dē-a) A rapid heartbeat or pulse rate.

Tactile disc Modified epidermal cell in the stratum basale of hairless skin that functions as a cutaneous receptor for discriminative touch. Also called a *Merkel's* (MER-kelz) *disc.*

Taenia coli (TĒ-nē-a KŌ-lī) One of three flat bands of thickened, longitudinal muscles running the length of the large intestine.

Target cell A cell whose activity is affected by a particular hormone.

Tarsal gland Sebaceous (oil) gland that opens on the edge of each eyelid. Also called **Meibomian** (mī-BŌ-mē-an) **gland.**

Tarsus (TAR-sus) A collective term for the seven bones of the ankle.

Tay-Sachs (TĀ-SAKS) **disease** Inherited, progressive neuronal degeneration of the central nervous system due to a deficient lysosomal enzyme that causes excessive accumulations of a lipid called ganglioside.

T cell A lymphocyte that can differentiate into one of six kinds of cells—killer, helper, suppressor, memory, amplifier, or delayed hypersensitivity—all of which function in cellular immunity.

Tectorial (tek-TŌ-rē-al) **membrane** A gelatinous membrane projecting over and in contact with the hair cells of the spiral organ (organ of Corti) in the cochlear duct.

Telophase (TEL-ō-fāz) The final stage of mitosis in which the daughter nuclei become established.

Temporomandibular joint (TMJ) syndrome A disorder of the temporomandibular joint (TMJ) characterized by dull pain around the ear, tenderness of jaw muscles, a clicking or popping noise when opening or closing the mouth, limited or abnormal opening of the mouth, headache, tooth sensitivity, and abnormal wearing of the teeth.

Tendon (TEN-don) A white fibrous cord of dense, regularly arranged connective tissue that attaches muscle to bone.

Tendon organ A proprioceptive receptor, sensitive to changes in muscle tension and force of contraction, found chiefly near the function of tendons and muscles. Also called a **Golgi** (GOL-jē) **tendon organ.**

Tendon reflex A polysynaptic, ipsilateral reflex that is designed to protect tendons and their associated muscles from damage that might be brought about by excessive tension. The receptors involved are called tendon organs (Golgi tendon organs).

Tenosynovitis (ten'-ō-sin-ō-VĪ-tis) Inflammation of a tendon sheath and synovial membrane at a joint.

Tentorium cerebelli (ten-TŌ-rē-um ser'-e-BEL-ē) A transverse shelf of dura mater that forms a partition between the occipital lobe of the cerebral hemispheres and the cerebellum and that covers the cerebellum.

Teratogen (TER-a-tō-jen) Any agent or factor that causes physical defects in a developing embryo.

Terminal ganglion (TER-min-al GANG-lē-on) A cluster of cell bodies of postganglionic parasympathetic neurons either lying very close to the visceral effectors or located within the walls of the visceral effectors supplied by the postganglionic fibers.

Testis (TES-tis) Male gonad that produces sperm and the hormones testosterone and inhibin. Also called a **testicle.**

Testosterone (tes-TOS-te-rōn) A male sex hormone (androgen) secreted by interstitial endocrinocytes (cells of Leydig) of a mature testis; controls the growth and development of male sex organs, secondary sex characteristics, spermatozoa, and body growth.

Tetanus (TET-a-nus) An infectious disease caused by the toxin of *Clostridium tetani,* characterized by tonic muscle spasms and exaggerated reflexes, lockjaw, and arching of the back. A smooth, sustained contraction produced by a series of very rapid stimuli to a muscle.

Tetany (TET-a-nē) A nervous condition caused by hypoparathyroidism and characterized by intermittent or continuous tonic muscular contractions of the extremities.

Tetralogy of Fallot (tet-RAL-ō-jē of fal-Ō) A combination of four congential heart defects: (1) constricted pulmonary semilunar valve, (2) interventricular septal opening, (3) emergence of aorta from both ventricles instead of from the left only, and (4) enlarged right ventricle.

Thalamus (THAL-a-mus) A large, oval structure located above the midbrain, consisting of two masses of gray matter covered by a thin layer of white matter.

Thalassemia (thal'-a-SĒ-mē-a) A group of hereditary hemolytic anemias.

Therapy (Ther-a-pē) The treatment of a disease or disorder.

Thermoreceptor (THER-mō-rē-sep-tor) Receptor that detects changes in temperature.

Thigh The portion of the lower extremity between the hip and the knee.

Third ventricle (VEN-tri-kul) A slitlike cavity between the right and left halves of the thalamus and between the lateral ventricles.

Thoracic (thō-RAS-ik) **cavity** Superior component of the ventral body cavity that contains two pleural cavities, the mediastinum, and the pericardial cavity.

Thoracic duct A lymphatic vessel that begins as a dilation called the cisterna chyli, receives lymph from the left side of the head, neck, and chest, the left arm, and the entire body below the ribs, and empties into the left subclavian vein. Also called the **left lymphatic** (lim-FAT-ik) **duct.**

Thorax (THŌ-raks) The chest.

Threshold potential The membrane voltage that must be reached in order to trigger an action potential (nerve impulse).

Threshold stimulus Any stimulus strong enough to initiate an action potential (nerve impulse). Also called a **liminal** (LIM-i-nal) **stimulus.**

Thrombin (THROM-bin) The active enzyme formed from prothrombin that acts to convert fibrinogen to fibrin.

Thrombocyte (THROM-bō-sīt) A fragment of cytoplasm enclosed in a cell membrane and lacking a nucleus; found in the circulating blood; plays a role in blood clotting. Also called a **platelet** (PLĀT-let).

Thrombophlebitis (throm'-bo-fle-BĪ-tis) A disorder in which inflammation of the wall of a vein is followed by the formation of a blood clot (thrombus).

Thrombosis (throm-BŌ-sis) The formation of a clot in an unbroken blood vessel, usually a vein.

Thrombus A clot formed in an unbroken blood vessel, usually a vein.

Thymus (THĪ-mus) **gland** A bilobed organ, located in the upper mediastinum posterior to the sternum and between the lungs, that plays a role in the immune mechanism of the body.

Thyroid cartilage (THĪ-royd KAR-ti-lij) The largest single cartilage of the larynx, consisting of two fused plates that form the anterior wall of the larynx. Also called the **Adam's apple.**

Thyroid colloid (KOL-loyd) A complex in thyroid follicles consisting of thyroglobulin and stored thyroid hormones.

Thyroid follicle (FOL-i-kul) Spherical sac that forms the parenchyma of the thyroid gland and consists of follicular cells that produce thyroxine (T_4) and triiodothyronine (T_3) and parafollicular cells that produce calcitonin (CT).

Thyroid gland An endocrine gland with right and left lateral lobes on either side of the trachea connected by an isthmus located in front of the trachea just below the cricoid cartilage.

Thyroid-stimulating hormone (TSH) A hormone secreted by the adenohypophysis (anterior lobe) of the pituitary gland that stimulates the synthesis and secretion of hormones produced by the thyroid gland.

Thyroxine (thī-ROK-sēn) (T_4) A hormone secreted by the thyroid that regulates organic metabolism, growth and development, and the activity of the nervous system.

Tic Spasmodic twitching made involuntarily by muscles that are ordinarily under voluntary control.

Tidal volume The volume of air breathed in and out in any one breath; about 500 ml in quiet, resting conditions.

Tinnitus (ti-NĪ-tus) A ringing, roaring, or clicking in the ears.

Tissue A group of similar cells and their intercellular substance joined together to perform a specific function.

Tissue factor (TF) A factor, or collection of factors, whose appearance initiates the blood clotting process. Also called **thromboplastin** (throm-bō-PLAS-tin).

Tissue plasminogen activator (t-PA) An enzyme that dissolves small blood clots by initiating a process that converts plasminogen to plasmin, which degrades the fibrin of a clot.

Tissue rejection Phenomenon by which the body recognizes the protein (HLA antigens) in transplanted tissues or organs as foreign and produces antibodies against them.

Tongue A large skeletal muscle covered by a mucous membrane located on the floor of the oral cavity.

Tonsil (TON-sil) A multiple aggregation of large lymphatic nodules embedded in mucous membrane.

Topical (TOP-i-kal) Applied to the surface rather than ingested or injected.

Torn cartilage A tearing of an articular disk in the knee.

Torpor (TOR-por) State of lethargy and sluggishness that precedes stupor, which precedes semicoma, which precedes coma.

Total lung capacity The sum of tidal volume, inspiratory reserve volume, expiratory reserve volume, and residual volume; about 6,000 ml.

Toxic (TOK-sik) Pertaining to poison; poisonous.

Toxic shock syndrome (TSS) A disease caused by the bacterium *Staphylococcus aureus,* occurring among menstruating females who use tampons and characterized by high fever, sore throat, headache, fatigue, irritability, and abdominal pain.

Trabecula (tra-BEK-yoo-la) Irregular latticework of thin plate of spongy bone. Fibrous cord of connective tissue serving as supporting fiber by forming a septum extending into an organ from its wall or capsule. *Plural,* **trabeculae** (tra-BEK-yoo-lē).

Trachea (TRĀ-kē-a) Tubular air passageway extending from the larynx to the fifth thoracic vertebra. Also called the **windpipe.**

Tracheostomy (trā-kē-OS-tō-mē) Creation of an opening into the trachea through the neck (below the cricoid cartilage), with insertion of a tube to facilitate passage of air or evacuation of secretions.

Trachoma (tra-KŌ-ma) A chronic infectious disease of the conjunctiva and cornea of the eye caused by *Chlamydia trachomatis.*

Tract A bundle of nerve fibers in the central nervous system.

Transcription (trans-KRIP-shun) The first step in the transfer of genetic information in which a single strand of a DNA molecule serves as a template for the formation of an RNA molecule.

Transfusion (trans-FYOO-shun) Transfer of whole blood, blood components, or bone marrow directly into the bloodstream.

Transient ischemic (is-KĒ-mik) **attack (TIA)** Episode of temporary focal, nonconvulsive cerebral dysfunction caused by interference of the blood supply to the brain.

Translation (trans-LĀ-shun) The construction of a new protein on the ribosome of a cell as dictated by the sequence of codons in messenger RNA.

Transplantation (trans-plan-TĀ-shun) The replacement of injured or diseased tissues or organs with natural ones.

Transvaginal oocyte retrieval Procedure in which aspirated secondary oocytes are combined with a solution containing sperm outside the body and then the fertilized ova are implanted in the uterus.

Transverse colon (trans-VERS KŌ-lon) The portion of the large intestine extending across the abdomen from right colic (hepatic) flexure to the left colic (splenic) flexure.

Transverse fissure (FISH-er) The deep cleft that separates the cerebrum from the cerebellum.

Transverse tubules (TOO-byools) (*T tubules*) Minute, cylindrical invaginations of the muscle fiber (cell) membrane that carry the muscle action potentials deep into the muscle fiber.

Trauma (TRAW-ma) An injury, either a physical wound or psychic disorder, caused by an external agent or force, such as a physical blow or emotional shock; the agent or force that causes the injury.

Tremor (TREM-or) Rhythmic, involuntary, purposeless contraction of opposing muscle groups.

Treppe (TREP-ē) The gradual increase in the amount of contraction by a muscle caused by rapid, repeated stimuli of the same strength.

Triad (TRĪ-ad) A complex of three units in a muscle fiber (cell) composed of a transverse tubule and the segments of sarcoplasmic reticulum on both sides of it.

Tricuspid (trī-KUS-pid) *valve* Atrioventricular (AV) valve on the right side of the heart.

Trigeminal neuralgia (trī-JEM-i-nal noo-RAL-jē-a) Pain in one or more of the branches of the trigeminal (V) nerve. Also called **tic douloureux** (doo-loo-ROO).

Trigone (TRĪ-gon) A triangular area at the base of the urinary bladder.

Triiodothyronine (trī-ī-od-ō-THĪ-rō-nēn) (T_3) A hormone produced by the thyroid gland that regulates organic metabolism, growth and development, and the activity of the nervous system.

Trochlea (TROK-lē-a) A pulleylike surface.

Trophoblast (TRŌF-ō-blast) The outer covering of cells of the blastocyst.

Tropic (TRŌ-pik) *hormone* A hormone whose target is another endocrine gland.

Trunk The part of the body to which the upper and lower extremities are attached.

Tubal ligation (lī-GĀ-shun) A sterilization procedure in which the uterine (Fallopian) tubes are tied and cut.

Tuberculosis (too-berk-yoo-LŌ-sis) An infection of the lungs and pleurae caused by *Mycobacterium tuberculosis* resulting in destruction of lung tissue and its replacement by fibrous connective tissue.

Tubular transport maximum (*Tm*)   The maximum amount of a substance that can be reabsorbed by renal tubules under any condition.

Tubular reabsorption   The movement of filtrate from renal tubules back into blood in response to the body's specific needs.

Tubular secretion   The movement of substances in blood back into filtrate in response to the body's specific needs.

Tumor (TOO-mor)   A growth of excess tissue due to an unusually rapid division of cells.

Tunica albuginea (TOO-ni-ka al'-byoo-JIN-ē-a)   A dense layer of white fibrous tissue covering a testis or deep to the surface of an ovary.

Tunica externa (eks-TER-na)   The outer coat of an artery or vein, composed mostly of elastic and collagenous fibers. Also called the *adventitia.*

Tunica interna (in-TER-na)   The inner coat of an artery or vein, consisting of a lining of endothelium, basement membrane, and internal elastic lamina. Also called the *tunica intima* (IN-ti-ma).

Tunica media (MĒ-dē-a)   The middle coat of an artery or vein, composed of smooth muscle and elastic fibers.

T wave   The deflection wave of an electrocardiogram that records ventricular repolarization (relaxation).

Twitch   Rapid, jerky contraction of a musle in response to a single stimulus.

Tympanic antrum (tim-PAN-ik AN-trum)   An air space in the posterior wall of the middle ear that leads into the mastoid air cells or sinus.

Tympanic (tim-PAN-ik) *membrane*   A thin, semitransparent partition of fibrous connective tissue between the external auditory meatus and the middle ear. Also called the *eardrum.*

Type II cutaneous mechanoreceptor   A receptor embedded deeply in the dermis and deeper tissue that detects heavy and continuous touch sensations. Also called an *end organ of Ruffini.*

Ulcer (UL-ser)   An open lesion of the skin or a mucous membrane of the body with loss of substance and necrosis of the tissue.

Umbilical (um-BIL-i-kal) *cord*   The long, ropelike structure, containing the umbilical arteries and vein that connect the fetus to the placenta.

Umbilicus (um-BIL-i-kus or um-bil-Ī-kus)   A small scar on the abdomen that marks the former attachment of the umbilical cord to the fetus. Also called the *navel.*

Upper extremity   The appendage attached at the shoulder girdle, consisting of the arm, forearm, wrist, hand, and fingers.

Uremia (yoo-RĒ-mē-a)   Accumulation of toxic levels of urea and other nitrogenous waste products in the blood, usually resulting from severe kidney malfunction.

Ureter (YOO-re-ter)   One of two tubes that connect the kidney with the urinary bladder.

Urethra (yoo-RĒ-thra)   The duct from the urinary bladder to the exterior of the body that conveys urine in females and urine and semen in males.

Urinalysis   The physical, chemical, and microscopic analysis or examination of urine.

Urinary (YOO-ri-ner-ē) *bladder*   A hollow, muscular organ situated in the pelvic cavity posterior to the symphysis pubis.

Urinary tract infection (*UTI*)   An infection of a part of the urinary tract or the presence of large numbers of microbes in urine.

Urine   The fluid produced by the kidneys that contains wastes or excess materials and is excreted from the body through the urethra.

Urobilinogenuria   The presence of urobilinogen in urine.

Urogenital (yoo'-rō-JEN-i-tal) *triangle*   The region of the pelvic floor below the symphysis pubis, bounded by the symphysis pubis and the ischial tuberosities and containing the external genitalia.

Urology (yoo-ROL-ō-jē)   The specialized branch of medicine that deals with the structure, function, and diseases of the male and female urinary systems and the male reproductive system.

Uterine (YOO-ter-in) *tube*   Duct that transports ova from the ovary to the uterus. Also called the *Fallopian* (fal-LŌ-pē-an) *tube* or *oviduct.*

Uterosacral ligament (yoo'-ter-ō-SĀ-kral LIG-a-ment)   A fibrous band of tissue extending from the cervix of the uterus laterally to attach to the sacrum.

Uterovesical (yoo'-ter-ō-VES-î-kal) *pouch*   A shallow pouch formed by the reflection of the peritoneum from the anterior surface of the uterus, at the junction of the cervix and the body, to the posterior surface of the urinary bladder.

Uterus (YOO-te-rus)   The hollow, muscular organ in females that is the site of menstruation, implantation, development of the fetus, and labor. Also called the *womb.*

Utricle (YOO-tri-kul)   The larger of the two divisions of the membranous labyrinth located inside the vestibule of the inner ear, containing a receptor organ for static equilibrium.

Uvea (YOO-vē-a)   The three structures that together make up the vascular tunic of the eye.

Uvula (YOO-vyoo-la)   A soft, fleshy mass, especially the V-shaped pendant part, descending from the soft palate.

Vacuole (VAK-you-ōl)   Membrane-bound organelle that, in animal cells, frequently functions in temporary storage or transportation.

Vagina (va-JĪ-na)   A muscular, tubular organ that leads from the uterus to the vestibule, situated between the urinary bladder and the rectum of the female.

Valvular stenosis (VAL-vyoo-lar STEN-Ō-sis)   A narrowing of heart valve, usually the bicuspid (mitral) valve.

Varicose (VAR-i-kōs)   Pertaining to an unnatural swelling, as in the case of a varicose vein.

Vasa recta (REK-ta)   Extensions of the efferent arteriole of a juxtaglomerular nephron that run alongside the loop of the nephron (Henle) in the medullary region.

Vasa vasorum (VĀ-sa va-SŌ-rum)   Blood vessels that supply nutrients to the larger arteries and veins.

Vascular (VAS-kyoo-lar)   Pertaining to or containing many blood vessels.

Vascular spasm   Contraction of the smooth muscle in the wall of a damaged blood vessel to prevent blood loss.

Vascular tunic (TOO-nik)   The middle layer of the eyeball, composed of the choroid, ciliary body, and iris. Also called the *uvea* (YOO-vē-a).

Vascular (venous) sinus   A vein with a thin endothelial wall that lacks a tunica media and externa and is supported by surrounding tissue.

Vasectomy (va-SEK-tō-mē) A means of sterilization of males in which a portion of each ductus (vas) deferens is removed.

Vasoconstriction (vāz-ō-kon-STRIK-shun) A decrease in the size of the lumen of a blood vessel caused by contracting of the smooth muscle in the wall of the vessel.

Vasodilation (vās'-ō-DĪ-lā-shun) An increase in the size of the lumen of a blood vessel caused by relaxation of the smooth muscle in the wall of the vessel.

Vasomotion (vāz-ō-MŌ-shun) Intermittent contraction and relaxation of the smooth muscle of the metarterioles and precapillary sphincters that result in an intermittent blood flow.

Vasomotor (vā-sō-MŌ-tor) *center* A cluster of neurons in the medulla that controls the diameter of blood vessels, especially arteries.

Vein A blood vessel that conveys blood from tissues back to the heart.

Vena cava (VĒ-na KĀ-va) One of two large veins that open into the right atrium, returning to the heart all of the deoxygenated blood from the systemic circulation except from the coronary circulation.

Venesection (vēn'-e-SEK-shun) Opening of a vein for withdrawal of blood.

Ventral (VEN-tral) Pertaining to the anterior or front side of the body; opposite of dorsal.

Ventral body cavity Cavity near the ventral aspect of the body that contains viscera and consists of a superior thoracic cavity and an inferior abdominopelvic cavity.

Ventricle (VEN-tri-kul) A cavity in the brain or an inferior chamber of the heart.

Ventricular fibrillation (ven-TRIK-yoo-lar fib-ri-LĀ-shun) Asynchronous ventricular contractions that result in cardiovascular failure.

Venule (VEN-yool) A small vein that collects blood from capillaries and delivers it to a vein.

Vermiform appendix (VER-mi-form a-PEN-diks) A twisted, coiled tube attached to the cecum.

Vermis (VER-mis) The central constricted area of the cerebellum that separates the two cerebellar hemispheres.

Vertebral (VER-te-bral) *canal* A cavity within the vertebral column formed by the vertebral foramina of all the vertebrae and containing the spinal cord. Also called the *spinal canal.*

Vertebral column The 26 vertebrae; encloses and protects the spinal cord and serves as a point of attachment for the ribs and back muscles. Also called the *spine, spinal column,* or *backbone.*

Vertigo (VER-ti-go) Sensation of spinning or movement.

Vesicle (VES-i-kul) A small bladder or sac containing liquid.

Vesicular ovarian follicle A relatively large, fluid-filled follicle containing an immature ovum and its surrounding tissues that secretes estrogens. Also called a *Graafian* (GRAF-ē-an) *follicle.*

Vestibular (ves-TIB-yoo-lar) *membrane* The membrane that separates the cochlear duct from the scala vertibuli.

Vestibule (VES-ti-byool) A small space or cavity at the beginning of a canal, especially the inner ear, larynx, mouth, nose, and vagina.

Villus (VIL-lus) A projection of the intestinal mucosal cells containing connective tissue, blood vessels, and a lymphatic vessel; functions in the absorption of the end products of digestion. *Plural, villi* (VIL-Ī).

Viscera (VIS-er-a) The organs inside the ventral body cavity. *Singular, viscus* (VIS-kus).

Visceral (VIS-er-al) Pertaining to the organs or to the covering of an organ.

Visceral effector (e-FEK-tor) Cardiac muscle, smooth muscle, and glandular epithelium.

Visceral muscle An organ specialized for contraction, composed of smooth muscle fibers (cells), located in the walls of hollow internal structures, and stimulated by visceral efferent neurons.

Visceral pleura (PLOO-ra) The inner layer of the serous membrane that covers the lungs.

Visceroceptor (vis'-er-ō-SEP-tor) Receptor that provides information about the body's internal environment.

Viscosity (vis-KOS-i-tē) The state of being sticky or thick.

Vital capacity The sum of inspiratory reserve volume, tidal volume, and expiratory reserve volume; about 4,800 ml.

Vital signs Signs necessary to life that include temperature (T), pulse (P), respiratory rate (RR), and blood pressure (BP).

Vitamin An organic molecule necessary in trace amounts that acts as a catalyst in normal metabolic processes in the body.

Vitiligo (vit-i-LĪ-go) Patchy, white spots on the skin due to partial or complete loss of melanocytes.

Vitreous (VIT-rē-us) *body* A soft, jellylike substance that fills the vitreous chambers of the eyeball, lying between the lens and the retina.

Vocal folds Pair of mucous membrane folds below the ventricular folds that function in voice production. Also called *true vocal cords.*

Volkmann's canal *See Perforating canal.*

Voltage-sensitive channel An ion channel in a plasma membrane composed of integral proteins that functions like a gate to permit or restrict the movement of ions in response to the voltage state of the membrane.

Vomiting Forcible expulsion of the contents of the upper gastrointestinal tract through the mouth.

Vulva (VUL-va) Collective designation for the external genitalia of the female. Also called the *pudendum* (poo-DEN-dum).

Wallerian (wal-LE-rē-an) *degeneration* Degeneration of the portion of the axon and myelin sheath of a neuron distal to the site of injury.

Wandering macrophage (MAK-rō-fāj) Phagocytic cell that develops from a monocyte, leaves the blood, and migrates to infected tissues.

Wart Generally benign tumor of epithelial skin cells caused by a virus.

Wave summation (sum-MĀ-shun) The algebraic addition of the excitatory and inhibitory effects of many stimuli applied to a nerve cell body. The increased strength of muscle contraction that results when stimuli follow in rapid succession. Also called *temporal summation.*

White matter Aggregations or bundles of myelinated axons located in the brain and spinal cord.

White matter tract The treelike appearance of the white matter of the cerebellum when seen in midsagittal section. Also called *arbor vitae* (AR-bōr VĒ-te). A series of branching ridges within the cervix of the uterus.

Wormian bone *See Sutural bone.*

Xiphoid (ZĪ-foyd) Sword-shaped. The lowest portion of the sternum.

Yolk sac An extraembryonic membrane that connects with the midgut during early embryonic development, but is nonfunctional in humans.

Zona fasciculata (ZŌ-na fa-sik'-yoo-LA-ta) The middle zone of the adrenal cortex that consists of cells arranged in long, straight cords and that secretes glucocorticoid hormones.

Zona glomerulosa (glo-mer'-yoo-LŌ-sa) The outer zone of the adrenal cortex, directly under the connective tissue covering, that consists of cells arranged in arched loops or round balls and that secretes mineralocorticoid hormones.

Zona pellucida (pe-LOO-si-da) Gelatinous glycoprotein layer internal to the corona radiata that surrounds a secondary oocyte.

Zona reticularis (ret-ik'-yoo-LAR-is) The inner zone of the adrenal cortex, consisting of cords of branching cells that secrete sex hormones, chiefly androgens.

Zygote (ZĪ-gōt) The single cell resulting from the union of a male and female gamete; the fertilized ovum.

Zymogenic (zī-mō-JEN-ik) *cell* One of the cells of a gastric gland that secretes the principal gastric enzyme precursor, pepsinogen. Also called a *peptic cell.*

Index

Page numbers followed by the letter E indicate terms to be found in exhibits.

A band, 144
Abdominal aorta, 319
Abdominal cavity, 8, 9
Abdominopelvic cavity, 8, 9, 10
Abdominopelvic diaphragm, 8, 164
Abdominopelvic quadrants
  left lower, 10, 12
  left upper, 9, 10, 12
  right lower, 12
  right upper, 9, 12
Abdominopelvic region, 8, 9, 12
Abduction, 136
Abnormal curve, 127
Abnormal uterine bleeding, 503
ABO blood grouping, 310–312
Abortion, 529
Abrasion, 93
Abscess, 362
Absorption, 400, 410, 417–419
  carbohydrate, 417
  electrolyte, 419
  lipid, 417
  protein, 417
  vitamin, 419
  water, 419
Accessory organs of skin
  ceruminous glands, 91
  hair, 90–91
  nail, 91–92
  sebaceous glands, 91
  sudoriferous glands, 91
Accessory sex glands, 490
Accessory structures of digestion, 400
  gallbladder, 414–415
  liver, 410–411, 414
  pancreas, 410
  salivary glands, 402–404
  teeth, 404
  tongue, 402
Accessory structures of eye, 249
Accommodation, 252
Acetabulum, 117
Acetylcholine (ACh), 146, 147, 221, 222E
Acetylcholinesterase (AchE), 235
Acetyl coenzyme A, 430
Achilles tendon. *See* Calcaneal tendon
Achondroplasia, 127
Achromatopsia, 265
Acid, 25–26
Acid–base balance, 25, 26, 478–480
Acid–base imbalances, 480–481
Acidosis
  metabolic, 481

respiratory, 481
  treatment of, 481
Acini, 410
Acne, 92
Acquired immune deficiency syndrome (AIDS),
    369–372
Acquired immunity
  artificially acquired active, 362
  artificially acquired passive, 363
  naturally acquired active, 362
  naturally acquired passive, 362
Acromegaly, 291
Acrosome, 489
Actin, 47, 144
Actin-binding site, 144
Active processes, 41–44, 44
Active site, 428
Active transport, 41–42, 44
Adam's apple. *See* Thyroid cartilage
Adaptation, 242
Addison's disease, 291
Adduction, 136
Adenitis, 373
Adenosarcoma, 56
Adenosine diphosphate
  chemical composition of, 32
  and electron transport chain, 430
  function of, 32
  and glycolysis, 430
  and Krebs cycle, 430
  and muscle contraction, 146–149
Adherence, phagocytic, 359
Adipocyte, 76
Adipose tissue, 766
ADP. *See* Adenosine diphosphate
Adrenal cortex, 281–284
Adrenal gland, 281–285
Adrenal medulla, 284–285
Adrenergic fiber, 235
Adrenocorticotropic hormone (ACTH), 275
Adrenogenital syndrome, 293
Adult connective tissue
  definition of, 76
  types of, 76–77
Aerobic respiration, 32, 149
Aerobics, 348
Afferent lymphatic vessel, 356
Afferent neuron, 191
A fiber, 197
Afterbirth, 524
Afterimage, 242
Agglutination, 310

Agglutinin, 310
Agglutinogen, 310
Aggregated lymphatic follicle, 416
Agnosia, 226
Agranular leucocyte, 302
Agraphia, 221
AIDS. *See* Acquired immune deficiency syn-
    drome (AIDS)
Alarm reaction of stress, 287
Albinism, 90
Albumin, 304, 465E
Albuminuria, 465E
Aldosterone, 282
Aldosteronism, 291
Alimentary canal. *See* Gastrointestinal (GI) tract
Alkalosis
  metabolic, 481
  respiratory, 481
  treatment of, 481
Allele, 527
Allergen, 372
Allergy. *See* Hypersensitivity
All-or-nothing principle, 150, 195
Alpha cell, 285
Alveolar-capillary (respiratory) membrane,
    382–383
Alveolar duct, 382
Alveolar macrophage, 382
Alveolar sac, 382
Alveolus
  of lung, 382
  of mammary gland, 497
  of mandible, 110
  of maxilla, 107
Alzheimer's disease (AD), 226
Amenorrhea, 503
Ametropia, 265
Amino acid, 31, 435
  essential, 435
  nonessential, 435
Amniocentesis, 519–520
Amnion, 514
Amniotic cavity, 514
Amniotic fluid (AF), 514
Amphiarthrosis, 131
Amplifier T cells, 365
Ampulla, of semicircular canal, 260
Amylase. *See* Pancreatic amylase; Salivary amy-
    lase
Anabolism
  of carbohydrates, 429–430
  definition of, 23, 428
  of lipids, 430–431, 433–434
  of proteins, 434–435

Anaerobic respiration, 32, 149, 429
Anal canal, 420
Analgesia, 226
Anaphase, 53
Anatomical position, 6
Anatomic dead space, 385
Anatomy, 3
Androgen, 284, 486
Anemia
  aplastic, 312
  hemolytic, 312
  hemorrhagic, 312
  nutritional, 312
  pernicious, 312
  sickle-cell, 312
Aneurysm, 349
Angina pectoris, 322
Angiocardiography, 330
Angiotensin I, 282, 456
Angiotensin II, 282, 456
Anion, 23
Ankylosis, 140
Annulus fibrosus, 111
Anopsia, 223E, 265
Anorexia nervosa, 422
Anosmia, 223E
ANS. See Autonomic nervous system
Antagonist, 154
Anterior cavity of eyeball, 251
Anterior corticospinal tract, 207E
Anterior pituitary, 274
Anterior root, 206E, 207
Antibody (Ab)
  agglutinin, 310
  and allergy, 372
  and autoimmune disease, 372
  cellular immunity and, 364–365
  classes of, 363
  definition of, 323
  formation of, 365, 368
  humoral activity and, 365, 368
  primary response and, 368
  structure of, 363
  and tissue rejection, 372
Anticoagulant, 308
Antidiuretic, 277
Antidiuretic hormone (ADH), 277–278
Antigen (Ag)
  agglutinogen, 310
  and allergy, 372
  and autoimmune disease, 372
  and cellular immunity, 363–365
  characteristics of, 363
  definition of, 363
  HLA, 302
  and humoral immunity, 364, 365, 368
  and tissue rejection, 372
  tumor-specific, 39
Antigen–antibody response, 303
Antigenic determinant site, 363
Anti-oncogenes, 56
Anuria, 457
Anus, 420
Aorta, 319, 541
Aortic body, 541
Aortic insufficiency, 330
Aortic semilunar valve, 319

Apex
  of heart, 317
  of lung, 380
Aphasia, 221
Apnea, 394
Apneustic area, 391
Aponeurosis
  definition of, 143
  galea aponeurotica, 158E
Appendicitis, 422
Appendicular skeleton, 103, 115–123
Appendix. See Vermiform appendix
Apraxia, 226
Aqueous humor, 251
Arachnoid, 204, 209
Arachnoid villi, 210
Arch of aorta, 319, 341
Arch of foot, 122–123
Areola, 497
Arousal, 215
Arrector pili muscle, 91
Arrhythmia, 330
Arteries
  abdominal aorta, 341
  afferent arteriole, 450
  arch of aorta, 319, 341
  ascending aorta, 319, 341
  coats of, 334
  common disorders of, 348–349
  common iliac, 341
  coronary, 322
  of coronary circulation, 322
  definition of, 334
  ductus arteriosus, 516
  of fetal circulation, 348, 516
  of hepatic portal circulation, 348
  histology of, 334
  pulmonary, 319, 342–343
  of pulmonary circulation, 322, 342–343
  pulmonary trunk, 319, 342
  renal, 450
  spiral arteriole, 496
  straight arteriole, 495
  of systemic circulation, 341–342
  thoracic aorta, 319, 341
  umbilical, 514
  vasoconstriction of, 334
  vasodilation of, 334, 360
Arteriole, 334
Arthritis
  definition of, 136
  gouty, 136
  osteoarthritis, 136
  rheumatoid (RA), 136
Arthrology, 131
Arthrosis, 140
Articular capsule, 131, 132
Articular cartilage, 98, 100, 132
Articular disc, 132
Articulations
  amphiarthrosis, 131
  ball-and-socket, 137
  cartilaginous, 131
  common disorders, 136
  definition of, 131
  diarthrosis, 132–133
  ellipsoidal, 134, 136
  fibrous, 131

  as fulcrums, 154
  functional classification of, 131
  gliding, 133
  gomphosis, 132
  hinge, 134
  medical terminology and conditions, 140
  movements at, 132–133
  pivot, 134
  saddle, 136
  structural classification of, 131
  suture, 131
  symphysis, 132
  symphysis pubis, 117, 132
  synchondrosis, 132
  syndesmosis, 132
  synovial, 132–137
Artificial kidney, 458
Artificial pacemaker, 323
Ascending aorta, 319, 341
Ascending colon, 420
Ascending limb of loop of nephron (Henle),
  450
Ascending tract, 204
Asphyxia, 396
Aspiration, 396
Association areas of cerebrum, 221
Association fiber, 217
Association (interneuron) neuron, 194
Astereognosis, 265
Asthma, bronchial, 380, 395
Astigmatism, 252
Astrocyte, 189E
Ataxia, 221
Atherosclerosis, 329
Athlete's foot, 93
Atlas, 113
Atom, 19, 21
Atomic number, 21
Atomic structure
  electron, 19
  neuron, 19
  nucleus, 19
  proton, 19
Atomic weight, 21
ATP. See Adenosine triphosphate
ATP-binding site, 144
Atrial diastole, 324
Atrial fibrillation, 330
Atrial flutter, 330
Atrial natriuretic factor (ANF), 287
Atrial systole, 324
Atrioventricular (AV) bundle, 322
Atrioventricular (AV) node, 322
Atrioventricular (AV) valve, 322
Atrium, 319
Atrophy, 57
Audiometer, 265
Auditory association area, 217
Auditory ossicle, 254
Auditory sensation, 254–261
Auditory tube, 254
Auricle, of ear. See Pinna
Auricle, of heart, 319
Autoimmune disease, 372
Autologous intraoperative transfusion (AIT),
  313
Autolysis, 46–47
Autonomic control, of heart, 375

Autonomic ganglia
  prevertebral, 234
  sympathetic trunk, 234
  terminal, 234
Autonomic nervous system (ANS), 188, 231–239
  activities of, 235
  comparison with somatic efferent, 231
  definition of, 231
  ganglia of, 232, 234, 235
  and heart control, 235
  neurotransmitters, 235
  parasympathetic division, 235
  postganglionic neuron, 235
  structure of, 232, 234–235
  sympathetic division, 235
  visceral efferent neurons of, 232
Autophagy, 46
Autoregulation, 338–339
Autorhythmicity, 180
Autosome, 528
Avascular, 62
Axial skeleton, 103–115
Axis, 113
Axon
  postganglionic, 232
  preganglionic, 232
Axon collateral, 189
Axon terminal, 189

Backbone. See Vertebral column
Ball-and-socket joint, 137
Baroreceptor, 375
Bartholin's gland. See Greater vestibular gland
Basal ganglion, 217
Basal metabolic rate (BMR), 442
Base
  of heart, 317
  inorganic, 25
  of lung, 380
Basement membrane, 62
Basilar membrane, 260
Basophil, 303
B cell, 364–365, 368
Bedsore. See Decubitus ulcer
Bell's palsy, 224E
Belly, of skeletal muscle, 154
Benign tumor, 53
Beri-beri, 438E
Beta cell, 285
Beta oxidation, 433
B fiber, 197
Bicuspid (mitral) valve, 319
Bile, 411, 414
Bile secretion, regulation of, 414
Bilirubin, 301, 466
Biliverdin, 301
Biopsy, 57
Bipolar neuron, 191, 251
Birth. See Parturition
Birth control (BC)
  barrier method, 525
  castration, 524
  cervical cap, 525
  chemical methods, 525
  coitus interruptus, 525
  condom, 525
  contraceptive sponge, 525

diaphragm, 525
  hormonal method, 524
  induced abortion, 525
  intrauterine device (IUD), 524–525
  laparoscopy, 524
  oral contraception, 524
  physiologic (natural) methods, 525
  rhythm method, 525
  sterilization, 524
  symptothermal method, 525
  tubal ligation, 524
  vasectomy, 524
Blackhead. See Comedo
Blastocoel, 512
Blastocyst, 512
Blepharitis, 265
Blind spot. See Optic disc
Blood
  coagulation, 305, 308
  common disorders of, 312–313
  formed elements, 299–304
  functions of, 298
  group, 310–312
  homeostasis of, 304–305, 308–310
  medical terminology and conditions, 313
  origin, 299
  physical characteristics of, 298
  plasma, 77, 304
Blood–brain barrier (BBB), 212
Blood–cerebrospinal fluid barrier, 209
Blood clot, 305, 308, 310
Blood clotting. See Blood coagulation
Blood coagulation
  definition of, 305
  extrinsic pathway of, 308
  factors in, 305
  fibrinolysis and, 308
  intrinsic pathway of, 308
  prevention of, 308, 310
  retraction and, 308
  stages of, 305
Blood colloidal osmotic pressure, 454
Blood flow, 336
Blood grouping (typing)
  ABO, 310–311
  Rh, 311–312
Blood hydrostatic pressure (BHP), 475
Blood osmotic pressure (BOP), 475
Blood plasma, 38, 304
Blood reservoir, 336
Blood sugar (BS)
  control of, 285–286
  homeostasis of, 13
Blood vessels
  arteries, 334
  arterioles, 334
  as blood reservoirs, 336
  capillaries, 344–345
  common disorders of, 348–349
  of coronary (cardiac) circulation, 322
  definition of, 334
  and exercise, 348
  of fetal circulation, 348
  of hepatic portal circulation, 348
  medical terminology and conditions, 349
  pressure within, 336–337
  of pulmonary circulation, 342–343
  of systemic circulation, 341–342

veins, 336
  venules, 335
BMR. See Basal metabolic rate
Body
  of femur, 118
  of humerus, 116
  of scapula, 115
  of sternum, 114
  of stomach, 407
  of uterus, 495
  of vertebra, 112
Body cavity
  abdominal, 8
  abdominopelvic, 8
  cranial, 8
  definition of, 8
  dorsal, 8
  nasal, 376
  oral, 401–405
  pelvic, 9
  pericardial, 9, 317
  pleural, 8, 380
  thoracic, 8
  ventral, 8
  vertebral, 8
Body heat
  loss of, 442
  measurement of, 437
  production of, 437, 442
  regulation of, 442
Body temperature, regulation by skin, 92
Bolus, 402
Bone marrow
  red, 96
  yellow, 98
Bones
  atlas, 113
  auditory ossicles, 254
  axis, 113
  calcaneus, 121
  capitate, 117
  carpus, 116
  cervical vertebrae, 113
  clavicle, 115
  coccyx, 113–114
  cranial, 105–107
  cuboid, 121
  ethmoid, 106–107
  facial, 107, 110
  femur, 118
  fibula, 121
  first (medial) cuneiform, 121
  fontanels of, 110
  frontal, 105
  hamate, 117
  hip, 117
  humerus, 116
  hyoid, 111
  incus, 254
  inferior nasal concha, 110
  lacrimal, 110
  as levers, 154
  lower extremity, 117–123
  lumbar vertebrae, 113
  lunate, 116
  malleus, 254
  mandible, 110
  maxilla, 107

Bones (*Continued*)
metacarpus, 117
metatarsus, 122
nasal, 107
navicular, of foot, 121
occipital, 105–10
organization into skeleton, 103E
palatine, 110
paranasal sinuses of, 110
parietal, 105
patella, 120
pectoral (shoulder) girdle, 115–116
phalanges, of fingers, 117
phalanges, of toes, 122
pisiform, 117
radius, 116
ribs, 114–115
sacrum, 113
scaphoid, 116
scapula, 115
second (intermediate) cuneiform, 121
skull, 103–111
sphenoid, 106
stapes, 254
sternum, 114
surface markings of, 102–103
sutures of, 103
talus, 121
tarsus, 121
temporal, 105
third (lateral) cuneiform, 121
thoracic vertebrae, 113
thorax, 114
tibia, 120
trapezium, 117
trapezoid, 117
triquetral, 117
types of, 96
ulna, 116
upper extremity, 116–117
vertebral column, 111–114
vomer, 110
zygomatic, 110
Bone tissue
common disorders of, 126–127
functions of, 96
growth of, 98–101
histology, 97–98
homeostasis of, 101–102
medical terminology and conditions, 127
ossification, 98–101
remodeling, 101
Bony labyrinth, 254
Bowman's capsule. *See* Glomerular capsule
Bowman's gland. *See* Olfactory gland
Brachial plexus, 209
Bradycardia, 339
Brain
association areas, 217
basal ganglia, 217
blood–brain barrier (BBB) of, 212
blood supply, 210, 212
brain stem, 212–214
brain waves, 221
cerebellum, 221
cerebrospinal fluid (CSF), 209–210
cerebrum, 203
common disorders of, 224–226

cranial meninges, 209
cranial nerves and, 221, 222E–224E
diencephalon, 214
hypothalamus, 214–215
lateralization and, 216–217
limbic system, 217
lobes of, 215–216
medical terminology and conditions, 226
medulla oblongata, 212–213
midbrain, 213–214
motor areas, 217
neurotransmitters, 221
parts of, 209
pons, 213
protection, 209
reticular formation, 212
sensory areas, 217
thalamus, 214
ventricles of, 209
white matter, 217
Brain death, 221
Brain lateralization, 216–217
Brain sand, 286
Brain stem, 212–214
Brain tumor, 225
Brain wave, 221
Breast. *See* Mammary gland
Breast cancer, 497, 506
Breathing. *See* Pulmonary ventilation
Breech presentation, 530
Broca's area of cerebrum. *See* Motor speech
area
Bronchi
alveolar duct, 382
bronchioles, 380
primary, 380
respiratory, 382
secondary, 380
terminal, 380
Bronchial tree, 380
Bronchiole, 380
Bronchitis, 395
Bronchogenic carcinoma, 395
Bronchoscopy, 380
Brunner's gland. *See* Duodenal gland
Buffer system, 26, 479–480
carbonic acid-bicarbonate, 479
hemoglobin, 479–480
phosphate, 479
protein, 480
Bulb of hair, 90
Bulbourethral gland, 490
Bulimia, 422
Bundle branch, 322
Bundle of His. *See* Atrioventricular (AV) bundle
Bunion, 123
Burn
classification of, 92
definition of, 92
Bursae, 132
Bursitis, 133, 136

Calcaneal tendon, 180
Calcaneus, 121
Calcification, 99
Calcitonin (CT), 281
Calcium
distribution of, 474

functions of, 474
regulation of, 474
Calyx, 449
Canaliculi, 77, 98
Canal of Schlemm. *See* Scleral venous sinus
Cancellous bone tissue. *See* Spongy bone
Cancer (CA)
antibodies and, 369
antigens and, 369
of brain, 225
of breasts, 497, 506
causes of, 56
cervical, 506
colorectal, 422
definition of, 53
of GI tract, 422
of lung, 395
malignant tumor, 53
metastasis, 53
prostate, 503
of skin, 90, 92
testicular, 503
types of, 56
of uterus, 496
Canker sore, 422
Capacitation, 512
Capillaries, 334–335
Capitate, 117
Capsular hydrostatic pressure, 454
Carbaminohemoglobin, 390
Carbohydrate
absorption of, 414
definition of, 27
digestion of, 416
disaccharide, 27
fate of, 429
functions of, 27
metabolism of, 429–430
monosaccharide, 27
polysaccharide, 27
structure of, 27
Carbon dioxide
and hemoglobin, 390
and respiratory control, 291, 294
transport of, 390
Carbonic acid–bicarbonate buffer system, 479
Carbon monoxide poisoning, 389–390
Carboxypeptidase, 410
Carbuncle, 93
Carcinogen, 56
Carcinoma, 56
Cardia, of stomach, 407
Cardiac arrest, 330
Cardiac center, 212, 325
Cardiac cycle, 324
Cardiac muscle tissue, 180
Cardiac notch, 382
Cardiac output (CO), 375
Cardiac tamponade, 317
Cardioacceleratory center (CAC), 375
Cardioinhibitory center (CIC), 325
Cardiology, 317
Cardiomegaly, 330
Cardiopulmonary resuscitation (CPR), 396
Cardiovascular system, 298–349
arteries, 334
blood, 298–305, 308–312
blood pressure, 336

capillaries, 334–335
circulatory routes, 341–345, 348
common disorders of, 312–313, 329–330, 348–349
coronary (cardiac) circulation, 322
definition of, 298
exercise and, 348
fetal circulation, 348
formed elements, 299
heart, 317–330
hepatic portal circulation, 348
interstitial fluid, lymph and, 298
medical terminology and conditions, 313, 330, 349
physiology of circulation, 336–339
plasma, 304
pulmonary circulation, 342–343
pulse, 339
systemic circulation, 341–342
veins, 336
Carotene, 90
Carotid body, 338
Carpus, 116
Carrier, 527
Cartilage
articular, 98, 100, 132
costal, 115
cricoid, 378
definition of, 76
thyroid, 377
types of, 76
Cartilaginous joint
definition of, 131
types of, 132
Cast, 466
Castration, 524
Catabolism
of carbohydrates, 429–430
definition of, 24, 25, 428
of lipids, 430–434
of proteins, 434–435
Catalyst, 428
Cataract, 251, 264
Cation, 23, 472
Cauda equina, 205
Caudate nucleus, 217
Cecum, 420
Cell
adipocyte, 76
alpha, 285
alveolar macrophage, 382
B, 364–365, 368
beta, 285
and cancer, 53, 56
cytoplasms of, 44
definition of, 3, 36
delta, 285
division of, 51–53
enteroendocrine, 407
gametes, 486
generalized animal, 36
hepatic, 410
inclusions of, 48
interstitial endocrinocyte, 486
medical terminology and conditions, 56–57
mucous, 407
organelles of, 44–47
parietal, 407

parts of, 36, 37E
peptic, 407
plasma membrane of, 36–44
spermatozoon, 489
stellate reticuloendothelial, 410
sustentacular, 485
T, 364–365
Cell body, of neuron, 189
Cell division
abnormal, 53, 56
definition of, 51
reproductive, 51
somatic, 51
types of, 51
Cell inclusions
definition of, 48
types of, 48
Cell junction, 62
Cell membrane. See Plasma membrane
Cellular immunity, 363–365, 368
Cellular respiration, 149
Cementum, 404
Center, of reflex arc, 207–208
Center of ossification, 99
Central canal
of bone, 77, 98
of spinal cord, 206
Central fovea, 251
Central nervous system (CNS), 204–223
brain, 209–217, 220–221
cerebrospinal fluid of, 209–210
common disorders of, 224–226
cranial nerves, 221, 223, 223E, 224E
definition of, 188
grouping of neural tissue, 204
medical terminology and conditions, 226
neurotransmitters, 199, 221
plexuses of, 209
protection and coverings, 204–206
reflexes and, 207
spinal cord, 204–206
spinal nerves, 208, 209
spinal tap, 205
tracts of, 206
Central sulcus, 216
Central vein, 410
Centriole, 47
Centromere, 53
Centrosome, 47
Cephalgia, 225
Cerebellar nucleus, 221
Cerebellar peduncle, 213, 221
Cerebellum, 221
Cerebral aqueduct, 209
Cerebral arterial circle, 210, 343
Cerebral circulation, 343, 348
Cerebral cortex, 215
Cerebral hemisphere, 215
Cerebral nucleus. See Basal ganglion
Cerebral palsy, 225
Cerebral peduncle, 213
Cerebral sulcus
central, 216
definition of, 216
lateral, 216
parietooccipital, 216
Cerebrospinal fluid (CSF), 209–210
Cerebrovascular accident (CVA), 224

Cerebrum
association areas, 221
basal ganglia, 217
brain lateralization, 216–217
brain waves, 221
convolutions, 215
cortex of, 215
fissures, 215
limbic system, 217
lobes, 215–216
motor areas, 217
sensory areas, 217
sulci, 215
white matter, 217
Cerumen, 91, 254
Ceruminous gland, 91, 254
Cervical cancer, 506–507
Cervical cap, 525
Cervical dysplasia, 506
Cervical enlargement, 206
Cervical plexus, 209
Cervical vertebrae, 111
Cervix, of uterus, 495
C fiber, 197
Chambers of heart, 319
Chancre, 503
Chemical bond
covalent, 23
hydrogen, 23
ionic, 22–23
Chemical compound, 22
inorganic, 25–26
organic, 27, 30–31
Chemical digestion, 400
Chemical element, 19
Chemical energy, 25
Chemical method, of contraception, 525
Chemical reaction
and covalent bonds, 23
decomposition, 24–25
and energy, 21, 25
and hydrogen bonds, 23
and ionic bonds, 22–23
synthesis, 23–24
Chemical symbol, 19
Chemical theory, of olfaction, 247
Chemoreceptor, 242, 338, 394
Chemosensitive area, 391
Chemotaxis, 361
Chest. See Thorax
Cheyne-Stokes respiration, 396
Chill, 444
Chlamydia, 503
Chloride
control of, 475
deficiency of, 475
distribution of, 475
functions of, 475
Cholecystitis, 422
Cholecystokinin (CCK), 287, 409, 409E
Cholesterol, 30, 329
Cholinergic fiber, 235
Chondrocyte, 76
Chondroitin sulfate, 61
Chondrosarcoma, 56
Chordae tendineae, 319
Chorion, 514
Chorionic villi, 514

Chorionic villus sampling (CVS), 520
Choroid, 249
Choroid plexus, 209
Chromatid, 53
Chromatin, 44, 51
Chromatophilic substance, 189
Chromosome, 44, 51
Chronic fatigue syndrome, 313
Chylomicron, 418
Chyme, 407
Chymotrypsin, 410
Cigarette smoking, 395
Cilia, 47
Ciliary body, 249
Ciliary muscle, 249
Ciliary process, 249
Circadian rhythm, 215
Circle of Willis. *See* Cerebral arterial circle
Circular fold, 416
Circulatory routes, 341–345, 348
Circumcision, 492
Circumduction, 136
Circumvalate papilla, 402
Cirrhosis, 422
Citric acid cycle. *See* Krebs cycle
Clavicle, 115
Clawfoot, 123
Cleavage, 512
Cleavage furrow, 53
Cleft lip, 107
Cleft palate, 107
Climacteric, 501
Climax. *See* Orgasm
Clinical perineum, 497
Clitoris, 496
Clot retraction, 308
CNA. *See* Central nervous system
Coagulation factors, 305
Coarctation of aorta, 329
Coccyx, 113–114
Cochlea, 260
Cochlear duct, 260
Coenzyme A (CoA), 430
Coitus. *See* Sexual intercourse
Coitus interruptus, 525
Cold sore, 93
Colitis, 422
Collagen, 87
Collagenous fiber, 61
Collecting tubule, 450
Colliculus, 213
Colon, 420
Colony-stimulating factor (CSF), 304
Color blindness, 254, 528
Color vision, 251
Colostomy, 422
Colostrum, 524
Column, 206
Coma, 215
Comedo, 93
Commissural fiber, 217
Common bile duct, 414
Common cold, 395
Common disorders
    of articulations, 136
    of brain, 224–226
    of the cardiovascular system, 312–313, 329–
        330, 348–349

of the digestive system, 421–422
of the endocrine system, 291–293
of the integumentary system, 92
of the lymphatic system and immunity, 369–
    371
of metabolism, 444
of muscle tissue, 183
of the reproductive system, 507
of the respiratory system, 395
of the skeletal system, 126–127
of special senses, 264–265
of the spinal cord, 224–226
of the urinary system, 466
Common hepatic duct, 410
Compact bone, 98
Compensation, 349, 480
Complement, 359
Complete blood count (CBC), 304
Complete tetanus, 151
Complete transection, 225
Compliance, 330, 385
Concentration gradient, 39
Condom, 525
Conduction, of heat, 442
Conduction myofiber, 322
Conduction pathway, 207
Conduction system, of heart, 322
Conductivity, 6
Cone, 251
Congenital heart defect, 329
Congestive heart failure (CHF), 330
Conjunctiva, 349
Conjunctivitis, 264
Connecting neuron. *See* Association neuron
Connective tissue
    classification of, 69
    definition of, 69
    types of, 69–77
Connective tissue proper, 76
Consciousness, 215
Constipation, 421
Constriction of pupil, 253
Contact inhibition, 53
Continuous conduction, 196
Contraception. *See* Birth control
Contraceptive diaphragm, 525
Contraceptive sponge, 525
Contractility, 6, 142
Contraction, muscular, 144–151, 154
Contraction period, 150
Convection, 442
Convergence, 253
Convolution, 215
Cooper's ligament. *See* Suspensatory ligament
    of the breast
Copulation. *See* Sexual intercourse
Corn, 93
Cornea, 249
Corneal transplant, 249
Corona radiata, 512
Coronary artery, 322
Coronary artery disease (CAD), 329
Coronary artery spasm, 329
Coronary (cardiac) circulation, 322
Coronary sinus, 319
Coronary sulcus, 319
Corpora cavernosa penis, 491
Cor pulmonale, 330

Corpus albicans, 501
Corpus callosum, 215
Corpuscle of touch, 87, 242
Corpus hemorrhagicum, 501
Corpus luteum, 493, 501
Corpus spongiosum penis, 491
Corpus striatum, 217
Cortex
    adrenal, 281–284
    of cerebellum, 221
    of cerebrum, 215
    of kidney, 449
    of ovary, 493
Corticosterone, 284
Corticotropin releasing hormone (CRH), 275
Cortisol (hydrocortisone), 284
Cortisone, 284
Costal cartilage, 45
Covalent bond, 23
Cowper's gland. *See* Bulbourethral gland
Crack, 200
Cranial bones, 105–107
Cranial cavity, 8
Cranial meninges, 209
Cranial nerve damage, 223E, 224E
Cranial nerves, 221, 223, 223E, 224E
Craniotomy, 127
Crenation, 41
Cretinism, 291
Cribriform plate, 106
Cricoid cartilage, 378
Crisis, 444
Crista
    of inner ear, 263
    of mitochondrion, 47
Crista galli, 106
Cross bridge, 144
Crossing-over, 489
Crown, of tooth, 404
Crypt of Lieberkühn. *See* Intestinal gland
Cryptorchidism, 485
Cuboid, 121
Culdoscopy, 507
Cuneiform bones, 121
Cupula, 263
Cushing's syndrome, 291
Cutaneous membrane, 81, 85–87, 90
Cutaneous sensation
    definition of, 242
    pain, 243, 246
    pressure, 243
    thermoreceptive, 243
    touch, 242
    vibration, 243
Cuticle, of nail, 92
Cyanosis, 313
Cyclic adenosine–3', 5'-monophosphate. *See*
    Cyclic AMP
Cyclic AMP, 272
Cyst, 93
Cystic duct, 410
Cystic fibrosis (CF), 444
Cystitis, 40
Cytochrome system. *See* Electron transport
    chain
Cytokinesis, 51, 53
Cytology, 36
Cytolysis, 359

Cytoplasm
  composition of, 44
  definition of, 44
  functions of, 44
Cytoskeleton, 47
Cytotoxic (killer) T cell, 303, 364

Daughter cell, 51
Deafness, 264
Deamination, 435
Decibel (dB), 260
Decubitus ulcer, 92
Decussation of pyramids, 212
Deep fascia, 143
Deep venous thrombosis (DVT), 349
Defecation, 421
Dehydration, 471
Dehydration synthesis, 27
Delayed hypersensitivity T cell, 365
Delirium, 226
Delta cell, 285
Dementia, 226
Demineralization, 126
Dendrite, 189
Dense body, 182
Dense bone tissue. See Compact bone
Dense (collagenous) connective tissue, 76
Dental caries, 421
Dental plaque, 421
Dentin, 404
Deoxygenated blood, 387
Deoxyribonuclease, 410, 416
Deoxyribonucleic acid (DNA)
  in cell division, 51
  chemical composition of, 31–32
  and chromatin, 51
  and chromosomes, 51
  functions of, 31–32
  and protein synthesis, 48, 50E
  structure of, 31–32
Depolarized membrane, 195
Depression, 137
Dermabrasion, 93
Dermal papilla, 87
Dermatology, 85
Dermis
  definition of, 87
  papilla of, 87
  receptors of, 87
  regions of, 87
Descending aorta, 341
Descending colon, 420
Descending limb of loop of nephron (Henle), 450
Descending tract, 204
Detached retina, 251
Development
  birth control, 524, 526E
  embryonic period, 513–516
  fertilization, 512
  fetal growth, 516, 519E
  gamete formation, 486, 492–493
  gestation, 517, 519
  hormones, 516–517
  implantation, 512–513
  lactation, 524
  medical terminology and conditions, 529–530

pregnancy, 512
  sexual intercourse, 511–512
Development anatomy, 512
Deviated nasal septum (DNS), 110
Diabetes insipidus, 291
Diabetes mellitus, 293
  type I, 293
  type II, 293
Diapedesis, 303, 361
Diaphragm
  abdominopelvic, 8, 164
  contraceptive, 525
Diaphysis, 98
Diarrhea, 421
Diarthrosis, 131
Diastole, 324
Diastolic blood pressure (DBP), 340
Diencephalon, 214
Differential white blood cell count (DIF), 304
Differentiation, 6
Diffuse lymphatic tissue, 353
Diffusion, 39–40
Digestion
  definition of, 27, 400
  in large intestine, 420–421
  in mouth, 405
  processes of, 400
  in small intestine, 416
  in stomach, 407–408
Digestive system, 400–423
  absorption by, 417
  common disorders of, 421–422
  definition of, 400
  general histology, 400–401
  medical terminology and conditions, 422–423
  organization of GI tract, 400–401
  organs of, 401–421
Diphtheria, 396
Diplegia, 225
Diploid cell, 486
Diplopia, 223E
Directional terms, 6, 6E
Direct transfusion, 313
Disaccharide, 27
Discriminative touch, 242
Dislocation, 136
Dissociation. See Ionization
Distal convoluted tubule, 450
Distress, 287
Diverticulitis, 422
Diverticulosis, 422
DNA. See Deoxyribonucleic acid
Dominant gene, 527
Dopamine (DA), 221
Dorsal body cavity, 8
Dorsal ramus, 209
Dorsiflexion, 137
Dual innervation, 231
Ductless glands. See Endocrine system
Duct of Wirsung. See Pancreatic duct
Ductus arteriosus, 516
Ductus (vas) deferens, 490
Ductus epididymis, 490
Ductus venosus, 516
Duodenal gland, 416
Duodenum, 415
Dura mater, 204, 209
Dynamic equilibrium, 262–263

Dynorphin, 222E
Dyskinesia, 265
Dyslexia, 225
Dysmenorrhea, 503
Dysphagia, 422
Dysplasia, 57
Dyspnea, 396
Dysuria, 467

Ear
  common disorders of, 264, 265
  and equilibrium, 261–263
  external, 254
  and hearing, 260–261
  internal, 254
  middle, 254
Eardrum. See Tympanic membrane
ECF. See Extracellular fluid
ECG. See Electrocardiogram
Ectoderm
  derivatives of, 514E
  development of, 514
Ectopic pregnancy (EP), 495
Eczema, 93
Edema, 358
EEG. See Electroencephalogram
Effector, 208
Efferent lymphatic vessel, 356
Efferent neuron, 191, 208
Ejaculation, 511
Ejaculatory duct, 490
Elastic cartilage, 77
Elastic connective tissue, 76
Elastic fiber, 61
Elasticity, 87, 142
Electrocardiogram (ECG or EKG)
  definition of, 322
  deflection waves of, 323
  use in diagnosis, 323
Electroconvulsive therapy (ECT), 226
Electroencephalogram (EEG), 221
Electrolyte, 472–475
  absorption by small intestine, 419
  concentration of, 472–473
  effects on osmosis, 472
  functions of, 474–475
Electromyography (EMG), 184
Electron transport chain, 430
Elephantiasis, 373
Elevation, 137
Ellipsoidal joint, 134, 136
Embolus, 310
Embryo
  definition of, 513
  development of, 513–516
  fertilization of, 512
  implantation of, 512–513
  membranes of, 514
  placenta of, 514
  primary germ layers of, 514E
Embryology, 513
Embryonic connective tissue, 69, 70E, 76
Embryonic membranes
  amnion, 514
  chorion, 514
Embryonic period
  definition of, 513
  embryonic membranes, 514

Embryonic period (*Continued*)
  placenta and umbilical cord, 514
  primary germ layers, 514E
Emission, 511
Emmetropic eye, 252
Emphysema, 395
Emulsification, 411
Enamel, 404
Endocardium, 320
Endochondral ossification, 99–101
Endocrine gland, 69, 270–286
Endocrine system, 269–293
  common disorders of, 291–293
  control of hormonal secretions, 272–274
  definition of, 269
  glands of, 270–286
  hormones of, 270–286
  mechanism of hormonal action, 270–272
  medical terminology and conditions, 293
  prostaglandins and, 272
  and stress, 287
Endocrinology, 269
Endocytosis, 42
  phagocytosis, 42
  pinocytosis, 42
  receptor-mediated, 42
Endoderm
  derivatives of, 514E
  development of, 514E
Endodontics, 404
Endolymph, 254
Endometriosis, 506
Endometrium, 495
Endomysium, 143
Endoplasmic reticulum (ER)
  agranular, 45
  granular, 45
End organ of Ruffini. *See* Type II cutaneous
  mechanoreceptor
Endorphin, 221
Endosteum, 98
Endothelial-capsular membrane, 450
Endothelium, 2, 334
Energy level, 21
Engram, 221
Enkephalin, 221
Enteric gastrin, 287, 409E
Enteroendocrine cell, 407
Enterokinase, 410
Enteroreceptor, 242
Enuresis, 467
Enzymes
  active site of, 428
  carboxypeptidase, 410
  chymotrypsin, 410
  definition of, 428
  deoxyribonuclease, 410, 416
  enterokinase, 410
  enzyme-substrate complex, 428
  function of, 428
  gastric lipase, 408
  lactase, 416
  maltase, 416
  nuclease, 416
  pancreatic amylase, 410
  pancreatic lipase, 410
  pepsin, 408
  peptidase, 416

renal erythropoietic factor (REF), 287, 301
  rennin, 408
  ribonuclease, 416
  salivary amylase, 405
  sucrase, 416
  trypsin, 410
Enzyme-substrate complex, 428
Eosinophil, 302
Ependyma, 189E
Epicardium, 317–319
Epidemiology, 56
Epidermis
  definition of, 86
  layers of, 87
  pigments of, 87, 90
Epididymis, 490
Epidural space, 204
Epiglottis, 377
Epilepsy, 225
Epimysium, 143
Epinephrine, 284
Epiphyseal line, 101
Epiphyseal plate, 101
Epiphysis, 98
Epiphysis cerebri. *See* Pineal gland
Episiotomy, 497
Epithelial membrane, 81
Epithelial tissue
  arrangement of layers, 62
  basement membrane of, 62
  cell shapes, 62
  classification, 62
  covering and lining, 62
  definition of, 62
  glandular, 69
  subtypes, 62–69
Equilibrium, 39, 261–263
ER. *See* Endoplasmic reticulum
Erection, 492, 511
Erythema, 93
Erythroblastosis fetalis. *See* Hemolytic disease
  of newborn
Erythrocyte
  anemia and, 312
  functions of, 301
  hematocrit, 302
  lifespan, 301
  number per cubic millimeter, 301
  production of, 301
  reticulocyte count, 302
  structure of, 301
Erythropoiesis, 301
Erythropoietin, 287, 301
Esophagus, 406–407
Essential amino acids, 435
Estrogens, 286, 499, 516
Ethmoid bone, 106–107
Eupnea, 384
Eustachian tube. *See* Auditory tube
Eustress, 287
Evaporation, 442
Excitability, 5–6, 142, 195
Excitatory transmission, 199
Exhaustion reaction, of stress, 291
Exocrine gland, 69, 270
Exocytosis, 42
Exophthalmic goiter, 291

Exophthalmos, 291
Expiration, 384
Expiratory reserve volume (ERV), 385
Extensibility, 87, 142
Extension, 134
External auditory meatus, 105
External ear, 254
External nares, 376
External (pulmonary) respiration, 376, 387–388
External sphincter, of urinary bladder, 461
External urethral orifice, 496
Exteroreceptor, 242
Extracellular fluid (ECF), 10, 471
Extracellular materials
  definition of, 61
  types of, 61
Extrapyramidal pathway, 207
Extrinsic blood clotting pathway, 308
Eyeball
  accessory structures of, 249
  afferent pathway to brain, 254
  common disorders of, 264–265
  structure of, 249–251
  and vision, 251–254
Eyebrows, 249
Eyelids, 249

Facet, 113
Facial bones, 107–110
Facilitated diffusion, 39–40
Facilitation, 199
Facultative water reabsorption, 457
Fallopian tube. *See* Uterine tube
False vocal cords, 378
Fascia
  deep, 143
  definition of, 143
  superficial, 143
Fasciculation, 183
Fasciculus, of skeletal muscle, 143
Fat. *See* Lipid
Fat-soluble vitamins, 437, 438E
Fat storage, 431
Fatty acid, 30, 433–434
Fauces, 402
Feces, 421
Feedback system
  and blood pressure, 13
  and blood sugar level, 13
  and body temperature, 92
  and hormones, 272–274
  input, 13
  negative, 13, 272–274
  output, 13
  positive, 13
Female infertility, 506
Female reproductive cycle (FRC), 498–502
Female reproductive system, 492–502
Femur, 118
Fenestra cochlea, 254
Fenestra vestibuli, 254
Ferritin, 301
Fertilization, of ovum, 512
Fetal alcohol syndrome (FAS), 528
Fetal circulation, 341, 348, 514, 516
Fetal growth
  changes associated with, 519E
  definition of, 516
Fetal period, 513, 516

Fetus
definition of, 513
development of, 519E
Fever, 362, 443–444
Fibrillation, 183
Fibrin, 308, 361
Fibrinogen, 304, 361
Fibrinolysis, 308
Fibroadema, 506
Fibroblast, 76
Fibrocartilage, 77
Fibrocyte, 76
Fibromyalgia, 183
Fibrosis, 183
Fibrous capsule, 132
Fibrous joints
definition of, 131
types of, 132
Fibrous pericardium, 317
Fibrous tunic, 249
Fibula, 120
Fight-or-flight response, 238, 287
Filiform papilla, 402
Filtrate, glomerular, 453
Filtration
definition of, 41, 339
glomerular, 453–457
Fimbriae, 494
First-degree burn, 92
First (medial) cuneiform, 121
First messenger, 272
First polar body, 494
Fissure, of brain, 215
Fixator, 154
Fixed macrophage, 359
Flaccid, 151
Flagellum, 47
Flat bone, 96
Flatfoot, 123
Flexion, 134
Flu. See Influenza
Fluid balance, 471
Fluid compartment, 471
Fluid intake, 471
Fluid output, 471
Fluids, body
aqueous humor, 251
balance of, 471
cerebrospinal, 209–210
edema and, 358
extracellular, 10, 471
intake of, 471
interstitial, 298, 353
intracellular, 471
lymph, 357
movement between interstitial and intracellular, 478
movement between plasma and interstitial,
475, 478
output of, 471
pericardial, 317
perilymph, 254
plasma, 304
synovial, 81, 132
Follicle-stimulating hormone (FSH), 275–276
Follicular cell, 279
Fontanel
anterior, 110

definition of, 110
posterior, 110
posterolateral, 110
Food intake, regulation of, 427
Foramen
definition of, 106
intervertebral, 111, 112
magnum, 106
obturator, 117
ovale, 516
sacral, 113
transverse, 112
vertebral, 112
Foramen ovale, of heart, 516
Foreskin. See Prepuce
Formed elements of blood, 299–304
Fornix, 496
Fossa ovalis, 516
Fourth ventricle, 209
Fractures
closed, 126
complete, 126
definition of, 126
open, 126
partial, 126
pathologic, 127
stress, 126
Freckles, 90
Free nerve ending, 242
Frontal bone, 105
Frontal eye field area, 220
Frontal plane, 6
Fulcrum, 154
Full-thickness burn. See Third-degree burn
Functional residual capacity, 386
Fundus
of stomach, 407
of uterus, 495
Fungiform papilla, 402
Furuncle, 93

Galea aponeurotica, 158E
Gallbladder (GB), 414–415
Gallstone, 422
Gamete, 486
Gamete formation, 486
Gamma aminobutyric acid (GABA), 221
Gamma globulin, 313
Ganglion
autonomic, 234
definition of, 204
posterior root, 207
prevertebral, 234
sympathetic trunk, 234
terminal, 234
Ganglion neuron, 251
Gangrene, 184
Gap junction, 180, 198, 322
Gastric absorption, 410
Gastric emptying, 409–410
Gastric gland, 407
Gastric inhibiting peptide (GIP), 287
Gastric juice, 359, 407
Gastric lipase, 408
Gastric secretion, regulation of, 408–409
Gastrin
enteric, 287, 409
stomach, 287, 409

Gastroenterology, 400
Gastrointestinal (GI) tract, 400
Gene
action of, 48
definition of, 31
structure of, 31
General adaptation syndrome (GAS), 287, 290–
291
Generalized animal cell, 36
General senses
cutaneous, 242–243, 246
pain, 243, 246
proprioceptive, 246
General sensory area. See Primary somesthetic
area
Generator potential, 241
Genetics, 525, 527–528
Genital herpes, 503
Genomic imprinting, 528
Genotype, 525, 527–528
Germinal epithelium, 493
Gestation, 517, 519
Giantism, 291
Gingivae, 404
Gland
adrenal (suprarenal), 281–285
bulbourethral, 490
ceruminous, 91
definition, 69
duodenal, 416
endocrine, 270–286
exocrine, 69, 270
gastric, 407
greater vestibular, 496
intestinal, 416
lacrimal, 249
mammary, 497–498
olfactory, 247
ovaries, 286
pancreas, 285–286, 410
parathyroids, 281
parotid, 403
pineal, 286
pituitary, 274–278
prostate, 490
salivary, 402–404
sebaceous, 91
seminal vesicles, 490
sublingual, 403
submandibular, 403
sudoriferous, 91
testes, 286, 485–486, 488–489
thymus, 286, 357
thyroid, 278–281
Glandular epithelium, 9
Glans
of clitoris, 496
of penis, 492
Glaucoma, 264
Glenoid cavity, 116
Glial cells. See Neuroglia
Gliding joint, 133
Gliding movement, 133
Globulin, 304
Globus pallidus, 217
Glomerular blood hydrostatic pressure, 454
Glomerular capsule, 450
Glomerular filtration, 453–457

Glomerular filtration rate (GFR), 457
Glomerulonephritis, 466
Glomerulus, 450
Glottis, 377
Glucagon, 285–286
Glucocorticoid, 284
Gluconeogenesis, 430
Glucose
    absorption of, 417
    alarm reactions and, 287
    anabolism of, 430
    catabolism of, 429
    and gluconeogenesis, 430
    and glycogenolysis, 430
    storage of, 430
Glycerol, 30, 433
Glycogen, 48, 430
Glycogenesis, 430
Glycogenolysis, 430
Glycolysis, 149, 430
Glycosuria, 457
Gnostic area, 217
Goblet cell, 62
Goiter, 291
Golgi complex, 46
Golgi tendon organ. See Tendon organ
Gomphosis, 132
Gonad, 485
Gonadocorticoid, 284
Gonadotropin releasing hormone (GnRH), 276
Gonorrhea, 502
"Goosebumps," 91
Gouty arthritis, 136
Graafian follicle. See Vesicular ovarian follicle
Graded potential, 200
Granstein cell. See Nonpigmented granular dendrocyte
Granular leucocytes, 302
Gray matter, 204
Greater sciatic notch, 117
Greater vesticular gland, 496
Growth, 6
Growth hormone inhibiting hormone (GHIH). See Somatostatin
Growth hormone releasing hormone (GHRH), 274
Gums. See Gingivae
Gustation, mechanism of, 248
Gustatory cell, 248
Gustatory hair, 248
Gustatory sensation, 247–248
Gynecology, 493
Gyrus. See Convolutions

Hair
    functions of, 90
    replacement of, 90
    structure of, 90
Hair cell, 261
Hair follicle, 90
Hair root, 90
Hair root plexus, 91, 242
Hair shaft, 90
Hamate, 117
Haploid cell, 486
Haploid chromosome number, 486
Hapten, 363
Hard palate, 107

Haustra, 420
Haustral churning, 420
Haversian canal. See Central canal
Haversian system. See Osteon
HDL. See High-density lipoprotein
Headache. See Cephalgia
Hearing, physiology of, 260–261
Heart
    angina pectoris and, 322
    atria of, 319
    autonomic control of, 325
    blood flow through, 323–324
    blood supply of, 222
    cardiac cycle of, 324
    cardiac output (CO) and, 325
    chambers of, 319
    chemicals and, 325
    common disorders of, 329–330
    conduction system of, 322
    electrocardiogram (ECG or EKG) of, 322
    emotions of, 329
    endocardium of, 330
    epicardium of, 317
    great vessels of, 319
    location, 317
    medical terminology and conditions, 330
    myocardial infarction (MI), 322
    myocardium of, 317
    pericardium, 317
    rate of, 324
    regulation of rate, 325, 329
    sex and age and, 329
    shock and, 340
    sounds of, 324
    temperature and, 325
    timing of, 324
    valves of, 319
    ventricles of, 319
    wall of, 317, 319
Heart attack, 322
Heartbeat, 324
Heart block, atrioventricular (AV), 330
Heartburn, 407
Heart sounds, 324
Heart valve, 319
Heat-losing center, 442
Heat-promoting center, 442
Heimlich (abdominal thrust) maneuver, 396
Helper T cell, 365
Hematocrit (Hct), 302
Hematology, 298
Hemiplegia, 225
Hemisection, 225
Hemochromatosis, 313
Hemocytoblast, 299
Hemodialysis, 458
Hemoglobin, 301, 388–389
Hemoglobin buffer, 479
Hemolysis, 41, 311
Hemolytic disease of newborn (HDN), 312
Hemophilia, 308, 529
Hemopoiesis, 96, 299
Hemoptysis, 396
Hemorrhage, 312
Hemorrhoid, 420
Hemosiderin, 301
Hemostasis, 304–305, 308–310
Heparin, 308

Hepatic cell, 410
Hepatic duct, 410
Hepatic portal circulation, 341
Hepatitis, 422
Hereditary traits in humans, 527E
Hermaphroditism, 507
Hernia, 423
Herniated (slipped) disc, 127
Heterozygous, 527
High altitude sickness, 387
High-density lipoprotein, 418
Hilus
    of kidney, 449
    of lung, 382
    of lymph node, 356
    of ovary, 493
Hinge joint, 134
Hip bones, 117
Hip joint, 117
Histamine, 360
Histology, 61
Hives, 93
HLA antigen, 302
Hodgkin's disease, 372
Homeostasis
    of blood pressure, 12–13
    of blood sugar level, 13
    and body temperature, 92
    of bone, 101
    definition of, 10
    and enzymes, 430
    and feedback system, 272–274
    kidneys and, 459
    of muscle tissue, 149–150
    stress and, 11
    and tissue inflammation, 360–362
Homologous chromosomes, 486, 525
Homozygous, 527
Horizontal plane, 6
Hormonal action, mechanism of, 270–272
Hormones
    activation of genes, 272
    adrenocorticotropic (ACTH), 275
    aldosterone, 282
    androgens, 284, 486
    antidiuretic (ADH), 277
    atrial natriuretic factor (ANF), 287
    and bone growth, 101
    calcitonin (CT), 281
    chemistry of, 270
    cholecystokinin (CCK), 287, 409E
    corticosterone, 284
    cortisol (hydrocortisone), 284
    cortisone, 284
    and cyclic AMP, 272
    definition of, 269
    enteric gastrin, 287, 409E
    epinephrine, 284
    erythropoietin, 287, 301
    estrogens, 286, 499, 516
    follicle-stimulating (FSH), 275–276
    glucagon, 285–286
    glucocorticoids, 284
    gonadocorticoids, 284
    human chorionic gonadotropin (hCG), 287, 501, 516
    human chorionic somatomammotropin (hCS), 287, 517

human growth (hGH), 274
inhibin, 286
insulin, 286
interaction with intracellular receptors, 272
interaction with plasma membrane receptors, 272
luteinizing (LH), 276
mechanism of action, 270–272
melanocyte-stimulating (MSH), 277
melatonin, 286
mineralocorticoids, 282
and negative feedback control, 272–274
norepinephrine (NE), 284
oxytocin (OT), 277, 524
parathyroid (PTH), 281
of pregnancy, 499, 516
progesterone (PROG), 287, 499, 516
prolactin (PRL), 276, 524
and prostaglandin, 272
relaxin, 286, 499
secretin, 287, 409E
stomach gastrin, 287, 409E
testosterone, 286, 489
thymosin, 286
thyroid, 279
thyroid-stimulating (TSH), 275
thyroxine (T_4), 286
triiodothyronine (T_3), 286
Horn, 204
Human chorionic gonadotropin (hCG), 287, 501, 516
Human chorionic somatomammotropin (hCS), 287, 517
Human growth hormone (hGH), 274
Human immunodeficiency virus (HIV), 370
Humerus, 116
Humoral immunity, 363–364
Hunger center, 427
Huntington's chorea, 226, 527
Hyaline cartilage, 76–77
Hyaluronic acid, 61
Hydrocephalus
    external, 210
    internal, 210
Hydrocortisone. See Cortisol
Hydrogen bond, 23
Hymen, 496
Hyoid bone, 111
Hypercalcemia, 474
Hyperextension, 134
Hyperkalemia, 474
Hypermagnesemia, 475
Hypermetropia, 252
Hypernatremia, 474
Hyperphosphatemia, 475
Hyperplasia, 57, 293
Hyperpolarization, 199
Hypersensitivity, 372
Hypertension, 348
Hypertonia, 151
Hypertonic solution, 41
Hypertrophy, 57
Hyperventilation, 394
Hypocalcemia, 474
Hypochloremia, 475
Hypodermic, 93
Hypoglycemia, 275
Hypokalemia, 474

Hypomagnesemia, 475
Hyponatremia, 474
Hypoparathyroidism, 291
Hypophosphatemia, 475
Hypophysis. See Pituitary gland
Hypoplasia, 293
Hypospadias, 507
Hypotension, 349
Hypothalamus, 214
Hypotonia, 151
Hypotonic solution, 41
Hypoxia, 301, 390

I band, 144
Ileocecal sphincter (valve), 420
Ileum, 415
Iliac crest, 117
Ilium, 117
Immunity
    acquired, 362–365, 368–369
    acquired immune deficiency syndrome (AIDS) and, 369–372
    allergy and, 372
    antibodies and, 363
    antigens and, 363
    autoimmune diseases and, 372
    and cancer, 369
    cellular, 363–364
    definition of, 362
    humoral, 363–364
    and skin, 368
    tissue rejection and, 372
Immunogenicity, 363
Immunoglobin. See Antibody
Immunologic escape, 369
Immunologic surveillance, 369
Immunologic tolerance, 372
Immunology, 362
Immunosuppressive drug, 372
Impetigo, 93
Implantation, of fertilized egg, 512
Impotence, 503
Incompetent valve, 330
Incomplete tetanus, 151
Incontinence, 464
Incus, 254
Indirect transfusion, 313
Induced abortion, 525
Infarction, 322
Infectious mononucleosis (IM), 312
Inferior nasal concha, 110
Inferior vena cava (IVC), 319
Infertility, 503
Inflammation. See Inflammatory response
Inflammatory bowel disease (IBD), 423
Inflammatory response
    definition of, 360
    stages of, 360–361
    symptoms of, 360
Inflation reflex, 391
Influenza, 395
Infundibulum
    of pituitary, 274
    of uterine (Fallopian) tube, 494
Ingestion
    digestive, 400
    phagocytic, 358

Inheritance
    of color-blindness, 528
    genotype, 525, 527–528
    phenotype, 525, 527–528
    of PKU, 525
    of sex, 528
    x-linked, 528
Inhibin, 286, 489, 517
Inhibiting hormone (or factor), 274
Inhibitory transmission, 199
Inner cell mass, 512
Inorganic compound, 25–26
Insertion of skeletal muscle, 154
Inspiration, 383–384
Inspiratory reserve volume (IRV), 385
Insula, 216
Insulin, 286
Integral protein, 36
Integration at synapses, 199
Integumentary system, 85–93
    common disorders of, 92
    definition of, 85
    medical terminology and conditions, 93
    and nonspecific resistance, 358
    skin, 85–87, 90
Interatrial septum, 319
Intercalcated disc, 180
Intercostal space, 115
Intercostal (thoracic) nerve, 209
Interferon (IFN), 359
Interleukin 1, 364
Interleukin 2, 365
Intermediate filaments, 47
Internal ear, 254, 260
Internal environment, 38, 298
Internal nares, 376
Internal sphincter, of urinary bladder, 461
Internal (tissue) respiration, 376, 388
Internuncial neuron. See Association neuron
Interphase, 51
Interstitial cell of Leydig. See Interstitial endocrinocyte
Interstitial endocrinocyte, 486
Interstitial fluid, 38, 298, 353
Interstitial fluid hydrostatic pressure (IFHP), 475
Interstitial fluid osmotic pressure (IFOP), 475
Interventricular septum, 319
Intervertebral disc, 111
Intervertebral foramen, 111
Intervillous space, 514
Intestinal gland, 416
Intestinal juice, 416
Intestinal secretion, regulation of, 416
Intracellular fluid (ICF), 10, 38, 471
Intramembranous ossification, 98
Intraocular pressure (IOP), 251
Intrapleural pressure, 384
Intrapulmonic pressure, 384
Intrauterine device (IUD), 524–525
Intravenous pyelogram (IVP), 467
Intrinsic blood-clotting pathway, 308
Intrinsic factor (IF), 302
Intubation, 379
Inversion, 137
Inverted image formation, 253
Ion
    anion, 23
    cation, 23

Ion channel, 194
Ionic bond, 22–23
Ionization, 25
Iris, 249
Irregular bone, 96
Irritable bowel syndrome (IBS), 223
Ischemia, 322
Ischium, 117
Island of Reil. *See* Insula
Islets of Langerhans. *See* Pancreatic islet
Isometric contraction, 151
Isotonic contraction, 151
Isotonic solution, 40–41

Jaundice, 414
Jejunum, 415
Joint kinesthetic receptor, 246
Joints. *See* Articulations
Juxtaglomerular cell, 450

Kaposi's sarcoma, 370
Keratin, 69, 86
Keratinization, 87
Keratinocyte, 86
Keratoplasty. *See* Corneal transplant
Keratosis, 93
Ketogenesis, 433
Ketone bodies, 423
Ketosis, 433, 466E
Kidney, 449–452
Kidney stone, 466E
Kilocalorie (kcal), 437
Kinesthetic sense. *See* Proprioception
Kinetic energy, 39
Kinocilium, 261
Knee jerk. *See* Patellar reflex
Krebs cycle, 430
Kupffer cell. *See* Stellate reticuloendothelial cell
Kyphosis, 127

Labia majora, 496
Labia minora, 496
Labor
  placental stage, 524
  stage of dilation, 520
  stage of expulsion, 520
Labyrinthine disease, 264
Labyrinth. *See* Internal ear
Laceration, 93
Lacrimal apparatus, 249, 358
Lacrimal bone, 110
Lacrimal gland, 249
Lactase, 416
Lactation, 497, 524
Lacteal, 401
Lacuna
  in bone, 77, 98
  in cartilage, 76
  definition of, 76
Lamella, 77, 98
Lamellated corpuscle, 87, 243
Langerhans' cell. *See* Nonpigmented granular
  dendrocyte
Language area of cerebrum, 220
Laparoscopy, 524
Large intestine, 419–421
Laryngitis, 379
Laryngopharynx, 377

Larynx, 377–379
Latent period, in muscle contraction, 150
Lateral cerebral sulcus, 216
Lateral malleolus, 121
Lateral ventricle, 209
LDL. *See* Low-density lipoprotein
Lens, 251
Lentiform nucleus, 217
Leucocyte
  agranular, 302
  and white blood cell differential count, 304
  functions of, 302–304
  granular, 302
  life span, 304
  number per cubic millimeter, 304
  production, 304
  structure of, 302
  types of, 302
Leucocytosis, 304
Leucopenia, 304
Leukemia, 313
Leukorrhea, 507
Lever, 154
Life processes, 5, 6
Ligaments, 76
Ligand, 42
Light touch, 342
Limbic system, 217
Liminal stimulus. *See* Threshold stimulus
Lingual frenulum, 402
Lingual tonsil, 357
Lipid
  absorption of, 418
  anabolism of, 434
  catabolism of, 431, 433–434
  definition of, 27
  digestion of, 416
  fate of, 430–431
  functions of, 27
  storage, 431
  structure of, 27
Lipogenesis, 434
Lipoprotein, 418
Lips, 401
Liver, 410–411, 414
Local disease, 56
Long bone, 96
Longitudinal arch, of foot, 122
Longitudinal fissure, 215
Long-term memory, 221
Loop of Henle. *See* Loop of the nephron
Loop of the nephron, 450
Loose connective tissue, 76
Lordosis, 127
Low-density lipoprotein, 418
Lower extremity, 117–123
Lubrication, during sexual intercourse, 511
Lumbar enlargement, 206
Lumbar plexus, 209
Lumbar vertebra, 111
Lumen, of blood vessel, 334
Lunate, 116
Lung, 380–383
Lunula, 92
Luteinizing hormone (LH), 276
Luxation. *See* Dislocation
Lyme disease, 136
Lymph, 357–358

Lymphadenopathy, 373
Lymphatic nodule, 353
Lymphatic organ, 353
Lymphatic system, 353–373
  and allergy, 372
  and autoimmune disease, 372
  common disorders of, 369–372
  definition of, 353
  and immunity, 362–365, 368–369
  lymphatic tissue, 356–357
  lymphatic vessels, 354–356
  lymph nodes, 356–357
  medical terminology and conditions, 373
  plan of lymph circulation, 357
  and tissue rejection, 372
Lymphatic tissue, 356–357
Lymphatic vessels
  afferent, 356
  efferent, 356
  lymph capillaries, 354
  right lymphatic duct, 357
  thoracic duct, 357
Lymph capillary, 354
Lymph circulation, 357–358
Lymphedema, 373
Lymph node, 356–357
Lymphocyte, 302
Lymphokine, 364
Lymphoma, 373
Lymph trunk, 357
Lysosome, 46–47
Lysozyme, 249, 359

Macrophage, 76, 359
  fixed, 359
  wandering, 359
Macula, 261
Macula densa, 450
Macula lutea, 251
Magnesium
  control of, 475
  deficiency of, 475
  distribution of, 475
  excess of, 475
  functions of, 475
Male infertility (sterility), 503
Male reproductive system, 485–492
Malignant melanoma, 90
Malignant tumor, 53
Malleus, 254
Maltase, 416
Mammary gland, 497–498
Mammography, 497
Mandible, 110
Mandibular fossa, 105
Manubrium, 114
Margination, 361
Marrow biopsy, 114
Mass peristalsis, 421
Mast cell, 76
Mastication, 405
Mastoid process, 105
Matrix
  of hair follicle, 90
  of mitochondrion, 47
  of nail, 92
Maxillae, 107
Maximal oxygen uptake, 150

Mechanical digestion, 400
Mechanoreceptor, 242
Medial malleolus, 120
Mediastinum, 9
Medical terminology and conditions
  associated with articulations, 140
  associated with brain, 226
  associated with the cardiovascular system,
    313, 330, 349
  associated with cells, 56–57
  associated with development, 529–530
  associated with the digestive system, 422–
    423
  associated with the endocrine system, 293
  associated with the integumentary system, 93
  associated with the lymphatic system and im-
    munity, 373
  associated with muscle tissue, 184
  associated with the reproductive systems,
    507
  associated with the respiratory system, 396
  associated with skeletal tissue, 127
  associated with special senses, 265
  associated with the urinary system, 467
Medulla
  adrenal, 284
  of brain, 212–213
  of kidney, 449
  of ovary, 493
Medulla oblongata
  cranial nerves of, 213
  pyramids of, 212
  reflex centers of, 212
  reticular formation of, 212
Medullary cavity, 98
Medullary rhythmicity area, 390
Meiosis
  meiosis I (reduction division), 488–489
  meiosis II (equatorial division), 489
Meissner's corpuscle. See Corpuscle of touch
Melanin, 48, 87
Melanocyte, 86, 87
Melanocyte-stimulating hormone (MSH), 277
Melanocyte-stimulating hormone releasing fac-
    tor (MRF), 277
Melatonin, 286
Membrane
  cutaneous, 81
  definition of, 81
  epithelial, 81
  mucous, 81
  pleural, 81, 380
  serous, 81
  synovial, 81
  types, 81
Membrane potential, 194–195
Membranous labyrinth, 254
Memory
  consolidation, 221
  engram of, 221
  long-term, 221
  short-term, 221
Memory B cell, 368
Memory consolidation, 221
Memory T cell, 368
Menarche, 498
Ménière's syndrome, 264

Meninges
  cranial, 209
  spinal, 204–205
Meningitis, 205
Meniscus. See Articular disc
Menopause, 501–502
Menstrual cycle
  hormonal control of, 499
  menstrual phase, 499–501
  ovulation, 501
  postovulatory phase, 501
  preovulatory phase, 501
Menstruation, 499–501
Merkel's disc. See Tactile disc
Mesencephalon. See Midbrain
Mesenchyme, 69
Mesoderm
  derivatives of, 514E
  development of, 514
Messenger RNA (mRNA), 48
Metabolic rate, 437, 442
Metabolism
  anabolism, 5, 428
  basal metabolic rate, 442
  and calories, 437
  of carbohydrates, 429–430
  catabolism, 5, 428
  common disorders of, 444
  definition of, 5, 427
  in lipids, 430–434
  and minerals, 436, 436E–437E
  and obesity, 444
  and proteins, 434–435
  and vitamins, 437, 438E–439E
Metacarpus, 117
Metaphase, 53
Metaplasia, 57
Metastasis, 53, 57, 357
Metatarsus, 122
Micelle, 417
Microfilament, 47
Microglia, 189E
Microtubule, 47
Microvillus, 62, 416
Micturition, 461
Midbrain
  cerebral peduncles, 213–214
  cranial nerves of, 214
  parts of, 213
Middle ear, 254
Midsagittal plane, 6
Milking, 339
Milk letdown, 524
Milk letdown reflex, 277
Milk secretion. See Lactation
Mineral, 436, 436E, 437E
Mineralocorticoid, 282
Minimal volume, 386
Minute volume of respiration (MVR), 385
Mitochondrion, 47
Mitosis, 52–53
Mitotic spindle, 53
Mitral valve. See Bicuspid valve
Mixed nerve, 209, 221
Mixing wave, 407
Modality, 242
Molecule, 21

Monocyte, 302
Monoplegia, 225
Monosaccharide, 27
Monounsaturated fat, 31
Mons pubis, 496
Morula, 512
Motion sickness, 264
Motor areas of cerebrum, 220
Motor cortex, 207
Motor end plate, 144
Motor neuron. See Efferent neuron
Motor pathways
  extrapyramidal, 207
  pyramidal, 207
Motor speech area, 220
Motor unit, 150
Mouth. See Oral cavity
Mucosa. See Mucous membrane
Mucous cell, 407
Mucous connective tissue, 76
Mucous membrane
  definition of, 81
  of GI tract, 400
  and nonspecific resistance, 358
Mucus, 48, 358
Multiple sclerosis (MS), 225
Multipolar neuron, 191
Multiunit smooth muscle tissue, 183
Mumps, 403
Muscle action potential, 143, 195
Muscle-building anabolic steroids, 144
Muscle contraction
  all-or-none principle, 150
  kinds of, 142
  motor unit, 150
  neuromuscular junction, 144, 146
  physiology of, 146–148
  role of ATP, 146–148
  role of calcium, 147
  role of phosphocreatine, 148
  sliding-filament mechanism, 144
  tone and, 151
Muscle fatigue, 150
Muscle fiber, 142, 180, 182–183
Muscles. See Skeletal muscle
Muscle spindle, 246
Muscle tissue
  all-or-none principle of, 150
  characteristics of, 142
  common disorders of, 183
  connective tissue components of, 143
  contraction of, 144–151, 154
  energy and, 148–149
  fascia of, 143
  fatigue of, 150
  functions of, 142
  heat production by, 150
  histology of, 144, 180, 182–183
  medical terminology and conditions, 184
  nerve and blood supply, 143–144
  oxygen debt and, 150
  physiology and contraction, 146–148
  types of, 142
Muscle tone, 151
Muscular atrophy, 154
Muscular dystrophies, 183
Muscular hypertrophy, 151

Muscularis
  definition of, 401
  of GI tract, 401
Muscular system
  characteristics of, 142
  common disorders of, 183
  connective tissue components of, 143
  contraction and, 144–151, 154
  definition of, 142
  fascia of, 143
  functions of, 142
  group actions of, 154
  histology of, 144, 180, 182–183
  insertions of, 154
  leverage and, 154
  medical terminology and conditions, 184
  naming skeletal muscles, 154
  nerve and blood supply of, 143–155
  origins of, 154
Mutation, 530
Myasthenia gravia (MG), 183
Myelin sheath, 191
Myeloma, 56
Myocardial infarction (MI), 322
Myocardium, 317
Myofibril, 144
Myofilament
  thick, 144
  thin, 144
Myoglobin, 144
Myogram, 150
Myology, 142
Myometrium, 495
Myoneural junction. See Neuromuscular junction
Myopathy, 184
Myopia, 252
Myosin, 144
Myosin-binding site, 144
Myositis, 184
Myotonia, 184
Myxedema, 291

Nail, 91–92
Nasal bone, 107
Nasal cavity, 376
Nasal septum, 110, 376
Nasopharynx, 377
Natural killer (NK) cells, 364
Nausea, 423
Navel. See Umbilicus
Navicular, of foot, 121
Neck, of tooth, 404
Necrosis, 57, 127
Negative feedback system, 13, 272–274
Neoplasm, 57
Nephrology, 449
Nephron, 450
Nephrotic syndrome, 466
Nerve
  abducens (VI), 223E
  accessory (XI), 224E
  cardiac, 325
  cochlear, 224E
  definition of, 204
  facial (VII), 224E
  glossopharyngeal (IX), 224E
  hypoglossal (XII), 224E

  intercostal, 208
  oculomotor (III), 223E
  olfactory (I), 223E
  opthalmic, 223E
  optic (II), 223E
  plexus, 209
  spinal, 209
  trigeminal (V), 223E
  trochlear (IV), 223E
  vagus (X), 224E, 325
  vestibular, 224E
  vestibulocochlear (VIII), 224E
Nerve action potential. See Nerve impulse
Nerve fiber
  A, 197
  adrenergic, 235
  B, 197
  cholinergic, 235
  C, 197
  definition of, 189
  postganglionic, 232
  preganglionic, 232
Nerve impulse
  all-or-none principle, 150, 195
  conduction across synapses, 197–200
  definition of, 195
  excitability, 195
  facilitation and, 199
  inhibitory transmission, 199
  integration at synapses, 199
  membrane potential, 194–195
  neurotransmitters, 221
  one-way conduction, 199
  saltatory conduction, 195–197
  speed of, 197
  summation and, 199
  synapse and, 197–200
Nervous system, 188–200
  autonomic, 231–239
  basal ganglia, 217
  blood supply of, 210, 212
  brain, 209–217, 220–221
  central, 188–200
  cerebrospinal fluid (CSF), 209–210
  common disorders of, 224–226
  cranial nerves, 221, 223
  definition of, 188
  electroencephalogram (EEG), 221
  excitatory transmission of, 199
  grouping of neural tissue, 204
  histology of, 188–191, 194
  inhibitory transmission of, 199
  medical terminology and conditions, 226
  memory and, 221
  meninges, 204–205, 209
  motor pathways of, 220
  nerve impulse of, 294–297
  neuroglia, 188, 189E
  neurons of, 189–191, 194
  neurotransmitters of, 221
  organization of, 188
  peripheral, 188
  plexuses, 209
  reflexes and, 207–208
  regeneration of, 200
  sensations and, 242–263
  sleep and wakefulness and, 215
  somatic, 188

  spinal cord, 204–208
  spinal nerves, 209
  synapses of, 197–200
Nervous tissue, 188–200
  all-or-nothing principle of, 150, 195
  classification of, 191, 194
  and conductivity, 195–200
  excitability of, 195
  excitatory transmission of, 195
  grouping of, 204
  histology of, 188–191, 194
  impulse speed of, 197
  inhibitory transmission of, 199
  and integration of synapses, 199
  nerve impulse and, 194–197
  neuroglia of, 188, 189E
  organization of, 188
  regeneration of, 200
  saltatory conduction and, 195–197
  synapses of, 197–200
Nervous tunic. See Retina
Net filtration pressure (NFP), 455, 475
Neuralgia, 226
Neuritis, 225
Neuroblastoma, 293
Neuroeffector junction, 198, 235
Neurofibral node, 189
Neurofibril, 189
Neuroglandular junction, 189
Neuroglia
  astrocytes, 189E
  ependyma, 189E
  microglia, 189E
  oligodendrocytes, 189E
Neurolemma, 191
Neurolemmocyte, 191
Neurology, 188
Neuromuscular junction, 144–146
Neuron
  afferent, 191
  all-or-none principle of, 195
  association, 191
  bipolar, 191, 251
  classification of, 191, 194
  and conductivity, 195–200
  definition of, 77, 189
  efferent, 191
  excitability, 195, 199
  ganglion, 257
  impulse speed of, 197
  inhibitory transmission of, 199
  and integration at synapses, 199
  membrane potential of, 194–195
  multipolar, 191
  nerve impulse and, 194–197
  neurotransmitters of, 198
  and one-way impulse conduction, 198
  photoreceptor, 251
  physiology of, 194–200
  postganglionic, 232
  postsynaptic, 198
  preganglionic, 232
  presynaptic, 198
  regeneration of, 200
  saltatory conduction and, 195–197
  structure of, 189–191
  synapses of, 197–200
  unipolar, 191

Neuropeptide, 221
Neurosecretory cell, 277
Neurotransmitter, 144, 198, 222E, 235
Neutrophil, 302
Nevus, 93
Night blindness, 253, 438E
Nipple, 497
Nissl body. *See* Chromatophilic substance
Nitrogen base, 31
Nociceptor, 242, 243
Node of Ranvier. *See* Neurofibral node
Nodule, 93
Nonpigmented granular dendrocyte, 86
Nonrapid eye movement (NREM) sleep, 215
Nonspecific resistance
  antimicrobial substances, 359
  chemical factors, 359
  fever, 362
  inflammation, 360–362
  mechanical factors, 358
  phagocytosis, 359–360
Norepinephrine (NE), 221, 232, 284–285
Nose, 376
Nostril. *See* External nares
Nuclear membrane, 44
Nuclease, 416
Nucleic acid
  chemical composition of, 31
  definition of, 31
  DNA, 31
  functions of, 31
  RNA, 31
  structure of, 31
Nucleolus, 44
Nucleotide, 44
Nucleus
  caudate, 217
  of cell, 44
  cerebellar, 221
  cerebral, 217
  corpus striatum, 217
  globus pallidus, 217
  lentiform, 217
  of nervous system, 204
  pulposus, 111
  putamen, 217
Nucleus pulposus, 111
Nutrient
  absorption of, 417–419
  and calories, 437
  carbohydrates, 429–430
  definition of, 427
  lipids, 430–434
  and metabolic rate, 442
  metabolism of, 427–435
  minerals, 436, 436E, 437E
  and obesity, 444
  proteins, 434–435
  vitamins, 437, 438E, 439E
  water, 471
Nystagmus, 222E, 265

Obesity, 444
Obligatory water reabsorption, 457
Obstetrics, 517
Obturator foramen, 117
Occipital bone, 105–106
Occlusion, 349

Oil gland. *See* Sebaceous gland
Olecranon, 116
Olfaction, theories of, 247
Olfactory bulb, 247
Olfactory cell, 246
Olfactory gland, 247
Olfactory hair, 247
Olfactory region, 247
Olfactory sensation, 246–247
Olfactory tract, 247
Oligodendrocyte, 189E
Oncogene, 56
Oncologist, 53
Oncology, 53
One-way impulse conduction,
  199
Oogenesis, 493–494
Oogonium, 493
Ophthalmology, 248
Opsonization, 359
Optic chiasma, 254
Optic disc, 251
Oral contraception (OC), 401–405
Orbit, 105
Organ, 3, 85
Organ of Corti. *See* Spiral organ
Organelle
  centriole, 47
  centrosome, 47
  cilium, 47
  definition of, 44
  endoplasmic reticulum, 45
  flagellum, 47
  Golgi complex, 46
  intermediate filament, 47
  lysosome, 46–47
  microfilament, 47
  microtubule, 47
  mitochondrion, 47
  nucleus, 44
  ribosome, 44–45
Organic compound, 27, 30–31
Organism, 5
Orgasm, 511
Origin of skeletal muscle, 154
Oropharynx, 377
Orthopedics, 96
Orthopnea, 396
Orthostatic hypotension, 349
Osmosis, 40
Osmotic pressure, 40, 454
Osmoreceptor, 278
Osseous tissue. *See* Bone tissue
Ossification
  definition of, 98
  endochondral, 99–101
  intramembraneous, 99
Osteitis, 127
Osteoarthritis, 136
Osteoblast, 97
Osteoblastoma, 127
Osteochondroma, 127
Osteoclast, 98, 101
Osteocyte, 77, 98
Osteogenic sarcoma, 56
Osteoma, 127
Osteomalacia, 126
Osteomyelitis, 126

Osteon, 77, 98
Osteoporosis, 126
Osteosarcoma, 127
Otalgia, 265
Otitis media, 265
Otolith, 261
Otolith membrane, 261
Otorhinolaryngology, 377
Otosclerosis, 265
Oval window. *See* Fenestra vestibuli
Ovarian cycle, 498
Ovarian cyst, 506
Ovarian follicle, 493
Ovary, 286, 493–494
Overhydration, 478
Ovulation, 495, 501
Ovum, 495
Oxidation. *See* Cellular respiration
Oxidation-reduction reaction, 428
Oxygen
  and carbon monoxide poisoning, 389–390
  and DPG, 389
  and hemoglobin, 388
  and pH, 388
  and temperature, 389
  transport of, 388–390
Oxygenated blood, 386
Oxygen debt, 150
Oxyhemoglobin, 388
Oxyphil cell, 281
Oxytocin (OT), 277, 524

Pacemaker. *See* Sinoatrial node
Pacinian corpuscle. *See* Lamellated corpuscle
Paget's disease, 126
Pain
  and enkephalin, 246
  phantom, 246
  referred, 246
Palatine bone, 110
Palatine tonsil, 357
Palpebrae. *See* Eyelids
Palpitation, 330
Pancreas, 285–286, 410
Pancreatic amylase, 410
Pancreatic duct, 410
Pancreatic islet, 285, 410
Pancreatic juice, 410
Pancreatic lipase, 410
Pancreatic secretion, regulation of, 410
Papilla
  dermal, 27
  of hair, 90
  of tongue, 402
Papillary duct, 450
Papillary muscle, 319
Pap smear, 495
Papule, 93
Parafollicular cell, 279
Paralysis, 185, 226
Paranasal sinus
  definition of, 110
  ethmoidal, 106–107
  frontal, 110
  maxillary, 107
  sphenoidal, 106
Paraplegia, 225
Parasagittal plane, 6

Parasympathetic division of ANS, 235
Parathyroid gland, 281
Parathyroid hormone (PTH), 281
Paraurethral gland, 496
Parent cell, 51
Parietal bone, 105
Parietal cell, 407
Parietal pleura, 380
Parietooccipital sulcus, 216
Parkinson's disease, 225
Parotid gland, 403
Paroxysmal tachycardia, 330
Partial antigen, 363
Partial pressure, 386
Parturition, 520
Passive processes, 39–41
Patella, 120
Patellar reflex, 208
Patent ductus arteriosus, 330
Pathogen, 358
Pectoral (shoulder) girdle, 115–116
Pelvic cavity, 9
Pelvic (hip) girdle, 117
Pelvic inflammatory disease (PID), 507
Pelvic inlet, 117
Pelvic outlet, 117
Pelvimetry, 117
Pelvis, 117
Penis, 491–492
Pepsin, 408
Pepsinogen, 408
Peptic cell, 407
Peptic ulcer, 421
Peptidase, 416
Peptide, 408
Peptide bond, 31
Perforating canal, 98
Pericardial cavity, 8, 317
Pericardial fluid, 317
Pericarditis, 317
Pericardium, 317
Perichondrium, 76, 99
Perikaryon. See Cell body, of neuron
Perilymph, 254
Perimysium, 143
Perineum, 496–497
Periodontal disease, 421
Periodontal ligament, 404
Periosteum, 98, 99
Peripheral nervous system (PNS), 188
Peripheral protein, 36
Peripheral resistance, 337
Peristalsis, 406, 416, 421
Peritoneum, 81
Peritonitis, 422
Peritubular capillary, 450
Perpendicular plate, 106
Perspiration, 91, 359
Peyer's patch. See Aggregated lymphatic follicle
PG. See Prostaglandins
pH
    and buffer systems, 25
    definition of, 25
    and kidney excretion, 457
    maintenance of, 26
    and respirations, 391
    scale, 25
Phagocytic vesicle, 42, 360

Phagocytosis
    adherence phase of, 359
    cells involved in, 359
    definition of, 42
    ingestion phase of, 360
    mechanism, 359
Phalanges
    of fingers, 116–117
    of toes, 121–122
Phantom pain, 246
Pharmacology, 56
Pharyngeal tonsil, 357
Pharynx, 376–377
Phenotype, 525, 527–528
Phenylketonuria (PKU), 444
Phlebitis, 349
Phosphate
    distribution of, 479
    functions of, 479
    regulation of, 479
Phosphate buffer system, 479
Phosphocreatine, 148
Phospholipid, 36
Phospholipid bilayer, 36
Photopigment, 253
Photoreceptor neuron, 251
Physiologic (natural) method of birth control, 525
Physiology, 3
Pia mater, 205, 209
Piezoelectric effect, 101
Pili. See Hair
Pineal gland, 286
Pinna, 254
Pinocytosis, 42
Pisiform, 117
Pitch, 379
Pituitary dwarfism, 291
Pituitary gland
    anterior lobe, 274
    common disorders of, 291
    hormones of, 274–278
    posterior lobe, 274
Pivot joint, 134
Placenta, 348, 514
Planes of body
    frontal, 6
    horizontal, 6
    midsagittal, 6
    parasagittal, 6
    sagittal, 6
Plantar flexion, 137
Plasma cell, 76, 304, 365
Plasma membrane
    definition of, 36
    functions of, 36
    movements of materials across, 38–44
    selective permeability of, 38–44
    structure of, 36
Plasma protein, 304
Plasmin, 308
Plasminogen, 308
Platelet. See Thrombocyte
Platelet plug formation, 305
Pleural cavity, 9, 380
Pleural membrane, 380
Pleurisy, 380

Plexus
    brachial, 209
    cervical, 209
    choroid, 209
    definition of, 209
    lumbar, 209
    sacral, 209
Pneumonia, 395
Pneumotaxic area, 391
Polarized membrane, 195
Poliomyelitis, 225
Polycythemia, 312
Polyneuritis, 438E
Polyp, 93
Polysaccharide, 27
Polyunsaturated fat, 30
Polyuria, 466
Pons
    cranial nerves of, 213
    structure of, 213
Positive feedback system, 13
Postcentral gyrus, 216
Posterior column pathway, 206
Posterior pituitary, 274
Posterior root, 207, 209
Posterior root ganglion, 207, 209
Postganglionic fiber, 232
Postganglionic neuron, 232
Postovulatory phase, 491
Postsynaptic neuron, 198
Potassium
    control of, 474
    deficiency of, 474
    distribution of, 474
    functions of, 474
Pott's disease, 127
Power stroke, 147
Precapillary sphincter, 335
Precentral gyrus, 216
Preeclampsia, 530
Preganglionic fiber, 232
Preganglionic neuron, 232
Pregnancy
    beginnings of organ systems during, 514–516
    birth control, 524–525
    definition of, 511
    development of blastocyst during, 512
    embryonic membranes during, 514
    embryonic period, 513–516, 514E
    fertilization of ovum, 512
    fetal growth, 516, 519E
    formation of morula, 512
    gestation, 517–519
    hormones of, 516–517
    implantation during, 512–513
    lactation, 497, 524
    parturition and labor, 520–521, 524
    placenta formation during, 514
    prenatal diagnostic techniques, 519–520
    sexual intercourse and, 511–512
    umbilical cord, 514
Pregnancy test, 516
Premenstrual syndrome (PMS), 503
Preovulatory phase, 501
Prepuce, 492
Presbyopia, 252, 265
Pressure sensation, 243
Presynaptic neuron, 198

Prevertebral ganglion, 232
Primary auditory area, 217
Primary bronchus, 380
Primary follicle, 500
Primary germ layers
    derivatives of, 514E
    development of, 514
Primary gustatory area, 217
Primary motor area, 217
Primary olfactory area, 217
Primary oocyte, 493
Primary ossification center, 99
Primary response, of antibody, 368
Primary somesthetic area, 217
Primary spermatocyte, 486
Primary visual area, 217
Prime mover (agonist), 154
Principal cell, 281
Proctology, 420
Progesterone (PROG), 286, 499, 516
Projection, 242
Projection fiber, 217
Prolactin (PRL), 276, 524
Prolactin inhibiting factor (PIF), 276
Prolactin releasing factor (PRF), 276
Pronation, 137
Properdin, 359
Prophase, 53
Proprioception, 242, 246
Proprioceptive sensation, 246
Proprioceptor, 242
Prostaglandins (PGs)
    functions of, 272
    and hormonal action, 272
    and inflammation, 360
Prostate cancer, 503
Prostate gland, 490
Protein
    absorption of, 416
    amino acids and, 31
    anabolism of, 434–435
    catabolism, 434–435
    definition of, 31
    digestion of, 416
    fate of, 434–435
    functions of, 31
    peptide bonds, 31
    synthesis of, 48
Protein buffer, 480
Protein synthesis, 48
Prothrombin, 305, 308
Prothrombin activator, 305, 308
Proto-oncogene, 56
Protraction, 137
Proximal convoluted tubule, 450
Pruritus, 93
Pseudopodia, 42, 358
Pseudostratified columnar epithelium, 62
Psoriasis, 92
Ptosis, 223E, 265
Puberty
    female, 498
    male, 489–490
Pubis, 117
Pudendum. See Vulva
Puerperal fever, 530
Puerperium, 524
Pulmonary artery, 319

Pulmonary capacity, 386
Pulmonary circulation, 341
Pulmonary edema, 395
Pulmonary embolism (PE), 349, 395
Pulmonary semilunar valve, 319
Pulmonary trunk, 319, 342
Pulmonary vein, 319, 341
Pulmonary ventilation, 383–385
Pulmonary volume, 385–386
Pulp, 404
Pulp cavity, 404
Pulse, 339
Pulse pressure, 340
Punnett square, 527
Pupil, 249
Purkinje fiber. See Conduction myofiber
Pus, 362
Pustule, 93
Putamen, 217
P wave, 322
Pyelitis, 466
Pyelonephritis, 466
Pyloric sphincter (valve), 407
Pylorus, of stomach, 407
Pyramid, of medulla, 212
Pyramidal pathway, 207
Pyruvic acid
    in glycolysis, 429
    in oxygen debt, 150
Pyuria, 465

QRS wave, 322
Quadrants, of abdominopelvic cavity, 9–10
Quadriplegia, 225

Radiation, 442
Radius, 116
Rales, 396
Rami communicantes, 209
Ramus, 209
Rapid eye movement (REM) sleep, 215
Raynaud's disease, 349
Reactivity, 363
Receptor
    complex, 242
    definition of, 208, 242
    for discriminative touch, 242
    for equilibrium, 261–263
    for gustation, 247
    for hearing, 254–260
    hormone, 270–272
    for light touch, 242
    for olfaction, 246–247
    for pain, 243, 246
    for pressure, 243
    for proprioception, 246
    simple, 242
    thermoreceptive, 243
    for vibration, 243
    for vision, 251
Receptor-mediated endocytosis, 42
Recessive gene, 527
Recruitment, 150
Rectum, 420
Red blood cell. See Erythrocyte
Red-green color blindness, 254
Red marrow, 96
Reduction, chemical, 428

Referred pain, 246
Reflex
    definition of, 208
    patellar, 208
    somatic, 208
    visceral autonomic, 208
    withdrawal, 208
Reflex arc, 208
Reflex center, 207
Refraction, 252
Refractory period, 195
Regeneration, 200
Regulating hormone (or factor), 274
Regulation of heart rate, 325, 329
Rejection, of tissue, 372
Relaxation period, 150, 324
Relaxin, 286, 499, 517
Releasing hormone (or factor), 274
Remodeling of bone, 101
Renal artery, 450
Renal autoregulation, 455
Renal capsule, 449
Renal column, 449
Renal erythropoietic factor (REF), 287, 301
Renal failure
    acute, 466
    chronic, 466
Renal papilla, 449
Renal pelvis, 449
Renal pyramid, 449
Renin, 282, 455
Renin-angiotensin pathway, 282, 457
Rennin, 408
Repolarized membrane, 195
Reproduction, 6, 485
Reproductive cell division, 51
Reproductive systems, 485–507
    birth control and, 524–525, 526E
    common disorders of, 502–503, 506–507
    definition of, 485
    female organs of, 492–502
    gamete formation and, 486, 493–494
    male organs of, 485–492
    medical terminology and conditions, 507
    pregnancy and, 512–517
    sexual intercourse and, 511–512
Residual body, 360
Residual volume, 386
Resistance, 337
    nonspecific, 358–362
    specific, 362–365, 368–369
Resistance reaction, of stress, 287, 290
Respiration, 5, 376
    air volumes exchanged, 385
    control of, 390–391, 394
    external (pulmonary), 387–388
    internal (tissue), 388
    and pH, 480
    pulmonary ventilation, 383–385
    and smoking, 395
    transport of gases, 388–390
Respiration, control of
    chemical, 391, 394
    nervous, 390–391
Respirator, 396
Respiratory bronchiole, 382
Respiratory center, 212, 390

Respiratory distress syndrome (RDS) of the new-born, 395
Respiratory failure, 395
Respiratory gases
   exchange of, 385–388
   transport of, 388–390
Respiratory system, 376–396
   air volumes exchanged, 385
   common disorders of, 395
   control of, 390–391, 394
   definition of, 376
   exchange of gases, 385–388
   expiration, 384
   external (pulmonary) respiration, 387–388
   inspiration, 383–384
   internal (tissue) respiration, 388
   medical terminology and conditions, 396
   organs of, 376–383
   pulmonary ventilation, 383–385
   smoking and, 395
   transportation of gases, 388–390
Resting membrane potential, 195
Rete testis, 490
Reticular activating system (RAS), 215
Reticular connective tissue, 215
Reticular fiber, 61
Reticular formation, 212
Reticulocyte count, 302
Retina, 249, 251
Retinal, 253
Retinal image formation, 251–258
Retinoblastoma, 56, 265
Retraction, 137
Retroperitoneal, 449
Reye's syndrome (RS), 226
Rh blood grouping, 311–312
Rheumatism, 136
Rheumatoid arthritis (RA), 136
Rheumatology, 140
Rhinitis, 396
Rhodopsin, 253
Rhythm method of birth control, 525
Rib
   false, 115
   floating, 115
   true, 115
Ribonuclease, 416
Ribonucleic acid (RNA)
   chemical composition of, 31
   comparison with DNA, 31
   functions of, 31, 48
   messenger, 48
   and nucleoli, 44
   and protein synthesis, 48
   ribosomal, 49
   transfer, 49
Ribosomal RNA (rRNA), 49
Ribosome
   definition of, 44
   and protein synthesis, 48
Rickets, 126
Right lymphatic duct, 357
Rigidity, 151
Rigor mortis, 148
Risk factors, heart disease, 395
RNA. See Ribonucleic acid
Rod, 251

Root
   of hair, 90
   of lung, 382
   of nail, 92
   of tooth, 404
Root canal, 404
Rotation, 134
Round window. See Fenestra cochlea
rRNA. See Ribosomal RNA
Rugae
   of stomach, 407
   of urinary bladder, 455
   of vagina, 496

Saccule, 254
Sacral canal, 114
Sacral hiatus, 114
Sacral plexus, 209
Sacrum, 113
Saddle joint, 136
Saliva
   composition of, 403
   secretion of, 403–404
Salivary amylase, 405
Salivary gland, 402–404
Salt, 26
Saltatory conduction, 196
Sarcolemma, 144
Sarcoma
   definition of, 56
   types of, 56
Sarcomere, 144
Sarcoplasm, 144
Sarcoplasmic reticulum, 144
Satiety center, 427
Saturated fat, 30
Scala tympani, 260
Scala vestibuli, 260
Scaphoid, 116
Scapula, 115
Schwann cell. See Neurolemmocyte
Sciatica, 225
Sclera, 249
Scleral venous sinus, 249
Scoliosis, 127
Scotopsin, 253
Scrotum, 485
Sebaceous gland, 91
Sebum, 91, 359
Secondary bronchus, 380
Secondary follicle, 500
Secondary oocyte, 494
Secondary ossification center, 100
Secondary response, 368
Secondary spermatocyte, 488
Second-degree burn, 92
Second (intermediate) cuneiform, 121
Second messenger, 272
Second polar body, 495
Secretin, 409
Secretion, 62
Segmentation, 416
Selective permeability
   definition of, 38
   factors determining, 38
Self excitation, 322
Sella turcica, 106

Semen, 491
Semicircular canal, 254
Semicircular duct, 254
Semilunar valve, 319
Seminal fluid. See Semen
Seminal vesicle, 490
Seminiferous tubule, 485
Sensations
   auditory, 254–263
   characteristics of, 241–242
   classifications, 242
   common disorders of, 264–265
   cutaneous, 242–243, 246
   definition of, 241
   equilibrium, 261–262
   general, 242–243, 246
   gustatory, 247–248
   medical terminology and conditions, 265
   olfactory, 246–247
   pain, 243, 246
   prerequisites for, 241
   pressure, 243
   proprioceptive, 246
   special, 246–263
   tactile, 242–243
   thermoreceptive, 243
   touch, 242
   vibration, 243
   visual, 248–254
Sensitization, 364
Sensory areas of cerebrum, 217–220
Sensory neuron. See Afferent neuron
Sensory pathway
   posterior column, 206
   spinothalamic, 206
Septal cells, 382
Septicemia, 313
Serosa. See Serous membrane
Serotonin (5–HT), 222E
Serous membrane
   definition of, 81, 401
   of GI tract, 401
   parietal layer of, 81
   visceral layer of, 81
Serous pericardium, 317
Sertoli cell. See Sustentacular cell
Serum, 305
Sesamoid bone, 96
Severe combined immunodeficiency (SCID), 372
Sex chromosomes, 528
Sexual intercourse, 511–512
Sexually transmitted disease (STD), 502
Sheath of Schwann. See Neurolemma
Shingles, 225
Shinsplints, 120
Shock, 340
Short bone, 96
Short-term memory, 221
Sigmoid colon, 420
Simple columnar epithelium, 62
Simple cuboidal epithelium, 62, 69
Simple epithelium
   definition of, 62
   types of, 62–69
Simple goiter, 291
Simple squamous epithelium, 62
Single binocular vision, 253

Sinoatrial (SA) node, 322
Sinusitis, 110
Skeletal muscle
  actions of, 158E
  adductor longus, 176E
  adductor magnus, 176E
  biceps brachii, 170E
  biceps femoris, 179E
  brachialis, 170E
  brachioradialis, 170E
  buccinator, 158E
  deltoid, 168E
  diaphragm, 180E
  epicranius, 158E
  erector spinae. *See* Skeletal muscle, sacrospi-
    nalis
  extensor carpi radialis longus, 172E
  extensor carpi ulnaris, 172E
  extensor digitorum, 172E
  extensor digitorum longus, 180E
  external intercostals, 180E
  external oblique, 162E
  flexor carpi radialis, 172E
  flexor carpi ulnaris, 172E
  flexor digitorum longus, 180E
  flexor digitorum profundus, 172E
  flexor digitorum superficialis, 172E
  frontalis, 158E
  gastrocnemius, 180E
  gluteus maximus, 176E
  gluteus medius, 176E
  gluteus minimus, 176E
  gracilis, 179E
  group actions of, 154
  hamstrings, 179E
  iliacus, 176E
  inferior oblique, 161E
  inferior rectus, 161E
  infraspinatus, 168E
  insertions of, 154
  internal intercostals, 164E
  internal oblique, 162E
  lateral rectus, 161E
  latissimus dorsi, 168E
  levator palpebrae superioris, 158E
  levator scapulae, 165E
  lever systems and, 154
  masseter, 160E
  medial rectus, 161E
  naming, 154
  occipitalis, 158E
  orbicularis oculi, 158E
  orbicularis oris, 158E
  origins of, 154
  palmaris longus, 172E
  pectineus, 176E
  pectoralis major, 168E
  pectoralis minor, 165E
  peroneus longus, 180E
  piriformis, 176E
  platysma, 158E
  pronator teres, 170E
  psoas major, 176E
  quadratus lumborum, 174E
  quadriceps femoris, 179E
  rectus abdominis, 162E, 174E
  rectus femoris, 179E
  rhomboideus major, 165E
  sarcospinalis, 174E
  sartorius, 179E
  semimembranosus, 179E
  semitendinosus, 179E
  serratus anterior, 165E
  soleus, 180E
  sternocleidomastoid, 174E
  subclavius, 165E
  subscalularis, 168E
  superior oblique, 161E
  superior rectus, 161E
  supinator, 170E
  temporalis, 160E
  tensor fasciae latae, 176E
  teres major, 168E
  tibialis anterior, 180E
  tibialis posterior, 180E
  transversus abdominis, 162E
  trapezius, 165E
  triceps brachii, 170E
  vastus lateralis, 179E
  vastus medialis, 179E
  zygomaticus major, 158E
Skeletal muscle tissue
  all-or-none principle of, 150
  connective tissue components of, 143
  contraction of, 144–151, 154
  definition of, 142
  energy and, 148–149
  fascia of, 143
  flaccidity of, 151
  histology of, 144
  hypertrophy of, 151
  motor unit of, 150
  nerve and blood supply of, 143–144
  physiology of contraction, 146–148
  sliding-filament mechanism of, 144
Skeletal system, 95–126
  appendicular skeleton of, 115–126
  arches of foot, 122–123
  axial skeleton of, 103
  common disorders of, 126–127
  comparison of female and male, 123
  definition of, 96
  divisions of, 103
  fontanels, 110
  fractures of, 126
  functions of, 96
  histology of, 97–98
  homeostasis of, 101–102
  hyoid, 111
  lower extremities, 117–123
  medical terminology and conditions, 127
  ossification, 98–10i
  paranasal sinuses of, 110
  pectoral (shoulder) girdles, 115–116
  pelvic (hip) girdle, 117
  remodeling of, 101
  skull, 103–111
  surface markings of, 102–103
  sutures, 103
  thorax, 114–115
  upper extremities, 116–117
  vertebral column, 111–114
Skene's gland. *See* Paraurethral gland
Skin
  acne of, 92
  and body temperature, 92
  burns of, 92
  cancer of, 90, 92
  color of, 87
  common disorders of, 92
  decubitus ulcer of, 92
  definition of, 85
  dermis of, 87
  epidermis, 86–87
  functions of, 85
  and immunity, 368
  mechanical factors of, 358
  structure of, 86–87, 90
  sunburn of, 92
  temperature regulation and, 92, 93
Skin cancer, 90, 92
Skull
  bones of, 103–111
  fontanels of, 110
  foramina of, 111
  paranasal sinuses of, 110
  sutures of, 103
Sleep
  and arousal, 215
  nonrapid eye movement (NREM), 215
  rapid eye movement (REM), 215
  and reticular activating system (RAS), 215
Sliding-filament mechanism, 144
Small intestinal absorption, 417–419
Small intestine, 415–419
Smegma, 507
Smoking and respiration, 395
Smooth muscle tissue, 180, 182–183
Sodium
  control of, 474
  deficiency of, 474
  distribution of, 474
  functions of, 474
Sodium-potassium pump, 194
Solitary lymphatic nodule, 416
Solute, 25
Solution, 25
Solvent, 25
Somatic cell division, 51–53
Somatic nervous system (SNS), 188
Somatomedin, 274
Somesthetic association area, 217, 220
Sound wave, 260
Spasm, 183
Spastic, 226
Spasticity, 151
Special movement, 137
Special senses
  auditory, 254–263
  equilibrium, 261–263
  gustatory, 247–248
  olfactory, 246–247
  visual, 248–254
Specific gravity, of urine, 464E
Spermatic cord, 490
Spermatid, 489
Spermatogenesis, 486–489
Spermatogonium, 486
Spermatozoon, 489
Sperm cell. *See* Spermatozoon
Spermiogenesis, 489
Sphenoid bone, 106
Sphygmomanometer, 340
Spina bifida, 127

Spinal column. *See* Vertebral column
Spinal cord
  column of, 206
  common disorders of, 224–226
  cross-sectional structure, 206
  functions of, 206–208
  general features, 205–206
  horns of, 204
  impulse conduction and, 206–207
  injury to, 224–225
  meninges, 204–205
  nerves of, 209
  protection of, 204–205
  reflex center, 207
  spinal tap, 205
  tracts of, 206–207
Spinal cord injury, 225
Spinal meninges, 204–205
Spinal nerves
  branches, 209
  definition of, 209
  names of, 209
  plexuses, 209
Spinal segment, 206
Spinal shock, 225
Spinal tap, 205
Spinothalamic pathway, 206
Spinous process, 112
Spiral organ, 254
Spirogram, 385
Spleen, 357
  removal of, 357
  rupture of, 357
Splenectomy, 357
Spongy bone, 98
Sprain, 136
Stapes, 254
Starling's law of the heart, 375
Starvation, 444
Static equilibrium, 261–262
Stellate reticuloendothelial cell, 410
Stereocilia, 261
Sterilization, 524
Sternum, 114
Stimulus, 195, 241
Stokes-Adams syndrome, 330
Stomach, 407–410
Stomach gastrin, 287
Strabismus, 223E, 265
Straight tubule, 490
Strain, 136
Stratified columnar epithelium, 69
Stratified cuboidal epithelium, 69
Stratified epithelium
  definition of, 69
  types of, 69
Stratified squamous epithelium, 69
Stratum basale, 86
Stratum basalis, 495
Stratum corneum, 87
Stratum functionalis, 495
Stratum germinativum, 87
Stratum granulosum, 87
Stratum lucidum, 87
Stratum spinosum, 87
Stress
  alarm and, 287
  and blood pressure, 12

and blood sugar level, 13
  definition of, 11, 287
  and disease, 291
  exhaustion reaction and, 290–291
  general adaptation syndrome (GAS), 287, 290–291
  and homeostasis, 12, 13
  resistance reaction and, 287, 290
  stressors and, 287
Stressor, 287
Stretch receptor, 391
Striae, 87
Stricture, 467
Stroke volume (SV), 375
Stroma, of ovary, 493
Styloid process, of temporal bone, 105
Subarachnoid space, 205
Subcutaneous (SC) layer, 76, 143
Sublingual gland, 403
Subluxation, 140
Submandibular gland, 403
Submucosa
  definition of, 401
  of GI tract, 401
Substance P, 222E
Substrate, 428
Subthreshold stimulus, 195
Sucrase, 416
Sucrose, 27
Sudden cardiac death, 330
Sudden infant death syndrome (SIDS), 395
Sudoriferous gland, 91
Sulcus
  of brain, 216
  coronary, 319
Summation, 199
Sunburn, 92
Superficial fascia. *See* Subcutaneous layer
Superior vena cava (SVC), 319
Supination, 137
Suppressor T cell, 365
Suprarenal (adrenal) gland. *See* Adrenal gland
Surface markings of bone, 102–103
Surfactant, 382, 385
Susceptibility, 358
Suspensory ligament of lens, 251
Suspensory ligament of the breast, 497
Sustentacular cell, 485
Sutural bone, 96
Suture
  coronal, 103
  definition of, 103, 131
  lambdoidal, 103
  sagittal, 103
  squamosal, 103
Swallowing. *See* Deglutition
Sweat gland. *See* Sudoriferous gland
Sympathetic chain, 235
Sympathetic division of ANS, 235
Sympathetic trunk ganglia, 234
Sympathomimetic, 284
Symphysis joint, 132
Symphysis pubis, 117, 132
Symptothermal method of birth control, 525
Synapse
  chemical, 198
  definition of, 198
  electrical, 198

excitatory transmission at, 199
  facilitation at, 199
  factors affecting transmission at, 199–200
  inhibitory transmission at, 199
  integration at, 199
  neurotransmitters at, 199
  one-way impulse conduction at, 199
  summation at, 199
Synaptic cleft, 144, 198
Synaptic end bulb, 144, 189, 198
Synaptic vesicle, 144, 189, 198
Synarthrosis, 131
Synchrondrosis, 132
Syncope, 349
Syndesmosis, 132
Synergist, 154
Synovial fluid, 81, 132
Synovial joint
  definition of, 132
  movements of, 133–137
  structure, 132
  types of, 133–137
Synovial (joint) cavity, 132
Synovial membrane, 81, 132
Synovitis, 140
Syphilis, 503
System
  cardiovascular, 5E, 298–313
  definition of, 5
  digestive, 5E, 400–421
  endocrine, 5E, 269–293
  integumentary, 5E, 85–92
  lymphatic and immune, 5E, 353–369
  muscular, 5E, 142–184
  nervous, 5E, 188–200
  reproductive, 5E, 485–507
  respiratory, 5E, 376–396
  skeletal, 5E, 95–126
  urinary, 5E, 449–466
Systemic circulation, 341, 343
Systemic disease, 56
Systemic lupus erythematosus (SLE), 92
Systole, 324
Systolic blood pressure (SBP), 340

Tachycardia, 339
Tactile disc, 87
Tactile sensation, 242
Taenia coli, 420
Talus, 121
Target cell, 270
Tarsus, 121
Taste bud, 247
Taste pore, 247
Tay-Sachs disease, 225
T cell, 364–365
  amplifier, 365
  cytotoxic (killer), 303, 364
  delayed hypersensitivity, 365
  helper, 364
  memory, 365
  suppressor, 365
Tectorial membrane, 254
Teeth
  deciduous, 404
  permanent, 404
Telodendrium. *See* Axon terminal
Telophase, 53

Temperature, regulation of, 442–444
Temporal bone, 105
Tendon, 76, 143
Tendon organ, 246
Tendon sheaths, 143
Teratogen, 528
Terminal bronchiole, 380
Terminal ganglion, 232
Tertiary bronchus, 380
Testes, 286, 485–489
Testicular cancer, 503
Testis determining factor (TDF), 528
Testosterone, 286, 489
Tetanus
    complete, 151
    incomplete, 151
Tetany, 291
Tetrad, 489
Tetralogy of Fallot, 330
Thalamus, 214
Thermoreceptive sensation, 243
Thermoreceptor, 92, 243
Third-degree burn, 92
Third (lateral) cuneiform, 121
Third ventricle, 209
Thoracic aorta, 319, 343E
Thoracic cavity, 9
Thoracic duct, 357
Thoracic vertebrae, 111
Thorax, 114–115
Threshold stimulus, 150, 195
Thrombin, 305, 308
Thrombocyte
    function of, 304
    number per cubic centimeter, 304
    production, 304
    structure of, 304
Thrombocytopenia, 312
Thrombophlebitis, 349
Thromboplastin. See Tissue factor (TF)
Thrombosis, 305
Thrombus. See Blood clot
Thymosin, 286
Thymus gland, 286, 357
Thyroid cartilage, 377
Thyroid follicle, 281
Thyroid gland, 278–281
Thyroid hormones, 280
Thyroid-stimulating hormone (TSH), 275
Thyroid storm, 293
Thyrotropin releasing hormone (TRH), 275
Thyroxine (T_4), 281
TIA. See Transient ischemic attack
Tibia, 120
Tic, muscular, 183
Tidal volume (TV), 385
Timing, of cardiac cycle, 324
Tinnitus, 265
Tissue
    connective, 69–77
    definition of, 3, 61
    epithelial, 62–69
    inflammation of, 360–361
    muscle, 77, 142–154
    nervous, 77, 188–200
    osseous, 77, 96–101
    rejection of, 372

    types of, 61
    vascular, 77, 298–313
Tissue factor (TF), 308
Tissue rejection, 372
Tongue, 402
Tonsils
    lingual, 357
    palatine, 357
    pharyngeal, 357
Tooth decay. See Dental caries
Torn cartilage, 132
Total lung capacity, 386
Touch sensation, 242
Toxic shock syndrome (TSS), 503
Trabecula, of spongy bone, 98
Trachea, 379
Tracheostomy, 379
Trachoma, 264
Tract
    ascending, 204
    definition of, 204
    descending, 204
    olfactory, 204
    optic, 244
Transcription, 48
Transferrin, 301
Transfer RNA (tRNA), 48
Transfusion, 313
    autologous, 313
    direct, 313
    exchange, 313
    indirect, 313
    reciprocal, 313
Transient ischemic attack, 225
Transitional epithelium, 69
Translation, 48
Transmission of nerve impulses, 194–200
Transplantation, 372
Transverse arch, of foot, 123
Transverse colon, 420
Transverse foramen, 113
Transverse tubule, 144
Trapezium, 117
Trapezoid, 117
Traveler's diarrhea, 423
Tremor, 183
Treppe, 183
Trichinosis, 184
Trichomoniasis, 503
Tricuspid valve, 322
Trigeminal neuralgia, 223E, 226
Trigone, 461
Triiodothyronine (T_3), 281
Triquetral, 116
Trophoblast, 512
Tropomyosin, 144
Troponin, 144
True vocal cords, 378
Trypsin, 410
Trypsinogen, 410
T tubule. See Transverse tubule
Tubal ligation, 524
Tuberculosis (TB), 395
Tubular maximum (Tm), 457
Tubular reabsorption, 457
Tubular secretion, 457–458
Tumor
    benign, 53

    of brain, 225
    of breasts, 497, 506
    of cervix, 506–507
    definition of, 53
    of GI tract, 422
    of lung, 395
    malignant, 53
    of skin, 90, 92
    of uterus, 495
Tumor-specific antigen, 372
Tunica albuginea, 485, 493
T wave, 322
Twitch contraction, 150
Tympanic antrum, 254
Tympanic cavity. See Middle ear
Tympanic membrane, 254
Type II cutaneous mechanoreceptor, 243
Typical vertebra, 112–113

Ulcer
    decubitus, 92
    peptic, 421
Ulna, 116
Umbilical cord, 524
Umbilicus, 524
Unipolar neuron, 191
Universal donor, 311
Universal recipient, 311
Upper extremity, 311
Urea, 465E
Uremia, 467
Ureter, 459–460
Urethra, 464, 490
Urethritis, 466
Urinalysis (UA), 464
Urinary bladder, 460–461, 464
Urinary system, 449–466
    blood supply, 450
    common disorders of, 466
    definition of, 449
    hemodialysis, 458
    medical terminology and conditions, 467
    nephron physiology, 450
    urine, 464
Urinary tract infection (UTI), 466
Urine
    abnormal constituents, 464, 465–466
    chemical composition, 464, 465
    factors affecting volume, 464
    physical characteristics, 464
Urobilinogen, 301, 466E
Urobilinogenuria, 466E
Urology, 449
Uterine cavity, 495
Uterine tube, 495
Uterus, 495
Utricle, 254
Uvula, 401

Vagina, 496
Vaginal orifice, 496
Vaginitis, 507
Valence, 21, 363
Valve, of heart, 319
Varicose vein (VV), 336
Vascular spasm, 304
Vascular tissue. See Blood
Vascular tunic, 249

Vas deferens. *See* Ductus deferens
Vasectomy, 490
Vasoconstriction, 334
Vasoconstrictor center. *See* Vasomotor center
Vasodilation, 334, 360
Vasodilator substance, 338
Vasomotion, 338
Vasomotor center, 212
Veins
　cardiac, 322
　central, 410
　coats of, 336
　coronary sinus, 319, 342
　definition of, 334
　ductus venosus, 516
　hepatic, 348
　hepatic portal, 341
　of hepatic portal circulation, 348
　inferior vena cava (IVC), 319, 342
　pulmonary, 319
　of pulmonary circulation, 342–343
　renal, 450
　superior vena cava (SVC), 319, 342
　of systemic circulation, 341
　umbilical, 516
　varicose, 336
Venereal disease. *See* Sexually transmitted disease
Venesection, 313
Ventral body cavity, 8
Ventral ramus, 209
Ventricle
　of brain, 209
　of heart, 319
Ventricular diastole, 324
Ventricular fibrillation (VF), 330
Ventricular systole, 324
Venule, 335
Vermiform appendix, 420
Vertebra
　cervical, 113
　coccyx, 113
　lumbar, 113

parts of, 112–113
　sacrum, 113
　thoracic, 113
　typical, 112–113
Vertebral body cavity, 8
Vertebral canal, 112
Vertebral column
　abnormal region, 127
　cervical region, 113
　coccygeal region, 113
　divisions of, 111
　herniated (slipped) disc, 127
　lumbar region, 113
　normal curves, 111–112
　sacral region, 113
　spina bifida, 127
Vertebral curve, 111–112
Vertebral foramen, 113
Vertebra prominens, 113
Vertigo, 265
Vesicular membrane, 254
Vestibule
　of internal ear, 254
　of vulva, 492
Vibration sensation, 243
Villus, 415
Viral encephalitis, 226
Virilism, 291
Viscera, definition of, 8
Visceral pericardium. *See* Epicardium
Visceral pleura, 380
Visual association area, 217
Visual sensation, 248–254
Vital capacity, 386
Vitamins
　absorption by small intestine, 419
　and bone growth, 101
　deficiencies of, 438E–439E
　definition of, 437
　fat-soluble, 437, 438E
　functions of, 438E–439E
　sources of, 438E–439E
　water-soluble, 437, 438E–439E

Vitreous body, 251
Vitreous chamber, 251
Voice production, 378
Volkmann's canal. *See* Perforating canal
Volkmann's contracture, 185–186
Vomer, 110
Vomiting, 409
Vulva, 496

Wandering (tissue) macrophage, 361
Wart, 93
Water
　absorption by small intestine, 419
　and body fluids, 471
　in chemical reactions, 25–26
　and heat, 26
　importance of, 25–26
　and lubrication, 26
　solvating property of, 25
Water intoxication, 478
Water-soluble vitamins, 437, 438E–439E
Wave summation, 151
Wharton's jelly. *See* Mucous connective tissue
Wheeze, 396
White blood cell. *See* Leucocyte
White coat hypertension, 349
White matter, 204, 217
Word blindness, 221
Word deafness, 221
Wormian bone. *See* Sutural bone
Wryneck, 184

Xiphoid process, 114
X-linked trait, 528

Yellow marrow, 98
Yolk sac, 514

Z line, 144
Zona pellucida, 512
Zygomatic arch, 105
Zygomatic bone, 110
Zygote, 486, 512